NATURAL SCIENCES IN AMERICA

THE ORIGIN OF THE FITTEST

and

THE PRIMARY FACTORS OF ORGANIC EVOLUTION

E[dward] D[rinker] Cope

ARNO PRESS
A New York Times Company
New York, N. Y. • 1974

Reprint Edition 1974 by Arno Press Inc.

Reprinted from a copy in
The Newark Public Library

NATURAL SCIENCES IN AMERICA
ISBN for complete set: 0-405-05700-8
See last pages of this volume for titles.

Manufactured in the United States of America

Library of Congress Cataloging in Publication Data

Cope, Edward Drinker, 1840-1897.
The origin of the fittest.

(Natural sciences in America)
Reprint of the author's The origin of the fittest; essays on evolution, published in 1887 by D. Appleton, New York, and of the author's The primary factors of organic evolution, published in 1896 by The Open Court Pub. Co., Chicago.

1. Evolution. 2. Paleontology. I. Cope, Edward Drinker, 1840-1897. The primary factors of organic evolution. 1974. II. Title. III. Series.

QH366.2.C66 1974 575 73-17813
ISBN 0-405-05729-6

THE

ORIGIN OF THE FITTEST

ESSAYS ON EVOLUTION

BY

E. D. COPE, A. M., Ph. D. (Heidelberg)

MEMBER OF THE UNITED STATES NATIONAL ACADEMY OF SCIENCES; CORRESPONDENT OF THE ROYAL BAVARIAN ACADEMY OF SCIENCES

NEW YORK
D. APPLETON AND COMPANY
1887

TO MY FRIEND

PROFESSOR ALPHEUS HYATT,

OF BOSTON, MASS.,

WITH WHOSE STUDIES MY OWN HAVE LONG BEEN PARALLEL,

AND WITH WHOSE CONCLUSIONS I HAVE LONG COINCIDED,

THIS WORK IS DEDICATED

BY

THE AUTHOR.

PREFACE.

The twenty-one essays which constitute the present volume represent the reflections which have suggested themselves to the author while engaged in special zoölogical and paleontological studies. While the original work of the author has been nearly confined to the vertebrata, his studies have taken in a wider range. These have convinced him that the conclusions derived from the investigations of the vertebrata are applicable to invertebrate animals and to plants.

In reaching conclusions, the author has endeavored to avoid, as much as possible, any bias due to the influence of any opinions whatsoever on evolution and allied subjects which were already in the field. He therefore avoided, for a time, reading the works of the masters of the subject, applying to them for confirmation or criticism only after the publication of his own results. It is therefore true that the generalizations contained in these essays have been worked out by the author from such material as has come under his own eyes, with little other aid. It has also followed that not a few of the conclusions he has reached were not new. On the other hand, some of the principles enunciated had not been clearly stated prior to the publication of these essays. In some cases the work accomplished has consisted in throwing well-known principles into accord with each other, as in the case of the laws of acceleration and retardation.

The earlier essays are the more hypothetical, and the later present more numerous demonstrations. The latter have resulted chiefly from the author's researches in the field of vertebrate paleontology, which have thrown the greatest possible light on the fact and method of evolution. For the history of this subject the reader is referred to the author's forthcoming "Manual of the Vertebrate Paleontology of North America"; and, for the more detailed work, to the author's publications in the "Final Reports of the United States Geological Surveys," under Wheeler and Hayden, and to the "Proceedings of the American Philosophical Society of Philadelphia."

The present essays are arranged (see table of contents) into four series, as follows: First, on General Evolution; second, on the Structural Evidences of Evolution; third, on Mechanical Evolution; fourth, on Metaphysical Evolution. In the first series the author's earlier essays are arranged. The general principles are here laid down or foreshadowed. The essays of the succeeding sections are occupied with the demonstration of these and other generalizations, so far as practicable. In each of the sections some essays will be found to be more, and others less, adapted to popular use. It is believed that the general reader can select a sufficient number of articles of minimum technicality to convey to his or her mind a sufficient idea of the views set forth. In Part I, articles first, fourth, and sixth are of this character. In Part II, the first, second, and third essays are the least technical. In Part III, the first article is the most popular, although the others are essential to an understanding of the doctrine of mechanical evolution. In Part IV, all are sufficiently popular for the reader who has some knowledge of mental science.

A historical synopsis of the essays may be now given. The attempt is made to point out the aim of each, with an indication of what may have been new in its contents. They are taken up in the order of date of publication:

II. The Origin of Genera. From the "Proceedings of the Philadelphia Academy of Natural Sciences" for October, 1868, and published separately by the author early in 1869. In this essay the following doctrines were taught:

First, that development of new characters has been accomplished by an *acceleration* or *retardation* in the growth of the parts changed. This was demonstrated by reference to a class of facts, some of which were new, which gave ground for the establishment of the new doctrine.

Second, that of *exact parallelism* between the adult of one individual or set of individuals, and a transitional stage of one or more other individuals. This doctrine is distinct from that of inexact parallelism which had already been stated by von Baer. And that this law expresses the origin of genera and higher groups, because—

Third, they can only be distinguished *by single characters* when all their representatives come to be known.

Fourth, that genera and various other groups have descended, not from a single generalized genus, etc., of the same group, but from corresponding genera of one or more other groups. This was called the doctrine of *homologous groups*.

Fifth, the doctrine that these homologous groups belong to different geological periods, and,

Sixth, to different geographical areas, which, therefore, in some instances, are,

Seventh, related to each other in a successional way like the epochs of geological time.

Of these doctrines it may be observed that the first and second are now the common property of evolutionists, and are recognized everywhere as matter of fact. The names which I selected to express them have, however, only come into partial use. The author believes that, although the doctrine was vaguely shadowed out in the minds of students prior to the publication of this essay, it had not previously been clearly expressed, nor been reduced to a demonstration. Of the truth of the doctrine the author is more than ever convinced, and he believes that paleontological discovery has demonstrated it in many instances, and that other demonstrations will follow. The fourth proposition (that of homologous groups) is now held as a hypothesis explaining the phylogeny of various groups of animals. For the descent of one homologous group from another, the term *polyphyletic* has been coined. It remains to be seen whether the doctrine is of universal application or not. That homologous groups belong to different geological horizons, as stated under the fifth head, has been frequently demonstrated since the publication of the essay. That the sixth proposition is true in a certain number of cases is well known, and it follows that the seventh proposition is also true in those cases. The latter hypothesis, which was originally advanced by Prof. Agassiz, is, however, only partially true, and the advance of paleontological study has not demonstrated that it has had a very wide application in geological time.

A proposition which was made prominent in this essay was, that the prevalence of non-adaptive characters, in animals, proves the inadequacy of hypotheses which ascribe the survival of types to their superior adaptation to their environment. Numerous facts of this kind undoubtedly indicate little or no activity of a selective agency in nature, and do point to the existence of an especial developmental force acting by a direct influence on growth. The action of this force is the acceleration and retardation appealed to in this paper. The force itself was not distinguished until the publication of the essay entitled "The Method of Creation" (No. V), where it was named growth-force, or bathmism. The energetic action of this force accounts for the origin of characters, whether adaptive or non-adaptive, the former differing from the latter in an intelligent direction, which adapts them to the environment. The numerous adaptive characters of animals had by that time engaged the attention of the author, and he found that they are even more numerous than the non-adaptive. Some of the latter were accounted for on the theory of the "complementary location of growth-force."

IV. The Hypothesis of Evolution, Physical and Metaphysical. "Lippincott's Magazine," Philadelphia, 1870; reprinted by Charles C. Chatfield & Co. New Haven, 1870.

This essay embraces a popular exposition of the principles maintained in the essay entitled the "Origin of Genera," with some conclusions derived from the general facts of anthropology. To this were added some facts in the evolution of human physiognomy and human character, which had not been previously thrown into harmony with the laws already set forth. Unfortunately, the author attempted to correlate these again with the theories of some theologians, and, in some instances, without success. A few paragraphs have been stricken from this part of the essay, and others are allowed to remain as illustrations of far-reaching hypotheses resting on little information.

V. The Method of Creation of Organic Types. From the "Proceedings of the American Philosophical Society," December, 1871; republished by McCalla & Stavely the same year. Read before the American Association for the Advancement of Science, at Detroit, August 27, 1871. Received the Walker prize of the Boston Society of Natural History.

In this essay were added to the preceding, the following hypotheses:

1. The law of repetitive addition, in which the structures of animals were shown to have originated from simple repetitions of identical elements.

2. The existence of an especial force which exhibits itself in the growth of organic beings, which was called growth-force, or bathmism.

3. That development consists in the location of this energy at certain parts of the organism.

4. That this location was accomplished by use or effort, modifying and being modified by the environment; or the doctrine of kinetogenesis.

5. That the location of this energy at one point causes its abstraction from other points, producing "complementary diminution" of force at the latter.

6. That the location of this energy, so as to produce the progressive change called evolution, is due to an influence called "grade influence."

7. That inheritance is a transmission of this form of energy, which builds in precise accord with the sources from which it is derived.

8. That this "grade influence" is an expression of the intelligence of the animal, which adapts the possessor to the environment by an "intelligent selection."

9. An attempt to account for the origin of "mimetic analogy" by "maternal impressions."

On these propositions, the following comments may be made: First, the law of repetitive addition is much like the law of rhythm previously proposed by Herbert Spencer.* Second, the force of growth, or bathmism, had already been called constructive force by Carpenter, who, however, did not treat of its evolutionary or "grade" characteristics. That such force exists there can be no doubt at the present time, but it may be that its varied aspects should each be considered a separate species of force (i. e., energy). Third, the relations of the energetic and static conditions of this force were considered, but were not sufficiently followed out to be clear. It is hoped that greater clearness has been attained by the omission of a few paragraphs and the insertion of an explanatory foot-note.

Fourth, that the location of this energy is due to the influence of use and effort. The doctrine of the development of parts of living beings by use, and their loss by disuse, is well known to have been put forth by Lamarck in 1809, who devotes one of the longest chapters (No. VII) of his "Philosophie Zoölogique" to its discussion. He did not, however, include the element of *effort*, prior to the appearance of any rudiment of an organ, in his hypothesis, which was proposed, so far as I know, for the first time in the present essay. The doctrine of use and disuse has been sustained

* "Principles of Biology."

by Spencer in his "Principles of Biology" as respects the effect of motion on structure in general,* and in the particular case of the origin of vertebræ.† Fifth, the complementary development of parts had also been pointed out by Herbert Spencer.‡

Seventh, the explanation of inheritance by the transmission of the type of growth-force possessed by one generation to another. This doctrine was subsequently announced by Haeckel, under the name of *perigenesis*,# and is the only good hypothesis yet proposed for the explanation of this phenomenon. Eighth, the theory of "intelligent selection," or the agency of the intelligence of a living being in directing its movements, and therefore its growth, although a plain and necessary consequence of the "law of use and effort," had not been, so far as I am aware, announced prior to the publication of this paper. This important theory at once opened the way for an investigation of the general relations of mind to evolution, which involved the question of the origin and development of mind itself. These questions were more fully discussed in the papers of Part IV, on "Metaphysical Evolution." Ninth, the origin of mimetic analogy. The explanation offered is almost necessary, if the doctrine of the influence of effort on structural growth be true.

I. Evolution and its Consequences. From the "Penn Monthly Magazine," Philadelphia, for May, July, and August, 1872.

This is a popular exposition, with elaboration, of the doctrines contained in the preceding essays. The evolution of mind is more fully stated, the hypothesis adopted being that proposed by Spencer in his "Principles of Psychology," published in 1855.

VII. The Homologies and Origin of the Types of Molar Teeth of the Mammalia Educabilia. From the "Journal of the Academy of Natural Sciences," Philadelphia, March 30, 1874.

The object of this paper was to show that the various more or less complex types of molar teeth displayed by the Ungulate Mammalia are referable to modifications of a primitive quadituberculur type, from which they were supposed to have been descended. The histories and homologies of the carnivorous dentition were not included in this paper. In order to complete the subject, I have inserted brief notes of the conclusions I have since attained in this field: first, as to the type of inferior sectorial teeth, in 1875; and, second, as to the superior molars, in 1883.

At the end of the paper a similar comparison between the feet of the same type of mammals is made, and general conclusions reached in the following language: "I trust that I have made it sufficiently obvious that the primitive genera of this division of mammals [Mammalia Educabilia = Unguiculata and Ungulata sensu latâ] must have been bunodonts with pentadactyl plantigrade feet."

---

* Vol. ii, p. 167. † Vol. ii, p. 195. ‡ "Principles of Biology."
# See Ryder, "American Naturalist," January, 1879.

The nearest approaches to a similar anticipation on the part of other naturalists, which I have been able to find, refer to the number of toes only, and are of restricted application. Thus Kowalevsky remarks ("Monographie der Gattung Anthracotherium, Palaeontographica," xxii, p. 1452): "So we can assume a tetradactyl foot as our point of departure, although it can not have the least effect on the result in case the original ungulate foot should have been pentadactyl. If I have set out with a tetradactyl foot it is simply because I wish to adhere so far as possible to facts." This was written August, 1873, but how soon thereafter it was printed I do not know. I did not meet with it until at least a year after the publication of my paper of March, 1874, cited. Secondly, Marsh, in writing on the genealogy of the horses ("American Journal of Science and Arts," March, 1874, p. 257), says: "A still older ancestor [of the horse], possibly in the Cretaceous, doubtless had five toes on each foot, the typical number in mammals." My paper was published during the same month as the above; but I communicated the substance of the generalization in question to the Philadelphia Academy the day it was read, November 18, 1873, which was published in the "Proceedings of the Society," January 13, 1874, p. 2.

XVIII. Consciousness in Evolution. From the "Penn Monthly Magazine," Philadelphia, July, 1875.

In this paper the doctrine of intelligent selection is analyzed, and the problem reduced to its essential—the relations of consciousness to matter. The doctrine of the origin of reflex and automatic acts from conscious states, or archaesthetism, is here first proposed. From the characters of protoplasm the inference is derived that that substance is not necessarily the only one capable of supporting consciousness. The author is not aware of any previous attempt to render these propositions probable.

VIII. The Relation of Man to the Tertiary Mammalia. From the "Penn Monthly Magazine," December, 1875. Read before the American Association for the Advancement of Science, at Detroit.

The fact that the hard tissues, and probably the digestive system, of man, are constructed on the type of the Mammalia of the early Eocene period is here pointed out for the first time.

III. The Theory of Evolution. Remarks made before the Academy of Natural Sciences of Philadelphia, February 22, 1876, and published in the "Proceedings of the Society," 1876, p. 15.

These remarks exhibit the correspondence between the evolutionary systems of Haeckel and of the writer, and combine them into a symmetrical whole.

XXI. The Origin of the Will. From the "Penn Monthly Magazine," Philadelphia, June, 1877.

In this paper the attempt is made to render the existence of freedom of will probable by a process of argument, and also to demonstrate its existence by another kind of argument. As is well known, there are two opposed doctrines respecting this important question. One of these, which has by far the larger number of adherents, is that the human mind embraces among its powers a freedom of will, or spontaneity in action. The other view is, that there is no such power, but that the actions are merely the necessary result of the strongest pressure of the strongest inducement or motive. The doctrine of evolution is known to lend support to the latter doctrine. In the present paper the attempt is made to prove that free will is a new power which supervenes on the process of evolution. This is done by assuming the existence of a class of acts for which the term *altruistic* is retained, which, undoubtedly, would require freedom for their performance. The only question here is, whether there be any such class of acts as are defined under the above name in this essay. If there be no such class of acts, the demonstration based on it falls to the ground; and the author is not at present sure whether there be such a class of acts or not.

The argument by which freedom of will is rendered probable is not open to any serious objection, and rests on the necessity for action which sometimes arises in cases where there is no experience or knowledge to serve as a determining motive or power. These cases are supposed to involve moral questions. The doctrine is then intermediate between the two opposing ones which have long divided the world of thought.* It permits of the development of a free will in previously automatic beings, as a phenomenon superposed on mental evolution. The argument demonstrates nothing more than that freedom is possible; a conclusion which is, however, important, since it shows that the position of the determinists is not impregnable.

XII. The Relation of Animal Motion to Animal Evolution. From "The American Naturalist," January, 1878. Read before the American Association for the Advancement of Science, at Nashville, August, 1877.

The effects of the actions of animals on their structure are considered, as in previous papers, and especial attention is paid to the influence of animals in the changes they produce in their environment.†

XIV. The Origin of the Specialized Teeth of the Carnivora. From "The American Naturalist," March, 1879.

The mechanical reasons for the changes in the dentition of carnivorous mammals during geologic time are pointed out for the first time.

VI. A Review of the Modern Doctrine of Evolution. From "The American Naturalist," March and April, 1880. A lecture delivered before the California Academy of Sciences, October 27, 1879. It is a general synopsis of views presented in preceding papers.

XV. On the Origin of the Foot-Structure of the Ungulates. From "The American Naturalist," April, 1881.

This essay embraces an explanation of the cause of the diminution of the number of digits in the diplarthrous ungulate mammals; of the reasons why some are even- and some odd-toed; and of the origin of some of the

---

* This view is adopted by President Noah Porter in one of his latest metaphysical works.

† A synopsis of the relations of animals to their environment is given by Spencer, "Principles of Biology," vol. i, p. 466.

articulations of the tarsus. These explanations, so far as based on paleontological grounds, were new at the time.

XVI. The Effect of Impact and Strains on the Feet of Mammalia. From "The American Naturalist," July, 1881.

The origin of the structures of all the articulations of the limbs of all the Mammalia are explained as the effects of impacts and strains. The demonstration is based largely on paleontological evidence, and is new.

XIX. On Archaesthetism. From "The American Naturalist," June, 1882.

This doctrine is discussed and illustrated on the basis laid down in the essay "Consciousness in Evolution," and a classification of theories of creation is presented.

IX. The Developmental Significance of Human Physiognomy. From "The American Naturalist," June, 1883.

In this paper suggestions presented in Art. IV, entitled "The Hypothesis of Evolution, Physical and Metaphysical," are developed, and the physical significance of the form-characters of men in general are considered. Nothing except reference to a few leading points of the subject had been published prior to this paper, so far as the author has been able to discover. In connection with a paper on the "Evolutionary Significance of Human Character," a foundation was laid for a scientific physiognomy.

X. The Evidence for Evolution in the History of the Extinct Mammalia. A lecture delivered before the American Association for the Advancement of Science, at Minneapolis, August, 1883.

In this paper are collected the evidences of descent displayed by the Mammalia (and in one instance by the Batrachia), as derived from the paleontological researches of the author. These had been in some points foreshadowed in the author's memoir on the homologies and origin of the structure of the molar teeth in the Mammalia Educabilia in 1874, which were here shown to have been realized by subsequent discovery, and a number of other evidences added. Restatements of the laws of kinetogenesis, and of the origin of morals, were made.

XVII. The Evolutionary Significance of Human Character. From the "American Naturalist," September, 1883.

The characters of the adult mind are compared with those of the child, and with those of the lower animals, and the direction of their evolution pointed out. Most of the propositions contained in this paper were new at the time of its publication.

XIII. The Trituberculae Type of Molar Teeth in the Mammalia. From the proceedings of the American Philosophical Society, 1883, page 324;* published in advance in the "Paleontological Bulletin," No. 37, January 2, 1884.

The origin of the ungulate molar tooth had already been traced to a quadritubercular type in paper No. VII (March and January, 1874). The

* Not published until 1884.

present paper shows that this type in the upper jaw is a derivative of a tritubercular type, while that of the lower jaw is a derivative of a quinquetubercular type, or a tritubercular type with a heel, which may support two additional tubercles. The tritubercular type was again traced to the simple cone. This generalization was new at the time of publication.

XX. Catagenesis. Vice-presidential address delivered before the biological section of the American Association for the Advancement of Science, Philadelphia, September 4, 1884.

The hypothesis of catagenesis which is put forth in this paper teaches that primitive energy was and is conscious, and that all unconscious forms of energy, whether "vital" or non-vital, have been derived from it by a process of retrograde metamorphosis. The first stage of this retrogression is the loss of consciousness, or cryptopnoÿ. Evidence for this kind of metamorphosis is derived from every-day experience, and from the designed character of automatic acts. That a form of energy is conscious is inferred from the nature of designed conscious acts of animals. The author had not met with any scientific statement of this theory prior to the preparation of this lecture.

XI. The Evolution of the Vertebrata, Progressive and Retrogressive. From the "American Naturalist," February, March, and April, 1885.

This paper sets forth the results of paleontological investigation of the Vertebrata, in a series of phylogenies. These are, first, the phylogeny of the classes; then the special phylogenies of their contents or of the orders. Here are introduced the newly-discovered relations of the Antiarcha and the Ichthyotomi to the fish-like vertebrates, and of the theromorphous reptiles to the Mammalia. Also, the ancestral relation of the Theromorpha to most other reptiles, and of the Condylarthra to the placental Mammalia, and to man. Many of the other results set forth in this essay are derived from the paleontological researches of the author. Some of them, especially the lemurine (condylarthrous) ancestry of the placental Mammalia, had been anticipated on theoretical grounds by Haeckel in his "History of Creation" ("Schöpfungsgeschichte"). Haeckel was very general in his proposition, and did not anticipate the details of the demonstration. My investigations enabled me to produce these, which bring out in a striking manner the sagacity of Prof. Haeckel. I consider further the question of degeneracy, and the significance of the phylogeny with reference to this subject is pointed out.

In conclusion, it may be said that the principal object which the author set before him, in the studies here recorded, has been the discovery of the laws of variation, or of the "Origin of the Fittest." These essays express the light which he has been able to obtain on this difficult question up to the present time. The results could be better and more briefly presented in a systematic form, but the author reserves this for a future occasion.

CONTENTS.

PART I.—GENERAL EVOLUTION.

PART II.—STRUCTURAL EVIDENCE OF EVOLUTION.

PART III.—MECHANICAL EVOLUTION.

PART IV.—METAPHYSICAL EVOLUTION.

LIST OF ILLUSTRATIONS.

Plates.

Wood-Cuts.

PART I.

GENERAL EVOLUTION.

I.

EVOLUTION AND ITS CONSEQUENCES.

THE broad theory of evolution includes the theories of development of the solar system and its members, as expressed by the nebular hypothesis; the theory of development of life by molecular movements consequent upon certain combinations of non-living matter; and, lastly, the theory of development of the species of animals and plants by descent, the later from the earlier, with accompanying change of form and character. It attempts nothing less than a history of the process of creation of the universe, so far as we can behold it; and is, therefore, an attempt to formulate the plans and thoughts of the Author of that universe. Hence, it is not surprising that it excites the interest of the best of men, especially as it is one of the results of the efforts of a class of these, crowning many centuries of labor and thought.

The object of the present essay is to discuss familiarly the latter of these theories of evolution, viz., that respecting the species of animals and plants. As all are aware, this mode of accounting for the creation of organized beings has attained especial prominence at the present time, and possesses more of interest to most readers because of its explanation of our own origin. Moreover, it rests on more indubitable evidence than the two other theories. The constitution and arrangement of the members of the solar system point to their origin by derivation from primal masses of vaporous matter through the mutual operation of the ordinary laws of attraction and motion. The *positioning* is precisely as it should be had such process taken place, but the process itself, that is, the change from type to type of celestial body, has not been observed.

The case is far different with the theory as regards organized or living beings. Not only are the mutual relations of animals and plants to each other such as should have resulted from a descent or development, but the changes from type to type have been actually observed, and in sufficient number to place the hypothesis on the basis of ascertained fact, as referring to a certain range of objects—say, in the case of the animal kingdom, to individuals distinguished by structural characters within the range of each of the three to six great primary divisions or "branches."

There are two totally distinct propositions involved in this question, which are confounded by the general public, and not unfrequently by students and writers on it. These are, first, the evidence which seems to prove that this evolution has taken place; secondly, the evidence as to the nature of the laws of its progress. A want of constant distinction between these views of the case has greatly obscured it and injured the evidence on one side or the other.

The evidence in favor of evolution is abundant, and is cited in fragments by various contemporary writers, foremost among whom, both in time and abundance of writings, comes Charles Darwin.

Much less has been done in explanation of the laws of evolution. Darwin and his immediate followers have brought out the law of "natural selection"; Spencer has endeavored to express them in terms of force; while Hyatt, Cope, Packard, and others have advanced the law of "acceleration and retardation."

In earlier days, when information was distributed slowly and books were few, it was long before any new truth or doctrine reached the majority of people, still less was adopted by the ruling classes. But the modern theory of evolution has been spread everywhere with unexampled rapidity, thanks to our means of printing and transportation. It has met with remarkably rapid acceptance by those best qualified to judge of its merits, viz., the zoölogists and botanists, while probably a majority of the public, in this region at least, profess to reject it. This inconsistency is due to two principal causes. In the first place, Darwin's demonstration contained in the "Origin of Species" extends little further than as stated in the title of his work. He proves little more than that species of the same genus or other restricted groups have had a common origin; and, further, his theory of natural selection is to the plainest understanding incomplete as an explanation of their "origin," as its author indeed freely allows. Besides, the unscientific world is particularly unreasonable on one point. Little knowing the slow steps and laborious effort by which any general truth is reached, they find in incompleteness ground of condemnation of the whole. Science is glad if she can prove that the earth stands on an elephant, and gladder if she can demonstrate that the elephant stands on a turtle; but, if she can not show the support of the turtle, she is not discouraged, but labors patiently, trusting that the future of discovery will justify the experience of the past.

If, then, some of the people find Mr. Darwin's argument incomplete, or in some points weak, it may be answered, so do the student classes, who, nevertheless, believe it. This is largely because Darwin's facts and thoughts repeat a vast multiplicity of experiences of every student, which are of as much significance as those cited by him, and which only required a courageous officer to marshal them into line, a mighty host, conquering and to conquer. These will slowly find their way into print, some in one country and some in another.

I. THE FACT OF EVOLUTION.

As to the truth of the theory, the proof has been stated in more than one form. The first and simplest, and essentially the central argument of Darwin, is as follows:

In every family or larger group of animals and plants there exists one or more genera in which the species present an aggregation of specific intensity of form; that is, that species become more and more closely related, and finally varieties of single species have to be admitted for the sake of obtaining a systematic definition or "diagnosis," which will apply to all the individuals. These varieties are frequently as well marked as the nearly-related species, so far as *amount* of difference is concerned, the distinction between the two cases being that in the varieties there is a gradation from one to the other; in the species, none. Nevertheless, between some of the varieties transitions may be of rare occurrence, and in the case of the "species" an intermediate individual or two may occasionally be found. Thus it is that differences, called varietal and specific, are distinguished by degree only, and not in kind, and are, therefore, the results of the operation of uniform laws. Yet, according to the old theory, the varieties have a common origin, and the species an independent one!

To find examples of what is asserted, it is only necessary to refer to the diagnostic tables and keys of the best and most honest zoölogists and botanists. It is true that these diagnoses are dry reading to the non-professional, yet they embrace nearly all that is of value in this part of biological science, and must be mastered in some department before the student is in possession of the means of forming an opinion. The neglect to do this explains why it is that, after all that has been written and said about *protean* species, etc., the subject should be so little understood.

It is true that in but few of these cases have the varieties been seen to be bred from common parents, a circumstance entirely owing to the difficulties of observation. The reasoning derived from the relations of differences appears to be conclusive as to their common origin, unless we are prepared to adopt the opposite view, that the varieties have originated separately. As these avowedly grade into individual variations, we must at once be led to believe that individuals have been created independently—a manifest absurdity.

But variations in the same brood have been found among wild animals; for example, both the red and gray varieties of the little horned owl (*Scops asio*) have been taken from the same nest.

As further examples of gradation between species and variety, found in nature, I only have to select those genera most numerous in species, and *best studied*. Among birds, *Corvus, Empidonax, Buteo, Falco*, etc. Reptiles, *Eutaenia, Anolis, Lycodon, Naja, Caudisona, Elaps, Oxyhrropus*, etc. Batrachia, *Rana, Hyla, Chorophilus, Borborocoetes, Amblystoma, Spelerpes*, etc. Fishes, *Ptychostomus, Plecostomus, Amiurus, Salmo, Perca*, and many others.

In all these groups of species, or "genera," it is impossible in some cases to determine what is variety and what species. This is notoriously the case with the salmon and trout (*Salmo*), for one of the greatest opponents of close division of species, Dr. Günther, of London, thought himself necessitated, a very few years ago, to name and describe half a dozen new species of trout from the lakes of the British Islands, and, from being a stanch supporter of the old view of distinct creations, was completely converted to evolutionism.

Such is one of the views which has forced conviction on the minds of thoroughly honest men who were not only desirous of knowing the truth, but were in many cases brought over from a position of strong opposition. But the earnest objector says, you have not after all shown me any real transitions from species to species; until that is done your development is but a supposition.

The all-sufficient answer to this statement is to be found in the imperfection of our system of classification. Thus, if we first assume, with the anti-developmentalist, that varieties have a common parentage, and species distinct ones, when intermediate forms connecting so-called species are discovered, we must confess ourselves in error, and admit that the forms supposed to have had a different origin really had a common one. Such intermediate forms really establish the connection between species, but the question is begged at once by asserting unity of species, and, therefore, of origin, so soon as the intermediate form is found; for, as before observed, it is not degree, but constancy of distinction, which establishes the species of the zoölogical systems. Transitions between species are constantly discovered in existing animals; when numerous in individuals, the more diverse forms are regarded as "aberrant"; when few, the extremes become "varieties," and it is only necessary to destroy the annectant forms altogether to leave two or more species. As the whole of a variable species generally has wide geographical range, the varieties coinciding with sub-areas, the submergence, or other change in the intervening surface, would destroy connecting forms, and naturally produce the isolated species.

Formerly naturalists sometimes did this in their studies. A zoölogist known to fame once pointed out to me some troublesome specimens which set his attempts at definition of certain species at defiance. "These," said he, "are the kind that I throw out of the window." Naturalists having abandoned throwing puzzling forms out of the window, the result of more honest study is a belief in evolution by nine tenths of them.

But, says the inquirer again, your variations and transitions are but a drop in the ocean of well-distinguished species, classes, etc. The permanent distinction of species is matter of every-day observation; your examples of changes are few and far between, and utterly insufficient for your purpose.

It is true that the cases of transition, intermediate forms, or diversity in the brood, observed and cited by naturalists in proof of evolution, are few compared with the number of well-defined, isolated species, genera, etc., known; though far more numerous than the book-student of natural history is apt to discover. But although the origin of most species by descent has not been observed, every one knows the worthlessness of argument based on a negative. Unless these cases exhibit opposing evidence of a positive character, they are absolutely silent witnesses.

He who cites them against evolution commits the error of the native of the Green Isle who testified at a murder trial. "Although the prosecuting attorney brought three witnesses to swear positively that they saw the murder committed, I could produce *thirty* who swore they did not see it done!"

By the inductive process of reasoning we transfer the unknown to the known, for it is the key of knowledge. It rests upon the invariability of Nature's operations under identical circumstances, and for its application merely demands that analysis and comparison shall fix that the nature of that of which something is unknown is identical with that of which the same thing is known. We then with certainty refer that which is known as an attribute of that object of which the same quality had been previously unknown. The following form exhibits its application to the question of evolution. As preliminary facts it may be assumed that:

1. Many species are composed of identical elemental parts which present minor differences.

2. Some of these differences have been seen to originate spontaneously from parents which did not possess them, or, what is the same thing, are known to exist in individuals whose parentage is identical with others which do not possess them.

3. The gradation of differences of the same elemental parts is one of degree only, and not of kind.

4. *Induction.*—Therefore all such differences have originated by a modification in growth, or have made their appearance without transmission in descent.

II. THE MANNER OF EVOLUTION.

In discussing this point, new evidence in favor of development must be produced, and some statements of the history of the opinion made.

The laws which are expressed by all that we find of structure in animals are four, viz.:

1. *Homology.*—This means that animals are composed of corresponding parts; that the variations of an original and fixed number of elements constitute their only differences. A part

large in one animal may be small in another, or *vice versa;* or complex in one and simple in another. The analysis of animals with skeletons or vertebrata has yielded several hundred original elements, out of which the 28,000 included species are constructed. Different this from the inorganic world, which can only claim about sixty-two elemental substances. The study of homologies is thus an extended one, and is far from complete at the present day.

2. *Successional Relation.*—This expresses the fact that species naturally arrange themselves into series in consequence of a mathematical order of excess and deficiency in some feature or features. Thus species with three toes naturally intervene between those with one and four toes. So with the number of chambers of the heart, of segments of the body, the skeleton, etc. There are greater series and lesser series, and mistakes are easily made by taking the one for the other.

3. *Parallelism.*—This states that while all animals in their embryonic and later growth pass through a number of stages and conditions, some traverse more and others traverse fewer stages; and that, as the stages are nearly the same for both, those which accomplish less resemble or are *parallel* with the young of those which accomplish more. This is the broad statement, and is qualified by the details.

4. *Teleology.*—This is the law of adaptation so much dwelt upon by the old writers, and admired in its exhibitions by men generally. It includes the many cases of fitness of a structure for its special use, and expresses broadly the general adaptations of an animal to its home and habits.

Of course, these laws must be all laws of evolution, if evolution be true. And such they are; but this is far from being perceived by some students, for some of them were in abeyance or neglect prior to the stimulus to thought caused by the appearance of the "Origin of Species."

Forty or fifty years ago Germany had been flooded with the writings of the "physiophilosophs." Oken and Goethe had obtained glimpses of the wonderful "unity in variety" expressed by the laws of homology. The latter saw vertebræ in the segments of the skull, and leaves in the floral organs of plants. He had found the magic wand, and many were the harmonious visions that delighted the laborious toilers among old bones and dusty skins; the patient haruspices saw omens in the intestines of birds and snakes, and he whose hours were spent over his lens ceased to be a mere wondermonger. But fashion is fashion, and always ends in absurdity and stagnation. The physiophilosophs became extravagant, and mistook superficial appearances for realities. They did not dream how misleading some of the resemblances between different elements, for example, of the skeleton may be, and for once German students did not analyze exhaustively. Cuvier laughed at these seekers for beauty, and confounded the true and the untrue in one condemnation. But the best men labored forward; errors began to be exposed, and soon a reaction set in. Another extreme followed, and the school of Müller, at Berlin, denied the meaning of these resemblances and ceased to see anything but differences. Minute and thorough investigation flourished in their hands, and the modern school of German anatomists has seen no superiors. So the theory of evolution found Berlin. The disfavor in which physiophilosophy was held secured to evolution a cold welcome, and it has been for Jena and other universities to give it its true impetus in Germany.

So it has been with the law of parallelism. Some of the physiophilosophs declared it, stating that the inferior animals were merely the repressed conditions of the higher. This view was taught by some men in high position in France. Their statements were, however, too broad and uncritical. The father of embryology, von Baer, of Koenigsberg, declared there was "*keine Rede*" of such theory, and Lereboullet stated "that it is founded on false and deceptive appearances." Even Professor Agassiz in our day has asserted that no embryonic animal is ever the same as the adult of another, though he also once informed the writer that the embryology of two nearly related species had never been studied and compared. This was subsequently done by Professor Hyatt, of Salem, for the nautilus and ammonite division of mollusks, and at about the same time by the writer, for many species of our native frogs and salamanders, and the result has been a complete clearing up of the confusion about parallelism, and the clear establishment of the law.

The results attained are these: The smaller the number of structural characters which separate the two species when adult, the more nearly will the less complete of the series be identical with an incomplete stage of the higher species. As we compare species which are more and more different, the more necessarily must we confine the assertion of parallelism to single parts of the animals, and less to the whole animal. When we reach species as far removed as man and a shark, which are separated by the extent of the series of vertebrated animals, we can only say that the infant man is identical in its numerous origins of the arteries from the heart, and in the cartilaginous skeletal tissue, with the class of sharks, and in but few other respects. But the importance of this consideration must be seen from the fact that it is *on single characters of this kind that the divisions of the zoölogist depend.* Hence we can say truly that one order is identical with an incomplete stage of another order, though the species of the one may never at the present time bear the same relation in their entirety to the species of the other. Still more frequently can we say that such a genus is the same in character as a stage passed by the next higher genus ; but when we can say this of species, then their distinction is almost gone. It will then depend on the opinion of the naturalist as to whether the repressed characters are permanent or not. Parallelism is then reduced to this definition : that each separate character of every kind, which we find in a species, represents a more or less complete stage of the fullest growth of which the character appears to be capable. In proportion as those characters in one species are contrasted with those of another by reason of their number, by so much must we confine our comparison to the characters alone, and the divisions they represent ; but when the contrast is reduced by reason of the fewness of differing characters, so much the more truly can we say that the one species is really a suppressed or incomplete form of the other. The denial of this principle by the authorities cited has been in consequence of this relation having been assigned to orders and classes, when the statement should have been confined *to single characters;* and divisions characterized by them. There seems, however, to have been a want of exercise of the classifying quality or power of "abstraction" of the mind on the part of the objectors. This faculty seems to be by no means so common as one would expect, judging from the systematic ideas of many.

To explain by a few examples selected at random : First, of species characters, I may cite the fact that all deer are spotted when young, and that some of the species of eastern and southern Asia retain the spotted coloration throughout life. All salamanders are uniform, often olive during a larval stage ; some species, and some individuals of other species, retain the color in maturity. To take a genus character : all the deer in the second year develop their first horn, which is unbranched and small, or a "spike." A genus of deer inhabiting South America never develops anything else. To take a character of higher grade: the exogenous plants usually present net-veined leaves, but the first pair, or those of the plumule, are of much simpler structure, being often parallel-veined; for example, the cucumbers and squashes. Now, the endogens usually produce nothing else than parallel-veined leaves, and no case is known where a plant bearing this type of leaf exhibits the net-veined type as its earliest growth.

But what do these facts mean? As in growth the genus characters usually appear last, I will suppose a case where one genus represents truly, or is identical with, the incomplete stage of another one.

In *A* we have four species whose growth attains a given point, a certain number of stages having been passed prior to its termination, or maturity. In *B* we have another series of four (the number a matter of no importance), which, during the period of growth, can not be distinguished by any common, i. e., generic character, from the individuals of group *A*, but whose growth has only attained to a point short of that reached by those of group *A* at maturity.

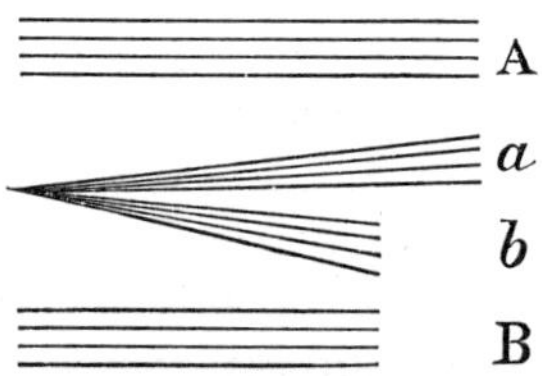

FIG. 1.

Here we have a parallelism, but no true evidence of descent. But if we now find a set of individuals belonging to one species (or, still better, the individuals of a single brood), and therefore held to have had a common origin or parentage, which present differences among themselves of the character in question, we have gained a point. We know in this case that the individuals, *a*, have attained to the completeness of character presented by group *A*, while others, *b*, of the same parentage have only attained to the structure of those of group *B*. It is perfectly obvious that the individuals of the first part of the family have grown further, and, therefore, in one sense faster, than those of group *b*. If the parents were like the individuals of the more completely grown, then the offspring which did not attain that completeness may be said to have been *retarded* in their development. If, on the other hand, the parents were like those less fully grown, then the offspring which have added something have been *accelerated* in their development. I claim that a consideration of the uniformity of nature's processes, or inductive reasoning, requires me to believe that the groups of species, that is, groups *A* and *B*, are also derived from common parents, and the more advanced have been *accelerated* or the less advanced *retarded*, as the case may have been with regard to the parents.

This is not an imaginary case, but a true representation of many cases which have come under notice. I can not repeat them here, but refer to the original memoirs, where they may be found.*

This is a simple statement of the law of "acceleration and retardation" of some American naturalists, which probably expresses better than any other the "manner of evolution," the proposition with which we started.

Hyatt thus defines it as seen in a group of ammonites which he studied: "The young of higher species are thus constantly accelerating their development, and reducing to a more and more embryonic condition the stages of growth corresponding to the adult periods of preceding or lower species." †

This form of demonstration of evolution is of far wider application than that which I first brought forward. ‡ In the latter case the induction may be limited to a certain range of variation, but the present law is as extensive as the organic world; that is, the "positioning" essential to it is found everywhere, from the lowest to the highest, and in characters from the least to the greatest in import.

Let an application be made to the origin of the human species. It is scarcely necessary to point out at the start the fact, universally admitted by anatomists, that man and monkeys belong to the same order of Mammalia, and differ in those minor characters, generally used to define a "family" in zoölogy.

Now, these differences are as follows: In man we have the large head with prominent forehead and short jaws; short canine teeth without interruption behind (above); short arms, and thumb

* See "Origin of Genera, and Method of Creation," Naturalists' Agency, Salem, Massachusetts; or McCalla & Stavely, 237 Dock Street, Philadelphia.

† "On the Parallelism between Stages in the Individual and those in the Group of the Tetrabranchiata." "Boston Society of Natural History," 4to, 1866, p. 203.

‡ It is quite misunderstood by Darwin, as will be sufficiently evident from the following quotation from the last edition of his "Origin of Species," 1872, p. 149: "There is another possible mode of transition, namely, through the acceleration or retardation of the period of reproduction. This has lately been insisted on by Prof. Cope and others in the United States." This has only been dwelt on as accounting for a very minor grade of differences seen in race and sex.

of hind foot not opposable. In monkeys we have the reverse of all these characters. But what do we see in young monkeys? A head and brain as large relatively to the body as in man; a facial angle quite as large as in many men, with jaws not more prominent than in some races; the arms not longer than in the long-armed races of men, that is, a little beyond half way along the femur. These observations are made on a half-grown *Cebus apella,* from Brazil, a member of a group more remote from men than are the Old World apes, yet with an unusually large facial angle. At this age of the individual the distinctive characters are therefore those of *homo,* with the exception of the opposable thumb of the hind foot, and the longer canine tooth; nevertheless, the canine tooth is shorter in the young than in the adult.

Now, in the light of various cases observed, where members of the same species or brood are found at adult age to differ in the number of immature characters they possess, we may conclude that man originated in the following way: that is, by a delay or retardation in growth of the body and fore limbs as compared with the head; retardation of the jaws as compared with the brain case, and retardation in the protrusion of the canine teeth. The precise process as regards the hinder thumb remains obscure, but it is probably a very simple matter. The proportions of the young *Cebus apella* enable it to walk on the hind limbs with great facility, and it does so much more frequently than an adult *C. capucinus* with which it is confined.*

The "retardation" in the growth of the jaws still progresses. Some of our dentists have observed that the last (3d) molar teeth (wisdom teeth) are in natives of the United States very liable to imperfect growth or suppression, and to a degree entirely unknown among savage or even many civilized races. The same suppression has been observed in the outer pair of superior incisors. This is not only owing to a reduction in the size of the arches of the jaws, but to successively prolonged delay in the appearance of the teeth. In the same way men, and the man-like apes, have fewer teeth than the lower monkeys, and these again fewer than the ordinary Mammalia, and this reduction has proceeded in relation to an enlargement of the upper part of the head and of the brain.

The *cause* of development may be next considered, and under

---

* The same relations of man to the anthropoid apes have been dwelt upon by Prof. C. Vogt.

this head may be discussed the natural selection of Wallace and Darwin and other propositions of similar import.

"Retardation" continued terminates in extinction. Examples of this result are common; among the best known are those of the atrophy of the organs of sight in animals inhabiting caves. It is asserted that the young of both the blind crawfish (*Orconectes pellucidus*) and the lesser blind fish (*Typhlichthys subterraneus*) of the Mammoth Cave possess eyes. If these statements be accurate, we have here an example of what is known to occur elsewhere, for instance, in the whalebone whales. In a fœtal stage these animals possess rudimental teeth like those of many other Cetacea when adult, which are subsequently absorbed. So also with the fœtal ox; the upper incisor teeth appear in a rudimental condition, but are very early removed. The disappearance of the eyes is regarded by Dr. Packard, with reason, as evidence of the descent of the blind forms from those with visual organs. I would suggest that the process of reduction illustrates the law of "retardation" accompanied by another phenomenon. Where characters which appear latest in embryonic history are lost, we have simple retardation, that is, the animal in successive generations fails to grow up to the highest point of completion, falling farther and farther back, thus presenting an increasingly slower growth in the special direction in question. Where, as in the presence of eyes, we have a character early assumed in embryonic life, retardation presents a somewhat different phase. Each successive generation, it is true, fails to come up to the completeness of its predecessor at maturity, and thus exhibits "retardation," but this process of reduction of rate of growth is followed by its termination in the part long before growth has ceased in other organs. This is an exaggeration of retardation, and means the early termination of the process of force-conversion, which has been previously diminishing steadily in activity.

Thus the eyes of the *Orconectes* probably exhibited for a time at maturity the incomplete character now found in the young, a retarded growth continuing to adult age, before the termination of growth was withdrawn by degrees to earlier stages. With this early termination of growth came the phase of atrophy, the incomplete organ being removed and its materials transferred to other parts through the greater activity of "growth-force." Thus, for the reduction of organs, we have "retardation"; but for their extinction, "retardation and atrophy."

III. ON THE CAUSES OF EVOLUTION.

1. *Inductive Reasoning.*

In the present investigation we are endeavoring to discover new principles, not to apply old ones. The work is similar to that which occupied Newton in his investigation of the law of attraction or gravitation. The process by which we arrive at general truths rests on the consideration of a sufficient number of observed facts, and the determined qualities which are common to all we regard as a law. This process requires for its proper conduct a careful analysis and discrimination of the nature of the objects considered; otherwise fallacy will result. As exact analysis is not always observed by the average mind, this inductive reasoning is not always successfully employed by it, nor understood when presented by others. In the deductive process it is more at home. With an ascertained principle given, as something like a staff for the mind to lean on, its application is not so difficult; but to such a mind induction presents an appearance of uncertainty and even of confusion, and these will certainly exist until order is evoked by the first step—classification. The theory of evolution has thus been charged with confusion, as though it asserted that which overthrew the order of nature. But the confusion only exists in the mind of such critics. The order of the creation is one of the foundation facts, and thus enters the inductive argument as one of its elements. That conclusion which is consistent with this order can not be regarded as its enemy.

2. *On Natural Selection.*

In endeavoring to assign a cause for the existence of the peculiar structures which define the divisions among animals, Messrs. Wallace and Darwin have proposed the now well-known law of natural selection. This states, that, inasmuch as slight variations appear continually in all species, it is evident that some will be more beneficial to the animal than others, in its exertions to supply itself with food, protect itself from enemies, the weather, etc. It then asserts that those whose peculiarities are beneficial will excel those less favored, in the successful use of their powers, and hence will live better, grow better, and increase more rapidly. That by the force of numbers, if not by direct conflict, they will ultimately supersede the weaker and destroy or drive them away.

Then, as there are many fields of action and possibilities of obtaining support in the world, that the weaker will first be driven to adopt such of these as their peculiarities may adapt them for, or not exclude them from. Thus all the positions in the world's economy are filled and the surplus destroyed. This is styled by Spencer the "survival of the fittest"; an expression both comprehensive and exact.

This doctrine is no doubt a true one, and has regulated the preservation of the variations of species, and assigned them their locations in the economy of nature. It was natural that this great law should have been brought out by such men as Darwin and Wallace, who are by nature much more of observers of life in the field, or out-door physiologists, than they are (or were) anatomists and embryologists. Their writings in their chosen field of the mutual relations of living beings in their search and struggle for means of existence are admirable, and almost unique, especially some of those of Darwin.

It is to be observed, however, that they both (especially Darwin) start with the variations observed. This is assumed at the outset, and necessarily so, for "selection" requires alternatives, and these are the product of variation. Great obscurity has arisen from the supposition that natural selection can *originate* anything, and the obscurity has not been lessened by the assertion often made that these variations are due to inheritance! What is inheritance but repetition of characters possessed by some (no matter what) ancestor; and if so, where did that ancestor obtain the peculiarity? The origin of variation is thus only thrown upon an earlier period.

Another reason why natural selection fails to account for the structures of many organic beings is the fact that in expressing "the survival of the fittest" it requires that the structures preserved should be especially useful to their possessors. Now, perhaps half of all the peculiarities of the parts of animals (and probably of plants) are of no use to their possessors, or not more useful to them than many other existing structures would have been. It fails to account for many characters which express the relations of *homology* and *parallelism,* and is almost confined in its exhibitions to features which express *teleology.* This objection has been insisted on by Kölliker, the writer, and by Mivart; and now Huxley, while defending Darwinism proper against the last-named author, says that "what the hypothesis of evolution wants is *a good theory of variation.*"

Plainly enough, then, nothing ever originated by natural selection, and as the present essay relates to the *origin* of types, little space can be given to its discussion; for natural selection, important though it be, is but half the question, and indeed the lesser half. It is to the great causative forces as are the gutters and channels which conduct the water in comparison with the pump and the man who pumps it.

3. *On Teleology.*

Two classes of structures have been alluded to: those which are useful to an animal, and those which are not useful; or the adaptive and non-adaptive. Nothing is better known than that animals are well adapted to their situations in the world, and for their needs as to supplying themselves with food, etc. Some part of every species is so constructed as to enable it to live under conditions where most other kinds of animals would perish. Thus the sea-rangers, among birds, as the great albatross, etc., possess long and pointed wings; while those that live in thickets or under cover have short, round ones, as grouse, woodcock, etc. Even our sparrows—those that love the bushes and swamps, as the song-sparrow (*Melospiza melodia*)—have short, rounded wings, while those that haunt trees have them sharper and pointed, as the chipping and tree sparrow (*Spizella socialis*, *pusilla*, etc.). Water-frogs have their feet webbed; land-frogs have small or no webs; while tree-toads possess sucker-like expansions of the ends of the toes, which secrete a glutinous fluid, by which they adhere to the trunks and leaves of trees. Finally, frogs that burrow have one or more of the bones of the base of the hind foot (*tarsus*) modified into a projecting blade, like that of a shovel; and, as they squat down, they literally sit into the ground, and are soon out of sight in the hole which they dig with these busy trowels.

Cave insects have long and delicate antennæ and limbs, exceeding those of their out-door relations by much. Moreover, their usual lack of eyes is a clear case of the reverse of adaptation, i. e., the absence of an organ where not needed.

Less attention has been directed to the non-adaptive characters, yet they are as numerous as the adaptive. I do not include under this head useless organs or parts only, but also those which are useful, but whose peculiarities do not relate to that use as advantageous to it.

Notable examples of this kind are to be found in the characters which distinguish all of the higher groups of animals and of plants among themselves. It is easier to ask than to answer what advantages the mammalian skeleton possesses over the reptilian that it should have superseded it. What end was served by aborting the coracöid bone, which in reptiles supports the shoulder-joint from behind, answering to the ischium of the pelvis? I do not know how to answer this question on a teleological basis, although it involves one of the principal characters of the class of reptiles. What mechanical end was gained by withdrawing the rib-bones of two cranial segments into the cavity of the ear, to become the hammer, anvil, and stirrup of the organ of hearing?* Was it to perfect the auditory faculty? Scarcely; for birds possess as refined and as musical an ear as any mammals, and appear to be superior to them in discriminating power, yet in them the hammer is the basal element of the lower jaw, and the anvil supports it, being entirely outside the cranial walls. See again one ground of distinction between reptiles and batrachians. The base of the brain-case in the former consists of an axis of bony segments developed in the primordial cartilage, while in the batrachian it is a single bony plate, formed by deposit in the membrane which originally bounded this cartilage. Who can assign any advantage of the one type above the other which can be looked upon as in any way related to the external needs of the animals of those classes?

Another example may be found in the ankle-joints of reptiles, birds, and mammals. In the first two the hinge is between the first and second series of tarsal bones; in the mammals, between the bones of the lower leg and the first row of tarsal bones. Something besides the superior mechanical advantages of the latter has given it predominance over the former.

To turn to the nautilus and ammonite types of Mollusca, we observe beautiful illustrations of all the laws already stated. As is well known, these shells have their tubular cavity divided by transverse partitions. In the nautilus these unite with the outer wall by a plain angle, but in various genera which lead toward the ammonites this margin becomes complicated. This results from an excessive growth of the peripheral part of the partition or septum, so that in order to confine it to the same space of contact it must be folded. This plication takes place in a symmetrical manner. The folded edge in *Aturia* forms a tongue-like loop on each

* The homologies here expressed have been rendered improbable by late investigation. The argument is, however, not affected.

side, on the outer surface. In *Goniatites* a fold is added on the back. In successive genera other main folds are added, the last appearing nearest the center of the coil. These then become complicated by subordinate undulations which in the more complex forms become the axes of a double row of new lobes and folds, the whole presenting symmetrical lobate outlines of much complexity and beauty.

But another series of changes accompanies those of the septa, and are entirely independent of them. These relate to the various degrees of winding of the shell. The early form with simple septa (*Orthoceras*) was straight, but others which succeeded began to turn their shell-making round an axis, thus commencing a partial coil. In some the coil was very open; in others it began close, and then ceased, the shell finishing straight. Others after such a course began to wind again, while some made a single turn near the middle. The most remote from the starting-point (*Ammonites*, sp.) made a tight and complete flat coil, while some, whose septa remained simple, did the same (*Nautilus*).

The direct uses of these various forms of septum and coil are simply inexplicable, and that one of them was any "fitter" to "survive" than another, by virtue of its usefulness, is for me more than doubtful.

I am tempted to continue this theme, for it might be prolonged indefinitely by any one familiar with the details of anatomical structure, but I will only repeat that the illustrations would be drawn chiefly from the characters of the classes, orders, families, and other higher groups.

It is not difficult to believe, in the case of the useful structures first cited, that the law of natural selection has had much, probably everything, to do with the preservation of the animals possessing them in the various localities to which they are adapted. But that it has had opportunity to direct the lines of progress in the second series is not likely. That it had nothing to do with the origin of either, is certain.

4. *On Growth-Force.*

Every change by complication of structure is by addition; every simplification is by subtraction. Every addition is a matter of growth, which is carried on by a process of nutrition. The inquiry respecting the origin of new forms centers itself at once on the history of growth and the influences affecting it.

If, as I suppose, these additions, either adaptive or non-adaptive, be produced by an *acceleration* * of growth, it is evident that the same immediate cause of that increase must be potent in both cases. That one of the "forces" is concerned in growth as well as in all the active animal and vegetable processes, is obvious to those who have carefully observed it. The fact that growth, like work, requires food for its progress and continuance, is reason enough for suspecting the existence of a force, and in some cases the relation between this force and other known forces may be measured.

Prof. Henry pointed out these facts many years ago, and illustrated them by observations on the growth of the potato and of the egg. The starch of the former, a complex "organic" chemical compound, weighs much more than the young shoot of cellulose, etc., into which it is converted by the process of growth, so that a portion of the substance of the tuber has evidently escaped in some other direction. This was found to be carbonic-acid gas and water, derived from the slow combustion of the starch, which, in thus "running down" from the complex organic state, to the more simple inorganic compounds, evolves an amount of force precisely equal in amount to the chemical force (chemism) requisite to bind together the elements in the new and complex substance cellulose.†

It is well known that substances differ in their capacities for giving out different kinds of force. This, of course, means their capacity for converting one kind of force into another. Thus, if glass be rubbed with silk, the motion is converted into electricity, while, if it be rubbed by the hand, heat is the principal result. In some cases chemical force, set free by decomposition, is converted into light; in others, heat; in others, to electricity, oftener to several at once. But one substance, so far as known, possesses the power of converting this chemical force or heat into growth-activity, and that is the material out of which the living parts of animals and plants are composed. This is a protein, a compound of carbon, oxygen, and nitrogen in the order of relative quantity, with a smaller proportion of hydrogen, the whole being often associated with still smaller quantities of sulphur

* For the definition of this term see first article, in May number of "Penn Monthly." (Antea, p. 11.)

† "Agricultural Report of Patent Office," 1857.

and phosphorus. In its mechanical aspect, as the material out of which structures and tissues are made, it is called protoplasm.

This substance exhibits two different phenomena of force-energy, viz., *motion* and *growth*. Motion is exhibited by contraction and expansion, but which is the active state and which the passive state is matter of question. Some physiologists regard "contractility" as the energetic state. Radcliffe believes that extension is the energetic condition, and contraction a rebound or discharge of the extending force. He finds dead protoplasm to be a dielectric, and believes that in life each muscular fibrilla acts like a Leyden jar. It has been demonstrated that the outer layer of the sheath of a muscular fibrilla is positively electrified in life, while the cut extremities are negative, and it is shown by Radcliffe that the inner side of the sheath becomes negatively electrified by induction. The attraction of the opposite electricities on opposite sides of the sheath compresses and elongates it, thus, according to his theory, producing muscular extension. The nervous cells and tubes he believes to act in the same way; the difference being that the walls and sheaths are in a state which prevents compression and extension. The phenomena of muscular extension and nervous tension he believes to be terminated by a discharge of the force, such as takes place in electrical fishes, but in much smaller quantity. Thus motive force resides as a form of electricity in protoplasm, and in highly organized animals is specialized into *neurism*.

As to growth-force, all its exhibitions may be reduced to cell-division, cell-nutrition, and cell-origin. Cell-division exhibits two prominent varieties. In both the cell nucleus first divides; in the ordinary mode, the cell-wall contracts at opposite points, forming approaching plications, which, when they meet, divide the cell. In the other mode or proliferation, the divisions of the nuclei approach the cell-walls, which bulge opposite to them, forming diverticula, which isolate themselves by opposing contractions at the base, which meet as before. The nature of the force thus displayed is as yet only speculative, and its demonstration will result largely from observation on cell-origin from homogeneous protoplasm. Radcliffe suggests an ingenious theory. He supposes that the protoplasm acquires an external layer differentiated from the internal mass by exposure to and contact with external substances, and that the electricity generated in the interior is distributed on the under surface of this stratum. That this induces opposite electricity on the external surface, which, as in the muscular cell, causes compression, and therefore extension of the stratum. This extends itself beyond its contents, which may be attenuated, filling the space, or contracting, forms the nucleus.

This view will not, however, account for nucleus and cell-division, which are the subsequent and principal exhibitions of growth-force. Perhaps the following suggestions may throw some light on it. Should a polar tendency appear in the nucleus, were the protoplasm of the proper viscidity, the poles being of like electric name, would tend to separate by repulsion, thus forming the hour-glass shape so well known, and afterward division. This nucleus division would inevitably be followed by division of the cell-wall, if its inner face were electrified in opposition to the nucleus. For the wall being attracted toward it equally all round, the separation of the two nuclei would be followed by an incurvature of the walls opposite the interval between the nuclei, in order that each should maintain a position equidistant from its center. Further divergence would result in the completion of the cell-wall of each, apart from the other. This process would be supported by nutrition of the cell, which proceeds by the passage through the cell-wall to the nucleus, of additional protoplasm from the blood. Of course, the prime question is as to whether polarity can appear in the nucleus. That growth is first polar appears probable on consideration of the globular and discoidal forms of the lower animals, and that this principle lies at the basis of the growth of the higher is rendered equally probable by the phenomena of symmetry of different kinds, as bilateral, anteroposterior, etc. The electricity supplied to the outer sheaths of muscular fibrillæ is, according to Dubois Reymond, positive. That covering the surface of the nucleus must be thought to be positive also, if it be effected in opposition to the inner wall of the cell-wall, which is known to be negative by experiment. But in that case it would be difficult to perceive why it should not fill the cell by attraction to the wall. If, on the other hand, it be a generator of negative electricity, it would retain its integrity and induce a temporary change in the wall. The transmission of protoplasm through the cell-wall to the nucleus would produce an accumulation of electricity to be discharged in extension, or a disturbance of equilibrium, to be followed by polarity and division; but the conditions under which these results would appear are, of course, unknown. The mechanical arrangement and condition of the parts would have much to do with it.

We are, however, here in the arcana of life, and the forms of law which rise before the mental vision are but as the statues of Memnon that greet the toiler on the river of mysteries ; they loom upon its banks in twilight, and the when, the how, and the wherefore remain unanswered. But the river of Africa is yielding her secrets, and, though the life that she nourishes may be the last she will give up, it is no less surely promised to the patient inquirer.

A great advance toward an explanation of the operation of this growth-force was made in the demonstration of the fact that its highest exhibitions are confined to the multiplication of cells by division of pre-existent ones, by contraction of their nuclei and walls in lines which finally meet.

The construction of additional parts consists solely of this cell-growth, but the character of the result is, of course, dependent on the position at which this addition takes place. It may be at the terminus of a limb to add another toe, or on the wall of an artery to add a valve. It may be in the brain to add a band of fibers, or on the edge of a muscle to extend its width.

That tissues are made of cells of original or altered shape, separate or confluent (flown together) is well known. That the arrangement of tissues into organs is due to the direction of this multiplication is also true. Thus a gland is a collection of follicles, each of which is a bagging of a plane tissue. This bagging is an exaggerated convexity, which is occasioned mechanically by excess of cell-growth at one spot on a uniform surface. Solid parts are all formed, in *the first instance*, of simple segments. These are parts produced in a straight rod by excessive elongation or growth ; the process being as before, the division of cells and distribution of homogeneous protein between them. All this is derived from the study of *homologies* combined with embryology, and the result is wonderful, and simplifies at once our ideas of the action of the growth-force. It is mainly a repetition of cell-division, the result as to structure being entirely dependent on the influences which *locate* its activity and regulate its amount.

Now, its amount will depend on the capacity of the existing organism to convert heat, etc., into it ; and the form, as to tissue, etc., in which it appears, depends no doubt on the complexity of the machines or organs of which the organism is composed. In the case of the higher organisms we have muscle, converting the results of nutrition into muscle ; skin and nails making more skin, etc. ; brain making more brain ; bone making more bone.

The prime question is the cause of the location of growth-force. Experiments on this point are greatly needed, and in their absence it will be necessary to take a wide survey of facts.

A given animal organism can only convert a given amount of force, and that capacity must remain uniform so long as the machine or structure remains the same. If, however, an addition to its work is developed in one quarter, a subtraction from some other region must take place if the whole amount remain the same. When, then, a useful organ is added, subtraction from some less important locality must result, and, as a consequence, the latter must become still less prominent in the general economy. Hence, the development of the useful class of organs already cited must always be accompanied by a corresponding disappearance of useless ones. This would be by reduction or retardation. But, in the case of the complex folds of the margins of the septa in the ammonite, useless parts are added by acceleration. Here a gradual increase in the amount of growth-force must be believed in.

The representation of simple growth-force, i. e., the result of bathmism, not specially located, is seen in general addition to size. In domestication careful feeding, associated with protection against all exhausting exposure or exertion, tends to this increase of size. This is well known to be the case in hogs and cattle and animals of quiet and easy life. The reverse conditions, as poor food, exposure, and *disproportionate* exertion, reduce the size ; witness the Maryland breed of cattle, and the semi-wild hogs of the Southern States. This matter is, however, nicely balanced with the production of motion, for if the latter be excessive, or the supply of material for consumption be insufficient, growth-force must be evolved in less amount.

It will be necessary here to state that the supply for the conversion of all the forces of the body is derived from the nutriment as circulated in the blood.

But the condition of plethora of growth-force is also the one highly favorable for the appearance of variations of structure, or the location of growth-force in new places. Domesticated animals are notorious for this variation, while the still more numerous examples of "protean" species in nature are always predominant forms, abundant in individuals and widely spread. For example, the wolf, the red fox, the red-tailed hawk, the garter-snake, the tiger, salamander, etc.

That the variability depends on a peculiar condition of the animals themselves, and not on domestication, excepting in so far as it produces these conditions, is plain, not only from the above facts, but from those observed in domestication. It is well known that while pigeons, fowls, cattle, dogs, etc., are variable, or "protean," the peafowl (*Pavo*) has maintained its specific characters with great accuracy during a period of domestication as long as that of the other species named. The same may be said of the guinea (*Numida*) and the turkey (*Meleagris*). These facts show that domestication is only a remote cause of variability.

5. *On the Location of Growth-Force.*

With the fact of increase of general growth-force before us, have we any others which can guide us in fixing on a cause for its special location? It is plain that the useful additions which have constituted certain genera, families, orders, etc., what they are, must have been produced as a consequence of the existence of a need for them; or, on the other hand, being created first, they must have sought for use, and found it. But what are the relative chances of truth for these two propositions? In the second case, admitting evolution as proved (see Part I), we perceive that an almost infinite chance exists against any usual amount of variation, as observed, producing a structure which shall be fit to survive in consequence of its superior adaptation to external circumstances. It would be incredible that a blind or undirected variation should not fail in a vast majority of instances to produce a single case of the beautiful adaptation to means and ends which we see so abundantly around us. The amount of attempt, failure, and consequent destruction, would be preposterously large, and in no wise consistent with the facts of teleology as we behold them.

What of the opposite view? We have in its support the well-known facts of determination of nutritive fluid during use of parts, not only for supplying fuel for the motions of use, but also for growth-force and material for the increase in bulk of the part used. Who has not remarked the large size of the hands of the laborer, and of the nails of the working-woman? Who can not remember some of the countless examples of certain modifications of form being associated with special excellence of use of the parts in different races of the same species of individuals of common parentage, showing that they must have grown with the history of those races? Who does not know the short, wide jaws of the bull-dog, with their oblique teeth, produced by the expansion of the zygomata to accommodate the huge temporal muscles so necessary for maintaining a firm hold of its enemy. Then the long and full nose of the hound, and its more extended turbinate bones—how closely is this connected with its developed scent; while the light muscular forms of the greyhound are undoubtedly necessary to its well-known speed. If it be said that these variations have not resulted in a single change of structure worthy of note by a systematist, we can point to the Japanese dog, where the excessive reduction of the jaws anteriorly has resulted in a total loss of some of the premolars and molars. The loss of molar and incisor teeth from shortening of the jaws, in the human species, has been already noted. The number of such instances is very great, but, as space to enumerate them fails, it is only necessary to add that they are characters of high importance in a systematic sense. Their importance will be more readily conceded in remembering the proposition, already stated, that species are simply isolated varieties, and of similar origin.

Confirmatory of this view are the facts already cited with reference to the relation of motive to growth-force. The force involved in both being seen to be similar, perhaps identical, the former represents its energetic state as discharge and motion, the other energetic without discharge, in the growth of cells. And whatever determines this force to a given part of the body must then probably result in both of its exhibitions, dependent on the kind of cells which receive it. As above remarked, the conditions which determine the result are unknown, except that the two kinds of muscular cells are the only ones in which elongation is strongly marked.

Another reason for believing in use as a cause of structural change is the manner in which the same useful structures have evidently appeared on totally distinct stems, as an evident adaptation to the same circumstances in which the different types have been equally placed. Thus the birds of prey possess the hooked, often toothed, beak, appropriate for tearing and destroying animals. Their stock is the same as that of the cuckoos and parrots, and even of the pigeons. The butcher-birds are of the division of songsters, not widely removed from the thrushes, and far enough from the raptores, yet the same hook and dentate bill reappears in them, as adapted to flesh-eating habits.

Among the reptiles, which were no doubt originally land animals, and derived from batrachians, we have a large number adapted to swimming in the ocean, and these not all of the same stock. Thus the *Plesiosauri* are crocodilian in relationship, while the *Pythonomorpha* of America were of the same that produced the snake and lizard. Again, the same modification appears in the Cetaceans, or whale, etc., among mammals which are primarily a terrestrial division.

It is a nice point of *phylogeny* (or the science of genealogy) to ascertain whether adaptive or strictly "bathmic" (or embryonic grade) characters came first in a time in a given group. Among frogs we have four divisions. One has an embryonic mouth and embryonic breast-bones; another, embryonic mouth and complete breast-bones; the third, embryonic breast-bone but complete mouth; and the fourth, with both complete. The first is the lowest and probably the oldest in time; agreeably to this supposition, it is distributed over the whole earth. The second is East Indian and African; the third and most extensive is Australian and American; while the fourth is confined to the Old World and North America. The second, third, and fourth divisions possess corresponding series of genera of different structure adapted to different modes of life. Some have shovels for burrowing, some webs for swimming, and some palettes for adhering to branches and leaves of trees. If these characters were first fixed, then those with the more perfect breast-bone and mouth are descendants of those with the less perfect; if, on the other hand, the conditions of mouth and breast-bone were first fixed, then each division thus defined in its special region was modified into the subdivisions, each adapted to a special mode of life. I have called these parallel divisions "homologous groups," and probably the origin of the embryonic modifications has sometimes come first, and sometimes the adaptive structures have preceded. Nevertheless, the lowest or most embryonic division will often have developed its own adaptive divisions, and each of these will agree in producing descendants which have advanced in the embryonic scale, and so produce homologous groups.

For these and other reasons it is concluded that the useful characters, defining natural divisions of animals, have been produced by the special "location of growth-force" by use. Useless ones have been produced by location of growth-force without the influence of use, or by its subtraction, due to a disturbance of equilibrium, consequent on the special location elsewhere.

But we go back to the origin of the question in investigating the action of growth-force undisturbed by the interference of locating influences. This is only to be ascertained by an examination of lower organisms in connection with the higher. A point that first strikes the student of higher animals is, that after he has proved the law of *homology* to be true, as regards different animals when compared, he further finds that the parts of the same animal are also homologous, that is, right hand and left, front and behind, bottom and top are also constructed on the same plan, so as to be reduced to the same elemental parts. This is described under the names of "anteroposterior" and "bilateral symmetry," etc.* These laws apply to all parts of the organism, and, though there appear to be exceptions, these are traceable to a common symmetrical type, the change wrought being one form of specialization. It is not surprising then that the lower we descend in the scale of life the less marked is the departure from the entirely symmetrical disposition of parts as regards the common center. This is beautifully seen in the *Radiata*, etc., where the jelly-fish, the star-fish, and the sea-urchin are familiar examples.

But these are complex compared with forms below them. The minute *Polycystina* construct shells of beautiful sculpture or tracery; but, though they present many forms, they may be all referred as modifications, to the symmetrically-marked disk. Of cellular animals, *Trichodiscus* represents a discoidal aggregate, and *Actinophrys* a globular mass of cells. *Monas* and its allies consist of one or several cells in globular form.

Lastly, in the first appearance of life in the most complex animal we have first a cell, which, then "segmenting," divides itself so as to become a globular mass of identical cells, arranged about the center.

From the cellular animal upward the operation of addition of parts is then one of *repetition*, and that centrifugally in the case of bodies which are developed in all planes, or peripherally in those which are in one plane.

Thus the complication of the edges of the septa of the ammonites is accounted for. First, the growing animal was supplied

* See Prof. B. Wilder, in Proceed. Boston Society Nat. Hist.

with an excess of growth-force. The type of the preceding generation being strictly adhered to according to the laws of inheritance, the only opportunity for its exhibition was necessarily *after* the inherited parts were completed by acceleration. The results could only, therefore, be exhibited on the edges of the new septa and free margin of the outer wall. As the older growths mechanically restrained the new septa from mere extension, folding or plication must have resulted from the crowding of additional surface into the small space. This folding would take the form of simple branching, and then "repetition" of the process by a refolding of the outlines of the first fold. The edge would thus have the digitate or pinnately branched appearance it is known to present.

Another case of folding is that of the brains of the higher mammals, including man. This was occasioned by excess of growth-force distributed to the circumference (from an inherited central region), exceeding in amount that of the space (skull) in which it was inclosed. Folding was necessary for its accommodation in the contracted space. This is a highly useful modification. Another case like that of the ammonite above cited, whose use is not so readily determined, but which illustrates excess of growth-force on some other account, is that of cotyledonary leaves in some plants. In these cases the growth of the leaf is accelerated, requiring plication for its accommodation.

We therefore have the two laws regulating the "location of growth-force" irrespective of use, viz.: 1, the centripetal or polar tendency; 2, the peripheral tendency; both more or less determined by the inherited central or first grown parts.

There is, however, another element which in animals disturbs the symmetrical direction of growth-force besides use, and which precedes use, i. e., effort. Use presupposes a part to use, and a simple part is sufficient for its influence, so that it be *usable;* but the first beginnings of few structures are usable. As Mivart has objected to natural selection, that the first rudiments of organs could never be sufficiently important to an animal to give it a preference in the struggle for existence, so it might be urged against the theory of use that rudiments are generally *useless*, and could not have been originated by use.

The effect of use is, however, twofold. The contact with objects used has some effect in stimulating nutrition, as well as the exertion of the muscles necessary to use. But determination of nutritive fluid is well known to be under the influence of nerve-force. How imagination stimulates secretion is seen in the familiar example of the flow of saliva in anticipation of food ; a very different example is the phenomenon of blushing under emotional stimulus. Nevertheless, it is not evident that growth can result with any such facility in a fully grown animal. It is thought that the effort becomes incorporated into the metaphysical acquisitions of the parent, and is inherited with other metaphysical qualities by the young, which during the period of growth is much more susceptible to modifying influences and is likely to exhibit structural change in consequence. Certain it is that acceleration ceases with growth, and, as the young of animals are not in complete relation with the surrounding world, the influences controlling it must be inherited. This consideration renders it doubly probable that the results of effort on the part of the parent appear in change of structure in the offspring.

Of course, immense numbers of cases of continued effort can be produced by the objector, in which no structural modification has resulted. There are various reasons why a modification should not take place. In the first place, the exertion of use must be great, habitual, and long-continued ; in the next place, abundant food must be at hand ; finally, growth-force must be to spare in the growing young, either from some less necessary part or by excess. Now, cases are probably not rare where none can be spared from another part without injuring the efficiency or viability of the animal ; hence, all such changed individuals would perish through some form of natural selection or disease.

Domesticated animals can be pointed out where effort and use have long been put forth in the service of man without changing structure. But such effort is not to be compared for a moment with that put forth by animals in a wild state, in seeking food or protection from enemies. The protection furnished by man, and consequent release from the struggle for existence, has reduced the chances for such variations greatly. Nevertheless, variations profitable to man have resulted ; witness the race-horse and carter.

In cases where one side of the body is used in excess of the other, unsymmetrical development would be counteracted by the law of polar or centrifugal growth, all that might be acquired by the one side being inherited by both. Even this original symmetry has, however, been overcome in some types, as in the flounders (in the jaws and teeth as used parts). This part of the subject is purely hypothetical, however probable, but the aspect of the discussion will be materially brightened to the reader, if he have previously adopted, with the author, the principle that evolution has been the mode of origin of the present life-forms of the earth.

Reverting now to those beings which are either in part or entirely destitute of the power of determining these movements, as plants, various causes present themselves as modifiers of the polar or centripetal activity of growth-force. In the case of a plant, the medium in which it grows modifies the result; for example, the downward growing axis or root differs much from the upward aërial portion or stem. Aquatic plants, supported beneath or on the surface of the water, lack the strong stems and braces necessary to air-growing plants; while of the latter, those with weak stems develop tendrils and other supports. In these cases physical laws have been the guides of growth-force. In the case of bright colors, which we know to be impossible in vegetation without sunlight, the influence is chemical.

The first physical law is, that growth-force, uninfluenced by inherited peculiarity,* or any stronger influence locating a nutritive fluid, must *develop extent in the direction of least resistance, and density on the side of greatest resistance,* when not too great. The illustration of this statement would be that a globular mass of cells brought to the point of junction of two media, as water and earth or air and earth, elongates in the direction of the medium presenting the least resistance, i. e., air. Thus a longitudinal development would originate from a centrifugal, and a repetition of the same process would produce branches. The reason why repetition should appear along the sides instead of as continued prolongation of the axis, arises probably from the difficulty of conveying fluid nutritive material far from the base as a source of supply, and the occurrence of various mechanical obstacles easy to be conceived. In low plants, where nutrition is absorbed by all parts of the axis, which branch, as the subterranean parts of fungi (*mycelium*), difference of local supply would produce an effect. In higher plants, where fluid nutriment is only introduced at one point (the root), and conveyed by special layers of cells or tubes, the difficulty of maintaining supply at a distance from its source would encourage lateral repetition or branching.†

* This is supposed to be due to atomic peculiarity of cell-substance.

† The great power exerted in a given direction by growth-force due to gradual

6. *On the Origin of Intelligence.*

Leaving this part of the subject, we approach one of higher interest, viz., the effects of the metaphysical or mental acquirements of animals on their exertions in effort and use. The growth of the mind in animals has, no doubt, followed the same laws obeyed by that of man; the difference being that the lower forms have remained permanently fixed in stages early passed by the lord of living beings. The foundation qualities from which all the phenomena of intellect may be derived are, the powers of retentiveness (memory) and of perceptions of resemblance and difference. These traits are well known to be possessed by many animals, and perhaps in some degree by all. Their possession will be modified by the power of exercising *attention,* which, in its turn, will depend on the sensitiveness of the animal to impressions—in other words, the ease with which consciousness may be aroused.

The origin of the disposition to take food will be the rudiment of all that appears as *will* in higher animals, and which, though supposed to guide, is the creature of so many *stimuli.* This origin is supposed by metaphysicians to be the result of education of the "spontaneous activity" of animals by their pleasures and pains.

The brain of man and of other animals is an organ which receives and retains pictures and impressions, both painful and pleasurable. The retention of these pictures is not a state of consciousness, but they may be brought into the consciousness according to the law of "contiguity," or association. That is, that the recurrence in the actual of some object or event, which was perceived on a former occasion, at or near the same time as another object or event not again repeated in the actual, will bring the latter before the consciousness. So, also, the revival of one such picture will bring within the mental vision others impressed on the mind at or near the same time as the first. These events may have been in the reality either painful or pleasurable. On the recurrence of circumstances which on a former occasion resulted in pain, the resuscitation of the mental picture, then impressed on

---

acquisition and inheritance is illustrated by the plantain, *Plantago major*. Although without axis, it has been observed by James C. Cresson, in the Philadelphia Park, and Alfred Cope, on his drive, to force itself through a solid bed of hard vulcanite pavement, several inches in thickness.

the memory, produces an anticipation of the pain, and the animal at once flies from the source of danger. So, also, with pleasurable objects, the resultant action being the reverse, or an attraction to the object. In both cases a previous experience of the relation between the object and the sensation of pleasure or pain must have been had.

There is, in addition, the power of determining differences and likenesses, by which contiguities or associations are originated in the mind, of a character different from that resulting from the relations of times of receipt of the impression. Such contiguities recall pictures to the consciousness in consequence of their resemblances in essential qualities.

On these two bases, together with the perceptive faculties, rest the complex phenomena of the animal and human minds. They are probably physiological functions of brain tissue, for the following reasons, among others:

1. Impressions are conveyed by physical means to the brain. The brain can only receive a given number at a time without exhaustion, and is prepared to receive more after being nourished. Impressions long forgotten are revived in certain states of disease. Impressibility and memory are most marked during growth, and diminish with age.

2. If impressions are physically produced and preserved in the brain, those made at or near the same time would remain close together in the brain, and this material proximity would be the "contiguity" according to which they would come before the consciousness.

3. Classification or appreciation of resemblances takes place unconsciously in the mind (by "unconscious cerebration"). Resemblances not seen at the time of impression suddenly flash into the consciousness on a subsequent revival of it. The "contiguity" to like things thus established may be more or less lasting in the mind than the contiguity of circumstances under which the impression was made.

4. As "contiguity" in time is believed to be revived by contiguity in location of impressions in the "perceptive" part of the brain, so "contiguity" of resemblance may be believed to depend on contiguity of location in the "reflective" part of the brain.

The retention of the contiguities of time and of resemblance constitute the basis of education of an animal, and its intelligence will be in proportion to the number and complexity of these acquisitions.

The lowest exhibitions of contiguity, or association in time, do not require a nervous system for their display. The *Protozoa*, which are without nervous system, exhibit its results in their determinate seizure of some small objects as food and rejection of others. The sea anemones (Actinia) display some preferences as to the substances to which they attach themselves. All power of taking food implies the retention of the impression of pleasure on first accidentally coming in contact with it. This power is then present in protoplasmic beings of the simplest type.

All the movements of animals have been shown to depend on a direction of this motive force, consequent on a necessity for avoiding pain and obtaining pleasure. It may be regarded, moreover, as a truth that heightened vitality or energetic conversion of force is always a state of pleasure, while depressed vitality is generally the cause (as well as a consequence) of pain. Hence the pleasurable nature of taking food, and the early education of an animal in the distinction between objects nutritive and non-nutritive.

It is well known, however, that food may be taken, and many or all other functions and acts be performed *automatically*, or in a state of unconsciousness. This is as much the case with the highest powers of thought (as in unconscious cerebration) as with the humblest acts which satisfy bodily wants. The question then arises whether these acts may not *arise* in a state of unconsciousness. So far as our own self-knowledge goes, we would reply in the negative. All intellectual functions are produced by education, and education involves consciousness at every step. Other habitual and automatic acts were originated consciously, but the contiguity of parts of the act becoming impressed on the brain, future repetitions of it are reflex or unconscious. We have seen that the development of the habits of animals is in strict obedience to the preference for pleasure and avoidance of pain. Pleasures and pains of course express sensations which involve consciousness. It then appears to me that, in the lowest animal, consciousness must be present at the time of origin of every habit, but that it may have been soon lost in each case, and the habit become automatic.

If this position be true, every subsequent addition to or change of habits must have been accompanied by a resuscitation of consciousness. But how is consciousness aroused in the cases which we can investigate—that is, in our own species? In the common case of sleep, a mere sound is sufficient to cause its return. In syncope and other forms of loss of consciousness, a sudden shock, as of cold water thrown on the face, will often restore it. In ordinary states, compulsion or pressure will produce a more intense degree of consciousness. Indeed, in many persons consciousness varies greatly under different influences; thus a dreamy state naturally follows a lack of nutrition of the brain, as during a fast, but the mind is readily aroused from it by a strong or sudden impression made on the senses or by an effort of will. It is, hence, to be believed that in animals whose acts may be largely automatic consciousness is aroused by exceptional sights, sounds, hunger, and other sensible impressions, the more readily in proportion to the development of the sensory ganglia of the brain. In the lowest animals, consciousness will be aroused with much greater difficulty, and life must be in part a blank, and in part a dream, at rare intervals illumined by gleams of consciousness produced under the influence of strong external stimulus. At these times of awakening we must believe that new movements are instituted, which will become habitual in proportion to the extent to which they are repeated.

From what is known of reflex actions, it is evident that movements may become habitual, and hence automatic, wherever cellular or gray nerve-substance exists. Thus, movements instigated by the spinal nerves derive their automatism from the spinal medulla. This axis no doubt derived the original nerve-force from the brain, but repetition of the act under the original stimulus confers the automatic power on the spinal cord or ganglionic center as certainly as on the brain. This must of necessity follow the complication of nervous structures by addition to the original center. Thus are explained the automatic movements of the frog's or centipede's legs in the well-known experiments, when separated from the brain or principal nervous center.

Greater difficulties are experienced in accounting for the origin of the automatic movements of plants. It is evident, however, that many of these, perhaps all, are not due to the energy of motive-force, but are phenomena of growth-force alone. Motion is produced by the change of direction imparted to growth-force by the influence of heat, light, contact, etc. This will largely follow as a consequence of the longer persistence of growth in plants than in animals. In some of the very lowest plants movement would seem to be similar to that seen in the lowest animals. The origin of intelligence is then to be found in the first traces of consciousness. But inquiry will go further in the search for truth. One school of physiologists believes consciousness to be due to certain molecular changes of nerve (i. e., brain) substance. No doubt these are essential to consciousness; yet it is, at least, utterly inconceivable that any molecular condition should be capable of recognizing either pleasure or pain. Here we find the first appearance of the "feelings," that part of the human mind which deserves to be called, *par excellence*, "the soul," and which is the "locator" of all the forces which go to build up the tissues, organs, and parts of the body; and hence the organic type of the intellect and even of the will. For if we find growth-force to be a correlative of physical force, and subject to motive-force for its "location," so, while we find that motive-force is in like manner but another equivalent of ordinary force, we find it also subordinate to this principle which "locates" it.

7. *On Intelligent Selection.*

We will now retrace our steps along the line of causes and effects, and see, if possible, the nature and results of "intelligent selection."

Protoplasm, homogeneous or cellular, discharges a force called "motive-force," or neurism, which it converts from surrounding forces by virtue of its molecular (chemical) or atomic constitution. Its spontaneous exhibitions as movements bring it into contact with surrounding objects, when, if sufficiently sensitive to impressions, consciousness of a pleasurable or painful character is aroused. Then in proportion to its sensitiveness, and the strength of the impression, the protoplasm *retains* the latter in an unconscious way, and, on the subsequent recurrence of the object to the consciousness, the pleasurable or painful attribute is also revived by the law of "contiguity." The "vital principle" interfering restrains the motive-force from exercise toward it, if it be painful, or releases it for motion toward it, if it be pleasurable. This quality of the vital principle is a power of choice, and, in so far as *consciously* exercised, is *will.** Nothing is better known than that acts originated consciously may, by repetition,

---

* For further development of this subject see Part IV of this book.

come to be exercised unconsciously, or as habits. The form these habits take, will depend on the opportunities and conditions offered, for instance, in the matter of food, by the surrounding order of things. Motive-force is, of course, only another name for effort and use; the former being a putting forth by means of *conducting* material or organs, without *executive* apparatus; the latter always requiring organs or parts by which to execute.

If, as previously assumed, effort and use locate growth-force, cellular structure will appear in the directions indicated by the activity of motive-force. It is converted by protoplasm from heat and chemism, or one only of these. That it does not require differentiated "motive-force" as a source, is proved by the growth of plants, which have no motion proper.

Growth-force, by its "repetitive" action,* creates organs. These at first will be extremely simple, but, as machines, at once increase the power of the animal to produce motive-force by conversion, whether the machine be a digestive apparatus for the appropriation of the material, or a mechanical one for the exercise of the force, the former necessarily preceding the latter in time.

With the increased power of assimilation (digestion) comes a larger amount of material for increased exhibition of growth-force, a part being burned or otherwise converted into the force, and a part remaining as the material from which the cells are constructed. In the latter part of the growth-period a considerable portion is usually consumed for motive-force.

In the history of the material environment, various changes of condition succeeded each other. Changes of level took place; waters were purified by precipitation of chemical compounds; fresh waters were established; the atmosphere deprived of various gases; new mineral, and especially vegetable, products took their appropriate places. All these offer a vast variety of food-supply and opportunity for the pleasurable discharge of motive-force, and, under the laws pointed out, efforts of animals were directed in various lines, as the conditions presented themselves. Thus executive organs were produced of varied character. Some acquired limbs and others wings for transportation from place to place. What a vast addition to their impressions must have been acquired by the first animals which could thus leave the place of their birth! How many new "contiguities" were established, and how

---

* See "Method of Creation," on the Law of Repetition.

many new habits originated! Look again at the acquisition of teeth. From a soft uniform diet the animal no doubt gradually learned to appropriate hard substances, and what a world of experience and consequent habit must have been at once placed in its way!

These acquisitions are of course mental, and include both kinds of contiguity, viz., that of succession in time, and that of association through resemblance. Animals choose between objects chiefly in accordance with the first mode, but are not lacking in the second quality. As an example of the latter, classification according to color is exhibited by some birds, which choose brightly colored objects and reject dull ones—the Australian bower-bird, for instance.

As is well known to metaphysicians, these acquisitions lead to "predication" and "forethought." Thus a hound becomes acquainted with the habit of a buck or rabbit in doubling, or in following a given circuit. The recurrence of the chase recalls the habit in consequence of contiguity of the impression of the former pursuit of the animal and the course it took. The hound then supposes or "predicates" that the deer will repeat the course. We know that some do so from the fact that they have been observed to cut off the curves in the animal's track, or to station themselves at the point where the deer, for instance, will pass. In the second act forethought is also involved. The hound sacrifices the lesser pleasure of the chase for the greater one of securing the prey. In forethought, experience having taught which circumstance results in greater and which in less pleasure, action is restrained in the presence of the second for the sake of procuring the first. Thus in ants, immediate pleasure suggests a life of labor enough for present wants, and ease for the remaining time; but the greater pleasure of existence during some time of scarcity has induced some of them to lay up a store, which has developed into the complete protection against winter they display in cold climates.

In the cases cited it is perfectly evident that the hound would never have learned to predicate had he not had limbs to bring him in contact (by pursuit) with the habits of the buck. Nor would the ant have learned to provide if it had not been furnished with the jaws necessary to the excavation of chambers and the carrying of food. And neither would have performed these mental acts had they not possessed nervous centers capable of sensation, retention, and classification. But the development of these qualities depended on the possession of the executive organs.

Thus intelligence of various degrees has resulted—first, from possession of executive organs; secondly, establishment of contiguities by their use; thirdly, from classifications based on contiguities.

There are two notable examples of the great advance in mental accomplishment inaugurated by the acquisition of an executive organ. The first is the acquisition of the hand by the monkey. Any one who has compared their manners with those of the species of unguiculate (clawed) animals—as, for example, the dog—must be struck by the immense advantage possessed by the former. To begin with, it renders the monkey an expert climber and catcher of moving objects. Then he greatly increases his habits of observation by the facility with which he brings objects before his eyes. The minuteness of his criticism is aided by his power of manipulating with his fingers. He learns the results of throwing. A cautious monkey, set on a strange animal or person, does not rush at them to bite like the dog, but takes the more refined method of throwing sticks and stones. In obtaining objects, if they be out of reach, a monkey uses a stick to draw them in with. In a *Cebus capucinus,* in my possession, a leather strap was preferred, and was thrown with great dexterity, and the object always secured by drawing it nearer in loops of the strap. If the strap were sometimes thrown out of reach, the monkey would use a stick or poker in order to obtain it again. All this would have been quite impossible without a hand. This facility has not been lost on the intelligence of the monkeys. Their curiosity is proverbial, and no animal can compare with them as successful thieves; moving in bands, with watchers stationed, and taking pains to store their booty in their cheek-pouches while they have opportunity, and not wasting time in eating exclusively.

Another instance is that of man. Here the acquisition consisted of vocal organs capable of speech. Any means of communicating and recording thoughts had the stupendous consequence of rendering the acquisitions of one man or class of men the common property of all. Another consequence, equally important, is the preservation of successive races of men from the necessity of passing through the experiences of those that preceded them, and the enabling them to commence their intellectual life nearly where the others left off. So great would necessarily be the location of activity in the brain, that its stages of growth would be the principal changes of structure to be witnessed in human history, together with those which should result from loss of growth-force in other regions by abstraction.

To return to the material aspect of the case. The discrimination between pleasure and pain locates motive-force, which is derived from without. Motion or use locates growth-force, also derived from without, and thus produces organs out of material derived from without. These organs diversify the directions of motion. From new movements arise new pleasures and pains, and motion is again "located" in its exhibitions in some particular directions, and restrained in others. These directions depend on external circumstances at the first. The determination of motion to certain lines locates growth-force in those lines, and new parts are produced, which are further executive organs and types of structure. We now repeat the circuit. New executive organs introduce new contiguities, the number depending on the general complication of the animal in connection with that of the organ, and new pleasures and pains result. The pleasures again determine activity, and, under the circumstances already mentioned, growth-force is again located.

In these facts lies the explanation of the principle of direction or location of executive structures about the head, termed by Prof. Dana "cephalization." The increasing demands of intelligence locate growth-force round its organ, the brain, etc., while such location reacts by furnishing means of increased activity of mind.

These conjunctions of growth with executive capacity, constitute a class of "expression points"; points attained without leaps, and abandoned without abruptness, but constituting great steps of progress, pregnant with future results.

The part played by "intelligent selection" remains to be considered. I have heretofore spoken only of variation of types, and not dwelt on their *persistence.* This is a most important consideration, for most men see a great deal more of the latter than the former. Species present their characters intact for long periods, and many structural peculiarities have had great range in time.

I recur to the origin of the "location of growth-force," i. e., the direction of motive-force. This is located by the appreciation of pleasure and pain. Now, every one knows that different animals of the same species vary exceedingly in their sensitiveness to exterior influences; that is, they differ in the degree in which the same object affects their *consciousness*. That which excites one is unnoticed by another; what fills one with alarm scarcely rouses the attention of another. It is well known also that differences in the power of retention are to be found in the same species of animals. The variations in persistence of memory are very great. While, therefore, one individual of a species will acquire a store of "contiguities," another of the same will possess but few. In other words, the degree with which objects are *impressed on* the consciousness, and the degree with which they are *returned to* the consciousness, vary greatly in the same species. Here we have the basis for the origin of totally different unconscious habits and reflex actions, and consequent divergences in the "location of growth-force."

Those in which these impressibilities are most highly developed will accumulate mental acquisitions most rapidly; in other words, they will be the most *intelligent* of their species. While others follow the old routine of once acquired and then inherited habits, those in whom consciousness most frequently recognizes events will *originate* new acts and new habits.

Intelligence is a conservative principle, and always will direct effort and use into lines which will be beneficial to its possessor. Here we have the source of the fittest—i. e., addition of parts by increase and location of growth-force, directed by the influence of various kinds of compulsion in the lower, and intelligent option among higher animals.

Thus, intelligent choice, taking advantage of the successive evolution of physical conditions, may be regarded as the *originator of the fittest*, while natural selection is the tribunal to which all the results of accelerated growth are submitted. This preserves or destroys them, and determines the new points of departure on which accelerated growth shall build.

If the above positions be true, we have here also the theory of the development of intelligence and of other metaphysical traits. In accordance with it, each trait appropriates from the material world the means of perpetuating its exhibitions by constructing its instruments. These react by furnishing increased means of exercise of these qualities, which have thus grown to their full expression in man.

II.

ON THE ORIGIN OF GENERA.

Introduction.—The present fragmentary essay is a portion of what other occupation has prevented the author from completing. It does not, therefore, amount to a complete demonstration of the points in question, but it is hoped that it may aid some in a classification of facts with a reference to their signification. When all the vast array of facts in possession of the many more learned than the writer are so arranged, a *demonstration* of the origin of species may be looked for somewhere in the direction here attempted to be followed.

Conclusions of any kind will scarcely be reached, either by anatomists who neglect specific and generic characters, or, secondly, by systematists who in like manner neglect internal structure. Such will never perceive the system of nature.*

ANALYSIS OF THE SUBJECT.

I. Relations of allied genera.
  First: in adult age.
  Second: in relation to their development.
    α. On exact parallelism.
    β. On inexact or remote parallelism.
    γ. On parallelism in higher groups.
    δ. On the extent of parallelisms.
II. Of retardation and acceleration in generic characters.
  First: metamorphoses in adult age.
    α. The developmental relations of generic and specific characters.

* It might seem incredible that either class should systematize with confidence, yet a justly esteemed author writes even at the present day, "However, there is scarcely a systematist of the present day who does not pay more or less attention to *anatomical* characters, in establishing the higher groups!" (The italics are our own.) As though a system were of any value which is not based on the *whole structure*, and as though *lower* groups were only visible in external characters; in a word, as though external (muco-dermal, dental, etc.) characters were not "anatomical"!

β. Probable cases of transition.
γ. Ascertained cases of transition.
Second: earlier metamorphoses.
δ. The origin of inexact parallelisms.

III. Relations of higher groups.
α. Of homologous groups.
β. Of heterology.
γ. Of mimetic analogy.

IV. Of natural selection.
α. As affecting class and ordinal characters.
β. As affecting family characters.
γ. As affecting generic characters.
δ. As affecting specific characters.
ε. On metaphysical species.

V. Of epochal relations.

The laws which have regulated the successive creation of organic beings will be found to be of two kinds, as it appears to the writer. The first, that which has impelled matter to produce numberless ultimate types from common origins; second, that which expresses the mode or manner in which this first law has executed its course, from its commencement to its determined end, in the many cases before us.

That a descent, with modifications, has progressed from the beginning of the creation, is exceedingly probable. The best enumerations of facts and arguments in its favor are those of Darwin, as given in his various important works, "The Origin of Species," etc. There are, however, some views respecting the laws of development on which he does not dwell, and which it is proposed here to point out.

In the first place, it is an undoubted fact that the origin of genera is a more distinct subject from the origin of species than has been supposed.

A descent with modification involves continuous series of organic types through one or many geologic ages, and the co-existence of such parts of such various series at one time as the law of mutual adaptation may permit.

These series, as now found, are of two kinds: the uninterrupted line of specific, and the same uninterrupted line of generic characters. These are independent of each other, and have not, it appears to the writer, been developed *pari passu*. As a general law, it is proposed to render highly probable that the same specific form has existed through a succession of genera, and perhaps in different epochs of geologic time.

With regard to the first law of development, as above proposed, no one has found means of discovering it, and perhaps no one ever will. It would answer such questions as this: What necessary coincidence of forces has resulted in the terminus of the series of fishes in the perches as its most specialized extreme; or, of the Batrachia, in the fresh-water frogs as its ultimum; or, of the thrushes, among birds, as their highest extreme: in a word, what necessity resulted in man as the crown of the mammalian series, instead of some other organic type? Our only answer and law for these questions must be, the will of the Creator.

The second law, of modes and means, has been represented to be that of natural selection, by Darwin. This is, in brief, that a disposition to a general variation on the part of *species* has been met by the greater or less adaptation of the results of such variation to the varying necessities of their respective situations. That the result of such conflict has been the extinction of those types that are not adapted to their immediate or changed conditions, and the preservation of those that are.

In determining those characters of plants and animals which constitute them what they are, we have, among others of higher import, those which constitute them species and those which constitute them genera. What we propose is: that, of the latter, comparatively very few in the whole range of animals and plants are *adaptations* to external needs or forces, and of the former a large proportion are of the same kind. How, then, could they owe their existence to a process regulated by adaptation?

Darwin is aware of these facts to some degree, but, as already said, he does not dwell on them. Where he does, he does not attempt to account for them on the principle of natural selection.

There are, it appears to us, two laws of means and modes of development: I. The law of acceleration and retardation. II. The law of natural selection.

It is my purpose to show that these propositions are distinct, and not one a part of the other: in brief, that, while natural selection operates by the "preservation of the fittest," retardation and acceleration act without any reference to "fitness" at all; that, instead of being controlled by fitness, it is the controller of fitness. Perhaps all the characteristics supposed to mark generalized groups from genera up (excepting, perhaps, families) have been evolved under the first mode, combined with some intervention of the second, and that specific characters or *species* have been evolved by a combination of a lesser degree of the first with a greater degree of the second mode.

I propose to bring forward some facts and propositions in the present essay illustrative of the first mode.

I. ON THE RELATIONS OF NEARLY ALLIED GENERA.

First. The writer's views of the relations of genera have already been given at the close of an "Essay on the Cyprinoid Fishes of Pennsylvania."* It is easy enough to define isolated genera which have few immediate affines; but among extensive series of related forms the case is different. One principle, however, pervades the conception and practice of all zoölogists and botanists, which few take pains to analyze or explain. It is simply that they observe a successional relation of groups, by which they pass from one type of structure to one or several other types, and the presence or absence of the steps in this succession they regard as definitions of the genera.

It is true that the reader will often find introduced into diagnoses of genera characters which indicate nothing of this sort. It is often necessary, indeed, to introduce characters which are not peculiar to the genus characterized, for the sake of distinguishing it from similar ones of *other series*, but this only in an imperfect state of the record. Moreover, the ability of the writer to distinguish genera being thus tested, he too often fails by introducing family and specific characters, or by indulging in an unnecessary redundancy. In general, it may be said that adjacent genera of the same series differ from each other by but a single character; and, generally, that the more remote differ by characters as numerous as the stages of their remove.

It is precisely as, among the inorganic elements, we pass from the electro-negative, non-oxidizing extreme of the halogens, with fluorine as the extreme, to the electro-positive, violently oxidizing extreme of the alkaline metals, whose extreme is potassium, by steps whose relative position is measured or determined first by these tests; and as these steps have each their included series of

* "Trans. Amer. Philos. Soc.," 1866, from "Proc. Acad. Nat. Sci.," Phil., 1859, p. 332.

bodies, characterized by their successive relations on the lower level of a subordinate range of characters. This principle is distinctly admitted by many zoölogists,* those who deny it generally failing to perceive it because they attempt to gauge a major scale by characters which are really the test of one or all of the subordinate or included scales. It holds true of most of the groups of organic beings; thus the class is a scale of orders, the order of tribes. I will not now say that the tribe is a scale of families, as the case is here much modified; but what is chiefly to be considered in this essay is, that the family is composed of one or several scales of genera.

Second. Now, the more nearly allied genera are, the more surely will these generic steps be found to fall into the direct line of the steps of the development of the highest, or that with the longest scale, the former being truly identical with the latter in generic characters. Less allied genera will offer an inexact or incomplete imitation of such identity—some additional character being present to disturb it. Such genus belongs to another series, characterized by the disturbing feature, whose members, however, bear to each other the relation claimed above for such.

The relation of genera, which are simply steps in one and the same line of development, may be called *exact parallelism*, while that of those where one or more characters intervene in the maturity of either the lower or higher genus, to destroy identity, may be called *incomplete parallelism.*

The latter relation has been dwelt on by von Baer, Agassiz, and other writers, but none have accepted the existence of *exact parallelism*, or seen its important relation to the origin of genera.

Third. That the lowest or most generalized terms or genera of a number of allied series will stand to each other in a relation of exact parallelism. That is, if we trace each series of a number, up to its lowest or most generalized genus, the latter together will form a series, similar in kind to each of the sub-series; i. e., each genus will be identical with the undeveloped conditions of that which progresses the farthest, in respect, of course, to the characters which define it as a series.

Those characters of the skeleton which we are accustomed to call embryonic are only so because they relate to the develop-

* Prof. Bronn, in his "Classen u. Ordnungen des Thierreiches," has everywhere a chapter on "Die aufsteigende Reihe" ("the ascending scale").

mental succession witnessed in animals at the present time. Characters not so called now were probably as much so at one period now passed. Hence embryonic characters of the bony system do not, as I have often had occasion to observe, characterize the types of the highest rank, but only subordinate divisions of them. Thus the elasmobranchs are probably repressed forms of groups of a really higher grade than the bony fishes, or teleostei, which may be known to us. In their early presence in the geologic series we have evidence of the first beginning of a higher type.

In the same manner it has been discovered that the molecular aggregation of the elementary substances does not characterize their highest or most distinct series, but rather the substances themselves within the group or family to which they belong. The gaseous, liquid, and solid molecular conditions, being characters distinguishing otherwise allied substances in the same way morphologically (we can not say yet developmentally), as the cartilaginous, osseous, and exostosed or dermosseous characters distinguish otherwise nearly allied genera.

The "family" group embraces one or many of such series. If we trace the series in several families to their simplest or most generalized terms or genera, and compare them, we will not find the relation to be one of exact parallelism in the series of the "order," so far as our present knowledge extends, but, in a developmental sense, one of divergence from the commencement.

If we could know the simplest known terms or family characters of a number of groups of families, or "orders," we would probably find them to represent a series of exact parallelism, though to find such simplest terms we must go far into past periods, since the higher the group the more extensive the range of its character, and the less likely to be found unmixed with additions and extensions, in modern times.

Finally, the series of classes is in the relation of the essential characters of the same, as expressed in their now extinct, most generalized and simple representatives, also one of "exact parallelism."

*a. Examples of Exact Parallelism.**

1. As an example we may take the genus Trachycephalus (*Batrachia anura*). Nearly allied to it is the genus Osteo-

* In generic series.

cephalus, which differs in the normal exostosis of the cranium not involving the derm, as in the former. Close to this is Scytopis, where the fully ossified cranium is not covered by an exostosis. Next below Scytopis is Hyla, where the upper surface of the cranium is not ossified at all, but is a membranous roof over a great fontanelle. Still more imperfect is Hylella,* which differs from Hyla in the absence of vomerine teeth. Now, the genus Trachycephalus, after losing its tail and branchiæ, possesses all the characters possessed by the genera Hylella and Hyla, either at or just before the mature state of the latter, as the ethmoid bone is not always ossified in advance of the parietals. It soon, however, becomes a Scytopis, next an Osteocephalus, and finally a Trachycephalus. It belongs successively to these genera, for an exhaustive anatomical examination has failed to reveal any characters by which, during these stages, it could be distinguished from these genera.

Now, it would be a false comparison to say that the young of Trachycephalus was identical with the genus Agalychnis, which in truth it resembles, because that genus is furnished with one other character—the presence of a vertical pupil—and belongs to another series in consequence, which is represented as yet, with our present imperfect knowledge—or perhaps imperfect fauna—by three genera only.

2. The lowest type of the near allies of our common freshwater frogs is the genus Ranula, where the prefrontal bones are narrow strips on each side the ethmoid cartilage ; the ethmoid cartilage itself entirely unossified above, and the vomerine teeth very few and on a small elevation. There are two species, *R. affinis* and *R. palmipes.*† The other species have the ethmoid cartilage ossified above, at least beneath the extremities of the frontoparietals.

Those of the latter most like Ranula possess the same type of narrow prefrontals, separated by a broad area of cartilaginous ethmoid, and fasciculi of teeth. Of this type is *Rana delalandii,* and probably *R. porosissima,* Steind., of the South Ethiopian region. Other species of the same type extend their vomerine patches into lines ; such are *R. mascariensis, R. fasciata, R. oxyrhynchus, R. grayi,* and other South African species.

* I refer to *H. carnea m.*, not having Reinhardt and Lütken's type of this genus.

† These species are now believed to be identical, but three others have been discovered (1886).

The prefrontals are subtriangular, and approach each other more or less in the numerous species of North America and of the Regio Palæarctica, while generally the vomerine teeth are in fascicles or very short series. In the Æthiopian *Rana fuscigula* the prefrontals unite on the median line, roofing over the ethmoid cartilage and reducing it, while the vomerine teeth are in very short lines.

In the species of the Palæotropical region, *Rana tigrina, R. vittigera, R. cyanophlyctis, R. grunniens, R. hexadactyla, R. corrugata, R. ehrenbergii, R. gracilis,* and the Æthiopian *R. occipitalis,* the prefrontals not only unite solidly (the suture remaining on the median line), but extend and closely fit to the frontoparietals. The vomerine patches have lengthened out into series.

Now, the young of the latter type of Rana (I take as an example the *R. tigrina,* one of the most abundant and largest of Indian frogs) presents the subtriangular prefrontals neither in contact with each other or with the frontoparietals, and the vomerine series is much reduced; in fact, it belongs in all respects to the Palæarctic group. I have not examined younger specimens, but have no doubt they are like those of the Palæarctic; which are, in their young stage, precisely of the type of the Æthiopian Rana, with fasciculate teeth like the young of those of the same region with teeth in series, since the prefrontals are still more reduced, becoming linear. Finally, the first stage of the Nearctic Rana, after losing the larval tail, is the genus Ranula, having linear prefrontals, minute vomerine teeth, and the ethmoid ring cartilaginous above.

These points of structure are of generic quality, but I have not regarded any group as sufficiently defined to be so regarded, except Ranula, as the adults of some species appear not to be constant in possessing them. Thus a very large *Rana catesbeyana* sometimes exhibits prefrontals in contact on the median line, while it is difficult to say whether *R. areolata* of North America is of the Nearctic type so much as of the Æthiopian. Nevertheless, the groups are generally quite geographically restricted.

3. A similar relation exists between the genera Hyperolius, Staurois, and Heteroglossa in respect to the prefrontal bones and the separation of the outer metatarsi, and—

4. Between Ixalus, Rhacophorus, and Polypedates also, in reference to vomerine teeth, bifurcation of last phalange, and dermoössification of the cranium.

5. When the larvæ of certain salamanders (Spelerpes) possess branchiæ, they also lack one digit of the hind foot, also the maxillary, nasal, and prefrontal bones, and exhibit a broad continuous palatopterygoid arch, in close contact with the parasphenoid. The proötic is separated from the exoccipital by a membranous space, and the exoccipitals themselves are not yet united above the foramen magnum. There is at the same time a series of splenial teeth. Both ceratohyals are confluent, the posterior is present, and there are but three superior hyoid arches. After they lose the branchiæ, the hinder foot, which has four toes only for a time, gradually adds another at first rudimental digit, in the Mexican species; in most North American species the fifth digit appears at an early larval stage. Five digits are finally present in all Spelerpes.

We have thus four combinations of the above characters, at different periods of the life history of certain (but not of all) of the species of Spelerpes. There exist four permanent series of species or genera, equivalent to these stages. The well-known "perennibranchiate" Necturus is nearly identical with the first, Batrachoseps with the second, the half-toed Spelerpes with the third, and the typical Spelerpes is the last.

In one character of generic value only do I find that Necturus differs from the early larval Spelerpes. It closes the premaxillary fontanelle with which it commences, by an approximation of the premaxillary spines, but not by a sutural union, as takes place in Amblystoma. It thus, in this one point, advances a stage beyond the condition to which Spelerpes attains, though it may be a question whether such a closure without union should not be classed among the specific characters by which *N. maculatus* differs from the young of the various Spelerpes, as they do from each other. Characters of the latter kind are the following: in *N. maculatus* the frontals are more deeply emarginate behind; it has little or no ala on the inferior keel of the caudal vertebræ, which is prominent in Spelerpes larvæ.

It may be that the parallelism in the case of Spelerpes is inexact by one character, and that a strictly developmental one; or it may be regarded otherwise.*

6. It is well known that the Cervidæ of the Old World devel-

* Necturus differs from these larvæ by another and more important character, viz., the presence of the os intercalare.

op a basal snag of the antler (see Cuvier, "Ossem. Fossiles"; Gray, "Catal. Brit. Mus.") at the third year; a majority of those of the New World (genera Cariacus, Subulo) never develop it except in "abnormal" cases in the most vigorous maturity of the most northern Cariacus (*C. virginianus*); while the South American Subulo retains to adult age the simple horn of the second year of Cervus.

Among the higher Cervidæ, Rusa and Axis never assume characters beyond an equivalent of the fourth year of Cervus. In Dama the characters are on the other hand assumed more rapidly than in Cervus, its third year corresponding to the fourth of the latter, and the development in after years of a broad plate of bone, with points, being substituted for the addition of the corresponding snags, thus commencing another series.

Returning to the American deer, we have Blastocerus, whose antlers are identical with those of the fourth year of Cariacus.

Now, individuals of the genus Cervus of the second year do not belong to Subulo, because they have not as yet their mature dentition. Rusa, however, is identical with those Cervi whose dentition is complete before they gain the antlers of the fifth year. When the first trace of a snag appears on one beam of *Cariacus virginianus*, the dentition includes the full number, but there remain $\frac{1}{3}$ milk molars much worn and ready to be shed. Perhaps the snag is developed before these are displaced. If so, the Cariacus is never a Subulo, but there can be little doubt that the young Blastocerus belongs to that genus before its adult characters appear.

7. Leidy states * that certain Perissodactyl remains, containing a foot of a horse, contained the teeth of a genus, Protohippus, which has the permanent teeth of Equus, and the deciduous dentition of Anchitherium. He observes: "The deciduous and permanent dentitions of both these genera are alike; therefore the new genus is in early life an Anchitherium, and later in life a true horse." This is therefore a case of exact parallelism, always providing that the Protohippus has not added to its immature equine characters others in other parts of the body, which invalidate the identity. In the latter case it will still be an interesting example of the "inexact parallelism." †

---

* "Proceed. Acad. Nat. Sci.," 1858, p. 7.

† This is not a proper example of parallelism if, as some zoölogists believe, the

8. It is well known that the Cephalopoda form a number of series of remarkable regularity, the advance being, in the first place, in the complication of the folds of the external margins of the septa, and, in the second place, in the degree of involution of one or both extremities of the shell to the spiral; third, in the position of the siphon.

Alpheus Hyatt, in an important essay on this subject,* points out that the less complex forms are in many cases identical with the undeveloped conditions of the more complex. He says: "There is a direct connection between the position of a shell, in the completed cycle of the life of this order, and its own development. Those shells occupying the extremes of the cycle" (in time), "the polar forms, being more embryonic than the intermediate forms.† The first epoch of the order is especially the era of rounded, and, in the majority of the species, of unornamented shells with simple septa; the second is the era of ornamentation, and the septa are steadily complicating; in the third the complication of the septa, the ornamentation, and the number of species, about twice that of any other epoch, all combine to make it the zenith of development in the order; the fourth is distinguishable from all the preceding as the era of retrogression in the form, and partially in the septa.

"The four periods of the individual are similarly arranged, and have comparable characteristics. As has been previously stated, the first is rounded and smooth, with simple septa; the second tuberculated, and the septa more complicated; the third was the only one in which the septa, form, and ornamentation simultaneously attained the climax of individual complication;

---

deciduous or temporary dentition is not a remnant of the primitive dentition, but is a later product of mammalian evolution. If it be a case of parallelism, it is inexact, because the genus Protohippus was discovered by the writer to have three toes, while Equus has but one.

* "Memoirs Boston Soc. Nat. Hist.," 1866, p. 193.

† He adds here: "Although in regard to geological sequence and structural position one of the extremes must be of higher geological rank." The "*highest*" extreme will be of higher geological rank, according to the complexity of structure and length of developmental scale, whether it come at the middle or end of the history of the class in time. If, as has been the case, so far as known, a decline has terminated the history of a class, its later forms are zoölogically *lower* than its older ones. Hence the adjective *high* is only appropriate to types of the latter kind, when used as synonymous with extreme.

the fourth, when amounting to anything more important than the loss of a few ornaments, was marked by a retrogression of the whorl to a more tabular aspect, and by the partial degradation of the septa."

I will here quote an entirely antagonistic statement of Bronn's,* as follows: "In the development of lamellibranchiate mollusks it is not possible to estimate the successional changes of one genus by those of another, though nearly related; so diverse are the most significant relations in the manner of progress among nearest allies. Therefore, embryologic indications are throughout useless in classification, and it is necessary to keep carefully separate the statements of observations on development of a given species, and not transfer such facts to the history of another species for the purpose of completing it. We can not even range these histories in conformity with family groups." For us this statement, though no doubt largely true, is an indication of imperfection—first, of knowledge of true affinities of recent, but especially of extinct adults, and, second, of imperfection of knowledge of development. The position appears to be based on negative evidence, while the opposing can and does stand on nothing but positive.

β. *Examples of the Inexact Parallelism.*

1. The genera of the batrachian family Scaphiopodidæ form a series of steps differing a little more than as repressions or permanent primary conditions in the development of the highest.† Thus two of the genera, which are North American, maintain their tubæ eustachii and tympanum through life, while three European lose them at an early period.‡ The three European genera also advance beyond the larval character of the American in the ossification of the basis of the xiphisternum into a broad style. Thus we have two series established, which differ only in the two characters named. Each shows its developmental steps in a similar manner, the European series extending further; thus:

---

* "Classen u. Ordnungen des Thierreichs," iii, p. 445.

† See "Journal Academy," Philadelphia, 1866, on Arcifera.

‡ According to Bruch and Tschudi, in Pelobates. I have found traces of the eustachian diverticula, in a tailed *Pelobates fuscus*, whose body measured 1 in. 4 lin., from Mus. Peabody Institute, Salem, Mass.

	European.	*North American.*
1.	Temporal fossa over arched. Cultripes.	* *
	Temporal roof not ossified.	
2.	Fronto-parietal bones ossified, involving derm. Pelobates.	Scaphiopus.
3.	Fronto-parietals ossified, distinct from derm. * * (Unknown.)	* *
4.	Fronto-parietals not ossified, distinct from derm. Didocus.	Spea.

In this case Didocus can not be said to be identical as a genus with an undeveloped stage of Cultripes, since, while the cranium of the latter is in the condition of Didocus, it bears a long tail, and the limbs are but little developed. Nor is Didocus identical with the undeveloped condition of Pelobates, since both cranium and limbs of the latter are developed before the tail is absorbed. Nor is Pelobates identical with the undeveloped condition of Cultripes, since, while the cranium of the latter is that of the former, the limbs and tail are still larval. The same relations exist between the other members of the family. The genus Scaphiopus is not an undeveloped form of Pelobates as to its auditory organs, for, when the latter is identical with the former in this respect, it bears otherwise entirely larval characters. Nor is Spea an arrested Scaphiopus, the relation being here precisely that between Didocus and Pelobates. Spea approaches more closely an arrested Didocus in all respects, but that when the latter possesses the auditory apparatus * of the former, it is a larva in limbs and tail, and that it loses this apparatus before reaching the other characters of Spea. The relations of these genera, as compared with those of the Trachycephalus, Cystignathidæ, and Bufo series, may be represented as follows: the lines represent the developmental scale of each.

This is an example of the simplest case of inexact parallelism, as distinguished from the exact parallelism or identity. As the fauna of the present period is but a fragment, so the simple inex-

* The possession of cavum tympani and tuba Eustachii in the undeveloped condition of this genus is only assumed from its close relation to Pelobates.

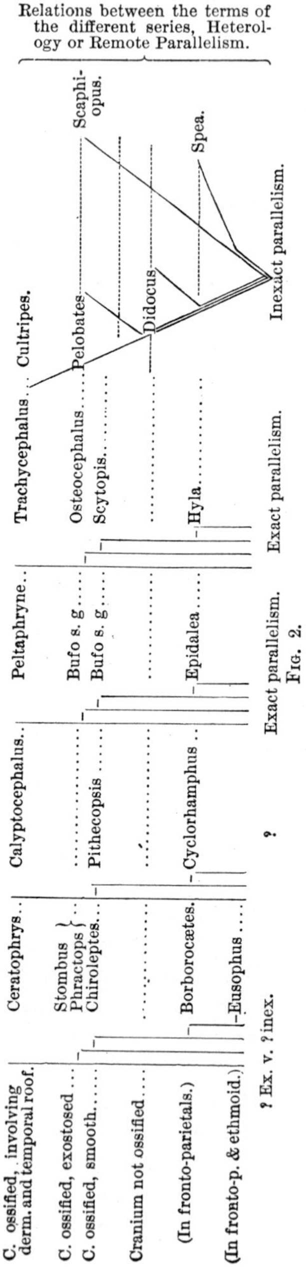

Fig. 2.

act is a more frequent relation than the exact, while the more complex inexact relation is still more common. The greater the inexactitude, the more frequently do such parallels occur till we have those of the most remote character, as, for instance, the parallelism between the different stages of the development of the mammal, in the structure of the heart and origins of the aorta, and the existing classes of vertebrates. The relation of these facts to the origin of genera will be noted hereafter.

It will be borne in mind that in the Scaphiopodidæ the generic types are identical for a long portion of their developmental history. (See figures on plate iv for the representation of these facts.)

2. In both Perissodactylous and Artiodactylous Mammalia, certain types develop their family character of canines at the earliest appearance of dentition, others not till a comparatively late period of life (Equus), and the extreme individuals never produce them.

3. Among cetaceans the genus Orca maintains a powerful and permanent series of teeth, which is an important generic character. In Beluga the series is shed in old age, in Globiocephalus, or the Caing whales, they are shed at middle age, while in the Balænidæ, of which the absence of teeth is an essential character, these organs are developed and absorbed *during fœtal life* (Eschricht). Though the condition of the teeth is not of systematic value in the two named intermediate genera, it is the important feature in the history of progress to such value.

4. Among the tortoises, the Testudinidæ rapidly extend the ribs into a carapace, which fits closely the marginal bones, while equally early in life the elements of the sternum unite together. This is also the case with most Emydidæ; among whose genera, however, we find the transitional scale. In Dermatemys and Batagur the carapace is very late in attaining its complete ossification, while the plastron is early finished. In Chelydra, on the other hand, while the carapace is even more slowly developed, the plastron is never free from its larval fontanelles. In the marine turtles neither plastron nor carapace is ever completed, while in the Trionychidæ the marginal bones are also entirely undeveloped.

In order that this last illustration be a true one for the theory in question, as applied to the *families*, these developmental characters should be the true distinctive features of these families respectively. This, as is well known, they are not. The Cheloniidæ are characterized by the form of their anterior limbs, which is in an adaptive structure, while the Testudinidæ similarly are distinguished by an extreme opposite modification of foot-structure, adapted to an extreme difference of habit. Here there is an example of the co-working of both laws. Nevertheless, we only claim at present to show the developmental relation of *genera* of the same family and the same series. This we see among the Emydidæ.

5. In the important character of the scutellation of the tarsi among the Passerine birds, the "boot" appears early in life in the highest Oscines, later in the lower, and does not appear at all in the majority. In respect to the still more important feature of the long posterior plates which appear very early in most Oscines, in the Myiadestes type * they appear late, the squamæ remaining long, while the Clamatores never develop the plates, not advancing beyond the infantile squamous stage.

6. It has been shown by Falconer that the genera of great Proboscidians form a remarkably regular and graded series, distinguished by their dentition. These are Dinotherium Kaup, Trilophodon Falc., Mastodon Cuv., Pentalophodon Falc., Stegodon Falc., Loxodon F. Cuv., and Elephas Linn. In the first there are but two cross crests on the third molars, and a pair of permanent mandibular tusks; in the second, three cross crests

* Baird, "Review of Birds of North America."

and mandibular tusks permanent in some males; in the third, four cross crests and the mandibular tusks variable; in the fourth, five cross crests on the third molar; tusks unknown. In Stegodon the mandibular tusks cease to appear, the crests of the third molar become more numerous, and embrace between them, in the bottoms of the valleys, a strong deposit of cementum. In Loxodon the crests have the whole interspaces filled with cementum, while the same thing holds in Elephas, with a greatly increased number of cross crests, which become vertical laminæ. The laminar character has become apparent from its rudimental condition in Stegodon.

Now, these are stages of development, though not in a continuous, single line. The shedding of the inferior tusks takes place earlier and earlier in the genera from Dinotherium, till they never appear in Stegodon. The molar teeth, it is well known, present, as they succeed each other from back to front, a regularly increasing number of transverse crests in the same species. Thus, in *Trilophodon ohioticus* the first molar presents but two, while the last presents six. The last molars of other genera present a very much increased number. What is it, then, but that the increased number of crests in the third molar, definitive of these genera, is an acceleration of growth; the fourth in Trilophodon is structurally third in Mastodon, and the fourth of Mastodon being third in Pentalophodon; the fourth of Pentalophodon becoming third in Stegodon, and so to the end? This is confirmed from the proved fact of the disappearance of the premolars. They are fewer in Trilophodon than in Dinotherium, and are soon shed; they are also early shed in Mastodon and Stegodon (*insignis Falc. Caut.*), and are not known to exist in the succeeding types; the acceleration of succession of teeth has caused them to be entirely omitted. The young tooth of Elephas, moreover, is represented by a series of independent parallel laminæ at first, which, when they unite, form a series of crests similar to the type of the genus Mastodon and others of the beginning of the series. The deposit of cementum takes place later, till the valleys are entirely filled up. Thus the relations of this part of the tooth-structure in the series are also those of the successional growth of those of Elephas, the extreme of the series.

It would be only necessary to show that two distinct conditions, in any of these respects, occurred among the different individuals of the same species of any of these genera, to render a hypothesis of evolution a demonstrated fact.*

It must be here observed that great size indicates little or nothing as to zoölogical rank. The greatest species are often not far removed in affinity from the least; thus there can be but little doubt that elephants are not far removed from the rodents, and the rhinoceros is near the cony. Indeed, in the same genus the most extraordinary diversity prevails, for we have a very small elephant of Malta, and in the Miocene of Maryland a fin-back whale not so large as the new-born young of the fin-backs now living. Hence the objection to the developmental hypothesis, based on the great size of the primal Selachians and Ganoids, has but little weight.

7. Rathke has shown that the arteria ophthalmica of the higher Ophidians is originally a branch of the arteria cerebralis anterior, and that it later forms a connection with the arteria facialis. This connection increases in strength, while the other diminishes, until finally its supply of blood is derived from the facialis instead of the cerebralis.

Rathke has also shown that the cerebral origin of this artery is continued through life in the three lowest suborders of the serpents, the Scolecophidia, Catodonta, and Tortricina; also in the next succeeding group, the Peropoda.

8. In most serpents the left lung is never developed; in such the pulmonary artery, instead of being totally wanting, remains as a posterior aorta bow, connected with the aorta by a ductus botalli; serpents without left lung being therefore identical in this respect with the embryonic type of those in which that lung exists.

9. Dr. Lespes states that the optic region of the brain of blind cave Coleoptera, examined by him is similar in structure to that in the blind larvæ of Coleoptera, whose imagos possess visual organs.

10. † Those Saurians (Uromastix, etc.), in which the premaxillary region is produced into a uniform cutting edge, are furnished during early stages with a series of premaxillary teeth, which become gradually fused and confluent with the alveolar margin. Hence other Acrodonts are equivalent, in this respect,

* This variability has since been shown to exist in species of the *M. angustidens* type (1886).

† See under section on Acceleration and Retardation.

to the young of Uromastix, etc. The same thing occurs among the Scaroid and Labroid fishes. In this most natural family we find the majority of generic forms provided with a normal complete dentition; in others (Chaerops, Xiphochilus, Pseudodax, etc.) the lateral teeth are gradually and normally replaced by a more or less cutting edge of the mandible; and finally, in the Scarina and Odacina, the entire mass of teeth and jaws are coalesced, forming a beak with sharp cutting edges, the single teeth being still visible in the true Scarus, while they have entirely disappeared in adult Pseudoscarus and Odax.* Thus, in dentition, the adult Scarus is identical with not fully developed Odax; Chaerops, with the teeth less confluent, equals a still younger stage of Odax, while those with distinct teeth are the same in this point as the embryos of the highest—Odax, etc. I venture to predict that here will be found a long series of *exact parallelism*, in which the different genera, resting exclusively on these dental characters, will be found to be *identical* generically with the various stages of the successively most advanced.

11. Prof. Agassiz states that the absence of ventral fins is characteristic of an embryonic condition of the Cyprinodont fishes. The genus Orestias does not progress beyond this stage in this one point. Probably the genus will be found which will only differ from Orestias in the presence of ventral fins. If so, Orestias will be identical with an imperfect stage of that genus, if, as will probably be the case, the fins appear in the latter, after other structures are fully completed.

γγ. *Parallelism in Higher Groups.*

It is not to be anticipated that the series of genera exhibiting exact parallelism can embrace many such terms, since comparatively few stages in the developmental condition of the same part, in the highest, would bring us back to a larval condition, which, as far as we yet know, has no *exact parallel* among existing genera. But it is to be believed that the lowest terms of a number of the most nearly allied of such series do of themselves form another series of exact parallelisms.

Thus exact parallelism between *existing* genera of mammals ceases, with all characters which are larval or fœtal, only prior to

---

* Günther on Hatteria, "Philosophical Transactions," 1867, ii. I had already noticed the peculiar development in Uromastix, but not published it.

the assumption of the adult dentition, since among the higher Mammalia at least we know of no genus which, however similar to undeveloped stages of the higher, never loses the milk dentition. It is nevertheless an important fact that among smooth-brained mammals, or many of them, but one tooth of the second series appears; and inasmuch as smooth-brained forms of the higher orders have become extinct, it is not too much to anticipate that a type of permanent milk dentition will be found among the extinct forms of the same high orders.

As an example of exact parallelisms in series of series, I select the following:

1. In the batrachian family Cystignathidæ there are six groups or sets of genera. In the highest of these we have an ossified cranium and xiphisternum—i. e., in the Cystignathi; in the Pleurodemæ the cranium is not ossified, thus representing the Cystignathi while incomplete; in the Criniæ the xiphisternum is cartilaginous, as well as the fronto-parietal region, being an equivalent of a still lower stage of the Cystignathi. From this simplest type we can find a rising series by a different combination of characters; thus the Ceratophydes add an osseous cranium to the incomplete xiphisternum, while two succeeding groups diverge from each other at the start, the Pseudes loosening the outer metatarsus in their development to maturity, while the Hylodes add by degrees a cross-limb to the last phalange. The Ceratophrydes and Criniæ are stages in the development of these; but neither one of them is a step in the development of the other. They are measured by adaptive characters purely.

2. The whole suborder of the Anurous Batrachia, to which the above family belongs, the Arcifera, differs from the suborder Raniformia by a character which distinguishes a primary stage of growth of the latter from its fully developed form. That is, the Raniformia present, at one period of their development, a pair of parallel or over-lapping curved cartilages, connecting the procoracoid and coracoid bones, which subsequently unite and become a single, slender median, scarcely visible rod, while the bones named expand and meet. The first condition is the permanent and sole systematic character of the Arcifera.*

Objection.—It may be objected, by those who have observed

* This may be readily understood by comparing my monograph of the Arcifera, "Jour. Ac. Nat. Sci.," Phil., 1866, with Duges's work, or Gegenbaur and Parker's memoir on the Shoulder-Girdle.

some of these developmental relations, that they are exhibited by certain single structures only, and not by whole organisms. These objectors must not forget that the distinctions of those groups, which alone we have in one geological period in a relation of near affinity, exists in *single characters only;* and that it is therefore infinitely probable that the higher groups, when we come to know their representatives with the same completeness, will prove to be separated by single characters of difference also.

3. The following table (pages 63 to 73) is introduced to illustrate the relations of groups higher than the preceding. This is largely measured by the circulatory system, not only as to the class relations, but also as regards orders. In its less central portions it is, however, definitive of families at times.* (See also Plate I.)

If the reader will compare the history of the development of vertebrates of any class or order, as those of Teleosts and the lizard by Lereboullet, of the snake and tortoise by Rathke and Agassiz, and of the bird and mammal by von Baer, he will find the most complete examples of the *inexact parallelism* of the lower types with the embryonic stages of the higher. A few points are selected as examples, from the histories included in a few of the columns of the table, and given at its end.

Similar parallels may be found to exist in the most beautiful manner between the adult anatomy and structure of the urogenital apparatus within each class of the series taken separately, as indicating ordinal relationship. This department is, however, omitted for the present.

As an example of the homologies derivable from the circulatory system, and of the use of the following table, I give the following relations between the types of the origins of the aorta.†

The single ventricle of Teleostei is no doubt homologous with that of Lepidosteus, and that of Lepidosiren. The *arteria vesicæ*

* This sketch is not nearly complete, but is published in hopes of its being useful to students. It is compiled from the works of Meckel, Rathke, Barkow, Müller, Hyrtl, Brücke, Stannius, and others, in connection with numerous dissections.

† Prof. Agassiz ("Contrib. Nat. Hist. U. S.," i, p. 285) states that the ventricle of the Testudinata "is not any more identical with the one ventricle of fishes than with the two ventricles of warm-blooded vertebrata; for in fishes we find only one vessel, the aorta, arising from it, while in turtles both the *aorta* and *arteria pulmonalis* start together from it." We think this statement, which, if true, is destructive to the asserted homologies of the circulatory system, can not be substantiated, for the reasons above given.

PLATE I.

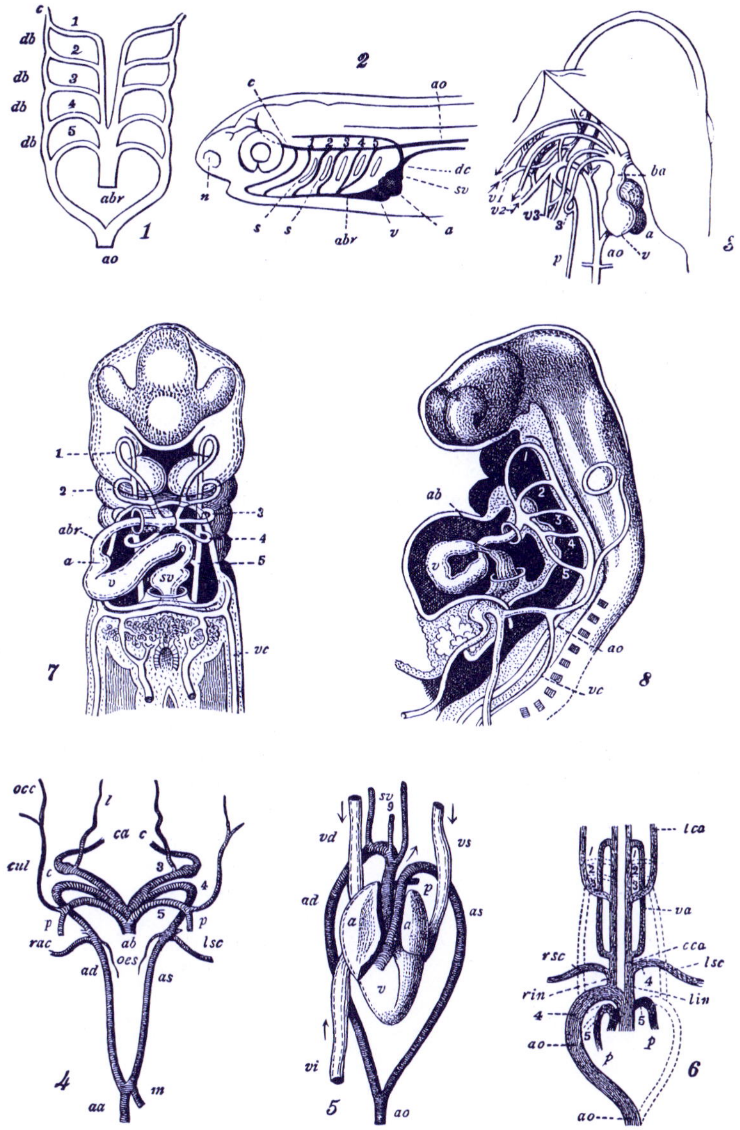

CENTRAL CIRCULATORY SYSTEM.

FIGS. 1, 2, fish. FIGS. 3, 4, batrachian. FIG. 5, reptile. FIG. 6, bird.
FIGS. 7, 8, human fœtus.

PLATE I.

Figures and Diagrams of the circulatory centers of Vertebrata. Copied from Gegenbaur and His, mostly enlarged.

FIG. 1. Diagram of the arterial arches of a Fish (species not given).

FIG. 2. Head of an embryonic Teleostean, with the rudiments of the vascular system (diagrammatic).

FIG. 3. Heart and arterial trunks of a larva of a Salamander.

FIG. 4. Arterial system of the Frog.

FIG. 5. Heart and arteries of an Ophidian (Boa).

FIG. 6. Diagram of the arterial system of the adult Fowl.

FIG. 7. Central circulatory system of Man; fœtus of the second month; front view.

FIG. 8. Same as Fig. 7, left side.

v, ventricle of heart.
a, auricle of heart.
s v, sinus venosus of heart.
b a, bulbus arteriosus of heart.
a b, aorta branchialis.
1, 2, 3, 4, 5, arteriæ branchiales.
v 1, 2, 3, venæ branchiales.
ao, aorta.
ad, aorta dextra (right).
a s, aorta sinistra (left).
p, arteria pulmonalis.
db, ductus botalli.
ca, arteria carotis.
cca, common carotid.
Rsc, *lsc*, right and left subclavian arteries.
Rin, *lin*, right and left innominate arteries.
cut, arteria cuticularis.
occ, arteria occipitalis.
v c, vena cardinalis.
v d, right vena cava.
v s, left vena cava.
vi, inferior (posterior) vena cava.
œs, œsophageal artery.
m, mesenteric artery.
s, spiracular or branchial fissure.
n, nostril.
dc, ductus cuvieri.
l, lingual artery.

natatoriæ, which is the homologue of the *A. pulmonalis* of air-breathers, issues in Lepidosteus from the last *vena branchialis*, thus receiving aërated blood from the gills. In Lepidosiren it issues from the point of junction of two gill-less and two gill-bearing *venæ branchiales*, thus receiving mixed blue and red blood, or blue blood altogether, when the branchiæ are not in functional activity. In Proteus it issues from the last vena branchialis, where it receives the *ductus botalli* of the preceding vein, which, when the gill is inactive, becomes a gill-less aorta-bow, which brings it only carbonized blood, which it readily aërates in the swim-bladder, now become a lung. The ventricle is homologous with the preceding. In salamanders, where the substitution of the accessory gill arches by the *ductus botalli*, converts the *arteriæ* and *venæ branchiales* into "aorta-bows," the *A. pulmonalis* is given off from the posterior bow, and receives henceforth mixed blood. In the Anura the origin is the same, but nearer the heart. In Cæciliidæ it approaches the heart so far as to issue from the extremity of the *bulbus arteriosus*, which is now divided by an incomplete septum, one half conveying blood to the *aorta-roots*, and the other to the *A. pulmonalis*. This septum was already preceded by a longitudinal valve with free margin in the Anura! As if to meet the coming event, a trace of ventricular septum appears at the apex within. There can now be no question of the homology of the ventricles of the gar, and of the Cæcilia. But we have next the true Reptilia. The *bulbus arteriosus* is split externally, as it already was internally, but it is first represented in most tortoises by an adherent portion, one half being the now, to this point, independent *arteria pulmonalis* and the other the nearly split aorta-roots. There can, I think, be little question of the exactitude of the homology throughout.

It is no less certain that the salamander * fulfills in its development the different stages to its permanent one, and is *identical* in each stage, in *respect to this point*, with the orders it represents at the time. This is true even of the long period during which it bears the long branchial appendages and contained arteries and veins which are not found in fishes; it is then like the Protopterus, which has hyoid venous arches and appendages of those arches at the same time. The tortoise † and Tropidonotus ‡ are also identical in their successive stages with the types already

* Amblystoma. † Agassiz. ‡ Rathke.

enumerated, the external or appendicular branchial vessels being omitted, as belonging to the special serial development of the line of air-breathing Anallantoidans. The division of the *bulbus arteriosus* into three instead of two may indicate a case of *inexact parallelism;* but, on the other hand, it may be that the pulmonary partition is completed a little before the aorta-root partition, thus passing through the batrachian permanent type. For explanations of inexactitude see under Part II. No doubt the batrachian type of *bulbus arteriosus* is passed by many serpents less extreme and specialized than the Tropidonotus.

The aortic and pulmonary divisions of the bulbus in the Cæcilia are not laterally placed, but one is dorsal and the other ventral, the one passing a little spirally to the right of the other. So the pulmonary division of the bulbus turns over to the right in the Anura. When the septum of the true reptiles appear, it rises on the anterior wall of the ventricle till it is seen, in Eunectes, to meet the partition between the *arteria pulmonalis* and *aorta-roots,* and we have at once the right and left ventricles of the bird and mammal structurally and functionally. Thus are the two ventricles of man the same as the one ventricle of the fish, merely divided by a septum.*

In the fissure of the aortic bulbus, in the reptiles, a spiral turn is again given, and in Testudo the one aorta-root issues behind the other. In the crocodile the turn is still greater, and the right aorta-root issues to the left of the left root, and vice versa. In the birds we have lost the left root, and parallelism ceases with this change. In the Mammalia the right root turns to the left, so that in the comparison of these classes the rule of von Baer above quoted is true; no mammal at present known is identical in a fœtal stage with any fully grown bird, but with a fœtus of the same, up to a certain point. But for both classes the parallelism of those below them holds true.

But it is with the exact parallelism or identity of genera that we have to do in the present essay. That being established, the inexact parallelism between the modern representatives of higher groups follows by a process of reduction.

---

* Agassiz, *l. c.*, denies the homology of the ventricles of the turtle and mammal; but it appears to me erroneously. He says: "The fact that the great blood-vessels (*aorta* and *art. pulmonalis*) start together from the cavum venosum seems to prove that the two cavities in the heart of turtles, which are by no means very marked, do not correspond to the two ventricles in Mammalia and birds."

	Heart sac.	*Auricles.*	*Ventricles.*	*Atrio ventric. valves.*	*Ventr. aorto-pulm., or aorto-bulbus valves.*	*Auriculo-sinous valves.*
Leptocardii	None	None	*One*, elongate	None.	None.	None.
Dermopteri	Present, att. to heart at *sinus & bran. trunk*	Present. Septum none. *Externally one.*	One, oval	*Orifice* 1; two valv's, membranous; no columnæ or chordæ.	*A.* To trunc's branch'l. Two semilunar.	Double.
Hyperotreti	Open into abd'm cav.	Poster. to ventricle.				
Hyperoarti	Clos'd in petromyzon.					
Elasmobranchii	Conn. by tube with abdominal cavity.	......	One trans.; (2) musc. reticul.; no column. carneæ.	" " two valves	*B.* To bulb's arterio's.	" with threads.
Plagiostomi	a. Bifurcate.					
Holocephali	b. Simple.					
Teleostei Apodes	Closed; con. w. heart by bands in muræna and some gadidæ.	Opposite ventr.	One, pyramid'l; (2) "	" " two valves	Two (4 in Orthagoriscus).	" (Four in Orthagoriscus); threads not constant.
Eventognathi	......	Opposite ventr.				
Heterosomata	......	Opposite ventr.				
In general	......	Dorsal anter.				
Ganoidei	......	Dors. ant. No sept.	(2) "			
Chondrostei	Conn. w. abd. by simple duct.	......	" with vasc'lar elevat's dors'ly & con. by vessels to pericard'm.	" " two valves; w. chordæ tendineæ.	......	One, annular, split, each half w. pockets.
Holostei	......	......	One, pyriform, (3) w. trabiculæ carneæ.	......	......	Double.
Lepidosteus	......	......	......	" " ⅔ the apert're bound'd by one valve.	None....	
Polypterus	Only a deep concav. toward abd. cav..					
Dipnoi	Closed	Dors. ant. An incomplete septum.	One, pyriform.	" " ; *one valve* bounded by chordæ cartilaginæ.	No semilunar.	None.
Batrachia Urodela	Closed.			*Two orifices.*		
Proteida	......	A complete septum.	......	" "		
Trachystomata	Closed; *att. by infer. 3d of post. face.*	......	......			
Trematodera	Closed; att. to apex in Amphiuma.	A complete septum.	......	" "		
Myctodera	......	A complete septum.				
Anura	......	A complete septum.	No rudiment'y sept. 1 & 2 " 3 & (4) a rudiment. sept. fr. apex.	" " approximate, small.	One at orifice of e. of the two primary div. of bulbus.	None in Pipa.
Gymnophiona	......	......				
Reptilia	Closed	A compl. sept. *Externally two.*	......	*Two orifices.*	*Three semil.* at ostium bulbi. ? valv. at m'th of bulb.	
Lacertilia	......	......	......	......	......	Two.
Amphisbænia						
Ophidia	......	......	Walls not thick; very coarse network in sin. arteriosus.	One. One.	aor. 2 semi. 2 semi. | p. 2 semil.	Two, semilun.

	Heart sac.	*Auricles.*	*Ventricles.*	*Atrio ventric. valves.*	*Ventr. aorto-pulm., or aorto-bulbus valves.*		*Auriculo-sinous valves.*
Testudinata........	Closed; att. by infer. 3d of post. face.	A complete sept....	" 1 & 2 " 3, 4, (5) walls very thick & spongy; ossiculum on anter. end, complete side of septum.	Two; the right rudimentary. One.	2 semi.; 2 semi.; inn'r of both on ossicu.	2; inner on ossiculum.	Two semil.
Crocodilia..........	" " "	A complete sept....	*Two*, sept. compl'te; 2 & 4 " occasion'l perforations near apex.	Two; the right muscular. One(? or 2).	2 semi.; 2 semi.	2 semil.	"
Aves................	Closed.........	A compl. sept. Externally slightly separated fr. ventricles.	1 " 2, musc. trebeculæ & column. carn'æ; septum convex into right ventr.	Two; right larg'r, musc. Two.	3 semil.	3 "	" inferior = valvula eustachii.
Cursores Apteryx.	....	....	....	Chordæ tendineæ sinistræ.			
Mammalia	Closed	A complete septum.	1 & 2 " walls more compact.	3 ("tricuspid"). 3 ("mitral").	3 "	3 "	One = v. eustachii; rudimentary.
Implacentialia	....	The right, small.	....	All membranous & bound by chordæ tendineæ; in monotremata; two in mitral, w. are att. to columnæ carneæ.			
Monotremata. ...	....	....	....				
Marsupialia......	....	No fossa ovalis on septum.					
Placentialia							
Halicore	....	....	Separated for some length.				

	Post-ventricular cav.	*Arteriæ branchiales.*	*Venæ branchiales.*	*Arteria pulmonalis.*	*Vena pulmonalis. At first v. vesicæ pneum.*	*Art. subclaviæ.*
Leptocardii	I. Series of lateral alternating bulbilli.	One *from* each *bulbillus.*	*Enter aorta direct;* All separate.	None (no swim-	bladder). None.	None.
Dermopteri	II. Truncus branchialis communis.	[A pair of ant. or false aor. bows; rudimentary in Hyperotreti.] *Fr. truncus;* 7 usually anter. somet. united, = bifurca. of truncus.	? united..........	"	....	"
Elasmobranchii	III. Bulbus arteriosus w. musc. coat & 2 or more rows valves.	*Fr. truncus,* 5; fr. bulbus; 2 ant. united.	Anterior united.	"		
Plagiostomi........						
Squali.........	....	Three post. sepa'te.				
Rajæ..........	....	Three post. united.				
Pristes..........	....	Three post. united.	....	....	....	From origin of aort.
Holocephali.						

Teleostei	IV. Bulbus arter. without musc. walls, no valves. (exc. Butyrinus.)	*Fr. truncus* which is fr. bulbus; 4 on each side (anter. wanting). a. Two ant. sepa'te. b. Two ant. united. 1. All to gills. * Post. sending art. to air-sac, = branch. operc.	*Enter aort.* by a. roots, i. e., by poster. trunks of circulus cephalicus, or direct when poster. to completion of circulus.	? Empties into portal, hepatic, or cardinal vein.	Into vena hemiazygia or from vertebralis posterior; often subdivides into rete, and unites.	Origin often unsymmetrical. Comm. trunk of anter. venæ branchiales, Esox.
Apodes Murænophis … Nematogn. Saccobranchus	…					From origin of aort.; Perca.
Class in general	…	** Post. not sending art. to air-sac, = bran. operc.	…	…	Into v. portæ in Lota.	Fr. each aorta-root; Gadus.
Ap. Symbranchus Monopterus	…	2. 4th no gill, $\therefore$ = aor. bow.	…	…	…	From origin of aorta, then dividing into 3; one branchialis, one *lateralis!* one dorsal. ascend.
Amphipnous	…	Five art. bran. ant. to air-sac. 1st & 4th no gills; vein from air-sac to 4th.				
Ganoidea	V. Bulbus arter. w. musc. walls and 3 or very many rows valv.	*Fr. truncus*, fr. bulbus; 5 on e. side; ant. is an opercular pseudo-branch'l and revehent art.	*Enter aorta* direct.			
Chondrostei		1. Rising from 2d art. branch.				
Spatularia	…	a. 1st true branch'l to 2d gill; 2d and 3d united.				
Accipenser	…	b. 1st true br. art. to 1st gill.				
Holostei	…	2. Rising fr. & = ant. fork of truncus; 1st true art. br. to 1st gill.				
Lepidosteus	…	…	…	From last vena branchialis.	*Enters hepatic vein.*	Very large; fr. aort., poster. to entrance of 3d vena. branch. communis.
Polypterus	…	…	…			
Amia	…	Four on e. side fr. truncus, 2 ant. unit'd.	…	? Empt. into vena cava ascend.	" vena cava ascendens.	
Dipnoi	VI. Bulb. arter. spiral. w. two elongate spir. valves.	*Fr. 2 primary heads of bulbus;* 5 on e. side; 1st rising fr. 2d; 2d & 3 gill-less; 5th to 2 gills.	*Enter two aorta-roots;* two become parts of aorta-bows.	From confluence of v. branchiales & aorta-bows.	*Enters auricula sinistra,* pass's vena cava & sinus, and has valv. tuberc. at ostium.	
Protopterus	…	a. Trunc. to ext'l gills.				
Lepidosiren	…	b. No " " "			"	
Batrachia	VII. Bulb. not muscular; a longitud. vlv.	…	…	…		
Urodela. Proteida	Bulb. narrow at beginning.	Prim'y div. of bulb. = 2 trunc. br. comm.: 3 Art. br. on each side, 2 post. united first.	Are both ven. branch and aorta-bows, from existence of 3 duct. botalli.	From poster. vena branch. " " "	…	From aorta.

5

	Post-ventricular cav.	*Arteriæ branchiales.*	*Venæ branchiales.*	*Arteria pulmonalis.*	*Vena pulmonalis. At first v. vesicæ pneum.*	*Art. subclaviæ.*
Trachystomata...	Bulb. twisted; one longit. elevat. (= septum) w. six grooves corr. to art. branch.	a, b. *From bulbus*; 3 on each side.	No ductus botalli.			
Trematodera.....	....	Become *aorta-bows* from want of gills. Four; ant. & post. conn. by duct's botalli; 2 median = aorta-root.		Fr. poster. aort.-bow.		
Myctodera	....	Four; ant. & post. conn. by duct's botalli; 2 median = aorta-root and with one ostium into bulbus art.		" " " giving branch's to œsoph. and pericardium.	....	" "
Anura.............	Contain. longit. septum w. a free edge: each half prolonged into 1 primary div. of bulbus.	Three; no ductus botalli; median = aort.-root.		Fr. poster. aort.-bow sends off ramus cutaneus in front of shoulder to back.		
Gymnophiona......	Elongate; div. anterior. into dorsal & ventral cham.; valves at origin.	One; by obliteration of 1st and last, *aorta-roots* only remain.		Fr. dorsal chamber of *bulbus arteriosus.*		
Reptilia..............	VIII. No proper bulbus; the aorta-roots and pulm. art. adherent.			From cavum venosum = (dextrum) of ventricle.		
Lacertilia in gen.	....	Two aort.-bows, fr. comm. ostium fr. cavum arteriosum (sinistrum).		....	"	From trun. subclav. comm.; fr. right aorta-root.
Varanidæ and Tupinambis	....	One aorta-bow, = aorta-root. " "				
Amphisbænia......	....	*One* " " " "				
Ophidia...........	....	" " "				
Testudinata.... ...	Aorta-roots and art. pulm. adherent and surr. by ring of musc. tissue at origins.			....	"	Each fr. innommin., fr. trun. innom. comm. from right aorta-root.
Crocodilia..........	Aorta-roots and pulm. art. adherent; right aorta-root from left ventr., left from right, commun. by a foramen above valves.			From right ventricle; attached.	"	Dextra fr. right aorta-root; sinist. fr. innom.
Aves.................	IX. Right aorta-root (fr. left ventr.) only = aorta.			Do. Free.	"	Sinistra very large; given from aorta immediately; first as innominata.
Mammalia	X. Right aorta-root (i. e., aorta) turning over the left bronchus to the left side.			Do. "	"	Dextra issues later; sometimes an innominata.

	Carotides.	Cerebrales.	Ophthalmicæ.	Vertebrales.	Intercostales.
Leptocardii.					
Dermopteri.					
Hyperotreti	1. Each carot. communis a continuation of a vessel connecting venæ branch. parallel to aorta. Divide to *c. ext.* and *int.* behind skull; *intern.* unite at base of cranium and receive vertebralis.	Median from conjunction of carot. intern.	......	Median, a continuation of aorta.	
Hyperoarti	2. Carot. communis from anter. venabranchialis. Carot. intern. do not unite.				
Elasmobranchii	3. No c. communis; posterior from 1st vena branch. or from united anterior ven. branchiales; united beneath cranium.	......	......	......	From aorta direct.
Plagiostomi	(a). Anter. carot. from pseudobranchia, i. e., vena branch. prima.	......	Fr. pseudobranchiæ.		
Holocephali	b. Ant. ("poster.?") from ven. branch. secunda.	......	Fr. carotis anterior.		
Teleostei	4. C. commun. from circulus cephalicus.	......	Fr. pseudobranch., ends in "glandula choroidalis."	......	Often break into retes.
Ganoidei	......	......	Fr. pseudobranch.		
Chondrostei	As in 3 (a).				
Holostei.					
Lepidosteus ..	5. Anter. carot. from aorta, anter. to entr. of 3d pair of v. branchiales; poster. carot. sending no branch to brain.	From near origin of aorta.	Fr. pseudobranch.		
Dipnoi	As in 3 (b)	......	Fr. carotis anterior.		
Batrachia.					
Proteida	As in 3: (c), the ?anter. ("post.") carot. from aorta-bow.				
Trachystomata.					
Trematodera	6 (a). Carotis comm. from anter aorta-bow. No glandula carotica.				
Myctodera	6 (b). A glandula carotica......	......	From aorta-root (most anteriorly).	......	In pairs from aorta.
Anura	6 (b). Carot. extern. cont. 1st as max. intern. and then ophthalm.	......	From carot. extern.	......	Small; from an "a. supravertebralis" run'g. ab. diapoph. on each side, ? homolog. w. intercostal. prim.?
Gymnophiona ..	7. Carotis commun. from anter. curve of each aorta-root.				
Reptilia	A. Carotis poster. connected w. its fellow by ramus communicans behind foramen magn.	B. Continuation of carotis interna. Circulus willisii compl'te.	From A. facialis (= carot. ext.) more or less connected with cerebr. anter.		

	Carotides.	*Cerebrales.*	*Ophthalmicæ.*	*Vertebrales.*	*Intercostales.*
Ophidia........	8. Both carot. commun. from right aorta-root direct. (a). Entirely separate. (b). Running together after issuing from a short tr. carot. impar. (c). Left carot. comm. rudimentary. (d). " " " absent.	....	....	One from trunc. arter. dexter.; divides; each half running beneath diapophyses & entering ramus com. of carotis internal; through this to basilaris.	Fr. trunc. art. dexter., fr. comm. truncs fr. aorta.
Peropoda, Angiostomata	....	....	Fr. cerebralis anter.	In Rhinophis, strong, like a cont. of aorta.	
Amphisbænia..	7. Each carot. comm. from anter. part of each aorta-root.		From ?		
Lacertilia	....	....	Fr. art. facialis....	Two from axillaris; each entering spinal canal and forming a supraspin. thro' those to basilaris.	Fr. subclaviæ, trunc. arterios. dexter, and aorta.
Varanidæ and Tupinambis	As in 7.				
In general....	9. Each carot. comm. a continuation of cardiac extrem. of anter. aorta-bow.				
Testudinata........	10. Each c. c. from innominata; fr. common innom.; from right aorta-root.	....	Fr. cerebralis anter.	" " " also connect w. cervicalis descend. from carot. extern.	In body and tail fr. an intercostal, prima continued fr. subcl. to iliaca, the "intern. recurrens." In neck from ram. comm. fr. carot. ext. to subclav.
Crocodilia............	11. (a). Car. comm. fr. truncus carot. impar, w. is from left innom.; or (b) c. com., each from an innom.; all from right aorta-root, or as in 10.				
Aves.................	B. *No occip. union of carotides posteriores:* one or both c. comm. run on sides of neck.	Fr. a union or communication of carot. int. w. = ramus com. of cerebralis anter. of mamm.; circ. willisii is abbreviated anteriorly. Cer. anter. goes to orbit or olfactory reg. Poster. to basilaris & spinalis anter.	External branch of carotis internus.	B. *Traverses cervical canal.*	Usually from aorta.
Insessores......... Passeres, Zygodactyli ¾, Syndactyli ½	(a). Car. comm. fr. tr. carotic. impar. fr. left innom. [vid. on Pitta sp. Tidjschr, Neederl. Ind.]			1. Not forming basilaris; conn. w. occipitalis. From truncus or carot. communis of its side.	
Synactyli, Zygodactyli cuculus and in general, Cursores, exc. Rhea, Natatores	(b). Carot. comm. from each innom. from aorta.	....	....	....	In Gallus, a few fr. intercost. prima; in Ciconia last from Epigastr., rest fr. prima.

Natatores Podiceps	As (a).....	.....	.....	.....	In Anas 5 or 6 from intercost. prima, 3 or 4 fr. vess. connect. in. terc. prima w. aorta.
Cursores Botaurus.	As in 11 (b)!				
" Phœnicopterus.	(c). Carot. comm. from right innom. from aorta.				
Mammalia	.....	Via canal. carot., or foramen jugulare.			
Implacentialia.....					
Monotremata, Phascolomys & Phascolarctus	d. One car. comm. fr. innom. dextra fr. aorta; one from aorta direct.	Ramus communic. betw. cerebr. anter. C. willisii alw'ys pres'nt.		3. Forming basilar.	
Marsupialia.....	c. Both c. c. fr. innom. dextr.; fr. aorta.	Somet. ramus comm. is between art. corporis callosi.			
Placent. Rodentia...	e. " " " " " except Muridæ. Cricetus, Dipus Meriones, which exhibit d.	Fr. carot. ext. via a. ophthalmic and foram. opht. in some Caviidæ; via max. intern. and foram. lac. anter. Hystrix.	Sciurus = intern. br. of ant. cerebr. (br. ext. leaves cran. and joins max. interna). Arctomys same, without extern. branch.	In Mustela fr. tr. communis vertebr. fr. aorta.	In Mustela, anterior fr. tr. com. verteb.
Proboscidia.....	f. From truncus car. impar.; fr. aorta direct; Elephas.				
Artiodactyla	g. From truncus car. impar.; from right innom. Sus.	From carot. ext. via max. intern. in Ovis.	Sus; from max. int. = carot. ext.	2. Connected with occipitalis; w. basilar by small branches.	
Perissodactyla	h. Each c. com. from right innom.; from innom. communis.				
Equus.....	i. From tr. carot. impar.; from r. innom.; from innom. communis.	.....	.....	As in 1.	
Cetacea, Halicore	As e.....	.....	.....	As in 3.	
Cetacea, Phocæna	As i.				
Cetacea, Delphinidæ	k. No carot. comm. Carot. intern. and ext. fr. innom.	.....	.....	4. Confluent w. occipital. and cervic. ascend. = "cervico-occip."	5. Anter. fr. mammar. intern.
Carnivora.....	As in e, except Phocidæ = d., also Brachypus, Dasypus and Cycloturus.	.....	.....	As in 2 in Canis, Mustela.	
Edentata.....					
Insectivora, Sorex	As in e.....	& in some rodents, enter'g via stapes!!			
Insectivora, Talpa	As in l.....				
Insectivora, Erinaceus	As in d.....				
Cheiroptera	l. C. comm. fr. each innom.; fr. aorta.				
Quadrumana	As in d.				
Homo.....	.....	.....	.....	.....	1st fr. subcl., rest of dors. div. from aorta.

	A. cœliaca.	*A. mesenterica.*	*A. iliaca.*	*Sinus venosus.*	*Venæ anonymæ.*	*v. Hemiazygeæ or Vertebrales poster.*
Leptocardi.						
Dermopteri.	From aorta.	...	...	a. Present; contractile.		I. Supravertebral. (a continuation of caudalis):
Hyperoptreti	...	...	...	...	I. One on left side...	On left side only.
Hyperoarti	...	...	...	...	II. Two, called "Trunci transversi." Receive subclaviæ, vertebrales and hemiazygeæ.	On both; equally large. (In Petromyzon *conn. w. subvert.* sac.)
Elasmobranchii	" Includes Mesenterica anterior.	A posterior only.	...	Present ...	...	II. Subvertebral.
Plagiostomi						a. Emptying into anonyma.
Squali	In Lamna sp. developes retes mirabiles near diaphragm.					* Of equal size. Receive renales revehentes.
Pristes						
Rajæ	[Mesent. ant. separate]	...	...	...	...	Conn. w. a subvert. sac.
Holocephali						
Teleostei	... [In Thynnus a subhepatic rete.] [In Gadus g. from aorta-root.]	...	...	Present ...	...	** The dexter larger.
Plectognathi	...	...	...	...	...	* Of equal size (Diodon).
Ganoidei						
Chondrostei	...	...	...	...	...	* Of equal size; breaking into spongy masses or retes in Accipenser.
Holostei						
Dipnoi	...	...	...	Present.		
Batrachia	...	...	...	"		
Proteida	" Gastrica separate from aorta.	Many, small ...	Symmetrical.			
Trachystomata						
Trematodera						
Myctodera	" Gastrica indepen.	Two, anter. & post.	Give femoralis and epigastrica.	...	...	Do not receive renal. revehs.
Anura	From left a.-root, including both mesenterics.	...	Rise at mid. of coccyx			aa. Emptying into iliaca. Two, equal.
Gymnophiona						
Reptilia	...	...	...	Present.		

Ophidia	" Four gastricæ.	" " "		Continuation of anonyma dextra and v. cava inf.	An. sinistra the continuation of jugularis only.	a. Emptying into Anonyma. One median.
Amphisbænia						
Lacertilia. (Psammosaurus	From aorta.)	United, fr. left a.-rt.				
Testudinata	A gastroepiploica instead.	One, fr. left a.-root.	Hypogastric & cruralis separate.	a. Present.		III. Above heads of ribs conn. iliaca and anonymæ.
Crocodilia	One, incl. gastr. and mesenterica; fr. left aorta-root.			?		II.?
Aves	" Excl. mesent.; fr. abdom. aorta.	Super. fr. aorta; inf. fr. caudalis s. sacra-media.	1. *None;* cruralis & hyp'g. separate; latter fr. caud. s. sacra med., not giving ischiadica, w. rises from aorta & is larger th. cruralis.	aa. Absent.	Coronaria enters an. sinistr. anonym. from union of vertebrales first, then jugularis, then subclavia.	III. a. Each unites w. ant. verteb. Comm. trunk then ent. anonyma.
Apteryx	" Fr. ab. diaphr'm.	Sup. fr. ab. diaphr.				
Mammalia	One fr. aorta.	Independent.	Cruralis and hypog. mostly separate.	aa		(II.) aa. Enter v. cava descend. or anter., or the two anonym.
Implacentialia		Infer. wanting	Ischiadicæ fr. aorta [(larger than cruralis) =hypog. of Meckel ?]		"	" }
Monotremata						
Marsupialia						" } 1. Two, equal.
Placentialia		Infer. present	Ischiad. fr. hypogastr., *exc.?*			
Rodentia		Super. wanting in Cavia.				" } Except ?
Proboscidia					Elephas	2. One, or dextra much larger.
Perissodactyla					III. An. sinist. enters dextra (coronaria ent. comm. tr.) Conn. by a cross trunk in Lepus.	"
Solidungula						"
Cetacea			Wanting in cruralis. Hyp'g. ending in retes in Manatus.			IV. a. Run in spinal canal; emerge, unite and enter anon. dext.
Carnivora						II.
Edentata						
Insectivora		Talpa, super. emptying in cœliaca.			Erinaceus & Sorex as II.	" In Talpa, 2, equal; one entering right atrium.
Cheiroptera		Vespertilio, super. emptying in cœliaca.			Same as in II.	" The right much larger, empt. into v. cav. desc.
Quadrumana			2. *Present.*			

	Vena caudalis.	(V. vertebralis et) V. Jugularis (extern.)	V. cava ascendens.	V. iliaca.	V. portæ.	V. hepaticæ.
Leptocardii	...	...	...	...	Contractile rhythmically.	
Dermopteri	A. Single ... a. Prolonged directly into hemiazygeæ.	A. *Does not take blood from brain.* Single (enters sinus venosus); on l'ft side.	None ...	...	Fr. aliment. canal and organs.	I. Two or three; emp. into sinus venosus.
Hyperotreti	...	(a). Verteb. sinister only (fr. skull).	...	...	(a). Contr. rhythm'y. Receives veins of abd. wall. Has a sac-like expansion.	
Hyperoarti	...	(b). Two vertebral. (from skull), mostly receive subclavia.	...	...	(b). Not contractile.	
Elasmobranchii	...	I. Double; one each side. Verteb. take blood from skull.	" ...	...		
Plagiostomi	Common intercost. trunks subdivided in kidneys.	...	...	...	Concealed in margin of spiral valve.	
Rajæ	...	...	...	...	Finally single.	Exhibit sinuses.
Holocephali						
Teleostei	b. Continued as renalis advehens; as renalis, receives com. trunks of 2 or 4 intercostals.	I, or single (entering sin.) in Cottus, Thynnus; often dilate into sinuses.	"[present in Perca, from spermatica and pneumatica; does not ramify in liver.]		Often rec. veins of walls of abdomen (homolog. of epigastrica). Often tr. to liver are very numerous, esp. Eventognathi.	II. Mostly one; often two; rarely three.
Ganoidea Chondrostei	...	...	None.		One trunk.	
Polyodon	...	...	...	...		
Accipenser	...	...	...	...	...	One.
Holostei						
Dipnoi	Divides to join renalis advehens.					
Batrachia	...	B. *Takes blood fr. brain.*	A. Present; formed fr. renal. revehentes. *Passes through liver.*	A. *Is a renal. adveh.*	From a median abdominalis anter., or epigastrica.	a. Empty into v. cava ascend.
Trachystomata						
Proteida						
Myctodera	c. Forks, each fork uniting w. cruralis.	...	...	1. a. From fork of caudalis & cruralis.		
Anura	None; crural. = renalis adv.	...	...	a. Forks of caudal. = a transv. anastomosis.	An origin fr. each iliaca.	
Gymnophiona						
Reptilia						

Ophidia	...	(I.) Dexter only into anonyma.	...	From fork of caud. (=? hypogas.) only on left; on right from caudalis and abdom. impar.	Is continuation of v. abdominalis w. rec. v. of aliment. canal & accom. organs; contains a long spiral valve; originates fr. iliaca dextra only.	
Amphisbænia.						
Lacertilia	...	...	...	...	An origin fr. each iliaca.	
Testudinata	B. Double; ab. diapophyses.	Fr. cranial sinuses.	...	Fr. caud. (*hypogas.?*) and cruralis.	Three; 1st from two symmet. v. abd. ant., 2 fr. do. abd. post. 3d fr. v. tract. intestinal.	b. Empty into atrium dextrum.
Crocodilia	A. Single; hypaxonic.	...	Formed from renal. revehs. and branches of iliacæ.	AA. *Is not renalis adv.* I. Sends off large ren. adv., takes up. ren. rev. [Previously conn. by anast.] continued as abd. anter.	Three; an abd. fr. each side fr. iliaca, & v. of aliment. canal & organs.	a.
Aves	Two, unite and continue into V. portæ.	A. Do not receive bl. fr. brain, or a small branch only. Not near course of carotids; an anastomosis nr. head; ∴ right is larger, or alone (as in Picidæ sp.) Subst. by verteb. Run in cervic. canal.	A. Fr. iliacæ; receiv's abdom. ant. after hepaticæ.	2. Fr. hypogas. and crur. [hypogas. a continuation of lat. fk. of common caudalis; as such perforates kidney.]	Comm. trunk from united caud., taking up v. tractus intest.; [∴ = mesenterica.] or former is a right v. port., the latter one or more v. p. sinistræ.	a. Usually 2, dext. & sinist. in liver.
Apteryx	...	...	...	Not imb. in kidney.		
Mammalia	Ent. v. cava ascend.	B. Take blood from brain; verteb. small, in cervic. canal.	B. Through groove of liver only.	2. No part. of hypogas. passes thr. kidn.	From splenic and sup. mesenter. only.	a.
Implacentialia.						
Placentialia.						
Rodentia and Pro	boscidia.					
Artiodactyla	...	...	...	...	Semilunar valves in Bos.	
Perissodactyla	...	...	...	...	Semilunar valves in Equus.	
Cetacea	In many into right iliaca.	...	With expansions in most divers.			
Carnivora	...	...	In Phoca inclosed by contractile muscular right at diaph'm.			
Edentata and Ins	ectivora.					
Cheiroptera.						
Quadrumana	...	C. Take no blood fr. brain; jugular. intern. do this.				

δ. *The Extent of Parallelism.*

Prof. de Serres and others have stated it as their belief that the lower "branches" of the animal kingdom are identical with the undeveloped forms of the higher; i. e., that the mollusk and articulate are not merely parallel with, but the same as the lower conditions of, the vertebrate. The works of various embryologists, as von Baer and Lereboullet, have shown this statement to be erroneous, "and founded on false and deceptive appearances." The embryos of the four great branches of the animal kingdom appear to be distinct in essential characters from their first appearance. But Lereboullet, who, in his prize essay, has compared with care the development of the trout, pike, and perch of the Teleosts, with that of a Lacerta among reptiles, has failed to point out characters by which the embryos of the two vertebrate classes essentially differ, for a considerable period. It is true that, as each and all of the species belong to widely different generic series, parallelism is of the kind to be called *inexact* or *remote*. But enough is known of embryology and paleontology to render it extremely probable that the historic predecessors, of the types whose embryology Lereboullet studied, formed a series of parallels of the kind termed in this essay *exact*.

Lereboullet states that a certain difference exists between the eggs of the fishes and those of the Lacerta. This is for us merely stating that the parents of the embryos differ, a fact which no one will contest. The same may be said of the elevated or depressed character of the surface of the vitellus on which the embryo reposes.

Secondly, after the appearance of the embryo the Lacerta is furnished with the amnios and allantois; the Teleost not. This is certainly neither a generic, ordinal, nor class character of the adult, for it is but temporary; therefore, in generic, ordinal, and class characters the embryos of the Teleost and Reptile are still identical. It is a physiological character, and not morphological, and therefore far the less likely to be a permanent one, even in embryos, under changed circumstances. The female of one of the species of Trachycephalus inverts the skin of the back at one season of the year to receive her eggs, because she can not lay them in the water; the other species of the genus do not. The next genus in direct morphological line possesses a single species whose female does the same for the same reason; but the relations of these species and genera are zoölogically the same as though this modification did not occur. Many such instances will occur to many naturalists. It is not pretended that they are as important as the presence of the allantois; but they constitute a character, no doubt, similar in kind, and entirely at the service of the needs of the great system of morphological succession. The same may be said of the vascular area of the Reptile.

Lereboullet concludes his summary of the differences between the Teleost and Reptile, up to the period of completion of the heart, by saying, "It is easy to perceive that all these differences, however important they may appear, are constituted by the accessory organs of the embryo, and do not modify the development of the latter, which progresses in reality exactly as in the fishes." He says the same previously, as to the relation of the same to the bird and mammal.

We have, then, in the embryos of the lower vertebrates at a certain time in the history of each, an "*exact parallelism*" or *identity* with the embryonic condition of the type which progresses to the next degree beyond it, and of all the other types which progress successively to more distant extremes.

We have, however, so far, every reason to suppose that the embryos of the other branches of animals never present an exact parallelism with those of the Vertebrata.*

The embryo of the fish and that of the reptile and mammal may be said to be generically, if not specifically, identical up to the point where preparation for the aërial respiration of the latter appears. They each take different lines at this point. The fish diverges from the course of the reptile, and proceeds to a different goal; the shark does the same, but proceeds a shorter distance; while the Dermopter scarcely leaves the point of departure. No doubt, there have been types which never left this point and *whose plan* of *circulatory system* is identical with that of the embryo reptile and mammal. *Such a type was only generically different from the reptile or mammal which had only taken the succeeding step*, provided other structures were not superadded.

By comparing the development of types of different classes in

* At about the time this was written the important papers of Haeckel on his admirable Gastræa theory were published, but had not reached the author. Haeckel shows the approximate identity of all the types of embryonic development. (Note, 1886.)

certain features which are only ordinal or generic in meaning, very erroneous conclusions may be reached by the inexact student as to the want of parallelism of classes to each other. Thus Rathke says of the development of the eye of the snake Tropidonotus, at a certain period, that it is far in advance of that of the mammal at the same stage.' Here, says the objector, is a case where their parallelisms do not coincide ; the mammal is really similar to a younger stage of the reptile.

But, in fact, the size of the eye is but a generic or family character ; if the development of the lemur had been compared with the snake, the mammal would have been found to be in advance ; of the mole, much farther behind. If the snake selected were the purblind Atractaspis, almost any mammal would have been in advance ; if, on the other hand, the great-eyed Dipsas, but few Mammalia would have been parallel to it.

In a word, to find *exact parallelism* it is necessary to examine the closest allies.

It is also of first importance to distinguish between the *existence* of generic or higher characters, and their condition under various circumstances of *individual life.* If a fœtal or larval character be conserved through the adult life of a type, it will be of course adapted to the functions of mature age. Thus the undeveloped character of the horns of the genus of deer, Rusa, are not accompanied with the marks of individual youth of the corresponding stage of Cervus ; its individuals are fully grown and *functionally* perfect. The species of Hyla are not small and incapable of self-preservation and reproduction, as is the corresponding stage of Trachycephalus ; they are functionally developed. The student need not be surprised, then, if, when identity or exact parallelism is asserted, he finds some differences dependent on age and adaptation, for if he be an anatomist he need not be informed that a morphological relation constitutes types what they are, not a physiological.

II. OF RETARDATION AND ACCELERATION IN GENERIC CHARACTERS.

First. Of adult metamorphosis :

The question has necessarily arisen, Have these remarkable relations between genera resulted from an arrangement of distinct generations according to a permanent scale of harmony, or have the same genetic series of individuals been made to assume the different positions at the same or different periods of the earth's history.*

Prof. Marcel de Serres proposed the theory of repressions of development to account for the existence of the lower groups of animals *as now existing*, an error easily exposed, as has been done by Lereboullet in his various important embryological writings. But little observation is sufficient to prove that a mammal is not a shark where it has five gill-arches or aorta-bows, nor a batrachian where it has three, or a reptile where it has the two aorta-roots. This has been already sufficiently pointed out by von Baer, who says there is "keine Rede" of such a theory as was afterward proposed by de Serres. Thus are true the rules propounded by this author.† 3. "Each embryo of a given animal type, instead of passing through the other given animal found, diverges still more from it." 4. "In the basis, therefore, the embryo of a higher animal type is never identical with an inferior type, but with the embryo only of the latter."

I think that I have already made some progress in proving that the near or true generic relationship is one of absolute developmental repression or advance. Paleontology shows that families and orders, as now existing, were preceded in time by groups which are synthetic or comprehensive, combining the common characters of modern generic series. This process of synthesis must, it is obvious, if continued, result in the near approximation of the single representatives of the now numerous and diverse groups. There is every reason to believe that a backward view through time will show this to have prevailed throughout the

---

* Some naturalists seem to imagine that the demonstration of the existence of intermediate types is only necessary to establish a developmental hypothesis. Thus Dr. Dohrn ("Ann. Magaz. N. Hist.," 1868), writing of his discovery of that most interesting genus, Eugereon, which combines characters of Neuroptera with those of Hemiptera, does not hesitate to say that it proves the truth of Darwin's theory. Now, it appears to me that a demonstration of the existence of a regularly graduated succession of types, from the monad to man, would be only the minor of a syllogism without its major, in evidence for development, so long as the proof of transition of one step into another is wanting; and the idea that such a discovery could establish a developmental theory is entirely unfounded. Indeed, the reasoning in which some indulge—if we dare so call the spurious article—based on this premise alone, is unworthy of science. The successional relation of types, though a most important element in our argument, has been long known to many who give no sanction to the idea of development.

† "Entwickclungsgeschichte," p. 224.

Vertebrata and other branches, as we already can in part prove. And I have no doubt that the synthetic types, which represent modern orders, have existed in a generic relationship subordinate to the plan of the synthetic class, and that the latter have existed as genera only, of the type of the great branch. This is not ideal. We only have to look to our extinct ganoids, Archegosaurs, Labyrinthodonts, Compsognathus, Archæopteryx, Ornithorhynchus, etc., to realize these facts.

The first genera then formed a scale of which the members were identical with the undeveloped stages of the highest, and each to each according to their position.

Such a series of antitypic groups having been thus established, our present knowledge will only permit us to suppose that the resulting and now existing kingdoms and classes of animals and plants were conceived by the Creator according to a plan of his own, according to his pleasure. That directions or lines of development toward these ends were ordained, and certain laws applied for their realization. That these laws are the before-mentioned law of RETARDATION AND ACCELERATION ; and law of NATURAL SELECTION.

The first consists in a continual crowding backward of the successive steps of individual development, so that the period of reproduction, while occurring periodically with the change of the year, falls later and later in the *life history* of the species, conferring upon its offspring features in advance of those possessed by its predecessors, in the line already laid down partly by a prior suppression on a higher platform, and partly, as above supposed, by the special creative plan. This progressive crowding back of stages is not, however, supposed to have progressed regularly. On the contrary, in the development of all animals there are well-known periods when the most important transitions are accomplished in an incredibly short space of time (as the passage of man through the stages of the aorta-bows, and the production of limbs in Batrachia anura) ; while other transitions occupy long periods, and apparently little progress is made.

The rapid change is called metamorphosis ; the intervening stages may be called larval or pupal. The most familiar examples are those which come latest in life, and hence are most easily observed, as in the insects and frogs. When, during the substationary period, the species reproduces, a constancy of type is the result ; when the metamorphosis only appears at the period of reproduction, a protean type is the result ; when the metamorphosis is crowded back to an earlier period of life, then we have another persistent type, but a new genus of a higher grade than its predecessor.

In reviewing many examples everywhere coming under the eye of the naturalist, it is easy to perceive what would constitute a *plastic* and what a *conserved* condition of generic, or even of specific form.

As one or more periods in the life of every species is characterized by a greater rapidity of development (or metamorphosis) than the remainder, so in proportion to the approximation of such a period to the epoch of maturity or reproduction, is the offspring liable to variation. During the periods corresponding to those between the rapid metamorphoses the characters of the genus would be preserved unaltered, though the period of change would be ever approaching.

Hence the transformation of genera may have been rapid and abrupt, and the intervening periods of persistency very long ; for it is ever true that the macrocosm is a parallel or repetition of the microcosm in matter and mind. As the development of the individual, so the development of the genus. We may add : so the development of the whole of organized beings.

These metamorphoses may be fitly compared to those in the molecular constitution of matter. The force of cohesion between the atoms of a vapor steadily increases with descending temperature, and in a regular ratio, till a given point is reached, when a sudden metamorphosis to a denser or liquid condition takes place. Nor have we reason to believe, with regard to many substances, that there is any parallel relation between the temperature and the molecular constitution before or after the metamorphosis takes place. So, the temperature continuing to descend, the molecular character of the liquid remains unchanged until, the *vis conservatrix* suddenly giving way at the ordained point, a soild is the result. Thus, while the change is really progressing, the external features remain unchanged at other than those points, which may be called *expression-points.*

Now, the *expression-point* of a new generic type is reached when its appearance in the adult falls so far prior to the period of reproduction as to transmit it to the offspring and to their descendants, until another *expression-point* of progress be reached.

Thus a developmental succession does not so obliterate the lines drawn around Nature's types as to render our system ineffectual as an expression of them.

The successional acceleration or retardation in metamorphosis may be best illustrated in certain tailless batrachians, by the following tables. These are taken, it will be remembered, from the Bufonidæ and Hylidæ as examples of "exact parallelism"; three are now added from the Ranidæ and Discoglossidæ. The case of "inexact parallelism" is that of the Scaphiopodidæ.

Whether they are cases of acceleration or retardation can only be determined by reference to the paleontology of the respective groups, or a careful comparison of times of metamorphosis. In the case of the Discoglossidæ I suspect it to be retardation, as the highest genus is extinct. The others I shall arrange with them for temporary convenience. Were I dealing with a group of Ganoids, I should imagine the process to be retardation, as this group is going out of existence. On the other hand, were they higher Oscine birds, we might imagine the case to be reversed.

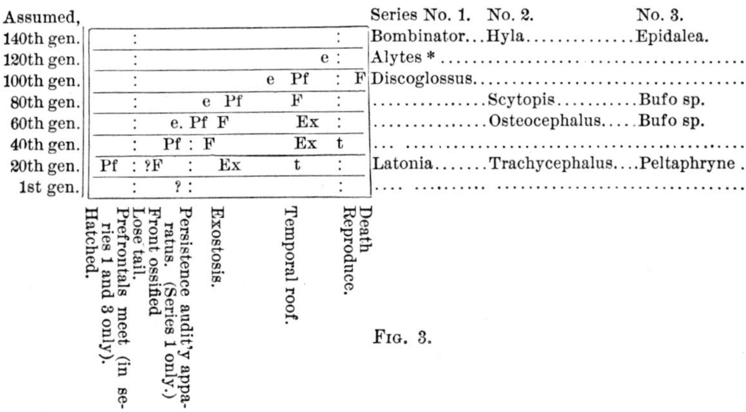

FIG. 3.

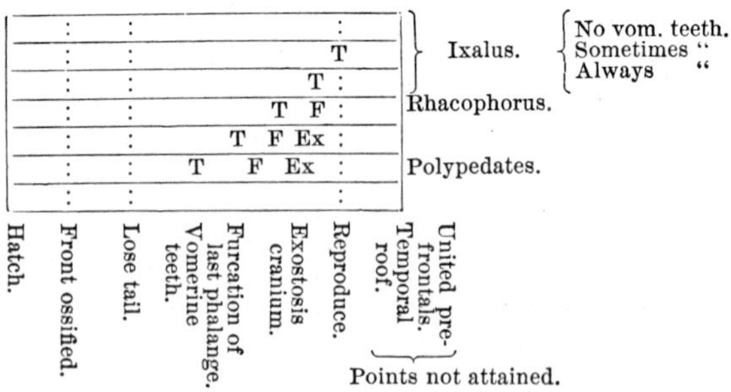

FIG. 4.

* A parotoid gland of small size is added here, but is not generic as compared with Bombinator, as the latter has collections of crypts on the same region and over the body.

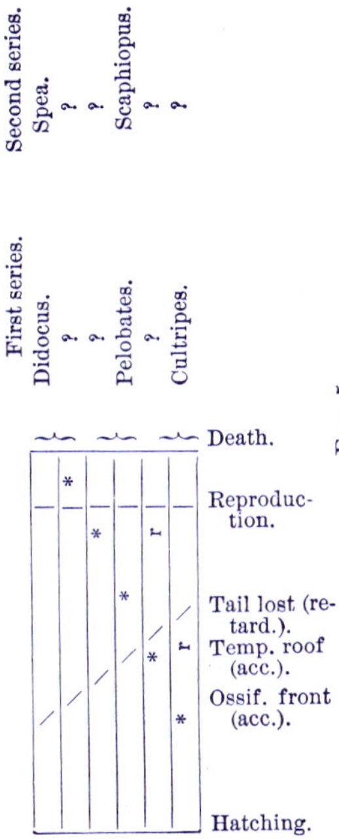

FIG. 5

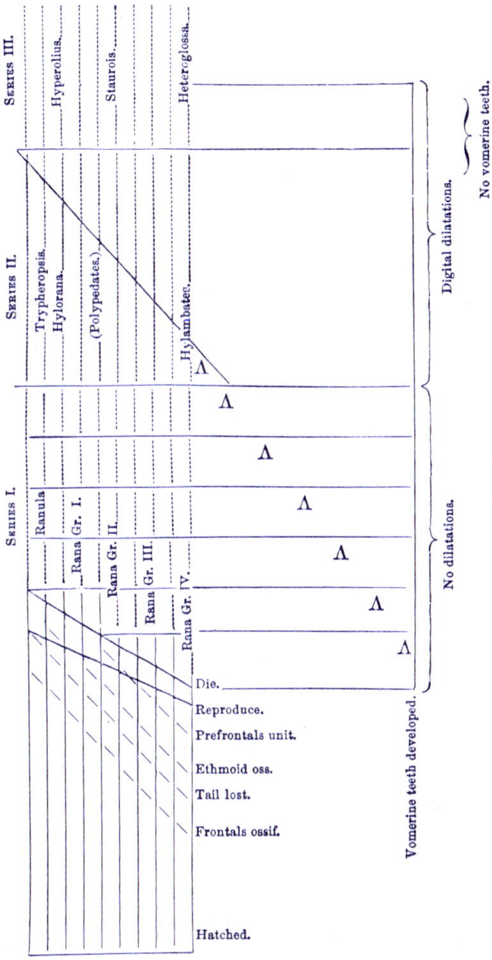

FIG. 6.

In the preceding diagrams each horizontal column represents the life history of the individuals of each genus. The line of dots, stars, etc., represents the same developmental stage of each, as it appears earlier or later in the life of the individuals. The point of crossing the breeding period is that at which the character is rendered permanent. When the change falls on this period the character is not generic, as in Ixalus, Fig. 4. The period of losing the tail, like that of breeding, is represented as occurring at nearly the same time in the history of every genus, as it is generally seasonal. Yet this is not always so, and, like the other characters, has most likely had its period of shifting. Compare difference of time of development, for instance, of the frontal and prefrontal bones in Figs. 3 and 6. The comparison of the adult stages of the less developed genera, at the tops of the columns, with the larval conditions of those more fully developed, may be traced in the absence of characters which appear in the latter. I have convinced myself of the accuracy of the above relations by the examination of many skeletons and wet preparations of adults and larvæ.

The diagrams * are representations of nature, and not ideal sketches. It is to be noted as remarkable that the advance throughout so many diverse groups is in the *same direction,* viz., to complete or excessive ossification of the cranium ; and this identity of progress might be readily shown by adding other characters, were it not that the tables would become too complex for convenience.

Has any such transition from genus to genus ever been seen to occur?

It must of course take place during the life of the individuals of a species, and probably at different times during the lives of different individuals, dependent on their relative vigor. In our view, ordinary metamorphosis is such a change, and we have stated its bearing in this form, that "every character distinguishing suborders, families, and genera is to be found among the individuals of some species, living or extinct, to mark new varieties or stages of growth."

a. The Developmental Relation of Generic to Specific Characters.

For the relation of the law of retardation and acceleration to specific characters we will look to development again. While the

* Notes on the diagrams : Fig. 4. Polypedates is here restricted to P. maculatus and P. quadrilineatus. The other species are referred to Rhacophorus, which has not hitherto rested on any proper basis ; the asserted character—the palmation of the hands — being one quite graduated from species to species among Hylæ. *Chiromantis*, Peters, is referred to the same, as its character is not strongly marked and is visible in other species. For similar reasons *Leptomantis* is referred to Ixalus.

Fig. 6. In each of series ii and iii, two series are mingled for the sake of comparing the structures of the prefontal bones. Thus Heteroglossa, Staurois, Hylorana and Trypheropsis are one series, and Hyperolius and Hylambates members of another.

young of Trachycephalus are successively different genera, they preserve most of their specific characters so as not to be mistaken. Agassiz says of the development of the North American turtles,* "I do not know a turtle which does not exhibit marked specific peculiarities long before its generic characters are fully developed." The same thing can be said of the characters of our salamanders, whose specific marks appear before their generic or even family characters. I suspect that this will be found to be a universal law.

It also follows, if a developmental process, as proposed, has existed, that at times *the change of generic type has taken place more rapidly than that of specific,*† *and that one and the same species (if origin be the definition) has, in the natural succession, existed in more than one genus.*

Apart from any question of origin, so soon as a species should assume a new generic character it ceases, of course, to be specifically the same as other individuals which have not assumed it. If supposed distinctness of origin be, however, a test of specific difference, we shall then have to contend with the paradox of the same species belonging to two different genera at one and the same time.

It follows, therefore, in our interpretation of nature, that groups defined by coloration alone are not to be regarded as genera, as is done by some ornithologists and entomologists. They are simply groups of species in which distinctive generic characters had not appeared up to the period of reproduction. Inasmuch as in development certain specific characters appear first, among them part or all of the coloration pattern, it is obvious that the latter do not belong to the generic category. The employment of such characters, then, in this sense, is only to commence reversing the terms generic and specific, and to inaugurate the process of regarding each species as type of a separate genus.

β. *Of Probable Cases of Transition.*

Thus the transition between the toothed and edentulous conditions in Cetacea takes place in the ordinary growth of the individuals of the genus Globiocephalus, and the transition between

---

* Contrib. "N. Hist. United States," i, p. 391. Note.

† See "Proceedings Academy," Philadelphia, 1867, p. 86, where I observe that generic characters are probably less inherent than specific.

the ossified and non-ossified types of Chelonia occurs during the life of the individuals of the genus Dermatemys.

But, in attempting to demonstrate this proposition, we must bring forward facts of another kind. The anti-developmentalists are accustomed to put such changes aside, as part of the necessary history of established types ; hence we will not appeal to such.

1. The frog *Ranula affinis*, of South America, was described by Peters as probably a climatal variety of European *Rana temporaria*. In this he is supported by the fact that the specific characters do not differ more than would characterize it as a local variety, were it an inhabitant of Europe. But I have found that it differs generically in the non-ossification of the ethmoid bone, as has been confirmed by Steindachner, and represents an embryonic condition of the same bone in Rana. It is in fact an undeveloped Rana. That this is a true genus is confirmed by many specimens, by additional species, and by the fact that the allied genus Trypheropsis, embracing three species in the same region, differs in the same way from the otherwise identical genus of the Old World, Hylorana.

2. The South African Saurians, *Chamæsaura anguina* and *Mancus macrolepis*, are very closely allied in specific characters in all respects, though distinct. They have one important ground of generic distinction : the latter has one pair of limbs less than the former. They are rudimental in Chamæsaura, and the disappearance in Mancus is but another step in the same direction. The difference in specific characters is of much less degree.

3. In the genus Celestus there are numerous species, which range from a slender snake-like form with weak limbs, to stouter, strong-limbed forms with a more saurian build. Among these the Haytian *C. phoxinus* is well distinguished by form and coloration. An allied genus from the same region is Panolopus, which in specific characters approaches the *C. phoxinus* very closely, much more so than any Celestus (one species possibly excepted). But in generic characters it is distinguished by the loss of all its toes and the non-separation of nine plates on the end of the muzzle. The genus Diploglossus, on the other hand, occupying a superior place on account of the division of the fronto-nasal into three, is, in specific characters (of *D. monotropis*) much closer to the stout Celesti than the species of the latter genus are among themselves.

4. The *Gronias nigrilabris* is a Silurid, which in specific char-

acters more nearly resembles the *Amiurus lynx,* than the latter does the *A. albidus* and many other species of the genus. The *A. lynx* is found in the same streams. The important generic character, the absence of eyes, is, however, its constant feature (in three specimens known to naturalists, others to fishermen).

5. The *Cinclidium maximum,* a large tree-toad of Brazil, resembles in all its characters the *Centrotelma geographicum.* The specific differences between them amount to almost nothing, but both sexes of the former grow larger and are furnished with a generic peculiarity in the addition of some phalanges to the thumb.

6. The *Oporornis agilis,* Baird, a North American bird of the Tanager family, resembles very closely, in form, color, and habits, the adjacent species of the adjacent genus Geothlypis. While its specific characters are thus very close to *Geothlypis tephrocotis,* it differs in the generic feature of a longer wing. By this it is associated, and properly so, with another species *O. formosus,* which has the general color and habits of species of Myiodioctes (*M. canadensis*), the next related genus.

7. The following fact I give on the authority of Prof. Leidy, who will publish it in his forthcoming work on the extinct Mammalia of Nebraska, etc.

Three species of Oreodon occur in the Miocene strata; they are a larger, a medium, and a small sized species. In the Pliocene beds above them they are represented by three species of Merychyus, which are in all respects known, identical specifically with the three preceding. Each one may thus be said to be more nearly allied to the species of the other genus than to its fellow of the same genus, in specific characters. But each, on the other hand, differs from each in generic characters. The teeth of Merychyus are more prismatic, have longer crowns and shorter roots, approaching the sheep, as Oreodon does the deer.*

---

* This phenomenon suggests an explanation on the score of adaptation, which the other cases do not. The existence during the later period of a tougher material of diet would increase the rapidity of wearing of the crown of the tooth, and require a longer crown and greater rapidity of protrusion. This necessitates a diminution of the basal shoulder and shortening of the roots, producing the prismatic form aforesaid. The deer browse on forest foliage, which is more tender, while the Cavicornia graze the grasses, which contain, as is known, a greater amount of silex; hence the more rapid attrition of the tooth.

This may have been the case with the two extinct genera; the different periods

8. The *Coreopsis discoidea T.* and *G.*, var. *anomala*, Gray, is, according to Mr. Aubrey Smith, much more nearly allied to *Bidens frondosa* than to other species of its own genus, and the latter is nearer to it than to other species of Bidens. It differs chiefly, if not altogether, in the generic character : the barbs of the achenia are directed upward ; those of the Bidens downward.

From these and many other such instances it may be derived : *That the nearest species of adjacent genera are more nearly allied in specific characters than the most diverse species of the same genus.*

9. While *Taxodium distichum* and *Glyptostrobus europæus*, conifers of North America and of Eastern Asia, respectively, are readily distinguished by generic peculiarities of their cones, in specific characters they appear to be identical.*

Confirmatory of this proposition is the statement of Parker : † "In tracing out the almost infinite varieties of the modifications of any one specific type of shelled Rhizopod, my friend Prof. Rupert Jones and I found that *like varieties of distinct species are much nearer in shape and appearance than unlike varieties of the same essential species.*" (It is not unlikely that species should here be read genus and variety species, though the latter may not fulfill the requirements in regard to distinctiveness observed among higher animals. In types like the Rhizopod, forms of this grade may not be really differentiated. Their enormous geographical range would suggest this, if nothing else.)

Objection.—A class of objectors to the preceding explanation of the relations in question will ascribe them to hybridization. They have already done so to considerable extent among the Teleosts (see the writings of von Siebold, Steindachner, and Günther). That hybrids exist in nature will be denied by none, but that they are usual or abundant is not a probable condition of a creation regulated by such order as ours is. The tendency to modify in given lines of generic series, if admitted, will account for many of the cases regarded as hybrids by the above authors, for it is to be remarked in many cases how the generic characters

during which they lived may have seen a change from forest to prairie. (It is not intended to suggest that the species of the two genera are necessarily of the same or any given number.)

* See Meehan, "Proc. Amer. Ass. Adv. Sci.," 1868. Newberry, "Ann. Lyc.," N. Y., 1868.

† "Transac. Zoöl. Soc.," London, 1864, p. 151.

are strikingly affected, and are chiefly used in guessing at the parentage. This is among Cyprinidæ so much the case that there is scarcely an example of a hybrid between two species of the same genus brought forward, but often between species of different genera.

γ. *Ascertained Cases of Transition.*

This naturally suggests that, in accordance with the theory of acceleration and retardation, a transition can take place in the life history of species. Have we any means of proving this suspicion ?

1. The genus Ameiva (Saurians of South America) has been composed of species of moderate size furnished with acutely tricuspid teeth. Teius, on the other hand, embraces very large species with the molars obtusely rounded and of the grinding type. These genera are generally held to be well founded at present. I find, however, that in *Ameiva pleii,* which is the largest species of the genus, in adults the greater part of the maxillary and mandibular teeth lose their cusps, become rounded, then obtuse, and finally like those of Teius. While young, they are true Ameivæ. Strangely enough the *A. pleii,* from Porto Rico, acquires but three such obtuse teeth when of the size of the other (St. Croix) forms. In youth the teeth of all are as in other Ameivæ. Here is a case of transition from one genus to another in the same species.

2. In the important characters of the possession of branchiæ, of maxillary bones, and of ossified vertebræ, the tailed Batrachia presents a series of a rising scale, measured by their successively earlier assumption. Thus *Salamandra atra* * produces living young, which have already lost the branchiæ ; *S. maculosa* living young with branchiæ ; Plethodon † produces young from eggs which bear branchiæ but a short time, and do not use them functionally ; *Desmoganthus nigra* uses them during a very short aquatic life ; *D. fusca* and other Salamanders maintain them longer ; while Spelerpes preserves them till full length is nearly reached. Finally, species of Amblystoma reproduce while carrying branchiæ, thus transmitting this feature to their young as an adult character. And it is a very significant fact that Spelerpes, which bears branchiæ longest, next to Amblystoma, is associated

---

* See Schreiber's "Isis," 1833, p. 527 ; Koeliker, "Zeitschr. f. wissensch. Zoölogie," ix, p. 464.

† Baird, "Iconographic Encyclopædia" ; Wyman, Cope.

in the same zoölogical region with a genus (Necturus) which differs from its four-toed form (Batrachoseps *) in nothing more than the possession of the osseous and branchial characters of its larva, in a permanent and reproducing condition. That this is a genus, to be one day converted into Batrachoseps by an acceleration of its metamorphosis, or that has been derived from it by the reverse process, I am much inclined to believe.† In support of this I quote the following examination into the time of change of the species of Amblystoma from my essay on that genus : ‡

"The great difference between the different species, and between individual species in this respect, may be illustrated by the following comparison between the size of the animals at the time of losing the branchiæ, so far as known, and that to which they ultimately attain.

Species.	Size at loss of branchiæ.		Average full size.	
	In.	Lines.	In.	Lines.
A. jeffersonianum,	1	5·75	6	
A. punctatum,	1	10	6·1	6
A. conspersum,	1	10·5	2	7·5
A. opacum,	2	2	3	9·5
A. texense,	2	1	?	
A. microstomum,	2	3·5	4	
A. talpoideum,	3 (perhaps too large).		3	9·5
A. paroticum,	3	7·5 (not smallest).	7·2	2·5
A. tigrinum,	{ 3 6	7 to 7	8 to 10	
A. mavortium,	{ 3 8	9·5 to 0	8	9
A. mexicanum,	? branchiæ persistent.		8	"

The last species, though not uncommon in collections, is not known to pass through its metamorphoses in its native country, but reproduces as a larva, and is therefore type of the genus *Siredon* of Wagler, Cuvier, Owen, and others. The larva of *A. mavortium* in like manner reproduces, but their offspring have in the Jardin des Plantes and at Yale College undergone an early metamorphosis. #

* See Cope, " Jour. Ac. Nat. Sci.," Phila., 1866.

† Necturus differs from all true Salamanders in the possession of the os intercalare of the skull; the parallelism is therefore inexact. (Note, 1886.)

‡ " Proceed. Academy," Phila., 1867.

\# Through the kindness of Prof. Duméril I have received both larvæ and adult of the species here noted, and observed by him. The larva is, as he states, *Siredon lichenoides* of Baird, while the adult is his *Amblystoma mavortium*, not *A. tigrinum* (= *luridum*), as also supposed by Duméril. (Note, 1886. I now believe these forms to belong to one species.)

Here is a case where all the species but two change their generic characters; one changes them or not, according to circumstances, and one does not change them at all. What are the probabilities respecting the change in the first set of species?

As we know from the experiments of Hogg, Duméril, and others that metamorphosis is greatly hastened or delayed by the conditions of temperature and light, what would not be the effect on such a protean species of a change of topographical situation, such as the elevation or depression of the land? And I have no hesitation in saying that if the peculiarities of series of individuals of *A. tigrinum* and *A. mavortium*, in the respects above enumerated, were permanent, they would characterize those series as species, as completely as any that zoölogists are accustomed to recognize. For the evidences on this head, see the discussions of those species in my monograph.

The experiments of Hogg, above alluded to, are as follows, as given by him in the "Annals and Magazine of Natural History."

He placed a number of impregnated ova of frogs in vessels arranged at regular distances from the light, in a cave. The lessening degrees of light were of course accompanied by a corresponding but much less rapid decline in temperature. The resulting effects on the metamorphosis may be tabulated as follows:

Mo.	day.	60°	56°	53°	51°
3	11	Egg.	Egg.	Egg.	Egg.
	20	Larva free,	*	*	*
	25	*	Larva free,	*	*
	31	*	*	Larva free,	Larva free.
4	10	Larva very large,	*	*	*
	22	Metam. complete,	Larva large,	Larva large,	Larva small.
8	11		Metam. complete,	*	*
	28			Metam. complt.	*
10	31				Metam. comp.

3. The reproduction of some species of insects before they complete their metamorphosis is a well-known fact, and it is particularly to the point that, in many of them, some individuals do attain to their full development, while the many do not. Westwood says,* "Two British species of this family (the Reduviidæ), *Prostemma guttula* and *Coranus subapterus*, are interesting on account of their being generally found in an undeveloped state, the latter being either entirely apterous or with the fore-wings rudimental, although occasionally met with having the fore-wings

* Uhler informs me that Amyot's asserted color characters are not reliable.

completely developed." "I think," says Spinola, "that the presence of wings and their development depends on the climate"; and, in speaking of *Oncocephalus griseus*, he says, "The influence of the northern climate appears to have arrested the development of the organs of flight. It will be seen that I have referred elsewhere that I have noticed that it is especially in hot seasons that certain species acquire wings, while the circumstance noticed respecting the ordinary occurrence of winged specimens of Microcœlia in the West Indies is confirmatory of the same opinion."

4. It is now known that certain Orthoptera do not get through their metamorphosis in time for the period of reproduction, and hence never, or in rare instances only, develop more than a short distance beyond the pupa state.

5. My friend P. R. Uhler tells me of an example among Hemiptera of the genus Velia. The species *V. rivulorum*, Fab., and *V. currens*,* of Europe, are only distinguished by the developmental feature of the presence of wings in one, and their absence in the other. Another species of the tropical region of the West Indies, *Halobates americanus*, Uhler, is furnished with wings, while its individuals, which occur abundantly in North America, have been generally supposed to lack them. Individuals, however, no doubt occur whose development is so far accelerated as to permit them to acquire wings before the period of reproduction, since one such has been found by Uhler.

These wing characters are in many cases generic, it appears to the writer; and the fact that they differ, without corresponding specific differences, is important evidence as to the origin of the genera.

6. The females of the Lepidopterous genus Thyridopteryx never develop beyond the pupa state, according to the same authority, before reproduction; they are reproducing pupæ, so far as the external characters concerned in metamorphosis go. In other words, the latter have been retarded, while the reproductive system and others have progressed. Now, generic characters are seen in the first, not in the last. The influence of the males is sufficient to prevent more than a part of the offspring from being retarded in the same manner.

I have selected a few of this class of facts which have come before my mind during the present writing, as drawn mainly from

* On "Insects," ii, p. 493.

my own experience. How many more of the same purport could be found by search through the great literature of science, or in the field of nature, may be readily imagined. I have no doubt that the field of entomology especially will furnish a great number of evidences of the theory of acceleration and retardation, especially among the insects with active pupæ.

Finally, having already stated the law according to which these processes naturally take place, I quote the following significant language of Hyatt in the above quoted essay on the Cephalopoda, as approaching nearer to the "law of acceleration and retardation" than anything I have found written. He says:

"In other words, there is an increasing concentration of the adult characteristics of lower species, in the young of higher species, and a consequent displacement of other embryonic features, which had themselves, also, previously belonged to the adult periods of still lower forms."

The preceding propositions have been formulated as follows, a few additions being now made:

I. That genera form series indicated by successional differences of structural character, so that one extreme of such series is very different from the other, by the regular addition or subtraction of characters, step by step.*

II. That one extreme of such series is a more generalized type, nearly approaching in characters the corresponding extreme of other series.

III. That the other extreme of such series is excessively modified and specialized, and so diverging from all other forms as to admit of no type of form beyond it.†

IV. That the peculiarities presented by such extremes are either only in part or not at all of the nature of adaptations to the external life of the type.‡

V. That rudimental organs are undeveloped or degraded conditions of the respective characters developed or obliterated in the extreme of the series.

VI. That the differences between genera of the same *natural* series are only in the single modifications of those characters which characterize the extreme of that series.

---

* St. Hilaire, Owen, Agassiz, Duméril.

† Dana on "Cephalization"; Leconte.

‡ Owen on "Cetacea," "Trans. Zoöl. Soc.," London, 1866, p. 44. Leconte on "Carabidæ," "Trans. Amer. Philos. Soc.," 1853, p. 364.

VII. That the relations of the genera of a primary series are those of the different steps in the development of the individuals of the extreme genus *ab ovo* (*von Baer, Agassiz*) (with sometimes the addition of special adaptive features ?).

VIII. That the presence, rudimental condition, or absence of a given generic character can be accounted for on the hypothesis of a greater rapidity of development in the individuals of the species of the extreme type, such stimulus being more and more vigorous in the individuals of the types as we advance toward the same, or by a reversed impulse of development, where the extreme is characterized by absence or "mutilation" of characters.

IX. And that, as the character of the genus at the period of reproduction of its species is that which is perpetuated ;

X. So the character of the genus has been first inferior, then protean, and then advanced, as the metamorphosis has been by a retrograde movement in time, posterior to, at, or anterior to the period of reproduction.

XI. That it therefore results that there is one primary structural type involved in such a series of species, which is made to present, at any given period in its geologic history, that appearance of succession of genera ordained by Creative Power.

δ. *On the Origin of Inexact Parallelism.*

The hypothesis can only be demonstrated in case of *exact parallelism.* If proved in these, it readily accounts for the cases of *inexact parallelism,* which are of course in any single period vastly in the majority. First take the case of *simple inexact parallelism.* A series of individuals of the genus Didocus undergo the metamorphosis of the cranial structure earlier and earlier in life, commencing by completing the ossification of the membrane of the fronto-parietal region in full age, until at last it becomes completed as early as the period of reproduction. Heretofore the adult offspring have appeared during a long period, invariably characterized by the larval cranium ; but like now producing like, this development springs into new power, and the offspring ossify the cranial bones far earlier than their immediate predecessors ; in a word, the genus Pelobates has been created ! At this state of progress Didocus is an undeveloped Pelobates.

Let us, however, suppose the "acceleration" of development of the cranial bones still to progress. The character appears now soon after the ordinary metamorphosis has been passed, and now a little before. The identity of Didocus with the undeveloped Pelobates is thereupon lost!

So may have been the relations between Pelobates and Cultripes. Pelobates was probably once identical with the undeveloped Cultripes; but the same acceleration has concentrated the characters more rapidly than the other larval stages, leaving Pelobates behind.

This I conceive to be the explanation of this relation: when the parallelism is inexact by two steps, as in Spea to Didocus, by the obliterated ear and ossified xiphisternum. The continued concentration of characters has been carried to earlier stages till the identity exists in the adult state of neither one, but at a period of larval life of both, shortly preceding the adult period of the lower. The relations between the Amblystomidæ and Plethodontidæ, which I have elsewhere * pointed out, have probably had their origin in this way.

If we attempt to prove the identity of the modern Mammalian fœtal circulation with that of the modern adult fish, we may find nearly an exact parallel in this respect, as it is the basis of class distinction; but in other respects the identity will not exist, rendering the parallel inexact or remote. The structure of the origins of the aorta is at one time identical with that of the shark, with one exception: in the former but four aorta-bows appear together; in the latter five. In the former the first disappears as the fifth comes into being. This is simply a continuation of *acceleration*. The first generalized representative of the Mammalia lost the first aorta-bow toward the latter part of its growth, and became the next genus in advance of the selachian. The fact that these bows do not appear exactly simultaneously, but rather successively, renders it necessary that in a regularly shortening period of possession of transitory characters, one such, as the existence of the first aorta-root, should vanish before the appearance of a permanent, the fifth, in the more specialized types, where acceleration reaches its maximum. This is indicated by the fact that in the Batrachia, where the acceleration has not attained so high a degree, the first and fifth aorta-bows co-exist for some time, though the first and second disappear before maturity.

So also with the splitting of the bulbus arteriosus. As in the Batrachia, the pulmonary *ductus communis* only is to be sepa-

* "Jour. Ac. Nat. Sci.," Phil., 1866, p. 100.

rated ; the remaining bulbus is divided by a long valve or incomplete septum, tracing the division of the aorta-roots. In the serpent (Rathke) this division is so accelerated as to appear at nearly the same time as the septum of the pulmonary duct. In the mammal, on the other hand, while the division of the aorta-root takes place as soon as in the last, the pulmonary septum is accelerated so as to appear long before the first named. Hence, in the septa in the serpent, the singular anomaly seems to present of the mammal passing through the Batrachian stage, while the serpent, a nearer relative, does not.* If, however, we take the less typical serpent, we will find the aortic septum to appear a little later, thus giving the Batrachian type, and if we reverse the order of time, so that the succession becomes one of retardations, we will find the same known ratio will bring us to an identity under all circumstances.

This, then, is the explanation of the divergence and want of "exact parallelism," which is observed in comparing the developmental histories of all types *not most closely allied. It has not,* according to our theory, *always been* a divergence, but was at a prior epoch in each case a relation of "exact parallelism," the lower type a repressed higher ; the former identical with one of the stages of the latter. But the process which has produced this relation, continued, has of necessity destroyed it, so that the exact parallelism has always been a temporary relation, and one shifting over the face of the system.

III. OF HIGHER GROUPS.

First. Comparison of the contemporary.

Having now admitted a developmental succession of genera, and, second, that this has progressed more rapidly at certain times in the earth's history than any modification of specific forms, the hypothesis already broached naturally comes up : *Has such transformation of types, generic or higher, taken place in any degree simultaneously, throughout a great number of species?* An affirmative answer to such a proposition is absolutely necessary to its acceptance as expressing the phenomena exhibited by geological succession of types. Let us try to answer the question put in a closer form. Have the same species been transferred from one

* This is the way indeed in which it is stated by Rathke, "Entwickelungsgeschichte der Natter," p. 164.

geologic epoch to another by a change of generic form ; and has not the genus been transferred from one epoch to another under change of ordinal type, and as a consequence the same species ?

As a reply I propose to render the affirmative of the first of these questions highly probable.

Paleontology only will be able to answer this question conclusively, though, as we have abundant evidence that the relations of species to genera and other higher groups were the same then as now, we may look to the present status as furnishing important evidence on the subject. We are turned at once to the probable history of development in the separate zoölogical areas of the earth's surface. The question may be asked, Are the present zoological regions on an equal plane as to the geologic relations of their faunæ, or are they related as the different subdivisions of a geologic period in time ?

I have on a former occasion asserted that the latter of these propositions was true.*

a. Of Homologous Groups.

Naturally following the admission of a developmental succession of organic beings is the question of its relation to the different surfaces of land and water on the earth. The following considerations bear on this subject.

Among the higher groups of animals can be detected series "homologous" on the same principle as the alcohols (? compound radicals) and their derivatives ; and the component types of each can be, and have been in many instances, shown to be "heterologous," as are the ethers, mercaptans, aldehydes, acids, etc. Among Mammalia two partly homologous series have been pointed out, Implacentialia and Placentialia ; possibly such are the types Altrices and Præcoces among Aves ; of a lesser grade in this class are the parallel series of Pullastræ and Gallinæ, of Clamatores and Oscines. Among tortoises I have alluded to the Pleurodira as compared with the remainder of the order, already parallelized by Wagler ; and, of lesser grades, the series among Lacertilia of Acrodonta and Iguania, parallelized by Duméril and Bibron, and of Teidæ and Lacertidæ, compared by Wiegmann. I have discovered a full parallelism between the Raniform and Arciferous Anura. It is carried out between the Characini and a group of remaining Physostomous Fishes, perhaps not yet well defined ; it

* On "Arciferous Anura," "Journ. Ac. Nat. Sci.," 1866, p. 108.

is exhibited between the orders Diptera and Hymenoptera among insects. None of these comparisons can be allowed, of course, without the most searching anatomical and embryological analysis.

This *heterology* is what Swainson and others called "analogy" as distinguished from affinity. It *generally* relates genera of different zoölogical regions. Mimetic analogy, on the contrary, relates genera of the same region; it is a superficial imitation which has occurred to critical biologists, and is of much interest, though as yet but little investigated. It has as yet been observed in external characters only, but occurs in internal also; it has been accounted for in the first case by the supposed immunity from enemies arising from resemblance to well-defended types. No such explanation will, however, answer in the latter case. I believe such coincidences express merely the developmental type common to many heterologous series of a given zoölogical "region"; this will be alluded to a few pages later.

We naturally inquire, Is there anything in the food, the vegetation, or the temperature to account for this apparent diversity in the different regions? Are there not carnivora, herbivora, seed-eaters, insectivores, and tree-climbers, where game and grass, seeds and insects and forests grow the world over? We answer undoubtedly there are, and these adaptations to food and climate are indeed as nothing in the general plan of creation, for every type of every age has performed these functions successively.

β. *Of Heterology.**

This relation will be exhibited by a few examples from groups known to the writer, commencing with the Batrachia anura.

	Raniformes.	Arciferi.
External metatarsal free.		
Aquatic.	Rana.	Pseudis.
Metatars. shovel.	Hoplobatrachus.	Mixophyes.
External metatarsal attached.		
	Feet webbed.	
Metatars. shovel.	Pyxicephalus.	Tomopterna.

* Some of the cases below cited as heterology I believe to be truly of this character; but some others are probably not such, but are merely series of genera presenting similar structural peculiarities as consequences of the operation of identical laws. I would place under this head, and withdraw from the homologous class, the families of *Lacertilia Leptoglossa*, *Diploglossa*, and *Typhlophthalmi*, those of the Old and New World *Quadrumana* and those of *Cephalopoda*. These distinct modes of origin of corresponding forms have been recently termed, by Lancaster, homogeny in the case of homologous groups, and homoplassy when the imitative types are on distinct lines. (Ed. 1886.)

PLATE II.

1

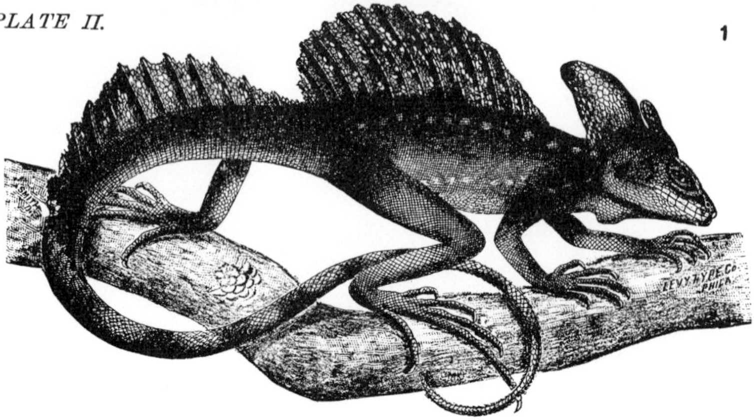

2

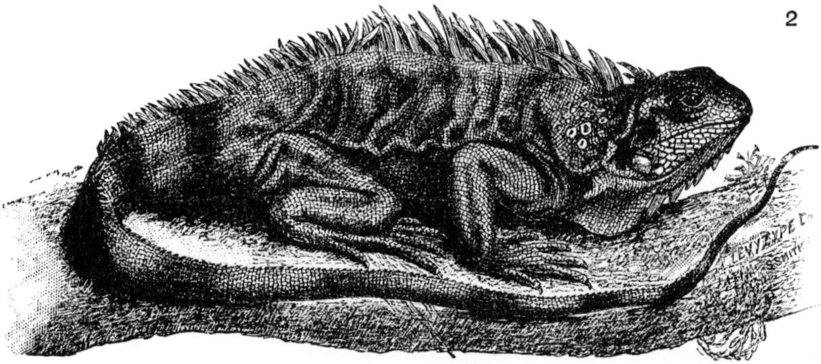

3

4

5

PLATE II a.

1

2

3

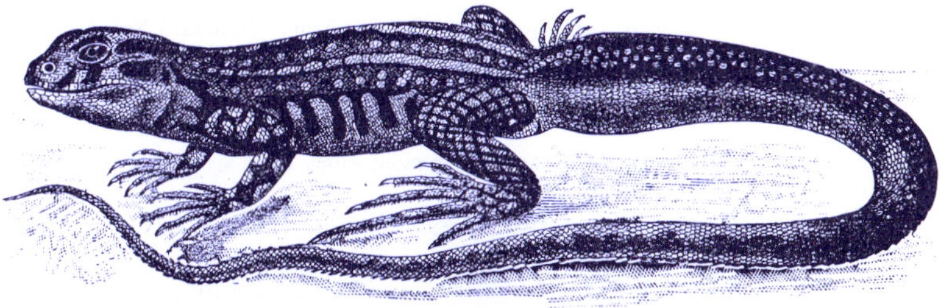

4

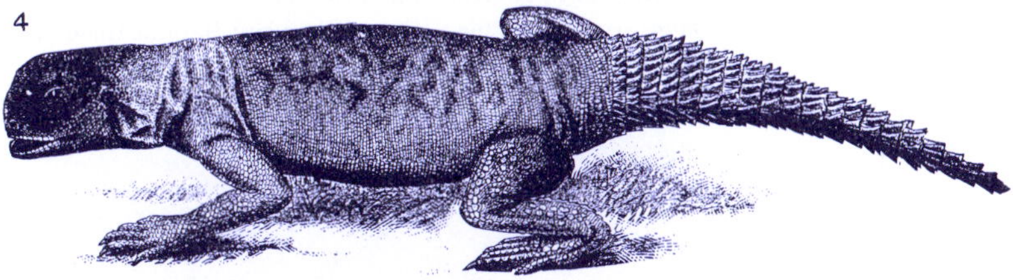

5

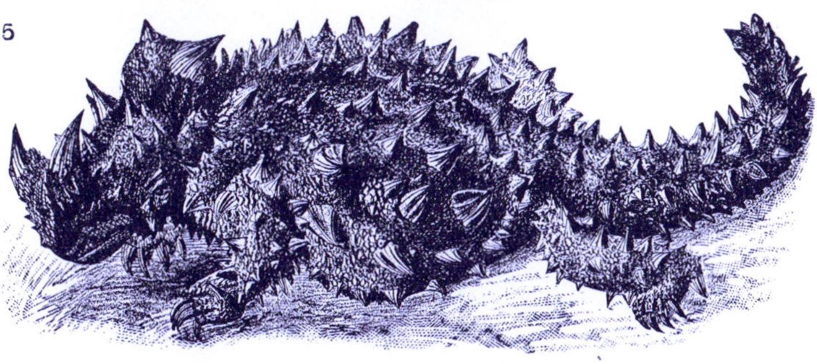

PLATE II.

Figures of Lizards of the family Iguanidæ to be compared with a homologous series of Agamidæ represented in Plate II*a*. The species are:

FIG. 1. *Basiliscus plumifrons* Cope, Costa Rica. From Cope.

FIG. 2. *Iguana tuberculata* Linn., South America. From the "Standard Natural History."

FIG. 3. *Crotaphytus wislizeni* B. G., Sonoran Region of North America. From Baird.

FIG. 4. *Phymaturus palluma* Grav., Chili. From Bell.

FIG. 5. *Phrynosoma cornutum* Harl., Texas. From "Standard Natural History."

PLATE II*a*.

Figures of Lizards of the family Agamidæ which form a homologous series with the Iguanidæ represented on Plate II.

FIG. 1. *Lophura amboinensis* Schloss., Amboina. From Wagler.

FIG. 2. *Physignathus mentager* Günther, Siam. From Günther.

FIG. 3. *Liolepis bellii* Gray, China. From Duméril and Bibron.

FIG. 4. *Uromastix spinipes* Cuv., Arabia. From Guérin.

FIG. 5. *Moloch horridus* Gray, Australia. From "Standard Natural History."

	RANIFORMES.	ARCIFERI.
Arboreal; vom. teeth.	Leptopelis.	Hyla.
" no " "	Hyperolius.	Hylella.
Subarboreal.	Hylambates.	Nototrema.
	Feet not webbed.	
Terrestrial.	Cassina.	Cystignathus.
" spurred,	Hemimantis.	Gomphobates.

Comparing the genera in a general physiological sense, we may parallelize further.

Aquatic, with digital dilatations.		
	Heteroglossa.	Acris.
Arboreal; cranium hyperostosed.	Polypedates.	Trachycephalus.
" cranium free.	Rhacophorus.	Hyla. Agalychnis.

The same kind of parallels exists between the primary groups of the Testudinata, as follows:

	CRYPTODIRA.	PLEURODIRA.
Five complete pairs of bones across the plastron.		
	Pleurosternidæ.	Sternothæridæ.
Four pairs of bones across plastron; not more than two phalanges on all toes.		
	Testudinidæ.	Pelomedusidæ.
Three phalanges on most digits;		
Zygomatic arch; no parieto-mastoid.		
	Emydidæ.	Podocnemididæ.
Temporal fossa over-roofed by parietal.		
	Macrochelys.	Podocnemis.
No zygoma; a parieto-mastoid arch.		
*	* *	Hydraspididæ.

If we compare the peculiarities of generic structure merely with reference to their adaptation to the animals' habits, we will see the following:

	CRYPTODIRA.	PLEURODIRA.
Feet reduced for terrestrial progress.		
	Testudinidæ.	Pelomedusidæ.
Feet normal.		
Anterior lobe of sternum movable.		
	Cistudo. Cinosternum.	Sternothærus.
Anterior lobe fixed.		
Neck very elongate.		
	Trionychidæ.	Chelodina.
Neck shorter; aquatic.		
Temporal fossa open.		
	Emydidæ in gen.	Hydraspididæ.
Temporal fossa over-roofed.		
	Cheloniidæ.	Podocnemis.

The parallels between the genera of the American Iguanidæ and the Old World Agamidæ are similarly quite close. They are shown on Plates II and II a.

	Iguanidæ.	Agamidæ.
Abdominal ribs.	Polychrus.	* *
No abdominal ribs.		
Ribs greatly prolonged into a lateral wing.	* *	Draco.
Ribs not prolonged.		
Arboreal types, generally compressed.		
A dorsal and caudal fin supported by bony rays.	Basiliscus (no fem. pores).	Lophura (pores).
No vertebral fin.		
	No femoral pores.	
Form slender, scales in equal series.	Calotes. Bronchocela.	Læmanctus.
Form elongate; eyebrows elevated, tail compressed.	Gonyocephalus.	Ophryoëssa.
Form stouter, scales less regular.	Hypsibates.	Tiaris.
	Femoral pores.	
Low crested; small hyoid disk.	Brachylophus.	Diporophora.
High crested; large hyoid disk.	Iguana.	Physignathus.
Tail with spinous whorls.	Cyclura.	* *
Terrestrial types of flattened form.		
	Femoral pores.	
Tail with whorls of spiny scales.	Hoplocercus.	Uromastix.
Tail long, simple; scales small.	Crotaphytus.	Liolepis.
Tail simple, scales large.	Sceloporus.	* *
	No femoral pores; preanal pores.	
Tail with whorls of spines.	* *	Stellio.
Tail simple, not elongate, ear open.	Proctotretus.	Agama.
	Neither femoral nor anal pores. Much flattened, tail short, scales irregular.	
Ear exposed.	Phrynosoma.	Moloch.
Ear concealed.	(Doliosaurus, s. g.)	Phrynocephalus. Megalochilus.

A similar parallel may be drawn between the American Teidæ, and the Old World Lacertidæ, and in fact between all the families of the Lacertilia Leptoglossa. I have added to these for comparison two families of the Typhlophthalmi. Each family embraces one or more series, and these exhibit a remarkable similarity in the relative development of the limbs and digits; among the higher groups the parallelisms lie in the arrangement—as greater or less separation—of the head shields. The Scincidæ are cosmopolite; the Gymnophthalmidæ, which have the eyelids of their fœtus, are Australian; the Sepsidæ, either larval or senile in head shields, are mostly Ethiopian.

Two pair limbs. I. Toes 5—5.	*Lacertidæ.*	*Teidæ.* *(no supranasals.)*	*Anguidæ.*	*Gymnophthalmidæ.*	*Scincidæ.*	*Sepsidæ.*	*Pygopodidæ.*	*Acontiadidæ & Anelytropidæ.*
A. Scales granular.	Lacerta.	Ameiva. Cnemidophorus.						
	Eremias.							
B. Scales imbricate.	Notopholis.							
Supranasals. Prefrontals, 2. (divided).	Gerrhosaurus.	Centropyx.	Diploglossus.	Morethia.	Euprepis. Eumeces.			
Prefrontals, 0.	Pleurostrichus.		Celestus.			Amphiglossus. Sphenops. Gongylus.		
No supranasals.	Platysaurus.... Cicigna.	Cercosaura. Iphisa. Ecpleopus. Lepidosoma.		Cryptoblepharus. Ablepharus.	Cyclodus. Hinulia. Mocoa.			
II. Toes 4—5.		Tretioscincus. (Acrantus.)		Menetia. Gymnophthalmus.	Campsodactylus. Ristella. Heteropus.			
III. Divided 3—3 to 2—3 times.	Saurophis.	Brachypus.	Sauresia.	Miculia. Blepharactisis.	Tetradactylus. Chiamela.	Sepsina. Anisoterma (2—4).		
IV. Divided 1—2 or 2—1 or 2—2.		Chalcis.		Lerista.	Anomalopus. Siaphus. Hemiergus. Chelomeles.	Seps. Heteromeles. Sphenocephalus. Sepomorphus.		Nessia.
V. One limb divided.		Ophiognomon.			Brachymeles. Rhodona.			
VI. Neither divided.	Cætia. Chamæsaura.		Panolopus.					Evesia.
One pair limbs.	Mancus.	Propus.	Pseudopus.			Scelotes.	Pygopus. Delma. Pletholax.	Dibamus.
No limbs.			Opheodes. Opheosaurus. Anguis. Opheomorus.		Soridia.		Aprasia.	Herpetosaura. Acontias. Typhlosaurus. Feylinia. Anelytropsis.

The first comparison of these groups was made by Wiegmann (Herpetologia Mexicana), who employed, however, only the Scincidæ and Lacertidæ, and could not include the many types made known since his day.

From the class Aves I have selected only the homologous series of the Clanatorial and Oscine Passeres. Naturalists more fully acquainted with the genera could probably increase the examples of heterology largely. Each group furnishes us with carnivorous, insectivorous, and frugivorous forms; each with walkers, climbers, and sedentary genera; each with butcher-birds, thrushes, warblers (not in song!), wrens, and fly-catchers. Each and all of these types are *teleologically* necessary to any country complete in the wealth of nature, and to each geological period.

CLAMATORES.		OSCINES.
I. Tree-climbers, with long hind toe and tail feathers stiffened and acute.		
Dendrocolaptidæ.		*Certhiidæ.*
II. Terrestrial in part, with the tertials as long as the primary quills.		
Geobatidæ.		*Motacillidæ.*
III. Tree-perchers with hooked bill, graduating from powerful to medium and slender.		
Formicariidæ		*Turdidæ.*
Thamnophilus.	Bill strongest, hooked.	*Lanius.*
Formicarius.	" moderate.	*Turdus.*
Formicivora.	" weak.	*Sylvia.*
Rhamphocœmus.	" slender (wrens).	*Troglodytes.*
IV. Fly-catchers with flat bill and weak legs; wait for their prey and take it on the wing.		
Tyrannidæ.		*Muscicapa* et aff.
V. Flat-billed berry and fruit eaters.		
Cotingidæ.		*Bombycillidæ.*

From the Mammalia the well-known series of the Marsupialia and Placentialia may be chosen.

PLACENTIALIA.		MARSUPIALIA.
I. Toes unguiculate, in normal number; sectorial teeth; i. e., one or more molars with one or no internal tubercles; canines strong:		
Carnivora.		*Sarcophaga.*
	I. Digitigrade. Toes 5—4. True molars $\frac{3}{3}$ (upper incisors more numerous in some).	
Amphicyon.		Thylacinus.
	II. Plantigrade; molars tubercular.	
	a. Posterior molars $\frac{4}{4}$.	
* *		Dasyuridæ.
	aa. Posterior molars $\frac{2}{2}$.	
Ursidæ.		* *

II. Toes unguiculate; molars with more than one row of pointed tubercles; canines weak or none; incisors large.

Insectivora.		*Entomophaga.*
	a. Tubercular molars $\frac{4}{4}$, toes 4—5. Tail naked.	
Gymnura.		Didelphys.
	Tail hairy.	
Cladobates.		Myoictis.

IV. Molars with transverse crests, no canines; tusk-like incisors; pairs of limbs of similar proportions.

Proboscidia.		*Diprotodontidæ.*
	Two inferior incisors; molars with two cross-crests; size huge.	
	a. Two rudimental lateral incisors above.	
* *		Diprotodon.
	aa. ? One pair of incisors only above; a trunk.	
Dinotherium.		

V. No canines; two pairs of cutting incisors.

	a. Three true molars.	
Rodentia.		* *
	aa. Four true molars.	
* *		*Rhizophaga.*

The parallels are in this case very imperfect in details, and but few worthy of the name can be made. They are, however, illustrative of a remote heterology, sufficiently remarkable to have claimed the notice of naturalists for many years.* I also have little doubt but that future paleontological discoveries will increase the number of parallels,† and bring to light truly heterologous generic terms of the Marsupial series. Predictions of this kind have been on many occasions fulfilled (e. g., some of D'Orbigny's among the Cephalopoda), and I look with confidence to the ultimate demonstration of that heterology here which has been already seen in the Batrachia and Reptilia.

The homologous groups of the Catarrhine and Platyrrhine Quadrumana are measured as follows:

	Catarrhini.	*Platyrrhini.*
Tailless.	Andropithecus.	*
	Simia.	*
	Hylobates.	*
Tail short.	Cynocephalus.	*
	Macacus.	Brachyurus.
Long tail.		
Thumb developed.		
	Cercopithecus.	Lagothrix.
		Mycetes.
Thumb rudimental.		
	Semnopithecus.	Brachyteles.
Thumb none.		
	Colobus.	Ateles.

* We owe very many observations on the Marsupials to Owen.

† The extinct carnivorous family of the Nimravidæ which forms a homologous series with the cats (Felidæ) was discovered by the writer some years after this was written. (Ed. 1886.)

I append two homologous series, represented by the Nautilea and the Ammonites of the Tetrabranchiate Cephalopoda, which are distinguished, the first by the simple septa and the siphon central or marginal ventral ; and the second by the complex and folded septa and siphon central or marginal dorsal. The parallelisms have been noted by Barrande, Bronn, and many conchologists, who can furnish a much more full table than the following, from the most recent sources :

NAUTILI.		AMMONITES.
	A. The shell straight, unwound.	
Orthoceras.		Baculites.
	B. The shell more or less curved or wound.	
	a. Simply curved.	
Cyrtoceras, Phragmoceras,	}	Toxoceras.
	aa. A more or less straight portion, folded on the remainder.	
	β. Folded portion in close contact with remainder.	
Ascoceras.		Ptychoceras.
	ββ. Folded portions not in contact.	
?	?	Hamites.
	aaa. One extremity spirally wound, the volutions not in contact.	
	β. Extremity of the shell prolonged beyond the wound portion.	
Lituites.		Ancyloceras.
	ββ. Extremity not prolonged in a line.	
	γ The spiral flat.	
Gyroceras.		Crioceras.
	γγ. The spiral elevated (heliciform).	
Trochocerus.		Turrilites.
	aaaa. Spiral turns of the shell in contact.	
	β. Extremity prolonged in line beyond the spiral.	
*	*	Scaphites.
	ββ. Extremity not prolonged beyond spiral.	
Nautilus.		Ammonites.

We may now consider the question of the origin of these higher groups. In the first place, we must lay down the proposition *that the characters which constitute groups "higher" in the comparison of rank* (we do not of course mean higher in the same line, as we say higher genus in a family, or higher order in a class) *are such solely from their being more comprehensive, or present throughout a greater range of species.*

What is true, therefore, in respect to characters of genera, is likely to be true in respect to characters of higher groups, such as we have been considering in the preceding pages. Believing, then, that a new genus has been established by the transition of a number of species of a preceding genus in order, without necessary loss of specific characters, I think the same process may have established the suborders and orders in question. That is, *that a large number of genera have near the same time, in past or pres-*

ent geological history, passed into another suborder or order by the assumption or loss of the character or characters of that to or from which they were transferred, and that without necessary loss of their generic characters.

I will cite a probable case of this kind, the facts of which I have already adduced.

It has been shown above that the genera of six of the families of the Batrachia anura form series characterized by the successive stages of ossification of the skull, terminating in a dermoössified condition, with over-roofed temporal fossæ. That in nearly all the other families similar relations between genera exist, but are nowhere carried so far. The character attained by all the first series is now only generic; but should all the genera of each of the six families assume this character in time, as is necessary in accordance with a development hypothesis, it would at once possess a new and higher importance, and would become ordinal or otherwise superior. It would define a series homologous with all those types which had not attained it. This character of the over-roofing of the temporal fossæ has actually attained a family significance among the Testudinata—e. g., as defining the marine turtles; and similar characters are found by Owen to characterize the Labyrinthodontian order of Batrachia.*

Agassiz has pointed out a similar and more extended case, in the Heterocercal and Homocercal ganoids. Had we not so many of the closest approximations between members of these groups, they would stand in the systems as two great homologous series, with their contained heterologous genera. As it is, these heterologous terms or genera are evidently so nearly allied that Agassiz, in the "Poissons Fossiles," has thought it best to arrange the latter together, thus instituting a system *transverse*, as it were, to the other. This may be necessary, since Kölliker points out transitional forms, and perhaps certain types may have begun to abandon the heterocercal form near the period of reproduction, producing offspring somewhat protean in character, preparatory to an earlier appearance and consequent permanence of the homocercal type. This is to be derived from the history of the metamorphosis of Amblystoma.

In the same manner the development of the convolutions of the brain does not define groups of the highest rank, since it pro-

* The roof here alluded to by Owen includes some two distinct bones not known in the arch of the Anura, and therefore different.

gresses chiefly during the later periods of embryonic life, and is therefore a "developmental character." Owen has endeavored to distinguish the primary divisions of Mammalia by the character of these convolutions, whereas they really define only the sub-groups of the orders. For we have Lissencephalous (smooth-brained) monkeys—certain lemurs—and smooth-brained Ruminants—i. e., the extinct Brachyodon and Anoplotherium, according to Lartet and Gratiolet. The lowest types of the existing smooth-brained Mammalia, including especially those with no or rudimental corpus callosum, the Marsupials, are also distinguished by the non-development of the deciduous teeth * (excepting one premolar). If now, through some topographical change, the whole series of Mammalia between the smooth-brained and convolute-brained were lost to us, as by the elevation of a region, and the absence of favorable localities or bodies of water for the preservation of their remains, we would have to study two homologous groups, with the heterologous terms of each corresponding with each other, as do now the genera of the Clamatores and Oscines of the Arcifera and Raniformia, etc.

In the same way the characters defining Implacental Mammalia will be found transitional in some type, and this great series, homologous with the Placentals, will have to be placed in closer connection, in its genera, with the series of the latter, with genera of the same, perhaps, now extinct.

γ. *Of Mimetic Analogy.*

It has been often remarked that the animals of the Equatorial Ethiopian region were very generally of smoky and black colors. This is remarkably the case, and the peculiarity of the genus Homo in this respect is shared by birds, reptiles, and fishes in a considerable degree. This can not be traced to the effect of torrid climate, for the same latitudes in India and the Malaysian Archipelago, and in South America, do not produce such colors.

The similarity in color of desert types has also been remarked. The gray sand-hue so well adapted for concealment is universal, with few variations, in the reptiles of the Tartar and Arabian

* This I have alluded to as the non-development of the permanent series; the homology of the dental system of Marsupials appears, however, to be with the latter and not with the milk series. See "Flower," "Trans. Roy. Soc.," 1867. (Wortman now denies the correctness of Flower's view, and demonstrates that the milk series is primitive. See "Encyclopædia of Dentistry," 1886. Ed. 1886.)

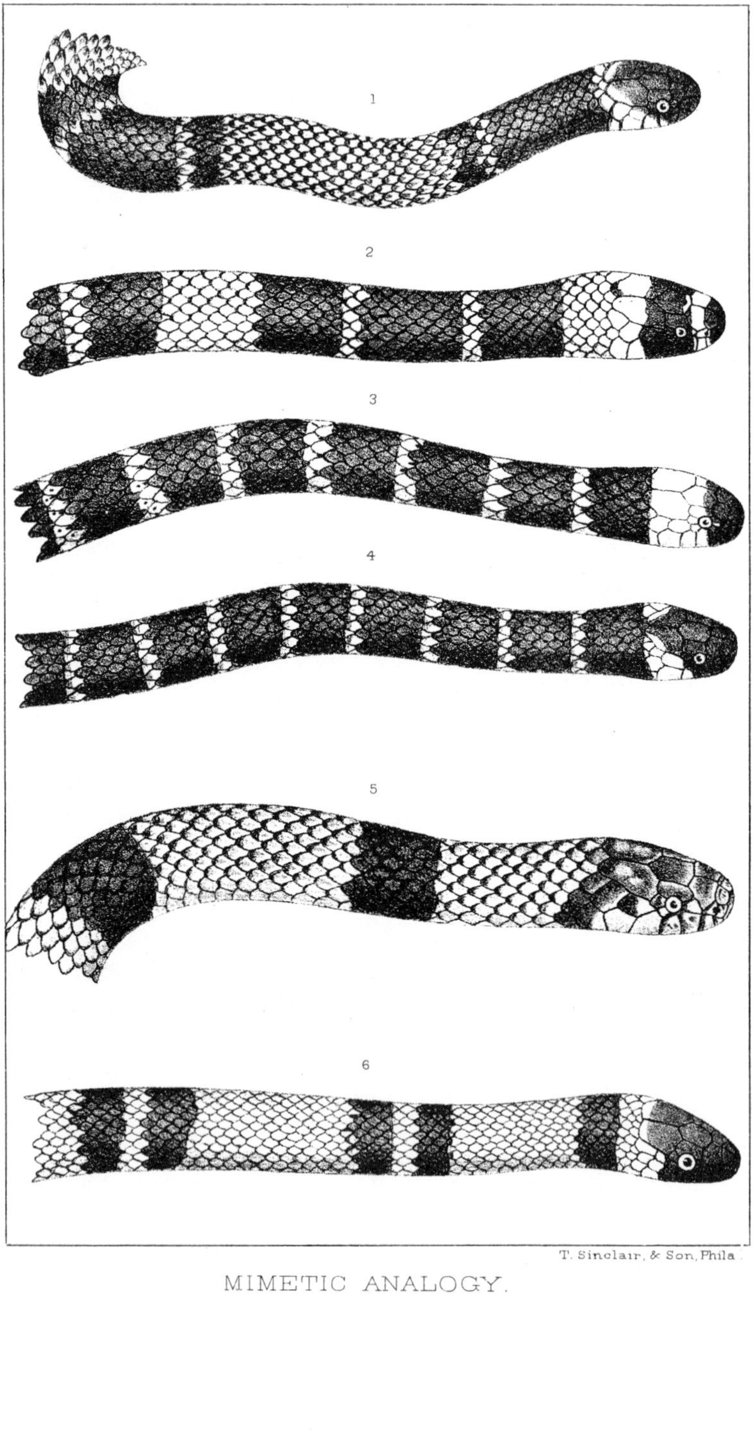

T. Sinclair, & Son, Phila.

MIMETIC ANALOGY.

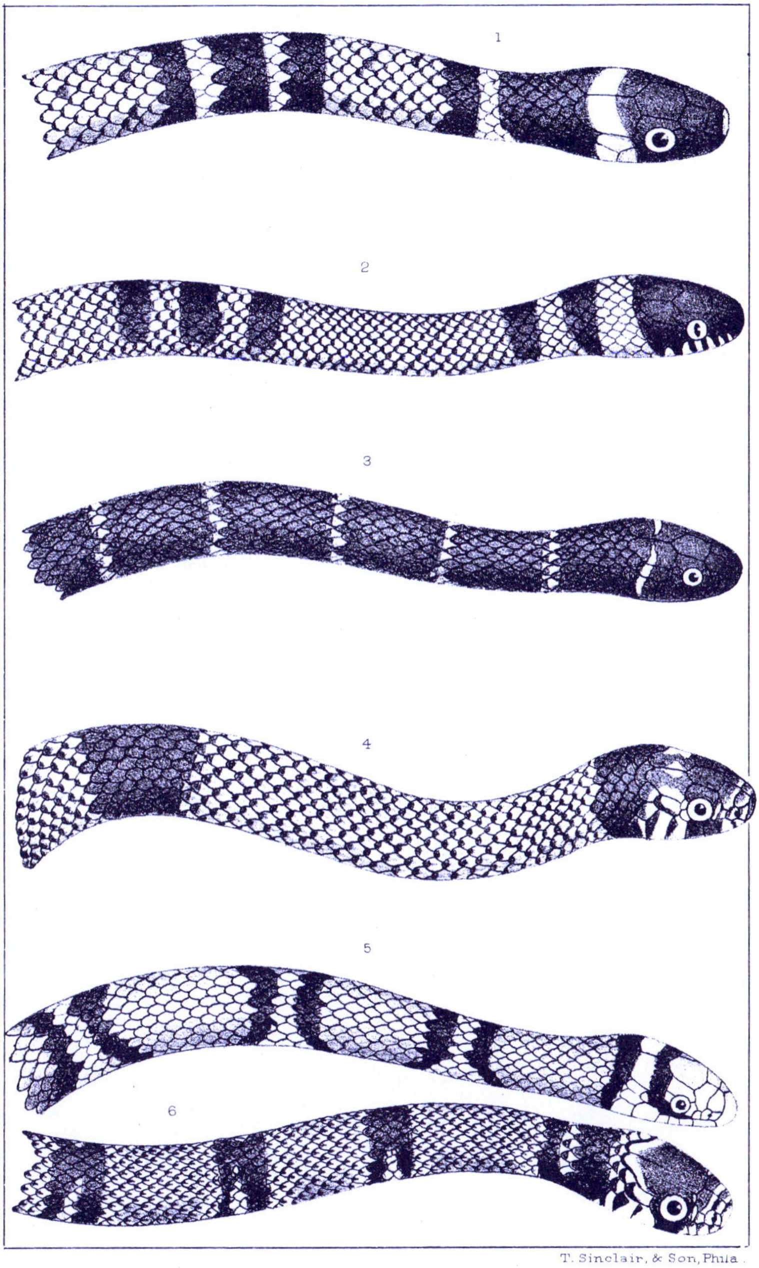

T. Sinclair, & Son, Phila.

MIMETIC ANALOGY.

PLATE III.	PLATE III*a*.
Fig. 1. *Elaps dumerili* Jan; New Granada.	Fig. 1. *Pliocercus elapoides* Cope; Mexico.
Fig. 2. *Elaps lemniscatus* L.; Brazil.	Fig. 2. *Oxyrrhopus trigeminus* D. and B.; Brazil.
Fig. 3. *Elaps semipartitus* D. and B; New Granada. Fig. 4. *Elaps psyche* Daud.; Brazil.	Fig. 3. *Pliocercus euryzonus* Cope; New Granada.
Fig. 5. *Elaps corallinus* Linn.; Brazil, Central America.	Fig. 4. *Erythrolamprus esculapii* L.; Brazil.
Fig. 6. *Ophibolus doliatus* Linn.; Southern United States to Central America.	Fig. 5. *Cemophora coccinea* Blum.; Southern United States. Fig. 6. *Erythrolamprus venustissimus* Neuw.; Brazil, Centr'l America.

The species placed opposite to each other present similar patterns of coloration. Figs. 3 and 4, Plate III, correspond with Fig. 3, Plate III*a*, and Fig. 6, Plate III, with Figs. 5 and 6, Plate III*a*. The species inhabit the same regions, more or less exactly, excepting the two Figs. 1. The *Elaps dumerili* would have been better replaced by a variety of the *Elaps fulvius* of Mexico, which has the red spaces black bordered; but a good colored plate was not accessible at the time the drawing was made. All the figures are copied from Jan's "Iconographic Générale des Ophidiens"; and the colors of such as I have not at hand for reference (*Elaps dumerili*, *Elaps semipartitus*, and *Elaps psyche*) are derived from the "Erpetologie Générale," and from Jan's "Prodrome." The species of Elaps are poisonous; the others are harmless.

deserts, the great Sahara and the sands of Arizona and California. There is also a tendency to produce spiny forms in such places; witness the Stellios and Uromastix and Cerastes of the Sahara, the Phrynosomas and horned rattlesnake of Southwestern America. The vegetation of every order, we are also informed, is in these situations extremely liable to produce spines and thorns.

The serpents of the Neotropical region furnish remarkable illustrations of mimetic analogy. All the species of the genera Elaps, Pliocercus, Erythrolamprus, and many of those of Oxyrrhopus, Ophibolus and Rhabdosoma are ornamented with black and yellow rings on a crimson ground. The species of all these genera are harmless, except in the case of Elaps, which is venomous. We may give for this genus, as the most varied, the following range of variation in coloration :

Pairs of black rings ;	*Single black rings, far apart.*	*Single black rings, very close.*
	Elaps corrallinus. b	Elaps mipartitus. d
	nigrocinctus. c	
	Pliocercus equalis. c	Pliocercus euryzonus. d
Opheomorphus mimus. d		
	Oxyrrhopus ?	Oxyrrhopus petolarius. d
Erythrolamprus venustissimus. a	Erythrolamprus albostolatus. b	Scolecophis zonatus. a
Ophibolus polyzonus. a		Leptognathus anthracops. a
Xenodon bicinctus. b		
Single black rings with faint laterals.	*Black rings in threes.*	*Single black rings about equal to intervals.*
Elaps fulvius.	Elaps lemniscatus. b	Elaps.
elegans. a		Pliocercus dimidiatus. a
Pliocercus elapoides. a		Catostoma semidoliatum. a
Procinura æmula. a	Oxyrrhopus trigeminus. b	Oxyrrhopus sebæ. d
	Pliocercus elapoides var. a	Ophibolus pyrrhomelas. h
		Chionactis occipitale. h
		Sonora semiannulata. h
		Contia isozona. h
		Chilomeniscus ephippicus. h

Species a, from Mexico and Central America. (Two species are now added, 1886.)
" b, " Brazil, Venezuela.
" c, " Central America.
" d, " western side of Andes.
" h, " Arizona and Sonora.

Many of the species in the same column are exceedingly similar, and some have little (perhaps nothing) to distinguish them but generic characters. The most similar are almost always from the same sub-region.* These facts are illustrated in Plates III and III a.

* Similar parallels exist between the Mexican species of Rhadinæa, Conophis, and Erythrolamprus (= Coniophanes). (Note, 1886.)

Similar analogies have been pointed out by Bates among the Lepidoptera of Brazil, and by Wallace among those of Borneo and Celebes, etc. I call attention to these authors here without copying them, as they will repay perusal in the originals.

A case of analogy which may belong to this class is that of the three genera Chelys among tortoises, Pipa among frogs, and Aspredo among Siluroid fishes, species of which inhabit at the same time the rivers of Guiana. The crania of these genera are similarly excessively flattened and furnished with dermal appendages, and their eyes are very minute. The singular similarity need only be mentioned to those familiar with these genera, to be recognized.

The bearing of the mimetic analogy, on the question of transition of types in the developmental hypothesis, is its demonstration of the independence of generic and specific characters of each other, which may suggest the possibility of the former being modified without affecting the latter.

These facts might have been introduced under Section II *a*, but they illustrate the general laws of the present section.

IV. OF NATURAL SELECTION.

a. As affecting Class and Ordinal Characters.

The second law which may be supposed to have governed a descent with modification, in the production of existing genera, is the force which the environment exercises in permitting or forbidding the existence or persistence of new forms. The forms which survive are supposed to have done so by virtue of their superior adaptation to their environment. This is the "natural selection" of Darwin.

That this law is subordinate to the one first propounded must, I think, be evident to any one who studies the assumed results of the workings of both, as seen in the characters of genera. It is sufficiently well known that the essential features of a majority of genera are not adaptive in their natures, and that those of many others are so slightly so, as to offer little ground for the supposition that this necessity has preserved them.

Both laws must be subordinate to that unknown force which determines the direction of the great series. If a series of suppressions of the nervous and circulatory systems of beings of common birth produced the "synthetic" predecessors of the classes of Vertebrata, the direction toward which the highest advanced, or its ultimate type, can be only ascribed as yet to the divine fiat. So far as we can see, there is no reason or law to produce a preference for this direction above any other direction.

If from these fixed bases descendants have attained to successive stations on the same line of progress, in subordinate features of the nervous and circulatory systems, constituting the "synthetic" predecessors of the orders in each class, the type finally reached seems to rest on no other basis than the pleasure of the Almighty.

β. *As affecting Family Characters.*

If from the single species generalizing a modern order we attempt to deduce synthetic predecessors of existing families, we find some difficulty, if we attempt to see in these stages a uniform succession of progress. A suppression of some features, and advance in others, in one and the same individual up to the period of reproduction, would produce offspring divergent from the start, and represent the relationship of families as we find them.

γ. *As affecting Generic Characters.*

If the extremes of our series of genera were characterized by structures particularly adapting them above all others to some contemporary necessity of existence, this second law, or Darwin's, might be regarded as primary. But the writer's experience of comparative anatomy has led him to believe that this is not the case, as expressed in Proposition IV, page 91.

This view had not been overlooked by Darwin, who, however, treats of it very briefly, and appears to attach it to the theory of adaptations, or modifications for a physiological purpose. He says, "Origin of Species," page 388 (Amer. edit., 1860): "We may extend this view to whole families, or even classes. The fore-limbs, which served as legs in the parent species, may become, by a long course of modification, adapted in one descendant to act as hands, in another as paddles, in another as wings; and on the above two principles—namely, of each successive modification supervening at a rather later age, and being inherited at a correspondingly late age—the fore-limbs in the embryos of the several descendants of the parent species will still resemble each other closely, for they will not have been modified. But in each individual new species the embryonic fore-limbs will differ greatly from the fore-limbs in the mature animal; the limbs in the latter have undergone much modification at a rather late period of life, and having thus been converted into hands, paddles, or wings." He then inclines to assign this change to the necessity of external circumstance. But such modification must be the same in kind as others, which the same hypothesis must explain, and of which the same author remarks (page 382) : "We can not, for instance, suppose that in the embryos of the Vertebrata the peculiar loop-like course of the arteries near the branchial slits are related to similar conditions in the young mammal, which is nourished in the womb of its mother, in the egg of the bird which is hatched in a nest, and in the spawn of a frog under water. We have no more reason to believe in such a relation than we have to believe that the same bones in the hand of a man, wing of a bat, and fin of a porpoise, are related to similar conditions of life. No one will suppose that the stripes on the whelp of the lion, or the spots on the young blackbird, are of any use to these animals, or related to the conditions to which they are exposed."

The law of natural selection, however, has no doubt been a very important agency in the production of organic types in different periods of the world's history ; but the part it has played in the determination of generic features would appear to have been very small.

In its first effect—that of producing a structure adapted for a particular purpose—it would seem to have acted differently to produce the same results, and hence not to have produced any of the more extended groups, as families, where hundreds of species are identical in a single feature. Witness the differences in diverse types of the tree-frogs, each type adapting its possessor to an aboreal life :

I. Claw-like, with globular base	HYLIDÆ.
	Leptopelis.
II. Simple, obtuse-depressed at tip	RANIDÆ. I *aa* and III *a.*
III. With a terminal transverse limb	RANIDÆ, *Hylarana* et aff.
	Callula.
	Brachymerus.
	Hylodes.
IV. Bifurcate	*Batrachyla.*
	Dendrobates.
	Polypedates.
	Rhacophorus.

The short foot of the Testudinidæ, where one row of phalanges is omitted, has been already alluded to. The gradual reduction of this set of bones, accompanying general modification of form in the increased convexity of dorsal region, as we leave the more aquatic and progress toward the terrestrial tortoises, would seem to be intimately connected with difference of habit. The increased convexity of carapace is an increased defense from falling objects—a danger to which land tortoises are far more subject than the aquatic. Another protection, not needed by water tortoises so much as by terrestrial, is the faculty of closing one or both free lobes of the plastron, as seen in the Cistudo, Sternothærus, etc., or of portions of the carapace, as in Pixys, Cinixys, etc. This might really have been produced by excessive tension on the sternal and pelvic muscles while young, and while the sutures were not fully interlocked. This, continued for a long time, might have produced the result. Yet it is not easy to see what protection the aquatic Kinosterna need in this respect, above the Emydes of the same countries. The backs of these genera are also as convex as are many of the terrestrial genera or Testudinidæ.

I can not better express my views than by quoting the following from the pen of the late Dr. Falconer. It is extracted from one of his essays on the Elephantidæ : *

"Each instance, however different from another, can be shown to be a term of some series of continued fractions. When this is coupled with the geometrical law governing the evolution of form, so manifest in shells of the Mollusca, it is difficult to believe that there is not in nature a deeper-seated and innate principle, to the operation of which natural selection is merely an adjunct.

"The whole range of the Mammalia, fossil and recent, can not furnish a species which has had a wider geographical distribution, and at the same time passed through a longer term of time, and through more extreme changes of climatal conditions than the mammoth.

"If species are so unstable and so susceptible of mutation through such influences, why does that extinct form stand out so signally a monument of stability? By his admirable researches and earnest writings, Darwin has, beyond all his contemporaries, given an impulse to the philosophical investigation of the most backward and obscure branch of the biological sciences of the day; he has laid the foundation of a great edifice; but he need not be surprised if, in the progress of erection, the superstructure is altered by his successors, like the Duomo of Milan, from the Roman to a different style of architecture.

---

* See writings of Hugh Falconer, vol. ii (ed. by Murchison).

"The inferences which I draw from these facts are not opposed to one of the leading propositions of Darwin's theory.

"With him I have no faith in the opinion that the mammoth and other extinct elephants made their appearance suddenly, after the type in which their fossil remains are presented to us. The most rational view seems to be, that they are in some shape the modified descendants of earlier progenitors. But if the asserted facts be correct, they seem clearly to indicate that the older elephants of Europe, such as *E. meridionalis* and *E. antiquus*, were not the stocks from which the later species, *E. primigenius* and *E. africanus* sprung, and that we must look elsewhere for their origin. The nearest affinity, and that a very close one, of the European *E. meridionalis*, is with the Miocene *E.* (Loxod.) *planifrons* of India, and of *E. primigenius* with the existing Indian species.

"Another reflection is equally strong in my mind, that the [theories of the origin of] species by 'natural selection,' or a process of variation from external influences, are inadequate to account for the phenomena. The law of Phyllotaxis, which governs the evolution of leaves around the axis of a plant, is nearly as constant in its manifestation as any of the physical laws connected with the material world."

δ. *As affecting Specific Characters.*

As I have hitherto attempted to prove that the higher grade of groups, or, in other words, the higher grade of characters, could not have had their origin through natural selection alone, though admitting it as a conserving or restricting principle, I now come to ground where natural selection must be allowed full sway. The "origin of species" is not the object of this essay, as a greater has gone before me, and has done a great deal toward showing that a selective power, dependent on adaptation and teleological relation, has favored or repressed, or even called into existence, the varied peculiarities that characterize species and races. I will therefore only refer to his well-known works on the "Origin of Species" and the "Modifications of Animals under Domestication."

I may add that it is within the range of possibility that that grade or kind of characters found to define the *family* group may be more or less the result of natural selection.

Acceleration and retardation are also far from excluded from the probable causes of specific characters. The species of many genera do exhibit a proportion of characters which are the successive stages of that one which progresses farthest, as the species of Amblystoma in the position of their teeth, nostrils, form of tail, and coloration ; of Hyla in form of vomerine teeth, etc. But the majority of specific characters are of divergent origin—are "morphic" as distinguished from developmental.

ε. *On Metaphysical Species.*

One of the arguments employed against the developmental hypothesis, in any form, is that that inherent "potentiality," which causes that like shall always produce like, is a metaphysical being, which can not be transformed, and which holds the structure which it vivifies as a material expression or stamp of itself, and which therefore can not be changed.

One expression of this inherent metaphysical specific individuality, if the term may be allowed, has been said to be the peculiar traits of the intelligence of species, their motions, voices, and instincts. But intelligence of all animals is susceptible of impressions, the lower the intelligence the less susceptible, and the more automatic. But, as we rise in the scale of animal being, this impressibility and capacity for education is undeniably exhibited by the dog, horse, and all the well-known domesticated companions of man. There can, in view of the capacities of Aves and Mammalia in these respects, be little doubt that all animals are educated by the "logic of events," that their intelligence, impressed by changed circumstances, can accommodate itself more or less to them, and that there is nothing in this part of their being opposed to the principle of "descent with modification."

There is another difficulty in the way of accepting metaphysical peculiarity or progenitiveness as isolating species. It is marked often strongly in races or varieties, which no one pretends to have had distinct origin. Here like produces like continually, though not persistently, but sufficiently to show that it resides in varieties of common origin. The isolation of allied species in fact depends, I believe, solely on the supremacy of the automatic over the intelligent spirit. When the intelligent rises above the bounds of nature, or the automatic, the mixture or separation of allied species depends merely on circumstances of necessity, determined by that intelligence.

But the metaphysical "potentiality" loses all basis, if the law of acceleration and retardation be true, for in accordance with it, at certain times, *like does not produce like.*

V. OF EPOCHAL RELATIONS, OR THOSE MEASURING GEOLOGIC TIME.

If it can be shown that groups having the developmental relation above insisted on are contemporaries, and if it can be shown that this relation is identical in kind with that which we regard as measuring the successions of geologic time, we will be led to doubt the existence of any very great interruptions in the course of this succession throughout geologic time. And if we can show that faunæ so related are more or less characteristic of distinct portions of the earth's surface, at the present time, we will be led to anticipate that contemporaneous faunæ in different regions, during geologic periods also, bore such a relation. If this proposition be true, we are led to the further conclusion, which is at variance with received canons, that identity of faunæ proves successional relation in time, instead of synchronism.* That this will ultimately be demonstrated appears highly probable to the writer, though, as yet, the evidence is but fragmentary.

If the relations expressed under the terms homology and heterology, taken together with the observations on metamorphosis, render it probable that a number of genera have reached their *expression-points*, or periods of metamorphosis, at near the same time in geologic history, an important point has been gained. If we can render it probable that a change in any organic character has been nearly simultaneous throughout a large extent of specific forms, the change becomes, on the latter account alone, of higher than generic value, but characteristic of such groups as Marsupialia, Clamatores, Acrodonta, Arcifera, Heterocerca, and the like.

We have here, also, an important element in the estimation of the value of apparent interruptions in the geological history of the life of the globe. These interruptions, it is true, are greater than any such theory as the present can bridge over; yet such a theory, if true, lessens their importance. They are in any case well accounted for on the theory of the existence of periods of elevation, during which the life of a given region is necessarily almost entirely lost to us, through lack of means of preservation of their remains.

* This view has been insisted on by Huxley.

We may also compare such extended metamorphoses with those of cosmical matter, such as when, in the course of ages, a primeval vapor has in a short time collapsed to the liquid form, or as when the vast of liquid in turn has shrunk to its solid condition; both alike for ages approaching their change, yet stationary in external relations till the moment of transition has arrived.

The following are the zoölogical relations of the groups already referred to:

The most generalized group of fishes of the Regio Neotropica is that of Characins. Its type, in respect to fin-structure, which is common to all the Malacopterygians, is that of an undeveloped stage of the Acanthopterygians, the adipose fin being an undeveloped cartilaginous fin, and the cartilaginous fin an undeveloped spinous fin.* It may be said to be the highest among Malacopterygians, if we look to the complete oviducts, opercula, jaws, etc.,† but it is the lowest as removed farthest from the extreme of Malacopterygian peculiarities, as being most generalized or embracing representatives of all the rest, and approaching nearest the types of the past—the Ganoids. For example, Butyrinus and Vastres may be compared with Amia. The family is distributed chiefly in the Southern Hemisphere.

The genus Orestias, which Agassiz says is characterized by a feature which exists in the immature state of all other Cyprinodonts—the absence of ventral fins—is only found in the Neotropical region.

Of the venomous serpents, the inferior group, the Proteroglypha, belong to the Southern Hemisphere, and the Australian and Neotropical regions almost exclusively embrace by far the greater proportion. Australia contains none other.

The Iguanian lizards are lower than the Acrodont, exhibiting a larval type of detention, and one characteristic of all lower Sauria and Batrachia. The only acrodont type of Ophiosaura (Trogonophis) is Old World.

The New World Teidæ have not the extent of ossified temporal roof that their representatives, the Old World Lacertidæ, have. So the chiefly Neotropical Anguidæ have the tongue part-

---

* Kner, "Ueber den Bau der Flossen."

† This is the correct view, for this family and the Siluroids are the most specialized of the Malacopterygian fishes. (1886.)

ly of papillose type of their Old World representatives, the Zonuridæ, and partly the smooth or scaly type of the cosmopolite Scincidæ, which are inferior to them.

The snake-like forms of the families of the Lacertilia Leptoglossa greatly predominate in the Southern Hemisphere; also those with undeveloped palpebræ.

The Neotropical type of Testudinata is quite coincident with the family Characinidæ in relations. It is, like it, largely distributed over the Southern Hemisphere, and like it may be regarded, in respect to its pelvic peculiarities, as higher than the remaining types, but in its generalized character and relationship to the past periods may be called lower.

The Neotropical type of Batrachia anura, that is, the Arcifera, is lower in developmental characters than the opposed series, the Raniformia; such of the latter as are found in its limits partake in some way of larval incompleteness. The Arcifera are chiefly distributed elsewhere in Australia, where no Raniformia exist.* Those genera of Old World Raniformia of the lowest or toothless group, which display the least development of the cranial bones, as Brachymerus and Breviceps, are of the Southern Hemisphere—South African.

The Pullastrine birds are a generalized group, inferior to the group opposed to them—the Gallinæ. Their typical forms, like the last, are distributed to the Neotropical and Australian regions: the outliers (pigeons) are not so numerously distributed in the other regions.

The Struthious birds, the most synthetic of the class, belong exclusively to the Southern Hemisphere; as is well known, they chiefly abound in Australia and its adjacent islands, with an abundant outlying type—the Tinamus—in South America.

The penguins, which only of all birds display the metatarsus nearly divided, inhabit the Antarctic regions and Cape Horn.

---

* The *Eucnemis bicolor*, Gray, would appear to be an exception, were its generic and subordinate affinities truly represented by its name. I have examined the type specimen through the kindness of Dr. Günther, and can state that it is not an Ixalus (=*Eucnemis*), and does not even belong to the Raniformia, but is an Arcifer of the family Hylidæ. If it be not a young Calamita or Hyla, it will be a Hylella near the *H. carnea* type. (Boulenger, in 1882, determined it to be a Hylella. Ed. 1886.)

Günther states that *Hylorana erythraea* has been found at the extreme northern point—Cape York—of Australia. If so, the case is parallel to the occurrences of the Raniform Ranula in northern South America.

The Clamatorial type of the Passeres exhibit larval characters in the non-development of the singing apparatus, and the scaled or nearly naked tarso-metatarsus. These are chiefly South American.

Of Mammalia, the placentals without enamel on their teeth, which, in this respect, never reach the full development of the class, whose dentition is also monophyodont, i. e., the Edentata, inhabit only the Southern Hemisphere, and almost altogether the Neotropical region. The implacental Mammalia, also (except in one tooth) monophyodont, which approach birds and reptiles in so many respects, are confined to the Southern Hemisphere, and chiefly, as all know, to Australia.

Of the Quadrumana, the Platyrrhine group is known to be inferior to the Catarrhines: the former presents an entirely embryonic condition of the *os tympanicum*, which is passed by the latter in early age; * it contains also the only clawed genus of the true monkeys. It is confined to the Neotropical realm. To Madagascar, also of the Southern Hemisphere, and nearest in many ways to the Neotropical, pertain the lowest families of the Quadrumana, the Lemuridæ and Chiromyidæ; the former presenting brains without convolutions, and approaching in many ways the Insectivora; the last imitating, at least, a Rodent.

There are also other reasons for the inferiority of South America. Its deer, which are few, include those which never produce more than the "dague," or the first horn of the northern Cervus, and also those which never get beyond the fourth step in the development of the lower group of *R. Nearctica*.

The Loricariidæ, of South America, I am informed by Prof. Agassiz, possess the fœtal pupil of the vertebrate type.

If we glance at Coleoptera we find the great predominance of the groups with undeveloped tarsus, the three- and four-jointed Trimera and Tetramera, and of the lower group with undeveloped sternum, † the Rynchophora, in the Neotropical region.

Among Lepidoptera it is known that the most gigantic of the species of the Neotropical region are Noctuidæ (Erebus, etc.), and that in that region this low type of the order reaches its greatest development. The largest forms of the Regio Nearctica, as well as Palæarctica, are representatives of the higher type of the Satur-

---

* See Dr. H. Allen, "Proceed. A. N. S.," Phila., 1867.

† Leconte, American Association, 1867.

niidæ (Atticus, Telea, etc.), while the largest and most powerful of this order in the Palæotropical (Indian) region are the Papilionid forms of Ornithoptera, etc., the generally admitted crown and head of all. Of course other types, both higher and lower, are largely developed in each and all of these regions, and the significance of the above facts is perhaps only to be seen when taken in connection with a large number of others pointing in the same direction.

Two or three comparisons of different faunæ may be brought forward finally. First, returning to the birds, a survey of some of the differences between the birds of Panama, Pennsylvania, and Palestine may be made.*

Tristram noticed three hundred and twenty-two species of birds within the range of the ancient territory of Palestine. Of these two hundred and thirty were land and ninety-two water birds, i. e., Natatores and the wading Cursores. Of the two hundred and thirty, seventy-nine are common to the British Islands, and thirty-six of them are found in China, but a small proportion extending their range to both these extremes. Of the water birds, which are always more widely distributed, fifty-five of the ninety-two are British and fifty-seven Chinese. Twenty-seven appear to be confined to Palestine, and to the immediately adjacent country; the largest of these is a crow.

Taking the two hundred and thirty land birds at a glance, we find the utter absence of so many of the well-known forms that enliven our grounds and forests. The absence of Tanagridæ (including Sylvicolidæ) and Icteridæ, changes the aspect of the bird-fauna at once. What have we here, then, of nine-quilled Oscines to enliven the meadows like our swarms of blackbirds, or fill the tree-tops and thickets with flutter like our wood-warblers? Nothing; for the twenty-four species of finches, Fringillidæ, will but balance our own, though the genera are all different but four, and they the most weakly represented by species. We must look to the higher series, the ten-quilled song-birds, for the missing rank and file. While a much larger extent of the Eastern United States possesses fifty species of these types, the little Palestine has already furnished a list of one hundred and twenty-eight.

First, of the crows, which verge nearest Icteridæ by the starlings, we have thirteen species against five in our district of the

* From the "American Naturalist," 1868, by the author.

United States, and not less than seven of the type genus Corvus, to our one common and two rare. Of these two are of the larger species, the ravens. If we turn to the cheerful larks, we find the proportion again the same; fifteen species for Palestine and one for the whole United States. One congener of our species occurs there; the other genera call to mind the African deserts and Russian steppes. Motacillidæ, again, ten to one against our fauna. *We* have two Tanagridæ to imitate them, besides the one true relative. In swallows we are about equal, and in the forest-haunting Paridæ —titmice and wrens—we exceed a little; but the comparison of Sylviidæ and Turdidæ is most striking. These highest of the bird series, especially made to gladden man's haunts with song, exceed in number all the other ten-quilled Oscines together inhabiting Palestine, amounting to seventy-five species. In our corresponding region of the United States nineteen species is the quantum. It is true no mocking-bird or wood-robin is known away from our shores, but Palestine has the nightingale, the black-cap, and the true warblers or sylvias, which, while they glean from shrub and tree their smallest insect enemies, as do our equally numerous small Tanagridæ, have much louder and sweeter voices.

Our solitary bluebird represents the long-winged Turdidæ; in the Holy Land there are twenty species corresponding, though none are of our genus. There are, indeed, but three genera of these two families common to both countries. One of these, Lanius, the butcher-bird, occurs here in one species, in Palestine in six.

Turning now to a lower series, we look in vain for Clamatorial perchers; that series which gives us the fierce king-bird and querulous pewee, and which peoples South America with thrush and warbler, and shrike and tree-creeper.

In taking a hasty glance over the lower groups, where the carotid arteries begin to be double, as the Syndactyli, we find Palestine too far from the tropics to present us with much array; but in the related zygodactyles our forest-crowned continent must claim great pre-eminence. It has but a solitary Picus, while we have eight in the immediate neighborhood of lat. 40°, in our Eastern States.

I will close with the birds of prey. Four swamp-hawks, eleven species of falcons, four kites, and eight native eagles, form a list unequaled in the annals of nobility by any land. There are together thirty-one species of Falconidæ, and of Vultures four. The eagles appear to be all common, among them the most magnificent birds of prey, the imperial and golden species of these creatures.

To the ornithologist, acquainted with the fauna of North America, it will thus be readily perceived that, in comparison, the ornis just examined, possesses more numerous representatives of the higher groups of the birds, and among lower groups possesses chiefly those of superior grade, or lacks them altogether. Let us, however, compare it with that of Central America, where varied surface and temperature offer even greater opportunity for variety, within quite as restricted an area.

The bird fauna has been found by Messrs. Sclater and Salvin to embrace about three hundred and eighty-five species, which is sixty-three more than were mentioned to occur in Palestine, which is open on three sides to the great continent. Eighty of three hundred and forty-eight land birds are characteristic of Central America; and those which find their kin limited to the Isthmus and adjoining regions of New Grenada and Equador amount to about seventy-five more. Twenty-seven is the number not known to extend beyond the boundaries of Palestine; as to the Middle States of our Union, not one species has been shown to be restricted within such narrow limits.

A single species occurs in Europe; this is the fish-hawk, an animal which combines the cosmopolite habit of the sea-bird with the powerful flight of the bird of prey. This is also the only species common to the Panama and Palestine catalogues.

The birds of prey are numerous—twenty-nine species. Among these there is no true eagle or falcon, and of the nineteen genera but four belong to the fauna of the Holy Land. There is but one species to represent the great grouse family, but, instead, three *families* of their South American imitators, the Pullastræ, instead of the one—that of the Pigeons—slimly represented in Palestine, and in North America as well.

Coming to the closer test of superiority, the Passeres—those delicate creatures, apparently so dependent on those laws which govern increase and provision, and so affected by the changes that man works in the face of Nature—what do we find? We count one hundred and six distinct species. There are none in Palestine. Of songsters, the Oscines, ninety-six species, await man's conquest of the wilderness, to increase in numbers and to display their gifts, while Palestine rejoices in a whole army of them. But the contrast is more remarkable if we analyze these forms. Of the

Isthmian Oscines, seventeen only hold the first rank, by virtue of their additional, the tenth primary quill, while this feature marks one hundred and twenty-eight species of Palestine. As we rapidly follow the line to the point where its extreme is manifested, in the family of the Thrushes or Turdidæ, Panama is left but two solitary pioneers of these songsters of the North, while seventy-five species represent the family in Palestine.

The comparison between different faunæ exhibits an apparent gradation in some other groups equally curious. Thus, the true Cyprinidæ in the Palæarctic region reach a great development, and produce the highest number of teeth on their pharyngeal jaws known, as well as attains the greatest bulk and importance. The number of these teeth is usually seven to five in the inner row; only two or three genera exhibit only four on both. In the Nearctic region the number of teeth is almost always 4—4, more rarely 4—5, and very seldom as high as 5—5. The species of the family are excessively numerous, but are, with scarcely any exception, of small size and weak organism. These statements apply to those of the eastern district of the region between the Rocky Mountains and the Atlantic. Similar types occur in the northern region of the Neotropical—Mexico, but in no great numbers. Farther south the family disappears, its place being supplied by the generalized family of Characinidæ.

I have already alluded to the great variety of the highest or pentamerous carnivorous beetles in the Palæarctic region. They are extremely abundant in the Nearctic, while the intermediate territory, the Sonoran and Mexican sub-districts, are the headquarters of the next lower form, the Tenebrionidæ, which have the tarsal joints 4—5. These give place in the Neotropical to the multitudes of the still lower series—those with the joints 4—4 and 3—3—Tetramera and Trimera.

The preceding comparisons indicate that an inherent difference between the types of a continent exists at the present time, though the difference is subordinated to a universal distribution of the higher groups throughout the earth. Has this state of things existed for any long period, or is it a result of different progress in the same group since the human period? This brings us necessarily to a consideration of the truths of paleontology, especially of the last periods, which have been already urged by Darwin. Thus the present fauna of Australia was preceded in the Post-Pliocene and Pliocene by forms possessing similar peculiarities, and belonging to the same classes. That is by herbivorous and carnivorous marsupials and monotremes, and by Varanid Sauria, all of greater size than their predecessors.

The same fact is well known of the Neotropical region, its present peculiar Edentata having been preceded by giants of the same type in the Post-Pliocene and Pliocene.

In the Nearctic region peculiar existing genera, as Procyon, Alces, Castor, Bos, Sciurus, Arctomys, Lepus, Ovibos, Sorex, Mephitis, Felis, Ursus, Menopoma, Aspidonectes, Crotalus, are represented by Post-Pliocene fossils.

The same occurs in the later Palæarctic formations, where Cervus, Bos, Canis, Mustelidæ, Insectivora, Vipera, Alytes, Triton, etc., are allied predecessors of existing types. In the Palæotropical area a wonderful development of Elephas and Gavialis preceded the same types of the present.

Prior to these faunæ another state of things has, however, existed. North America has witnessed a withdrawal of a Neotropical fauna, and the Palæarctic the retreat of an Ethiopian type. During the Post-Pliocene in North America, Neotropical genera were to Nearctic as 12 to 29, as the record now stands. In the Pliocene beds of Pikermi (Greece) antelopes, giraffes, rhinoceros, hippopotamus, huge manis,* monkeys, monitors, and other genera and species of African relationship are the prevailing forms.

Still earlier, a strong mingling of Nearctic, and more of Neotropical types, abounded in the Palæarctic. The genera Chelydra, Andrias, Podocnemis,† Platemys, Caviiform, Psammoryctid and Hesperomys-like Rodentia, Opossums, and Raccoon-like Carnivora.

We have, then, three important terms from which to derive a theory of the creation: 1, the existing six faunæ bear in many of their parts developmental relations to one another; 2, they were preceded immediately by faunæ similar to them in each case, but more remotely by faunæ like that now next lower.

On the whole, there can be no doubt of the truth of the generalization: *That the Southern Hemisphere is a geologic stage behind the Northern Hemisphere in progress*, on account (1) of its perfection in types extinct in the Northern, and (2) inferiority in modern types prevalent in the Northern.

In order, however, to demonstrate this point more fully, let us

---

* Ancylotherium, Macrotherium.

† *P. bowerbankii* (*Platemys*, Owen). *P. lævis* (*Emys*, Owen).

examine to what extent the higher types exist in the Southern, and lower or ancient in the Northern.

The Percoid fishes and their allies have Australian and South American representatives in their fresh waters, but they are as mere outliers of the great mass in the Northern Hemisphere. The higher type of venomous serpents (Solenoglypha) occur in both the Ethiopian and Neotropical regions, but they preponderate in the Northern Hemisphere. The higher group of the Saurians (the Acrodonta) abounds in the Ethiopian and Australian regions; they are as abundant in the Indian and Palæarctic regions of the Northern Hemisphere. In the Southern, also, by Uromastix and the Rynchocephalia, they approach nearest the ancient types of the Dicynodontia and the Crocodilia. Lacertidæ, and not Teidæ, occur in the Ethiopian; but they are but a proportion of the whole, which chiefly exists in the Nearctic.

Raniform, and not Arciferous Anura, populate South Africa; they, however, form but a small proportion compared with the great series of the Nearctic, Palæarctic, and Palæotropical regions. It is, however, superior in Anura to the Nearctic, taken by itself.

Rasorial birds, and not Pullastræ, are the food species of South Africa; but they do not compare in abundance or size with those of the three regions just mentioned.

Moreover, but few Clamatores exist in either Australia or Ethiopia. The Oscine types are abundant; nevertheless, they can not be compared in relative abundance with those of the northern regions. It must also be remembered that the migratory capabilities of birds render them less expressive of the true nature of any fauna.

The higher family of the Quadrumana, the Simiidæ, replaces in Africa the Cebidæ of the Neotropical; they are, however, most abundant in the Palæotropical region, in the other hemisphere.

There are two ancient or inferior types of the Northern Hemisphere: First, its fishes, the Sturgeons of the Nearctic and Palæarctic, and the Gars of the Nearctic.* The latter only have representatives in the Southern Hemisphere, Polypterus and Calamoichthys in Africa, and so may be said to be equally distributed; but the former are confined to the north. We do not know, however, whether they are of a modern or an ancient type, nor do we

* Subsequent investigations have proved that Polypterus and Calamoichthys are of much more ancient type than gars and sturgeons. (Ed. 1886.)

know of extinct sturgeons in the Southern Hemisphere. Indeed, the Ganoid series is not well defined or known as yet. If, as Agassiz states, the Siluroids pertain to it, it is cosmopolitan, though least represented in the Palæarctic.

Second, the Tailed or Urodele Batrachia. This order, entirely characteristic of the Northern Hemisphere, is a group which combines characters of Anura with those of the ancient forms, and possesses in its Nearctic types many of low development. The Gymnophidia of the Southern Hemisphere can not be considered inferior to them. In the possession of this group the Northern Hemisphere presents its first element of inferiority.

The preceding comparisons indicate also the relations of the regions proper to each other. It is obvious enough that the Ethiopian is much superior to the two others of the Southern Hemisphere. As to the Australian and Neotropical, the former must still be regarded as probably the most ancient, though possessing at the same time a much stronger admixture of northern forms. I have already presented the relations, with the inferior forms of each, thus: *

R. Australis.—Inferior in possessing Monotrematous and Marsupial Mammalia, Pullastriform and Struthious Birds, Serpentiform Pleurodont Lacertilia, Arciferous Batrachia, Pleurodire Tortoises, its Elapid venomous snakes, and the whole Flora, according to Unger.

R. Neotropica.—Marsupial and Edentate Mammalia, Inferior Rodentia and Quadrumana, Pleurodire Tortoises, Pleurodont Lacertilia, Arciferous Batrachia, Clamatorial and Pullastriform Birds, Characin and Erythrinid Malacopterigii.

Conclusions.

The following may be looked upon as conclusions which have been indicated in the preceding pages:

I. Species have developed from pre-existent species by an inherent tendency to variation, and have been preserved in given directions and repressed in others by the operation of the law of natural selection (Darwin).

II. Genera have been produced by a system of retardation or acceleration in the development of individuals; the former on pre-established, the latter on preconceived lines of direction. Or,

* "Jour. Acad. Nat. Sci.," Philada., 1866, p. 109.

in other words, that while nature's series have been projected in accordance with the law of acceleration and retardation, they have been limited, modified, and terminated by the law of natural selection, which may itself have operated in part by the same law.

III. The processes of development of specific and generic characters have not proceeded *pari passu*, transitions of the one kind not being synchronous with transitions of the other; and that, therefore, species may be transferred from one genus to another without losing their specific characters, and genera from order to order without losing their generic characters.

IV. And as the heterologous terms of the peculiar homologous groups present an "inexact parallelism" with each other; and as types related by inexact parallelism are each among themselves exact parallels in separate series, whose earliest members present "exact parallelisms" with each other, it follows—

V. That the heterologous terms or genera in the later series are modified descendants of those of the earlier series; in other words, that certain groups higher than genera are produced from others of similar high value by "descent with modification."

VI. That the result of such successional metamorphoses will be expressed in geologic history by more or less abrupt transitions or expression-points, rather than by uniformly gradual successions.

Of course, under the conclusion stated in Proposition II, the genus Homo has been developed by the modification of some preexistent genus. All his traits which are merely functional have, as a consequence, been produced during the process. Those traits which are not functional, but spiritual, are of course amenable to a different class of laws, which belong to the province of religion.

III.

ON THE THEORY OF EVOLUTION.

At a meeting of the Philadelphia Academy of Natural Sciences, held February 22, 1876, Prof. Cope gave a history of the progress of the doctrine of evolution of animal and vegetable types. While Darwin has been its prominent advocate within the last few years, it was first presented to the scientific world, in a rational form, by Lamarck, of Paris, at the commencement of the present century. Owing to the adverse influence of Cuvier, the doctrine remained dormant for half a century, and Darwin resuscitated it, making important additions at the same time. Thus Lamarck found the variations of species to be the primary evidence of evolution by descent. Darwin enunciated the law of "natural selection" as a result of the struggle for existence, in accordance with which "the fittest" only survive. This law, now generally accepted, is Darwin's principal contribution to the doctrine. It, however, has a secondary position in relation to the *origin* of variation, which Lamarck saw, but did not account for, and which Darwin has to assume in order to have materials from which a "natural selection" can be made.

The relations exhibited by fully grown animals and plants with transitional or embryonic stages of other animals and plants, had attracted the attention of anatomists at the time of Lamarck. Some naturalists deduced from this now universally observed phenomenon that the lower types of animals were merely repressed conditions of the higher, or, in other words, were embryonic stages become permanent. But the resemblances do not usually extend to the entire organism, and the parallels are so incomplete that this view of the matter was clearly defective, and did not constitute an explanation. Some embryologists, as Lereboullet and Agassiz, asserted that no argument for a doctrine of descent could be drawn from such facts.

The speaker, not adopting either view, made a full investigation into the later embryonic stages, chiefly of the skeleton of the Batrachia, in 1865, and Prof. Hyatt, of Salem, Mass., at the same time made similar studies in the development of the Ammonites and Nautili. The results as bearing on the doctrine of evolution were published in 1869 (in "The Origin of Genera"). It was there pointed out that the most nearly related forms of animals do present a relation of repression and advance, or of permanent embryonic and adult type, leaving no doubt that the one is descended from the other. This relation was termed *exact parallelism.* It was also shown that, if the embryonic form were the parent, the advanced descendant was produced by an increased rate of growth, which phenomenon was called *acceleration;* but that if the embryonic type were the offspring, then its failure to attain to the condition of the parent is due to the supervention of a slower rate of growth; to this phenomenon the term *retardation* was applied. It was then shown that the *inexact parallelism* was the result of *unequal* acceleration or retardation; that is, acceleration affecting one organ or part more than another, thus disturbing the combination of characters which is necessary for the state of *exact parallelism* between the perfect stage of one animal and the transitional state of another. Moreover, acceleration implies constant addition to the parts of an animal, while retardation implies continual subtraction from its characters, or atrophy. He had also shown ("Method of Creation," 1871) that the additions either appeared as *exact repetitions* of pre-existent parts, or as *modified repetitions,* the former resulting in simple, the latter in more complex organisms.

Prof. Haeckel, of Jena, has added the keystone to the doctrine of evolution in his gastræa theory. Prior to this generalization it had been impossible to determine the true relation existing between the four types of embryonic growth, or, to speak otherwise, than that they are inherently distinct from each other. But Haeckel has happily determined the existence of identical stages of growth in all of the types of eggs, the last of which is the *gastrula;* and beyond which the identity ceases. Not that the four types of gastrula are without difference, but this difference may be accounted for, on plain principles. In 1874, Haeckel, in his "Anthropogenie," recognizes the importance of the irregularity of time of appearance of the different characters of animals, during the period of growth, as affecting their permanent structure. While maintaining the view that the low forms represent the transitional stages of the higher, he proceeds to account for the want of exact correspondence exhibited by them at the present time, by reference to this principle. He believes that the relation of parent and descendant has been concealed and changed by subsequent modifications of the order of appearance of characters in growth. To the original, simple descent he applies the term *palingenesis ;* to the modified and later growth, *cœnogenesis.* The causes of the change from palingenesis to cœnogenesis he regards as three, viz., acceleration, retardation, and heterotopy.

It is clear that the two types of growth distinguished by Prof. Haeckel are those which had been pointed out by Prof. Cope in "The Origin of Genera," as producing the relations of "exact" and "inexact parallelism"; and that his explanation of the origin of the latter relation by acceleration or retardation is the same as that of the latter essay. The importance which he attaches to the subject was a source of gratification to the speaker, as it was a similar impression that led to the publication of "The Origin of Genera" in 1869.

It remains to observe that the phenomena of exact parallelism or palingenesis are quite as necessarily accounted for on the principle of acceleration or retardation as are those of inexact parallelism or cœnogenesis. Were all parts of the organism accelerated or retarded at a like rate, the relation of exact parallelism would never be disturbed ; while the inexactitude of the parallelism will depend on the number of variations in the rate of growth of different organs of the individual, with additions introduced from time to time. Hence it may be laid down that *synchronous acceleration* or *retardation* produces exact parallelism, and *heterochronous acceleration* or *retardation* produces inexact parallelism.

In conclusion, it may be added that acceleration of the segmentation of the protoplasma or animal portion of the primordial egg, or retardation of segmentation of the deutoplasma or vegetative half of the egg, or both, or the same relation between the growth of the circumference and center of the egg, has given rise to the four types which the segmentation now presents.

An analysis of the laws of evolution may be tabulated as follows :

		Exact parallelism, the product of *Palingenesis*, which is *synchronous*....	*Inexact parallelism*, the product of *Cœnogenesis*, which is *heterochronous*..
acceleration, which proceeds by	Exact repetition....................	*	*
	Modified repetition.................	*	*
	Heterotopy.........................	*	*
retardation, which proceeds by	Exact atrophy.....................	*	*
	Inexact atrophy (or senility)*........	*	*

* So called by Prof. Hyatt.

IV.

ON THE HYPOTHESIS OF EVOLUTION:

PHYSICAL AND METAPHYSICAL.

"Man shall not live by bread alone, but by every word that proceedeth out of the mouth of God shall man live."

THERE is apparently considerable repugnance in the minds of many excellent people to the acceptance, or even consideration, of the hypothesis of development, or that of the gradual creation by descent, with modifications from the simplest beginnings, of the different forms of the organic world. This objection probably results from two considerations: first, that the human species is certainly involved, and man's descent from an ape asserted; and, secondly, that the scheme in general seems to conflict with that presented by the Mosaic account of the Creation, which is regarded as communicated to its author by an infallible inspiration.

As the truth of the hypothesis is held to be infinitely probable by a majority of the exponents of the natural sciences at the present day, and is held as absolutely demonstrated by another portion, it behooves those interested to restrain their condemnation, and on the other hand to examine its evidences, and look any consequent necessary modification of our metaphysical or theological views squarely in the face.

The following pages state a few of the former; if they suggest some of the latter, it is hoped that they may be such as any logical mind would deduce from the premises. That they will coincide with the spirit of the most advanced Christianity, I have no doubt; and that they will add an appeal through the reason to that direct influence of the Divine Spirit which should control the motives of human action, seems an unavoidable conclusion.

I. PHYSICAL EVOLUTION.

It is well known that a species is usually represented by a grea number of individuals, distinguished from all other similar associ ations by more or less numerous points of structure, color, size, etc., and by habits and instincts also, to a certain extent; that the individuals of such associations reproduce their like, and can not be produced by individuals of associations or species which present differences of structure, color, etc., as defined by naturalists; that the individuals of any such series or species are incapable of reproducing with those of any other species, with some exceptions; and that in the latter cases the offspring are usually entirely infertile.

The hypothesis of Cuvier assumes that each species was created by divine power as we now find it at some definite point of geologic time. The paleontologist holding this view sees, in accordance therewith, a succession of creations and destructions marking the history of life on our planet from its commencement.

The development hypothesis states that all existing species have been derived from species of pre-existent geological periods, as offspring or by direct descent; that there have been no total destructions of life in past time, but only a transfer of it from place to place, owing to changes of circumstance; that the types of structure become simpler and more similar to each other as we trace them from later to earlier periods; and that finally we reach the simplest forms consistent with one or several original parent types of the great divisions into which living beings naturally fall.

It is evident, therefore, that the hypothesis does not include change of species by hybridization, nor allow the descent of living species from any other *living* species: both these propositions are errors of misapprehension or misrepresentation.

In order to understand the history of creation of a complex being, it is necessary to analyze it and ascertain of what it consists. In analyzing the construction of an animal or plant we readily arrange its characters into those which it possesses in common with other animals or plants, and those in which it resembles none other: the latter are its *individual* characters, constituting its individuality. Next, we find a large body of characters, generally of a very obvious kind, which it possesses in common with a generally large number of individuals, which, taken collectively, all men are accustomed to call a species; these characters we consequently name *specific*. Thirdly, we find characters, generally in parts of the body which are of importance in the activities of the animal, or which lie in near relation to its mechanical construction in details, which are shared by a still larger number of individuals than those which were similar in specific characters. In other words, it is common to a large number of species. This kind of character we call *generic*, and the grouping it indicates is a genus.

Farther analysis brings to light characters of organism which are common to a still greater number of individuals; this we call a *family* character. Those which are common to still more numerous individuals are the *ordinal:* they are usually found in parts of the structure which have the closest connection with the whole life-history of the being. Finally, the individuals composing many orders will be found identical in some important character of the systems by which ordinary life is maintained, as in the nervous and circulatory: the divisions thus outlined are called *classes*.

By this process of analysis we reach in our animal or plant those peculiarities which are common to the whole animal or vegetable kingdom, and then we have exhausted the structure so completely that we have nothing remaining to take into account beyond the cell-structure or homogeneous protoplasm by which we know that it is organic, and not a mineral.

The history of the origin of a type, as species, genus, order, etc., is simply the history of the origin of the structure or structures which define those groups respectively. It is nothing more nor less than this, whether a man or an insect be the object of investigation.

EVIDENCES OF DERIVATION.

a. Of Specific Characters.

The evidences of derivation of species from species, within the limits of the genus, are abundant and conclusive. In the first place, the rule which naturalists observe in defining species is a clear consequence of such a state of things. It is not amount and degree of difference that determine the definition of species from species, but it is the *permanency* of the characters in all cases and under all circumstances. Many species of the systems include varieties and extremes of form, etc., which, were they at all times distinct, and not connected by intermediate forms, would be estimated as species by the same and other writers, as can be easily seen by reference to their works.

Thus, species are either "restricted" or "protean," the latter embracing many, the former few variations; and the varieties included by the protean species are often as different from each other in their typical forms as are the "restricted" species. As an example, the species *Homo sapiens* (man) will suffice. His primary varieties are as distinct as the species of many well-known genera, but can not be defined, owing to the existence of innumerable intermediate forms between them.

As to the common origin of such "varieties" of the protean species, naturalists never had any doubt; yet when it comes to the restricted "species," the anti-developmentalist denies it *in toto.* Thus the varieties of most of the domesticated animals are some of them known—others held with great probability to have had a common origin. Varieties of plumage in fowls and canaries are of every-day occurrence, and are produced under our eyes. The cart-horse and racer, the Shetland pony and the Norman, are without doubt derived from the same parentage. The varieties of pigeons and ducks are of the same kind, but not every one is aware of the extent and amount of such variations. The varieties in many characters seen in hogs and cattle, especially when examples from distant countries are compared, are very striking, and are confessedly equal in degree to those found to *define* species in a state of nature: here, however, they are not *definitive.*

It is easy to see that all that is necessary to produce in the mind of the anti-developmentalist the illusion of distinct origin by creation of many of these forms would be to destroy a number of the intermediate conditions of specific form and structure, and thus to leave remaining definable groups of individuals, and therefore "species."

That such destructions and extinctions have been going no ever since the existence of life on the globe is well known. That it should affect intermediate forms, such as bind together the types of a protean species as well as restricted species, is equally certain. That its result has been to produce *definable* species can not be denied, especially in consideration of the following facts: Protean species nearly always have a wide geographical distribution. They exist under more varied circumstances than do individuals of a more restricted species. The subordinate variations of the protean species are generally, like the restricted species, confined to distinct subdivisions of the geographical area which the whole occupies. As in geological time changes of level have separated areas once continuous by bodies of water or high mountain-ranges, so have vast numbers of individuals occupying such areas been destroyed. Important alterations of temperature, or great changes in abundance or character of vegetable life over given areas, would produce the same result.

This part of the subject might be prolonged, were it necessary, but it has been ably discussed by Darwin. The *rationale* of the "origin of species," as stated by him, may be examined a few pages farther on.

β. *Of the Characters of Higher Groups.*

a. Relations of Structures.—The evidences of derivative origin of the structures defining the groups called genera, and all those of higher grade, are of a very different character from those discussed in relation to specific characters; they are more difficult of observation and explanation.

First: It would appear to be supposed by many that the creation of organic types was an irregular and capricious process, variously pursued by its Author as regards time and place, and without definite final aim; and this notwithstanding the wonderful evidences we possess, in the facts of astronomy, chemistry, sound, etc., of his adhesion to harmonious and symmetrical sequences in his modes and plans.

Such regularity of plan is found to exist in the relations of the great divisions of the animal and vegetable kingdoms as at present existing on the earth. Thus, with animals we have a great class of species which consists of nothing more than masses or cells of protoplasmic matter, without distinct organs; or the Protozoa. We have then the Cœlenterata (example, corals), where the organism is composed of many cells arranged in distinct parts, but where a single very simple system of organs, forming the only internal cavity of the body, does the work of the many systems of the more complex animals. Next, the Echinodermata (such as star-fish) present us with a body containing distinct systems of organs inclosed in a visceral cavity, including a rudimental nervous system in the form of a ring. In the Molluscs to this condition is added further complication, including extensions of the nervous system from the ring as a starting-point, and a special organ for a heart. In the Articulates (crabs, insects) we have like complications, and a long distinct nervous axis on the lower surface of the body. The last branch or division of animals is considered to be higher, because all the systems of life organs are most complex or specialized. The nervous ring is almost obliterated by a great enlargement of its usual ganglia, thus become a brain, which is succeeded by a long axis on the upper side of the body. This and other points define the Vertebrata.

Plans of structure, independent of the simplicity or perfection of the special arrangement or structure of organs, also define these great groups. Thus the Protozoa present a spiral, the Cœlenterata a radiate, the Echinodermata a bilateral radiate plan. The Articulates are a series of external rings, each in one or more respects repeating the others. The Molluscs are a sac, while a ring above a ring, joined together by a solid center-piece, represents the plan of each of the many segments of the Vertebrates which give the members of that branch their form.

These bulwarks of distinction of animal types are entered into here simply because they are the most inviolable and radical of those with which we have to deal, and to give the anti-developmentalist the best foothold for his position. I will only allude to the relations of their points of approach as these are affected by considerations afterward introduced.

The Vertebrates approach the Molluscs closely at the lowest extreme of the former and higher of the latter. The lamprey eels of the one possess several characters in common with the cuttle-fish or squids of the latter. The amphioxus is called the lowest Vertebrate, and, though it is nothing else, the definition of the division must be altered to receive it; it has no brain!

The lowest forms of the Molluscs and Articulates are scarcely distinguishable from each other, so far as adhesion to the "plan" is concerned, and some of the latter division are very near certain Echinodermata. As we approach the boundary-lines of the two lowest divisions, the approaches become equally close.

More instructive is the evidence of the relation of the subordinate classes of any one of these divisions. The conditions of those organs or parts which define classes exhibit a regular relation, commencing with simplicity and ending with complication; first associated with weak exhibitions of the highest functions of the nervous system—at the last displaying the most exalted traits found in the series.

For example: In the classes of Vertebrates we find the lowest nervous system presents great simplicity—the brain can not be recognized; next (in lampreys), the end of the nervous axis is subdivided, but scarcely according to the complex type that follows. In fishes the cerebellum and cerebral hemispheres are minute, and the intermediate or optic lobes very large: in the reptiles the cerebral hemispheres exceed the optic lobes, while the cerebellum is smaller. In birds the cerebellum becomes complex and the cerebrum greatly increases. In mammals the cerebellum increases in complexity or number of parts, the optic lobes diminish, while the cerebral hemispheres become wonderfully complex and enlarged, bringing us to the highest development, in man.

The history of the circulatory system in the Vertebrates is the same. First, a heart with one chamber, then one with two divisions: three divisions belong to a large series, and the highest possess four. The origins of the great artery of the body, the aorta, are first five on each side: they lose one in the succeeding class in the ascending scale, and one in each succeeding class or order, till the Mammalia, including man, present us with but one on one side.

From an infinitude of such considerations as the above, we derive the certainty that the general arrangement of the various groups of the organic world is in scales, the subordinate within the more comprehensive divisions. The identification of all the parts in such a complexity of organism as the highest animals present is a matter requiring much care and attention, and constitutes the study of homologies. Its pursuit has resulted in the demonstration that every individual of every species of a given branch of the animal kingdom is composed of elements common to all, and that the differences which are so radical in the higher groups are but the modifications of the same elemental parts, representing completeness or incompleteness, obliteration or subdivision. Of the former character are rudimental organs, of which almost every species possesses an example in some part of its structure.

But we have other and still more satisfactory evidence of the meaning of these relations. By the study of embryology we can prove most indubitably that the simple and less complex are inferior to the more complex. Selecting the Vertebrates again as an example, the highest form of mammal—e. g., man—presents in his earliest stages of embryonic growth a skeleton of cartilage, like that of the lamprey; he also possesses five origins of the aorta and five slits on the neck: both which characters belong to the lamprey and the shark. If the whole number of these parts does not co-exist in the embryonic man, we find in embryos of lower forms, more nearly related to the lamprey, that they do. Later in the life of the mammal but four aortic origins are found, which arrangement, with the heart now divided into two chambers, from a beginning as a simple tube, is characteristic of the class of Vertebrates next in order—the bony fishes. The optic lobes of the human brain have also at this time a great predominance in size—a character above stated to be that of the same class. With advancing development the infant mammal follows the scale already pointed out. Three chambers of the heart and three aortic origins follow, presenting the condition permanent in the Batrachia; and two origins, with enlarged cerebral hemispheres of the brain, resemble the reptilian condition. Four heart-chambers, and one aortic-root on each side, with slight development of the cerebellum, follow, all characters defining the crocodiles, and immediately precede the special conditions defining the mammals. These are, the single aorta-root from one side, and the full development of the cerebellum; later comes that of the cerebrum also in its higher mammalian and human traits.

Thus we see the order already pointed out to be true, and to be an ascending one. This is the more evident as each type or class passes through the conditions of those below it, as did the mammal; each scale being shorter as its highest terminus is lower. Thus the crocodile passes through the stage of the lamprey, the fish, the batrachian, and the reptile proper.

b. In Time.—We have thus a scale of relations of existing forms of animals and plants of a remarkable kind, and such as to stimulate greatly our inquiries as to its significance. When we turn to the remains of the past creation, preserved to us in the deposits continued throughout geologic time, we are not disappointed, for great light is at once thrown upon the subject.

We find, in brief, that the lowest division of the animal kingdom appeared first, and long before any type of a higher character was created. The Protozoön, Eozoön, is the earliest of animals in geologic time, and represents the lowest type of animal life now existing. We learn also that the highest branch appeared last. No remains of Vertebrates have been found below the lower Devonion period, or not until the Echinoderms and

Molluscs had reached a great pre-eminence. It is difficult to be sure whether the Protozoa had a greater numerical extent in the earliest periods than now, but there can be no doubt that the Cœlenterata (corals) and Echinoderms (crinoids) greatly exceeded their present bounds in Paleozoic time, so that those at present existing are but a feeble remnant. If we examine the subdivisions known as classes, evidence of the nature of the succession of creation is still more conclusive. The most polyp-like of the Molluscs (Brachiopoda) constituted the great mass of its representatives during Paleozoic time. Among Vertebrates the fishes appear first, and had their greatest development in size and numbers during the earliest periods of the existence of the division. Batrachia were much the largest and most important of land animals during the Carboniferous period, while the higher Vertebrates were unknown. The later Mesozoic periods saw the reign of reptiles, whose position in structural development has been already stated. Finally, the most perfect, the mammal, came upon the scene, and in his humblest representatives. In Tertiary times Mammalia supplanted the reptiles entirely.

Thus the structural relations, the embryonic characters, and the successive appearance in time of animals coincide. The same is very probably true of plants.

That the existing state of the geological record of organic types should be regarded as anything but a fragment is, from our standpoint, quite preposterous. And, more, it may be assumed with safety that when completed it will furnish us with a series of regular successions, with but slight and regular interruptions, if any, from the species which represented the simplest beginnings of life at the dawn of creation, to those which have displayed complication and power in later or in the present periods.

For the labors of the paleontologist are daily bringing to light structures intermediate between those never before so connected, and thus creating lines of succession where before were only interruptions. Many such instances might be adduced : two might be selected as examples from American paleontology ; * i. e., the

* Prof. Huxley, in the last anniversary lecture before the Geological Society of London, recalls his opinion, enunciated in 1862, that "the positively-ascertained truths of Paleontology" negative "the doctrines of progressive modification, which suppose that modification to have taken place by a necessary progress from more to less embryonic forms, from more to less generalized types, within the limits of the

near approach to birds made by the reptiles Lælaps and Megadactylus, and the combination of characters of the old genera Ichthyosaurus and Plesiosaurus in the Polycotylus of Kansas.

We had no more reason to look for intermediate or connecting forms between such types as these than between any others of similar degree of remove from each other with which we are ac-

---

period represented by the fossiliferous rocks; that it shows no evidence of such modification; and, as to the nature of that modification, it yields no evidence whatsoever that the earlier members of any long-continued group were more generalized in structure than the later ones."

Respecting this position, he says: "Thus far I have endeavored to expand and enforce by fresh arguments, but not to modify in any important respect, the ideas submitted to you on a former occasion. But when I come to the propositions respecting progressive modification, it appears to me, with the help of the new light which has broken from various quarters, that there is much ground for softening the somewhat Brutus-like severity with which I have dealt with a doctrine for the truth of which I should have been glad enough to be able to find a good foundation in 1862. So far indeed as the Invertebrata and the lower Vertebrata are concerned, the facts and the conclusions which are to be drawn from them appear to me to remain what they were. For anything that as yet appears to the contrary, the earliest known Marsupials may have been as highly organized as their living congeners; the Permian lizards show no signs of inferiority to those of the present day; the labyrinthodonts can not be placed below the living salamander and triton; the Devonian ganoids are closely related to polypterus and lepidosiren."

To this it may be replied: 1. The scale of progression of the Vertebrata is measured by the condition of the circulatory system, and in some measure by the nervous, and not by the osseous: tested by this scale, there has been successional complication of structure among Vertebrata in time. 2. The question with the evolutionist is, not what types have persisted to the present day, but the order in which types appeared in time. 3. The Marsupials, Permian saurians, labyrinthodonts, and Devonian ganoids are remarkably generalized groups, and predecessors of types widely separated in the present period. 4. Prof. Huxley adduces many such examples among the mammalian subdivisions in the remaining portion of his lecture. 5. Two alternatives are yet open in the explanation of the process of evolution: since generalized types, which combine the characters of higher and lower groups of later periods, must thus be superior to the lower, the lower must (first) be descended from such a generalized form by degradation; or (second) not descended from it at all, but from some lower contemporaneous type by advance; the higher only of the two being derived from the first-mentioned. The last I suspect to be a true explanation, as it is in accordance with the homologous groups. This law will shorten the demands of paleontologists for time, since, instead of deriving all Reptilia, Batrachia, etc., from common origins, it points to the derivation of higher Reptilia of a higher order from higher Reptilia of a lower order, lower Reptilia of the first from lower Reptilia of the second; finally, the several groups of the lowest or most generalized order of Reptilia from a parallel series of the class below, or Batrachia.

quainted. And inasmuch as almost all groups, as genera, orders, etc., which are held to be distinct, but adjacent, present certain points of approximation to each other, the almost daily discovery of intermediate forms gives us confidence to believe that the pointings in other cases will also be realized.

γ. *Of Transitions.*

The preceding statements were necessary to the comprehension of the supposed mode of metamorphosis or development of the various types of living beings, or, in other words, of the single structural features which define them.

As it is evident that the groups of highest rank have had their origin in remote ages, cases of transition from one to the other by change of character can not be witnessed at the present day. We therefore look to the most nearly related divisions, or those of the lowest rank, for evidence of such change.

It is necessary to premise that embryology teaches that all the species of a given branch of the animal kingdom (e. g., Vertebrate, Mollusc, etc.) are quite identical in structural character at their first appearance on the germinal layer of the yolk of the parent egg. It shows that the character of the respective groups of high rank appear first, then those of less grade, and last of all those structures which distinguish them as genera. But among the earliest characters which appear are those of the species, and some of those of the individual.

We find the characters of different *genera* to bear the same relation to each other that we have already seen in the case of those definitive of orders, etc. In a natural assemblage of related genera we discover that some are defined by characters found only in the embryonic stages of others, while a second will present a permanent condition of its definitive part, which marks a more advanced stage of that highest. In this manner many stages of the highest genus appear to be represented by permanent genera in all natural groups. Generally, however, this resmblance does not involve an entire identity, there being some other immaturities found in the highest genus at the time it presents the character preserved in permanency by the lower, which the lower loses. Thus (to use a very gross illustration) a frog at one stage of growth has four legs and a tail; the salamander always preserves four legs and a tail, thus resembling the young frog. The latter is, however, not a salamander at that time, because, among other things, the skeleton is represented by cartilage only, and the salamander's is ossified. This relation is therefore an imitation only, and is called *inexact parallelism.*

As we compare nearer and nearer relations—i. e., the genera which present fewest points of difference—we find the differences between undeveloped stages of the higher and permanent conditions of the lower to grow fewer and fewer, until we find numerous instances where the lower genus is exactly the same as the undeveloped stage of the higher. This relation is called that of *exact parallelism.*

It must now be remembered that the permanence of a character is what gives it its value in defining genus, order, etc., in the eyes of the systematist. So long as the condition is permanent no transition can be seen; there is therefore no development. If the condition is transitional, it defines nothing, and nothing is developed; at least, so says the anti-developmentalist. It is the old story of the settler and the Indian: "Will you take owl and I take turkey, or I take turkey and you owl?"

If we find a relation of *exact parallelism* to exist between two sets of species in the condition of a certain organ, and the difference so expressed is the only one which distinguishes them as sets from each other—if that condition is always the same in each set—we call them two genera: if in any species the condition is variable at maturity, or sometimes the undeveloped condition of the part is persistent and sometimes transitory, the sets characterized by this difference must be united by the systematist, and the whole is called a single genus.

We know numerous cases where different individuals of the same species present this relation of *exact parallelism* to each other; and, as we ascribe common origin to the individuals of a species, we are assured that the condition of the inferior individual is, in this case, simply one of repressed growth, or a failure to fulfill the course accomplished by the highest. Thus, certain species of the salamandrine genus Amblystoma undergo a metamorphosis involving several parts of the osseous and circulatory systems, etc., while half grown; others delay it till fully grown; one or two species remain indifferently unchanged or changed, and breed in either condition, while another species breeds unchanged, and has never been known to complete a metamorphosis.

The nature of the relation of *exact parallelism* is thus explained to be that of checked or advanced growth of individuals having a common origin. The relation of *inexact parallelism* is readily explained as follows: With a case of *exact parallelism* in the mind, let the repression producing the character of form B parallelize the latter with a stage of form A in which a second part is not quite mature: we will have a slight want of correspondence between the two. Form B will be immature in but one point, the incompleteness of A higher being seen in two points. If we suppose the immaturity to consist in a repression at a still earlier point in the history of the higher, the latter will be undeveloped in other points also: thus, the spike-horned deer of South America have the horn of the second year of the North American genus. They would be generically identical with that stage of the latter, were it not that these still possess their milk dentition at two years of age. In the same way the nature of the parallelisms seen in higher groups, as orders, etc., may be accounted for.

The theory of homologous groups furnishes important evidence in favor of derivation. Many orders of animals (probably all, when we come to know them) are divisible into two or more sections, which I have called *homologous*. These are series of genera or families, which differ from each other by some marked character, but whose contained genera or families differ from each other in the same points of detail, and in fact correspond exactly. So striking is this correspondence that were it not for the general and common character separating the homologous series, they would be regarded as the same, each to each. Now, it is remarkable that, where studied, the difference common to all the terms of two homologous groups is found to be one of *inexact parallelism*, which has been shown above to be evidence of descent. Homologous groups always occupy different geographical areas on the earth's surface, and their relation is precisely that which holds between successive groups of life in the periods of geologic time.*

In a word, we learn from this source that distinct geologic epochs co-exist at the same time on the earth. I have been forced to this conclusion † by a study of the structure of terrestrial life, and it has been remarkably confirmed by the results of recent deep-sea dredgings made by the United States Coast Survey in the Gulf Stream, and by the British naturalists in the North Atlantic. These have brought to light types of Tertiary life, and of even

---

* The extinct family of the Nimravidæ, which is homologous with the existing family of Felidæ, has been discovered since this was written. (Ed. 1886.)

† "Origin of Genera," pp. 70, 77, 79.

the still more ancient Cretaceous periods, living at the present day.* That this discovery invalidates in any wise the conclusions of geology respecting lapse of time is an unwarranted assumption that some are forward to make. If it changes the views of some respecting the parallelism or co-existence of faunæ in different regions of the earth, it is only the anti-developmentalists whose position must be changed.

For, if we find distinct geologic faunæ, or epochs defined by faunæ, co-existing during the present period, and fading or merging into one another as they do at their geographical boundaries, it is proof positive that the geologic epochs and periods of past ages had in like manner no trenchant boundaries, but also passed the one into the other. The assumption that the apparent interruptions are the result of transfer of life rather than destruction, or of want of opportunities of preservation, is no doubt the true one.

δ. *Rationale of Development.*

a. In Characters of Higher Groups.—It is evident in the case of the species in which there is an irregularity in the time of completion of metamorphosis, that some individuals traverse a longer developmental line than those which remain more or less incomplete. As both accomplish growth in the same length of time, it is obvious that it proceeds with greater rapidity in one sense in that which accomplishes most; its growth is said to be accelerated. This phenomenon is especially common among insects, where the females of perfect males are sometimes larvæ or nearly so, or pupæ, or lack wings or some character of final development. Quite as frequently, some males assume characters in advance of others, sometimes in connection with a peculiar geographical range.

In cases of *exact parallelism* we reasonably suppose the cause to be the same, since the conditions are identical, as has been shown; that is, the higher conditions have been produced by a crowding back of the earlier characters and an acceleration of growth, so that a given succession in order of advance has extended over a longer range of growth than its predecessor in the same allotted time. That allotted time is the period before maturity and reproduction, and it is evident that as fast as modifications or characters should be assumed sufficiently in advance of that period, so certainly would they be conferred upon the offspring by

* Most of the deep-sea forms are, however, degenerate forms of existing orders. (Ed. 1886.)

reproduction. The *acceleration* in the assumption of a character, progressing more rapidly than the same in another character, must soon produce, in a type whose stages were once the exact parallel of a permanent lower form, the condition of *inexact parallelism.* As all the more comprehensive groups present this relation to each other, we are compelled to believe that *acceleration* has been the principle of their successive evolution during the long ages of geologic time.

Each type has, however, its day of supremacy and perfection of organism, and a retrogression in these respects has succeeded. This has no doubt followed a law the reverse of acceleration, which has been called *retardation.* By the increasing slowness of the growth of the individuals of a genus, and later and later assumption of the characters of the latter, they would be successively lost.

To what power shall we ascribe this acceleration, by which the first beginnings of structure have accumulated to themselves through the long geologic ages complication and power, till from the germ that was scarcely born into a sand-lance, a human being climbed the complete scale, and stood easily the chief of the whole?

In the cases of species, where some individuals develop farther than others, we say the former possess more growth-force, or "vigor," than the latter. We may therefore say that higher types of structure possess more "vigor" than the lower. This, however, we do not know to be true, nor can we readily find means to demonstrate it.

The food which is taken by an adult animal is either assimilated, to be consumed in immediate activity of some kind, or stored for future use, and the excess is rejected from the body. We have no reason to suppose that the same kind of material could be made to subserve the production of force by any other means than that furnished by a living animal organism. The material from which this organism is constructed is derived first from the parent, and afterward from the food, etc., assimilated by the individual itself so long as growth continues. As it is the activity of assimilation directed to a special end during this latter period which we suppose to be increased in accelerated development, the acceleration is evidently not brought about by increased facilities for obtaining the means of life which the same individual possesses as an adult. That it is not in consequence of such increased facilities possessed by its parents over those of the type preceding it seems equally improbable when we consider that the characters in which the parent's advance has appeared are rarely of a nature to increase those facilities.

The nearest approach to an explanation that can be offered appears to be somewhat in the following direction :

There is every reason to believe that the character of the atmosphere has gradually changed during geologic time, and that various constituents of the mixture have been successively removed from it, and been stored in the solid material of the earth's crust in a state of combination. Geological chemistry has shown that the cooling of the earth has been accompanied by the precipitation of many substances only gaseous at high temperatures. Hydrochloric and sulphuric acids have been transferred to mineral deposits or aqueous solutions. The removal of carbonic-acid gas and the vapor of water has been a process of much slower progress, and after the expiration of all the ages a proportion of both yet remains. Evidence of the abundance of the former in the earliest periods is seen in the vast deposits of limestone rock ; later, in the prodigious quantities of shells which have been elaborated from the same in solution. Proof of its abundance in the atmosphere in later periods is seen in the extensive deposits of coal of the Carboniferous, Triassic, and Jurassic periods. If the most luxuriant vegetation of the present day takes but fifty tons of carbon from the atmosphere in a century, per acre, thus producing a layer over that extent of less than a third of an inch in thickness, what amount of carbon must be abstracted in order to produce strata of thirty-five feet in depth ? No doubt it occupied a long period, but the atmosphere, thus deprived of a large proportion of carbonic acid, would in subsequent periods undoubtedly possess an improved capacity for the support of animal life.

The successively higher degree of oxidization of the blood in the organs designed for that function, whether performing it in water or air, would certainly accelerate the performances of all the vital functions, and among others that of growth. Thus it may be that *acceleration* can be accounted for, and the process of the development of the orders and sundry lesser groups of the Vertebrate kingdom indicated ; for, as already pointed out, the definitions of such are radically placed in the different structures of the organs which aërate the blood and distribute it to its various destinations.

But the great question, What determined the direction of this acceleration ? remains unanswered. One can not understand why more highly oxidized blood should hasten the growth of partition of the ventricle of the heart in the serpent, the more perfectly to separate the aërated from the impure fluid ; nor can we see why a more perfectly constructed circulatory system, sending purer blood to the brain, should direct accelerated growth to the cerebellum or cerebral hemispheres in the crocodile.

b. In Characters of the Specific Kind.—Some of the characters usually placed in the specific category have been shown to be the same in kind as those of higher categories. The majority are, however, of a different kind, and have been discussed several pages back.

The cause of the origin of these characters is shrouded in as much mystery as that of those which have occupied the pages immediately preceding. As in that case, we have to assume, as Darwin has done, a tendency in Nature to their production. This is what he terms "the principle of variation." Against an unlimited variation the great law of heredity or atavism has ever been opposed, as a conservator and multiplier of type. This principle is exemplified in the fact that like produces like—that children are like their parents, frequently even in minutiæ. It may be compared to habit in metaphysical matters, or to that singular love of time or rhythm seen in man and lower animals, in both of which the tendency is to repeat in continual cycles a motion or state of the mind or sense.

Further, only a proportion of the lines of variation is supposed to have been perpetuated, and the extinction of intermediate forms, as already stated, has left isolated groups or species.

The effective cause of these extinctions is stated by Darwin to have been a "natural selection"—a proposition which distinguishes his theory from other development hypotheses, and which is stated in brief by the expression, "the survival of the fittest." Its meaning is this : that those characters appearing as results of this spontaneous variation which are little adapted to the conflict for subsistence, with the nature of the supply, or with rivals in its pursuit, dwindle and are sooner or later extirpated ; while those which are adapted to their surroundings, and favored in the struggle for means of life and increase, predominate, and ultimately become the centers of new variation. "I am convinced," says Darwin, "that natural selection has been the main, but not exclusive, means of modification."

That it has been to a large extent the means of preservation of those structures known as specific, must, I think, be admitted. They are related to their peculiar surroundings very closely, and are therefore more likely to exist under their influence. Thus, if a given genus extends its range over a continent, it is usually found to be represented by peculiar species—one in a maritime division, another in the desert, others in the forest, in the swamp or the elevated areas of the region. The wonderful interdependence shown by Darwin to exist between insects and plants in the fertilization of the latter, or between animals and their food-plants, would almost induce one to believe that it were the true expression of the whole law of development.

But the following are serious objections to its universal application:

First: The characters of the higher groups, from genera up, are rarely of a character to fit their possessors especially for surrounding circumstances; that is, the differences which separate genus from genus, order from order, etc., in the ascending scale of each, do not seem to present a superior adaptation to surrounding circumstances in the higher genus to that seen in the lower genus, etc. Hence, superior adaptation could scarcely have caused their selection above other forms not existing. Or, in other words, the very differences in structure which indicate successional relation, or which measure the steps of progress, seem to be equally well fitted for their surroundings.

Second: The higher groups, as orders, classes, etc., have been in each geologic period alike distributed over the whole earth, under all the varied circumstances offered by climate and food. Their characters do not seem to have been modified in reference to these. Species, and often genera, are, on the other hand, eminently restricted according to climate, and consequently vegetable and animal food.

The law of development which we seek is indeed not that which preserves the higher forms and rejects the lower after their creation, but that which explains why higher forms were created at all. Why in the results of a creation we see any relation of higher and lower, and not rather a world of distinct types, each perfectly adapted to its situation, but none properly higher than another in an ascending scale, is the primary question. Given the principle of advance, then natural selection has no doubt modified the details; but in the successive advances we can

10

scarcely believe such a principle to be influential. We look rather upon a progress as the result of the expenditure of some force fore-arranged for that end.

It may become, then, a question whether in characters of high grade the habit or use is not rather the result of the acquisition of the structure than the structure the result of the encouragement offered to its assumed beginnings by use, or by liberal nutrition derived from the increasingly superior advantages it offers.

ε. *The Physical Origin of Man.*

If the hypothesis here maintained be true, man is the descendant of some pre-existent generic type, the which, if it were now living, we would probably call an ape.

Man and the chimpanzee were in Linnæus's system only two species of the same genus, but a truer anatomy places them in separate genera and distinct families. There is no doubt, however, that Cuvier went much too far when he proposed to consider Homo as the representative of an order distinct from the Quadrumana, under the name of Bimana. The structural differences will not bear any such interpretation, and have not the same value as those distinguishing the orders of Mammalia; as, for instance, between Carnivora and bats, or the cloven-footed animals and the rodents, or rodents and edentates. The differences between man and the chimpanzee are, as Huxley well puts it, much less than those between the chimpanzee and lower Quadrumana, as lemurs, etc. In fact, man is the type of a family, Hominidæ, of the order Quadrumana, as indicated by the characters of the dentition, extremities, brain, etc. The reader who may have any doubts on this score may read the dissections of Geoffroy St. Hilaire, made in 1856, before the issue of Darwin's "Origin of Species." He informs us that the brain of man is nearer in structure to that of the orang than the orang's is to that of the South American howler, and that the orang and howler are more nearly related in this regard than are the howler and the marmoset.

The modifications presented by man have, then, resulted from an acceleration in development in some respects, and retardation perhaps in others. But until the *combination* now characteristic of the genus Homo was attained, the being could not properly be called man.

And here it must be observed that as an organic type is characterized by the co-existence of a number of peculiarities which have been developed independently of each other, its distinctive features and striking functions are not exhibited until that co-existence is attained which is necessary for these ends.

Hence, the characters of the human genus were probably developed successively; but few of the indications of human superiority appeared until the combination was accomplished. Let the opposable thumb be first perfected, but of what use would it be in human affairs without a mind to direct? And of what use a mind without speech to unlock it? And speech could not be possible though all the muscles of the larynx but one were developed, or but a slight abnormal convexity in one pair of cartilages remained.

It would be an objection of little weight could it be truly urged that there have as yet no remains of ape-like men been discovered, for we have frequently been called upon in the course of paleontological discovery to bridge greater gaps than this, and greater remain, which we expect to fill. But we *have* ape-like characters exhibited by more than one race of men yet existing.

But the remains of that being which is supposed to have been the progenitor of man may have been discovered a short time since in the cave of Naulette, Belgium, with the bones of the extinct rhinoceros and elephant.

We all admit the existence of higher and lower races, the latter being those which we now find to present greater or less approximations to the apes. The peculiar structural characters that belong to the negro in his most typical form are of that kind, however great may be the distance of his remove therefrom. The flattening of the nose and prolongation of the jaws constitute such a resemblance; so are the deficiency of the calf of the leg, and the obliquity of the pelvis, which approaches more the horizontal position than it does in the Caucasian. The investigations made at Washington during the war with reference to the physical characteristics of the soldiers show that the arms of the negro are from one to two inches longer than those of the whites: another approximation to the ape. In fact, this race is a species of the genus Homo, as distinct in character from the Caucasian as those we are accustomed to recognize in other departments of the animal kingdom; but he is not distinct by isolation, since intermediate forms between him and the other species can be abundantly found.

And here let it be particularly observed that two of the most prominent characters of the negro are those of immature stages of the Indo-European race in its characteristic types. The deficient calf is the character of infants at a very early stage; but, what is more important, the flattened bridge of the nose and shortened nasal cartilages are universally immature conditions of the same parts in the Indo-European. Any one may convince himself of that by examining the physiognomies of infants. In some races—e. g., the Slavic—this undeveloped character persists later than in some others. The Greek nose, with its elevated bridge, coincides not only with æsthetic beauty, but with developmental perfection.

This is, however, only "*inexact* parallelism," as the characters of the hair, etc., can not be explained on this principle *among existing races*. The embryonic characters mentioned are probably a remnant of those characteristic of the primordial race or species.

But the man of Naulette, if he be not a monstrosity, is a still more distinct and ape-like species. The chin, that marked character of other species of men, is totally wanting, and the dentition is quite approximate to the man-like apes, and different from that of modern men. The form is very massive, as in apes. That he was not abnormal is rendered probable by approximate characters seen in a jaw from the cave of Puy-sur-Aube, and less marked in the lowest races of Australia and New Caledonia.

As to the single or multiple origin of man, science as yet furnishes no answer. It is very probable that, in many cases, the species of one genus have descended from corresponding species of another by change of generic characters only. It is a remarkable fact that the orang possesses the peculiarly developed malar bones and the copper color characteristic of the Mongolian inhabitants of the regions in which this animal is found, while the gorilla exhibits the prognathic jaws and black hue of the African races near whom he dwells. This kind of geographical imitation is very common in the animal kingdom.

II. METAPHYSICAL EVOLUTION.

It is infinitely improbable that a being endowed with such capacities for gradual progress, as man has exhibited, should have been full-fledged in accomplishments at the moment when he could first claim his high title, and abandon that of his simian ancestors. We are, therefore, required to admit the growth of human intelligence from a primitive state of inactivity and absolute ignorance; including the development of one important mode of its expression—speech; as well as that of the moral qualities, and of man's social system—the form in which his ideas of morality are first displayed.

The expression "evolution of morality" need not offend, for the question in regard to the *laws* of this evolution is the really important part of the discussion, and it is to the opposing views on this point that the most serious interest attaches.

a. *Development of Intelligence.*

If the brain is the organ of mind, we may be surprised to find that the brain of the intelligent man scarcely differs in structure from that of the ape. Whence, then, the difference of power? Though no one will now deny that many of the Mammalia are capable of reasoning upon observed facts, yet how greatly the results of this capacity differ in number and importance from those achieved by human intelligence! Like water at the temperatures of 50° and 53°, where we perceive no difference in essential character, so between the brains of the lower and higher monkeys no difference of function or of intelligence is perceptible. But what a difference do the two degrees of temperature from 33° to 31° produce in water! In like manner the difference between the brain of the higher ape and that of man is accompanied by a difference in function and power, on which man's earthly destiny depends. In development, as with the water, so with the higher ape; some Rubicon has been crossed, some floodgate has been opened, which marks one of Nature's great transitions, such as have been called "expression-points" of progress.

What point of progress in such a history would account for this accession of the powers of the human intelligence? It has been answered, with considerable confidence—the power of speech. Let us picture man without speech. Each generation would learn nothing from its predecessors. Whatever originality or observation might yield to a man would die with him. Each intellectual life would begin where every other life began, and would end at a point only differing with its original capacity. Concert of action, by which man's power over the material world is maintained, would not exceed, if it equaled, that which is seen among the bees; and the material results of his labors would not extend beyond securing the means of life and the employment of the simplest modes of defense and attack.

The first men, therefore, are looked upon by the developmentalists as extremely embryonic in all that characterizes humanity, and they appeal to the facts of history in support of this view. If they do not derive much assistance from written history, evidence is found in the more enduring relics of human handiwork.

The opposing view is, that the races which present or have presented this condition of inferiority or savagery have reached it by a process of degradation from a higher state—as some believe, through moral delinquency. This position may be true in certain cases, which represent perhaps a condition of senility, but in general we believe that savagery was the condition of the first man, which has in some races continued to the present day.

β. Evidence from Archæology.

As the object of the present essay is not to examine fully into the evidences for the theories of evolution here stated, but rather to give a sketch of such theories and their connection, a few facts only will be noticed.

Improvement in the Use of Materials.—As is well known, the remains of human handiwork of the earliest periods consist of nothing but rude implements of stone and bone, useful only in procuring food and preparing it for use. Even when enterprise extended beyond the ordinary routine, it was restrained by the want of proper instruments. Knives and other cutting implements of flint still attest the skill of the early races of men from Java to the Cape of Good Hope, from Egypt to Ireland, and through North and South America. Hatchets, spear-heads and ornaments of serpentine, granite, silex, clay slates, and all other suitable rock materials, are found to have been used by the first men, to the exclusion of metals, in most of the regions of the earth.

Later, the probably accidental discovery of the superiority of some of the metals resulted in the substitution of them for stone as a material for cutting implements, Copper—the only metal which, while malleable, is hard enough to bear an imperfect edge —was used by succeeding races in the Old World and the New. Implements of this material are found scattered over extensive regions. So desirable, however, did the hardening of the material appear for the improvement of the cutting edge, that combinations with other metals were sought for and discovered. The alloy with tin, forming bronze and brass, was discovered and used in Europe, while that with silver appears to have been most readily produced in America, and was consequently used by the Peruvians and other nations.

The discovery of the modes of reducing iron ores placed in the hands of man the best material for bringing to a shape convenient for his needs the raw material of the world. All improvements in this direction made since that time have been in the quality of iron itself, and not through the introduction of any new metal.

The prevalent phenomena of any given period are those which give it its character, and by which we distinguish it. But this fact does not exclude the co-existence of other phenomena belonging to prior or subsequent stages. Thus during the many stages of human progress there have been men more or less in advance of the general body, and their characteristics have given a peculiar stamp to the later and higher condition of the whole. It furnishes no objection to this view that we find, as might have been anticipated, the stone, bronze, and iron periods overlapping one another, or men of an inferior culture supplanting in some cases a superior people. A case of this kind is seen in North America, where the existing "Indians," stone-men, have succeeded the mound-builders, copper-men. The successional relation of discoveries is all that it is necessary to prove, and this seems to be established.

The period at which the use of metallic implements was introduced is unknown, but Whitney says that the language of the Aryans, the ancestors of all the modern Indo-Europeans, indicates an acquaintance with such implements, though it is not certain whether those of iron are to be included. The dispersion of the daughter races, the Hindoos, the Pelasgi, Teutons, Celts, etc., could not, it is thought, have taken place later than 3000 B. C.—a date seven hundred years prior to that assigned by the old chronology to the Deluge. Those races co-existed with the Egyptian and Chinese nations, already civilized, and as distinct from each other in feature as they are now.

Improvement in Architecture.—The earliest periods, then, were characterized by the utmost simplicity of invention and construction. Later, the efforts for defense from enemies and for architectural display, which have always employed so much time and power, began to be made. The megalithic period has left traces over much of the earth. The great masses of stone piled on each other in the simplest form in Southern India, and the circles of stones planted on end in England at Stonehenge and Abury, and in Peru at Sillustani, are relics of that period. More complex are the great Himyaritic walls of Arabia, the works of the ancestors of the Phœnicians in Asia Minor, and the titanic workmanship of the Pelasgi in Greece and Italy. In the iron age we find granitic hills shaped or excavated into temples ; as, for example, everywhere in Southern India. Near Madura the circumference of an acropolis-like hill is cut into a series of statues in high relief, of sixty feet in elevation. Easter Island, composed of two volcanic cones, one thousand miles from the west coast of South America, in the bosom of the Pacific, possesses several colossi cut from the intrusive basalt, some in high relief on the face of the rock, others in detached blocks removed by human art from their original positions and brought nearer the sea-shore.

Finally, at a more advanced stage, the more ornate and complex structures of Central America, of Cambodia, Nineveh and Egypt, represent the period of greatest display of architectural expenditure. The same amount of human force has perhaps never been expended in this direction since, though higher conceptions of beauty have been developed in architecture with increasing intellectuality.

Man has passed through the block-and-brick building period of his boyhood, and should rise to higher conceptions of what is the true disposition of power for "him who builds for aye," and learn that "spectacle" is often the unwilling friend of progress.

No traces of metallic implements have ever been found in the salt-mines of Armenia, the turquoise-quarries in Arabia, the cities of Central America, or the excavations for mica in North Carolina, while the direct evidence points to the conclusion that in those places flint was exclusively used.

The simplest occupations, as requiring the least exercise of mind, are the pursuit of the chase and the tending of flocks and herds. Accordingly, we find our first parents engaged in these occupations. Cain, we are told, was, in addition, a tiller of the ground. Agriculture in its simplest forms requires but little more intelligence than the pursuits just mentioned, though no employment is capable of higher development. If we look at the savage nations at present occupying nearly half the land surface of the earth, we shall find many examples of the former industrial condition of our race preserved to the present day. Many of them had no knowledge of the use of metals until they obtained it from civilized men who visited them, while their pursuits were and are those of the chase, tending domestic animals, and rudimental agriculture.

γ. *Development of the Fine Arts.*

If we look at representation by drawing or sculpture, we find that the efforts of the earliest races of which we have any knowledge were quite similar to those which the untaught hand of infancy traces on its slate or the savage depicts on the rocky faces of hills. The circle or triangle for the head and body, and straight lines for the limbs, have been preserved as the first attempts of the men of the stone period, as they are to this day the sole representations of the human form which the North American Indian places on his buffalo robe or mountain precipice. The stiff, barely outlined form of the deer, the turtle, etc., are literally those of the infancy of civilized man.

The first attempts at sculpture were marred by the influence of modism. Thus the idols of Coban and Palenque, with human faces of some merit, are overloaded with absurd ornament, and deformed into frightful asymmetry, in compliance with the demand of some imperious mode. In later days we have the stiff, conventionalized figures of the palaces of Nineveh and the temples of Egypt, where the representation of form has somewhat improved, but is too often distorted by false fashion or imitation of some unnatural standard, real or artistic. This is distinguished as the day of archaic sculpture, which disappeared with the Etruscan nation. So the drawings of the child, when he abandons the simple lines, are stiff and awkward, and but a stage nearer true representation ; and how often does he repeat some peculiarity or absurdity of his own !

The introduction of the action and pose of life into sculpture was not known before the early days of Greece, and it was there that the art was brought to perfection. When art rose from its mediæval slumber, much the same succession of development may be discovered. First, the stiff figures, with straightened limbs and cylindric drapery, found in the old Northern churches—then the forms of life that now adorn the porticoes and palaces of the cities of Germany.

δ. *Rationale of the Development of Intelligence.*

The history of material development shows that the transition from stage to stage of development, experienced by the most perfect forms of animals and plants in their growth from the primordial cell, is similar to the succession of created beings which the geological epochs produced. It also shows that the slow assumption of main characters in the line of succession in early geological periods produced the condition of inferiority, while an increased rapidity of growth in later days has resulted in an attainment of superiority. It is not to be supposed that in "acceleration" the period of growth is shortened; on the contrary, it continues the same. Of two beings whose characters are assumed at the same rate of succession, that with the quickest or shortest growth is necessarily inferior. "Acceleration" means a gradual increase of the rate of assumption of successive characters in the same period of time. A fixed rate of assumption of characters, with gradual increase in the length of the period of growth, would produce the same result—viz., a longer developmental scale and the attainment of an advanced position. The first is in part the relation of sexes of a species; the last of genera, and of other types of creation. If from an observed relation of many facts we derive a law, we are permitted, when we see in another class of facts similar relations, to suspect that a similar law has operated, differing only in its objects. We find a marked resemblance between the facts of structural progress in matter and the phenomena of intellectual and spiritual progress.

If the facts entering into the categories enumerated in the preceding section bear us out, we conclude that in the beginning of human history the progress of the individual man was very slow, and that but little was attained to; that, through the profitable direction of human energy, means were discovered from time to time by which the process of individual development in all metaphysical qualities has been accelerated; and that up to the present time the consequent advance of the whole race has been at an increasing rate of progress. This is in accordance with the general principle, that high development in intellectual things is accomplished by rapidity in traversing the preliminary stages of inferiority common to all, while low development signifies sluggishness in that progress, and a corresponding retention of inferiority.

How much meaning may we not see, from this standpoint, in the history of the intelligence of our little ones! First they crawl, they walk on all-fours; when they first assume the erect position they are generally speechless, and utter only inarticulate sounds. When they run about, stones and dirt, the objects that first meet the eye, are the delight of their awakening powers; but these are all cast aside when the boy obtains his first jackknife. Soon, however, reading and writing open a new world to him; and, finally, as a mature man he seizes the forces of Nature, and steam and electricity do his bidding in the active pursuit of power for still better and higher ends.

So with the history of the species: first, the quadrumane; then the speaking man, whose humble industry was, however, confined to the objects that came first to hand, this being the "stone age" of pre-historic time. When the use of metals was discovered, the range of industries expanded wonderfully, and the "iron age" saw many striking efforts of human power. With the introduction of letters it became possible to record events and experiences, and the spread of knowledge was thereby greatly increased, and the delays and mistakes of ignorance correspondingly diminished in the fields of the world's activity.

From the first we see in history a slow advance as knowledge gained by the accumulation of tradition and by improvements in habit based on experience; but how slow was this advance while the use of the metals was still unknown! The iron age brought with it not only new conveniences, but increased means of future progress; and here we have an acceleration in the rate of advance. With the introduction of letters this rate was increased manifold, and in the application of steam we have a change equal in utility to any that has preceded it, and adding to the possibilities of future advance in many directions. By it power, knowledge, and means of happiness were to be distributed among the many.

The uses to which human intelligence has successively applied the materials furnished by Nature have been—first, subsistence and defense; second, the accumulation of power in the shape of a representative of that labor which the use of matter involves—in other words, the accumulation of wealth. The possession of this power involves new possibilities, for opportunity is offered for the special pursuits of knowledge and the assistance of the weak or undeveloped part of mankind in its struggles.

Thus, while the first men possessed the power of speech, and could advance a little in knowledge through the accumulation of the experiences of their predecessors, they possessed no means of accumulating the power of labor, no control over the activity of numbers—in other words, no wealth.

But the accumulation of knowledge finally brought this advance about. The extraction and utilization of the metals, especially iron, formed the most important step, since labor was thus facilitated and its productiveness increased in an incalculable degree. We have little evidence of the existence of a medium of exchange during the first or stone period, and no doubt barter was the only form of trade. Before the use of metals, shells and other objects were used; remains of money of baked clay have been found in Mexico. Finally, though in still ancient times, the possession of wealth in money gradually became possible and more common, and from that day to this avenues for reaching this stage in social progress have ever been opening.

But wealth merely indicates a stage of progress, since it is but a comparative term. All men could not become rich, for in that case all would be equally poor. But labor has a still higher goal; for, thirdly, as capital, it constructs and employs machinery, which does the work of many hands, and thus cheapens products, which is equivalent in effect to an accumulation of wealth to the consumer. And this increase of power may be used for the intellectual and spiritual advance of men, or, otherwise, at the will of the men thus favored. Machinery places man in the position of a creator, operating on Nature through an increased number of "secondary causes."

Development of intelligence is seen, then, in the following directions: First, in the knowledge of facts, including science; second, in language; and, as consequences of these, the accumulation of power by development—first, of means of subsistence; and, second, of mechanical invention; and, third, in the apprehension of beauty.

Thus, we have two terms to start with in estimating the beginning of human development in knowledge and power: first, the primary capacities of the human mind itself; second, a material world, whose infinitely varied components are so arranged as to yield results to the energies of that mind. For example, the transition-points of vaporization and liquefaction are so placed as to be within the reach of man's agents; their weights are so fixed as to accord with the muscular or other forces which he is able to exert; and other living organizations are subject to his convenience and rule, and not, as in previous geological periods, entirely beyond his control. These two terms being given, it is maintained that the present situation of the most civilized men has been attained through the operation of a law of mutual action and reaction—a law whose results, seen at the present time, have depended on the acceleration or retardation of its rate of action; which rate has been regulated, according to the degree in which a third great term, viz., the law of moral or (what is the same thing) true religious development, has been combined in the plan. What it is necessary to establish in order to prove the above hypothesis is—

I. That in each of the particulars above enumerated the development of the human species is similar to that of the individual from infancy to maturity.

II. That from a condition of subserviency to the laws of matter, man's intelligence enables him, by an accumulation of power, to become in a sense independent of those laws, and to pursue a course of intellectual and spiritual progress.

III. That failure to accomplish a moral or spiritual development will again reduce him to a subserviency to the laws of matter.

This brings us to the subject of moral development. And here I may be allowed to suggest that the weight of the evidence is opposed to the philosophy, "falsely so called," of necessitarianism, which asserts that the first two terms alone were sufficient to work out man's salvation in this world and the next; and, on the other hand, to that anti-philosophy which asserts that all things in human progress, intellectual and moral, are regulated by immediate divine interposition instead of through instrumentalities. Hence, the subject divides itself at once into two great departments—viz., that of the development of mind or intelligence, and that of the development of morality.

That these laws are distinct there can be no doubt, since in the individual man one of them may produce results without the aid of the other. Yet it can be shown that each is the most invaluable aid and stimulant to the other, and most favorable to the rapid advance of the mind in either direction.

III. SPIRITUAL OR MORAL DEVELOPMENT.

In examining this subject, we first inquire (Section a) whether there is any connection between physical and moral or religious development; then (β), what indications of moral development may be derived from history. Finally (γ), a correlation of the results of these inquiries, with the nature of the religious development in the individual, is attempted. Of course in so stupendous an inquiry but a few leading points can be presented here.

If it be true that the period of human existence on the earth has seen a gradually increasing predominance of higher motives over lower ones among the mass of mankind, and if any parts of our metaphysical being have been derived by inheritance from pre-existent beings, we are incited to the inquiry whether any of the moral qualities are included among the latter; and whether there be any resemblance between moral and intellectual development.

Thus, if there have been a physical derivation from a pre-existent genus, and an embryonic condition of those physical characters which distinguish Homo—if there has been also an embryonic or infantile stage in intellectual qualities—we are led to inquire whether the development of the individual in moral nature will furnish us with a standard of estimation of the successive conditions or present relations of the human species in this aspect also.

α. *Relations of Physical and Moral Nature.*

Although, *cæteris paribus*, men are much alike in the deeper qualities of their nature, there is a range of variation which is best understood by a consideration of the extremes of such variation as seen in men of different latitudes, and women and children.

(*a*) *In Children.*—Youth is distinguished by a peculiarity, which no doubt depends upon an immature condition of the nervous center concerned, which might be called *nervous impressibility*. It is exhibited in a greater tendency to tearfulness, in timidity, less mental endurance, a greater facility in acquiring knowledge, and more ready susceptibility to the influence of sights, sounds, and sensations. In both sexes the emotional nature predominates over the intelligence and judgment. In those years the *character* is said to be in embryo, and theologians, in using the phrase, "reaching years of religious understanding," mean that in early years the religious *capacities* undergo development coincidentally with those of the body.

(*b*) *In Women.*—If we examine the metaphysical characteristics of women, we observe two classes of traits—namely, those which are also found in men, and those which are absent or but weakly developed in men. Those of the first class are very similar in essential nature to those which men exhibit at an early stage of development. This may be in some way related to the fact that physical maturity occurs earlier in women.

The gentler sex is characterized by a greater impressibility, often seen in the influence exercised by a stronger character, as well as by music, color, or spectacle generally; warmth of emotion, submission to its influence rather than that of logic; timidity and irregularity of action in the outer world. All these qualities belong to the male sex, as a general rule, at some period of life, though different individuals lose them at very various periods. Ruggedness and sternness may rarely be developed in infancy, yet at some still prior time they certainly do not exist in any.

Probably most men can recollect some early period of their lives when the emotional nature predominated—a time when emotion at the sight of suffering was more easily stirred than in maturer years. I do not now allude to the benevolence inspired, kept alive, or developed by the influence of the Christian religion on the heart, but rather to that which belongs to the natural man. Perhaps all men can recall a period of youth when they were hero-worshipers—when they felt the need of a stronger arm, and loved to look up to the powerful friend who could sympathize with and aid them. This is the "woman stage" of character: in a large number of cases it is early passed; in some it lasts longer; while in a very few men it persists through life. Severe discipline and labor are unfavorable to its persistence. Luxury preserves its bad qualities without its good, while Christianity preserves its good elements without its bad.

It is not designed to say that woman in her emotional nature does not differ from the undeveloped man. On the contrary, though she does not differ in kind, she differs greatly in degree, for her qualities grow with her growth, and exceed in *power* many fold those exhibited by her companion at the original point of departure. Hence, since it might be said that man is the undeveloped woman, a word of explanation will be useful. Embryonic types abound in the fields of nature, but they are not therefore immature in the usual sense. Maintaining the lower essential quality, they yet exhibit the usual results of growth in individual characters; that is, increase of strength, powers of support and protection, size and beauty. In order to maintain that the masculine character coincides with that of the undeveloped woman, it would be necessary to show that the latter during her infancy possesses the male characters predominating—that is, unimpressibility, judgment, physical courage, and the like.

If we look at the second class of female characters—namely, those which are imperfectly developed or absent in men, and in respect to which man may be called undeveloped woman—we note three prominent points: facility in language, tact or finesse, and the love of children. The first two appear to me to be altogether developed results of "impressibility," already considered as an indication of immaturity. Imagination is also a quality of impressibility, and, associated with finesse, is apt to degenerate into duplicity and untruthfulness—a peculiarity more natural to women than men.

The third quality is different. It generally appears at a very early period of life. Who does not know how soon the little girl selects the doll, and the boy the toy-horse or machine? Here man truly never gets beyond undeveloped woman. Nevertheless, "impressibility" seems to have a great deal to do with this quality also.

Thus the metaphysical relation of the sexes would appear to be one of *inexact parallelism*, as defined in Section I. That the physical relation is a remote one of the same kind, several characters seem to point out. The case of the vocal organs will suffice. Their structure is identical in both sexes in early youth, and both produce nearly similar sounds. They remain in this condition in the woman, while they undergo a metamorphosis and change both in structure and vocal power in the man. In the same way, in many of the lower creation, the females possess a majority of embryonic features, though not invariably. A common example is to be found in the plumage of birds, where the females and young males are often undistinguishable.* But there are a few points in

---

* Meehan states that the upper limbs and strong laterals in Coniferæ and other trees produce female flowers and cones, and the lower and more interior branches the male flowers. He calls the former condition one of greater "vigor," and the latter one of "weakness," and argues that the vigorous condition of growth produces females, and the weaker males. What he points out, however, is in harmony with the position here maintained—namely, that the female characters include more of those which are embryonic in the males than the male characters include of those

the physical structure of man also in which the male condition is the immature one. In regard to structure, the point at which the relation between the sexes is that of *exact parallelism,* or where the mature condition of the one sex accords with the undeveloped condition of the other, is when reproduction is no longer accomplished by budding or gemmation, but requires distinct organs. Metaphysically, this relation is to be found where distinct individuality of the sexes first appears; that is, where we pass from the hermaphrodite to the bisexual condition.

But let us put the whole interpretation on this partial undevelopment of woman.

The types or conditions of organic life which have been the most prominent in the world's history—the Ganoids of the first, the Dinosaurs of the second, and the Mammoths of the third period—have generally died with their day. The line of succession has not been from them. The law of anatomy and paleontology is, that we must seek the point of departure of the type which is to predominate in the future, at lower stages on the line, in less decided forms, or in what, in scientific parlance, are called generalized types. In the same way, though the adults of the tailless apes are in a physical sense more highly developed than their young, yet the latter far more closely resemble the human species in their large facial angle and shortened jaws.

How much significance, then, is added to the law uttered by Christ!—"Except ye become as little children, ye can not enter the kingdom of heaven." Submission of will, loving trust, confiding faith—these belong to the child: how strange they appear to the executing, commanding, reasoning man! Are they so strange to the woman? We all know the answer. Woman is nearer to the point of departure of that development which outlives time and peoples heaven; and if man would find it, he must retrace his steps, regain something he lost in youth, and join to

which are embryonic in the female: the female flowers are the product of the younger and more growing portions of the tree—that is, those last produced (the upper limbs and new branches)—while the male flowers are produced by the older or more mature portions—that is, lower limbs or more axial regions. Further, we are not accustomed to regard the condition of rapid growth as that of great vigor in animals, but rather ascribe that quality to maturity, after such growth has ceased.

Meehan's observations coincide with those of Thury and others on the origin of sexes in animals and plants, which it appears to me admit of a similar explanation.

the powers and energies of his character the submission, love and faith which the new birth alone can give.

Thus the summing up of the metaphysical qualities of woman would be thus expressed : In the emotional world, man's superior ; in the moral world, his equal ; in the laboring world, his inferior.

There are, however, vast differences in women in respect to the number of masculine traits they may have assumed before being determined into their own special development. Woman also, under the influence of necessity, in later years of life, may add more or less to those qualities in her which are fully developed in the man.

The relation of these facts to the principles stated as the two opposing laws of development is, it appears to me, to be explained thus : First, that woman's most inherent peculiarities are *not* the result of the external circumstances with which she has been placed in contact, as the *conflict theory* would indicate. Such circumstances are said to be her involuntary subserviency to the physically more powerful man, and the effect of a compulsory mode of life in preventing her from attaining a position of equality in the activities of the world. Second, that they *are* the result of the different distributions of qualities as already indicated by the *harmonic theory* of development ; that is, of the unequal possession of features which belong to different periods in the developmental succession of the highest. There is then another beautiful harmony which will ever remain, let the development of each sex be extended as far as it may.

(*c*) *In Men.*—If we look at the male sex, we shall find various exceptional approximations to the female in mental constitution. Further, there can be little doubt that in the Indo-European race maturity in some respects appears earlier in tropical than in northern regions ; and though subject to many exceptions, this is sufficiently general to be looked upon as a rule. Accordingly, we find in that race—at least in the warmer regions of Europe and America—a larger proportion of certain qualities which are more universal in women ; as greater activity of the emotional nature when compared with the judgment ; an impressibility of the nervous center, which, *cœteris paribus*, appreciates quickly the harmonies of sound, form, and color ; answers most quickly to the friendly greeting or the hostile menace ; is more careless of consequences in the material expression of generosity or hatred, and more indifferent to truth under the influence of personal relations.

The movements of the body and expressions of the countenance answer to the temperament. More of grace and elegance in the bearing marks the Greek, the Italian, and the Creole, than the German, the Englishman, or the Green Mountain man. More of vivacity and fire, for better or for worse, is displayed in the countenance.

Perhaps the more northern type left all that behind in its youth. The rugged, angular character which appreciates force better than harmony, the strong intellect which delights in forethought and calculation, the less impressibility, reaching stolidity in the uneducated, are its well-known traits. If there be in such a character less generosity and but little chivalry, there is persistency and unwavering fidelity, not readily obscured by the lightning of passion or the surmises of an active imagination.

All these peculiarities appear to result, *first*, from different degrees of quickness and depth in appreciating impressions from without; and, *second*, from differing degrees of attention to the intelligent judgment in consequent action. (I leave conscience out, as not belonging to the category of inherited qualities.)

The above observations have been confined to the Indo-European race. It may be objected to the theory that savagery means immaturity in the senses above described, as dependent largely on "impressibility," while savages in general display the least "impressibility," as that word is generally understood. This can not be asserted of the Africans, who, so far as we know them, possess this peculiarity in a high degree. Moreover, it must be remembered that the state of indifference which precedes that of impressibility in the individual may characterize many savages; while their varied peculiarities may be largely accounted for by recollecting that many combinations of different species of emotion and kinds of intelligence go to make up the complete result in each case.

(*d*) *Conclusions.*—Three types of religion may be selected from the developmental conditions of man: first, an absence of sensibility (early infancy); second, an emotional stage more productive of faith than of works; thirdly, an intellectual type, more favorable to works than faith. Though in regard to responsibility these states may be equal, there is absolutely no gain to laboring humanity from the first type, and a serious loss in actual results from the second, taken alone, as compared with the third.

These, then, are the *physical vehicles of religion*—if the phrase may be allowed—which give character and tone to the deeper spiritual life, as the color of the transparent vessel is communicated to the light which radiates from within.

But if evolution has taken place, there is evidently a provision for the progress from the lower to the higher states, either in the education of circumstances ("conflict") or in the power of an interior spiritual influence ("harmony"), or both.

β. Evidence derived from History.

We trace the development of Morality in—first, the family, or social order; second, the civil order, or government.

Whatever may have been the extent of moral ignorance before the Deluge, it does not appear that the earth was yet prepared for the permanent habitation of the human race. All nations preserve traditions of the drowning of the early peoples by floods, such as have occurred frequently during geologic time. At the close of each period of dry land, a period of submergence has set in, and the depression of the level of the earth, and consequent overflow by the sea, has caused the death and subsequent preservation of the remains of the fauna and flora living upon it, while the elevation of the same has produced that interruption in the process of deposit in the same region which marks the intervals between geologic periods. Changes in these respects do not occur to any very material extent at the present time in the regions inhabited by the most highly developed portions of the human race; and as the last which occurred seems to have been expressly designed for the preparation of the earth's surface for the occupation of organized human society, it may be doubted whether many such changes are to be looked for in the future. The last great flooding was that which stratified the drift materials of the north, and carried the finer portions far over the south, determining the minor topography of the surface and supplying it with soils.

The existence of floods which drowned many races of men may be considered as established. The men destroyed by the one recorded by Moses are described by him as exceedingly wicked, so that "the earth was filled with violence." In his eyes the Flood was designed for their extermination.

That their condition was evil must be fully believed if they were condemned by the executive of the Jewish law. This law, it will be remembered, permitted polygamy, slavery, revenge, aggressive war. The Jews were expected to rob their neighbors, the

Egyptians, of jewels, and they were allowed "an eye for an eye and a tooth for a tooth." They were expected to butcher other nations, with their women and children, their flocks and their herds. If we look at the lives of men recorded in the Old Testament as examples of distinguished excellence, we find that their standard, however superior to that of the people around them, would ill accord with the morality of the present day. They were all polygamists, slaveholders, and warriors. Abraham treated Hagar and Ishmael with inhumanity. Jacob, with his mother's aid, deceived Isaac, and received thereby a blessing which extended to the whole Jewish nation. David, a man whom Paul tells us the Lord found to be after his own heart, slew the messenger who brought tidings of the death of Saul, and committed other acts which would stain the reputation of a Christian beyond redemption. It is scarcely necessary to turn to other nations if this be true of the chosen men of a chosen people. History, indeed, presents us with no people prior to, or contemporary with, the Jews who were not morally their inferiors.*

If we turn to more modern periods, an examination of the morality of Greece and Rome reveals a curious intermixture of lower and higher moral conditions. While each of these nations produced excellent moralists, the influence of their teachings was not sufficient to elevate the masses above what would now be regarded as a very low standard. The popularity of those scenes of cruelty, the gladiatorial shows and the combats with wild beasts, sufficiently attests this. The Roman virtue of patriotism, while productive of many noble deeds, is in itself far from being a disinterested one, but partakes rather of the nature of partisanship and selfishness. If the Greeks were superior to the Romans in humanity, they were apparently their inferiors in the social virtues, and were much below the standard of Christian nations in both respects.

Ancient history points to a state of chronic war, in which the social relations were in confusion, and the development of the useful arts was almost impossible. Savage races, which continue to this day in a similar moral condition, are, we may easily believe, most unhappy. They are generally divided into tribes, which are mutually hostile, or friendly only with the view of injuring some other tribe. Might is their law, and robbery, rapine,

* The evidence on this point being very imperfect, judgment may be properly suspended. (Ed. 1886.)

and murder express their mutual relations. This is the history of the lowest grade of barbarism, and the history of primeval man so far as it has come down to us in sacred and profane records. Man as a species first appears in history as a sinful being. Then a race maintaining a contest with the prevailing corruption and exhibiting a higher moral idea is presented to us in Jewish history. Finally, early Christian society exhibits a greatly superior condition of things. In it polygamy scarcely existed, and slavery and war were condemned. But progress did not end here, for our Lord said, "I have yet many things to say unto you, but ye can not bear them now. Howbeit, when he, the spirit of truth, is come, he will guide you into all truth."

The progress revealed to us by history is truly great, and if a similar difference existed between the first of the human species and the first of whose condition we have information, we can conceive how low the origin must have been. History begins with a considerable progress in civilization, and from this we must infer a long preceding period of human existence, such as a gradual evolution would require.

γ. *Rationale of Moral Development.*

I. *Of the Species.*—Let us now look at the moral condition of the infant man of the present time. We know his small accountability, his trust, his innocence. We know that he is free from the law that when he "would do good, evil is present with him," for good and evil are alike unknown. We know that until growth has progressed to a certain degree he fully deserves the praise pronounced by our Saviour, that "of such is the kingdom of heaven." Growth, however, generally sees a change. We know that the buddings of evil appear but too soon; the lapse of a few months sees exhibitions of anger, disobedience, malice, falsehood, and their attendants—the fruit of a corruption within not manifested before.

In early youth it may be said that moral susceptibility is often in inverse ratio to physical vigor. But with growth the more physically vigorous are often sooner taught the lessons of life, for their energy brings them into earlier conflict with the antagonisms and contradictions of the world. Here is a beautiful example of the benevolent principle of compensation.

1. *Innocence and the Fall.*—If physical evolution be a reality, we have reason to believe that the infantile stage of human mor-

als, as well as of human intellect, was much prolonged in the history of our first parents. This constitutes the period of human purity, when we are told by Moses that the first pair dwelt in Eden. But the growth to maturity saw the development of all the qualities inherited from the irresponsible denizen of the forest. Man inherits from his predecessors in the creation the buddings of reason—he inherits passions, propensities, and appetites. His corruption is that of his animal progenitors, and his sin is the low and bestial instinct of the brute creation. Thus only is the origin of sin made clear—a problem which the pride of man would have explained in any other way had it been possible.

But how startling the exhibition of evil by this new being as compared with the scenes of the countless ages already past! Then the right of the strongest was God's law, and rapine and destruction were the history of life. But into man had been "breathed the breath of life," and he had "become a living soul." The law of right, the Divine Spirit, was planted within him, and the laws of the beast were in antagonism to that law. The natural development of his inherited qualities necessarily brought him into collision with that higher standard planted within him, and that war was commenced which shall never cease "till he hath put all things under his feet." The first act of man's disobedience constituted the Fall, and with it would come the first *intellectual* "knowledge of good and of evil"—an apprehension up to that time derived exclusively from the divinity within, or conscience.*

2. *Free Agency.*—Heretofore development had been that of physical types, but the Lord had rested on the seventh day, for man closed the line of the physical creation. Now a new development was to begin—the development of mind, of morality, and of grace.

On the previous days of Creation all had progressed in accordance with inevitable law apart from its objects. Now, two lines

* In our present translation of Genesis, the Fall is ascribed to the influence of Satan assuming the form of the serpent, and this animal was cursed in consequence, and compelled to assume a prone position. This rendering may well be revised, since serpents, prone like others, existed in both America and Europe during the Eocene epoch, five times as great a period before Adam as has elapsed since his day. Clark states, with great probability, that "serpent" should be translated monkey or ape—a conclusion, it will be observed, exactly coinciding with our inductions on the basis of evolution. The instigation to evil by an ape merely states inheritance in another form. His curse, then, refers to the retention of the horizontal position retained by all other quadrumana, as we find it at the present day.

of development were at the disposal of this being, between which his *free will* was to choose. Did he choose the courses dictated by the spirit of the brute, he was to be subject to the old law of the brute creation—the right of the strongest and spiritual death. Did he choose the guidance of the Divine Guest in his heart, he became subject to the laws which are to guide—1, the human species to an ultimate perfection, so far as consistent with this world; and, 2, the individual man to a higher life, where a new existence awaits him as a spiritual being, freed from the laws of terrestrial matter.

The charge brought against the theory of development, that it implies a necessary progress of man to all perfection without his co-operation—or *necessitarianism,* as it is called—is unfounded.

The free will of man remains the source alike of his progress and his relapse. But the choice once made, the laws of spiritual development are apparently as inevitable as those of matter. Thus men whose religious capacities are increased by attention to the Divine Monitor within are in the advance of progress—progress coinciding with that which in material things is called the *harmonic.* On the other hand, those whose motives are of the lower origin fall under the working of the law of *conflict.*

The lesson derivable from the preceding considerations would seem to be "necessitarian" as respects the whole human race, considered by itself; and I believe it is to be truly so interpreted. That is, the Creator of all things has set agencies at work which will slowly develop a perfect humanity out of his lower creation, and nothing can thwart the process or alter the result. "My word shall not return unto me void, but it shall accomplish that which I please, and it shall prosper in the thing whereto I sent it." This is our great encouragement, our noblest hope—second only to that which looks to a blessed inheritance in another world. It is this thought that should inspire the farmer, who, as he toils, wonders, "Why all this labor? The Good Father could have made me like the lilies, who, though they toil not, neither spin, are yet clothed in glory; and why should I, a nobler being, be subject to the dust and the sweat of labor?" This thought should enlighten every artisan of the thousands that people the factories and guide their whirling machinery in our modern cities. Every revolution of a wheel is moving the car of progress, and the timed stroke of the crank and the rhythmic throw of the shuttle are but the music the spheres have sung since time began. A new significance then appears in the prayer of David : "Let the beauty of the Lord our God be upon us, and establish thou the work of our hands upon us : the work of our hands, O Lord, establish thou it." But, beware of the catastrophe, for "He will sit as a refiner"; "The wheat shall be gathered into barns, but the chaff shall be burned with unquenchable fire." If this be true, let us look for—

3. *The Extinction of Evil.*—How is necessitarianism to be reconciled with free will ? It appears to me, thus : When a being whose safety depends on the perfection of a system of laws abandons the system by which he lives, he becomes subject to that lower grade of laws which govern lower intelligences. Man, falling from the laws of right, comes under the dominion of the laws of brute force ; as said our Saviour, "Salt is good, but if the salt have lost his savor, it is thenceforth good for nothing but to be cast forth and trodden under foot of men."

In estimating the practical results to man of the actions prompted by the lower portion of our nature, it is only necessary to carry out to its full development each of those animal qualities which may in certain states of society be restrained by the social system. In human history those qualities have repeatedly had this development, and the battle of progress is fought to decide whether they shall overthrow the system that restrains them, or be overthrown by it.

Entire obedience to the lower instincts of our nature insures destruction to the weaker, and generally to the stronger also. A most marked case of this kind is seen where the developed vices of civilization are introduced among a savage people—as, for example, the North American Indians. These seem in consequence to be hastening to extinction.

But a system or a circuit of existence has been allotted to the civil associations of the animal species man, independently of his moral development. It may be briefly stated thus : Races begin as poor offshoots or emigrants from a parent stock. The law of labor develops their powers, and increases their wealth and numbers. These will be diminished by their various vices ; but, on the whole, in proportion as the intellectual and economical elements prevail, wealth will increase — that is, they accumulate power. When this has been accomplished, and before activity has slackened its speed, the nation has reached the culminating point, and then it enters upon the period of decline. The restraints imposed by economy and active occupation being removed, the beastly traits find in accumulated power only increased means of gratification, and industry and prosperity sink together. Power is squandered, little is accumulated, and the nation goes down to its extinction amid scenes of internal strife and vice. Its cycle is soon fulfilled, and other nations, fresh from scenes of labor, assault it, absorb its fragments, and it dies. This has been the world's history, and it remains to be seen whether the virtues of the nations now existing will be sufficient to save them from a like fate.

Thus the history of the animal man in nations is wonderfully like that of the type or families of the animal and vegetable kingdoms during geologic ages. They rise, they increase, and reach a period of multiplication and power. The force allotted to them becoming exhausted, they diminish and sink and die.

II. *Of the Individual.*—In discussing physical development, we are as yet compelled to restrict ourselves to the evidence of its existence and some laws observed in the operation of its causative force. What that force is, or what are its primary laws, we know not.

So, in the progress of moral development, we endeavor to prove its existence and the mode of its operation, but why that mode should exist, rather than some other mode, we can not explain.

The moral progress of the species depends, of course, on the moral progress of the individuals embraced in it. Religion is the sum of those influences which determine the motives of men's actions into harmony with the divine perfection and the divine will.

Obedience to these influences constitutes the practice of religion, while the statement of the growth and operation of these influences constitutes the theory of religion, or doctrine.

The Divine Spirit planted in man shows him that which is in harmony with the Divine Mind, and it remains for his free will to conform to it or reject it. This harmony is man's highest ideal of happiness, and in seeking it, as well as in desiring to flee from dissonance or pain, he but obeys the disposition common to all conscious beings. If, however, he attempts to conform to it, he will find the law of evil present, and frequently obtaining the mastery. If now he be in any degree observing, he will find that the laws of morality and right are the only ones by which human society exists in a condition superior to that of the lower animals, and in which the capacities of man for happiness can approach a state of satisfaction. He may be then said to be "awakened" to the importance of religion. If he carry on the struggle to attain to the high goal presented to his spiritual vision, he will be deeply grieved and humbled at his failures; then he is said to be "convicted." Under these circumstances the necessity of a deliverance becomes clear, and is willingly accepted in the only way in which it has pleased the Author of all to present it, which has been epitomized by Paul as "the washing of regeneration and renewal of the Holy Spirit through Jesus Christ." Thus a life of advanced and ever-advancing moral excellence becomes possible, and the man makes nearer approaches to the "image of God."

Thus is opened a new era in spiritual development, which we are led to believe leads to an ultimate condition in which the nature inherited from our origin is entirely overcome, and an existence of moral perfection entered on. Thus, in the book of Mark the simile occurs: "First the blade, then the ear, after that the full corn in the ear"; and Solomon says that the development of righteousness "shines more and more unto the perfect day."

δ. *Summary.*

If it be true that general development in morality proceeds in spite of the original predominance of evil in the world, through the self-destructive nature of the latter, it is only necessary to examine the reasons why the excellence of the good may have been subject also to progress, and how the remainder of the race may have been influenced thereby.

The development of morality is then probably to be understood in the following sense: Since the Divine Spirit, as the prime force in human progress, can not in itself be supposed to have been in any way under the influence of natural laws, its capacities were no doubt as eternal and unerring in the first man as in the last. But the facts and probabilities discussed above point to development of *religious sensibility*, or capacity to appreciate moral good, or to receive impressions from the source of good.

The evidence of this is supposed to be seen in—*first*, improvement in man's views of his duty to his neighbor; and, *second*, the substitution of spiritual for symbolic religions: in other words, improvement in the capacity for receiving spiritual impressions.

What the primary cause of this supposed development of religious sensibility may have been, is a question we reverently leave untouched. That it is intimately connected in some way with, and in part dependent on, the evolution of the intelligence, appears very probable ; for this evolution is seen—*first*, in a better understanding of the consequences of action, and of good and of evil in many things ; and, *second*, in the production of means for the spread of the special instrumentalities of good. The following may be enumerated as such instrumentalities :

1. Furnishing literary means of record and distribution of the truths of religion, morality, and science.

2. Creating and increasing modes of transportation of teachers and literary means of disseminating truth.

3. Facilitating the migration and the spread of nations holding the highest position in the scale of morality.

4. The increase of wealth, which multiplies the extent of the preceding means.

And now, let no man attempt to set bounds to this development. Let no man say even that morality accomplished is all that is required of mankind, since that is not necessarily the evidence of a spiritual development. If a man possess the capacity for progress beyond the condition in which he finds himself, in refusing to enter upon it he declines to conform to the divine law. For "from those to whom little is given, little is required, but from those to whom much is given, much shall be required."

V.

THE METHOD OF CREATION OF ORGANIC FORMS.

CHAPTER I—ON THE LAW OF ACCELERATION AND RETARDATION.—Nature of law of Natural Selection. Two kinds of evidence. Illustration. Examples from Cervidæ, Helicidæ, insects and men.

CHAPTER II—THE LAW OF REPETITIVE ADDITION.—Segment and cell-repetition. Illustration from limbs and vertebral column. A, On segment addition; definitions. On repetition in bilateral and antero-posterior symmetry; in structure of compound teeth; in segments of Articulata; limbs of Reptilia; brain of lamprey. B, On cell-repetition; simple segment a repetition of cells; simple diverticulum the same. The cell theory; the nucleated cell. C, Synthesis of repetition. From unicellular to multicellular animals; simple repetition to compound repetition; *Actinia*, *Lepidosiren*, *Ichthyosaurus*, *Plesiosaurus*, *Tænia;* the heart; mammalian teeth. D, On growth-force; relation to other forces; definition. E, Direction of repetition, its location, centrifugal and longitudinal; movements longitudinal. Inheritance; its relation to growth-force.

CHAPTER III—THE LAW OF USE AND EFFORT.—Points to be investigated. A, On the location of growth-force. Relation of effort to use. Rudimental characters. Examples of growth under influence of physical laws; Examples of colors under influence of light. Use and disuse of gills. Rattlesnake; horned animals. Teeth of ruminants. B, Change in amount of growth-force. Local increase of growth-force. Convoluted structures; brain, teeth, cotyledons. Absolute loss of growth-force. Teeth and toes of Ruminants; incisors of Rodents.

CHAPTER IV—ON GRADE INFLUENCE.—A, On the nature of Grade Influence or Bathmism. Definitions. In plants; in animals. Increase in time of Bathmism and growth-force. Vital forces and vital influences. Thought-force. Origin of Bathmism in time. B, Physiological origin of Bathmism. Function of nervous system in force-conversion. Automatic and habitual movements. Effect on nervous system.

CHAPTER V—INTELLIGENT SELECTION.—Development of intelligence. Stimuli to use. Compulsion, Choice; Bees, Food, Rattlesnake; Change of color; Mimetic analogy; Examples. Development of character.

IN the present state of biological science, essays like the present can only be tentative in so far as they treat of the laws of evolution. Nevertheless the present time is pre-eminently one of generalization in this field, and properly so. Facts have been accumulating for a long period, and are now sufficiently numerous to yield important results, under proper classification and induction. Darwin led the way in this work, and the development hypothesis is regarded as demonstrated by most biologists. The discussion of the laws of its progress involves a multitude of subordinate hypotheses. In the following essay, these are arranged under five prominent heads, viz.: 1. The law of Acceleration and Retardation; 2. The law of Repetitive Addition; 3. The law of Use and Effort; 4. The law of Grade Influence; 5. The law of Intelligent Selection. Of these, the first and second are regarded by the author as demonstrated, the third and fourth as only reduced to a partial demonstration, while the fifth is a consequence of the third, and stands or falls with it.

The discussion of this subject divides itself into two parts, viz.: a consideration of the proof that evolution of organic types or descent with modification has taken place; and, secondly, the investigation of the laws in accordance with which this development has progressed. As the latter involves the use of the evidence included in the former, I will not devote a special chapter to the proof for evolution.

The influences and forces which have operated to produce the type structures of the animal kingdom have been plainly of two kinds: 1. *Originative; 2. Directive.* The prime importance of the former is obvious; that the latter is only secondary in the order of time or succession is evident from the fact that it controls the preservation or destruction of the results or creations of the first, and thus furnishes the bases of the exhibitions of the originative forces in the production of the successive generations of living beings.

Wallace and Darwin have propounded as the cause of modification in descent their law of natural selection. This law has been epitomized by Spencer as the "survival of the fittest." This neat expression no doubt covers the case, but it leaves the origin of the fittest entirely untouched. Darwin assumes a "tendency to variation" in nature, and it is plainly necessary to do this, in order that materials for the exercise of a selection should exist. Darwin and Wallace's law is, then, only restrictive, directive, conservative, or destructive of something already created. I propose then to seek for the originative laws by which these subjects are furnished—in other words, for the causes of the origin of the fittest.

It has seemed to the author so clear from the first as to require no demonstration, that natural selection includes no *actively* progressive principle whatever; that it must first wait for the development of variation, and then, after securing the survival of the best, wait again for the best to project its own variations for selection. In the question as to whether the latter are any better or worse than the characters of the parent, natural selection in no wise concerns itself.

I. ON THE LAW OF ACCELERATION AND RETARDATION.

There are two modes of demonstration of evolution, both depending on direct observation. One of these has been successfully presented by Darwin. He has observed the origin of varieties in animals and plants, either in the domesticated or wild states, and has shown, what had been known to many, the lack of distinction in the grades of difference which separate varieties and species. But he has also pointed out that species (such, so far, as distinctness goes) have been derived from other species among domesticated animals, and he infers by induction that other species, whose origin has not been observed, have also descended from common parents. So far I believe his induction to be justified; but when from this basis evolution of divisions defined by important structural characters, as genera, orders, classes, etc., is inferred, I believe that we do not know enough of the uniformity of Nature's processes in the premises to enable us to regard this kind of proof as conclusive.

I therefore appeal to another mode of proving it, and one which covers the case of all the more really structural features of animals and plants.

It is well known that in both kingdoms, in a general way, the young stages of the more perfect types are represented or imitated with more or less exactitude by the adults of inferior ones. But a true identity of these adults with the various stages of the higher has, comparatively, rarely been observed. Let such a case be supposed.

In *A* we have four species whose growth attains a given point, a certain number of stages having been passed prior to its termination or maturity. In *B* we have another series of four (the number a matter of no importance), which, during the period of growth, can not be distinguished by any common, i. e., generic character, from the individuals of group *A*, but whose growth has only attained to a point short of that reached by those of group *A* at maturity. Here we have a parallelism, but no true evidence of descent. But if we now find a set of individuals belonging to one species, or, still better, the individuals of a single brood, and therefore held to have had a common origin or parentage, which present differences among themselves of the character in question, we have gained a point. We know in this case that the individuals, *a*, have attained to the completeness of character presented by group *A*, while others, *b*, of the same parentage have only attained to the structure of those of group *B*. It is perfectly obvious that the individuals of the first part of the family have grown further, and, therefore, in one sense faster, than those of group *b*. If the parents were like the individuals of the more completely grown, then the offspring which did not attain that completeness may be said to have been *retarded* in their development. If, on the other hand, the parents were like those less fully grown, then the offspring which have added something have been *accelerated* in their development.

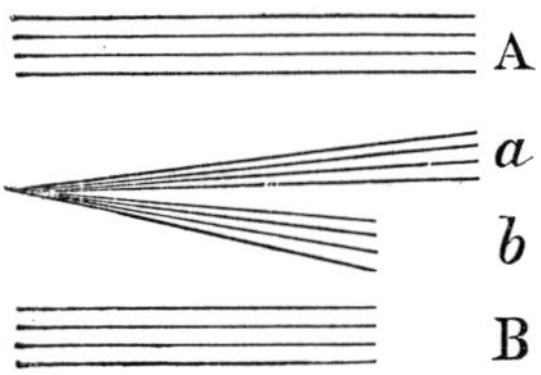

Fig. 7.

I claim that a consideration of the uniformity of Nature's processes, or inductive reasoning, requires me (however it may affect the minds of others) to believe that the groups of species, whose individuals I have never found to vary, but which differ in the same point as those in which I have observed the above variations, are also derived from common parents, and the more advanced have been *accelerated* or the less advanced *retarded*, as the case may have been with regard to the parents.

This is not an imaginary case, but a true representation of many which have come under observation. The developmental resemblances mentioned are universal in the animal, and probably in the vegetable kingdoms, approaching the exactitude above depicted in proportion to the near structural similarity of the species considered.

Example 1.—It is well known that the *Cervidæ* of the Old World develop a basal snag of the antler (see Cuvier, "Ossemens Fossiles," and Gray, "Cat. British Museum") at the third year; a majority of those of the New World (genera Subulo, Cariacus) never develop it except in abnormal cases in the most vigorous maturity of the most northern Cariacus (*C. virginianus*), while the South American Subulo retains to adult age the simple horn or spike of the second year of all *Cervidæ*.

Among the higher *Cervidæ*, Rusa and Axis never assume characters beyond an equivalent of the fourth year of Cervus. In Dama the characters are, on the other hand, assumed more rapidly than in Cervus, its third year corresponding to the fourth of the latter, and the development in after years of a broad plate of bone, with points being substituted for the addition of the corresponding snags, thus commencing another series which terminates in the great fossil elk, Megacerus.

Returning to the American deer, we have Blastocerus, whose antlers are identical with the fourth year of Cariacus. Corresponding with the Dama-Megacerus type of the Old World, we have the moose (Alces) developing the same palmate horn on the basis of Cariacus (i. e., without eye-snag).

Example 2.—I select the following series, embracing the majority of the genera of the North American Helicidæ.*

1. Turns of spire very few; wide umbilicus; shell thin, with thin lips.. *Binneya*.
2. Turns few, but more; rest as above.......................... *Vitrina*.
3. Turns still more numerous; rest as above..................... *Hyalina*.
4. As No. 3, but lip thickened inside.......................... *Hygromia*.
5. Coiled; *umbilicus closed;* lip thickened inside and out.. *Tachea* and *Pomatia*.
6. Same, with a parietal tooth................................ *Mesodon*.
7. Same, with parietal and two interior lip teeth............. *Isognomostoma*.

* * Recommencing at No. 4. All with *open umbilicus*.

5. As No. 4, but lip thickened in and out...................... *Arionta*.
6. Same as No. 5, but with parietal tooth...................... *Polymita*.
7. Same, with both parietal and lip teeth...................... *Triodopsis*.

The successional relation of these genera may be represented in such a diagram as this:

	Umbilicus open.					Umbilicus closed.	
7	*						*
6		*				*	
5			*		*		
4				*			
3				*			
2				*			
1				*			

* See Tryon, "Terrestrial Mollusca of the United States." Probably other (e. g., dental) characters distinguish some of these genera, but the above furnishes the history of one set of characters.

In the history of the growth of the genera Isognomostoma and Triodopsis, the extreme forms of the two series, it is well known that at first the coils of the shell are extremely few, as in Binneya; and that, like it, it is very thin and with a delicately thin edge; that the turns increase successively in number, as in Vitrina and Hyalina; and that, finally, the lip thickens, as in Hygromia. Then the umbilicus may close, as in Tachea, or (in Triodopsis) remain open, as in Arionta. In either case a tooth is soon added on the body-whorl (Polymita, Mesodon), and, finally, the full maturity of the shell is seen in the added teeth of the inside of the lip-margin. How many of the stages of the genera *Triodopsis* and *Mesodon* are identical with the genera of the series which represent them, I leave to more thorough conchologists, but that some now exhibit and all have once presented illustrations of the relation of exact parallelism, I can not doubt.

Example 1.—An abundant race of the American deer, *Cariacus virginianus*, exists in the Adirondack region of New York, in which the development of the antlers never progresses beyond the spike stage of the second year. Therefore, some individuals of this species belong to Cariacus and some to Subulo.*

Example 2.—A large part of the individuals of the common snail, *Mesodon albolabris*, never develop the tooth of the body-whorl, characteristic of the genus whose definition has to be modified to retain them.

Example 3.—Many individuals of *Triodopsis tridentata* from eastern North Carolina occur without the lip-teeth, characteristic of the genus *Triodopsis*. Hence these specimens, though of common origin with others of the species, must be referred to another genus.

Example 4.—Structural characters are known in many, if not all, species which are said to be "inconstant," being present or absent indifferently, thus being useless for definition. They may be rudimental when present or considerably developed. The presence or absence of wings in some species of insects may be cited; also the presence of generic characters in the male sex of many Coleoptera and their absence in the females. The characters of

* Since the above was written the facts as stated have been denied. But the author has seen specimens of *Cariacus virginianus* and *C. macrotis* in which the horns had assumed the characters of the genus Alces, throwing these individuals into that genus. See "American Naturalist," 1883–'84. (Ed. 1886.)

males, females, workers and soldiers in bees and ants may be added. All these facts belong to the same category as those cited among deer and mollusks, and have a similar explanation.

Example 5.—It does not seem to be the law in "retardation" that parallelisms exhibited by the series in its rise to its highest point of development should retrace the steps by which it attained it, and that "exact parallelisms" should be exhibited in a reversed order. Parallelisms, it is true, are exhibited; but so far as I have observed always "inexact," often in a high degree. A marked case of retardation occurs in the dental development of a number of persons who have come under my observation in the neighborhood of Philadelphia. It is not very uncommon to find persons in whom the third molars in both jaws are incomplete as to number, one, two, three, or all, being deficient. It is still more common for them to be incompletely covered by the enamel layer, and to become in consequence so worthless as to require early removal. I am acquainted with two families in which the absence of the exterior upper incisor on each side is common. In one of these the second and third generation have inherited it from the mother's side, and it now characterizes many of the children. The significance of this modification will be best understood by examining the dental structures of the *Quadrumana* in general. Commencing with the highest family and its abnormal dentition, we have:

		Incisors.	Canines.	Premolars.	Molars.
Hominidæ,..	Abnormal.	$\frac{1}{2}$	$\frac{1}{1}$	$\frac{2}{2}$	$\frac{2}{3}$—$\frac{2}{2}$
	Normal.	$\frac{2}{2}$	$\frac{1}{1}$	$\frac{2}{2}$	$\frac{3}{3}$
Simiidæ................		$\frac{2}{2}$	$\frac{1}{1}$	$\frac{2}{2}$	$\frac{3}{3}$
Cebidæ................		$\frac{2}{2}$	$\frac{1}{1}$	$\frac{3}{3}$	$\frac{3}{3}$
Lemuridæ..............		$\frac{2}{3}$	$\frac{1}{1}$	$\frac{3}{3}$—$\frac{2}{3}$	$\frac{3}{3}$
Mammalia, Normal.....		$\frac{3}{3}$	$\frac{1}{1}$	$\frac{4}{4}$	$\frac{3}{3}$

In this table we see a decline in the number of teeth of the higher groups. Thus, the premolars are one less than the normal number in the whole order, and they lose one in each jaw in the Old World apes, and man. The molars maintain the normal number throughout, but the third in both jaws is in the *Simiidæ* reduced by the loss of a fifth or odd tubercle, thus becoming four-lobed. In the upper jaw, this is first lost in the Semnopithecus; in the lower, in the next highest genus Cercopithecus. In Homo its appearance is "retarded," the interval between that event and the protrusion of the second molar—six to ten years—being relatively greater than in any genus of *Quadrumana*. Its absence is then the result of continued retardation, not of a new and adaptive suppression, and is of direct systematic zoölogical value.

In the incisors a reduction is also plainly visible, as we pass from the most completely furnished mammals to the genus Homo. One from the upper jaw is first lost, then, in the *Cebidæ*, one from the lower also. The number remains the same through the *Simiidæ* and normal *Hominidæ*, but, in the abnormal cases cited, the process of reduction is continued and another incisor from each side disappears. That this also is truly "retardation" is evident from the fact, that the exterior incisor is the last developed, being delayed in ordinary growth a year later than those of the inner pair. The same retardation is seen in the quadrumane *Cheiromys* (the *aye-aye*), and the whole order *Rodentia*. In the latter, the rare presence of the reduced second incisors, as in *Lepus*, shows a less degree of this modification. This retardation is also of systematic importance, and, should either of the characters described be constant in any of the species of the genus *Homo*, would at once entitle it to new generic rank. The very frequent absence of the posterior molars (wisdom teeth) has been recently found to characterize a race in India. Should this peculiarity prove constant, this race would with propriety be referred to as a new genus of *Hominidæ*, as we have many cases of very similar species being referred to different genera. It is altogether probable that such will, at some future time, be the condition of some race or races of men.*

I am now disposed to regard the above as the method of production, not only of generic but of all other, including specific characters. It would appear that, by excessive acceleration or retardation, some of the characters of a series may be skipped; but observations are not conclusive on this point, since very close examination is necessary for the appreciation of very transitory embryonic conditions.

II. ON THE LAW OF REPETITIVE ADDITION.

The origin of new structures, which distinguish one generation from those which have preceded it, I have stated to take place under the law of *acceleration*. As growth (creation) of parts usually ceases with maturity, it is entirely plain that the process of acceleration is limited to the period of infancy and youth in all animals. It is also plain that the question of growth is one of

---

* The preceding section is merely an abbreviation, with new illustrations, of the propositions brought forward in the writer's "Origin of Genera," 1868, where a considerable extension of the subject will be found.

nutrition, or of the construction of organs and tissues out of protoplasm.

The construction of the animal types may be referred to two kinds of increase—the addition of identical segments and the addition of identical cells. The first is probably to be referred to the last, but the laws which give rise to it can not now be explained. Certain it is that segmentation is not only produced by addition of identical parts, but also by subdivision of a homogeneous part. In reducing the vertebrate or most complex animal to its simplest expression, we find that all its specialized parts are but modifications of the segment, either simply or as sub-segments of compound but identical segments. Gegenbaur has pointed out that the most complex limb with hand or foot is constructed, first, of a single longitudinal series of identical segments, from each of which a similar segment diverges, the whole forming parallel series, not only in the oblique transverse, but generally in the longitudinal sense. Thus, the limb of the Lepidosiren represents the simple type, that of the Ichthyosaurus a modification. In the latter, the first segment only (femur or humerus) is specialized, the other pieces being undistinguishable. In the Plesiosaurian paddle the separate parts are distinguished, the ulna and radius well marked, the carpal pieces hexagonal, the phalanges defined, etc.

As regards the whole skeleton, the same position may be safely assumed. Though Huxley may reject Owen's theory of the vertebrate character of the segments of the brain-case, because they are so very different from the segments in other parts of the column, the question rests entirely on the definition of a vertebra. If a vertebra be a segment of the skeleton, of course the brain-case is composed of vertebræ ; if not, then the cranium may be said to be formed of "sclerotomes," or some other name may be used. Certain it is, however, that the parts of the segments of the cranium may be now more or less completely parallelized or homologized with each other, and that, as we descend the scale of vertebrated animals, the resemblance of these segments to vertebræ increases, and the constituent segments of each become more similar. In the types, as Amphioxus, etc., where the greatest resemblance is seen, segmentation of either is incomplete, for they retain the original non-osseous basis. Other animals which present cavities or parts of a solid support are still more easily reduced to a simple basis of segments, arranged either longitudinally (worm) or centrifugally (star-fish, etc.).

DEFINITIONS.

a. The succession of construction of parts of a complex, was originally a succession of identical repetitions; and grade influence merely determined the number and location of such repetitions.

β. *Acceleration* signifies addition to the number of those repetitions during the period preceding maturity, as compared with the preceding generation, and *retardation* signifies a reduction of the numbers of such repetitions during the same time.

γ. The successive additions now characterizing the growth of the highest animals are not exact repetitions of segments at this time, because of influences brought to bear on cell-nutrition during long periods. The nature of these influences is made the subject of another section.

In the endeavor to prove these positions, I will produce evidence, first, that some simpler animals grow according to the principle of modified repetitive addition, and that traces of it are to be observed in the most complex; second, that every addition to structure which has resulted in the complexity of the higher animals was originally a repetition of a pre-existent structure.

Detailed explanations of the law of repetitive addition are attempted in the following pages, under two heads—segment-repetition, and cell-repetition.

A. On Segment-Repetition.

This is everywhere seen in the construction of animals and plants. Double bilateral symmetry may serve as one example of repetition in growth.

a. Bilateral symmetry. Anatomists have little difficulty in determining the bilateral symmetry in most animals—that is, the homologies of the parts on opposite sides of the median line. It might be almost asserted that it was a necessity of organization; but, when we observe the growth of many plants, we are undeceived. And though bilateral symmetry in the *Cœlenterata* and many *Articulata* is perfect, yet in higher animals it is more or less departed from. In the Vertebrata the *Amphioxus* is almost completely bilaterally symmetrical. In the fishes, the digestive system is the only one which does not conform to it; while in the birds the reproductive system is atrophied on one side. In the serpents the respiratory and part of the circulatory are similarly modified; and in the Mammalia the digestive and circulatory systems have both become unsymmetrical; and the cranium, even, in the Cetacea.

If evolution be true, the unsymmetrical forms have descended from the symmetrical, and the asymmetry being thus not inherited, is the result of laws which have interfered with the original tendency to bilateral repetition.

Many cases of bilaterally symmetrical diseases have been enumerated by physiologists, and I will select as an example one which has come under my observation. They were those of two boys who had had that disease involving the muco-dermal system called varicella, while the crowns of the successional incisor teeth were still inclosed in the mucous capsules of the alvéolar walls. The deposit of phosphate of lime forming their surfaces was interrupted by the disease of the tissue, and the result was a surface pitted, or sculptured intaglio fashion. The sculpture of the two incisors of the right side was precisely imitated by those of the left in reversed order, even in minute details, which were numerous, thus producing a result not displeasing to the eye. This has been observed on two distinct occasions some years apart.

Another interesting example of bilaterally symmetrical disease is recorded in a paper on "A Case of Universal Hyperostosis, etc.," by Drs. Mears, Keen, Allen, and Pepper.* They describe the skeleton of a boy of fourteen which displayed an extraordinarily exostosed condition, the bones themselves remaining in the condition known as osteoporosis. They describe the uniform repetition of the abnormal growths of one side on the other in the following language (p. 22):

"Comparing the two sides externally, not only is there no difference in the extent and character of the disease, but there is the most remarkable symmetry of the corresponding diseased bones, which may be traced even into details. The disease begins and ends on both sides at corresponding points, it changes in character from simple porosity to the growth of osteophytes at corresponding points; if, on one side, the posterior part of the bone is most diseased, the same is true of the other side; if the osteophyte growth is continuous or interrupted on one bone (fibula, Fig. 18), it is so on the opposite one; if one is unusually diseased at a tendinous or aponeurotic insertion, so is its mate; if a groove or a variation in

---

* See "Proceed. Amer. Philos. Soc.," 1870, p. 19.

color exist on the one side, the same will be found on the other side; even of single marked spiculæ of bone the same may be said, so that a description of one side will answer for both, minute differences being noted as they occur."

b. Anteroposterior symmetry.

That this is an absolute law of creation will be less readily admitted than in the case of double bilateral symmetry, since the exceptions appear to be so universal. Nevertheless, I believe it to be as much a part of the law of repetitive nutrition as the other. The anteroposterior homologies even of the human skeleton have been largely demonstrated, but, as usual, we must appeal to the lower forms for a clear view of it. In the rudimental skeletal axis we find such symmetry almost perfect in the *Amphioxus,* but in no other vertebrate. In limbs we have it clearly indicated in the Reptilian order *Ichthyopterygia,* and in the Piscine order *Dipnoi,* where the anterior and posterior are scarcely or not at all distinguishable. In the scapular and pelvic arches we find it also approximated in the first-named orders.

In the nervous system it also exists approximately in the *Amphioxus.* It is not seen in any vertebrate, and in but few other animals, in the digestive system, but it appears to exist in some lower Articulata in both the respiratory and circulatory systems.

c. As illustrations of exact repetition involving large portions of the organism the higher Polyps may be cited, which differ from the lower chiefly by the addition of similar septa and similar tentacles. Examples of repetition of nearly the whole organism may be found in many Entozoa, as *Tænia,* where the cephalic segment only differs from the others, the remainder or proglottides being alike. The most entire repetition of structure is seen in *Vibrio,* where the segments are all alike, there being none representing a head.

d. As an example in special details of structure, the pelvic arch of *Ichthyosaurus,* when first created, was a repetition of the scapular, and the hind limb, of the fore limb. The segments of the limbs of the *Dipnoi* are mere repetitions, the later created of the earlier. The special parts of the pes and manus of *Ichthyosaurus* are simply repetitive efforts of growth-force joined with a diminishing amount. The addition of a digit, often distinguishing one genus of Salamanders or Saurians from another, is evidence of a similar repetitive effort. The low mammal *Ornithorhynchus* possesses but a single tooth in each jaw; the simple teeth of armadillos and cetaceans, increasing as they have done from a simpler commencement, present examples of repetitive acceleration of growth-force.

e. Complication of a single element of repetition is accomplished apparently by a double repetition. This is best understood by the consideration of the transition from simple to complex teeth. In the cetaceans this occurs in the Squalodonts ; the cylindric incisors are followed by flattened ones, then by others grooved on the fang, and then by two-rooted, but never double-crowned teeth. This is the result of anteroposterior repetitive acceleration of the simple cylindric dental type of the ordinary toothed cetacean.

Another mode of dental complication is by lateral repetition. Thus, the heel of the sectorial tooth of a Carnivore is supported by a fang alongside of the usual posterior support of a premolar, and is the result of a repetitive effort of growth-force in a transverse direction. More complex teeth, as the tubercular molars, merely exhibit an additional lateral repetition, and sometimes additional longitudinal ones. As is well known, the four tubercles of the human molar commence as similar separate knobs on the dental papilla.

The above are cited as examples to explain the meaning of the proposition. When fuller demonstration is desired, a greater number might be given.

B. On Cell-Repetition.

That each additional act of creation in growth was originally identical with one which preceded it, and therefore an *exact repetition* in its character and results, is proved by the following considerations.

It has been already determined by the study of homologies that all organs and parts of an organism can be referred to an original simple archetype.

The question then remains as to whether the first element, or lowest term, of a given organized part is essentially a new structure, or whether it be a repetition of some previously existing one. It may be asserted that the simplest expressions which shall cover all organs are the solid segment, and the hollow sack, and tube. For example, we have already noted that the ultimate element of the limb is the first segment of the single ray of Lepidosiren. Is this short cartilaginous cylinder (which probably represents the fore limb of some undiscovered member of the Dipnoi) a result of the repetition of a pre-existent structural element? This is no doubt the case, for, as will be shown beyond, cartilage, though the least cellular of all the tissues, is formed originally by cell repetition or division. Again, the ultimate lobules of the most complex gland are but repetitions of the diverticula of the simply branched, and each of the latter repetitions of the simple *cul-de-sac,* which has its origin in a convexity of an originally plane surface. This convexity is again the result of repetition of cells or cell-division, whereby their number is increased and the surface rendered convex.

We are thus, in both the solid segment and hollow sack, brought down to cell-repetition. Thus it is with organs, as with entire animals, in which, following the line of simplification, we reach at last forms composed of cells only (*Actinophrys,* e. g.) and then the unicellular (*Amœba*).

If this be the origin of organs, the question whether repetitive growth has constructed tissues remains for consideration.

In growth, each segment—and this term includes the parts of a complex whole or parts always undivided (as the jaw of a whale or the sac-body of a mollusk)—is constructed, as is well known, by cell-division. In the growing fœtus the first cell divides its nucleus and then its whole outline, and this process, repeated millions of times, produces, according to the cell theory, all the tissues of the animal organism or their bases, from first to last. That the ultimate or histological elements of all organs are produced originally by repetitive growth of simple nucleated cells, with various modifications of exactitude of repetition in the more complex, is taught by the cell theory. The formation of some of the tissues is as follows:

First Change.—Formation of simple nucleated cells from homogeneous protoplasm or the cytoblastema.

Second.—Formation of new cells by division of nucleus and body of the old.

Third.—Formation of tissues by multiplication of cells with or without addition of intercellular cytoblastema.

A. In connective tissue, by slight alteration of cells and addition of cytoblastema.

B. In blood, by addition of fluid cytoblastema (fibrin) to free cells (lymph-corpuscles), which in higher animals (vertebrates) develop into blood-corpuscles by loss of membrane, and by cell-development of nucleus.

C. In muscles, by simple confluence of cells end to end, and mingling of contents (Kölliker).

D. Of cartilage, by formation of cells in cytoblast which break up, their contents being added to cytoblast; this occurring several times, the result being an extensive cytoblast with few and small cells (Vogt). The process is here an attempt at development with only partial success, the result being a tissue of small vitality.

Even in repair-nutrition, recourse is had to the nucleated cell. For Cohnheim first showed that if the cornea of a frog's eye be scarified, repair is immediately set on foot by the transportation thither of white or lymph or nucleated corpuscles from the neighboring lymph-heart. This he ascertained by introducing aniline dye into the latter. Repeated experiments have shown that this is the history in great part of the construction of new tissues in the adult man.

Now, it is well known that the circulating fluid of the fœtus contains for a period only these nucleated cells as corpuscles, and that the lower vertebrates have a greater proportion of these corpuscles than the higher, whence probably the greater facility for repair or reconstruction of lost limbs or parts enjoyed by them. The invertebrates possess only nucleated blood-corpuscles.

C. Synthesis of Repetition.

That growth-force is capable of exhibiting great complexity of movement with increase in amount, will now be shown. That this quality of complication is one of its distinguishing features will appear plain.

The simplest forms of life, as stated by Haeckel, are simply homogeneous drops of protoplasm (*Protamœba*). These only grow by ordinary accretion, and display a form of self division or reproduction which is the simplest possible—i. e., the bisection of the mass by contraction at opposite points.

The next grade of animal type is represented by the nucleated cell. This is simple in *Amœba*, complex in *Actinophrys*, etc. With such forms as the latter, cell-growth begins, and its development is accomplished, by cell-division. This is simple repetition of ultimate parts. In the growth of all higher types we have nothing more than this, but following a law of *complex repetition.* Thus in the growth of the parts of an archetypal vertebral column or an archetypal limb, we have the repetition of cell-growth till the first segment is formed, when it ceases at that point, and repeats the process again, forming another segment like the first; repetition within repetition. So with the construction of muscular tissue; first, the nucleated cell repeated in a series, whose adjacent walls disappear, and whose cell-contents flow together, thus forming a fibrilla; then a repetition of the same process, forming a second fibrilla; and so on to the completion of thousands of them in fasciculi.

Let us then trace the series of repetitions and duplicated and still more complex repetitions, seen in following up animal forms from their archetypes.

In the simplest repetition of cell-growth in a longitudinal direction we have *Vibrio;* in the centrifugal, *Actinophrys.* The former may be represented by a line of simple dots, thus: Fig. 1.

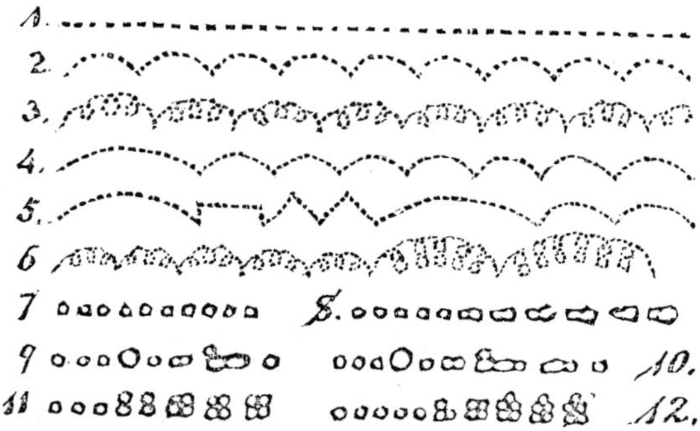

In a complex repetition we rarely have the same degree of complication in each repeated part. We have it centrifugally almost perfect in a Cœlenterate (*Actinia*), and linearly in some of the lower vermes. An archetype of the latter kind might be represented thus: Fig. 2. In a more complex form, as of the proglottides of *Tænia,* thus: Fig. 3. The same might represent an archetypal vertebrate.

If now we attempt to express the complication of an organ by modified repetition of once identical parts, the history of extremities will serve us. Thus the limb of Lepidosiren, which is composed of identical segments, may be thus represented: Fig. 2. Each longitudinal segment of the limb of Ichthyosaurus may be similarly represented with a modification, in size only, of the proximal or humerus; thus: Fig. 4. But in *Plesiosaurus* an important series of changes of shape (but not in complexity) appears, which may be represented thus: Fig. 5; the first being humerus, second ulna, third and fourth carpals (tarsals), the last phalanges, which are first specialized in this genus.

By far the most usual modification is, however, complication by duplicated and triplicated and still more highly multiplied repetition in some segments of the archetype, and its omission in other segments. Thus in the *Tænia*, the cephalic segments are much modified, and the nature of the repetition might be thus expressed : Fig. 6 ; the simpler segments representing the body segment, the two complex representing those of the head. In each, it will be observed, the complication is represented by loops of similar form, and each loop of dots which represent the cells in the first linear (Fig. 1) arrangement.

A somewhat similar figure might represent the nature of the complication in the Myriapod. In the insect the additional complications of the thoracic segments would alter the diagram near the middle.

In the vertebrate cranium a somewhat similar diagram might be used, except that the modification of the segments or vertebræ, as compared with the segments of the vertebral column, is not by repetition with modification of the parts of each segment, but rather by modification of the forms of the parts of the segments. The basicranial segments thus compare with the dorsal vertebræ as the segments of the limb of *Plesiosaurus* do with those of *Ichthyosaurus*.

The above considerations have reference to repetition of parts in a linear direction. Centrifugal repetition is seen in the addition of chambers to the heart, by the subdivision in the earliest stages into auricle and ventricle in the linear direction, considered in connection with the earlier division of each in the transverse direction by the growth of partitions. This mode of repetitive addition is not readily represented by diagram.

A good example of repetitive addition, in both linear and transverse direction, may be found in the successive complication of tooth structure seen in Mammalia. In the dolphin, the dental series may be represented thus : Fig. 7 ; in the squalodon thus : Fig. 8 ; in the cat : Fig. 9 ; in the dog : Fig. 10 ; in man : Fig. 11 ; in some Insectivora : Fig. 12.

The circles represented here are each a simple cusp.

In conclusion, the directions of repetitive growth may be tabulated as follows : The types to the left represent the original ; to the right, the derivative :

- Centrifugal.
  - Longitudinal antero-posterior and bilateral.
    - More bilaterally symmetrical.
      - Only bilateral.
    - More antero-posteriorly.
      - Only antero-posterior.
  - Centrifugal.
    - In plane.
    - In globe.

D. On Growth-Force.

From such examples as those that precede, but more especially from the last, it seems necessary to believe that there resides in organized matter, and in its most unmodified representative, the nucleated cell, an affection which displays itself in repetition. This phenomenon reduced to its lowest terms, may mean cell-division only, but the proof is only clear in cases of growth proper. This affection displays itself in very slow or more rapid repetitions —cell-division in growth occurring rapidly, while its recurrences at rutting seasons in the development of horns, feathers, etc., are separated by long intervals of time. In acceleration these repetitions occur with increased rapidity, i. e., in the adding of more structures during the same growth periods, while in low types its repetitions are few and therefore slow.

What is the relation of cell-division to the forces of nature, and to which of them as a cause is it to be referred, if to any? The animal organism transfers solar heat and the chemism of the food (protoplasm) to correlated amounts of heat, motion, electricity, light (phosphorescence), and nerve force. But cell-division is an affection of protoplasm distinct from any of these; although addition to homogeneous lumps or parts of protoplasm (as in that lowest animal, *Protamœba* of Hæckel) should prove to be an exhibition of mere molecular force, or attraction, cell-division is certainly something distinct. It looks like an exhibition of another force, which may be called *growth-force.* It is correlated to the other forces, for its exhibitions cease unless the protoplasm exhibiting it be fed.

Professor Henry pointed out many years ago that this must be the case, basing his belief on the observed phenomena of growth in the potato, and in the egg. The starch of the potato weighs much more than the young shoot of cellulose, etc., into which it has been converted by growth-activity, so that a portion of the substance of the tuber has evidently escaped in some other direction. This s shown to be carbonic acid gas and water, derived from the slow ombustion of the starch, which in thus running down from the omplex organic state to the more simple inorganic compounds, volves an amount of force precisely equal in amount to the chemial force (or chemism) requisite to bind together the elements in he more complex substance.*

Carpenter also states that in his opinion the growth of the Fungi s produced by a force liberated by the retrograde metamorphosis f their food, which is of an organic character (i. e., humus). 'his metamorphosis consists, as in the tuber, in the production of arbonic acid gas and water, and a force equivalent to the chemism 'hich had bound them in the former complex union.† But in igher forms of vegetable life, and in growth that follows germinaion, the plant must appropriate carbon from the carbonic acid of he atmosphere. The decomposition of the binary compound which sets free its oxygen) liberates the chemical force which had reviously maintained the compound (or an equivalent force), which Ienry regards as furnishing the growth-force, which produces the lant. Carpenter derives but a portion of the force in this way, btaining the greater part from the heat of the sun. To this source lso he looks for the heat necessary to the construction of coldlooded animals ; while in warm-blooded animals, the retrograde ıetamorphosis or running down of the material (protoplasm) of he food furnishes a requisite amount of heat.

Growth-force we may then regard as potential in organized tisıe, and as energetic during growth.‡ Our present knowledge

---

* " Agricultural Report of the Patent Office," 1857.

† " Correlation of Physical and Vital Forces," 1864 (" Quarterly Jour. of Science ").

‡ Bathmism or growth-force must be static or potential in each unit or plastid ell) of a living organism, the type differing with each organic species. When it is excess in a given locality it becomes energetic, and builds tissue in various forms. portion of this energy is, in this process, developed as molar motion of nutritive aterial, and is accompanied in the completed structure by the ordinary cohesive ergy, by which the newly-built material maintains its attachment as a whole and its parts. If living tissue be constructed, a portion of the excess remains as the atic energy of the plastids or cells of the new part, as it did of the old. The ermatozoid is highly endowed with static bathmism, and communicates it to the male ovum. The mingling of the two elements in the presence of nutritious maial presents an excess, and form-building results. Its activity will regulate subquent new growth, by giving the motion of nutritive material its proper direcon. When the tissue dies, this energy must leave behind a dynamic equivalent, it what this may be is as yet a mystery. (Ed. 1886.)

only permits us to believe that other force is only converted int it under the influence of pre-existent life, but of the real cause o this conversion we are as ignorant as in the case of the physica forces.

In the animal organism, different tissues display different de grees of "vitality." The most vital display cell-organization an its derivative forms, while the least so, approach nearer to homo geneity. As organized tissue is the machine for converting vita forces, we may believe that less growth-force is potential as such i cartilage than in muscle, for it is formed by a retrograde process by which cells once formed are mostly lost, and the contents forn the intercellular, nearly structureless mass characteristic of thi tissue. Growth-force must be here liberated in some other form perhaps the mere cohesive force of the former or "dead" inter cellular substance.

The higher vitality we may believe to result from the greate perfection of the more complex *machine* as a force converter, a compared with the inefficiency of the more simple.

E. On the Direction of Repetition.

It has been already pointed out that *growth-force* exhibits itsel in cell or segment repetition. The forms in which it thus display itself may be briefly considered. The approximate cause is treate of in the next chapter; but enough may be shown here to indicat that duplication and complex duplication is the law of growth-force and that therefore this process must always follow *an increase i amount* in any given locality.

The size of a part is then dependent on the amount of cell-divis ion or growth-force, which has given it origin, and the numbe and shape of segments is due to the same cause. The whole ques tion, then, of the creation of animal and vegetable types is reduce to one of the *amount and location of growth-force.*

Repetition is of two kinds, *centrifugal* and *longitudinal.* A an example of the former, the genus *Actinophrys* has been cited where the animal is composed of cells arranged equidistally aroun a common center. The arrangement in this type may be dis coidal or globular, providing no definite axis be discoverable. A an example of *longitudinal* repetition, *Vibrio,* and numerou cellular plants may be cited, where the arrangement is in a singl line.

In by far the greater number of animals these kinds of repeti ive structure co-exist. The *longitudinal* is, however, predomi-ant in the *Vertebrata*, *Mollusca*, and *Articulata*, while the *cen-rifugal* is greatly developed in the *Cœlenterata* and *Radiata*. In one but the simplest forms are either of these modes to be found lone.

The *centrifugal* repetition, or addition, more nearly resembles he mode of aggregation of atoms in inorganic or crystalline odies, and hence may be regarded as the inferior manifestation. t implies that growth-force in this case conforms to a law of olarity in exhibiting itself at equal distances from a center—hich is allied to ordinary molecular force, and independent of he localizing influences of which higher organisms seem capable. n centrifugal animals, then, the latter evidently plays an in-erior part. In Cœlenterates and Radiates, however, the body ossesses a short longitudinal axis, in some (*Asterias*) very short, n others (*Holothuria*) more elongate. The amount of com-lication of centrifugal growth greatly exceeds the complication n a longitudinal direction in all of these animals except the *Iolothurida*.

It is now important to observe that great numbers of centrifu-al animals are sedentary or sessile; while the longitudinal are agrant, moving from place to place. Many of the centrifugal nimals which wander, do not do so in the direction of their axis, ut sideways (*Medusæ*). It is also proper to notice that not only ne movements of the muscles but also the direction taken by ne food is in the long axis. It is therefore to be concluded that n longitudinal animals growth-force has assumed a more truly nimal type, and that this tendency has predominated over the olar or molecular tendency.

In most longitudinal animals, however, certain lateral portions, mbs, etc., extend on each side of the axis; and were the space narked by their extremities, and the extremities of the axis, filled, e would have the outline of a centrifugal animal.

Before discussing the influences which have increased and lo-ated growth-force, it will be necessary to point out the mode in hich these influences must necessarily have effected growth. cceleration is only possible during the period of growth in ani-nals, and during that time most of them are removed from the nfluence of physical or biological causes, either through their idden lives or incapacity for the energetic performance of life unctions. These influences must, then, have operated on the parents, and become energetic in the growing fœtus of the nex generation. However little we may understand this mysteriou process, it is nevertheless a fact. Says Murphy, "There is no ac which may not become habitual, and there is no habit that ma not be inherited." Materialized, this may be rendered—there i no act which does not direct growth-force, and therefore there i no determination of growth-force which may not become habitual there is, then, no habitual determination of growth-force whic may not be inherited ; and, of course, in a growing fœtus become at once energetic in the production of new structure in the direc tion inherited, which is acceleration.

But if the forces converted into growth-force are derived fro without the animal organism, whence and what the agency b which the acceleration or retardation of the latter is inherite from the parent? A few suggestions only on this head can b made in the fourth section.

III. THE LAW OF USE AND EFFORT.

Up to this point we have followed paths more or less distinctl traced in the field of nature. The positions taken appear to m either to have been demonstrated or to have a great balance o probability in their favor. In the closing part of this paper shall indulge in more of hypothesis than heretofore.

Since repetitive addition only produces identical results i archetypes, and each effort produces results more and more unlik its predecessor as structure becomes specialized, it becomes im portant to examine into the influences which have originall modified the repetitive efforts successively, producing structur more or less different in detail in the second generation fro those of the parents, in acceleration, or the reverse, in retarda tion.

Going further back, the question arises, why a simple exhib tion of repetition (e. g., cell-division) should be converted into complex or duplicated repetition (e. g., jointed ray). This, it ha already been stated, is one consequence of increased amount o the growth-force.

We then seek explanation of the main question, as to what d termines the location of this additional or new growth-forc (Div. A.)

Lastly, why the total amount of this force should change in given individual or part of an individual. (Div. B.)

A. On the Location of Growth-Force.

What are the influences locating growth-force? The only efficient ones with which we are acquainted, are, first, physical and chemical causes; second, use; and I would add a third, viz.: effort. I leave the first, as not especially prominent in the economy of type-growth among animals, and confine myself to the two following. The effects of use are well known. We can not use a muscle without increasing its bulk; we can not long use the teeth in mastication without inducing a renewed deposit of dentine within the pulp-cavity to meet the encroachments of attrition. The hands of the laborer are always larger than those of men of other pursuits. Pathology furnishes us with a host of hypertrophies, exostoses, etc., produced by excessive use, or necessity for increased means of performing excessive work. The tendency, then, induced by use in the parent, is to add segments or cells to the organ used. Use thus determines the locality of new repetitions of parts already existing, and determines an increase of growth-force at the same time, by the increase of food always accompanying increase of work done, in every animal.

But supposing there be no part or organ to use. Such must have been the condition of every animal prior to the appearance of an additional digit or limb or other useful element. It appears to me that the cause of the determination of growth-force is not merely the irritation of the part or organ used by contact with the objects of its use. This would seem to be the remote cause of the deposit of dentine in the used tooth; in the thickening epidermis of the hand of the laborer; in the wandering of the lymph-cells to the scarified cornea of the frog in Cohnheim's experiment. You can not rub the sclerotica of the eye without producing an expansion of the capillary arteries and corresponding increase in the amount of nutritive fluid. But the case may be different in the muscles and other organs (as the pigment cells of reptiles and fishes) which are under the control of the volition of the animal. Here, and in many other instances which might be cited, it can not be asserted that the nutrition of use is not under the direct control of the will through the mediation of nerve force. Therefore I am disposed to believe that growth-force may be, through the motive force of the animal, as readily determined to a locality where an executive organ does not exist, as to the first segment or cell of such an organ already commenced, and that therefore effort is, in the order of time, the first factor in acceleration.

Addition and subtraction of growth-force, in accordance with the modes pointed out below, account for the existence of many characters which are not adaptive in their nature.

Acceleration under the influence of effort accounts for the existence of rudiments of organs in process of development, while rudiments of organs in process of extinction are results of retardation, occasioned by absolute or complementary loss of growth-force. Many other characters will follow, at a distance, the modification resulting from the operation of these laws.

Examples of the Influence of Physical Causes.—This is nowhere better seen than in the case of coloration, which requires the light of the sun for its production. The most striking examples of this are seen in the colorless surface of animals inhabiting the recesses of caves, as the blind craw-fish and the *Amblyopsis*, etc. If evolution be true, these have descended from more highly colored progenitors. The flat fishes, also (*Pleuronectidæ*), as is known, swim on one side in adult age, but many of them are hatched symmetrical fishes, or nearly so, one eye rotating from one side to the other by a twisting of the cranial bones. It is thus probable that they have descended from symmetrical fishes, which were similarly colored on both sides. Now the lower side is colorless, the upper retaining often brilliant hues. The influence of sunlight is thus as distinctly discoverable among animals as among plants, where it has been generally accepted as a principle of vegetable physiology.*

Examples of the Effects of Effort and Use.—a. The Respiratory and Circulatory System of Vertebrates.—It is well known that the succession of classes of Vertebrates is measured first by their adaptation to aëration in water, and then by their successive departures from this type in connection with the faculty of breathing air. The same succession of structure is traversed by the embryos of the vertebrates, the number of stages passed being measured by the final status of the adult. This transition takes place in the Batrachia later in development than in any other class. Now, it is well known that the transition or metamorphosis may be delayed

* In this and similar cases, care must be taken not to misunderstand the writer by supposing him to mean that in *each generation separately* the peculiar coloration is the result of changed exposure to light. The evolutionist will understand that the effect of such influence increases with succeeding generations by the addition to inherited character of the effect of immediate external cause.

or encouraged by suppression of use of the branchial and encouragement of use of the pulmonary organs, or the reverse.

The aquatic respiration of tadpoles may be indefinitely prolonged by preventing their access to the surface, and it is known that in nature the size or age of the larva at time of metamorphosis may vary much in the same species. If perennibranchiates (*Siren*, e. g.) are deprived of their branchiæ, they will aërate blood by the lungs exclusively, and there is no reason to doubt that by use of these, and disuse of the branchiæ, aërial respiration might become the habit of the animal. It is also easy to perceive that geologic changes would bring about a necessity for precisely this change of habit. This occurred in the period of the coal measures, where large fresh-water areas were desiccated, and it was precisely at this period that many air-breathing Batrachians originated and had a great development.

β. The rattle of the Rattlesnake.—Nearly all the larger harmless snakes which live on the ground have a habit of throwing the end of the tail into violent vibrations when alarmed or excited, with the view of alarming a supposed enemy. Among *Coronelline* snakes, *Ophibolus triangulus* possesses it; among the water snakes, *Tropidonotus sipedon*. In the typical Colubrine group the black snake, *Bascanium constrictor*, is an example; *Pityophis sayi* also shakes the tail violently. The copperhead (*Ancistrodon contortrix*) and the moccasin (*A. piscivorus*) (fide Günther) have the habit in a marked degree. Among the rattlesnakes it is a means of both warning and defense, in connection with the rattle which they carry.

In the structure of the end of the tail of harmless snakes, we see a trace of the first button of the rattle in a horny cap that covers the terminal vertebræ. In the venomous genera, it is conspicuous in *Lachesis* especially, reaching a considerable length and having a lateral groove. In the plate-headed rattlesnakes (*Crotalus*) this corneous cap is inflated into a button with lateral groove, and in some of them possesses only one or two buttons or joints. In the perfected rattlesnakes (*Caudisona*) not only are the segments numerous and inflated, but a number of the terminal caudal vertebræ are greatly enlarged vertically, and co-ossified into a mass.* This is important from the fact that the rattlesnakes are the most spe-

* See good figures of this structure in "Zeitschr. f. wissensch. Zoologie," viii, Tab. 12.

cialized of all snakes, standing at the head of the order, and as such, on the principle of *acceleration,* present the greatest amount of *grade-nutrition.*

Now it appears to me that the constant habit of violent vibration in a part tends to determine an increased amount of nutritive fluid to it, in other words to localize growth-nutrition, and when this has attained complex repetition or grade-nutrition, to result in new grade-structure. (The segments of the rattle being nearly all alike, it is a case of simple repetition.) This view appears to be as reasonable as that generally entertained with regard to the cause of spavin in the horse's leg. Here, owing to excessive use, exostoses appear on the bones surrounding the tibio-tarsal articulation. As to the reason of the structure in question not appearing in forms lower in the scale than the rattlesnake, it is explained below, if the law of accumulation of grade-nutrition be true. (See Sec. B.) This is, that repetition (or acceleration) is only possible where the animal has an excess of growth-force at its disposal, or can abstract it from some portion which is unused or useless.

γ. On horns.—The possession of horns on the posterior part of the cranium, as defenses against enemies, is a character found in many distinct types of animals. (Herbivora have no (dental) weapons and need horns.) It is seen in the *Batrachia Stegocephala* in the extinct genus *Ceraterpeton;* among *Anura* it is approached by *Triprion* and *Hemiphractus.* Among Reptilia it is well marked in *Phrynosoma,* a Lacertilian genus. In *Mammalia* the *Artiodactyla Ruminantia* are the horned animals of the order. We have opportunities of observing the habits of these representatives of the Frogs, the Lizards, and the Mammals.

In the first case, any one who has kept ordinary toads and tree-toads in confinement, is aware that when attacked and unable to escape, they defend themselves by presenting the top of the head forward and using it as a shield. Now I have already pointed out* that in both toads, tree-toads, and frogs there are natural series of genera, measured by the degree of ossification of the superior cranial walls, the longest being that of the Hylidæ, which embraces six terms, viz. : *Hylella, Hyla, Scytopis, Osteocephalus, Trachycephalus,* and *Triprion.* The two last have the head thoroughly shielded, and *Hemiphractus* has projecting angles which appear in some South American forms, described by M.

* "Origin of Genera," 1868, p. 14. This work, article I.

Espada, to be developed into short horns. That this excessive ossification is associated with the habit of protecting the whole body with the front seems likely.

In the case of *Phrynosoma* we know that precisely the same habit is associated with the presence of the sharp horns; and that some genera without horns possess it also. *Phrynosoma* is an exceptionally sluggish genus in a family of most active forms, and must necessarily resort to this mode of defense more than they.

In the case of Ruminants, we also know that defense is accomplished by throwing the head down with the horns thrown forward. But this is not confined to this group. That generalized suborder, the *Artiodactyla Omnivora*, represented by the hog, which were no doubt the genetic predecessors of the Ruminants in time, also throw the head down in defense in the same way, having thus a manner totally distinct from that seen in the *Carnivora.* The latter show their *teeth* and often crouch preparatory to a leap.

These cases present so constant an association between habit and use that, admitting evolution, we are compelled to believe that the structure has given rise to the habit or the habit to the structure. In the former case we have to suppose, with the author of "Natural Selection," that among the many spontaneous variations rudimental horns occasionally appeared, and that their possessors, being thus favored in the struggle for existence, were preserved and multiplied; while those not favored dwindled, and were ultimately nearly all extirpated or starved. The question of origin is here left to chance, and Alfred Bennett has made a mathematical estimate of the chances of any particular profitable variation occurring among the great number of possibilities of the case. This has shown the chance to be so excessively small as to amount in most cases to a great improbability.

If we turn to the probabilities of such structure having arisen through the selection of that mode of defense by the animal, we find them greatly increased. The position occupied by the horns, in all the animals described, is that which is at once brought into contact with an enemy in conflict, and as sport among animals is a gentle imitation of conflict, the part would be constantly excited in sport as well. With an excess of growth-nutrition, our knowledge of the effects of friction on the epidermis, and of excessive ligamentous strain and inflammation on bone (e. g., spavin in horses), as well as of abnormal exostoses in general, would warrant us in the belief that the use of the angles of the parts in questio in these animals would result in a normal exostosis, of a simpl kind in the frogs, or as horn cores in the Ruminantia.* As to th sheathing of the cores in the *Bovidæ*, and nakedness in the *Cer vidæ*, it is in curious relation to their habitat and to their habits The epidermis and derm would of course share in the effects o friction. In the Bovidæ which dwell in treeless plains, or fee on the grasses in great part, the development of these coverings o the horn cores into a horny sheath would naturally meet wit no interruption. In the case of the deer, which mostly live i forests or browse on trees, constant contact with the latter woul prevent the healthy growth of the dermal covering, and it woul be liable to injury or constant excoriation by the animals them selves on the branches of trees, etc. This we know to be the pres ent habit of the deer as regards the dermal covering of the horns.† I have elsewhere pointed out the similar connection between th dental structure and habitat among the oxen and the deer. Th former, eating the harder grasses, are provided against the conse quent rapid attrition of the tooth by a prismatic form, whic allows of more prolonged growth and more rapid protrusion The deer, in accordance with their foliage-eating habits, do no wear the crown of the tooth with such rapidity. Long-continue protrusion is not so necessary, hence the teeth are more distinctl rooted and have a prominence or shoulder, distinguishing th body of the crown.

B. Change in Amount of Growth-Force.

1. *Absolute increase of Growth-Force.*—As every type has ha its period of greatest development in numbers, size, and complica tion of structure, the present law indicates as an explanation, culmination of the process of conversion of growth-force from it energetic to its potential state in tissue. The cause is primaril the increased exercise of effort and use, which, while effecting conversion, increases the capacity of the organs by which furthe conversion is effected.

---

* The now well-known fact that all cranial ossification was primitively derma is confirmatory of the idea that its appearance was due to moderate friction of th skin on resistant bodies. (Ed. 1886.)

† Palæontological studies go far to show that the origin of the shedding of th deer's horn was due to the loss of their dermal covering, and that this was caused b violent use. (1886.)

2. *Local increase of Growth-Force.*—Examples of a local increase of this kind are probably to be seen in convoluted organs; as the convolutions of the brain in higher Mammalia; the convolutions of the tooth-dentine of the Labyrinthodont Batrachia; the same phenomenon in the cotyledons or plumule of some seeds. In these cases the superficial area of the parts is excessively developed, and the inclosing organs not being proportionately enlarged, a convolution necessary follows. In the first case, the skull; in the second, the alveolus; in the third case, the seed-envelope, restrain the expanse of the contained part, which would otherwise follow increase of growth-force.

3. *Absolute loss of Growth-Force.*—This will follow defective nutrition, produced by inability of the animal to obtain heat and food requisite to that end. This is supposed to be due, according to the view hereafter proposed, primarily to deficiency of intelligence, in failing to adapt habits to changed physical circumstances, and secondarily to the unfavorable influence of such changed circumstances. The extinction of highly specialized types, which has closed so many lines of animal types, will be accounted for by their less degree of plasticity and want of capacity for change under such changed circumstances. Such changes consist of modified topography and temperature, with irruptions of many new forms of life by migration. The less developed forms would be most likely to experience modification of structure under a new order of things, and paleontology teaches that the predecessors of the characteristic types of one period were of the less specialized forms of that period which went before.

Thus is explained the fact that, in following out the line of succession of animal forms, we have constantly to retrace our steps from specialized extremes (as osseous fishes, tailless Batrachia, song birds, etc.) to more generalized or simple forms, in order to advance beyond.

4. *The complementary diminution* of growth-nutrition follows the excess of the same in a new locality or organ, of necessity, if the whole amount of which an animal is capable be, as I believe, fixed. In this way are explained the cases of retardation of character seen in most higher types. The discovery of truly complementary parts is a matter of nice observation and experiment. Perhaps the following cases may be correctly explained.

A complementary loss of growth-force may be seen in absence of superior incisor teeth and digits in ruminating Mammalia, where excessive force is evidently expended in the development of horns, and complication of stomach and digestive organs. The excess devoted to the latter region may account for the lack of teeth at its anterior orifice, the mouth ; otherwise, there appears to be no reason why the ruminating animals should not have the superior incisors as well developed as in the odd-toed (Perissodactyl) Ungulates, many of which graze and browse. The loss to the osseous system in the subtraction of digits may be made up in the development of horns and horn-cores, the horn sheath being perhaps the complement of the lost hoofs. It is not proposed to assert that similar parts or organs are necessarily and in all groups complementary to each other. The horse has the bones of the feet still further reduced than the ox, and is nevertheless without horns. The expenditure of the complementary growth-force may be sought elsewhere in this animal. The lateral digits of the *Equidæ* are successively retarded in their growth, their reduction being marked in *Hippotherium,* the last of the three-toed horses ; it is accompanied by an almost coincident acceleration in the growth-nutrition of the middle toe, which thus appears to be complementary to them.

The superior incisors of the *Artiodactyla* disappear coincidentally with the appearance of horns, which always exist in the toothless division of the order, except in some very small antelopes (*Cephalophus*, etc.) where the whole amount of growth-force is small. Possibly the superior incisors and horns are complementary here. The retardation in development of the teeth in the higher apes and men, as compared with the lower apes, is coincident with the increase of number of brain convolutions. That this is not necessarily coincident with reduction of teeth in other groups is plainly proved by the rodents and *Chiromys*, where the loss of many teeth is complementary to the great size of the incisors of the middle pair. But in man there is no complementary increase of other teeth, and the reduction is no doubt due to contraction of the jaws, which is complementary to increase in other parts of the cranium, in both apes and men.

I am confident that the origin and loss of many structures may be accounted for in this way, and the correlation of parts to each other be *measured* accurately.

Objection.—The first one which arises is that which the author of the "Vestiges of Creation" made against Lamarck's theory of a similar kind, i. e., that by assuming that effort, use, and physical causes have originated modifications of structure, we give the adaptive principle too much to do. I have made the same objection to the theory of natural selection. It is true that an application to a purpose is involved in the present theory of the "location of growth-force"; but in point of fact, a large number of non-adaptive characters are accounted for by it. These are the rudimental and transitional ones which mark the successive steps preliminary to the completion of an adaptive structure; second, those produced by deficiency of growth-force in less favored regions of the body; and third and fourth, phenomena consequent on general deficiency and excess of growth-force.

And it may be said in conclusion that if the three principles, or if use especially, should be found to be inadequate to the service here demanded of them, it may be at least said that they, or the last named, constitute the only controllers of growth-force to any degree at all with which we are acquainted.

IV. ON GRADE-INFLUENCE.

The object of the present section is the attempt to discuss how the influence of effort and use on the parent is placed in a position to be inherited by the offspring.

A. Of the Nature of Grade-Influence.

In the first place, it is necessary to note the definition and character of grade-influence.

a. Growth-force uninfluenced by grade-influence simply adds tissue either (*a*) in enlarging size, or (*b*) in replacing waste. It does this by repeating the cell, by division, in localities which have already assumed their specific form. This form of growth-force may persist throughout life, but with diminished energy in age.

β. Grade-influence directs growth-force in building up the tissues into organs, and constructs the parts of the body successively to completion, the result expressing the type or grade of the animal or plant. Its energy terminates with maturity, except in cases of periodical reproduction of sexual ornaments of the male (birds, deer), where it continues throughout life, appearing at regular intervals.

But it has occurred in acceleration that instead of a simple repetition of the ultimate histological element of an organism, in adding to its amount, it adds a completely organized part of the structure, as a tube, a phalange, a digit, a limb or an arch; an ocellus or a tooth. For instance, in the genus *Amblystoma,* one section possesses four phalanges on the longest digit; another section exhibits but three. In the species *A. mavortium,* some individuals have the small number of phalanges, but the majority possess the larger number. As all are of common parentage, a whole phalange has been lost or added. The explanation of this phenomenon is essential to the comprehension of the *origin* of type structures.

* In plants, growth-nutrition continues throughout life, but in the higher plants it is more active during the earlier years in perennial species, addition to size becoming less and less marked with increasing age. Grade-nutrition also persists throughout life, but is chiefly active during a short period only of every year, or during flowering and fruiting. Not only in the production of the reproductive organs, but also in the yearly additions to other typical parts of the plant, grade-nutrition is active.

** In animals, growth-nutrition is more active in the early stages of life, but is continued throughout in the lower divisions; in the highest, it is also continued throughout life, but there is a greater contrast between its results during youth, when nearly the whole size is attained, and during age, where the additions are much less.

Grade-nutrition is, on the other hand, entirely confined to infancy and youth, except in those low animals which produce their reproductive organs periodically (some *Entozoa,* etc.), where it may be said to be in nearly the same condition as in plants.

γ. While the amount of simple growth-force, represented in adult living animals, has varied very irregularly throughout the animal kingdom, there being large and small in every division, it would seem to have accumulated, on the whole, with the rising scale of animal types. Thus the lower or *Protozoa* are the smallest; *Radiates* are next in size; *Molluscs* and *Articulates* reach nearly the same maximum, which exceeds that of the *Radiates,* and falls far below that of the *Vertebrates.* Among the last the Mammalia have attained as large if not larger size than any of the other orders (e. g., *Cetacea*). This is, however, not necessary to the history of evolution.

That an increased amount of grade growth-force has been constantly rendered potential during the advance of time is clear, if the preceding inferences be true. It is also evident that some individuals have accumulated it more rapidly than others, if all alike originated from the simplest forms known to us. Multitudes have remained in the earliest stages (*Protozoa*) of the whole series, or of their own special series (*Lingula*), forming "persistent types"; or taken directions which rendered them incapable of expansion beyond a certain point without exhaustion or death; for example, complicated types, as *Ammonitidæ*. The quadrumanous animal, which was the progenitor of man, may thus be believed to have acquired a higher capacity of this accumulation than his contemporaries.

Assuming the nucleated cell to be the ultimate element of organic tissue, there are two types of life in which grade-influence has not appeared, viz.: unicellular animals and plants, and living forms composed of homogeneous protoplasm. In the latter neither grade-influence nor animal growth-force is potential; in the former, simple growth-force only. It is therefore apparent that grade-influence has been developed in the organism itself; perhaps this may have been in the plant, through the modified influence of external physical causes; in the animal, if our inductions as to use and effort be true, under the influence of the activities of the parent, which determined a structural change either in itself or in its offspring. The possibilities of this origin are considered in the next section.

δ. The *Location of Growth-Force* proceeds under the direction of what Prof. Henry calls "Vital Influence." With this author I discard the use of the term "Vital Force," what was originally understood by that term being a complex of distinct ideas. The *Vital forces* are (nerve-force) *Neurism*, (growth-force) *Bathmism*, and (thought-force) *Phrenism*.* All these are supposed to be correlated to the Physical Forces, but are under direction and control of the *Vital principle* which locates their action, etc., just as molecular or *atomic constitution* determines the locality and character of the physical forces. The laws of the vital principle and of atomic constitution also determine the nature of the conversion of one force into another. Now, since physical and vital forces are correlated and convertible, the close relationship of the two controlling principles becomes obvious and suggestive of their identity.

---

* The objection of President Barnard to thought being an exhibition of a force, is that "thought can not be measured." This objection does not take into consideration the two-fold nature of thought. The *amount* of thought can most assuredly be measured, the *quality* of the thought can not.

Dr. Carpenter, in describing the correlation of physical and vital forces, defines the difference of organic species to be similar to that prevailing between different chemical bodies (the latter depending on different molecular and atomic constitution), which leads them "*to behave* differently" from each other under similar circumstances. This may be more fully expressed by saying that different species possess different capacities for the location of the conversion of the physical forces into growth-force. A "descent with modifications" contemplated by a process of evolution, signifies a progressive change in this capacity. *Acceleration* means an increase in this capacity; *retardation* a diminution of it. *Grade-influence* means the influence which produces this change of capacity.

B. The Origin of Grade-Influence.

Living protoplasm can convert heat and nutriment into growth-force without the agency of the nervous system. This is proved by the nutrition of the *Protozoa* and *Cœlenterata,* and from experiments on the muscles of frogs, etc. In the latter case, as is well known, the nerve may be divided, and the muscle retain its size if a current of electricity be passed through it, thus sustaining the nutrition. As the presence and structure of the nervous system is in relation to the specialization of animal structure in other respects, it is very probable that the nervous system is in *higher* animals the agent of the *location* of growth-force. In the lowest it is not affected by any such means. As the nervous system is the instrument of the metaphysical peculiarities of the animal (emotions, choice, etc.), we may conclude that, in the lower animals, location of growth-force is influenced by necessity without choice; in the higher by necessity with choice.

The impulses derived from the nervous system, it is known, may be reflex or automatic in answer to stimuli from without. They may become so, also, after having been originated consciously or by effort of will. In the case of habits, frequent exercise of choice has so impressed the nervous system as to result in its repetition of effort, often in opposition to changed choice.

The influence of effort in muscular action, through the nervous system, appears to be, first, to enable it to convert heat to nerve-force, and, then, to conduct nerve-force to the involuntary muscles, or those controlling circulation, where it is converted into motion, which thus controls nutrition through circulation. The nervous system, like others, develops in capacity with use, hence probably nerve-tissue converts heat * into nerve-force as muscular tissue converts heat into motion. In other words, by repetition, the capacity of the nervous system for this conversion of heat is known to increase. As the amount of heat converted is in proportion to the amount of appropriate nerve-tissue (see above), it is evident that use and effort increase the amount of nerve-tissue.

The phenomena of thought render the same modification of structure probable. Effort in the direction of thought is supposed to convert heat into thought-force. Inasmuch as the more intelligent animals possess the highest development of cerebral hemispheres, it is highly probable that brain substance converts heat into growth-force also, which produces tissue of its own kind precisely as muscle does.

As different parts of the nervous centers subserve different purposes, the development of these parts must proceed approximately under the influence of special kinds of effort and use. Where, as in the adult, heat is converted into growth-force in the tissues to a very limited extent, if the above principles be true, the conversion of heat by the nervous system into nerve growth-force and tissue is, on the other hand, not terminated.

Capacity for effecting conversion of force is regarded, as above pointed out, as dependent on molecular constitution. Hence we conclude that change in that capacity on the part of the nervous system involves a molecular change in its constitution.

Now, we know physical and metaphysical peculiarities of parents to be inherited by offspring ; hence, no doubt, the nervous structure determinative of growth-force is inherited. This will then control the localities of special conversion of heat, etc. (from the mother), into growth-force, in accordance with the structure of the parent, and the more decidedly, as its own increase progresses.

The result will be *acceleration,* or construction of tissues and organs in excess of those of the parent, if the effort or use devoted to a nerve or organ be represented in the nerve-center of the parent by a greater amount of force-converting tissue than is

* The supposition that this and other forms of vital energy are derived from metamorphosis of heat, was in accordance with views held at the time this essay was written (1871). The much more probable hypothesis is that such energy is derived from retrograde metamorphosis of nutritive material. (Ed. 1886.)

necessary when inherited in the fœtus for the construction (by conversion) of tissues and organs like those of the parent.

That this is a partial explanation of inheritance, is rendered probable from the fact that the types of structure presented by the nervous centers express the grade of the animals possessing them far more nearly than those of any other organ or set of organs. If the brain, like other organs, develops by intelligent use, it can not be doubted that this relation of its development to grade is not accidental, but that grade-structure * is an expression of its capacities, physical and mental.

V. ON INTELLIGENT SELECTION.

As neither use nor effort can be ascribed to plants, and as we know that their life history is much more dependent on their surroundings than is that of animals, we naturally look to the physical and chemical causes as having a prime influence in the origination of their type-structures. Without greater familiarity with the subject, I will not attempt to say how far the various degrees of growth-force possessed by parent plants, located under the influence of meteoric and other surroundings, and preserved, destroyed or restricted by natural selection, may account for the characters of their successors of the present period. But other agencies similar to use, that is, automatic movements, may be also introduced as an element in the argument. The movements of tendrils seeking for support may be here considered, and, as Dr. Asa Gray has pointed out, have consequences similar to those of use in animals. When the tendril seizes a support, growth-force is located at the point of contact, for the tendril increases considerably in thickness.

Among animals of the lowest grade, movement must be quite similar to those of plants, or automatic from the start, and not even at the beginning under the influence of will. Evidence of will is, however, soon seen in the determinate movements of many of the Protozoa in the seizing of food. With *will* necessarily appears a *power of choice*, however limited in its lowest exhibitions, by the lack of suggestive metaphysical qualities, or the fewness of alternatives of action presented by surrounding circumstances, to animals of low and simple organism. We can, however, believe

* Grade growth-force is not regarded here or elsewhere as a simple form of energy, but as a class of energies, which are the resultants of the interference of mind (i. e., consciousness) with simple growth-force. (Ed. 1886.)

that the presence of greater or less number of external facilities for action characterize different situations on the earth's surface, as well as that greater and less metaphysical capacity for perceiving and taking advantage of them must exist in different individuals of every species of animal, however low, which possesses consciousness and will. These qualities must, of course, influence effort and use to the advantage of the animal, or the reverse.

Effort and use have very various immediate stimuli to their exertion.

Use of a part by an animal is either compulsory or optional. In either case, the use may be followed by an increase of nutrition under the influence of reflex action or of direct volition.

A compulsory use would naturally occur in new situations which take place apart from the control of the animal, where no alternatives are presented. Such a case would arise in a submergence of land where land-animals might be imprisoned on an island or in swamps surrounded by water, and compelled to assume a more or less aquatic life. Another case, which has also probably often occurred, would be when the enemies of a species should so increase as to compel a large number of the latter to combat who had previously escaped it.

In these cases, the structure produced would be necessarily adaptive. But the effect would sometimes be to destroy or injure the animals (retard them) thus brought into new situations and compelled to an additional struggle for existence, as has, no doubt, been the case in geologic history.

Direct compulsion would also exist where alternatives should be presented by nature, but of which the animal would not be sufficiently intelligent to take advantage.

Most situations in the struggle for existence afford alternatives, and the most intelligent individuals of a species will take advantage of those most beneficial. Nevertheless, it is scarcely conceivable that any change or increase of effort, or use, could take place apart from compulsion derived from the relation of external circumstances, as a more or less remote cause.

Preservation, with modifications, would most probably ensue when change of stimulus should occur gradually, though change of structure might occur abruptly, under the law of *expression points*.*

* See "Origin of Genera," p. 38.

Choice is influence not only by *intelligence*, but by the *imagination* and by the *emotions*.

Intelligence is a conservative principle, and always will direct effort and use into lines which will be beneficial to its possessor. Here we have the source of the fittest—i. e., addition of parts by increase and location of growth-force, directed by the will—the will being under the influence of various kinds of compulsion in the lower, and intelligent option among higher animals.

Thus, intelligent choice, taking advantage of the successive evolution of physical conditions, may be regarded as the *originator of the fittest*, while natural selection is the tribunal to which all the results of accelerated growth are submitted. This preserves or destroys them, and determines the new points of departure on which accelerated growth shall build.

The influences locating growth-force may be tabulated as follows:

DIVISION.	INFLUENCE.			
Plants.	Physical and chemical. + ? ?			
Plants with mechanical movements; animals with indeterminate movements.	"	+ use		
Animals with determinate movements or will,* but no intelligence.	"	"	+ effort under compulsion.	
Animals with will and less intelligence.	"	"	" "	+ choice.
Animals with more intelligence.	"	"	" "	+ intelligent choice.

As examples of intelligent selection, the modified organisms of the varieties of bees and ants must be regarded as striking cases. Had all in the hive or hill been modified alike, all soldiers, neuters, etc., the origin of the structures might have been thought to be compulsory; but varied and adapted as the different forms are to the wants of a community, the influence of intelligence is too obvious to be denied.

The selection of food offers an opportunity for the exercise of

---

* Will must be here regarded as the answer to stimulus, without any element of freedom. See Article XX, on the "Origin of the Will." (Ed. 1886.)

intelligence, and the adoption of means for obtaining it still greater ones. It is here that intelligent selection proves its supremacy as a guide of use, and consequently of structure, to all the other agencies here proposed. The preference for vegetable or for animal food determined by the choice of individual animals among the omnivores, which were, no doubt, according to the paleontological record, the predecessors of our herbivores, and perhaps of carnivores also, must have determined their course of life and thus all their parts, into these totally distinct directions. The choice of food under ground, on the ground, or in the trees would necessarily direct the uses of organs in the appropriate directions respectively.

In the selection of means of defense a minor range of choice is presented. The choice must be limited to the highest capabilities of the animal, since in defense, these will, as a general thing, be put forth. This will, however, not be necessarily the case, but will depend in some measure on the intelligence of the animal, as we readily observe in the case of domesticated species.

In the case of the rattlesnake, already cited, the habit of rapid vibration of the tail appears to me to be the result of choice, and not of compulsion. For the cobra of India, for the same purpose, expands the anterior ribs, forming a hood, which is a very different habit.* Here are two alternatives, from which choice might be made, and violent hissing is a third, which the species of the colubrine genus *Pityophis* have adopted to some purpose. As to the benefit of the rattle, it no doubt protects the animal from all foes other than man ; but is rather a disadvantage as regards the latter, being by a beautiful turn of events a protection to the higher animal.

On the principle of natural selection it might be supposed that the harmless snakes which imitate the Crotalus for the sake of defense were preserved ; but if the above explanation of the origin of the habit in the latter be true, the second explanation is not valid. (Since in time the harmless snakes preceded the rattlesnakes. Ed. 1886.)

The power of metachrosis, or of changing the color at will, by the expansion under nerve-influence of special pigment cells, exists in most *Reptilia*, *Batrachia*, and fishes. It is then easy to believe that free choice should, under certain circumstances, so habitually

* The North American Heterodons possess a similar habit. (Ed. 1886.)

avoid one or another color as to result finally in a loss of the power to produce it.

Thus, it appears to be a fact that not only are species of fishes which dwell in the mud of darker hues than those that inhabit clear water, but that individuals of the same species differ in a similar manner in relation to their habitats; those that live in impure or muddy waters having darker tints than those of clear streams.

Land animals present equally abundant and remarkable imitations of the objects or substances on which they live. This is well known in insects and spiders, which look like sticks or leaves, or the flowers on which they feed. It is seen in reptiles, which in very many cases can voluntarily assume the hue of leaf, stone, or bark, or have constantly the gray color of their native desert sands.

These cases are largely selective or optional in their origin, for though metachrosis is also induced by some external stimulus, as an enemy or a food animal, yet other means of escaping the one and procuring the other are generally open.

These facts pave the way for a consideration of the phenomenon of mimetic analogy, which, though well known to naturalists, may be illustrated by the following new facts:

On the plains of Kansas there is a species of *Mutilla* whose abdomen and thorax are colored ochraceous, or brown-yellow, above. A spider of the genus *Salticus* is equally abundant, and is almost precisely similar in the color of the upper surfaces, so much so as to deceive any but a most careful observer. The *Mutilla* being a well-armed insect, and a severe stinger, there can be no doubt that the *Salticus* derives considerable immunity from enemies from its resemblance.

On the same plains, the *Crotalus confluentus*, or prairie rattlesnake, abounds. It is an olive gray, with a series of transverse brown dorsal spots, and two rows of smaller lateral ones. The head exhibits a number of brown and white bands. The prairie *Heterodon* (*H. nasicus*) possesses not only the same tints but the same pattern of coloration, and at a short distance can not be distinguished from it.

In consequence, as one may justly say, this species is, with the rattlesnake, the most common serpent of the plains, as it shares, no doubt, in the protection which the armature of the *Caudisona* gives its possessor. This is in accordance with the views of Wallace and Bates.

A curious case occurred to me in four species of fishes, which I took in a small tributary of the Yadkin River, in Roane County, N. C. Among several others, there were varieties of the widely distributed species *Chaenobryttus viridis*, *Hypsilepis analostanus*, and *Ptychostomus pidiensis* (each representing a different family), which differ from the typical form of each in the same manner, viz.: in having the back and upper part of the sides with longitudinal black lines, produced by a line along the middle of each scale. This peculiarity I have not observed in these species from any other locality. Until I had examined them I thought them new species.

The only other species presenting such marking in the Yadkin River is the large perch, the *Roccus lineatus*. According to the theory of natural selection a resemblance to this well-armed species might be of advantage to the much weaker species in question; yet the same species co-exist in other rivers without presenting the same mimicry.

It is difficult not to urge the importance of the causes already regarded as efficient in the origination of structure in the present branch of the subject also. We are especially disposed to call in use and effort here, after noticing how much more distinctly change of color is under the control of the animal, than change of shape. It must, however, be borne in mind that similar resemblances exist among plants; though, as Professor Dyer shows, a large majority of these cases occur in species of different floral regions. Thus in this case, as in those of structure already cited, we appeal first to physical laws in the lowest beings, but with the increasing interference of use, effort, and intelligence, as we rise in the scale. Thus it is that, in the Vertebrates generally, the mimetic resemblances are found in species of the *same region*, where only an intelligent or emotional agency could be illustrated. If among animals as low as butterflies the influence of intelligence be denied, that of admiration for the beauty, or fear of the armature, of the predominant species imitated, would appear to be sufficient to account for the result. Admiration and fear are possessed by animals of very low organization, and, with the instincts of hunger and reproduction, constitute the most intense metaphysical conditions of which they are capable. But our knowledge of this branch of the subject is less than it ought to be, for animals possess many mental attributes for which they get little credit.

It appears to be impossible to account for the highest illustrations of mimetic analogy in any other way, the supposition of Wallace that such forms must be spontaneously produced, and then preserved by natural selection, being no explanation. It has been shown by Bennett that the chances of such modification arising out of the many possibilities are exceedingly small.

If the above positions be true, we have here also the theory of the development of intelligence and of other metaphysical traits. In accordance with it, each trait appropriates from the material world the means of perpetuating its exhibitions by constructing its instruments. These react by furnishing increased means of exercise of these qualities, which have thus grown to their full expression in man.

VI.

A REVIEW OF THE MODERN DOCTRINE OF EVOLUTION.*

THE doctrine of evolution of organic types is sometimes appropriately called the doctrine of derivation, and its supporters, derivatists. This is because it teaches the derivation of species, genera, and other divisions, from pre-existent ones, by a process of modification in ordinary descent by reproduction. The opposite or creativist doctrine teaches that these forms were created as we see them to-day, or nearly so ; and that the natural divisions and species of organic beings have never been capable of change, the one into the other.

I. THE EVIDENCE FOR EVOLUTION.

The reasons which induce me to accept the derivatist doctrine, and to reject the creational, fall under the two heads of probabilities and conclusive evidence. The probabilities are cumulative in their pointings, and strengthen that part of the evidence which is, to my mind, conclusive. The reasons why derivation is probable are the successional relation of increment or decrement of structure, observed in :

1. Systematic relation (taxonomy) ; 2. Embryonic growth (embryology) ; 3. In geologic time (paleontology) ; 4. And in the coincidence in the successions seen in Nos. 1, 2, and 3.

The fact that it is necessary to arrange animals in an order corresponding with the phases of their embryonic history is remarkable ; but the further fact, shown by paleontology, that the same succession marked the ages of past time, at once brings evolution within the limits of strong probability. Nevertheless, all this might have been a mere system, without transitions between its

---

* Abstract of a lecture delivered before the California Academy of Sciences, Oct. 27, 1879.

members; organic types might have been created unchangeable, but presenting the mutual relations in question. But if transitions among these members can be shown to take place, then indeed the phenomena mentioned received a sufficient explanation. They are seen to be the necessary relations of the parts of a shifting scene of progression and retrogression; they express combinations of structure, which, though often long enduring, are, nevertheless, not perpetual, but give way to other combinations to be in their turn dissolved. Now, if there is anything well known in nature, it is that there are divisions of various ranks in the vegetable and animal kingdoms, whose contents present variations of structure which are confessedly additions to or subtractions from the characters of ancestors, which have appeared during ordinary descent. The protean species, genera, etc., are well known to biologists, and every naturalist who admits varieties, sub-species, sub-genera, etc., admits derivation so far as they are concerned. The facts of variation, including "sporting," etc., are notorious, not only among domesticated, but also in wild animals and plants. The facts have led some persons to suggest that species have been produced by evolution from a single specific center, but that the genus and other comprehensive divisions are unchangeable. But I think I have shown, in a paper entitled "The Origin of Genera,"* that the structural characters which define genera, and even higher divisions, are subjects of variation to as great an extent as are the less profound specific characters; and, moreover, that the evidence of derivation which they present is singularly clear and conclusive. The changes of both genus and species character are always of the nature of additions to or subtractions from those of one generation displayed by their descendants. As such, they form the closing chapters of the embryonic or growth-history of the modified generation.

In order to explain more fully the application of the above statements, I introduce a few examples selected from the subjects of my studies. Their number might be indefinitely extended. I first cite the genera of the tailless *Batrachia Anura* (frogs, toads, etc.), whose relations are very simple and clear, and show the parallelism between adult structure and embryonic succession. See above, 1 and 2.

The greater number of *Batrachia Anura* fall into two divis-

---

* Philadelphia, 1869. "Proceedings Academy Natural Sciences, 1868."

ions, which differ only in the structure of the lower portion of their scapular arch, or shoulder-girdle. In the one the opposite halves are capable of movements which contract or expand the capacity of the thorax; in the other the opposite halves abut against each other so as to be incapable of movement, thus preserving the size of the thoracic cavity. But during the early stages, the frogs of this division have the movable shoulder-girdle which characterizes those of the other division, the consolidation constituting a modification superadded in attaining maturity. Furthermore, young *Anura* are toothless, and one section of the species with embryonic shoulder-girdle never acquire teeth. So here we have a group which is imperfect in two points instead of one. This is the tribe *Bufoniformia;* the tribe with teeth and embryonic shoulder-girdle is called the *Arcifera,* and that which is advanced in both these respects is the *Raniformia.* Now the frogs of each of these divisions present nearly similar scales of development of another part of the skeleton, viz., the bones of the top of the skull. We find some in which one of these bones (ethmoid) is represented by cartilage only, and the frontoparietals and nasals are represented by only a narrow strip of bone each. In the next type the ethmoid is ossified; in the next, we have the frontoparietal completely ossified, and the nasals range from narrow strips to complete roofs; in the fourth station on the line, these bones are rough, with a hyperostosis of their surfaces; and in the next set of species this ossification fills the skin, which is thus no longer separable from the cranial bones; in the sixth form the ossification is extended so as to roof in the temporal muscles and inclose the orbits behind, while in the rare seventh

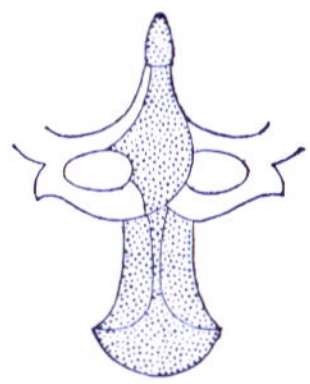

Fig. 8.

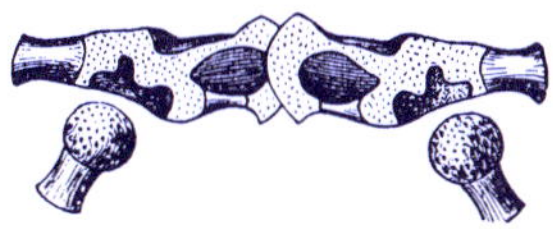

Fig. 9.

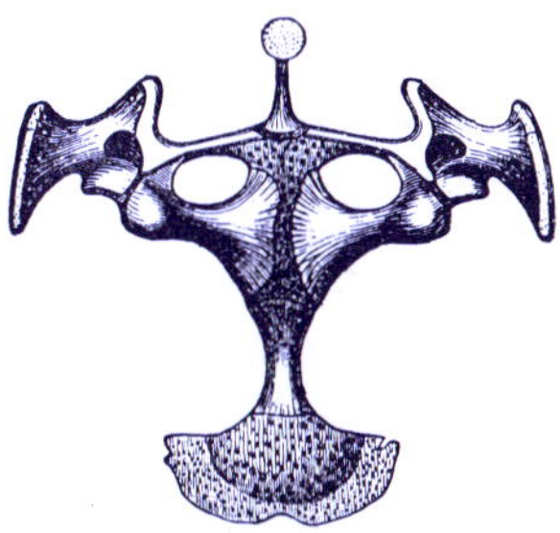

Fig. 10.

SHOULDER-GIRDLES OF "ANURA."

Fig. 8, of the Arciferous type (*Scaphiopus holbrooki*). Fig. 9, *Rana temporaria*, tadpole with budding limbs. Fig. 10, do. adult. Figs. 9 and 10 from Parker.

and last stage, the tympanum is also inclosed behind by bone. Now all of these types are not found in all of the families of the *Anura*, but the greater number of them are. Six principal families, four of which belong to the *Arcifera*, are named in the diagram below, and three or four others might have been added. I do not give the names of the genera which are defined as above described, referring to the explanation of the cuts for them, but indicate them by the numbers attached in the plate, which correspond to those of the definitions above given. A zero mark signifies the absence or non-discovery of a generic type.

	Sternum embryonic.				Sternum complete.
	Bufoniformia.	Arcifera.			Raniformia.
	Bufonidæ.	Scaphiopidæ and Pelobatidæ.	Cystignathidæ.	Hylidæ.	Ranidæ.
1—	1	0	1	1	0
2—	2	2	2	2	0
3—	3	0	3	3	3
4—	4	4	4	4	4
5—	5	5	0	5	5
6—	6	6	6	6	0
7—	7	0	0	0	0

It is evident, from what has preceded, that a perfecting of the shoulder-girdle in any of the species of the Bufoniform and Arciferous columns would place it in the series of *Raniformia*. An accession of teeth in a species of the division *Bufoniformia* would make it one of the *Arcifera*; while a small amount of change in the ossification of the bones of the skull would transfer a species from one to another of the generic stations represented by the numbers of the columns from one to seven.

There are few groups where this law of parallelism is so readily observed among contemporary types as the *Batrachia*, but it is none the less universal. The kind of parallelism usually observed is that in which there is only a partial resemblance between adults of certain animals and the young of others. This has been termed "inexact parallelism," and the relation is presented by forms not very nearly phylogenetically related. The more remote the phylogenetic lines of two types, the more "inexact" will their parallelism be. It was once a question whether any parallelism can be traced between the members of the five or six primary divisions of animals, and in my essay on the "Origin of Genera," I was compelled to state that there was then "no evidence of the community of origin of these divisions." Since that time, Haeckel has published his "Gastræa Theory." This is a grand generalization from the facts of embryology, which shows the community in type of the early stages of all animals, and the similarity of the phases which they present during a part of their larval life. The exceptions to this law which have been observed will probably be explained, as have been those which have been urged against the law of homologies in anatomy.

The paleontology of the *Batrachia anura* is largely unknown, so we must look elsewhere for proof of the truth of the fourth proposition, viz., that the successional relation in embryology corresponds with that shown by paleontology to have existed in geologic time.

For this purpose I select one of the most complete series known to paleontology; that of the camels or *Camelidæ*, whose remains are found abundantly in various parts of our country. The succession of the known genera is seen in the structure of the bones of the feet, and of the superior incisor and premolar teeth. The metatarsal and metacarpal bones are or are not co-ossified into a cannon bone; the first and second superior incisor teeth are present, rudimental or wanting, and the premolars number from four to one. The relations which these conditions bear

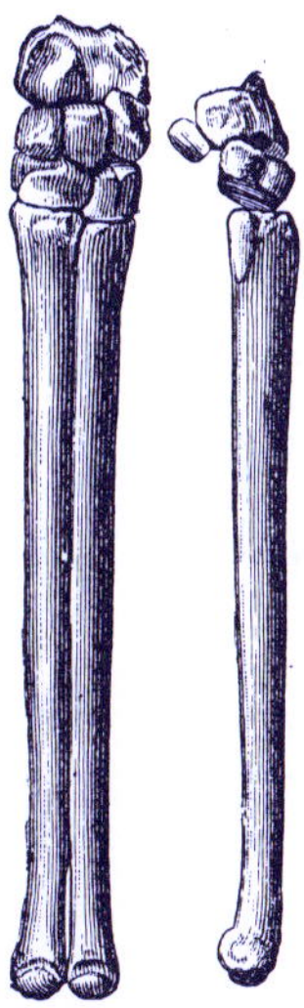

FIG. 11.

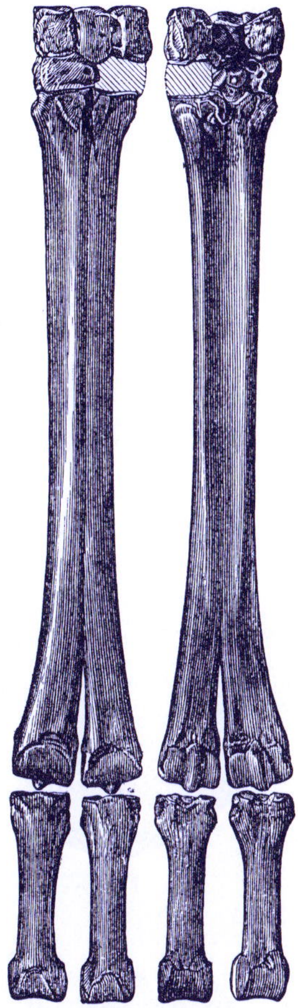

FIG. 12.

FIG. 11.—*Poëbrotherium vilsoni* carpus and metacarpus with end of radius, three fifths natural size. Original; from White River Miocene of Colorado.

FIG. 12.—Carpus and metacarpus of *Procamelus occidentalis*, about two fifths natural size. Original; from report of G. M. Wheeler, U. S. Expl. Surv. W. of 100th Mer., vol. iv, pt. ii. From New Mexico.

PLATE IV.

FIG. 1.

FIG. 2.

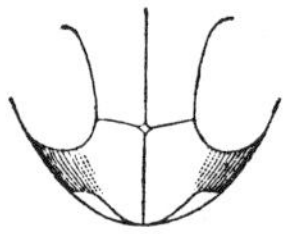

FIG. 3[3].

FIG. 3, wanting.

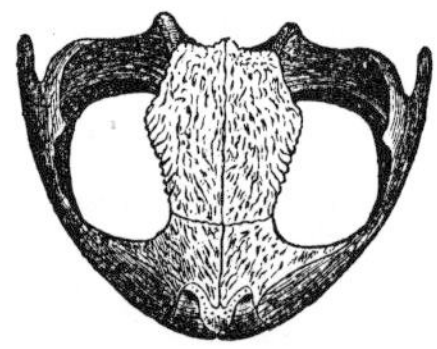

FIG. 5.

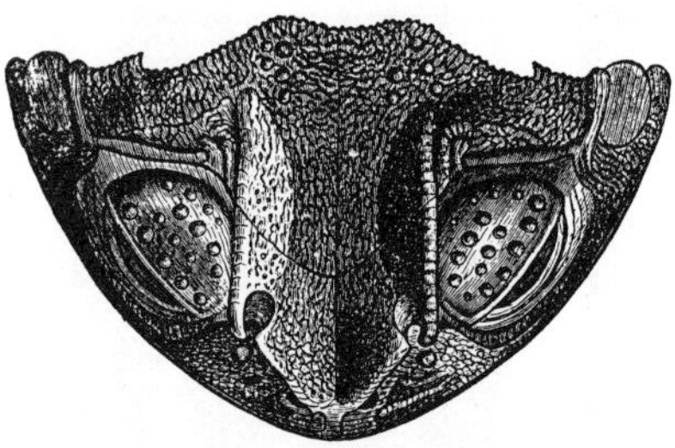

FIG. 6.

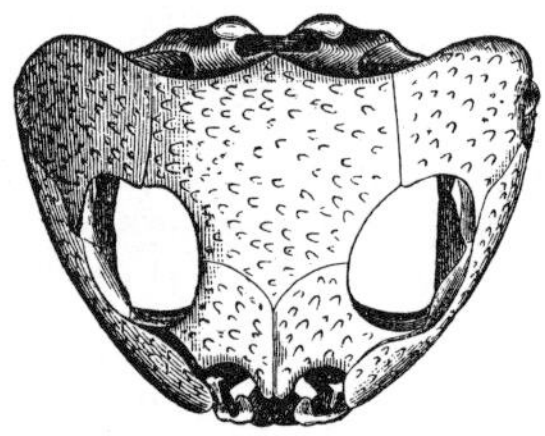

FIG. 6.

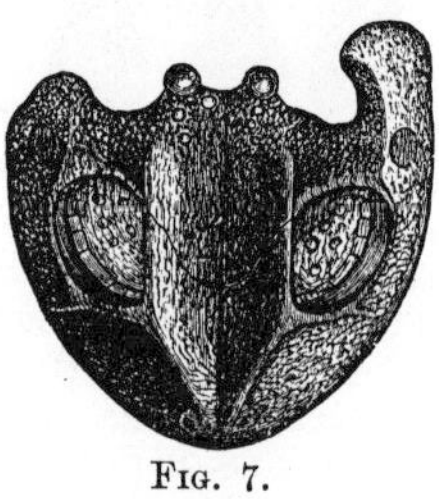

FIG. 7.

BUFONIDÆ.

FIG. 7, wanting.

SCAPHIOPIDÆ AND *PELOBATIDÆ.*

PLATE V.

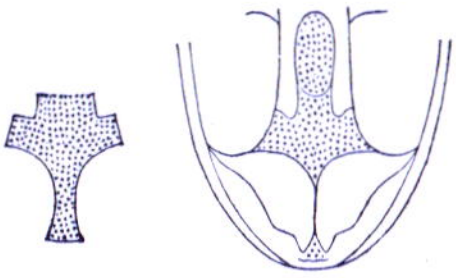

FIG. 1.

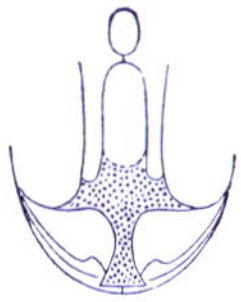

FIG. 1.

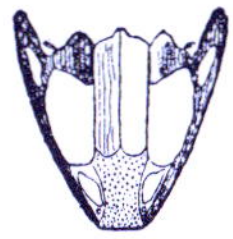

FIG. 3 1.

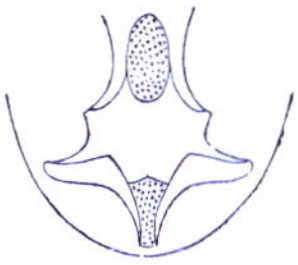

FIG. 2.

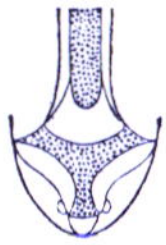

FIG. 2.

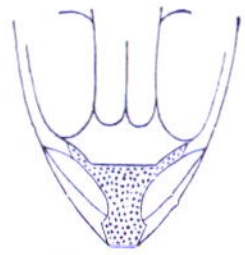

FIG. 3.

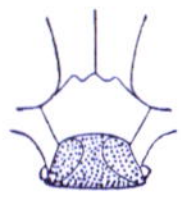

FIG. 3.

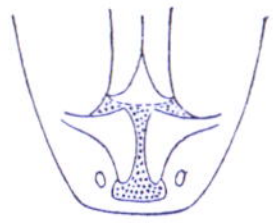

FIG. 31.

FIG. 21.

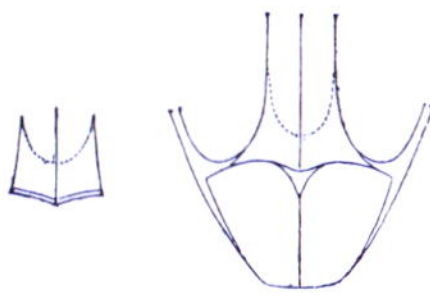

FIG. 33.

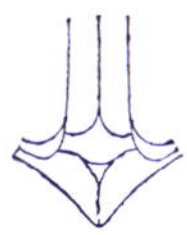

FIG. 32.

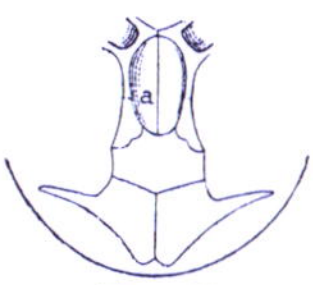

FIG. 32.

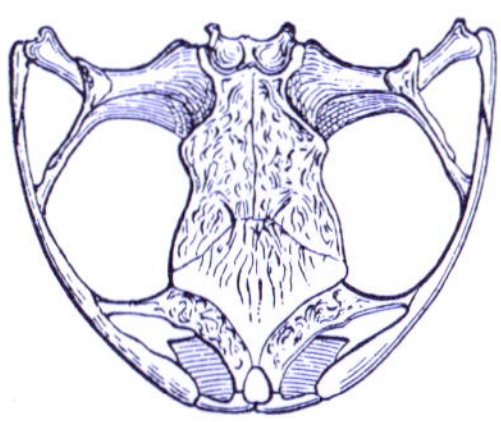

FIG. 4.

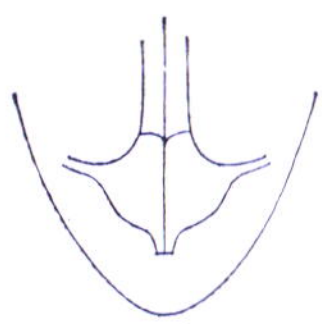

FIG. 33.

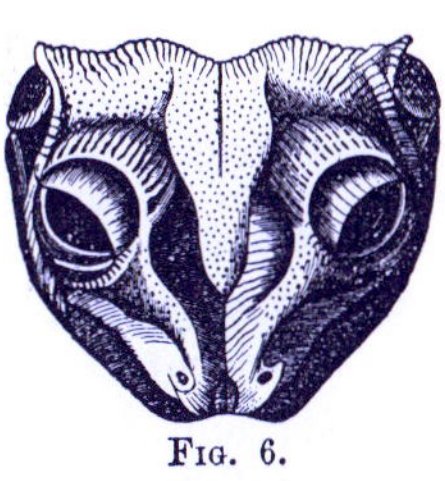

FIG. 6.

HYLIDÆ.

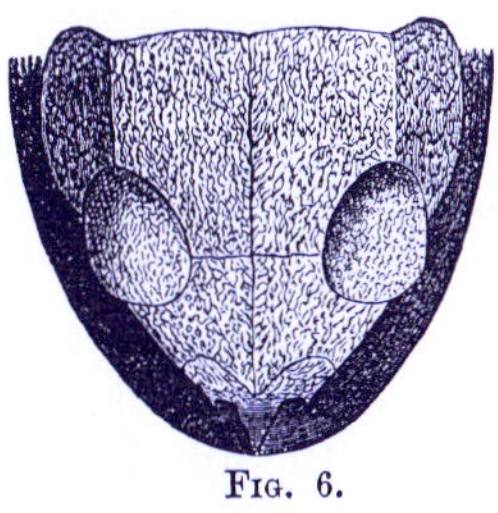

FIG. 6.

CYSTIGNATHIDÆ.

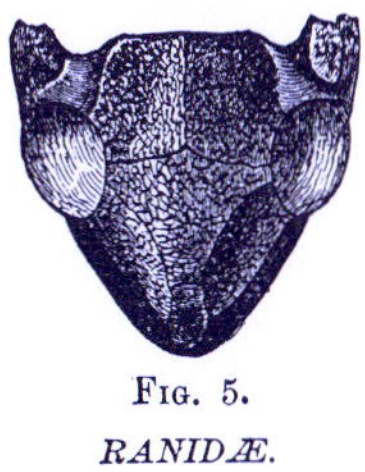

FIG. 5.

RANIDÆ.

EXPLANATION OF CUTS OF CRANIA OF ANURA.

The numbers in each column correspond with the types of ossification mentioned in the text, and are the same as those in the table of families given in the same connection. The power numbers attached to Fig. 3, represent the degree of ossification of the nasal bones, except the $^{-1}$, which signifies unossified ethmoid. Most of the cuts are original.

BUFONIDÆ.—Fig. 2, anterior part of skull of *Chelydobatrachus gouldi* Gray, from Australia. Fig. 3, do. of *Schismaderma carens* Smith, S. Africa. Fig. 6, top of head of *Peltaphryne peltacephala* D. and B., Cuba. Fig. 7, top of head of *Otaspis empusa* Cope, Cuba.

SCAPHIOPIDÆ AND PELOBATIDÆ.—Fig. 2, diagram of top of cranium of *Didocus calcaratus* Micahelles, Spain. Fig. 5, skull of *Scaphiopus holbrooki* Harl., United States. Fig. 6, skull of *Cultripes provincialis*, from France, after Dugès.

HYLIDÆ.—Fig. 1, *Thoropa miliaris* Spix., Brazil. Fig. 2, *Hypsiboas doumerci* D. and B., Surinam. Fig. 2^1, *Hypsiboas punctatus* Schn., Brazil. Fig. 3^2, *Scytopis venulosus* Daudin, Brazil. Fig. 6, *Trachycephalus geographicus* D. and B., Brazil, after Steindachner.

CYSTIGNATHIDÆ.—Fig. 1, *Eusophus nebulosus* Gir., Chili. Fig. 2, *Borborocœtes tasmaniensis* Gthr., Tasmania. Fig. 3, *Elosia nasus* Licht., Brazil. Fig. 3^3, *Hylodes oxyrhynchus* D. and B., W. Indies. Fig. 4, *Grypiscus umbrinus* Cope, Brazil. Fig. 6, *Calyptocephalus gayi* D. and B., Chili.

RANIDÆ.—Fig. 3^{-1}.—*Ranula chrysoprasina* Cope, Costa Rica. Fig. 3, *Rana oxyrhyncha* Sund., S. Africa. Fig. 3^1, *Rana clamata* Daud., N. America. Fig. 3^2, *Rana agilis* Thomas. Fig. 3^3, *Rana hexadactyla* Less., India. Fig. 4, *Polypedates quadrilineatus* D. and B., Ceylon.

---

to geologic time are displayed in the following table, commencing with the lowest horizon:

	No cannon bone.	Cannon bone present.			
	Incisor teeth present.		Incisors one and two wanting.		
		4 premolars.	3 prem'rs.	2 prem's.	1 prem'r.
Lower Miocene.	*Poëbrotherium.*				
Upper Miocene.		*Protolabis.* *Procamelus.*	*Pliauchenia.*		
Pliocene and Recent.				*Camelus.*	*Auchenia.*

This table shows that geological time has witnessed, in the history of the *Camelidæ*, the consolidation of the bones of the feet and a great reduction in the numbers of the incisor and premolar teeth. The embryonic history of these parts is as follows: In the 'œtal state all the *Ruminantia* (to which the camels belong) have ,he cannon bones divided as in *Poëbrotherium ;* they exhibit also ncisor teeth, as in that genus and *Protolabis*. Very young recent

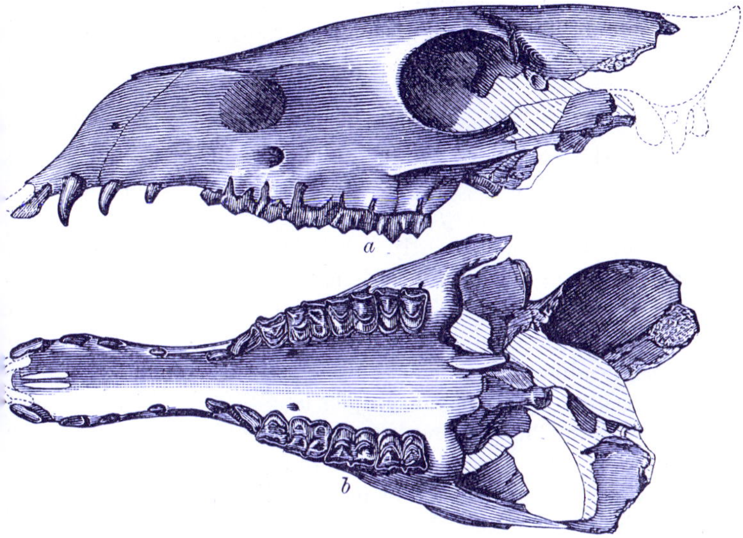

FIG. 13.—*Protolabis transmontanus* Cope, skull. *a*, in profile; *b*, from below, one hird (linear) natural size. Original; from vol. iv, Report U. S. Geol. Surv., under '. V. Hayden. Ticholeptus bed of Oregon.

·amels have the additional premolar of *Pliauchenia*. They shed his tooth at an early period, but very rarely a camel is found n which the tooth persists. The anterior premolar of the normal

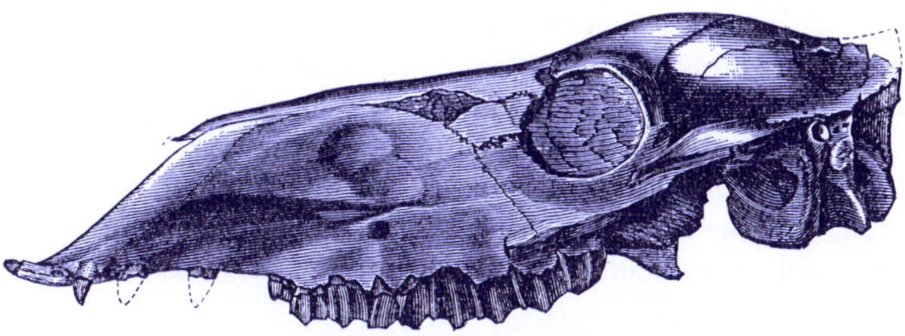

FIG. 14.—*Procamelus occidentalis* Leidy, profile of skull, about two sevenths natu-ıl size. Original; from report of Capt. G. M. Wheeler, U. S. Geol. Geog. Surv. W. ·f 100th Mer. Vol. iv, pt. ii. From Loup Fork bed of New Mexico.

'amelus is in like manner found in the young lama (*Auchenia*), ·ut is shed long before the animal attains maturity. I may add hat in some species of *Procame*... *s* caducous scales of enamel

and dentine in shallow cavities represent the incisive dentition o
Protolabis.

It remains to show that characters of the kind above men tioned are sometimes inconstant; that they may or may not ap pear in individuals of a species. Under such circumstances it i evident that their origin does not imply any break in the line o descent.

First, as to a family character. It is well known that the dee differ from the giraffes in the presence of a burr or ring of osseou excrescences surrounding the base of the horn. Now, in the ex tinct tertiary genus *Cosoryx* there are three species which posses or lack this burr indifferently. Why some individuals should and others should not possess it, is not known.*

Second, as to a generic character. The genus *Canis* (dog) i defined by the presence of two tubercular molars in the inferio series. The allied genus *Thous* possesses three such teeth, whil *Icticyon* has but one. Now examples of *Canis familiaris* (domes tic dog) with but one tubercular molar are not rare, while an in dividual with three is occasionally found.

To take another case. The normal dentition of *Homo* (man is, on each side, incisors, 2; canine, 1; premolars, 2; molars, 3 It is very common to find in the higher races, individuals wh have molars only two in one or both jaws; and the absence o the external incisors of the upper jaw is almost as frequently me with. Here we have two new generic variations in one and th same species.

In specific characters variations are most familiar. Thus, th young of deer are generally spotted, and the adults are nearl uniform in coloration. Some deer (as the *Axis*) retain the spotte coloration throughout life, while an occasional spotted individua of unicolor species is a violation of specific character by a failur to develop. The larvæ of some salamanders are of uniform col oration, and the adults spotted. The unicolor adults of the sam species, not uncommonly met with, present examples of the sam kind of variation.

Any biologist can select hundreds of similar cases from hi special department of study.

---

* The explanation I have offered is that the horn was stripped of its integument to the position of the burr by the animals in fighting. The condition of the speci mens renders this hypothesis probable. See "Report U. S. Expl. Surv. W. 100t Mer.," Capt. G. M. Wheeler, vol. iv, pt. ii.

II. THE LAWS OF EVOLUTION.

Having reviewed the reasons why the doctrine of evolution should be received as truth, I desire to give attention to the laws which may be made out by reference to its phenomena. Progress in this direction is difficult, owing to the natural impediments in the way of studying the history of the growth of living beings. We will, however, commence by examining more fully the phenomena with which we have to deal.

It is well understood that the world of animal life is a nicely adjusted equilibrium, maintained between each individual and its environment. This environment exerts forces both purely physical, and those exercised by other animals. Animals antagonize each other in procuring food, whether that food consist of vegetation or of other animals, but in the latter case the conflict is more severe. A similar competition exists among male animals in the matter of reproduction. These exhibitions of energy constitute the struggle for existence, which is the daily business of the living world. It is well understood that in this struggle the individuals best provided with means of self-preservation necessarily survive, while the weak in resources must disappear from the scene. Hence those which survive must display some especial fitness for existence under the circumstances of their environment, whatever they may be. So the "survival of the fittest" is believed to be a law of evolution, and the process by which it is brought about has been termed "natural selection." The works of Darwin and others have satisfied biologists that this is a *vera causa*.

Before the excellence of a machine can be tested, it must exist, and before man or nature selects the best, there must be at least two to choose from as alternatives. Furthermore, it is exceedingly improbable that the nicely adapted machinery of animals should have come into existence without the operation of causes leading directly to that end. The doctrines of "selection" and "survival" plainly do not reach the kernel of evolution, which is, as I have long since pointed out, the question of "the origin of the fittest." The omission of this problem from the discussion of evolution, is to leave Hamlet out of the play to which he has given the name. The law by which structures originate is one thing; those by which they are restricted, directed, or destroyed, is another thing.

There are two kinds of evolution, progressive and retrogressive; or, to use expressions more free from objection, by addition of parts, and by subtraction of parts. It is further evident that that animal which adds something to its structure which its parents did not possess, has grown more than they; while that which does not attain to all the characteristics of its ancestors has grown less than they. To express the change in the growth-history which constitutes the beginning of evolution, I have employed the terms "acceleration and retardation." Generally these expressions are literally exact, i. e., there is an increased rate of growth in evolution by addition, and a decreased rate in evolution by subtraction; but this is not always the case, for some divisions of animals have increased the length of their growth-period without reference to evolution in structure. The terms express the phenomena figuratively, where not exact in the sense of time, and I believe they are sufficiently clear. The origin of the fittest is then a result of either acceleration or retardation. It is easy to perceive that a character which makes its appearance in a parent before or near to the breeding season is likely to be transmitted to its descendants; so also a character which is lost near this time is likely to be wanting from the offspring. The causes of acceleration and retardation may next claim attention.

It is well known that the decomposition of the nutritive fluids within living animals gives rise, in the appropriate tissues, to exhibitions of different kinds of forces. These are, motion in all classes; heat in some only; in a still smaller number, electricity and light; in all, at certain times, growth-force or bathmism; in many, phrenism or mental or thought-force. These are all derived from equivalent amounts of chemical force which are liberated by the dissolution of protoplasm. This organic substance, consisting of CHON, undergoes retrograde metamorphosis, being resolved into the simpler CO_2, HO, etc., and necessarily liberates force in the process. None of the functions of animal life can be maintained without supplies of protoplasm. We have here to do with bathmism. It consists of the movement of material to, and its deposition in, certain definite portions of the growing egg, or fœtus, as the case may be. It is different in its movements in every species, and its direction is probably the resultant of a number of opposing strains. In the simplest animals its polar equilibrium is little disturbed, for these creatures consist of nearly globular masses of cells. As we ascend the scale a greater and more marked interference becomes apparent; radiated animals display energy in a number of radiating lines rather than in the spaces between them ; and in longitudinal animals, a longitudinal axis exceeds all others in extent and importance. In the highest animals its results are much more evident at one extremity of the axis (head) than at the other, and the diverging lines are reduced to four (the limbs). In each species the movements of this force are uniform and habitual, and it is evident that the habit is so deeply seated that only a very strong dynamic interference can modify or divert it. The interfering forces are probably all those transmissible through living tissue, and especially molar force. Thus every species has its own specific kind of bathmic force.

The characters of living beings are either adaptive or non-adaptive ; they are either machines especially fitted to meet the peculiarities of their environment, or they are not. Among the latter may be ranged rudimental structures and also many others of no sufficient use. They are all due either to excess or defect of growth-force ; they are either consequences of a removal of nutritive material to other portions of the body ; or they are due to an excess of such material which renders an organ or part useless through disproportionate size. Of the former class may be cited the absence of the tail in some monkeys and birds ; also of the teeth in some Cetaceans ; of the latter kind are the enormous tusks of the mammoth and the recurved superior canines of the babyrussa. The change of destination of this material has been probably due to the construction of adaptive machines whose perfection from time to time has required the use of larger and larger proportions of force and material.

In considering the origin of adaptive structures, two alternative propositions are presented to us. Did the occasion for its use follow the appearance of the structure, or did the need for the structure precede its appearance ? The following answer to the question has always been the most intelligible to me. Animals and plants are dependent for existence on their environment. It is an every-day experience that changes in environment occur without any preparation for them on the part of living things. If the changes are very great, death is the result. It is evident that the influence of environment is brought to bear on life as it is, or has been, and that special adaptations to it on their part must follow, not precede, changes of climate, topography, population, etc. We have another important consideration to add to this one, viz., the well-known influence of use, i. e., motion, on nutrition. Exercise of an organ determines nutritive material to it, and the nervous or other influence which does this, equally determines nutritive material to localities in the body to which an effort to move is directed, whether an executive organ exist there or not. The habit of effort or use determining the nutritive habit must be inherited, and result in the growing young, in additional structure. Change of structure, denied to the adult on account of its fixity, will be realized in the growing or plastic condition of fœtal or infant life. The two considerations here brought forward lead me to think that the cause of acceleration, in many adaptive structures, is environment alone, or environment producing movements, which in turn modify structure. The character of the stimulus in the successive grades of life may be expressed by the following table, passing from the lowest to the highest:

1. Passive or motionless beings:
   by climate and food only.
2. Movable beings:
   by climate, food and motion.
   By motion either:
   a, unconscious, or *
   aa, conscious, which is,
   b, reflex, or
   bb, directed by desire without ratiocination, or
   bbb, by desire directed by reason.

The only general rules as to the direct influence of motion on structure which can be laid down at present are two, viz., That density of tissue is in direct ratio to pressure, up to a certain point; † and that excess of growth-force, in a limited space, produces complications of the surfaces stimulated.‡ These and other laws, yet unknown, have probably led the changes expressed by evolution, while many others have followed the disturbance of equilibrium which they have produced.

I here allude incidentally to the question of transmission or inheritance. It has been maintained above that the bathmic force of each species is different from that of all other species. This force is characteristic of some unit of organization of living be-

---

* Movements coming under this head are often called reflex.

† See "Penn. Monthly," 1872; this work Art. I.

‡ "Method of Creation," Philadelphia, 1871.

ings; and this probably consists of several molecules. This unit has been termed, by Haeckel, the plastidule. The transmission of the bathmic force of one generation to another would be effected by the transmission of one or more living plastidules; and this is probably precisely what is accomplished in reproduction. The *Dynamic Theory* of reproduction I proposed in 1871,* and it has been since adopted by Haeckel under the name of perigenesis. I compared the transmission of bathmic force to that of the phenomenon of combustion, which is a force conversion transmitted from substance to substance by contact. The recent observations of Hertwig, Bütschli and others, confirm this view. The theory of pangenesis, devised to explain the phenomenon of reproduction, is to my mind quite inadequate.

III. METAPHYSICS OF EVOLUTION.

I enter here upon a wide field, over which I can only skim on an occasion like the present. The subject has been already introduced by reference to consciousness as modifying movement; of course, then, if movement modify structure, the latter is influenced by consciousness. The word consciousness was then, and is now, used in its simplest sense, viz.: as synonymous with physical sensibility. Its lowest and most usual exhibition is the sense of touch; the special senses, taste, sight, etc., are higher forms, while thoughts and desires are organized products of the same raw material. Consciousness can not be denied to many of the inferior animals; indeed, if we grant it to any, we must admit that it is displayed at times by even the lowest *Protozoön.* That these humble creatures should possess it, is apparently quite as probable as that the very similar bioplasts of the brain of man should be its seat.

Consciousness alone is not a sufficient basis for the development of mind. For this, one more element is necessary, and that is, *memory.* Impressions made by the environment are registered, and soon cease to be present in consciousness. Under the influence of association the impressions return to consciousness. Associations are those of place, of the order of time, and of similarity or difference in various qualities, as size, color, or any other physical features. Experiences of these qualities are to all conscious beings either painful, indifferent, or pleasurable. When associa-

---

* "Method of Creation," 1871.

tion requires, events, objects, or characteristics, are returned to consciousness in the order in which they cohere most firmly in the mind, which may or may not be that in which they entered it. The liking for or dislike to the object, are equivalent to an attraction to or repulsion from it. Thus experience is begotten : as its material increases, new combinations are formed, new relations observed, and in the highest types of mind, laws are discovered. No one can deny memory to animals ; it is the medium of their education by man, and has been as well the means of their education by nature. Impressions cause a re-arrangement of certain elements of structure which give the form to consciousness when it arises again. It is also probable that these arrangements are not the same as those which represent classifications and conclusions, but that nevertheless the arrangement or organization of these is determined by the simpler arrangements caused by perceptive stimuli. Experience produces these combinations in the bioplastic aggregations of all animals, be they in the form of ganglia, brains, or less specialized forms. Nowhere in the human organism are the effects of effort and use so strikingly witnessed as in the increase of brain power ; and familiarity with the education of the lower animals shows that this is the case with them also, though in a lesser degree than in man.

If, then, we grant the propositions, first, that effort and use modify structure ; and second, that effort and use are determined by mind in direct ratio to its development, we are led to the conclusion that evolution is an outgrowth of mind, and that mind is the parent of the forms of living nature. This is, however, to reverse a very usual evolutionary hypothesis, viz.: that mind is the product and highest development of the universe of matter and force. The contradiction is, however, not so absolute as at first appears. By mind, as the author of the organic world, I mean only the two elements, consciousness and memory. But it is the view of some thinkers that consciousness is a product ; that it is not only a correlative of force, but a kind of force. To the latter theory I can not subscribe ; when it becomes possible to metamorphose music into potatoes, mathematics into mountains, and natural history into brown paper, then we can identify consciousness with force. The nature of consciousness is such as to distinguish it from all other thinkable things, and it must be ranged with matter and force as the third element of the universe.

It is true that unconsciousness does not imply absence of life as generally understood. A majority of the processes of life are performed unconsciously by living creatures; mind itself being no exception to this rule. There is another class of acts whose performance produces sensation, but consciousness is not concerned in them as an immediate cause. Therefore, it is a common endeavor to associate reflex and unconscious acts with the molecular movements of inorganic and non-living substances. But the one great difficulty in making this identification has never been surmounted. This is the different nature of the movements in the two cases. In non-living matter they are simply polar, nothing more. In living beings they display design. Perhaps I use the word "design" in a new sense, but the expression is nevertheless appropriate. What I mean is, that the movements of living things have direct reference to consciousness, to the satisfaction of pleasures, and to the avoidance of pains. The *molecular* movements within animals of the simplest class are the digestion of food and the elaboration of the materials of reproduction. The *molar* movements of the simplest animals are to enable them to escape the pains of hunger and celibacy. Moreover there is reason why the movements of living beings display design. We all know the nature of habits; how they are performed unconsciously, and as automatically as digestion itself. But did any one ever know of a habit in an animal, whose origin he could trace, which has been formed in unconsciousness? According to our knowledge, habits are always the result of stimuli which are consciously felt, and which cause by repetition or through reminiscence a repetition of the resulting movement. After a sufficient number of repetitions such an act becomes a habit, i. e., is performed automatically, or without the intervention of effort, and frequently without consciousness. It thus becomes a part of the character of the individual or species. This common phenomenon is explained by the hypothesis that an organization of the centers controlling action is caused by the efforts of the animal under the stimulus, and that finally a machine is constructed which determines the nature of the force expended, without further mental exertion of the individual. Such a process is education, and the result is an addition to the stock of faculties already on hand. Thus is explained the vast number of automatic and unconscious activities displayed by animals; to the same source, I believe, the common reflex acts may be traced; it even appears to me probable that the organic functions in general have had the same origin.* While these latter have mostly long since passed beyond the control of the mind, portions of the uro-genital functions still linger within the confines of its jurisdiction. Thus have consciousness and mind endowed living nature with useful functions; and this, which may be called the *Theory of Endowment,* accounts for the element of design which is so puzzling when seen in unconscious and reflex acts.

As it has been maintained above, that structure is the effect of the control over matter exercised by mind, it is evident that the evolution of mind must be directly followed by corresponding development of organism. The science of paleontology substantiates this theory in a wonderful manner. But the animal mind being generally occupied with simple functions, its expressions in structure are usually nothing more than the progressive creation of improved instruments for obtaining food, resisting climate, escaping enemies, and reproducing their kind. The struggles of animals have been seen on this platform, and mind has only been necessary to aid in accomplishing the ends above mentioned. Wonderfully effective machines for grinding, cutting, seizing, and digging; for running, swimming, and flying have been produced. The development of mind proper must appear in the size and structure of the brain; and though the history of the latter in past ages must always remain, in large part, hidden from us, it is known that in the former respect there has been great progress made in various lines of animals. Now the line which has carried brain to its present development in man, the *Quadrumana,* has been deficient in special mechanical excellencies of the kind enumerated above. Perhaps primitive inferiority in these many respects has kept the *Quadrumana* under greater mental tension, and compelled them to exercise caution in their acts, and give that opportunity to thought which was less demanded in the case of other animals. Furthermore, if they are less specialized in their mechanism than most other *Mammalia,* they are less restricted by it to peculiar modes of life. They are more versatile, and more capable of the adoption of new habits as a consequence. And here we have a glimpse of a most important principle in evolution, which is the keynote to its method; this is what I have called the *The Doctrine of the Unspecialized.*

---

* "Consciousness in Evolution." "Penn Monthly," 1875.

Paleontology shows that the succession of living types has not been in a single straight line. It has been in many divergent lines, and a large number of them have not continued to the present time. The history of life has been well compared to a tree with divergent branches, many of which do not reach the elevation of the summit. Furthermore, in the many cases in which we can trace the lower lines to the present period, it is evident that in their present condition they could not have given rise to the higher forms. Each line, in fact, has developed to an extreme of specialization of structure, which it would seem is incapable of modification in any direction very divergent from that which it has already taken. Much less have such specialized types been able to survive the environment for which they were designed; with important changes in that respect they have perished. A few examples will serve to illustrate my meaning. The direction of development has been from fishes, through *Batrachia* and reptiles, to birds and mammals. But we can not derive any living type from the osseous fishes of the present or past ages (*Teleostomi*): to find the origin of *Batrachia*, we must pass below these to more generalized and older forms, the *Dipnoi*, a class whose position in the system was for years a controverted point. We can not obtain *Mammalia* from any of the existing types of reptiles, but we must go back to the Permian period, and trace their outlines in the *Theromorpha* of that day. In spite of the prophetic resemblance of these remarkable animals, they are inferior to later Reptilia in the structure of their vertebral column, and display resemblance to some of their immature stages, as well as to those of the *Mammalia*. Among mammals we can not derive monkeys from *Carnivora* or *Ungulata*, nor the latter from each other, but can only trace their close approximation in the Bunotherian types of the Lower Eocene. So with the great divisions of *Ungulata; Proboscidians, Hyrax*, and the even- and odd-toed orders must all be traced to the unspecialized *Amblypoda*, with small brains and five-toed plantigrade feet, as their ancestors.* It is easy to perceive that the generalization and plasticity of all these forms has furnished the ground of their ancestral relation.

We are now in a position to comprehend more clearly the general nature of evolution. The doctrine of the unspecialized

* See the origin of types of *Mammalia educabilia*, "Journal Academy," Philadelphia, 1874.

teaches that the perfection produced by each successive age has not been the source or parent of future perfection. The types which have displayed the most specialized mechanism have either passed away, or, undergoing no change, have witnessed the progress and ultimate supremacy of those which were once their inferiors. This is largely true of animals which have attained great bulk. Like those with perfected weapons, they have ever been superior to the attacks of other animals in their day, and doubtless led, so long as food abounded, lives of luxurious indolence. With change or diminution of food, such huge beasts would be the first to succumb, and it is a fact that no type of land animals has maintained great size through many geologic changes. It is true that all of the lines of ancestry of the existing higher *Mammalia,* as the subdivisions of the *Carnivora, Ungulata,* and *Quadrumana,* which we know in detail, commenced with types of small size and correspondingly little muscular power.

Some important conclusions may be derived from what has preceded. It seems that evolution has witnessed a continual running down of types to their great specialization or extinction. That many types have arisen in weak and small beginnings, but that the conflict with more powerful forms has developed some qualities in which they sooner or later excelled, and which formed the basis of their future superiority and persistence. That while this has probably been the true cause of the origin of the many admirable mechanical adaptations displayed by animals, it is preeminently true of the development of mind. That the reason why progress has reached its limit in the lines of greatest specialization, has probably been the removal of the occasion of its original cause, i. e., active exercise in the struggle for existence. This explanation is suggested by the remarkable degradation which is witnessed in animals whose mode of life relieves them from the necessity of working for a livelihood, e. g., the parasites and sessile animals whose young are free. Some of these creatures, on assuming their parasitic life, lose the semblance of even the order to which their young belong. The primary stages of various plants move actively through the water like the lowest forms of animals, and their sessile adult condition must be looked upon as a degeneration. It is well known that the endeavor to relegate the lowest forms of life to the two kingdoms of animal and vegetable, has been generally abandoned. The great vegetable kingdom probably exhibits a life degraded from more animal-like beginnings. Animal irritability and mobility have been lost, and their own consciousness must be entirely eliminated from the question of the origin of the many later and specialized types of plants. But I venture here the hypothesis that the consciousness of plant-using animals, as insects, has played a most important part in modifying the structure of the organs of fructification in the vegetable kingdom.* Certain it is that insects have been effective agents in the preservation of certain forms of plants. I would suggest whether the mutilations and strains they have for long periods inflicted on the flowering organs, may not, as in some similar cases in the animal kingdom, have *originated* peculiarities of structure.

Evolution of living types is then a succession of elevations of platforms on which succeeding ones have built. The history of one horizon of life is, that its own completion but prepares the way of a higher one, furnishing the latter with conditions of a still further development. Thus the vegetable kingdom died, so to speak, that the animal kingdom might live; having descended from an animal stage to subserve the function of food for animals. The successive types of animals have first stimulated the development of the most susceptible to the conflict of the struggle for existence, and afterward furnished them with food. Doubtless, in the occupation of the world's fields, the easiest and nearest at hand have been first occupied, and successively those which were more difficult. The digging animals are generally those which first abandoned the open field to more courageous or stronger rivals; and they remain to this day generally of low type compared with others of their classes (e. g., *Monotremata*, *Rodentia*, *Insectivora*). All occupations have been filled before that one which requires the greatest expenditure of energy, i. e., mental activity. But all other modes of life have fallen short of this one in giving the supremacy over nature.

Automatism then represents a condition of "lapsed intelligence" and diminished life. The unconscious automatism of animals is a condition of still greater lapse. On the contrary, sensibility is the condition of development, and the susceptibility and impressibility which is the extreme reverse of automatism is the especial character of youth. Here the "doctrine of the unspecialized" finds justification again.

---

* This opinion has also been expressed by Hermann Müller.

What the future has in store for us in the history of inorganic force and its results we can not now foresee, but I call attention in this connection to the important part played by life in the distribution of minerals. It has long been known that the carbon of the earth's crust was once in a living state, and it is admitted that the limestone once circulated in the fluids of animals. We have recently been compelled to believe that siliceous rocks are composed of the consolidated shells of minute plants, which they have elaborated from the water of the ocean. Silver and gold are segregated and deposited by seaweeds. The principal rock material, whose relation to life has not been ascertained, is alumina. How far the processes which now characterize dead matter were once related to life is a problem for the future.

IV. THE MORALS OF EVOLUTION.

The doctrines of the struggle for existence and survival of the fit in human life have a two-fold application. The relative proportions in which these applications are made will depend on the moral development of .him who makes them. Moral density and intellectual stupidity (often nearly allied) will see in these two laws only the struggle for material power, and the survival of the strongest. They will hardly urge in these days, as they would infallibly have done had they lived a few centuries ago, that the strongest means the hardest hitter, or the most successful assassin, but they will probably believe that this pre-eminent position belongs to the most wealthy. From a purely dynamical standpoint this position is correct, yet it might be a useful question for such advocates to consider why it is that physical oppression and assassination should be less successful avenues to power than they once were.

There are two reasons why man does not grant the first place in his esteem to physical force. The first principles of morals are acquired in the struggle for existence. The idea of *meum* and *tuum* was speedily developed so soon as men associated together; and the habit of justice has doubtless been formed by the insistence of every man on his own rights, and by the power of combinations of men to control those who may from superior strength or other cause seek to violate the rights of property. Thus law originated, and from the earliest history of the race to the present day it has educated the barbarous and semi-barbarous to civilization. It is then easy to perceive that man gives the highest place in his affections to the *most just;* but there is yet another reason why this should be the case.

The reproductive instinct in the lower animals has developed into social affections, and these form a part of the character of the higher animals and, in an especial degree, of man. The sentiments of sympathy and benevolence are probably outgrowths of the same. While the rational faculties are concerned in the *knowledge* of right, these sentiments are a source of the *love* of right. This disposition is trusted by men as leading to the *practice* of right, in cases where the power to enforce it is not immediately present. The struggle for existence then among men ranges all the way from a rivalry of physical force to a rivalry for the possession of human esteem and affection. The robber and assassin of the lowest human races are represented by the slanderer and defamer in the higher. The ultimate prosperity of the just, asserted and foretold by prophets and poets, is but a forecast of the doctrine of the survival of the fittest. The unjust are sooner or later eliminated by men from their society, either by death, seclusion, or ostracism.

But the organized moral qualities can not normally transcend in power, as motives of human action, those which secure man's physical preservation. Lines of men in whom the sympathetic and generous qualities predominate over the self-preservative, must inevitably become extinct. Evolution can produce no higher development of the race (whatever may sometimes appear in individuals) than an equivalency in these two classes of forces. Beyond this the organization of the social faculties of the brain must always be repressed in the race, so that we can only expect to attain an equilibrium between them and the more purely selfish ones, as the very highest result of unassisted evolution. In this position the judgment is suspended between the opposing classes of motives; and it must ever remain doubtful in general as to whether resulting action will be just and right, or the reverse. I exclude from this question those generous acts which do not appear to the actor to conflict with self-interest. These may be termed *sympathetic* acts, and are quite distinct from the altruistic.* The sympathetic actions are seen at times in most animals. The altruistic acts, on the other hand, are those that express what is usually called "moral principles." Such acts may often coincide

* "On the Origin of the Will." "Penn Monthly," 1877.

with the interest of the actor, but so long as they do not appear to him to do so, they are altruistic. It is part of the doctrine of evolution that habits will ultimately disappear on the removal of their stimulating cause. The moral nature originated, and has been maintained, through the pressure of the fear of consequences. The removal of this pressure, through the acquisition of power, would then ultimately result in the diminution or loss of the moral nature, through disuse. The abuses of power are well known. This appears to be all that evolution can do for us in the production of the moral nature. So it would appear that no organized faculty of *self-sufficient altruistic justice* can be derived by the process of mental evolution. The result is rather a continued struggle between justice and injustice. It is, then, evident that any power which shall cause the permanent predominance of the just over the selfish faculties must be derived from without.

After we omit from customary religion, cosmogony, which belongs to science, and theogony, which belongs to the imagination, we have left an art which has for its object the development and sustentation of good works or morals among men. If the teachers and professors of this art produce the results in this direction at which they aim, their great utility must be conceded by all. Their method has the advantage over that of the law, in being of the character of inducements supplied before action, instead of pains and penalties inflicted after action. They strive to *originate* good conduct, rather than to punish bad conduct. They are working on the side of the *originative* force in development, rather than the destructive; the "*origin of the fittest,*" rather than the "*survival of the fittest.*" Whether man possesses the spontaneous power called "free will" or not, the work of supplying inducements for good conduct is most useful to society. But religion, as generally understood, pre-supposes free will; and the definition of the word responsibility implies its existence. The question as to the presence of such a faculty is an interesting one, and will now be briefly considered.

The well-known doctrine of necessity leaves no place for free will. All acts are the consequences of motives, and are the outcome of a balancing of interests. The heaviest side of the account determines action. Our physical necessities supply the motives for most of our activities; our pursuit of food and clothing is of necessity, and no condition is free from it. Evolution supports and explains this doctrine, as can readily be perceived. It derives our instincts from an ancestry whose daily occupation has been their gratification. But it has been shown above that this development does not supply the motives of an independent morality.*

The direction of action under stimulus is determined by intelligence, which is, as has been above maintained, the product of experience. Intelligence is organized or classified knowledge, and directs the activities set on foot by the likes and dislikes, that is, the affections. *When there is knowledge, there is no necessity for spontaneous action or free will,* since action is determined by the organization of the mind. Even if the mind is conscious of insufficient knowledge, an inducement to seek knowledge is supplied, and according to the result of investigation will be the direction of knowledge.

But we are here brought to face the case where knowledge can not be or is not obtained. This is the condition of the two questions of the practice of morals, and the nature of the future life. The evolution of mind consists of a continual advance from the known into the unknown, and a transfer of the unknown to the known. So long as there is any inducement to progress of this kind, and nature responds to inquiry, development will go on. Although it is true that it is only among men, and but few men at that, that the pursuit of knowledge is an occupation, most men add to their stock incidentally as they pursue other avocations. The knowledge of right and the inducements to its practice are learned in their every-day intercourse, so far as it can be acquired. But knowledge in these directions soon attains its limit, and accordingly, development dependent on knowledge must cease. If any further progress in practical morals is to be made, some new force must intervene at this point.

Here is the opportunity for the appearance of will or spontaneity ; here it is at least needed. I am willing to believe that it may appear at this point, and that so long as we have to face the unknown in moral progress, so long it will remain. As a force it must be equivalent of other forces, but as a form of consciousness it is a new element of mind. As represented in new molecular organization, it may always continue, even after much of the unknown may have been conquered, and a stationary period may have ensued. Such an accession to character would be a fitting

* "On the Origin of the Will." "Penn Monthly," 1877.

crown of evolution, and a justification of this labor of the ages. If a true factor in human development, it might be compared, in the creation of character, to the apical bud of a growing tree. As the part pre-eminently living, it leads the growth of the trunk and branches. They all follow of necessity the path it has marked out. Under its lead they are successively formed, become fixed, and finally decay.

PART II.

THE STRUCTURAL EVIDENCE OF EVOLUTION.

VII.

ON THE HOMOLOGIES AND ORIGIN OF THE TYPES OF MOLAR TEETH OF THE MAMMALIA EDUCABILIA.

I. THE TYPES OF MAMMALIAN MOLARS.

It has been already stated * that the transition from simple to complex teeth is accomplished by repetition of the type of the former in different directions. "In the cetaceans this occurs in the Squalodonts; the cylindric incisors are followed by flattened ones, then by others grooved in the fang, and then by two-rooted, but never by double-crowned teeth. This is the result of antero-posterior repetitive acceleration of the simple cylindric dental type of the ordinary toothed cetacean. Another mode of dental complication is by lateral repetition. Thus the internal heel of the superior sectorial tooth of a carnivore is supported by a fang alongside of the usual posterior support of a premolar, and is the result of a repetitive effort of growth-force in a transverse direction. More complex teeth, as the tubercular molars, merely exhibit an additional lateral repetition, and sometimes additional longitudinal ones. As is well known, the four tubercles of the human molar commence as similar separated knobs on the [primitive] dental papilla."

Accordingly, the simple tubercle may be regarded as the least specialized form of tooth. It may be low and obtuse, as in the *Chiromys*, or the walrus; more elevated and conic, as in the dol-

---

* "Method of Creation," p. 10. Philadelphia, 1871.

phins; or truncate, as in sloths and some rodents. The form is complicated in two ways, viz., either by the folding of the sides, as in *Glyptodon* and many rodents, as *Arvicola, Castor, Lepus,* etc.; or by the development of tubercles on the crown, as in *Mus, Dicotyles, Homo,* etc. Upon this basis are constructed the more complex types of teeth exhibited by the various families of *Ungulata* and some *Rodentia*, as has been pointed out in the following language: "The genus *Uintatherium** has been shown to be a Proboscidian, which combines some important features of the *Perissodactyla* with those of its own order. . . . The number of such characters was shown to be somewhat increased in *Bathmodon,* which therefore stands still nearer to the common point of departure of the two orders. This point is to be found in types still nearer the clawed orders (*Unguiculata*) in the number of their digits (4, 5), and in which the transverse and longitudinal crests of the molar teeth are broken up into tubercles more or less connected, either type of dentition [i. e., Proboscidian or Perissodactyle] being derived according as such tubercles are expanded in the transverse or longitudinal directions." †

As is well known, the crowns of the superior molar teeth of the higher *Mammalia* are supported on three roots, two of which are external, and the third internal. The corresponding inferior molars are supported on two roots, and are therefore, in so far, less complex. But these two roots usually support four tubercles, two to each, while the roots of the superior molars support directly but one each. Hence, as Prof. Harrison Allen remarks, the crown of the inferior molar is more complex than that of the superior. ‡

In tracing the degrees of complication of the crowns of the superior molars of *Mammalia* from the simple cone of the Cetacean, the first step is seen in the *Squalodon* and teeth of similar character. In these there are two roots, antero-posteriorly related. In the Squalodont *Portheodon gervaisii* (*Squalodon* Auct.) the third root is present. In Mammalia with but three tubercles in the crown, e. g., *Carnivora,* the inner root has much the form of one of the external ones. In many of those with four tubercles of the crown, e. g., *Quadrumana,* the form of the internal root is

---

* *Eobasileus* in the original. (Ed. 1886.)

† "On the Primitive Types of Mammalia Educabilia," published by E. D. Cope, May 6, 1873. "Proceedings of the American Philosophical Society," 1873, p. 224.

‡ "Dental Cosmos," December, 1874.

not modified ; but in the Lophodont types it is materially altered. In *Hyracodon, Rhinocerus, Anchitherium, Oreodon, Cervus,* and others, it is flattened, with the long diameter anteroposterior, and is grooved on the inner face, so as to assume the form, as it does the function, of two roots conjoined. In the temporary dentition of *Sus, Merychyus,* etc., the inner root is divided, so that the superior molar has four distinct roots.

The proper homologizing of the various forms of dental structure of the Ungulates with each other, and with the primitive types of tubercular teeth, is entirely essential to their intelligent classification, and therefore comprehension of their mode of origin. In order to lay a foundation for this work, I define the four types as follows, giving the subdivisions of the first two in brief, and discussing those of the third and fourth more fully afterward :

Division I. *Haplodont* type ; the crown undivided or simple.

a. Crown low, obtuse; *Cetacea* (*Beluga*), *Carnivora* (*Rosmarus*).

b. Crown elevated, acute ; *Cetacea* (*Delphinus*) ; canine teeth in general.

c. Crown truncate ; *Edentata* (*Bradypus*) ; *Rodentia* (*Geomys, Dipodomys*).

FIG. 15.—*Globicephalus.*

Division II. *Ptychodont* type ; the crown folded on the sides ; the folds frequently crossing the crown.

a. Sides only folded ; *Rodentia* (*Arvicola Castor*) ; *Edentata* (*Glyptodon*).

b. Summit of crown also folded ; *Rodentia* (*Lepus, Chinchilla*).

FIG. 16.—*Jaculus hudsonicus.*

Division III. *Bunodont* type ; crown supporting tubercles.

a. Tubercles few, opposite ; *Ungulata, Achænodon, Dicotyles, Elotherium* ; *Carnivora, Procyon* ; *Rodentia, Heliscomys.*

b. Tubercles few, alternate ; *Hyopsodus.*

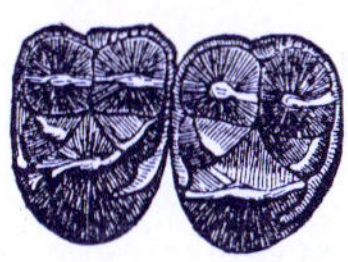

FIG. 17.—*Leptochœrus.*

c. Tubercles numerous, irregular ; *Mastodon, Phacochœrus.*

Division IV. *Lophodont* type ; the summit of the crowns thrown into folds of transverse or longitudinal direction. Higher Ungulates.

This division embraces the many types observed in the Ungulates, some Rodents, and possibly Carnivora. Inasmuch as the teeth of the maxillary and dentary (mandibular) bones do not always conform to the same type (e. g., *Symborodon, Equus*), it will be necessary to consider them separately. Besides the difference in type, they differ in their relative development in width in the more specialized forms; thus, in *Homo, Mus, Mastodon,* and such genera, the molars of both jaws are identical; in *Palæotherium, Eobasileus, Tapirus,* etc., and most *Carnivora,* the superior are the wider, the inferior narrower, appropriately to the greater slenderness of the mandibular bone. The latter, or anisognathous type, may then be regarded as the more specialized.

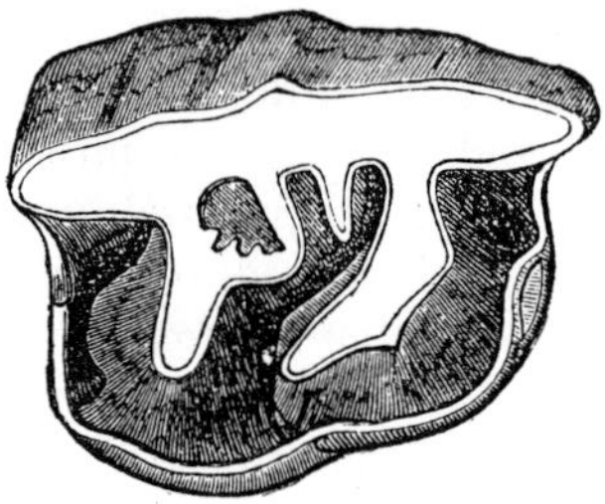

FIG. 18.—*Rhinocerus* (temporary).

The Bunodonts, except some *Carnivora,* are all of the former or isognathous type; among Lophodonts the few Rodents, the *Dinotheriidæ,* and Elephas are isognathous, while all of the other *Proboscidia,* the Perissodactyla, and *Ruminantia* are anisognathous. Examples may be selected as follows:

Isognathous; Bunodonts: *Homo, Dicotyles, Sus, Mastodon, Elotherium;* Lophodonts: *Dinotheriidæ, Elephas; Rodentia, Sciuridæ.*

Anisognathous; Bunodonts: *Macacus, Lemuroidea, Procyon;* Lophodonts: *Anoplotherium, Hyopotamus, Oreodon, Ruminantia, Perissodactyla, Bathmodon, Uintatherium.*

II. THE SUBORDINATE TYPES OF BUNODONTS.

1. *The Maxillary Teeth.**

It is apparent that the type of superior molar tooth which predominated during the Puerco epoch was triangular or tritubercular; that is, with two external and one internal tubercles.† Thus, of sixty-seven species of placental Mammalia of which the superior molars are known, all but four have three tubercles of the crown,

* This division is inserted from the "Proceedings of the American Philosophical Society," for December, 1883, for the sake of supplying an omission in the original memoir. (Ed. 1886.)

† See "American Naturalist," April, 1883, p. 407.

and of the remaining sixty-five, all are triangular, excepting those of three species of Periptychus, and three of Conoryctes, which have a small supplementary lobe on each side of the median principal inner tubercle.

This fact is important as indicating the mode of development of the various types of superior molar teeth, on which we have not heretofore had clear light. In the first place, this type of molar exists to-day only in the insectivorous and carnivorous Marsupialia; in the Creodonta, and the tubercular molars of such Carnivora as possess them (excepting the plantigrades). In the Ungulates its persistence is to be found in the molars of the Coryphodontidæ of the Wasatch, and Dinocerata of the Bridger Eocenes. In later epochs it is occasionally seen only in the last superior molar.

It is also evident that the quadritubercular molar is derived from the tritubercular by the addition of a lobe of the inner part of a cingulum of the posterior base of the crown. Transitional states are seen in some of the Periptychidæ (*Anisonchus*), and in the sectorials of the Procyonidæ.

2. *The Mandibular Teeth.*

The tritubercular or triangular superior molar is associated with a corresponding form of the anterior part of the inferior molar. This kind of inferior molar I have called the tubercular sectorial, and is very variable as to the degree of development of the sectorial cutting edge. The anterior triangle is formed by the connection by angle or crest, of the median and anterior internal crests with the anterior external. Its primitive form is seen in Didelphys, Pelycodus, Pantolambda and the Amblypoda generally; in Centetes and Talpa; and in its sectorial form, in Stypolophus and Oxyæna, etc.

The * tubercular molar of some *Viverridæ*, and among the extinct forms especially the *Didymictis protenus*, Cope, present a similar structure to that just described. This furnishes a ready explanation of the tooth immediately in advance, which is the primitive form of sectorial tooth characteristic of primitive Carnivora. The three anterior tubercles are largely developed, stand-

* The remainder of this section (2) is taken from the writer's "Synopsis of the Vertebrata of the Eocene of New Mexico," p. 800. Publication of the U. S. G. G. Survey, W. 100th Meridian, 1875. (Ed. 1886.)

ing at opposite angles of a triangular space; the outer and anterior cusps are the most elevated, and the ridge which connects them is now a cutting blade. The posterior portion of the tooth does not share in this elevation, and its two tubercles are in some genera obsolete, and in others replaced by an elevation of one margin, which leans obliquely toward the middle of the crown. In *Mesonyx* this is represented by a median longitudinal crest. If the two tubercles of the posterior part of this tooth (which may be termed a *tubercular sectorial*) are elevated and acute, we have the molar of many recent and extinct *Insectivora;* if the same portion (now called a *heel*) is much reduced, we have the type of *Oxyæna* and *Stypolophus.* In the *Canidæ* the three anterior tubercles are much less elevated than in the genera above named; the external is much the larger, and the anterior removed farther forward so as to give the blade a greater antero-posterior extent. The heel is large and without prominent tubercles. In the *Mustelidæ* the inner of the two median cusps is often reduced to a rudiment, or is entirely wanting, and the heel is large. The lower sectorial of the *Hyænidæ* has no inner tubercle, and the heel is much reduced. In some of the saber-toothed tigers the heel remains as a mere rudiment, while in the true cats it has entirely disappeared, and the carnassial tooth remains perfected by subtraction of parts, as a blade connecting two subequal cusps. The *Hyænodontidæ,* as is known, possess three carnassial teeth without inner tubercles. The history of this form is as yet uncertain, as it was evidently not derived from contemporary forms of the Eocene with tubercular sectorials.

The development of the inferior carnassial dentition has thus been accomplished by the subtraction of the inner and posterior cusps, so that of the original four of the quinquetuberculate molar but a *single one,* i. e., the anterior external, remains.

III. THE SUBORDINATE TYPES OF LOPHODONTS.

1. *The Maxillary Teeth.*

In the essay already quoted * the following remarks (page 7) explain the relation between the Bunodont genera and several of the Lophodont types of superior molar teeth: "In the superior

* "Primitive Types of Mammalia Educabilia," May, 1873, and Hayden's "Report on Geological Survey of Montana, Wyoming, etc.," 1873, p. 646.

molar series the flattening of the outer tubercles may proceed so far as to produce, on wearing, a confluence of the [resulting] crescentoid surfaces. . . . In both *Palæosyops* and *Hyrachyus* these tubercles of the upper molars are confluent into two Vs (more or less open when unworn). In the former the inner tubercles retain their primitive conic tubercular form, but in *Palæotherium, Rhinocerus, Lophiodon, Hyrachyus,* and *Tapirus* they elongate transversely so as to meet the corresponding outer tubercles (now crests) forming the familiar cross-crests of those genera. If alternate, the oblique crests of *Palæotherium ;* if opposite, the cross-crests of *Tapirus.* If, on the other hand, the inner tubercles flatten like the outer, we have, on wearing, the quadricrescentoid [Selenodont] type of the *Ruminantia* and *Anoplotherium.* In the Quadrumanous families, including man, the primitive quadrituberculate type of molars is preserved."

Four types of Lophodont dentition are included in the above discussion, and three others may be added. They belong to two series, viz., those in which the crests represent the modification of opposite tubercles, and those where the tubercular elements of the crown are alternate. These series may be called the *Amœbodont* (alternate), and *Antiodont* (opposite), and the component types are :

ANTIODONTS.—*Selenodont* (Ruminants) ; *Tapirodont* (*Tapir, Rhinocerus*) ; *Trichechodont* (Manati, Elephant).

AMŒBODONTS.—*Palæotheriodont* (Palæotherium, etc.) ; *Symborodont* (Palæosyops Symborodon, etc.) ; *Bathmodont* (Bathmodon, etc.) ; *Loxolophodont* (Uintatherium, etc.).

These types are defined as follows :

1. ANTIODONTS.

Selenodonts.—The tubercles separate or united at their angles, much elevated, narrow crescentic in section, separated by deep valleys.

To this group belong the molars of the Ruminants, the *Tragulidæ,* the *Oreodontidæ,* and *Merycopotamidæ ;* of Perissodactyles the *Equidæ.*

Tapirodonts.—The outer tubercles longitudinally compressed, subcrescentic in section ; the inner transversely compressed, continued as transverse ridges to the end or middle of the corresponding exterior crests.

Rhinocerus, Tapirus, Hyrachyus, Lophiodon, and *Hyrax* represent this type ; the last molar of *Lophiodon* appears to be *Amœbodont.*

Trichechodonts.—Tubercles confluent into two or more transverse crests.

The *Dinotherium* represents this form, so does *Elephas, Trichecus* (the manati), and a number of the larger *Marsupialia.*

2. Amœbodonts.

Palæotheriodonts.—External tubercles longitudinal, subcrescentic in section ; the inner united with them by transverse oblique crests.

Palæotherium and *Anchitherium* present this type, which only differs from the Tapirodont in the alternation of the opposing tubercles.

Symborodonts.—External tubercles longitudinally compressed and subcrescentic in section ; the inner independent and unaltered, i. e., conic.

To this group are to be referred the types of *Palæosyops, Menodus,* and *Symborodon.* They, of all *Lophodonts,* approach nearest to the *Bunodonts.*

Bathmodonts.—The posterior pair of tubercles approximated and connected, together compressed and subcrescentoid in section ; the anterior outer connected with the anterior inner by an oblique crest forming a V with the preceding.

Bathmodon and *Metalophodon.*—The homologies of the crests are difficult to make out ; the subcrescentic crest of the second molar may include only the posterior outer tubercle, and thus be entirely homologous with the posterior crescent of Palæosyops. In support of this view we have the structure of the premolars, where it becomes the only external crescent, while the anterior transverse crest turns round on its inner side, supporting the inner anterior tubercle of the tooth. [Note to this edition : The latter view is the correct one, as the posterior inner tubercle is wanting.]

Loxolophodonts—Anterior inner tubercle connected with the two external by oblique crests ; the posterior inner tubercle rudimental or wanting.

Uintatherium and probably *Tillotherium* represent this group, both being like those of the last, extinct genera from the Eocene of Wyoming.

2. *The Mandibular Teeth.*

The types of structure are less numerous than those of the maxillary teeth, since I am only acquainted with six. Still more distinctly than those of the upper jaw do they represent the types of opposite or alternating tubercles, or the antiodont and amœbodont. The essential principles of modification are the same as in the maxillaries, and they correspond with them as to genera, as follows :

Inferior.	Superior.
ANTIODONT.	
Selenodont.	Selenodont.
Hyracodont.	Tapirodont pt.
Trichecodont.	Tapirodont pt. / Trichecodont.
AMŒBODONT.	
Symborodont.	Palæotheriodont. / Symborodont.
Hippodont.	Selenodont pt.
Loxolophodont.	Bathmodont. / Loxolophodont.

The characteristics of these groups are as follows :

A. ANTIODONTS.

Selenodonts.—Constructed, typically, like the upper molars. *Ruminantia.*

Hyracodonts.—External tubercles compressed longitudinally and crescentic in section, continuous by a cross-crest with the corresponding tubercle of the inner side.

The animals which possess this type of teeth are the *Rhinocerus*, *Hyracodon*, and the *Hyrax ;* it is nearly approached by some of the *Hyrachyi*. It corresponds in structure with the true tapirodont arrangement of the maxillary teeth ; but many of the Tapirodonts have the Trichecodont type of mandibular teeth.

Trichecodonts.—Definition the same as for the maxillary teeth. *Tapirus, Lophiodon, Hyrachyus, Dinotherium, Elephas, Trichecus* (the manati), and the Kangaroos and their extinct allies belong here.

B. AMŒBODONTS.

Hippodonts.—In the horses the maxillary molars are constructed on an opposite crested basis, while the mandibulars represent an alternate crested type. This is not constituted as in the next form, by a union of alternating tubercles, but as in the *Selenodonts* by the special development of each crest into a crescent extended antero-posteriorly. As alternating, the inner crescents stand at the apices of the outer, and are connected with them. In *Anchitherium* the inner are so reduced as to constitute a condition intermediate between the *Hippodont* and *Symborodont* types.

Symborodonts. — The alternating tubercles connecting by oblique ridges which form together two Vs.

To this type is to be referred a great number of *Perissodactyles*, e. g., *Anchitherium*, *Palæotherium*, *Palæosyops*, *Menodus*, *Symborodon*, *Anchippodus*.

Loxolophodonts.—Alternate cusps connected by two cross ridges from the outside forward and one from the inside forward; from which result an oblique posterior cross ridge, and a V opening inward.

Here are *Eobasileus* and *Bathmodon*; the last molar of the latter having the anterior ridge of the V quite low.

3. Comparison of the Opposing Series.

In review, the above types of molar dentition may be classified as follows, with reference to the amount or complication of the modification of the tubercular type. The orders of *Ungulata* with which they correspond, are also given.

α. Both inner and outer tubercles crescentoid.

Selenodonts.	*Ruminantia.* (*Anoplotherium*, *Oreodon*, *Hyopotamus*). *Perissodactyla* (*Equidæ*).

β. External tubercles only crescentoid; the inner transverse or tubercular.

Tapirodonts. Palæotheriodonts. Symborodonts.	*Perissodactyla* in gen.

γ. Neither kind of tubercles crescentoid, but united in pairs.

Trichecodonts. *Proboscidia.*

It may be added that the groups arranged under *γ* are the only ones in which the types of crests of the superior and inferior molars are fundamentally simple and alike. Thus in the group *α*, tubercles of both upper and lower series are modified independently to produce the type; in group *β*, the tubercles of the upper series are modified independently of each other, while those of the inferior series unite, in order to produce the result; in division γ the tubercles of both jaws unite entirely across the crown, without any distinction between those of the outer and inner sides. Thus the molar type of dentition of the *Proboscidians* is the most generalized among the Lophodonts, resembling in this respect the type of construction of the feet.

IV. THE ORIGIN OF THE MODERN TYPES OF BUNODONT AND LOPHODONT DENTITION.

The four types of molar dentition, the Haplodont, Ptychodont, Bunodont, and Lophodont, are by no means sharply defined, but pass into each other by insensible gradations at many points. The addition of cusps to the simple cone has been accomplished by the appearance of folds at the base of the crown, which have increased in size so as to resemble the primitive cusp to which they are attached. As already pointed out, the first additional cusp in the transverse direction appears on the inner side of the first, or anterior cusp. The last one was apparently the inner posterior. I first consider the carnassial molars.

1. *The Carnassial Bunodonts.*

In tracing the appearance of the different types of superior sectorial in time, the following facts are patent: In the Eocene genera *Mesonyx, Amblyctonus, Oxyæna, Stypolophus,** etc., the inner tubercle is much more largely developed than in any existing *Carnivora*, with the exception of some of the *Viverridæ*. In all, the sectorial crowns are less efficient as shear-like cutters than in *Canidæ, Mustelidæ, Hyænidæ*, and *Felidæ*, on account of the shortness of the posterior blade and the slight compression of the anterior cone. On the other hand, in all of them the number of teeth presenting the carnassial or three-tubercled type is greater than in existing genera, the posterior upper molars in the latter being either narrowly tubercular or wanting. The genus *Hyænodon*, while agreeing with the others in this respect, differs from them in the great reduction of the inner cusp, which is obsolete even in the representative of the true carnassial tooth, although the inner root remains in all. Hence this type of dentition might be regarded as the most specialized of those of the *Carnivora*,

---

* In the original, the synonym *Prototomus* is used here.

but for the fact of the imperfection of the shear blades in all except the last. This is seen in the large and prominent anterior cusp and very short posterior blade.

In all recent and most Miocene *Carnivora,* the sectorials are reduced to one in each jaw, the exception in the latter period being the genus *Hyænodon,* which did not continue later. The posterior carnassials of the Eocene forms disappear, and in the most specialized recent families *Hyænidæ, Mustelidæ,* and *Felidæ* have scarcely any representatives. The shortening of the series appears in the premolars as well, until we have the formula of the *Felidæ,* P. m. 2 ; M. 2.

In the inferior dentition the same process may be observed in the successional modifications of the entire series. The Eocene forms of carnivora frequently display more numerous sectorial teeth (such as they are) than any of the existing families. The important change, which is clearly indicated, is the progressive extinction of the genera with numerous sectorial teeth, accompanying the increasing specialization of the sectorial tooth in the genera which remain. In other words, the numerous types of digitigrade carnivora which have survived are those developing but one sectorial tooth (whose earliest representative is *Didymictis*). The increased perfection of the sectorial tooth has been associated with a reduction in the number of other molars, first, posterior, then anterior to it, which reduction has been accompanied by an increased relative size of the sectorial. By this process concentration of the carnassial function has been gained, and increased robustness of the jaws, by progressive shortening. The slender form of the rami of the Eocene genera and *Hyænodon* are much less efficient in functional use than the stout jaws of existing *Mustelidæ, Hyænidæ,* and *Felidæ.*

2. *The Lophodonts.*

Transitions between the Bunodonts and Lophodonts are very obvious, so much so as to lead to the belief that the several subdivisions of the Lophodonts represent modifications of corresponding types of Bunodonts, and that the two are partially "homologous groups." Both present corresponding Amœbodont and Antiodont types ; as an example of the former kind of Bunodont, the mandibular molars of the genus *Hyopsodus* may be cited ; of the latter kind, the same of the genus *Achænodon,* both the earliest, or Eocene genera. It remains to indicate the intermediate forms, if any there be, which give color to the supposition that the various divisions of Lophodonts have descended from Bunodont predecessors. Here, then, I mention a fact of prime importance, i. e., that in America, at least, no Selenodonts are known from formations of older age than Miocene; while the greatest development of Bunodonts is in the beds of the next older epoch, the Eocene.

The special forms of Lophodonts may be separately considered as follows:

First, as to the opposite and alternate types, or the Antiodont and Amœbodont. They pass into each other by many intermediate conditions among the Bunodonts, as in *Notharctus*, *Limnotherium*, etc., of the Wyoming Eocene. There is reason, also, to believe that this has been the case with some of the Lophodonts after they had left the bunodont stage behind. Thus *Equus* is an antiodont as to its upper molars, but has been probably derived from *Palæotheriodont* ancestors, which are amœbodont; this is rendered especially probable by the fact that the mandibular teeth are of the amœbodont division (hippodont). It is also highly probable that the antiodont genus *Tapirus*, though so near to *Palæotherium*, was derived from an antiodont Bunodont. Hence, while the discrimination between opposite and alternate types is in some cases most radical, in others its importance is but slight.

I. Antiodonts: Bunodont type *Achænodon* (Fig. 19).

1. *Selenodont* type approximated by the bunodont *Hippopotamus*, where the tubercles are compressed, thus: the intervening

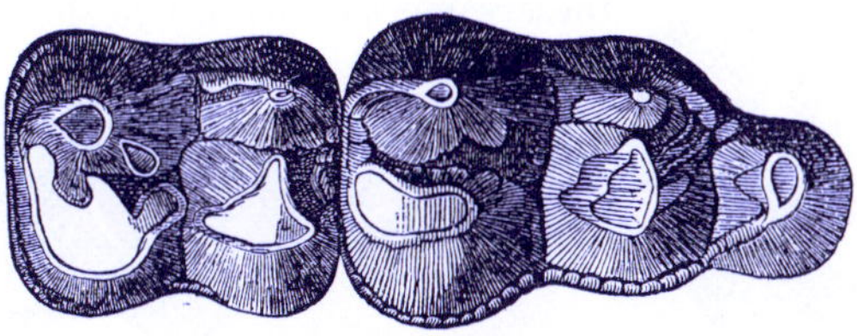

Fig. 19.—*Achænodon insolens.*

valleys are deepened, and the cusps wear readily into separate crescents. Another intermediate form is seen in the genus *Anthracotherium*, where the tubercles of the mandibular teeth are compressed, while they remain conic (Fig. 21); selenodont forms of *Omnivora* present us with near approaches to these Bunodont genera. Thus in *Hyopotamus* and *Anoplotherium*, the crowns, when unworn, present four principal tubercles, which are openly V-shaped in section, and which are separated by open valleys. The latter are deeper than those of the truly *Bunodont* genera, but much shallower than those of the typical Selenodonts. In *Oreodon* the valleys are somewhat deepened and the crescents elevated, while in the deer the same infolding is carried still further. In the *Cavicornia* the type reaches its fullest expression in the loss of the shoulder at the base of the crown, the great elongation of the latter, and correspondingly deep infolding of the terminal valleys.

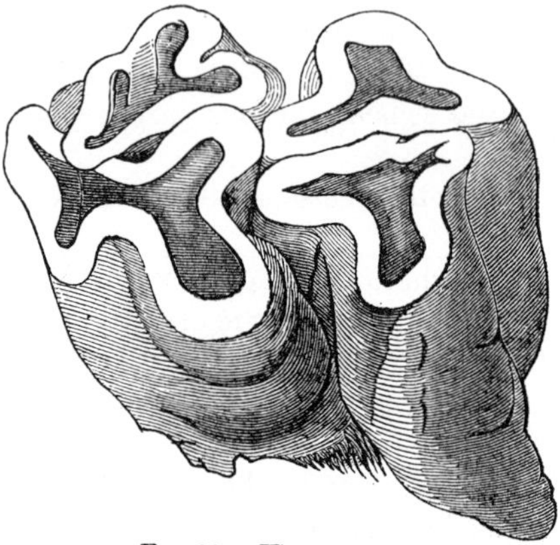

Fig. 20.—*Hippopotamus.*

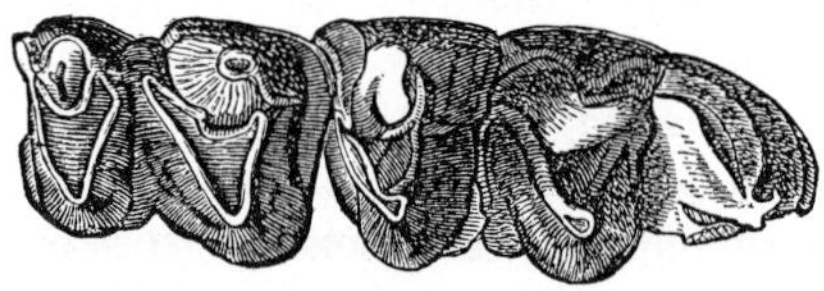

Fig. 21.—*Hyopotamus velaunus.*

Professor Lartet* states that the most ancient deer have very short-crowned molars, and the depressions on the surface are so shallow that the bottom is always visible, while in the *Cervidæ* of the more recent Tertiary periods, and especially the Plistocene and living species, these same cavities are so deep that whatever be the state of attrition, the bottom can not be seen. This, he says, is a perfectly reliable rule for distinguishing the ancient from the more modern forms of deer, and can be applied to other animals as well as the Cervidæ.

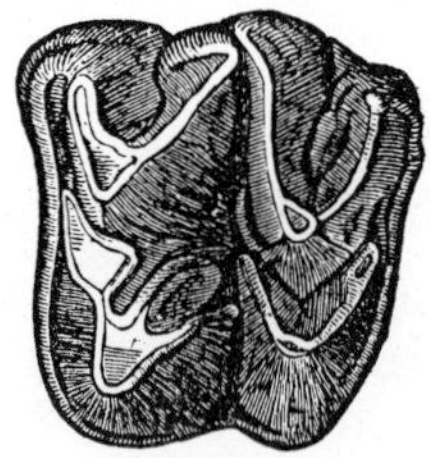

Fig. 22.—*Hyopotamus americanus.*

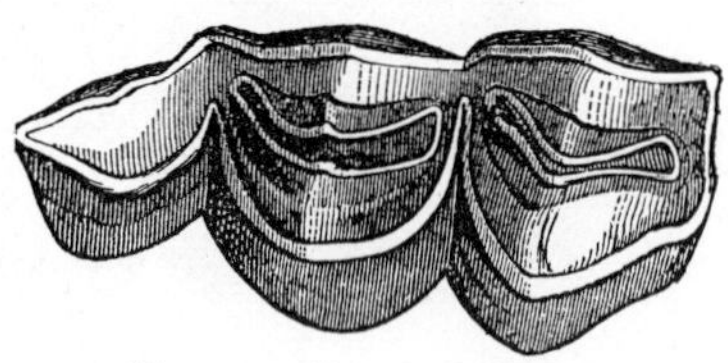

Fig. 23.—*Procamelus robustus.*

* "Comptes Rendus," 1868, p. 1119.

The writer nearly contemporaneously* recalled the observa-ons of Leidy that the teeth of the Oreodont "*Merychyus* are ore prismatic, have larger crowns and shorter roots, approaching e sheep, as *Oreodon* does the deer." Now *Oreodon* is Miocene, id *Merychyus* Pliocene. It was then ob-rved: "This phenomenon suggests an ex-anation on the score of adaptation which e other cases do not. The existence dur-g the later period of a [hard]er material diet would increase the rapidity of wear-g of the crown of the tooth, and require a nger crown and greater rapidity of protru-on. This necessitates a diminution of the isal shoulder and shortening of the roots, producing the prismatic rm aforesaid."

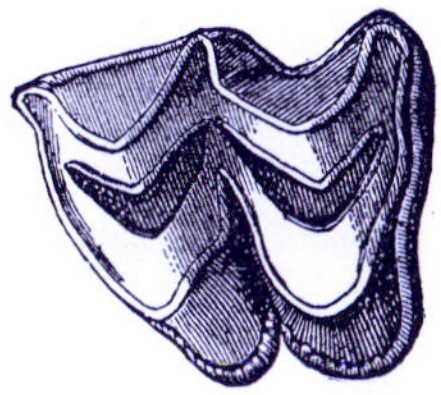

Fig. 24.—*Merychyus major.*

These observations render it highly probable that the seleno-ont molar is produced by a modification of the antiodont buno-ont molar. Also, that the manner of the change has been by nstant acceleration of growth of the folds of the tooth upward id perhaps downward in its long axis; and an acceleration in the ngthening of the crown.

2. *Tapirodonts.*—This form is so nearly similar to the Palæo-eriodont that any series annectant between the latter and the unodonts will render very probable such a connection for the apirodonts also. Indeed, it is clear that the same evidence will be sufficient in both cases, since the pre-molars and last molar of *Lophiodon* are amœbodont, like *Palæotherium*. In point of fact, however, the tubercles of the molars of *Achænodon* are partly united in transverse pairs, while there are intermediate tubercles connecting the opposite cones in some molars *Elotherium*. These structures foreshadow this group as well the Trichechodont.

Fig. 25.—*Tapirus.*

3. *Trichecodonts.*—The *Mastodons* and Elephants form a most mplete series between this form and the Bunodonts, as has been inted out by Falconer. In this series, the transverse rows or irs of tubercles, as well as the crests, may be few or many. Thus

* "Proceedings of the Academy of Natural Sciences of Philadelphia," 1868, 274. "Origin of Genera," p. 44.

in *Elotherium, Halitherium,* and *Trichechus* they are few ; *Trachytherium* and *Dinotherium* more numerous ; in *Mastodo Stegodon,* and *Elephas,* mo numerous. The tubercles a united into serrated cros crests in *Halitherium,* the e tinct sea-cow ; in *Trachyth rium,* another fossil ally of t Manati, the tubercles are n united. The succession fro *Mastodon* to *Elephas* may represented by the accompan ing figures : Fig. 26 (fro Cuvier) is a molar of *M. a gustidens,* where, beside the principal tubercles, numerous less ones appear. Fig. 27 represents *Mastodon ohioticus,* in whic

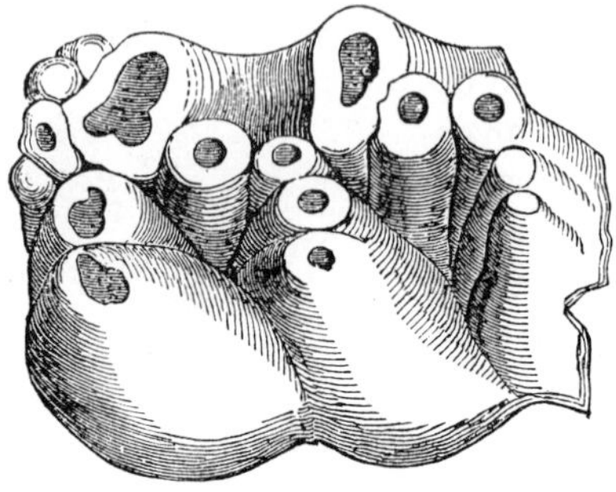

Fig. 26.—*Mastodon angustidens.*

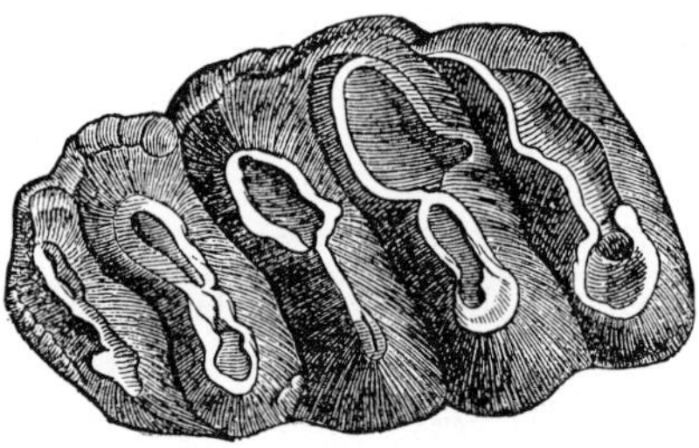

Fig. 27.—*Mastodon ohioticus.*

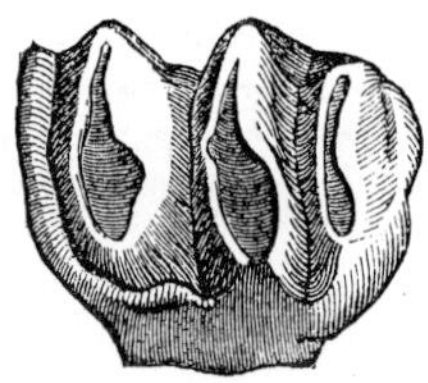

Fig. 28.—*Dinotherium giganteu*

the opposite tubercles are nearly united into transverse crest In *Dinotherium* (Fig. 28) and *Elephas* (Fig. 29) the union complete. The relation of these genera has been described as one of "inexact parallelism" ; a condition supposed by the writer to depend on modification in descent under the law of acceleration. The language used is :* "The young tooth of *Elephas,* moreover, is represented by a series of independent parallel laminæ at first, which, when they unite, form a series of cres

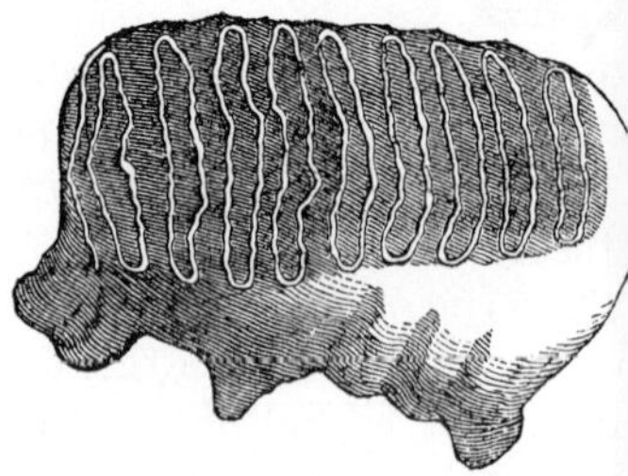

Fig. 29.—*Elephas indicus.*

* "Origin of Genera," p. 17.

milar to the type [i. e., pattern] of the genus *Mastodon* [*Stego-n*] and others of the beginning of the series. The deposit of mentum takes place later, till the valleys are entirely filled). Thus the relations of this rt of the tooth structure in e series are also those of the ccessional growth of *Elephas* the extreme of the series."

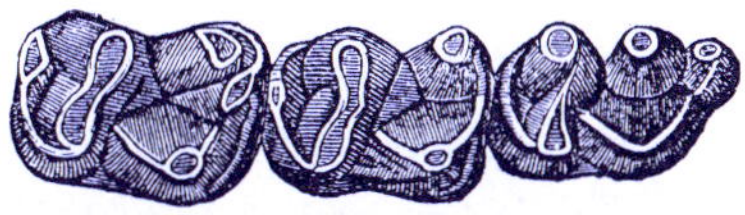

FIG. 30.—*Microsyops elegans.*

The transition from the inodont type to the lophodont in the mandibular dentition is en in the Eocene genera *Microsyops* and *Limnotherium*, where e opposite cones are connected by a low cross-crest.

II. AMŒBODONTS; Bunodont type *Hyopsodus*.*

4. *Symborodonts.*—In *Hyopsodus* the exterior cones are already mewhat excavated on the inner side, so that a section of each is mewhat triangular. It is obvious that but little more compres-on and curvature are required to produce the type of *Palæosyops*,

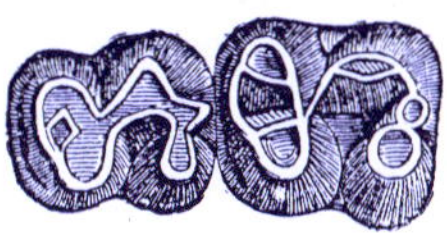

FIG. 31.—*Hyopsodus.*

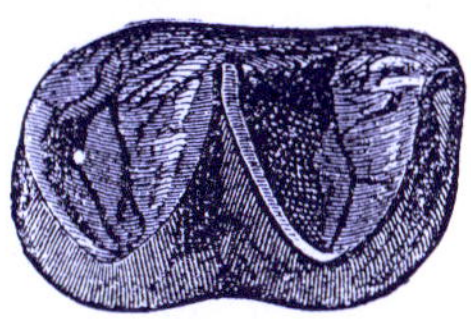

FIG. 33.—*Anchippodus.*

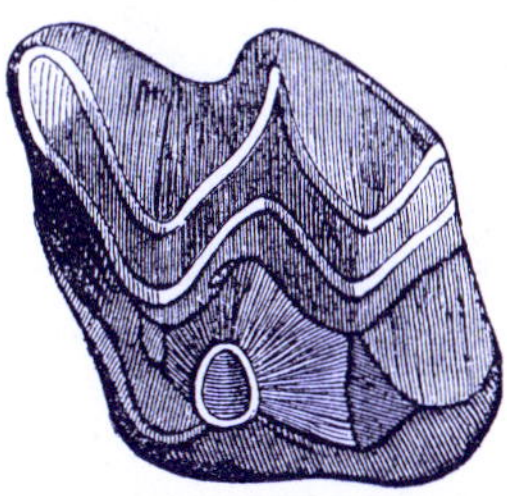

FIG. 32.—*Palæosyops lævidens.*

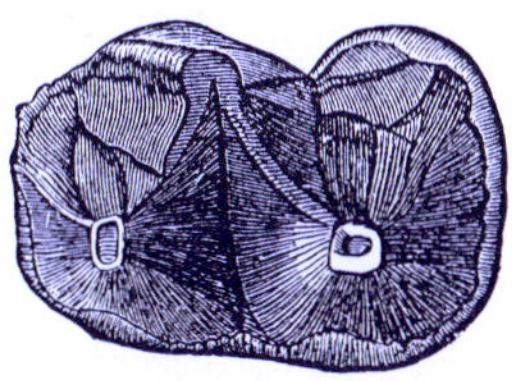

FIG. 34.—*Palæosyops.*

c. (Fig. 32). The angles of the outer cones in *Hyopsodus* are so slightly produced as low ridges to the bases of the alternating bercles of the opposite side: the elevation of these ridges is only

* I originally ("On the Primitive Types of Mammalia Educabilia," p. 9; "Hay-n's Geolog. Survey Terr.," 1872, 1873, p. 648) selected *Pliolophus* (*Oligotomus* Cope) this position, but as it has a considerable diastema, it is better exchanged for *yopsodus*, where the dental series is uninterrupted.

17

necessary to produce the two Vs of the mandibular dentition *Palæotherium*, *Palæosyops* (Fig. 34), *Symborodon*, *Anchippod* (Fig. 33), and all their allies (Fig. 35, *Palæotherium*).

5. *Palæotheriodont* type.—Immediately following the form the *Palæosyops* molar we have that of *Hipposyus*, where inte mediate tubercles stand b tween the inner subconic ar the exterior longitudinal cre centoid tubercles. They a compressed so as to be tran verse, and only need mo complete connection with t adjacent tubercles to give t oblique transverse ridges *Anchitherium*, *Palæotheriu* and *Hyracodon*, *Rhinoceru* etc. *Hipposyus* was origi ally compared with *Anchitherium* by Dr. Leidy, and the writer ignorance of his language remarked :* "An interesting annecta form is seen in *Lambdotherium*† *procyoninum*, where the two inte mediate tubercles which separate the inner cones from the out Vs in *Limnohyus* are so developed as to constitute parts of an i complete pair of transverse ridges which disappear in front of the bases of the outer Vs. They represent the oblique crests of *Palæotherium* and *Anchitherium*, and thus the genus *Lambdotherium* furnishes a station on the line from *Palæosyops* to the horses." Contemporaneously and quite indepen ently Prof. Marsh expressed similar views ‡ as to its affinities.

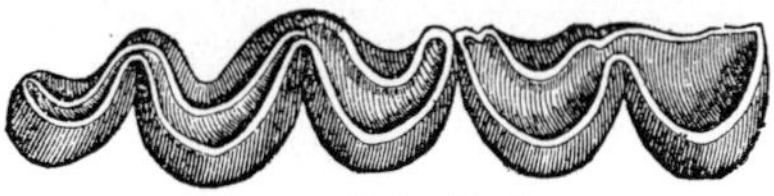

FIG. 35.—*Palæotherium*.

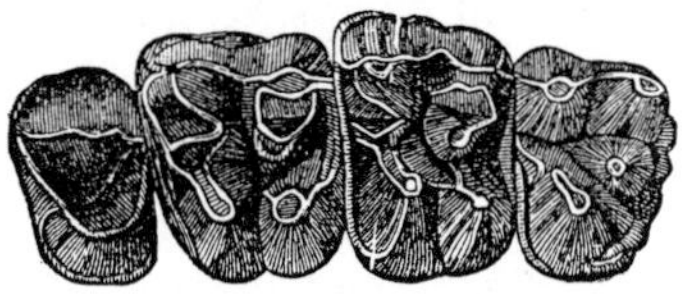

FIG. 36.—*Hipposyus*.

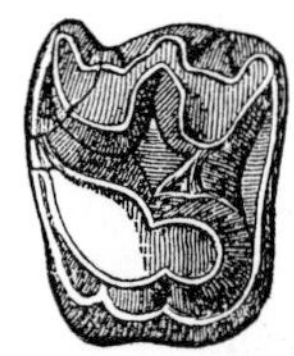

FIG. 37.—*Hipposyus*.

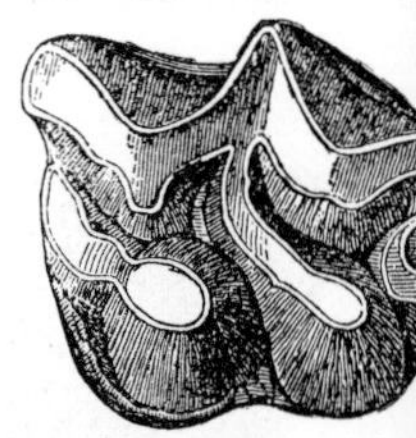

FIG. 38.—*Hypohippus*.

A greater longitudinal extent of these ridges or longitudin expansion of the tubercles in the molars in both jaws, the obliq connections being still retained, gives the type of *Equus* (Fig. 39 The elevation of the tubercles and deepening of the valleys giv

---

* Hayden's "Geol. Surv. Montana," etc., 1873, p. 647.

† This species was in the original erroneously called *Orohippus procyonin* (Ed. 1886.)

‡ "Amer. Journ. Sci. Arts," 1873, p. 407.

us the Selenodont type of superior molars again in this genus; while the lower molars only differ from that type in having the crescents alternate instead of opposite, forming the Hippodont pattern (Fig. 40). There can be little doubt that the line of the horses comes through *Hipposyus** from the Bunodonts, rather than through *Palæotherium*, as has been suggested by some writers.

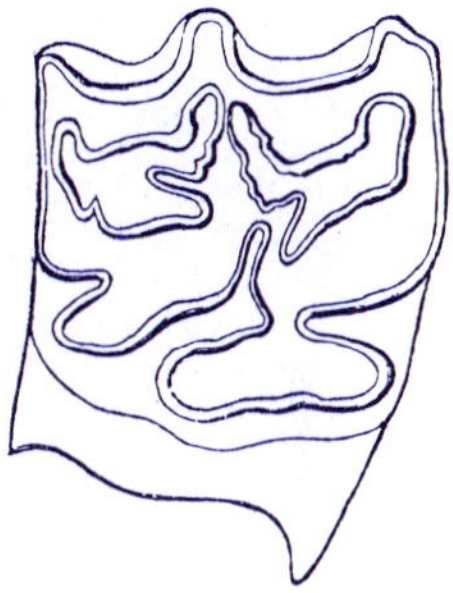

FIG. 39.—Superior molar of *Equus*.

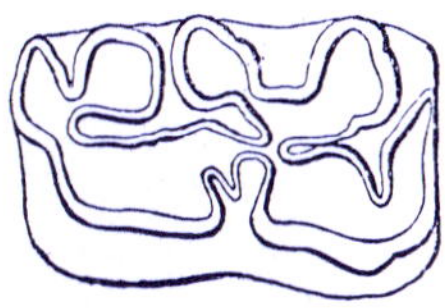

FIG. 40.—Inferior molar of *Equus*.

6. *Bathmodont* type.—I know of no genus which by its intermediate structure connects this type of molar with the Amœbodont form of Bunodonts.† Such will doubtless be discovered, for it is impossible that the upper molar of *Bathmodon* could have been produced by the modification of any known Palæotheriodont, the type which it most nearly resembles. The structure of the feet of the animal forbids any such supposition. Such intermediate types would have, firstly, the (oblique) crests more nearly equal in length and similar in direction. A depression of the crests and indication of diagonal ridges connecting the tubercles in the opposite direction would produce an approach to a W, and the form of *Hyopsodus*. That this was probably the history of this curious type is rendered probable by the form of the mandibular teeth, which exhibit two of the intermediate stages above anticipated. Thus the last inferior molar exhibits two obliquely transverse crests of subequal length, with rudimental oblique or diagonal ridges connecting them. In the median lower molars one of the latter is developed, giving a V, as in *Palæotherium*, but the posterior one is undeveloped, leaving only the original oblique cross-crest.‡

7. The *Loxolophodonts*.—Like the preceding group I know of no type connecting this form with the Bunodont, but anticipate the discovery of a type with a rudimental posterior V on the

---

* I have since shown that *Hyracotherium* and *Pliolophus*, which have very similar dentition, are the types which occupy this position. (Ed. 1886.)

† Such has since been discovered in the genus *Pantolambda* (Cope). (Ed. 1886).

‡ See "Proceed. Amer. Philosophical Society," Sept., 1872, "On the Dentition of *Metalophodon*."

upper molars,* which shall connect it with the W-shaped type proposed above as the probable predecessor of *Bathmodon*. Or, a pair of oblique parallel crests with rudimental diagonals like the posterior lower molars of *Bathmodon* may intervene between this form and the Bunodont. This is, however, not probable in view

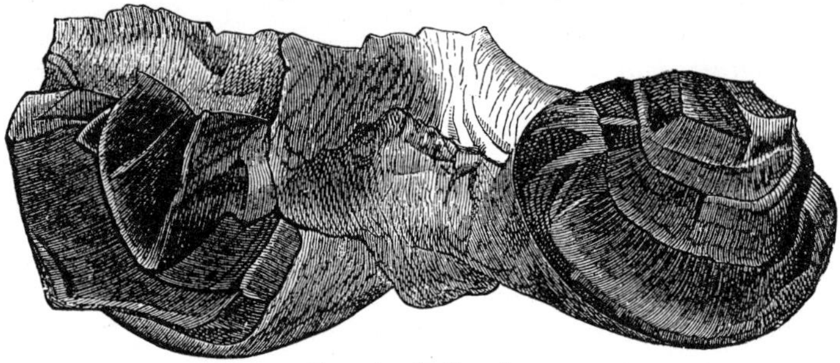

Fig. 41.—*Bathmodon*.

of the diagonal crest of the upper molars (Fig. 42, *Uintatherium robustum*), and especially if the parallel with the type of the lower molars is kept up. These are like those of *Bathmodon*, ex-

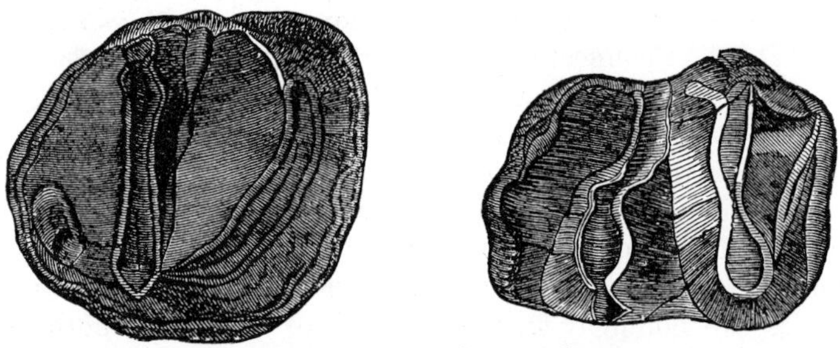

Figs. 42 and 43.—*Uintatherium robustum*.

cept that the type of the middle molars of the latter is continued to the posterior end of the series in *Uintatherium* (Fig. 43); that is, the last molar of the latter consists of a V and an oblique cross-crest.

V. RELATIONS OF THE TYPES OF DENTITION TO TYPES OF FOOT STRUCTURE.

I hope that I have succeeded in showing that the Bunodont and Lophodont types of dentition form two homologous series,

* This has since been discovered in Pantolambda (Cope). (Ed. 1886.)

similar to those already indicated among *Batrachia, Anura, Cephalopoda,* etc.* That this relation indicates descent of the corresponding terms of the one series from those of the other has also been rendered highly probable. This conclusion has also been previously stated as a theorem,† as follows: "V* The heterologous terms or genera in the later series are modified descendants of those of the earlier series"; in other words, that certain groups higher than genera are produced from others of a similar high value by "descent with modification."

As already pointed out, the Bunodont primary genera belong to the older geologic epoch of the Eocene, while most of the derivative ones belong to later periods. Some were contemporary with the primary forms, but doubtless have descended from pre-existent members of the same type as yet unknown to us. The genus *Achænodon,* Cope, is especially generalized in three respects: (1) the simplicity of the construction of its molars; (2) the same simplicity of the premolars, which are without inner or posterior lobes; (3) the absence of all diastemata and consequent continuity of the dental series. Hence it may be regarded as more primitive than *Palæochærus, Chæromorus, Dicotyles,* or *Elotherium,* in all which there are marked diastemata. The two series may then be arranged as follows: with the understanding that in some cases names of genera used represent rather family groups, in which the special generic lines have not yet been made out.‡

The following table has been already published in its essential features in the "Report on Geological Survey of the Territories," 1873, p. 648.#

It remains now to ascertain whether the genealogical or taxonomic relations expressed by the teeth coincide with those derived from the other diagnostic regions of the body. First of these must be selected, as of chief importance, the limbs and feet.

---

* See "Origin of Genera," p. 53. † *Loc. cit.* p. 79.

‡ Since the following table was published it has become probable that Achænodon is an unguiculate and a flesh-eater. It has also been discovered that the genus Phenacodus possesses the characters which give it the place as the ancestor of all the lines in the table. (Ed. 1886.)

It was previously published in a separate form in "On the Primitive Types of Mammalia Educabilia," May 6, 1873. An error occurs in this edition in the reversal by a lapsus calami of the positions of the types *Omnivora* and *Anoplotherium.* It is also important to note, that in the "Report G. S.," p. 645, where it is stated that "during the Eocene they (the orders) were in process of differentiation," etc., *Mammalia Educabilia* and not *Lissencephala,* are referred to.

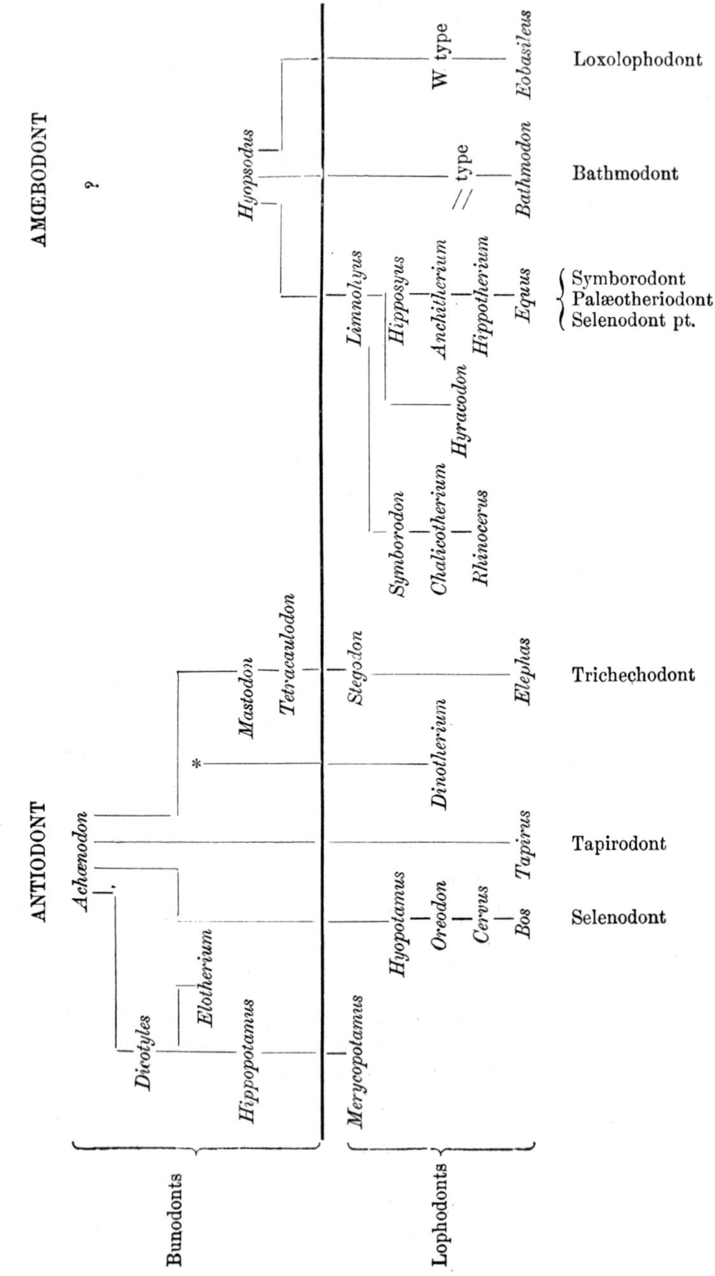
AMŒBODONT
?
Hyopsodus
W type
Eobasileus
Loxolophodont
type
Bathmodon
Bathmodont
Limnohyus
Hipposyus
Anchitherium
Hippotherium
Equus
Symborodont
Palæotheriodont
Selenodont pt.
Hyracodon
Symborodon
Chalicotherium
Rhinocerus
Mastodon
Tetracaulodon
Stegodon
Elephas
Trichechodont
*
Dinotherium
ANTIODONT
Achænodon
Tapirus
Tapirodont
Hyopotamus
Oreodon
Cervus
Bos
Selenodont
Elotherium
Dicotyles
Hippopotamus
Merycopotamus
Bunodonts
Lophodonts

We may look on the *Artiodactyla*—even-toed or cloven-footed mammals—as one of the most homogeneous groups in the class, not only in respect to the structure of the extremities, but also in that of the cranium, vertebræ, etc. But here we have both *Bunodont* and *Selenodont* types of molars. The *Perissodactyla*, as defined by the feet, axis, palate, etc., present us with the *Symborodont, Palæotheriodont, Tapirodont*, and *Selenodont* types of dentition in the superior series, and the *Hippodont, Hyracodont, Palæotheriodont*, and *Trichechodont* types in the mandibular teeth. The pentadactyle, plantigrade type, for which I have used the name *Proboscidia*, presents us with the *Trichechodont, Bathmodont*, and *Loxolophodont* types of molar structure. Among Sirenians, as defined by the marked peculiarities of the entire skeleton, we have the great differences in dentition presented by *Halicore* and *Trichecus*, the former being *Haplodont*, the other *Trichechodont*. Finally, the Marsupial group is unquestionably well defined, and here *Phascolomys* is *Ptychodont ; Didelphys, Bunodont ; Petaurus* between *Bunodont* and *Selenodont*, and the Kangaroos and their gigantic extinct allies the *Diprotodontidæ, Trichechodont.*

It is thus evident that the molar types are everywhere subordinated to those which we call ordinal ; therefore in the case of the placental mammals, and especially those with complex folding of the cerebral hemispheres, to the types of construction of the feet. As to the modifications presented by the canine and incisor teeth, these exist within a still more narrow range of variation ; for instance, in the allied genera *Equus, Rhinaster*, and *Symborodon ; Sus* and *Phacochærus ; Bathmodon* and *Unitatherium*, and others.

It is thus probable that modifications in the three points of structure considered were introduced in the following order :

First, Of the feet.

Second, Of molar type.*

Third, of the relations of canines and incisors.

With regard to the significance of the three types of feet, *Proboscidian, Perissodactyle*, and *Artiodactyle*, it has been already remarked : "It is to be observed that the lines of *Ungulata*,

---

* In the case of the *Marsupialia* the relation of the dental and extremital types may be reversed. Thus we have pentadactyle plantigrade forms (Opossums) and (nearly) didactyle digitigrade forms (*Macropus*) in the same order. Also *Halmaturus* and *Diprotodon*, both Trichechodonts, differ in the type of feet, as do the carnivorous *Didelphys* and *Thylacinus*, both *Bunodonts.*

Quadrumana, and *Carnivora* originate in plantigrade types, a state of things quite predominant among the lower series or *Lissencephala* (smooth brains). It is universal in *Edentata* and very usual in *Rodentia* and *Insectivora.* The lower forms of *Marsupialia* and all of the *Monotremes* present it. In the *Marsupials, Rodents, Ungulates,* and *Carnivores,* we have series whose highest expression is in the most highly digitigrade genera." * To this it may be added that the lower terrestrial vertebrates are plantigrade, with some exceptions. Thus in some Anurous *Batrachia* there is a partial digitigradism; the only digitigrade *Reptilia* are some *Dinosauria,* especially such carnivorous forms as *Lælaps;* all birds are digitigrade. The digitigrade modification evidently has reference to speed in running, or projectile force in leaping.

The connecting points between the different types of foot-structure among the *Mammalia Educabilia* are as obvious as in the case of the types of molar structure. Examples may be adduced as follows:

Carnivora.—In all of the genera of the Eocene Carnivora which I have had the opportunity of examining, excepting *Mesonyx,* namely, *Amblyctonus, Oxyæna, Prototomus, Didymictis,* the tibio-astragalar articulation is of a primitive character. The astragalus is flat, and the applied surfaces are nearly a plane, and without the pulley-shaped character seen in existing Carnivora; as dogs, cats, and in a less degree in the bears and in other Mammalia with specialized extremities, as *Perissodactyla, Artiodactyla,* etc. The simplicity of structure resembles, on the other hand, that found in the opossum and various *Insectivora, Rodentia,* and *Quadrumana,* and in the *Proboscidia,* most of which have the generalized type of feet. The structure indicates that the carnivorous genera named were plantigrade—a conclusion which is in conformity with the belief already expressed that the Mammalia of the Eocene exhibit much less marked ordinal distinction than do those of the Miocene or the recent periods. It is, indeed, questionable whether some of the genera here included in the Carnivora are not gigantic *Insectivora,* since the tibio-tarsal articulation in many, the separation of the scaphoid and lunar bones in *Mesonyx,* the form of the molars, and the absence of incisor teeth in some, are all characteristic of the latter rather than the former order.

Artiodactyla.—Approximations to the *Perissodactyla* are to be

* "Mammalia Educabilia," p. 8; Hayden's "Geological Survey," 1873, p. 647.

seen in *Hippopotamus* in the increase in development of the lateral or first and fourth digits, thus equaling the number in the fore-foot of *Tapirus* and *Menodus*, though preserving the equality of the two median digits. But an inequality of these digits appears in the genera *Anoplotherium* and *Cœnotherium*, as has been stated * in the following language: "In *Anoplotherium secundarium* the digit *ii* is developed in each foot, though not nearly so long as *iii*, which is nearly symmetrical in itself. There is an approach to the same structure in the manus of *Cœnotherium*." The only approximation to the Proboscidian type is to be seen in the shortening of the metapodial bones in *Hippopotamus*, a point of very inconsiderable value.

Perissodactyla.—Approximation to the preceding order is made in the anterior foot of *Menodus*, in which, according to Marsh, there are four toes of nearly equal size.† Approximation to the *Proboscidia* ‡ is seen in *Symborodon*, where the cuboid facet of the astragalus is rather larger than in *Rhinoceros*, and developed much as in *Bathmodon*; the small third trochanter of the femur is also much like that in *Bathmodon*. The osseous horn-cores may be compared with those on the front of *Loxolophodon*. The knee was probably free from the integument of the abdomen, as in Proboscidians. In all other respects there is no approximation to this order.

Proboscidia.—The approximations to other orders in the structure of the feet are only to be seen in the Eocene genera *Bathmodon* and *Eobasileus*. The latter, or its ally *Uintatherium*, presents, according to Marsh, but four toes on the hind foot; the anterior has five. In the former point we have a resemblance to *Hippopotamus*, but one of little significance, in view of the radical differences between the two in the form of the astragalus, calcaneum, and cuboid bones. The former is essentially Proboscidian in all respects, with the addition of a cuboid facet alongside of and behind the navicular, as in *Symborodon*; thus constituting a Perissodactyle character, but leaning to the forms of that order which betray probably the closest, though slight, approach to the omnivorous division of the *Artiodactyla*. Thus, while the *Uintatheriidæ* present the Proboscidian type of feet and molar dentition,

* Huxley, "Anatomy of the Vertebrated Animals," p. 321.

† "American Journal of Science and Arts," 1873, p. 486.

‡ This should be Amblypoda, not separated from the Proboscidia at the time this was written. (Ed. 1886.)

if they present any ordinal characters resembling those of the Artiodactyla, they are equally shared by certain extinct *Perissodactyla.*

From the hints above furnished, we may regard the succession of modifications of foot-structure to be nearly as follows :

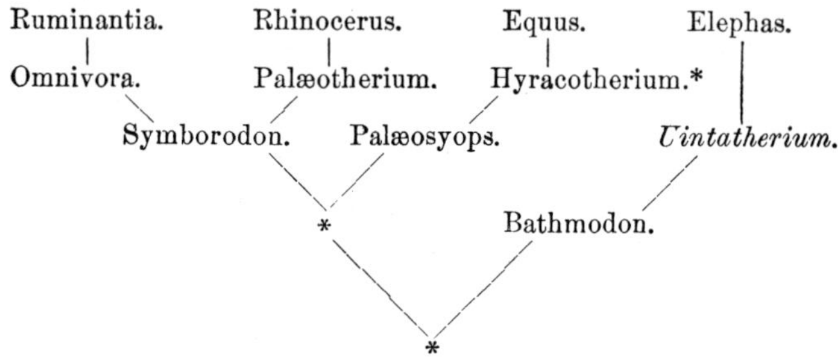

VI. THE ANCESTRAL TYPE OF MAMMALIA EDUCABILIA.

I trust that I have made it sufficiently obvious that the primitive genera of this division of mammals must have been Bunodonts with pentadactyle plantigrade feet. It therefore follows that *Elephas* was not the descendant of *Eobasileus* nor *Bathmodon* in a direct line, but from some common ancestor with tubercular teeth, through *Mastodon.* We may anticipate the discovery of such a genus, and believe that it will not be widely removed from the Eocene *Hyopsodus,* or perhaps *Achænodon.* This will, then, be the primitive ungulate.

But it will be more than this ; it can not be far removed from the primitive carnivore and the primitive quadrumane. The Carnivora are all modified bunodonts, and the lower forms (*Ursus Procyon,* e. g.) are pentadactyle and plantigrade. As to the *Quadrumana,* man himself is a pentadactyle plantigrade bunodont. This view has been already expressed, as follows : "The type of *Tomitherium,* already described, evidently stands between lemurine monkeys and such small allies of *Palæotheriidæ* with conic tubercular teeth (*Oligotomus, Orotherium,*† etc.), and which abound in the Eocenes of Wyoming. . . . The dentition of the two types is, indeed, but little different in the Quadrumanous and

* This was called Hipposyus in the original essay—a name which really applies to a different type. (Ed. 1886.)

† Both these are names for Pliolophus, a close allay of Hyracotherium. (Ed. 1886.)

Ungulate types respectively, being a continuous series of I. 1 or ; C. 1 ; P. m. 3—4 ; M. 3 ; the canines but moderately developed."* Such a hypothetical type might be expressed by the name *Bunotheriidæ,* with the expectation that it will present subordinate variations in premolar, canine, and incisor teeth. The premolars might be expected to differ in the degree of development of the internal lobes, the canine in its proportions, and the incisors in their number.

In respect to the limbs proper, neither the *Quadrumana* nor *Carnivora* attain to the specialization seen in the *Artiodactyla* and *Perissodactyla,* for the ulna and fibula are never atrophied nor co-ossified with the radius and tibia, but are always distinct and free ; the only modification of structure in these points being the slight one involved in developing the rotary capacity seen in the higher monkeys.

Thus the human series preserves in its feet, limbs, and dentition, more of the characteristics of the primitive *Bunotherium* than any other line of descent of the *Mammalia Educabilia.* It even exhibits a retrogression, in the transition from the anisognathous *Tomitherium* to the genus *Homo,* where the teeth in the two jaws are exactly alike, as well as in the resumption of the continuity of the dental series after the diastema had prevailed among the higher monkeys. In one respect it has steadily advanced, viz., in the number of convolutions and extent of the cerebral hemispheres and relative size of the brain as a whole.

NOTE (Ed. 1886).—As remarked in a previous note, the discovery of the general characters of the genus Phenacodus in 1881, more than six years after the publication of this paper, demonstrated the truth of the hypothesis here proposed, viz.: that the ancestor of the Mammalia Educabilia was a pentadactyle plantigrade bunodont. The numerous genera and species allied to Phenacodus have been placed in a suborder Condylarthra. See "American Naturalist," 1884, 790, for an illustrated article on this group.

---

* Hayden's "Geological Survey of Montana," etc., 1872, p. 645.

VIII.

THE RELATION OF MAN TO THE TERTIARY MAMMALIA.*

IN order to prove the affirmative of a doctrine of evolution b descent of the existing types of living beings, two proposition must be established. The first that a relation of orderly succes sion of structure exists, whicl corresponds with a succession ii time. Second, that the term (species, genus, etc.) of this suc cession actually display transition or connection by intermediat forms, whether observed to aris in descent, or to be of such vari etal character as to admit of n other explanation of their origi than that of descent.

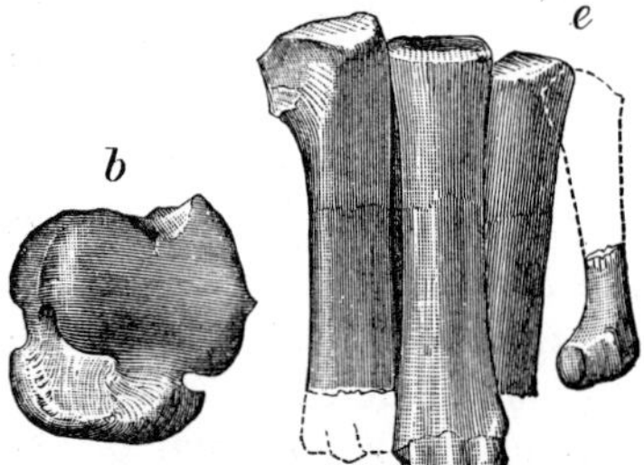

FIG. 44.—*Periptychus rhabdodon*, part of posterior foot, two thirds natural size; *b*, astragalus from above, showing flat face; *e*, metatarsals, the first lost, showing plantigrade foot. Original; from Puerco Epoch of New Mexico.

In the field of paleontology i is quite possible to demonstrat the first of these propositions, while the proof of the second i necessarily restricted to the observation of variations and the dis covery of connecting forms which destroy the supposed definition of species, genera, etc. The conditions are more favorable for th investigation of animals of the higher types than of those of th lower. Their late origin insures to us the opportunity of discover of their ancestry far more certainly than in the case of the lower whose beginnings are lost in the remote past, and belong to period whose deposits have undergone physical changes, or have been en tirely removed and redeposited elsewhere, thus insuring the de struction of the fossil remains once contained in them. The serie of the tertiary *Mammalia* is becoming more complete through th

* Read before the American Association for the Advancement of Science, a Detroit, 1875, under another title.

ecent explorations in the West, and the results are embraced in he forthcoming quarto eports of the Hayden and Vheeler U. S. Geological, tc., surveys of the Terri- ories. An abstract of ome of these is given in he present essay.

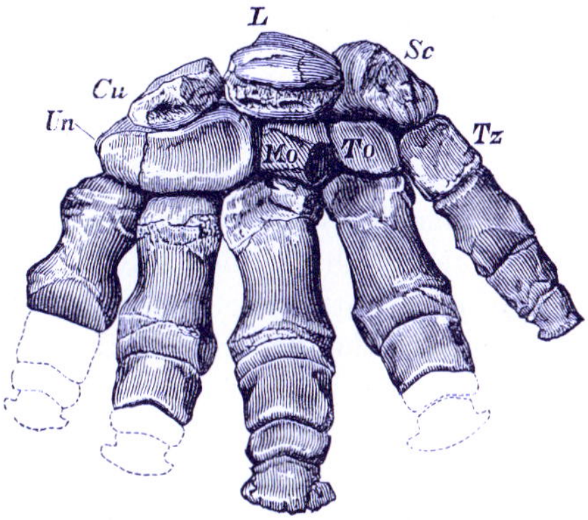

FIG. 45.—*Coryphodon elephantopus*, anterior foot from above, one third natural size. From Lower Eocene of New Mexico. The cuneiform (*Cu*) is injured. (Original.)

The primary forms of he Mammalia repose in great measure on the struc- ure of the feet. Those of he teeth are also very sig- ificant, but present a greater number of varia- ions among animals other- wise nearly related. The steology of the feet of re- ent land mammals falls into several categories. These may be called the plantigrade, many-toed type ; the car- nivorous type ; the horse type, and the ruminant (e. g., ox) type. The lower vertebrates, as sala- manders, lizards, etc., dis- play the simplest form of feet, having usually five toes, with numerous sepa- rate bones of the palm and the sole, which they apply to the ground in progression. The many- toed or multidigitate type of mammalian foot most nearly resembles this condition, but dif- fers in the points of dif- ference which are com- mon to all Mammalia. In the hind foot a succession of forms leads from this general-

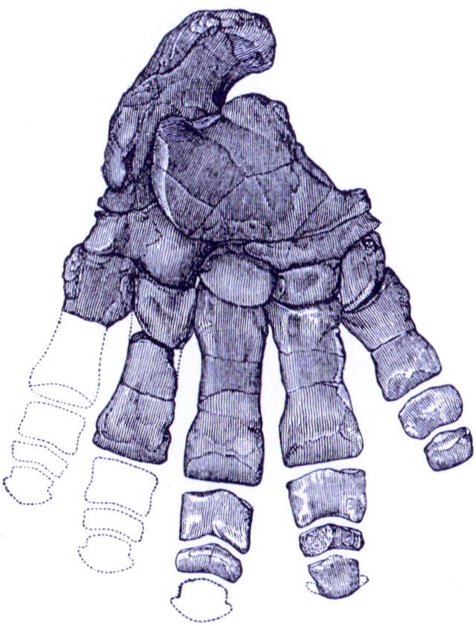

FIG. 46.—Posterior foot of *Coryphodon elephan- topus*. From Lower Eocene of New Mexico. (Orig- inal.)

ized type to the extreme specializations observed in the horse an the ox.

The modifications are as follows: The hind foot is compose of two rows of tarsal bones, of which the second is followed b the long metatarsal bones, from which the bones of the toes origi nate. The second segment of the hind leg is composed of tw bones, tibia and fibula, which in the salamander, etc., have a sub equal union with the foot. In some multidigitates, as the genu *Coryphodon*, both these bones articulate with the two bones of the first row of the tarsus, and one (fibula) is the smaller of the two. In many higher forms they articulate with but one of these tarsal bones, viz., the *astragalus*, with which they form a perfect hinge joint; the other tarsal bone of the first row is the *calcaneum* or heel-bone. In *Coryphodon* the astragalus and the applied leg-bone (tibia) are nearly flat, offering an extremely imperfect hinge for the foot, and the heel-bone (calcaneum) is exceedingly short. The animal plainly walked on the entire sole of the foot, and must have had an awkward gait, from the slight power of flexing the ankle-joint. From this point to the horse on one side, and to the ox on the other, we have a line of succession of intermediate forms. And before describing them, I may state that the *Coryphodon* is one of the oldest known Mammalia, its remains having been found in the Lower Eocene Tertiary of New Mexico and Wyoming, while the ox and horse are extremely modern animals, their advent on the earth having preceded that of man by but one geological period.

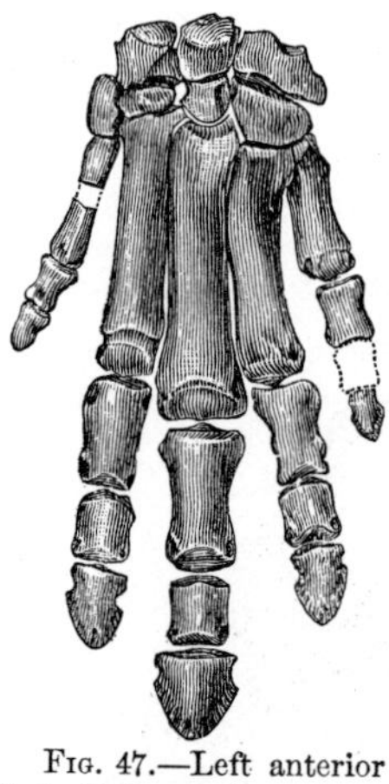

FIG. 47.—Left anterior foot of *Phenacodus primævus*, ⅛ nat. size. (Orig.)

FIG. 48.—Left posterior foot of *Phenacodus primævus*, ⅛ nat. size. (Orig.)

The most perfect ankle-joint is that of the ruminating animals. The astragalus presents a deeply grooved segment of a pulley; an angulated pulley, face downward to the rest of the foot, and a

PLATE VI.

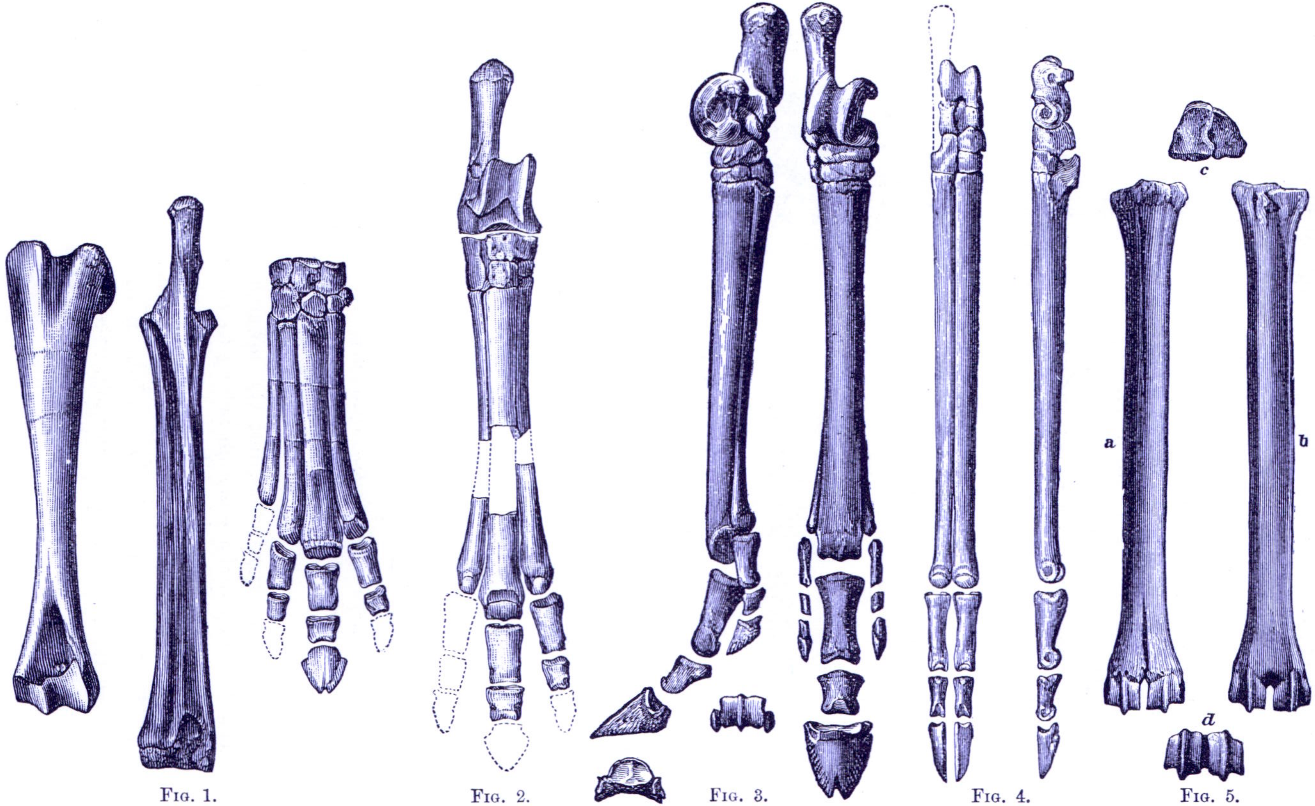

Fig. 1. Fig. 2. Fig. 3. Fig. 4. Fig. 5.

Fig. 1, *Hyracotherium venticolum*, fore leg. Fig. 2, do. hind foot, two thirds. Fig. 3, *Protohippus sejunctus*, hind foot, one half. Fig. 4, *Poëbrotherium* ***labiatum***, hind foot. Fig. 5, *Cosoryx furcatus*, cannon-bone of hind foot. All original.

smooth convexity to the hollow of the applied heel-bone behind. No such astragalus has ever been found in the Lower Eocene formations of America; animals bearing it in a less perfect stage appear in the next higher period, the Miocene; but it is not until the Pliocene and modern times that they abound. In the *Hippopotamus* foot, we have an example of the less perfect astragalus of this type of animals. The pulley surfaces are flatter and less deeply grooved.

In the horse, the upper surface of the astragalus forms as perfect

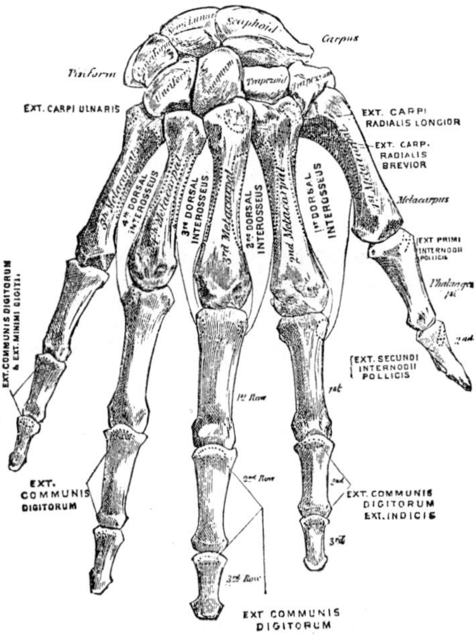

FIG. 49.—Anterior foot of *Homo sapiens* from above, one third natural size. From Allen's Anatomy.

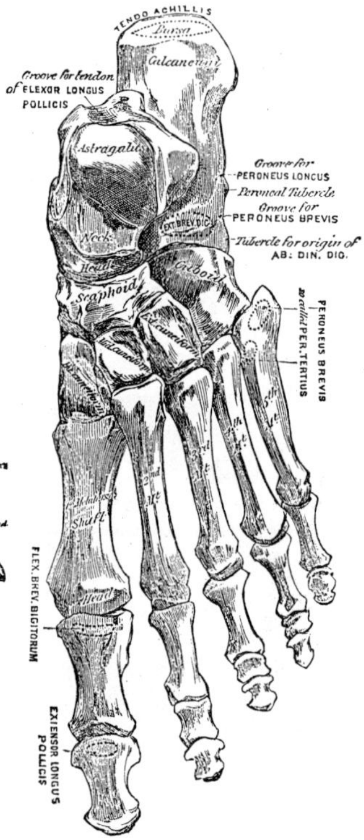

FIG. 50.—Posterior foot of do. from above, one third natural size. From Allen's Anatomy.

a joint as in ruminants, but the lower and hinder faces present the flattened surfaces which belong to the many-toed Mammalia. The lower face especially is mainly occupied by one large facet, instead of the two-faced pulley of the ruminants. In the rhinoceros a later facet is more distinct, while in the Miocene *Menodus* the second facet is larger, resembling, except in the still convex tibial articulation, the structure of that of the primitive *Coryphodon*.

In the heel-bone we have a succession from the short and flat form of *Coryphodon* to the long and slender one of the horse and ruminants; the increase in length being associated with the elongation of the bones of the toes, and the assumption of the digitigrade type from the plantigrade. The mammals of the Lower Eocene* exhibit a greater percentage of types that walk on the entire sole of the foot, while the succeeding periods exhibit an increasing number of those that walk on the toes, while the hoofed animals and *Carnivora* of recent times nearly all have the heel high in the air, the principal exceptions being the elephant and the bear families.

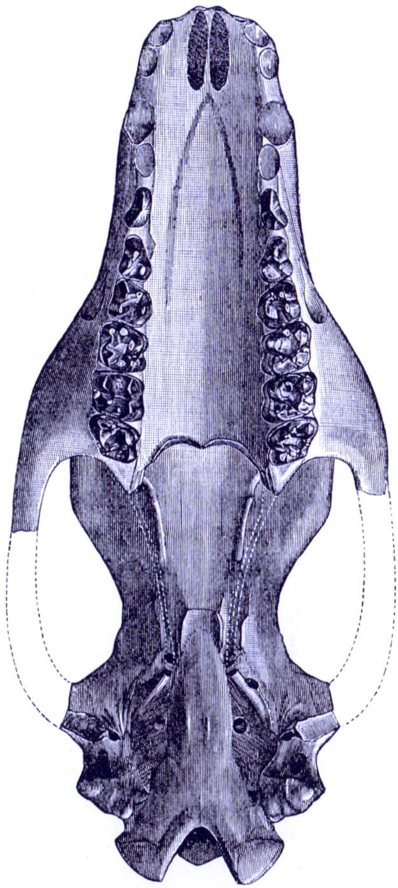

FIG. 51.—*Phenacodus primævus* skull, one third natural size, from below, showing quadritubercular true molars. From specimen figured on plate.

A most noticeable succession is seen in the diminution of the number of toes. In the series leading to the horse, the ox, and the hyæna and cat, this reduction proceeds by the loss of a toe from the one side or the other, until in the ruminants but two are left, and in the horse but one. The series extending from the primitive Eocene types with five digits, to the existing reduced forms, is most complete, although a few of the New Mexican Eocene genera themselves probably exhibit but four digits on one or both pairs of feet. The presence of the rudiments

* *All* the Mammalia of the Puerco fauna (which was unknown at the time this lecture was delivered) are plantigrade. (Ed. 1886.)

of the lost lateral digits is constantly observed, and when these disappear it is to be finally replaced by the rudiments of the adjoining toes in process of similar reduction. The bones of the second row of the tarsus which are in connection with the toes are not reduced so rapidly as the toes themselves; hence, the bones of the toes, in order to maintain the fit of the parts, increase in width, and consequently in strength. As is well known, in the horse the single toe is as stout as several united toes of lower forms, and the two toes of the ruminants have their basal segments (metatarsals) united into a stout solid mass, the cannon-bone. At the same time several of the small bones of the second tarsal row become coössified, so that we have, in the ruminants and horse, the greatest consolidation of structure, connected in the former with the most elegant mechanism. It is scarcely necessary to add that, in the various cases of coössification and consolidation described, the fœtus displays the original elements separated.

In the fore limb the same successional reduction in the number of toes may be traced as I have described in the hind foot; but, as the bones of the palm differ from those of the sole, the successional modification of these is also characteristic. The bones of the second row of the carpus are four in number, but as the toes are reduced, in the lines of the hoofed animals, the inner (trapezium) is soon dropped, and the second (trapezoides) becomes united with the third (magnum). In the carnivorous order, the trapezoides is always separate, but the inner pair of bones of the first row (scaphoid and the lunar) become consolidated into a single mass, although their original distinctness is easily determined by examination of the fœtus.

The two bones of the leg which articulate with the foot and hand, exhibit a succession of changes of relation in progress toward the more specialized types. In the *Coryphodon* and *Uintatherium* of the Eocene, each of these bones has considerable share in the articulation; but as we rise in the series, the surface of attachment of the lesser bones, the fibula in the foot, and the ulna in the hand, becomes successively smaller, until in the ruminants the fibula is almost obliterated, its distal end remaining as a small tuberosity coössified to the side of the end of the tibia. In the same manner the articular end of the ulna in the fore leg is successively reduced, until this bone also becomes a thin strip coössified to the lower side of the radius, with no distinct termination,

PLATE VII.

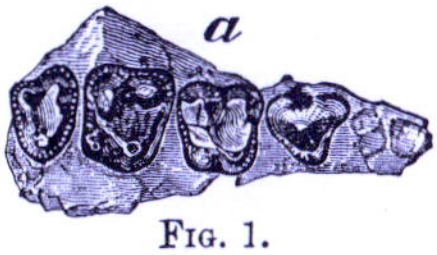

FIG. 1.

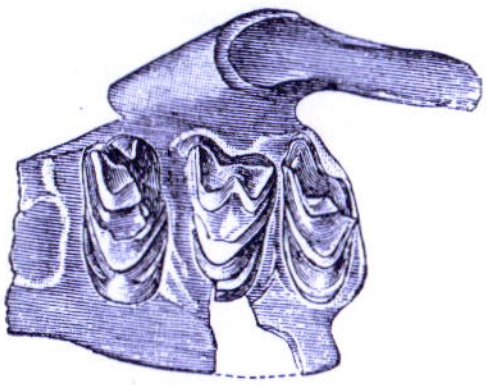

FIG. 2.

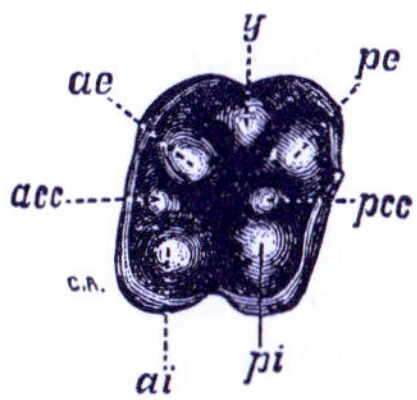

FIG. 3.

FIG. 4.

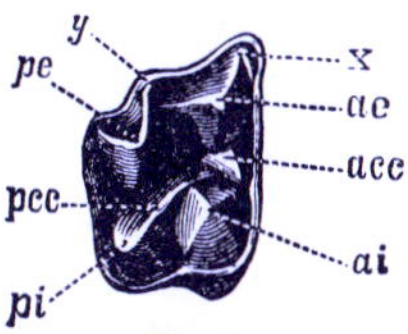

FIG. 5.

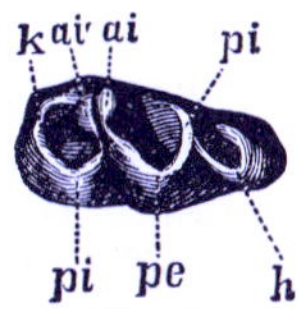

FIG. 6.

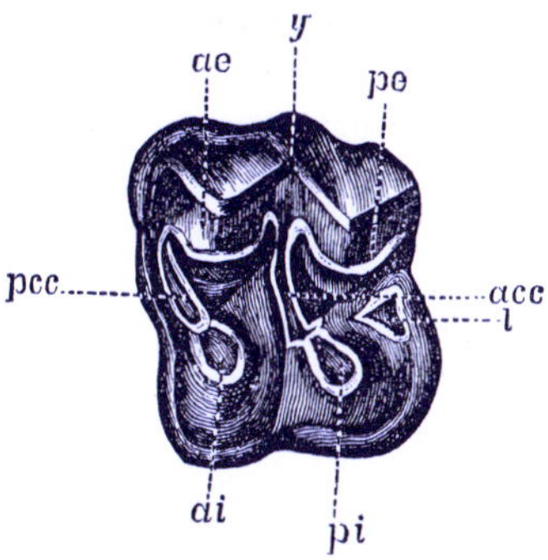

FIG. 7.

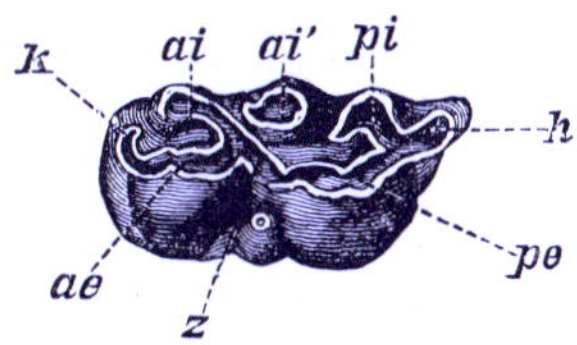

FIG. 8.

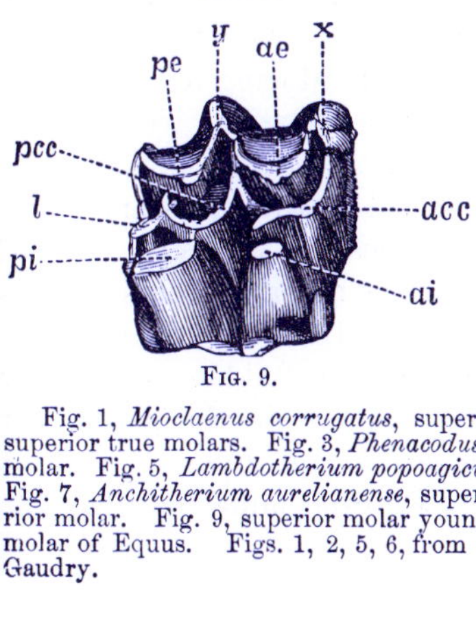

FIG. 9.

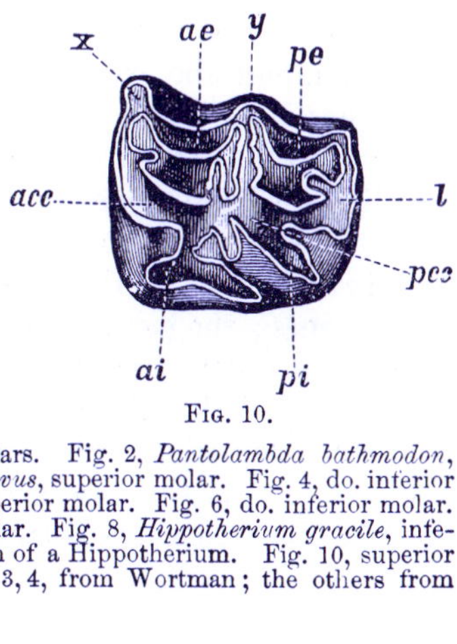

FIG. 10.

Fig. 1, *Mioclaenus corrugatus*, superior molars. Fig. 2, *Pantolambda bathmodon*, superior true molars. Fig. 3, *Phenacodus primævus*, superior molar. Fig. 4, do. inferior molar. Fig. 5, *Lambdotherium popoagicum*, superior molar. Fig. 6, do. inferior molar. Fig. 7, *Anchitherium aurelianense*, superior molar. Fig. 8, *Hippotherium gracile*, inferior molar. Fig. 9, superior molar young crown of a Hippotherium. Fig. 10, superior molar of Equus. Figs. 1, 2, 5, 6, from Cope; 3, 4, from Wortman; the others from Gaudry.

in both the horse and the ox. In the fœtus these bones are well distinguished.

The dentition tells as clearly as possible the same story. Here, again, as I have pointed out in a paper on the "Homologies and Origin of the Types of Dentition of the *Mammalia Educabilia*," the most specialized forms of dental structure are presented by the horse, the ox, and the tiger. But they are all modifications of a single type of tooth, viz., an oval crown supporting four tubercles on the summit, in the lower jaw, and three or four in the upper jaw. In the lower cutting molar of the cat but one of these tubercles remains, forming with another in front of it a double shear blade, whose development may be traced from its earliest beginnings in the genera of the Eocene. In the odd-toed forms (tapir, rhinoceros, etc.) the tubercles become connected transversely, forming cross-crests, and the outer ones are generally flattened on the outer side. In the horse the tubercles have a very complex form, and the spaces between them filled by a peculiar substance, the cementum. In the ruminants the tubercles come to have a crescent-shaped section and are drawn out to an enormous length, forming a prismatic tooth : here, also, the intervening deep valleys are filled with cementum. In the third series, that of the elephant, the original tubercles (permanently separate in the mastodons) are connected into cross-crests, which are drawn out to a great length, and as in the other series are supported by a deposit of hard cementum in the intervening valleys. The transitions between these and the primitive four-tubercled molar are numerous and direct.

There is not now opportunity to consider the question of transition from type to type by descent, further than to indicate by a few examples the manner in which it has evidently occurred. This has been by unequal growth of parts during fœtal life, according to the laws of acceleration and retardation. The union of the two basal bones of toes into a single one (the "cannon-bone") in the ruminants, is accomplished by the more and more rapid completion of the process of ossification in the growth of those bones ; the confluence of the various carpal and tarsal bones in various orders has the same history. In many genera it has been observed that the milk dentition has resemblances to other and older dentitions, which entirely disappear in the permanent teeth. This is the case with *Coryphodon* and *Equus;* while it may be observed in the kitten, whose sectorial milk-tooth has the

PLATE VIII.

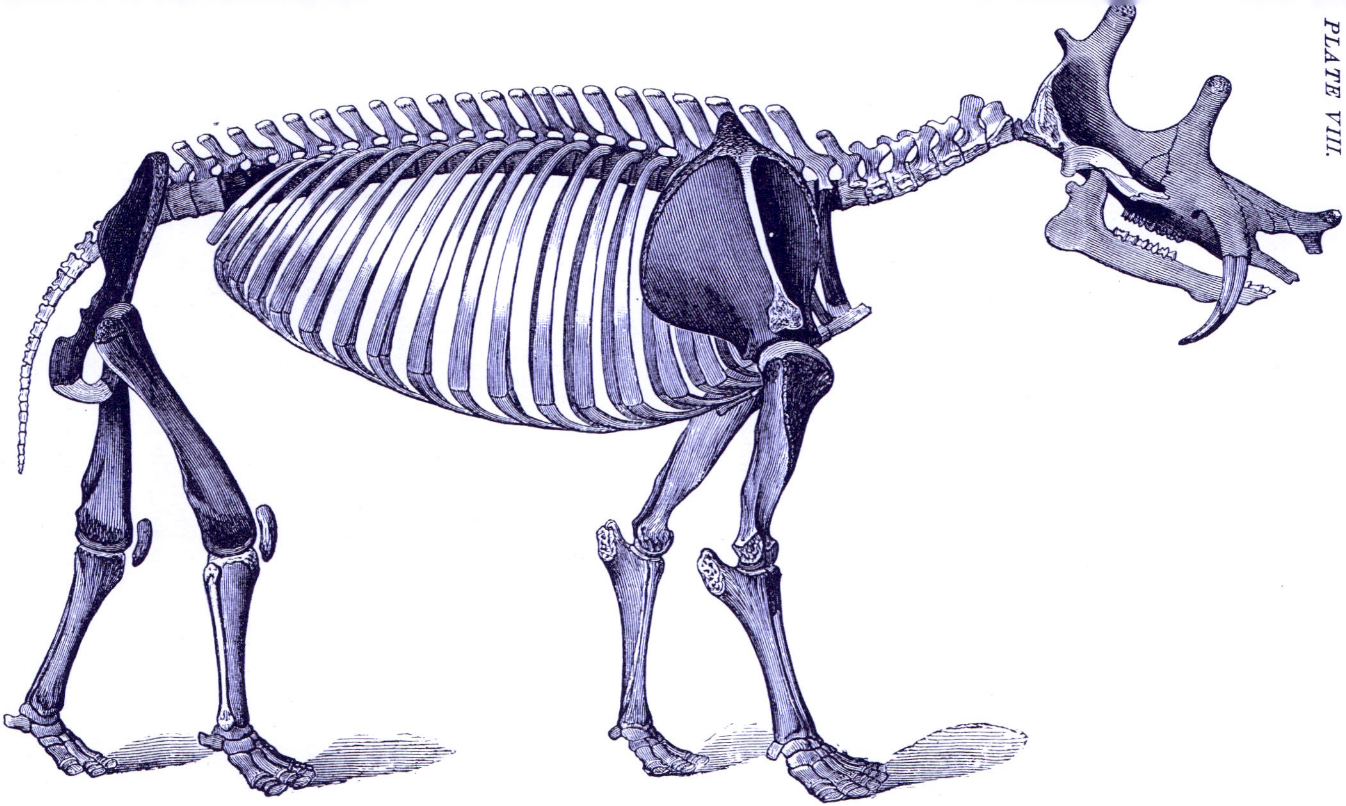

Uintatherium cornutum (Cope), restored to $^1/_{27}$ natural size. The darkly shaded portions are those in my possession; the feet and caudal and cervical vertebræ are restored after Marsh (*U. mirabile*); the lower jaw after Osborn. Original; from "American Naturalist," 1882, page 1029.

heel belonging to all the lower and primitive Carnivora which is wanting from the sectorial of the adult cat. Moreover, in complex teeth, the different stages of wear represent primitive conditions of the same animal, elsewhere preserved to us in extinct adult genera and species.

Forms which violate the definitions of the orders above given are also well known. Thus *Mesonyx,* with the skull and dentition of a carnivore, has the separate scaphoid and lunar bones and flat claws or hoofs of an ungulate. *Calamodon* has the molars of an ungulate, the incisors of a rodent, and claws resembling somewhat those of a carnivore. *Uintatherium* has feet partly like an elephant, and teeth of a tapir.

I have referred the Mammalia with five and four toes, the full number of distinct tarsal and carpal bones, which form interlocking series, and in which the tread is plantigrade, to a new order called the *Amblypoda.* These represent the primitive type of the higher *Mammalia* with convoluted brains, etc., but present much variation in the constitution of the teeth. From the latter have come off not only the three hoofed-orders represented by the elephant, the horse, and the ox, but the origin of the *Carnivora* and *Quadrumana* is scarcely distinguishable from it, if at all; while there is little doubt that the natatory *Sirenian* order was derived from it by a process of degradation, chiefly of the extremities, in connection with the assumption of an aquatic life. The manatee, of which a fine example may now be seen in our zoölogical gardens, represents this division.

The relation of man to this history is highly interesting. Thus in all general points his limbs are those of the primitive type so common in the Eocene. He is plantigrade, has five toes, separate carpals and tarsals; short heel, rather flat astragalus, and neither hoofs nor claws, but something between the two. The bones of the forearm and leg are not so unequal as in the higher types, and remain entirely distinct from each other, and the ankle-joint is not so perfect as in many of them. In his teeth his character is thoroughly primitive. He possesses, in fact, the original quadrituberculate molar with but little modification. His structural superiority consists solely in the complexity and size of his brain. The forms of the quadrumanous order, while agreeing with each other in most respects, display the greatest range of brain structure, and show that while they have made but little progress since the Eocene in perfection of organization of the skeleton, they

PLATE IX.

FIG. 1.

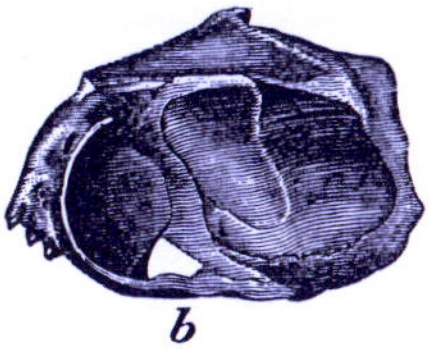

FIG. 2.

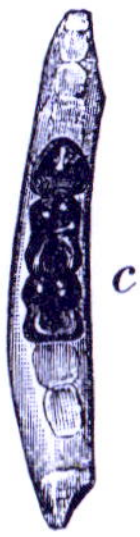

FIG. 7.

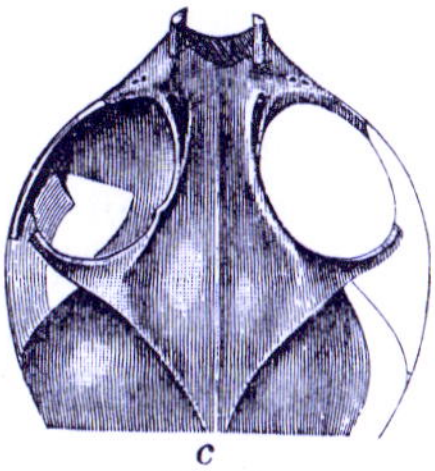

FIG. 3.

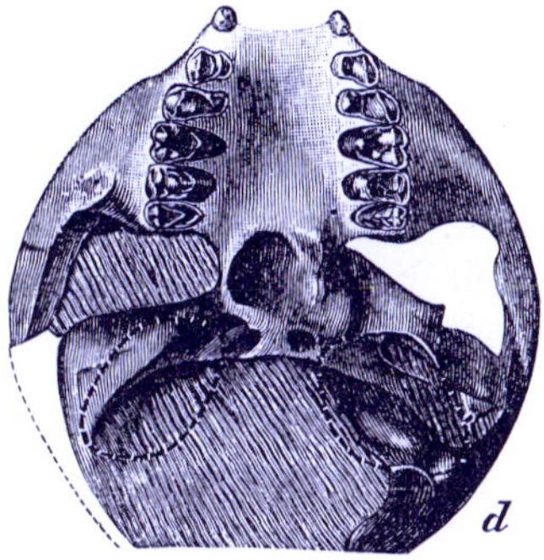

FIG. 4.

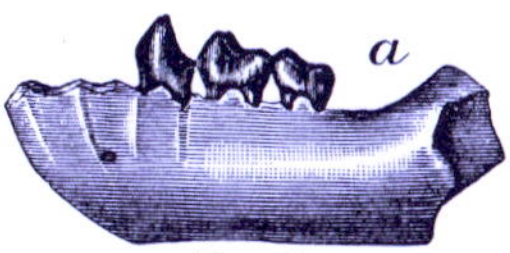

FIG. 5.

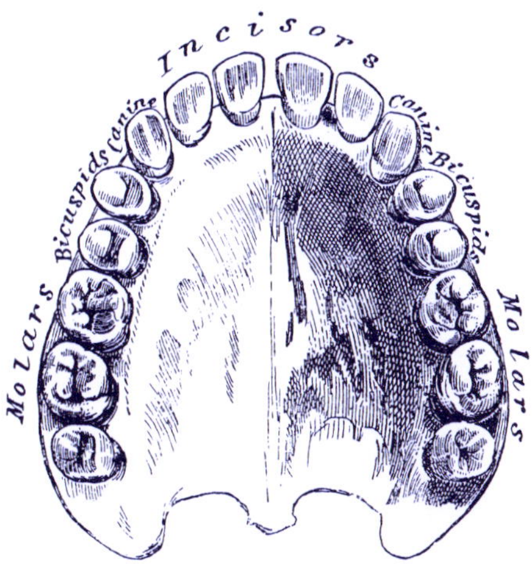

FIG. 8.

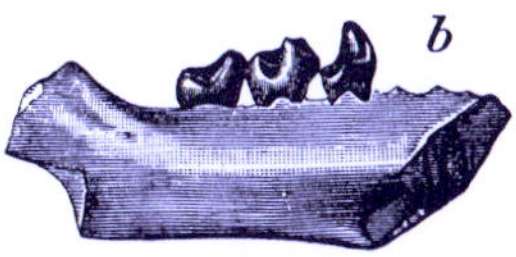

FIG. 6.

COMPARISON BETWEEN THE DENTITION OF THE LEMUR ANAPTOMORPHUS AND MAN.

Fig. 1, skull of *Anaptomorphus homunculus*, natural size. Fig. 2, same, oblique view, displaying the large cerebral hemispheres; compare Plate XIV for relatively small size in a contemporary (Fig. 3) or a successor in time (Fig. 2). Fig. 3, superior view of skull, natural size. Fig. 4, inferior view, $^3/_2$ natural size. Figs. 5, 6, and 7, left branch of lower jaw of *Anaptomorphus æmulus*, twice natural size; 5, from left side; 6, inner side; 7, from above. Fig. 8, superior dentition of *Homo sapiens* (from Allen), natural size.

accomplished a much greater work, the evolution of the human brain and its functions.

A very important lesson is derived from these and kindred facts. The monkeys were anticipated in the greater fields of the world's activity by more powerful rivals; the ancestors of the ungulates held the fields and the swamps, and the Carnivora, driven by hunger, learned the arts and cruelty of the chase. The weaker ancestors of the *Quadrumana* possessed neither speed nor weapons of offense and defense, and nothing but an arboreal life was left them, where they developed the prehensile powers of the feet. Their digestive system unspecialized, their food various, their life the price of ceaseless vigilance, no wonder that that quality of inquisitiveness and wakefulness was stimulated and developed which is the condition of progressive intelligence. So "the race has not been to the swift nor the battle to the strong"; the "survival of the fittest" has been the survival of the most intelligent, and natural selection proves to be, in its highest animal phase, intelligent selection.

IX.

THE DEVELOPMENTAL SIGNIFICANCE OF HUMAN PHYSIOGNOMY.*

THE ability to read character in the form of the human face and figure is a gift possessed by comparatively few persons, although most people interpret, more or less correctly, the salient points of human expression. The transient appearances of the face reveal temporary phases of feeling which are common to all men; but the constant qualities of the mind should be expressed, if at all, in the permanent forms of the executive instrument of the mind, the body. To detect the peculiarities of the mind by external marks has been the aim of the physiognomist of all times; but it is only in the light of modern evolutionary science that much progress in this direction can be made. The mind, as a function of part of the body, partakes of its perfections and its defects, and exhibits parallel types of development. Every peculiarity of the body has probably some corresponding significance in the mind, and the causes of the former are the remoter causes of the latter. Hence, before a true physiognomy can be attempted, the origin of the features of the face and general form must be known. Not that a perfect physiognomy will ever be possible. A mental constitution so complex as that of man can not be expected to exhibit more than its leading features in the body; but these include, after all, most of what it is important for us to be able to read from a practical point of view.

The present essay will consider the probable origin of the structural points which constitute the permanent expression. These may be divided into three heads, viz.: (1) Those of the general form or figure; (2) those of the surface or integument of the body with its appendages; and (3) those of the forms of the

---

* Abstract of a lecture delivered before the Franklin Institute of Philadelphia, January 20, 1881, in exposition of principles laid down in "The Hypothesis of Evolution," New Haven, 1870, p. 31.

head and face. The principal points to be considered under each of these heads are the following:

I. The General Form.

1. The size of the head.
2. The squareness or slope of the shoulders.
3. The length of the arms.
4. The constriction of the waist.
5. The width of the hips.
6. The length of the leg, principally of the thigh.
7. The sizes of the hands and feet.
8. The relative sizes of the muscles.

II. The Surfaces.

9. The structure of the hair (whether curled or not).
10. The length and position of the hair.
11. The size and shape of the nails.
12. The smoothness of the skin.
13. The color of the skin, hair, and irides.

III. The Head and Face.

14. The relative size of the cerebral to the facial regions.
15. The prominence of the forehead.
16. The prominence of the superciliary (eyebrow) ridges.
17. The prominence of the alveolar borders (jaws).
18. The prominence and width of the chin.
19. The relation of length to width of skull.
20. The prominence of the malar (cheek) bones.
21. The form of the nose.
22. The relative size of the orbits and eyes.
23. The size of the mouth and lips.

The significance of these, as of the more important structural characters of man and the lower animals, must be considered from two standpoints, the paleontological and the embryological. The immediate paleontological history of man is unknown, but may be easily inferred from the characteristics displayed by his nearest relatives of the order Quadrumana. If we compare these animals with man, we find the following general differences. The numbers correspond to those of the list above given.

I. As to General Form.—(3) In the apes the arms are longer; (8) the extensor muscles of the leg are smaller.

II. As to Surface.—(9) The body is covered with hair which is not crisp or woolly; (10) the hair of the head is short; (13) the color of the skin, etc., is dark.

III. As to Head and Face.—(14) The facial region of the skull is large as compared with the cerebral; (15) the forehead is

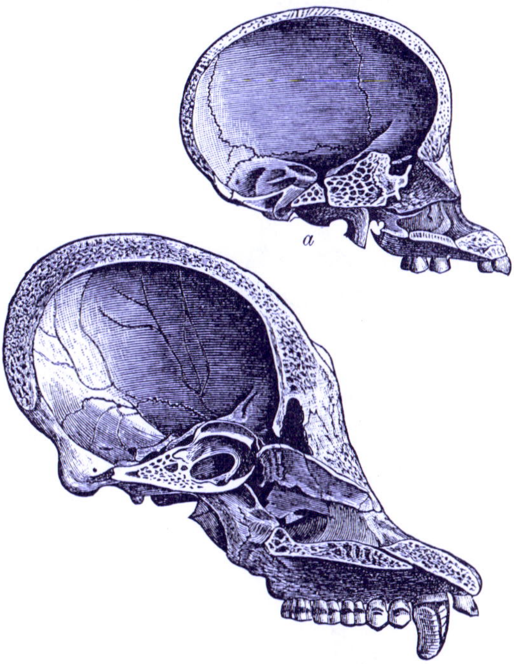

FIG. 52.—Section of skull of adult orang-outang (*Simia satyrus*). *a*, section of skull of young orang, showing relatively shorter jaws and more prominent cerebral region.

not prominent, and is generally retreating; (16) the superciliary ridges are more prominent; (17) the edges of the jaws are more prominent; (18) the chin is less prominent; (20) the cheek-bones are more prominent; (21) the nose is without bridge, and with short and flat cartilages; (22) the orbits and eyes are smaller (except in Nyctipithecus); (24) the mouth is small and the lips are thin.

It is evident that the possession of any one of the above characteristics by a man approximates him more to the monkeys, so far as it goes. He retains features which have been obliterated in other persons in the process of evolution.

In considering the physiognomy of man from an embryological standpoint, we must consider the peculiarities of the infant at birth. The numbers of the following list correspond with those already used (Fig. 53).

I. As to the General Form.—(1) The head of the infant is relatively much larger than in the adult; (3) the arms are relatively longer; (4) there is no waist; (6) the legs, and especially the thighs, are much shorter.

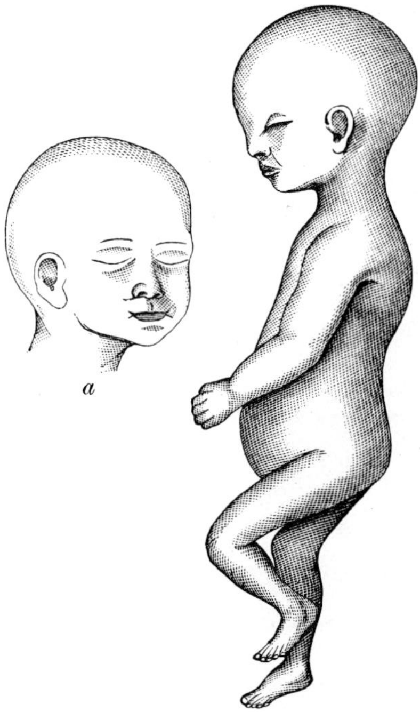

Fig. 53.—Figure of infant at birth; *a*, front of face.

II. As to the Surfaces.—(10) The body is covered with fine hair, and that of the head is short.

III. The Head and Face.—(14) The cerebral part of the skull greatly predominates over the facial; (16) the superciliary ridges are not developed; (17) the alveolar borders are not prominent; (20) the malar bones are not prominent; (21) the nose is without bridge and the cartilages are flat and generally short; (22) the eyes are larger.

It is evident that persons who present any of the characters cited in the above list are more infantile or embryonic in those respects than are others; and that those who lack them have left them behind in reaching maturity.

We have now two sets of characters in which men may differ from each other. In the one set the characters are those of monkeys, in the other they are those of infants. Let us see whether there be any identities in the two lists, i. e., whether there be any of the monkey-like characters which are also infantile. We find the following to be such:

I. As to General Form.—(3) The arms are longer.

II. Surface.—(10) The hair of the head is short, and the hair on the body is more distributed.

III. As to Head and Face.—(21) The nose is without bridge, and the cartilages are short and flat.

Three characters only out of twenty-three. On the other hand, the following characters of monkey-like significance are the opposites of those included in the embryonic list: (14) The facial region of the skull is large as compared with the cerebral; (15) the forehead is not prominent; (16) the superciliary ridges are more

FIG. 54.

FIG. 55.

FIG. 54.—Portrait of a girl at five years of age. FIG. 55.—Portrait of the same at seventeen years, showing the elongation of the facial region, and less protuberance of the cerebral.

prominent; (17) the edges of the jaws are more prominent. Four characters, all of the face and head. It is thus evident that in attaining maturity man resembles more and more the apes in some important parts of his facial expression.

It must be noted here that the difference between the young and embryonic monkeys and the adults is quite the same as those just mentioned as distinguishing the young from the adult of man (Figs. 1–2). The change, however, in the case of the monkeys is greater than in the case of man. That is, in the monkeys the jaws and superciliary ridges become still more prominent than in

PLATE X.

Esequibo Indian women, showing the following peculiarities: deficient bridge of nose, prognathism, no waist, and deficiency of stature through short femur. From photographs by Endlich.

ıan. As these characters result from a fuller course of growth rom the infant, it is evident that in these respects the apes are nore fully developed than man. Man stops short in the development of the face, and is in so far more embryonic.* The prominent forehead and reduced jaws of man are characters of "retardation." The characters of the prominent nose, with its elevated bridge, is a result of "acceleration," since it is a superaddition to he quadrumanous type from both the standpoints of paleontology nd embryology.† The development of the bridge of the nose is no doubt directly connected with the development of the front of he cerebral part of the skull and ethmoid bone, which sooner or ater carries the nasal bones with it.

If we now examine the leading characters of the physiognomy f three of the principal human sub-species, the Negro, the Mon-

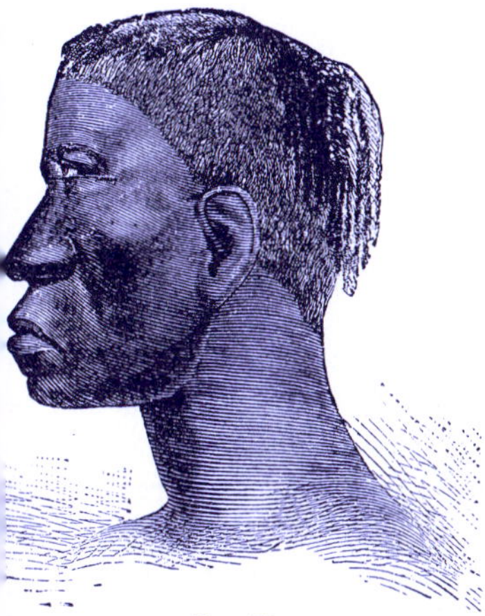

FIG. 56.

FIG. 57.

FIG. 56.—Profile of a Luchatze negro woman, showing deficient bridge of nose and hin, and elongate facial region and prognathism. FIG. 57.—Face of another Luchatze, howing flat nose, less prognathism and larger cerebral region. From Serpa Pinto.

olian, and the Indo-European, we can readily observe that it is n the two first named that there is a predominance of the quadumanous features which are retarded in man ; and that the emryonic characters which predominate, are those in which man is

* This fact has been well stated by C. S. Minot, in the "Naturalist" for 1882, . 511.

† See Cope, "The Hypothesis of Evolution," New Haven, 1870, p. 31.

accelerated. In race description the prominence of the edges o the jaws is called prognathism, and its absence orthognathism The significance of the two lower race characters, as compare with those of the Indo-European, is as follows :

Negro.—Hair crisp (a special character), short (quadrum. accel.) ; prognathous (quadrum. accel.) ; nose flat, without bridg (quadrum. retard.) ;* malar bones prominent (quadrum. accel.) beard short (quadrum. retard.) ; arms longer (quadrum. accel.) extensor muscles of legs small (quadrum. retard.).

Mongolian.—Hair straight, long (accel.) ; jaws prognathou (quadrum. accel.) ; nose flat or prominent, with or without bridge malar bones prominen (quadrum. accel.) ; bear none (embryonic) ; arm shorter (retard.) ; extenso muscles of leg ("calf" smaller (quad. retard.).

Fig. 58.—Portrait of Satanta, a late chief of the Kiowas (from the Red River of Texas), from a photograph. The predominance of the facial region, and especially of the malar bones, and the absence of beard, are noteworthy.

Indo-European.—Hai long (accel.) ; jaws orthognathous (embryonic retard.) ; nose (generally prominent with bridg (accel.) ; malar bones reduced (retard.) ; bear long (accel.) ; arms shorte (retard.) ; extensor muscles of the leg large (accel.)

The Indo-Europea race is then the highest b virtue of the acceleratio of growth in the development of the muscles b which the body is maintained in the erect positio (extensors of the leg), an in those important elements of beauty, a well developed nose and beard

* In the Bochimans, the flat nasal bones are coössified with the adjacent elements as in the apes (Thulié).

PLATE XI.

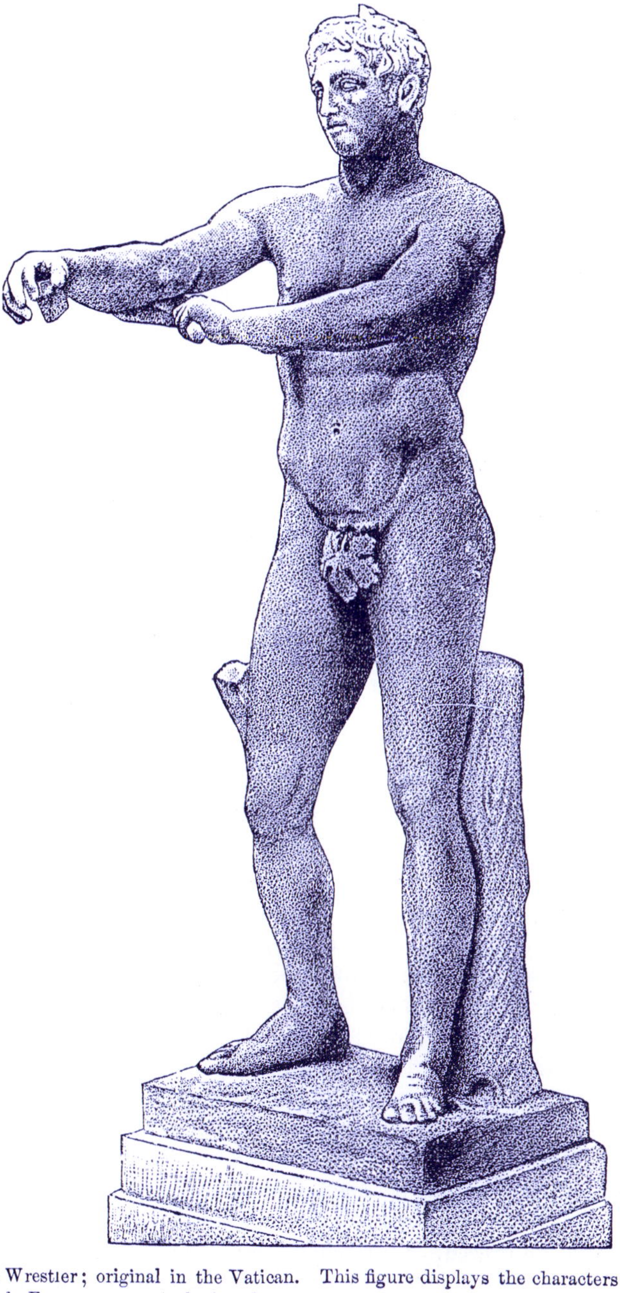

The Wrestler; original in the Vatican. This figure displays the characters of the male Indo-European, except the beard.

It is also superior in those points in which it is more embryonic than the other races, viz., the want of prominence of the jaws and cheek-bones, since these are associated with a greater predominance of the cerebral part of the skull, increased size of cerebral hemispheres, and greater intellectual power.

A comparison between the two sexes of the Indo-Europeans expresses their physical and mental relations in a definite way. I select the sexes of the most civilized races, since it is in these, according to Broca and Topinard, that the sex characters are most pronounced. They may be contrasted as follows. The numbers are those of the list on page 282, already used. I first consider those which are used in the tables of embryonic, quadrumanous, and race characters :

MALE.	FEMALE.
I. The General Form.	
2. Shoulders square.	Shoulders sloped.
4. Waist less constricted.	Waist more constricted.
5. Hips narrower.	Hips wider.
6. Legs longer.	Legs shorter (very frequently).
8. Muscles larger.	Muscles smaller.
II. The Integuments, etc.	
10. More hair on body, that of head shorter; beard.	Less hair on body, that of head longer no beard.
12. Skin rougher (generally).	Skin smoother.
III. The Head and Face.	
16. Superciliary ridges more prominent.	Superciliary ridges low.
22. Eyes often smaller.	Eyes often larger.

The characters in which the male is the most like the infant are two, viz., the narrow hips and short hair. Those in which the female is most embryonic are five, viz., the shorter legs, smaller muscles, absence of beard, low superciliary ridges, and frequently larger eyes. To these may be added two others not mentioned in the above lists ; these are (1) the high-pitched voice, which never falls an octave as does that of the male ; and (2) the structure of the generative organs, which in all Mammalia more nearly resemble the embryo and the lower Vertebrata, in the female than in the male. Nevertheless, as Bischoff has pointed out, one of the most important distinctions between man and the apes is to be found in the external reproductive organs of the female.

From the preceding summary sketch the reader will be able to explain the meaning of most of the peculiarities of face and form which he will meet with. Many persons possess at least one quadrumanous or embryonic character. The strongly convex upper lip frequently seen among the lower classes of the Irish is a modified quadrumanous character. Many people, especially those of the Sclavic races, have more or less embryonic noses. A retreating chin is a marked monkey character. Shortness of stature is mostly

FIG. 59.—Australian native (from Brough Smyth), showing small development of muscles of legs, and prognathism.

due to shortness of the femur, or thigh ; the inequalities of people sitting are much less than those of people standing. A short femur is embryonic ; so is a very large head. The faces of some people are partially embryonic, in having a short face and light lower jaw. Such faces are still more embryonic when the forehead and eyes are protuberant. Retardation of this kind is most frequently

PLATE XII.

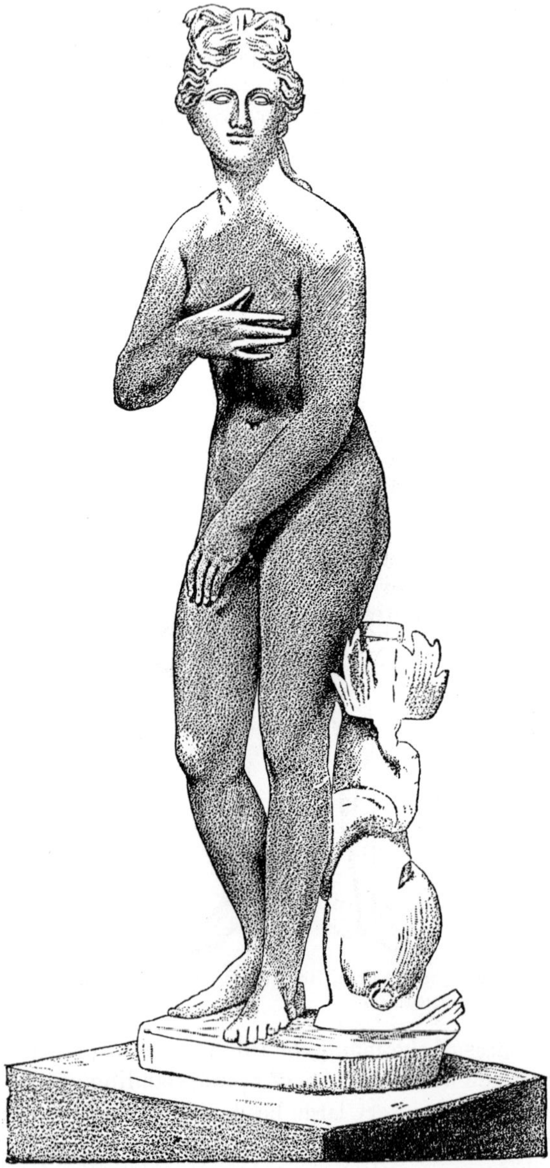

The Venus of the Capitol (Rome). The form and face present the characteristic peculiarities of the female of the Indo-European race.

seen in children, and more frequently in women than in men. The length of the arms would appear to have grown less in comparatively recent times. Thus the humerus in most of the Greek statues, including the Apollo Belvidere, is longer than those of modern Europeans, according to a writer in the "Bulletin de la Société d'Anthropologie" of Paris, and resembles more nearly that of the modern Nubians than any other people. This is a quadrumanous approximation. The miserably developed calves of many of the savages of Australia, Africa, and America, are well known. The fine swelling gastrocnemius and soleus muscles characterize the highest races, and are most remote from the slender shanks of the monkeys. The gluteus muscles developed in the lower races as well as in the higher, distinguish them well from the monkeys with their flat posterior outline.

It must be borne in mind that the quadrumanous indications are found in the lower classes of the most developed races. The status of a race or family is determined by the percentage of its individuals who do and do not present the features in question. Some embryonic characters may also appear in individuals of any race, as a consequence of special circumstances. Such are, however, as important to the physiognomist as the more normal variations.

Some of these features have a purely physical significance, but the majority of them are, as already remarked, intimately connected with the development of the mind, as an effect or necessary coincidence. I will examine these relations in a future article.

X.

THE EVIDENCE FOR EVOLUTION IN THE HISTORY OF THE EXTINCT MAMMALIA.*

THE subject to which I wish to call your attention this morning requires neither preface nor apology, as it is one with the discussion of which you are perfectly familiar. I bring it before the general session of the Association in view of the fact that you are all familiar with it in a general way, and that it probably interests the members of sections who do not pursue the special branch to which it refers, as well as those who do : also, since it has been brought before us in various public addresses for many years, during the meetings of this Association, I thought it might be well to introduce it at this meeting, in order that we might not omit to have all the sides of it presented.

The interests which are involved in it are large : they are chiefly, however, of a mental and metaphysical character ; they do not refer so much to industrial and practical interests, nor do they involve questions of applied science. They involve, however, questions of opinion, questions of belief, questions which affect human happiness, I venture to say, even more than questions of applied science ; certainly, which affect the happiness of the higher grades of men and women more than food or clothing, because they relate to the states of our mind, explaining as they do the reasons of our relations to our fellow beings, and to all other things by which we are surrounded, and the general system of the forces by which we are controlled. So it has always appeared to me : hence I have selected the department of biology for study, and have taken a great interest in this aspect of it.

The doctrine of evolution, as taught by the biologists of today, has several stages or parts of its presentation. First, the foundation principle is this : that the species of animals and of

---

* An address delivered before the American Association for the Advancement of Science, 1883.

plants, the species of organic beings, as well as the various natural divisions into which these organic beings fall, have not always been as we see them to-day, but they have been produced by a process of change which has progressed from age to age through the influence of natural laws; that, therefore, the species which now exist are the descendants of other species which have existed heretofore, by the ordinary processes of reproduction; and that all the various structures of organic beings, which make them what they are, and which compel them to act as they now act, are the result of gradual or sudden modifications and changes during the periods of geologic time. That is the first phase or aspect which meets the naturalist or biologist.

Another phase of the question relates to the origin of that life itself which is supposed to inhabit or possess organic beings. There is an hypothesis of evolution which derives this life from no-life, which derives vitality from non-vitality. This is another branch of the subject, to which I can not devote much attention to-day.

There is still another department of the subject, which relates to the origin of mind, and which derives the mental organization of the higher animals, especially of man, from pre-existent types of mental organization. This gives us a genealogy of mind, a history of the production or creation of mind, as it is now presented in its more complex aspects as a function of the human brain. This aspect of the subject is, of course, interesting; and upon it I can touch with more confidence than upon the question of the origin of life.

Coming now to the question of the origin of structures, we have by this time accumulated a vast number of facts which have been collated by laborious and faithful workers, in many countries and during many years; so that we can speak with a good deal of confidence on this subject also. As to the phenomena which meet the student of zoölogy and botany at every turn, I would merely repeat what every one knows—and I beg pardon of my biological friends for telling them a few well-known truths, for there may be those present who are not in the biological section—the phenomena which meet the student of biology come under two leading classes. The first is the remarkable fidelity of species in reproducing their like. "Like produces like," is the old theorem, and is true in a great many cases; just as coins are struck from the die, just as castings are turned out from a common mold. It is one of the most wonderful phenomena of nature, that such complex organisms, consisting of so many parts, should be repeated from age to age, and from generation to generation, with such surprising fidelity and precision. This fact is the first that strikes the student of these sciences. The first impression of the ordinary person would be, that these things must continue unchanged. When I began to study zoölogy and botany, I was surprised to find there was a science of which I had no conception, and that was this remarkable reproduction of types one after another in continued succession. After a man has had this idea thoroughly assimilated by his honest and conscientious studies, he will be again struck with another class of facts. He will find, not unfrequently, that this doctrine does not apply. He will find a series of facts which show that many individuals fail to coincide with their fellows precisely, the most remarkable variations and the most remarkable half-way attitudes and double-sided aspects occurring; and he will come to the conclusion, sooner or later, that like does not produce like in some animals with the same precision and fidelity with which it is accomplished in other animals. So that we have these two classes of facts—the one relating to, and expressing, the law of heredity; the other which expresses the law of variation. I should not like to say which class of facts is the most numerously presented to the student. In the present fauna we find many groups of species and varieties; but exactly how many species we have, how many genera and families we have, we can not definitely state. The more precise and exact a person is in his definition and in his analysis, the more definite his science becomes, and the more precise and scientific his work. Biology is a science of analysis of forms. What the scales are to the chemist and the physicist, the rule and measure are to the biologist. It is a question of dimension, a question of length and breath and thickness, a question of curves, a question of crooked shapes or simple shapes—rarely simple shapes, mostly crooked, generally bilateral. It requires that one should have a mechanical eye, and should have also something of an artistic eye, to appreciate these forms, to measure them, and to be able to compare and weigh them.

Now, when we come to arrange our shapes and our measurements, we find, as I said before, a certain number of identities, and a certain number of variations. This question of variation is so common and so remarkable, that it becomes perfectly evident to the specialist in each department, that like does not at all times produce like. It is perfectly clear, and I will venture the assertion that nearly all the biologists in this room will bear me witness, that variability is practically unlimited in its range, and multiplied in the number of its examples. That is to say: species vary by adding or by failing to retain certain characteristics; and generic and other characters are found to appear or disappear in accordance with some law to be discussed farther on.

I believe that this is the simplest mode of stating and explaining the law of variation: that some forms acquire something which their parents did not possess; and that those which acquire something additional have to pass through more numerous stages than their ancestors; and those which lose something pass through fewer stages than their ancestors; and these processes are expressed by the terms "acceleration" and "retardation."

Of course we are met with the opposite side of the case—the law of heredity. We are told that the facts there are not accounted for by any law of evolution; that we can not pass from one class of facts to the other class of facts; that the law of the one class is not that of the other. Here is a question of rational processes, of ordinary reason. If the rules of chemistry are true in America, I imagine they are true in Australia and Africa, although I have not been there to see. If the law of gravitation is effective here, I do not need to go to Australia or New Zealand to ascertain whether it be true there. So, if we find in a group of animals a law sufficient to account for their creation, it is not necessary to know that others of their relatives have gone through a similar process. I am willing to allow the ordinary practical law of induction, the practical law of inference, to carry me over these gaps, over these interruptions. And I state the case in this way, because it is just here that some people differ from me, and it is just here that I say the simple question of rationality comes in. I can not believe that Nature's laws are so dissimilar, so irregular, so inexact, that those which we can see and understand in one place are not true in another; I also believe that the question of geological likelihood is similar to the question of geographical likelihood. If a given process be true in one of the geological periods, it is true in another; if it be true in one part of the world, it is true in another; because I find interruptions in the series here, it does not follow that there need be interruptions clear through from age to age. The assumption is on the side of the man who asserts that transitions have not taken place between forms which are now distinct.

We are told that we find no sort of evidence of that transition in past geological periods ; we are assured that such changes have not taken place ; we are even assured that no such sign of such transition from one species to another has ever been observed—a most astonishing assertion to make *to* a biologist, or *by* a biologist ; and such persons have even the temerity to cite such a special case as that between the wolf and the dog. Many of our domestic dogs are nothing but wolves, which have been modified by the hand of man to a very slight extent indeed. Many dogs, in fact nearly all dogs, are descendants of wild species of various countries, and are variously modified.

To take the question of the definition of species. Supposing we have several species well defined, say four or five. In the process of investigation we obtain a larger number of individuals, many of which betray characters which invalidate the definitions. It becomes necessary to unite the four or five species into one. And so, then, because our system requires that we shall have accurate definitions (the whole basis of the system is definitions ; you know the very comprehension of the subject requires definitions), we throw them all together, because we can not define all the various special forms as we did before, until we have but one species. And the critic of the view of evolution tells us, "I told you so ! There is but one species, after all. There is no such thing as a connection between species : you never will find it." Now, how many discoveries of this kind will be necessary to convince the world that there are connections between species ? How long are we to go on finding connecting links, and putting them together, as we have to do for the sake of the definition, and then be told that we have, nevertheless, no intermediate forms between species ? The matter is too plain for further comment. We throw them together, simply because our definitions require it. If we knew all the known individuals which have lived, we should have no species, we should have no genera. That is all there is of it. It is simply a question of a universal accretion of material, and the collection of information. I do not believe that the well-defined groups will be found to run together, as we call it, in any one geological period, certainly in no one recent period. We recognize, however, in looking backward, that they converge to a wonderful extent : one group has diverged at one period, and another one has become diversified in a different period; and so each one has its history, some beginning farther back than others, some reaching far back beyond the very beginning of the time when fossils could be preserved. I call attention to this view because it is a very easy matter for us to use words for the purpose of confusing the mind; for, next to the power of language to express clear ideas, is its power of expressing no ideas at all. As we all know, we can *say* many things which we can not *think*. It is a very easy thing to say twice two are equal to six, but it is impossible to think it.

I would cite what I mean by variations of species in one of its phases: I mention a genus of snakes, Ophibolus, which is found in the United States. If we take the species of this snake-genus as found in the Northern States, we have a good many species well defined. If we go to the Gulf States, and examine our material, we see we have certain other species well defined, and they are very nicely distinguished. If, now, we go to the Pacific coast, to Arizona and New Mexico, we shall find another set of species well defined indeed. If we take all these different types of our specimens of different localities together, our species, as the Germans say, all tumble together: definitions disappear, and we have to recognize, out of the preliminary list of thirteen or fourteen, only four or five. That is simply a case of the kind of fact with which every biologist is perfectly familiar.

When we come to the history of the extinct forms of life, it is perfectly true that we can not observe the process of descent in actual operation, because, forsooth, fossils are necessarily dead. We can not perceive any activities, because fossils have ceased to act. But, if this doctrine be true, we should get the series, if there be such a thing; and we do, as a matter of fact, find longer or shorter series of structures, series of organisms proceeding from one form into another form, which are exactly as they ought to be if this process of development by descent had taken place.

I am careful to say this; because it is literally true, as we all must admit, that species must fall into some kind of order or other. You could not collect bottles, you could not collect old shoes, but you could make some kind of a serial order of them. There are, no doubt, characters by which such and such shoes could be distinguished from other shoes, these bottles from other bottles. But if serial order does not prove evolution, as is too often assumed, we have in recent forms of life in zoölogy and

PLATE XIII.

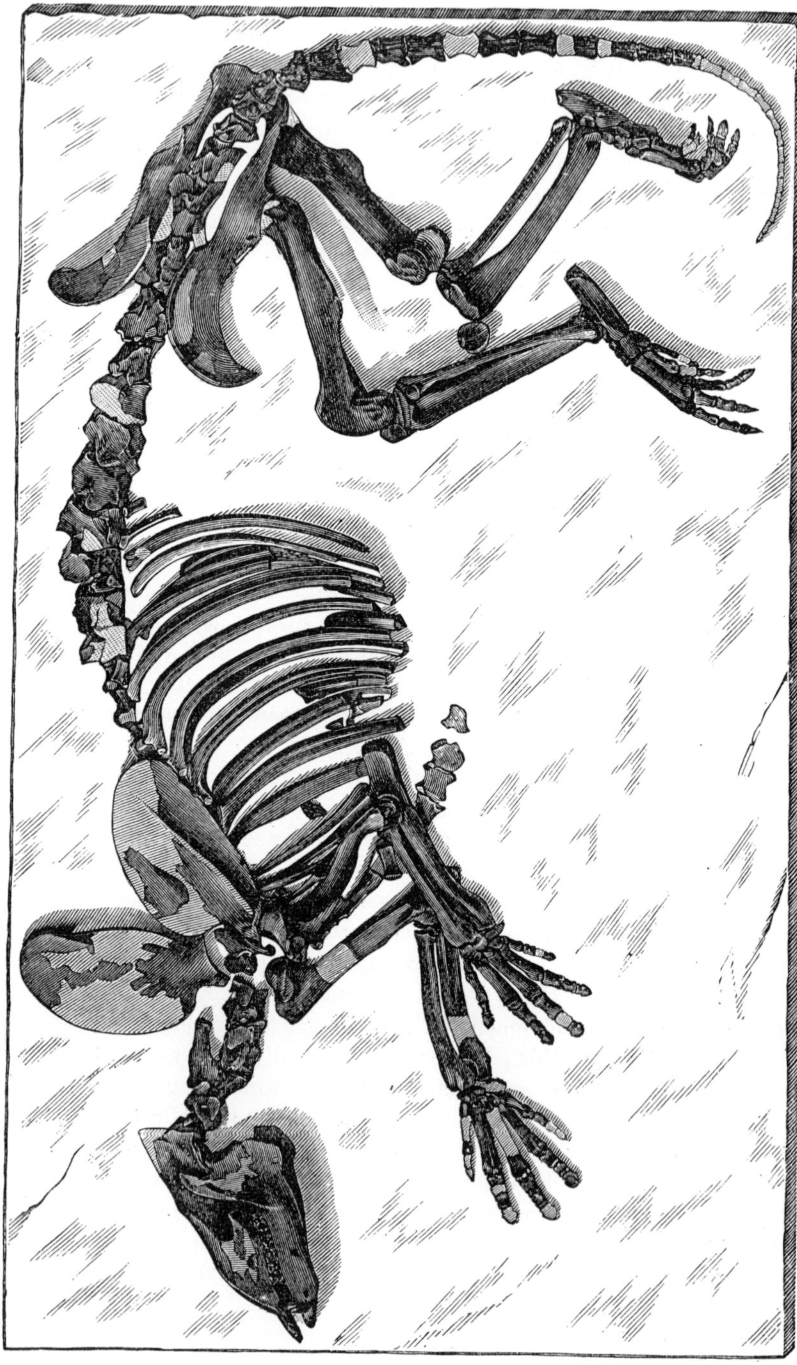

Phenacodus primævus (Cope), one seventh natural size. From specimen in coll. Cope.

botany, irrefragable proofs of the metamorphoses, and transformations, and changes of the species, in accordance with the doctrine with which we commenced.

We now come to the second chapter of our subject. With the assumption, as I take it already satisfactorily proved, of species having changed into others, in considering this matter of geological succession or biological succession, I bring you face to face with the nature and mode of the change; and hence we may get a glance perhaps at its laws.

I have on the board a sketch or table which represents the changes which took place in certain of the Mammalia. I give you a summary of the kind of thing which we find in one of the branches of paleontology. I have here two figures, one representing a restoration, and the other an actual picture, of two extinct species that belong to the early Eocene periods. One represents the ancestor of the horse line, Hyracotherium, which has four toes on his anterior feet, and three behind; and the other (Plate XIII) a type of animal, Phenacodus, which is antecedent to all the horse series, the elephant series, the hog, the rhinoceros, and all the other series of hoofed animals. It has five toes on all the feet. Each presents us with the primitive position in which their series first come to our knowledge in the history of geological time.

I have also arranged here a series of some leading forms of the three principal epochs of the Mesozoic times, and six of the leading ones of the Tertiary time. I have added some dates to show you the time when the faunæ which are entombed in those beds were discovered, in the course of our studies; and you will easily see how unsafe it is to say that any given type of life has never existed, or even to assert that such and such a form is unknown; and it is still more unsafe, I think, to assert that any given form of life properly defined, or that a specific intermediate form of life, will not be found. I think it is much safer to assert that such and such intermediate forms will be found. I have frequently had the pleasure of realizing anticipations of this kind. I have asserted that certain types would be found, and they have been found. You will see that I attend to the matter of time because there have been a great many things discovered in the last ten or fifteen years in this department. With these forms I give the date of the discovery of the fauna in which they are embraced.

Here we have the White River fauna discovered in 1856; then we skip a considerable period of time, and the next one was in 1869, when the Cretaceous series was found. Six or seven Cretaceous faunæ have been discovered. Then we have the Bridger fauna in 1870, the Wasatch fauna in 1874. Next we have, in 1877, the Equus beds and the fauna which they embrace, part of which was also found in 1878. The Permian fauna, which is one of the last, in 1879; and the latest, the Puerco, which gives the oldest and ancestral types of the modern forms of Mammalia, was only found in 1881. When I first commenced the study of this subject, about 1860, there were perhaps two hundred and fifty species known. There are now somewhere above 1,000, and we are augmenting them all the time. I have found many myself: if they were distributed through the days of the year, I think in some months I should have had several every day. You see then that the accessions to knowledge which are constantly being made make it unsafe to indulge in any prophecies; as, for instance, that, because such and such things have not been found, therefore they can not be; for we find such and such things really have been and really are discovered.

The successive changes that we have in the Mammalia have taken place in the limbs, feet, teeth, and brain, and the vertebral column. The parts which present us with the greatest numbers of variations are those in which many parts are concerned, as in the limbs and feet. In the Lower Eocene (Puerco), the toes were 5–5. In the Loup Fork fauna, some possess toes but 1–1. Prior to this period no such reduction was known, although in the Loup Fork fauna a very few species remained 5–5. Through this entire series we have transitions steady and constant, from 5–5, to 4–5, to 4–4, to 4–3, to 3–3, to 2–2, to 1–1. In the Puerco period there was not a single mammal of any kind which had a good ankle-joint; which had an ankle-joint constructed as ankle-joints ought to be, with tongue and groove. The model ankle-joint is a tongue-and-groove arrangement. In this period they were nearly all perfectly flat. As time passes on we get them more and more grooved, until in the Loup Fork fauna and the White River fauna they are nearly all grooved. The soles of the feet, in the Puerco fauna, are all flat; but in the Loup Fork fauna the soles of the feet are in the air, and the toes only are applied to the ground, with the exception of the line of monkeys, in which the feet have not become erect on the toes, and the elephant, in which the feet are nearly flat also, and the line of bears, where they are also flat. As regards the angulation between the small bones of the palm and of the sole,

Formation.	No. of toes.	Feet.	Astragalus.	Carpus and tarsus.	Ulno-radius.	Superior molars.	Zygapophyses.	Brain.
Pliocene	1–1 2–2	Digitigrade. (Plantigrade.)	Grooved. (Flat.)	Interlocking. (Opposite.)	Faceted.	4-tubercles, crested and cemented	Doubly involute. Singly involute.	Hemispheres larger, convoluted.
Upper Miocene....... (Loup Fork.)	3–3 4–4 (5–5)							
Middle............... (John Day.)	2–2 3–3 4–4	Digitigrade.	Grooved.	Interlocking.	Faceted. Smooth.	4-tubercles, and crested.	Singly involute. Doubly involute.	Hemispheres larger, convoluted.
Lower................ (White River.)	3–3 4–3 4–4	Digitigrade. Plantigrade.	Grooved.	Interlocking.	Smooth. Faceted.	4-tubercles, and crested.	? Singly involute.	Hemispheres small, and larger.
Eocene........... Upper............. (Bridger.)..........	3–3 4–3 4–5 5–5	(Digitigrade.) Plantigrade.	Grooved. (Flat.)	Opposite. Interlocking.	Smooth.	4-tubercles. 3-tubercles, and crested.	Singly involute. Plane.	Hemispheres small.
Middle............... (Wasatch.)	4–3 4–5 5–5	Plantigrade. (Digitigrade.)	Flat. (Grooved.)	Opposite. Interlocking.	Smooth.	4-tubercles. 3-tubercles, a few crested.	Plane. Singly involute.	Hemispheres small; mesencephalon sometimes exposed.
Lower................ (Puerco.)	5–5	Plantigrade.	Flat.	Opposite.	Smooth.	3-tubercles. (4-tubercles), none crested.	Plane.	Mesencephalon exposed; hemisphere small and smoother.

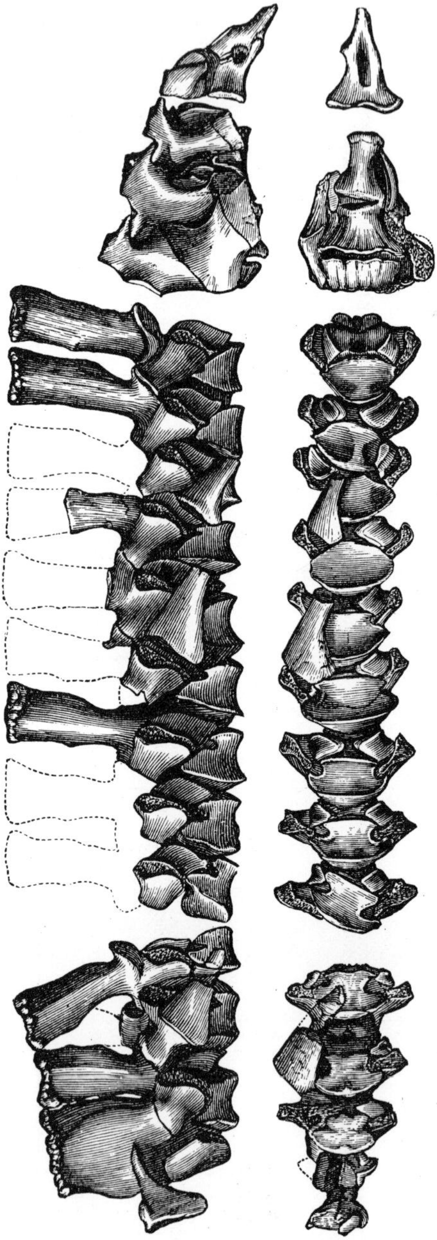

FIG. 60.—Vertebral column of *Eryops megacephalus*, one fourth natural size. *a*, profile; *b*, from below. From the Permian epoch of Texas.

there is not a single instance in which those bones are locked in the Lower Eocene, as they are in the later and latest Tertiary. So also with the articulation of the toes with the foot.

When we come to the limbs, the species of the Puerco fauna have short legs. They have gradually lengthened out, and in the late periods they are nearly all relatively long.

Coming to the vertebræ as a part of the osseous system, I mention the zygapophyses, or anteroposteriorly directed processes, of which the posterior looks down and the anterior looks up. They move on each other, as the vertebral column bends from side to side. In the lower forms of vertebrates they are always flat, and in the hoofed mammals of the Puerco period they are all flat. In the Wasatch period we get a single group in which the articulation, instead of being perfectly flat, becomes rounded; in the later periods we get them very much rounded; and, finally, in the latest forms, we get the double curve and the locking process in the vertebral column, which, as in the limb, secures the greatest strength with the greatest mobility.

In the first stages of the growth of the spinal column, it is a notochord, or a cylinder of cartilage or softer material. In later stages the bony deposit is made in its sheath until it is perfectly segmented. Now, all the Permian land-animals, reptiles and batrachians, retain this notochord with the elements of osseous vertebræ, in a greater or less degree of completeness. There are some in South Africa, I believe, in which the ossification has come clear through the notochord; but they are few. In this characteristic the Permian appears almost, perhaps absolutely, peculiar as

Fig. 61.—Sleeve of a coat showing folds produced by lateral flexure which leaves interspaces similar to the segments of a rhachitomous vertebra. Thus, *i* represents intercentrum; *p*, pleuro-centrum; and *n*, neurapophysis.

regards land-animals. There is something to be said as to the condition of the column from a mechanical standpoint, and it is this: that the chorda exists, with its osseous elements disposed about it; and in the Permian batrachians, equally related to salamanders and frogs, these osseous elements are arranged in the sheath or skin of the chorda; and they are in the form of regular concave segments, very much like such segments as you can take from the skin of an orange—but parts of a cylinder, and having greater or less dimensions according to the group or species. Now, the point of divergence of these segments is on the side of the column. The contacts are placed on the side of the column where the segments separate—the upper segments rising and the lower segments coming downward. To the upper segments are attached the arches and their articulations, and the lower segments are like the segments of a cylinder. If you take a flexible cylinder, and cover it with a more or less inflexible skin or sheath, and bend that cylinder sidewise, you of course will find that the wrinkles or fractures of that part of the surface will take place along the line of the shortest curve, which is on the side; and, as a matter of fact, you have breaks of very much the character of the segments of the Permian Batrachia. It may not be so symmetrical as in the actual animal, for organic growth is symmetrical so far as not interfered with; for, when we have two forces, the one of hereditary growth, and the other of change or alteration, and they contend, you will find in the organic being a quite symmetrical result. That is the universal rule. In the cylinder bending both ways, of course the shortest line of curve is right at the center of the side of that cylinder, and the longest curve is of course at the summit and base, and the shortest curve will be the point of fracture. And that is exactly what I presume has happened in the case of the construction of the segments of the sheath of the vertebral column, by the lateral motion of the animal in swimming, and which has been the actual cause of the disposition of the osseous material in its form. I have gone beyond the state of the discussion in calling attention to one of the forces which have probably produced this kind of result. That is the state of the vertebral column of many of the Vertebrata of the Permian period.*

I go back to the Mammalia and call attention to the teeth. The ordinary tooth of the higher type of the Mammalia, whether hoofed or not, with some exceptions, is complex with crests or cusps. By cutting the complex grinding surfaces, we find they have been derived by the infolding of extensions of four original cusps or tubercles. They have been flattened, have been rendered oblique, have run together, have folded up, have become acute, have descended deeply, or have lifted themselves, so that we have teeth of all sorts and kinds, sometimes very elegant, and often-

* Note (Ed. 1886).—Some further elucidation of this point is necessary, since the fishes have not produced this kind of segmentation by the lateral motions of the vertebral column in swimming; at least, such segmentation is not yet known among them, but rather simple discoidal segmentation. The rhachitomous segmentation above described would result from a greater flexure than that required for the propulsion of a fish through the water. This increased flexure was no doubt the concomitant of the acquisition of a terrestrial mode of life by the early batrachians. Progression on land by an animal with weak limbs requires much greater flexure of the column than the act of swimming by the use of the caudal part of the body as in fishes. One can easily convince himself of this by comparing the movements of a fish in the water with those of a salamander on land. The snakes, where limbs are wanting, show an even stronger flexure in progression.

times very effective in mechanism. In many primary ungulates, the primitive condition of four conical tubercles is found. In passing to older periods we find the Mammalia of the Puerco period, which never have more than three principal tubercles, with the exception of three or four species. In the succeeding periods, however, they get the fourth tubercle on the posterior side. Finally, you get a complicated series of grinding or cutting apparatus, as the case may be.

Last, but not least, we take the series of the brain. No doubt the generalization is true, that the primitive forms of Mammalia had small brains with smooth hemispheres; later ones had larger brains with complex hemispheres. In general the Carnivora have retained a more simple form of brain, while herbivorous animals have retained a more complicated type of brain. The lowest forms of Mammalia display the additional peculiarity of having the middle brain exposed; and the hemispheres or large lobes of the brain, which are supposed to be the seat of the mental phenomena, are so reduced in size at the back end, that you see the middle brain distinctly, though it is smaller than in reptiles and fishes. (See Plate XIV.)

It is beyond the possibility of controversy, that these series have existed, that they have originated in simplicity and have resulted in complication; and the further induction must be drawn, that the process of succession has been toward greater effectiveness of mechanical work. There are also cases of degradation, as in the growing deficiency in dentition in man. There is no doubt that a large number of people are now losing their wisdom-teeth in both jaws.

We are now brought to the question of the relations which mind bears to these principles. The question as to the nature of mind is not so complex as it might seem. There is a great deal of it, to be sure; but on examination it resolves itself into a few ultimate forms. An analysis reduces it to a few principal types or departments—the departments of the intelligence and of the emotions (with their primary simpler forms, likes and dislikes), and the will, if such there be. These three groups, proposed by Kant, are well known, and are adopted by many metaphysicians; and they stand the scrutiny of modern science as applied to both men and the lower animals. But the question of the material of the mind, the original raw stuff out of which mind was made, is one which is claiming attention now from biologists, as it always has

PLATE XIV.

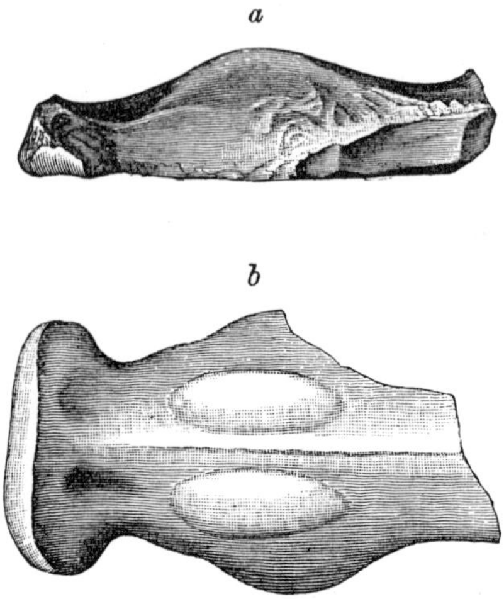

FIG. 1.

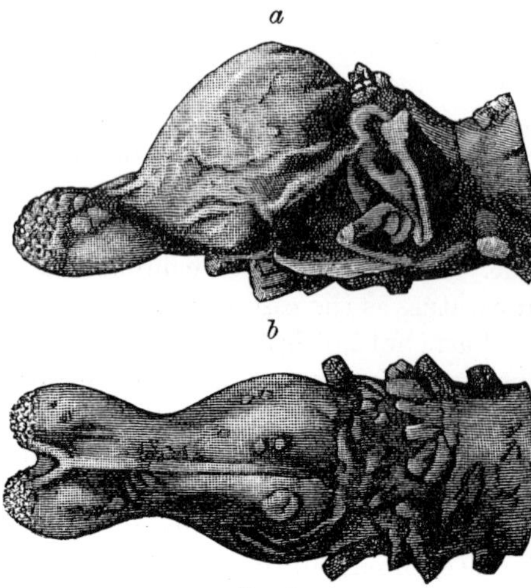

FIG. 2.

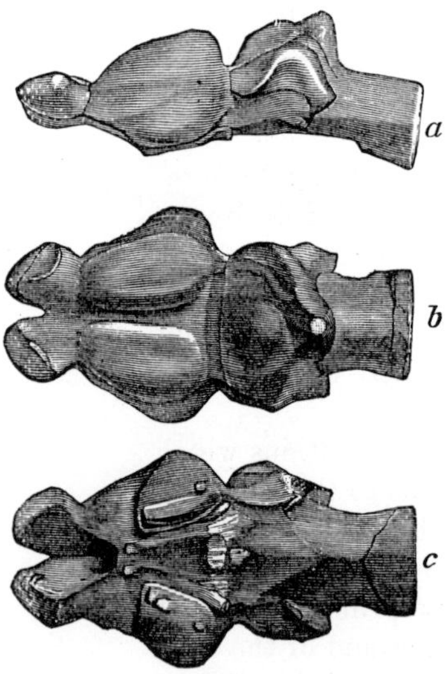

FIG. 3.

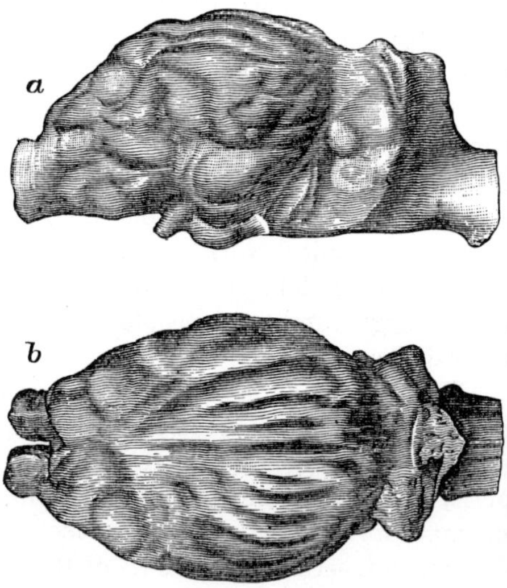

FIG. 4.

FIGURES OF CASTS OF BRAIN-CHAMBERS OF EXTINCT MAMMALIA.

Fig. 1, *Periptychus rhabdodon*, cast of roof of brain case, showing middle-brain hemispheres and part of olfactory lobes, natural size. Original; from Puerco bed of New Mexico. Fig. 2, *Uintatherium mirabile*, one third natural size, from Marsh. From Bridger bed of Wyoming. Fig. 3, *Phenacodus primævus*, one half natural size. Original; from Wasatch bed of Wyoming. Fig. 4, *Procamelus occidentalis*, one half natural size. Original; from Loup Fork bed of New Mexico. Figs. *a*, profile; *b*, above; *c*, below.

done from physiologists proper, and physicians. This is sensibility, mere simple sensibility, unmodified sensation, or consciousness. Sensibility, in connection with memory, is sufficient for the accomplishment of wonderful results. It is only necessary to impress the sensibility with the stimuli which this world affords, whether from the outside or the inside, to have the record made, and to have the record kept. Among wonderful things this is perhaps the most wonderful : that a given form of matter should be able to retain a record of events, a record which is made during a greater or less degree of sensibility ; which is retained in a state of insensibility ; and is finally returned to the sensibility by some curious process of adhesion, as the results of impresses which are found on the material tissue concerned.

And these simple elements of mind are found in animals. No zoölogist who has perception or honesty, nor any farmer or breeder, nor any person who has charge of animals in any way, can deny sensibility to all the lower animals at times. The great stumbling-block in the way of the thinker in all this field, is the great evanescence of this sensibility ; the great ease with which we dissipate it. The readiness with which we can deprive a fellow-being of his sense is a stumbling-block in more ways than one. While it is a question of the greatest difficulty, nevertheless, like other departments of nature, doubtless it will ultimately be explained by the researches of physiologists. I only need to call attention to consciousness as an important factor in evolution.

We now approach the question of the origin of organic machines with fresh resources. Did the consciousness of the animal find his structure made ready to hand, or did he, under the dire stimuli of necessity, produce through ages these modifications in his own structure ? We are told by some of our friends that law implies a lawgiver, that evolution implies an evolver : the next question is, Where is the lawgiver ? where is the evolver ? where are they located ? This question is best answered, as it appears to me, as follows. In the first place, I may say, it is distinctly proved in some directions, that the constant applications of force or motion in the form of strains, in the form of impacts and blows, upon any given part of the animal organism, do not fail to produce results in change of structure. I believe the changes in the ungulates, to which I have called your attention, are the result of strains and impacts, precisely as I have shown you the manner of the fracture of the vertebral column of the primitive vertebrates of the Permian period. This would require long discussion to render clear; nevertheless, I venture to make the assertion that this series of structures is the result of definite and distinct organic forces, directed to special ends. We have yet to get at the conflicting forces which have produced the results we see. Mechanical evolution will give us a good deal to do for some time to come. Of course, if motion has had any effect in modifying structure, it behooves us to investigate those conditions which give origin to motion in animals. First in order come the sensibilities of the animal, which we have traced to simple consciousness; and stimuli, upon notice of which he immediately begins to move. The primary stimulus of all kinds of motion is necessarily touch. If a stone falls upon the tail of some animal which has a tail, he immediately gets out of that vicinity. If a jelly-fish with a stinging apparatus runs across an eel which has no scales, the eel promptly moves. External applications of unpleasant bodies will always cause an animal to change his location. Then he is constantly assaulted by the dire enemy of beasts, hunger, an instinct which is evidently universal, to judge from the actions of animals. This seems to have fashioned, in large part, all forms of life from the least to the greatest, from the most unorganized to the most complex. Each exercised itself for the purpose of filling its stomach with protoplasm. Then come the stimuli which appeal to a sense allied to that of touch, changes of temperature. No animals like to be too cold or too hot; and when the temperature is disagreeable, the tendency is to go away from that locality. Among primary instincts must be included that of reproduction. After that comes the sensation of resistance, or, carried to a high degree, of anger: when an animal's interests are interfered with, its movements restricted, the most energetic displays are prompted. So, you see, it is a matter of necessity that mental phenomena lie at the back of evolution, provided always that the connecting link of the argument—that motion has ever affected structure—be true. That is a point which, of course, admits of much discussion. I have placed myself on the affirmative side of that question; and, if I live long enough, I expect to see it absolutely demonstrated.

Of course the development of mind becomes possible under such circumstances. It is not like a man lifting himself up by his boots, which it would be were there no such thing as memory. But with that memory which accumulates, which formulates, first habits and then structures, especially in the soft, delicate nervous tissue, the development of the function of the mind, as well as the machinery of the mind, becomes perfectly possible. We develop our intellect through the accumulation of exact facts; through the collation of pure truth, no matter whether it be a humble kind of truth—as the knowledge of the changes of the seasons, which induces some animals to lay up the winter's store; whether it be knowledge of the fact that the sting of the bee is very disagreeable; or knowledge of the fact (of which the wild ox, no doubt, is thoroughly aware) that the teeth of the wolf are not pleasant to come in contact with; or whether it be the complex knowledge of man. When the cerebral matter has become larger and more complex, it receives and retains a much greater number of impressions, and the animal becomes a more highly educated being.

As regards the department of emotions or passions, they are also much stimulated by the environment. Animals which live in a state of constant strife, naturally have their antagonistic passions much developed; while amiable, sympathetic sentiments are better and more largely produced by peace-loving animals. Thus it is that the various departments of the mind have the beautiful results which we now find in the human species.

There are some departments of the mind which some of our friends decline to admit having had such an origin. The moral faculty, for instance, is excepted by many from this series. But the reasons why they object to its production in this way are, to my mind, not valid. The development of the moral faculty, which is essentially the sense of justice, appears to them not to fall within the scope of a theory of descent or of evolution. It consists of two parts. First is the sentiment of benevolence, or of sympathy with mankind, which gives us the desire to treat them as they should be treated. It is not sufficient for justice that it is unmixed mercy or benevolence, which is sometimes very injurious, and very often misplaced. It requires, in the second place, the criticism of the judgment, of the mature intellect, of the rational faculty, to enable the possessor to dispose of his sentiments in the proper manner. The combination of rational discrimination and judgment, with benevolence, constitutes the sense of justice, which has been derived, no doubt, as a summary of the development of those two departments of the mind—the emotions and the intellect.

It is said that a sense of justice could not be derived from no sense of justice; that it could not have been derived from the state of things which we find in the animals, because no animal is known to exhibit real justice; and that objection is valid as far as it goes. I suspect that no animal has been observed to show a true sense of justice. That they show sympathy and kindness, there is no question; but when it comes to real justice, they do not display it. But do all men display justice? Do all men *understand* justice? I am very sure not. There are a good many men in civilized communities, and there are many tribes, who do not know what justice is. It does not exist as a part of every mental constitution. I never lived among the Bushmen, and do not know exactly what their mental constitution is; but in a general way the justice of savages is restricted to the very smallest possible circle—that of their tribe or of their own family. There is a class of people who do not understand justice. I do not refer to people who know what right is, and do not do it; but to the primitive state of moral character, in which, as in children, true justice is unknown. I call attention to the fact, because some of our friends have been very much afraid that the demonstration of the law of evolution, physical and metaphysical, would result in danger to society. I suspect not. The mode in which I understand this question appears to me to be beneficial to society, rather than injurious; and I therefore take the liberty of appending this part of the subject to its more material aspect.

I refer to another topic, that is to the nature of life, and the physical basis of life. The word "life" is so complex that it is necessary to define it, and so to define it away that really the word "life" does not retain its usual definition. Many phenomena of life are chemical, physical, mechanical. We have to remove all these from consideration, because they come within the ordinary laws of mechanical forces; but we have a few things left which are of a different character. One is the law of growth, which is displayed in the processes of embryonic succession; secondly, the wonderful phenomena of sensibility. Those two things we have not yet reduced to any identity with the ordinary laws of force, though we know of their dynamic equivalency. In the phenomena of embryology the phenomena of evolution are repeated, only concentrated in the early stages through which animals have to pass. So whatever explains the general phenomena of evolution explains the phenomena of embryology.

What is the nature of physical sensibility? In this planet, it is found residing only in one form of matter, which has a slightly varied chemical constitution, namely, protoplasm, so called from a physical standpoint. Now, this world, as you all know, has passed through many changes of temperature. Its early periods, it is probable, were so very hot that protoplasm had a very poor chance. Again, can we assume for a moment that this little speck in the great universe is the only seat of life? I suppose scarcely any scientific man will venture to do so. If, therefore, life exists in other planets, worlds, and systems, does it necessarily occupy bodies of protoplasm in those different, remote spheres? It would be a great assumption. It is altogether improbable. The certainty is, that in those planets which are in proximity to the sun's heat there could be no protoplasm. Protoplasm in the remote planets would be a hard mineral, and near the sun it would be dissipated into its component gases. So that, if life be found in other parts of this universe, it must reside in some different kind of material. It is extremely probable that the physical conditions that reside in protoplasm might be found in other kinds of matter. It is in its chemical inertness, and in its physical constitution, that its adaptation to life resides; and the physical constitution necessary for the sustentation of life may be well supposed to exist in matter in other parts of the universe. I only say the door is open and not closed: any one who asserts that life can not exist in any other material basis than protoplasm is assuming more than the world of science will permit him to assume. And that it is confined to this single planet, among the great systems of the universe —that assumption will not for a moment be allowed. Therefore the subject is one which allows us a free field for future investigation: it is by no means closed in the most important laws which it presents to the rational thinker. I hope, also, if the evidence in favor of this hypothesis of the creation of living forms be regarded as true, that no one will find in it any ground for any very serious modification of existing ideas on the great questions of right and wrong, which have been long since known by men as a result of experience, and without any other scientific demonstration whatsoever.

XI.

ON THE EVOLUTION OF THE VERTEBRATA, PROGRESSIVE AND RETROGRESSIVE.

I. PRELIMINARY.

In attempting to ascertain the course of evolution of the Vertebrata, and to construct phylogenetic diagrams which shall express this history, among the difficulties arising from deficient information one is especially prominent. As is well known, there are many types in all the orders of the Vertebrata which present us with rudimentary organs, as rudimental digits, feet or limbs, rudimental fins, teeth, and wings. There is scarcely an organ or part which is not somewhere in a rudimental and more or less useless condition. The difficulty which these cases present is, simply, whether they be persistent primitive conditions, to be regarded as ancestral types which have survived to the present time, or whether, on the other hand, they be results of a process of degeneration, and therefore of comparatively modern origin. The question, in brief, is, whether these creatures presenting these features be primitive ancestors or degenerate descendants.

In the first place, let us define the meaning of the word degenerate. This must be done first from a structural or anatomical standpoint. Degeneracy may be defined as a loss of parts without corresponding development of other parts. All animals are degenerate in some respect or another, as, for instance, the Mammalia in the small size of the pineal gland and of the coracoid bone; so that degeneracy, as a whole, can only be affirmed where the sum of the subtractions is greater than the sum of the additions. Function of the parts must, however, be consulted in this matter. We naturally regard sensibility as the highest of animal functions, and mind as the highest form of sensibility. Therefore development of organs of sensibility and sense and mind, constitutes a better claim of progress than development of stomach or of skin. Since motion is under the direction of sensibility, organs of movement have much to do with the question. When perfection in this respect conflicts with perfection of brain, in evidence of position, we naturally give the preference to the latter in deciding. Thus the ruminating mammals are much superior to man in the structure of their feet, teeth, and stomach, yet we properly assign the higher position to the Quadrumana and to man, on account of the superior complication of their brain-structure.

Paleontology has proved,* what had been already surmised, that the development of animal organisms has been on lines of increasing specialization of parts. That is, in lines of increasingly perfect adaptations of structures to ends, or functions. In certain series of animals we witness steadily increasing perfection of mechanisms of the limbs for running; in others for digging; in others for flying. In the teeth we find increasing perfection of machines for grinding, for cutting, or for seizing. In the brain the specialization has evidently been toward increased acuteness of perception, increased energy of action, and increased intelligence. Specialization does not, however, necessarily imply progressive development. Adaptation may be to a parasitic or a sessile mode of life. Such adaptation is often displayed in a very special modification of parts, as in the anterior limbs of some of the parasitic Crustacea; in the mouth parts of some Arachnida; in the feet of the sloth, and in the jaws of the ant-eaters.

Embryology has furnished, and will furnish, many important hints and demonstrations as to the true meaning of the rudimentary condition or absence of parts, and thus indicate the phylogenetic connections of animals. Thus the origin of the Tunicata from primitive vertebrate-like forms would probably never have been suspected but for embryological studies; and the origin of the very peculiar order of Rotifera has been explained in like manner. But embryology has its limitations, for the transitional characters presented by embryos are only partially of the nature of a record of the structures which belonged to their ancestors in successive geological ages, and are frequently special adaptations to the necessities of their embryonic life. Such are the statoblasts which are present in fresh-water sponges and Polyzoa, and wanting in the marine forms; and the allantois and placenta of

---

* Cfr. "On the Evidence for Evolution in the History of the Extinct Mammalia," "Proc. Amer. Assoc. Adv. Sciences" for 1883.

Vertebrata. In a number of groups the embryo seems to have been more susceptible to the influence of the environment than the adults.* It results that in many cases the phylogeny can only be determined by the discovery and investigation of the ancestors themselves, as they are preserved in the crust of the earth. In all cases this discovery confirms and establishes such definite conclusions as may be derived from embryology. It is also clear that on the discovery of phylogenetic series it becomes at once possible to determine the nature of defective types. It becomes possible to ascertain whether their rudimental parts represent the beginnings of organs, or whether they are the result of a process of degeneration of organs once well developed.

A great deal of light has been happily thrown on this question, as regards the Vertebrata, by the recent work done in North American paleontology. The lines of descent of many of the minor groups have been positively determined, and the phylogenetic connections of most of the primary divisions or classes have been made out. The result of these investigations has been to

---

* A remarkable instance of this state of things appears in the history of the evolution of the insects. It is quite impossible to understand this history without believing that the larval and pupal states of the highest insects are the results of a process of degeneracy which has affected the middle periods of growth but not the mature results. The earliest insects are the Orthoptera, which have active aggressive larvæ and pupæ, undergoing the least changes in their metamorphosis (Ametabola), and never getting beyond the primitive mandibulate condition at the end. The metamorphosis of the jawed Neuroptera is little more marked, and they are one of the oldest orders.

The highest orders with jaws undergo a marked metamorphosis (Coleoptera, Hymenoptera), the Hymenoptera even requiring artificial intervention in some instances to make it successful. Finally, the most specialized orders, the suctorial Diptera and Lepidoptera, especially the latter, present us with very unprotected more or less parasitic stages, both active and inactive. These animals have evidently degenerated, but not so as to prevent their completing a metamorphosis necessary for purposes of reproduction. As is well known, many imagines (Saturniidæ, Œstridæ) can perform no other function, and soon die, while in some Diptera the incomplete larvæ themselves reproduce, so that the metamorphosis is never completed.

This history is parallel to that proposed by Dohrn to account for the origin of the Ammocœtes larval stage of the Marsipobranchii. He supposes this form to be more degenerate than its probable ancestral type in the ancestral line of the Vertebrata, as it is inferior to its own adult. An inactive life in mud is supposed by Dohrn to have been the effective cause. An inactive life on the leaves of plants, or in dead carcases, has probably been the cause of the same phenomenon in the Lepidoptera and Diptera.

prove that the evolution of the Vertebrata has proceeded not only on lines of acceleration, but, to a much greater extent than has been heretofore suspected, on lines of retardation.* That is, that evolution has been not only progressive, but at times retrogressive. This is entirely in accord with the views derived by Dohrn from embryology,† who, however, wrote only of the origin of the Vertebrata as a whole and not of its divisions, excepting only the Leptocardii and Marsipobranchii, that is, of the sand-lance and the lampreys and hags. The demonstration of such relations for the higher Vertebrata is now done nearly for the first time.‡

Omitting from consideration the two classes above mentioned, whose remains have not yet been certainly found in a fossil state, there remain the following: the Pisces, Batrachia, Reptilia, Aves, and Mammalia.

The Mammalia have been traced to the theromorphous reptiles through the Monotremata. The birds, some of them at least, appear to have been derived from the Dinosaurian reptiles. The reptiles, in their primary representative order, the Theromorpha, have been probably derived from the rhachitomous Batrachia.# The Batrachia have originated from the subclass of fishes, the Dipnoi, though not from any known form. I have shown that the true fishes or Teleostomi have descended from an order of sharks,‖ the Ichthyotomi, which possess characters of the Dipnoi also. The origin of the sharks remains entirely obscure, as does also that of the Pisces as a whole. Dohrn believes the Marsipo-

---

* See "Origin of Genera," E. D. Cope, Philadelphia, 1868, where these terms are introduced.

† See "Der Ursprung der Wirbelthiere u. d. Princip des Functionwechsels," Leipsic, 1875.

‡ "On the Phylogeny of the Vertebrata," Cope, "Amer. Naturalist," Dec., 1884. I here remark that my researches have now, as I believe, disclosed the ancestry of the mammals, the birds, the reptiles, and the true fishes, or Teleostomi, including the special phylogenies of the Batrachia and Reptilia, and some of the Mammalia. See the following references: "American Naturalist," 1884, p. 1136; "Proceedings Academy Philadelphia," 1867, p. 234; "Proceedings American Philosoph. Society," 1884, p. 585; "American Naturalist," 1884, p. 27; "Proceedings American Association for the Advancement of Science," xix, 1871, p. 233; "Proceedings American Philosophical Society," 1882, p. 447; "American Naturalist," 1884, pp. 261 and 1121; "Report U. S. Geol. Survey W. of 100th Mer.," G. M. Wheeler, 1877, iv, ii, p. 282.

Through the batrachian order Embolomeri. (Ed. 1886.)

‖ "Proceedings Am. Phil. Soc.," 1884, p. 585.

branchii to have acquired its present characters by a process of degeneration. The origin of the Vertebrata is as yet entirely unknown, Kowalevsky deriving them from the Ascidians, and Semper from the Annelida. The above results I have embodied in the following phylogenetic diagram :

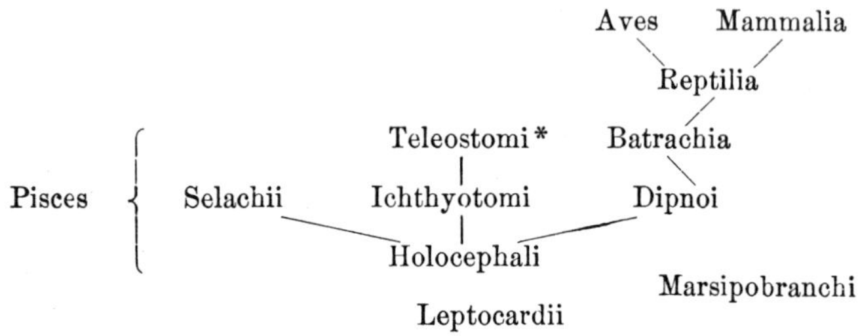

Accepting this phylogeny, it becomes possible to determine the course of development, first, of the whole series ; and, secondly, of the contents of each class taken by itself. I will first consider the direction of the evolution of the Vertebrata as a whole.

II. THE VERTEBRATE LINE.

The Vertebrata exhibit the most unmistakable gradation in the characters of the circulatory system.† It has long been the custom to define the classes by means of these characters, taken in connection with those of the skeleton. Commencing in the Leptocardii with the simple tube, we have two chambers in the Marsipobranchii and fishes ; three in the Batrachia and Reptilia ; and four in the Aves and Mammalia. The aorta-roots commence as numerous pairs of branchial arteries in the Leptocardii ; we see seven in the Marsipobranchi, five in the fishes (with number reduced in some) ; four and three in Batrachia, where they generally cease to perform branchial functions ; two and one on each side in Reptilia ; the right-hand one in birds, and the left-hand one in Mammalia. This order is clearly an ascending one throughout. It consists of, first, a transition from adaptation to an aquatic, to an aërial respiration ; and, second, an increase in the power to aërate and distribute a circulating fluid of increased quantity,

* In the original I used the name Hyopomata for this division, but Owen's name Teleostomi is prior. (Ed. 1886.)

† See "Origin of Genera," 1868, p. 20, for a table of the characters of the circulatory system.

nd of increased calorific capacity. In other words, the circula-on passes from the cold to the hot blooded type coincidentally 'ith the changes of structure above enumerated. The accession f a capacity to maintain a fixed temperature while that of the ırrounding medium changes, is an important advance in animal ɔonomy.

The brain and nervous system also display a general progress-'e ascent. Leaving the brainless Leptocardii, the Marsipo-ranchs and fishes present us with small hemispheres, larger ptic lobes, and well-developed cerebellum. The hemispheres re really larger than they appear to be, as Rabl Rückard has ıown * that the supposed hemispheres are only corpora striata. ut the superior walls are membranous, and support on their in-ernal side only a layer of epithelial cells, as in the embryos of ther Vertebrata, instead of the gray substance. So that, al-ıough we find that the cerebellum is really smaller in the atrachia and most Reptilia than in the fishes, the better de-elopment of the hemispheres in the former gives them the pre-ninence. The Elasmobranchii show themselves superior to any of the fishes in the large size of their corpora restiformia ıd cerebellum. The Reptilia constitute an advance on the Ba-'achia. In the latter the optic thalami are, with some excep-ons, of greater diameter than the hemispheres, while the reverse generally true of the reptiles. The crocodiles display much ıperiority over the other reptiles in the larger cerebellum, with ıdimental lateral lobes. The great development of the hemi-heres in birds is well known, while the general superiority of ıe brain of the living Mammalia over all other vertebrates is lmitted.

The consideration of the successive relations of the skeleton in ıe classes of vertebrates embraces, of course, only the characters hich distinguish those classes. These are not numerous. They nbrace the structure of the axis of the skull; of the ear-bones; the suspensors of the lower jaw; of the scapular arch and ıterior limb, and of the pelvic arch and posterior limb. Other ıaracters are numerous, but do not enter into consideration at ıis time.

The persistence of the primitive cartilage in any part of the :eleton is, embryologically speaking, a mark of inferiority.

* "Biologisches Centralblatt," 1884, p. 449.

From a physiological or functional standpoint it has the sam significance, since it is far less effective both for support and fc movement than is the segmented osseous skeleton. That this i a prevalent condition of the lower Vertebrata is well knowr The bony fishes and Batrachia have but little of the primitiv cartilage remaining, and the quantity is still more reduced in th higher classes. Systematically, then, the vertebrate series is i this respect an ascending one. The Leptocardii are membra nous; the Marsipobranchii and most of the Elasmobranch cartilaginous; the other Pisces and the Batrachia have th basicranial axis cartilaginous, so that it is not until the Reptili are reached that we have osseous sphenoid and presphenoid bone such as characterize the birds and mammals. The vertebra column follows more or less inexactly the history of the base c the skull, but its characters do not define the classes.

As regards the suspensor of the lower jaw, the scale is in th main ascending. We witness a gradual change in the segmenta tion of the mandibular visceral arch of the skull, which clearl has for its object such a concentration of the parts as will produc the greatest effectiveness of the biting function. This is accom plished by reducing the number of the segments, so as to brin the resistance of the teeth nearer and nearer to the power, that i the masseter and related muscles, and their base of attachment the brain-case. This is seen in bony Vertebrates in the reductio of the segments between the lower jaw proper and the skull, fror four to none. In the fishes we have the hyomandibular, the sym plectic, the inferior quadrate, and the articular. In the Ba trachia, reptiles and birds, we have the quadrate and articula only, while in the Mammalia these elements also are wanting.

The examination of the pectoral and pelvic arches reveals successive modification of the adaptation of the parts to the me chanical needs of the limbs. In this regard the air-breathin types display wide diversity from the gill-bearing types or fishes In the latter, the lateral elements unite below without the inter vention of a median element or sternum, while in the former th sternum, or parts of it, is generally present. Either arrange ment is susceptible of much mechanical strength, as witness th Siluroid fishes on the one hand, and the mole on the other. Th numerous segments of the fishes' pectoral arch must, however, b an element of weakness, so that from a mechanical standpoint i must take the lowest place. The presence of sternal elements ith both clavicle, procoracoid, and coracoid bones on each side, ives the Reptilia the highest place for mechanical strength. The oss of the coracoid seen in the tailed Batrachia, and loss of coraoid and procoracoid in the Mammalia, constitute an element of eakness. The line is not then one of uniform ascent in this respect.

The absence of pelvis, or its extremely rudimental condition in shes, places them at the foot of the line in this respect. The orward extension of the ilium in some Batrachia and in the Mammalia, is to be compared with its backward direction in Reptilia, nd its extension both ways in the birds. These conditions are ll derived by descent from a strictly intermediate position in the atrachia and Reptilia of the Permian epoch. The anterior diection must be regarded as having the mechanical advantage ver the posterior direction, since it shortens the vertebral column nd brings the posterior nearer to the anterior feet. The prevlence of the latter condition in the Mammalia enables them to and clear of the ground, while the Reptilia move with the abdoen resting upon it. As regards the inferior arches of the pelvis, he Mammalia have the advantage again, in the strong bony meian symphysis connecting the ischium and pubis.* This charcter, universal among the land Vertebrata of the Permian epoch, as been lost by the modern Reptilia, and birds, and is retained nly by the Mammalia. So the lines, excepting the mammalian, ave *degenerated* in every direction in the characters of the pelvis.

The limbs of the Pisces are as well adapted to their environent as are those of the land Vertebrata; but, from an embryologal standpoint, their structure is inferior. The primitive rays re less modified in the fin than in the limb; and limbs themelves display a constantly increasing differentiation of parts, comencing with the Batrachia and ending with the Mammalia. The details of these modifications belong to the history of the ontents of the classes, however, rather than to the succession of he Vertebrata as a whole.

In review, it may be said that a comparison of the characters hich define the classes of the Vertebrates shows that this branch f the animal kingdom has made with the ages successive steps f progress from lower to higher conditions. This progress has ot been without exception, since, as regards the construction of

* This is an advantage as a protection during gestation.

the scapular arch, the Mammalia have retrograded from the rep
tilian standard as a whole.

In subsequent pages I shall take up the lines of the classe
separately.

III. THE LINE OF THE UROCHORDA.

Embryological evidence leads us to anticipate that the prim
tive Vertebrata possessed nothing representative of the vertebrat
skeleton beyond a chorda dorsalis. Above this axis should lie th
nervous chord, and below it the nutritive and reproductive sy
tems and their appendages. Such a type we have in its simple
form in the Branchiostoma, the representative of the division
the Acrania. In the animals of this division the mouth and an
have the usual vertebrate position, at opposite ends of the bod
cavity. The Tunicata (formerly referred to the Mollusca) are no
known to present a still more primitive type of Vertebrata,
which the name of Urochorda has been given. These curiou
frequently sessile creatures, have a vertebrate structure during th
larval stage, which they ultimately lose. They have the necessa
chorda, and nervous axis with a brain, and a cerebral eye. The
have at this time a tail, and are free-swimming; a peculiarit
which a few of them retain throughout life (Appendicularia).
They differ from the Acrania in the positions of the extremiti
of the alimentary canal. The mouth is on the top of the anteri
end of the animal, and is supposed by some anatomists to repr
sent an open extremity of the pineal gland of other Vertebrata
while the tract represented by this body, the third ventricle of th
brain, and the pituitary body of the Craniata, are the remains
the primitive œsophagus of the Urochorda. The anus in th
adult tunicates is either dorsal, or it opens into the body-cavity,
in the young larvæ. In Appendicularia it is ventral (Gegenbaur

The history of the Tunicata can not be traced by paleontolo
gists as yet, owing to the absence of hard parts in their structur
The evidence of embryology has, however, convinced phylogenis
that the ancestors of this class resembled their larvæ, and that the
have as a whole undergone a remarkable degeneracy. They ha
passed from an active, free life to a sessile one, and have lost th
characters which pertain to the life of vertebrates generally.

It was to have been anticipated, however, that all of these a

* See Lankester on "Degeneration," "Nature Series," 1880.

cestral Tunicata did not undergo this degenerative metamorphosis, for it is to such types that we must look for the ancestors of the other Vertebrata, the Acrania and the Craniata. And here paleontology steps in and throws new light on the question. I have pointed out briefly, in the "American Naturalist,"* that a second order must be added to the Urochorda, viz., the Antiarcha, in which the anus presents the same position as in the Acrania, at the posterior end of the body, while an orifice of the upper surface represents the mouth of the Tunicata. To this order is to be referred the family of the Pterichthyidæ, of which the typical genus,

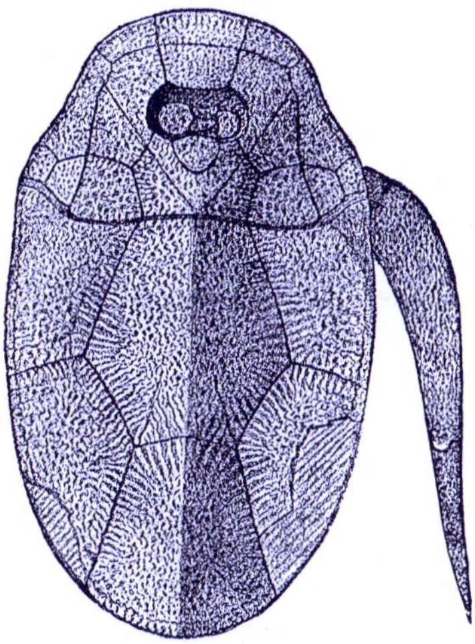

FIG. 62.

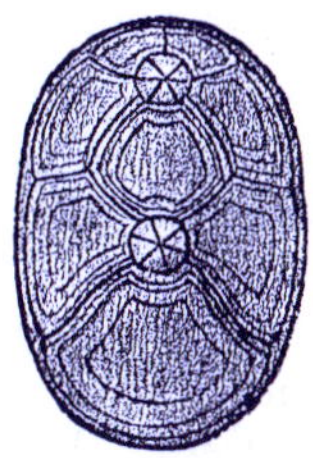

FIG. 63.

Fig. 62. *Bothriolepis canadensis* Whiteaves, from above, half size of a small specimen. The valve of the dorsal mouth, or notostome, is broken. Fig. 63. *Chelyosoma maclovianum* Brod. & Sow., $^4/_3$ natural size, from Point Barrow, Alaska.

Pterichthys, is a well-known form of the Devonian period. This genus retained its tail, which was the cause, in connection with the presence of lateral fin-like appendages, of its having been supposed to be a fish, by Agassiz, Hugh Miller, and others. It is possible that the American *Bothriolepis canadensis* lost its tail, as in the majority of Urochorda. The tunicate which approaches nearest to the Antiarcha is the Arctic Chelyosoma.

From the Antiarcha to the Acrania and Craniata, then, the line is an ascending one.

---

* March, 1885, p. 289.

IV. THE LINE OF THE PISCES.

The fishes form various series and subseries, and the tracing of all of them is not yet practicable, owing to the deficiency in our knowledge of the earliest or ancestral forms. Thus the origins of the four subclasses, Holocephali, Dipnoi, Elasmobranchii, and Teleostomi, are lost in the obscurity of the early Palæozoic ages.

A comparison of the four subclasses just named shows that they are related in pairs. The Holocephali and Dipnoi have no distinct suspensory segment for the lower jaw, while the Elasmobranchii and Hyopomata have such a separate element. The latter, therefore, present one step in the direction of complication beyond the former; but whether the one type is descended from the other, or whether both came from a common ancestor or not, is unknown. If one type be derived from the other, it is not certain which is ancestor, and whether the process has been one of advance or retrogression. The fauna of the Permian epoch throws some light on the relations of these subclasses in other respects. The order of the Ichthyotomi,* while belonging technically to the Elasmobranchi, presents characters of both the Dipnoi and the Teleostomi. It is so near to the Dipnoi in the characters of the skull that nothing save the presence of a free suspensor of the lower jaw prevents its entering that subclass. It indicates that the one of these divisions is descended from the other, or both from a common division which may well be the group Ichthyotomi itself. In case the Elasmobranchi have descended from the Ichthyotomi, they have undergone degeneracy, as the Ichthyotomi have a higher degree of ossification and differentiation of the bones of the skull. If they descended from a purely cartilaginous type of Dipnoi, they have advanced, in the addition of the free hyomandibular. If the Dipnoi have descended from either division, they have retrograded, in the loss of the free hyomandibular. As regards the Teleostomi, we have a clear advance over the other subclasses in the presence of the maxillary arch and the opercular apparatus.

Too little is known of the history of the subclasses, excepting the Teleostomi, for us to be able to say much of the direction of

---

* See "Palæontological Bulletin," No. 38, E. D. Cope, 1884, p. 572, on the genus Didymodus.

the descent of their contained orders. On the sharks some light is shed by the discovery of the genus Chlamydoselachus Garman,* which is apparently nearly related to the Cladodonts of the Devonian seas. This genus has more numerous branchial slits than all but two of the genera of existing sharks, and it differs from all but these two in having a more perfect articulation between the tooth-bearing bones and the cranium. Of the Teleostomi a much clearer history is accessible. It has three primary divisions or tribes which differ solely in the structure of the supports of the fins. In the first division, the Crossopterygia, the anterior limbs have numerous basilar bones which are supported on a peduncle of axial bones. The posterior limbs are similar. In the second division, or Chrondrostei (the sturgeons, etc.), the posterior limb remains the same, while the anterior limbs have undergone a great abbreviation in the loss of the axial bones and the reduction of the number and length of the basilar bones. In the third group, or Actinopteri,† both limbs have undergone reduction, the basilar bones in the posterior fin being almost all atrophied, while those of the fore limb are much reduced in number.

The phylogeny of these tribes is not easy to make out at present. The descent has been, no doubt, in the order named in time, but the starting-point is yet uncertain. Thus the Chondrostei appear later in time than either of the other tribes, a history which probably only represents our ignorance. The characters of the genus Crossopholis Cope, from the American Eocene, strongly suggest that the existing forms have descended from scaled ancestors. The Crossopterygian fore limb, with its arm-like axis, tells of the origin of the first limbed vertebrates, the Batrachia, whose skull-structure, however, only permits their derivation from the Dipnoi or Holocephali. As the former subclass has the Crossopterygian fin-structure, we can safely regard them as the ancestors of the Batrachia, while the Crossopterygia are a side-line from a similar type, probably the Ichthyotomi, because these have a free suspensor of the lower jaw. But of the structure of the fins of the Ichthyotomi unfortunately we know nothing. If this position be true, then the successive derivation of the Chondrostei and the Teleostomi in one line is rendered probable. The modification of structure has consisted in the contraction of the

* "Proceedings American Assoc. Adv. Sci.," 1884.

† Partly agrees with the Teleostei of Müller, but includes many of his Ganoidea.

supporting elements of the pectoral and ventral fins by the reduction of their numbers and length. According to paleontological history, however, the tribe of Teleostomi with most contracted fins, Actinopteri, appeared in the Coal Measures (Paleoniscidæ), or very soon after the Crossopterygia in the Devonian.

The descent of the fishes in general has witnessed, then, a contraction of the limbs to a very small compass, and their substitution by a system of accessory radii. This has been an ever-widening divergence from the type of the higher Vertebrata, and from this standpoint, and also a view of the "loss of parts without complementary addition of other parts," may be regarded as a process of degradation.

Taking up the great division of the Actinopteri, which embraces most of the species of living fishes, we can trace the direction of descent largely by reference to their systematic relations when we have no fossils to guide us.

The three subtribes adopted by Jordan represent three series of the true fishes which indicate lines of descent. The Holostei include the remainder of the old ganoids after the subtraction of the Crossopterygia and the Chondrostei. They resemble these forms in the muscular bulbus arteriosus of the heart and in the chiasm of the optic nerves. Both of these characters are complexities which the two other divisions do not possess, and which, as descendants coming later in time, must be regarded as inferior, and therefore to that extent degenerate. Of these divisions the Physostomi approach nearest the Holostei, and are indeed not distinctly definable without exceptions. The third division, or Physoclysti, shows a marked advance beyond the others in: (1) the obliteration of the primitive trachea, or ductus pneumaticus, which connects the swim-bladder and œsophagus; (2) the advance of the ventral fins from the abdomen forward to the throat; (3) the separation of the parietal bones by the supraoccipital; (4) the presence of numerous spinous rays in the fins; and (5) the roughening of the edges of the scales, forming the ctenoid type. There are more or less numerous exceptions to all of these characters. The changes are all further divergencies from the other vertebrate classes, or away from the general line of ascent of the vertebrate series taken as a whole. The end gained is specialization; but whether the series can be called either distinctively progressive or retrogressive, is not so clear. The development of osseous spines, rough scales, and other weapons of defense, together with the generally superior energy and tone which prevail among the Physoclysti, characterize them as superior to the Physostomi, but their departure from the ascending line of the Vertebrata has another appearance.

The descent of the Physoclystous fishes has probably been from Holostean ancestors, both with and without the intervention of Physostomous forms. This is indicated by increase in the number of basilar bones in the fins of families which have pectoral ventral fins, as in the extinct genus Dorypterus.*

The Physostomi display three or four distinct lines of descent. The simplest type is represented by the order Isospondyli, and paleontology indicates clearly that this order is also the oldest, as it dates from the Trias at least. In one line the anterior dorsal vertebræ have become complicated, and form an interlocking mass which is intimately connected with the sense of hearing. This series commences with the Characinidæ, passes through the Cyprinidæ, and ends with the Siluridæ. The arrangements for audition constitute a superadded complication, and to these are added in the Siluroids defensive spines and armor. Some of this order, however, are distinctly degenerate, as the soft purblind Ageniosus, and the parasitic Stegophilus and Vandellia, which are nearly blind, without weapons, and with greatly reduced fins.

The next line (the Haplomi, pike, etc.) loses the precoracoid arch and has the parietal bones separated, both characters of the Physoclysti. This group was apparently abundant during the Cretaceous period, and it may have given origin to many of the Physoclysti.

Another line also loses the precoracoid, but in other respects diverges totally from the Physoclysti and all other Physostomi. This is the line of the eels. They next lose the connection between the scapular arch and the skull, which is followed by the loss of the pectoral fin. The ventral fin disappeared sooner. The palatine bones and teeth disappear, and the suspensor of the lower jaw grows longer and loses its symplectic element. The opercular bones grow smaller, and some of them disappear. The ossification of most of the hyoid elements disappears, and some of their cartilaginous bases even vanish. These forms are the marine eels or Colocephali. The most extraordinary example of speciali-

* See "Proceeds. Amer. Assoc. Adv. Science," 1878, p. 297.

zation and degeneracy is seen in the abyssal eels of the family Eurypharyngidæ. Here all the degenerate features above mentioned are present in excess, and others are added, as the loss of ossification of a part of the skull, almost total obliteration of the hyoid and scapular arches, and the semi-notochordal condition of the vertebral column, etc.

The Physoclysti nearest the Physostomi have abdominal ventral fins, and belong to several orders. It is such types as these that may be supposed to have been derived directly from Holostean ancestors. They appear in the Cretaceous period (Dercetidæ), along with the types that connect with the Physostomi (Haplomi). Intermediate forms between these and typical Physoclysti occur in the Eocene (Trichophanes, Erismatopterus), showing several lines of descent. The Dercetidæ belong apparently to the order Hemibranchi, while the Eocene genera named belong apparently to the Aphododiridæ, the immediate ancestor of the highest Physoclysti, the Percomorphi. The order Hemibranchi is a series of much interest. Its members lose the membrane of their dorsal spinous fin (Gasterosteidæ), and then the fin itself (Fistularia, Pegasus). The branchial apparatus has undergone, as in the eels, successive deossification (by retardation), and this in direct relation to the degree with which the body comes to be protected by bony shields, reaching the greatest defect in the Amphisilidæ. One more downward step is seen in the next succeeding order of the Lophobranchii. The branchial hyoid apparatus is reduced to a few cartilaginous pieces, and the branchial fringes are much reduced in size. In the Hippocampidæ the caudal fin disappears and the tail becomes a prehensile organ by the aid of which the species lead a sedentary life. The mouth is much contracted and becomes the anterior orifice of a suctorial tube. This is a second line of unmistakable degeneracy among true fishes.

The Physoclysti with pectoral ventral fins present us with perhaps ten important ordinal or subordinal divisions. Until the paleontology of this series is better known, we shall have difficulty in constructing phylogenies. Some of the lines may, however, be made out. The accompanying diagram will assist in understanding them.

The Anacanthini present a general weakening of the organization in the less firmness of the osseous tissue and the frequent reduction in the size and character of the fins. The caudal ver tebræ are of the protocercal type. As this group does not appear early in geological time, and as it is largely represented now in the abyssal ocean fauna, there is every reason to regard it as a degenerate type.* The scyphobranch line presents a specialization of the superior pharyngeal bones, which is continued by the Haplodoci (Batrachidæ). This can not be called a degenerate line, although the fin-rays are soft. The Heterosomata (flounders) found it convenient to lie on one side, a habit which would appear to result from a want of motive energy. The fins are very inefficient organs of movement in them, and they are certainly no rivals for swift-swimming fishes in the struggle for existence, excepting as they conceal themselves. In order to see the better while unseen, the inferior eye has turned inward, i. e., upward, and finally has penetrated to the superior surface, so that both eyes are on one side. This peculiarity would be incredible if we did not know of its existence, and is an illustration of the extraordinary powers of accommodation pos-

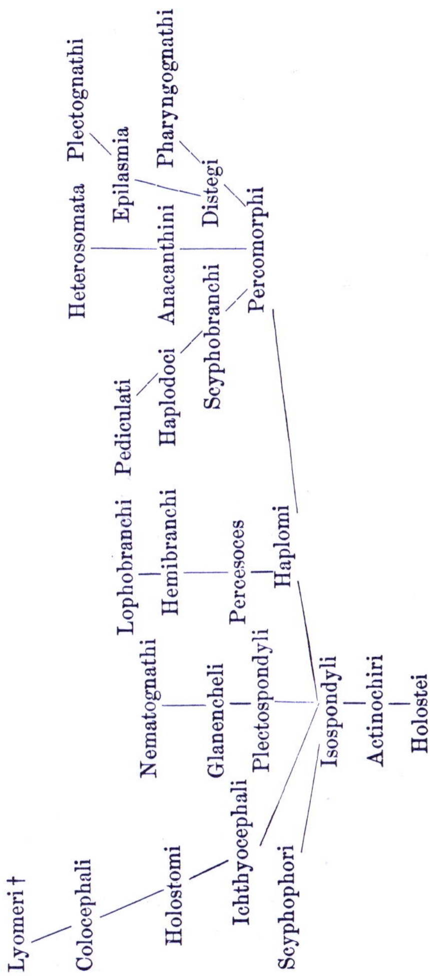

* The general characters of the deep-sea fish-fauna are those of degeneracy. (Ed. 1886.)

† This order was proposed by Gill for the Eurypharyngidæ, and is now added. (Ed. 1886.)

sessed by nature. The Heterosomata can only be considered a degenerate group.

The double bony floor of the skull of the Distegous percomorph fishes is a complication which places them at the summit of the line of true fishes. At the summit of this division must be placed the Pharyngognathi, which fill an important rôle in the economy of the tropical seas, and the fresh waters of the Southern hemisphere. By means of their powerful grinding pharyngeal apparatus they can reduce vegetable and animal food inaccessible to other fishes. The result is seen in their multifarious species and innumerable individuals decked in gorgeous colors, and often reaching considerable size. This is the royal order of fishes, and there is no reason why they should not continue to increase in importance in the present fauna.

Very different is the line of the Plectognathi. The probable ancestors of this division, the Epilasmia (Chætodontidæ, etc.), are also abundant in the tropical seas, and are among the most brilliantly colored of fishes. One of their peculiarities is seen in a shortening of the brain-case and prolongation of the jaws downward and forward. The utility of this arrangement is probably to enable them to procure their food from the holes and cavities of the coral reefs among which they dwell. In some of the genera the muzzle has become tubular (Chelmo), and is actually used as a blow-gun by which insects are secured by shooting them with drops of water. This shortening of the basicranial axis has produced a corresponding abbreviation of the hyoid apparatus. The superior pharyngeal bones are so crowded as to have become a series of vertical plates like the leaves of a book. These characters are further developed in the Plectognathi. The brain-case is very small, the face is very elongate, and the mouth is much contracted. The bones surrounding it in each jaw are co-ossified. The axial elements (femora) of the posterior fins unite together, become very elongate, and lose the natatory portion. In one group (Orthagoriscidæ) the posterior part of the vertebral column is lost and the caudal fin is a nearly useless rudiment. In the Ostraciontidæ (which may have had a different origin, as the pharyngeal bones are not contracted) the natatory powers are much reduced, and the body is inclosed in an osseous carapace so as to be capable of very little movement. The entire order is deficient in osseous tissue, the bones being thin and weak. It is a marked case of degeneracy.

There are several evident instances of sporadic degeneracy in other orders. One of these is the case of the family of the Icosteidæ, fishes from deep waters off the coast of California. Although members of the Percomorphi, the skeleton in the two genera Icosteus and Icichthys is unossified, and is perfectly flexible. Approximations to this state of things are seen in the parasitic genus Cyclopterus, and in the ribbon-fishes, Trachypteridæ.

Thus nearly all the main lines of the Physoclysti are degenerate; the exceptions are those that terminate in the Scombridæ (mackerel), Serranidæ, and Scaridæ (Pharyngognathi).

V. THE LINE OF THE BATRACHIA.

We know Batrachia first in the Coal Measures. They reach a great development in the Permian epoch, and are represented by large species in the Triassic period. From that time they diminish in numbers, and at the present day form an insignificant part of the vertebrate fauna of the earth. The history of their succession is told by a table of classification such as I give below:

I. Supraoccipital, intercalary and supratemporal bones present. Propodial bones distinct.

Vertebral centra, including atlas, segmented, one set of segments together supporting one arch.. *Rhachitomi.*

Vertebræ segmented, the superior and inferior segments each complete, forming two centra to each arch.. *Embolomeri.*

Vertebral centra, including atlas, not segmented, one to each arch..... *Stegocephali.*

II. Supraoccipital and supratemporal bones wanting. Frontal and propodial bones distinct.

a. An os intercalare.

A palatine arch and separate caudal vertebræ........................ *Proteida.*

aa. No os intercalare.

A maxillary arch; palatine arch imperfect; nasals, premaxillaries and caudal vertebræ distinct.. *Urodela.**

No maxillary or palatine arches; nasals and premaxillary, also caudal vertebræ, distinct.. *Trachystomata.*

III. Supraoccipital, intercalare and supratemporal bones wanting. Frontals and parietals connate; propodial bones and caudal vertebræ confluent.

Premaxillaries distinct from nasals; no palatine arch; astragalus and calcaneum elongate, forming a distinct segment of the limb.................. *Anura.*

The probable phylogeny of these orders as imperfectly indicated by paleontology is as follows:

---

* Probably includes the Gymnophiona.

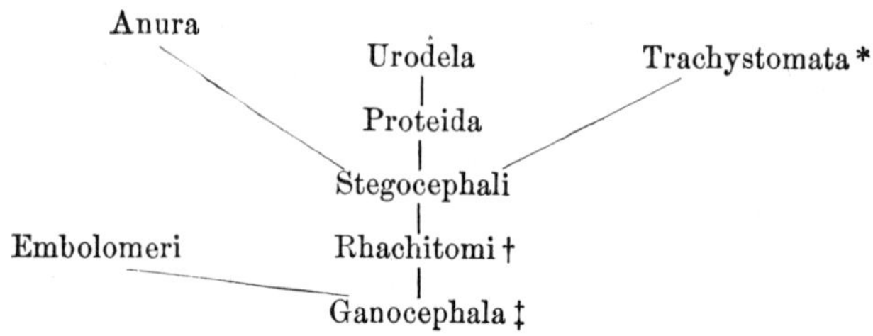

An examination of the above tables shows that there has been in the history of the Batrachian class a reduction in the number of the elements composing the skull, both by loss and by fusion with each other. It also shows that the vertebræ have passed from a notochordal state with segmented centra, to biconcave centra, and finally to ball-and-socket centra, with a great reduction of the caudal # series. It is also the fact that the earlier forms (those of the Permian epoch) show the most Mammalian characters of the tarsus and of the pelvis. The latter forms, the salamanders, show a more generalized form of carpus and tarsus and of pelvis also. In the latest forms, the Anura, the carpus and tarsus are reduced through loss of parts, except that the astragalus and calcaneum are phenomenally elongate. We have then, in the Batrachian series, a somewhat mixed kind of change ; but it principally consists of concentration and consolidation of parts. The question as to whether this process is one of progression or retrogression may be answered as follows : If degeneracy consists in "the loss of parts without complementary addition of other parts," then the Batrachian line is a degenerate line. This is only partly true of the vertebral column, which presents the most primitive characters in the early, Permian, genera (Rhachitomi). If departure from the nearest approximation to the Mammalia is degeneracy, then the changes in this class come under that head. The carpus, tarsus, and scapular and pelvic arches of the Rachitomi are more Mammalian than are those of any of their successors. ‖

---

* The Trachystomata probably came from the Urodela by a process of degeneracy. See "American Naturalist," Dec., 1885. (Ed. 1886.)

† Includes the Eryopidæ.

‡ Includes Trimerorhachidæ and Archegosauridæ ; and is distinguished from the Rhachitomi only by the presence of a single and cotyloid articulation of the skull with the atlas.

\# This reduction extends to the dorsal series as well. (Ed. 1886.)

‖ This should read, than their latest, or anurous successors. (Ed. 1886.)

There are several groups which show special marks of degeneracy. Such are the reduced maxillary bones and persistent gills of the Proteïda; the absence of the maxillary bones and the presence of gills in the Trachystomata; the loss of a pair of legs and feebleness of the remaining pair in the sirens; and the extreme reduction of the limbs in Amphiuma. Such I must also regard, with Lankester, the persistent branchiæ of the Siredons. I may add that in the brain of the Proteïd Necturus the hemispheres are relatively larger than in the Anura, which are at the end of the line.

It must be concluded, then, that in many respects the Batrachia have undergone degeneracy with the passage of time.

VI. THE REPTILIAN LINE.

As in the case of the Batrachia, the easiest way of obtaining a general view of the history of this class is by throwing their principal structural characters into a tabular form. As in the case of that class, I commence with the oldest forms and end with the latest in the order of time, which, as usual, corresponds with the order of structure. I except from this the first order, the Ichthyopterygia, which we do not know prior to the Triassic period:*

A. Extremities not differentiated in form beyond proximal segment.

I. Os quadratum immovably articulated to squamosal, etc.

Tubercular and capitular rib-articulations present and distinct. . . 1. *Ichthyopterygia.*

AA. Elements of extremities differentiated.

II. Os quadratum immovably articulated; capitular and tubercular rib-articulations distinct. Archosauria.

Pubis and ischium united, and with little or no obturator foramen; one posterior cranial arch; limbs ambulatory; a procoracoid. . . . . 2. *Theromorpha.*

Ischium and pubis distinct, the latter directed forward, backward, or downward; two posterior cranial arches; limbs ambulatory; no procoracoid 3. *Dinosauria.*†

Ischium and pubis united; two postcranial arches; anterior limbs volant 4. *Ornithosauria.*

III. Os quadratum closely united to cranial arches; but one rib-articulation. Synaptosauria.

---

* Generally similar to the system published by me in "Proceedings Amer. Ass. Adv. Science," xix, p. 233.

† This definition includes the Crocodilia in the Dinosauria, as it is absolutely connected with the typical Dinosaurs by the Opisthocœla (Sauropoda Marsh).

Distinct hyposternal and postabdominal bones; ribs joining each two vertebræ, and generally forming a carapace; one posterior cranial arch......5. *Testudinata*

Hyposternal and postabdominal bones not distinct; two posterior cranial arches ribs attached to one vertebra; a sternum; ? no procoracoid 6. *Rhynchocephalia*

Hyposternal and postabdominal bones not distinct; two posterior cranial arches ribs attached to one centrum; no sternum;* a procoracoid..7. *Sauropterygia.*

IV. Os quadratum attached only at the proximal extremity, and more or less movable; ribs with one head. Streptostylica.†

Brain-case membranous in front of proötic bone; trabecula not persistent 8. *Lacertilia.*

Brain-case with osseous walls anterior to proötic; a scapular arch and sternum 9. *Pythonomorpha.*

Brain-case with osseous walls anterior to proötic; no scapular arch nor sternum; trabecular grooves of sphenoid and presphenoid bones...........10. *Ophidia.*

An inspection of the characters of these ten orders, and their consideration in connection with their geological history, will give a definite idea as to the character of their evolution. The history of the class, and therefore the discussion of the question, is limited in time to the period which has elapsed since the Permian epoch inclusive, for it is then that the Reptilia enter the field of our knowledge. During this period but one order of reptiles inhabited the earth, so far as now known, that of the Theromorpha. The important character and rôle of this type may be inferred from the fact that they are structurally nearer to both the Batrachia and the Mammalia than any other, but present characters which render it probable that all the other reptiles, with possibly the exception of the Ichthyopterygia, derived their being from them. The phylogeny may be thus expressed :

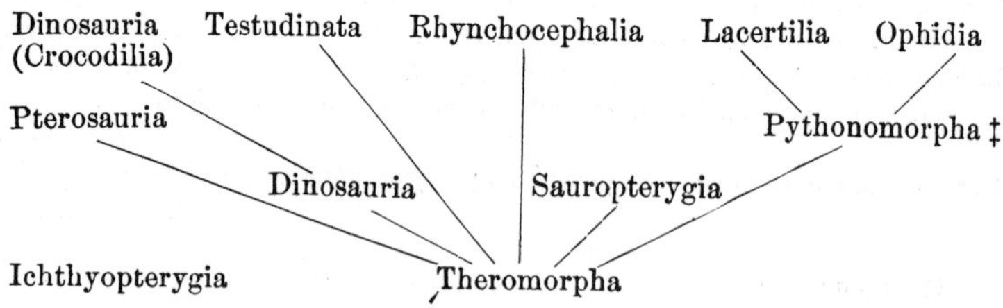

---

* Episternum present.

† It is quite possible that the three divisions of this head form one natural order, the Streptostylica, or Squamata.

‡ Some unknown type of this order will represent the ancestor of the Ophidia, while the Lacertilia may have come directly from the Theromorpha. (Ed. 1886.)

In the first place, this line departs with lapse of time from the primitive and ancestral order, the Theromorpha, in two respects: First, in the loss of the capitular articulation of the ribs, and, second, in the gradual elongation and final freedom of the suspensory bone of the lower jaw (the os quadratum). In so departing from the Theromorpha, it also departs from the mammalian type. The ribs assume the less perfect kind of attachment which the mammals only exhibit in some of the whales, and the articulation of the lower jaw loses in strength, while it gains in extensibility, as is seen in the development of the line of the eels among fishes. The end of this series, the snakes, must therefore be said to be the result of a process of creation by degeneration, and their lack of scapular arch and fore limb and usual lack of pelvic arch and hind limb, are confirmatory evidence of the truth of this view of the case.

Secondly, as regards the ossification of the anterior part of the brain-case. This is deficient in some of the Theromorpha, the ancestral order, which resemble in this, as in many other things, the contemporary Batrachia. Some of them, however (Diadectidæ), have the brain completely inclosed in front. The late orders mostly have the anterior walls membranous, but, in the streptostylicate series at the end, the skull in the snakes becomes entirely closed in front. In this respect, then, the latter may be said to be the highest or most perfect order.

As regards the scapular arch, including the sternum, no order possesses as many elements as thoroughly articulated for the use of the anterior leg as the Permian Theromorpha, though the coracoid is of reduced size. In all the orders there is loss of parts, excepting only in the Ornithosauria and the Lacertilia. In the former the adaptation is to flying. The latter retain nearly the Theromorph type, enlarging the coracoid. An especial side development is the modification of abdominal bones into two peculiar elements to be united with the scapular arch into a plastron, seen in the Testudinata. In this part of the skeleton the orders are generally degenerate, the last one, the Ophidia, especially so.

The pelvic arch has a more simple history. Again, in the Theromorpha we have the nearest approach to the Mammalia. The only other order which displays similar characters is the Ornithosauria (Dimorphodon, according to Seeley). In the Dinosauria we have a side modification which is an adaptation to the erect or bipedal mode of progression, the inferior bones being thrown backward so as to support the viscera in a more posterior position. This is an obvious necessity to a bipedal animal where the vertebral column is not perpendicular, as in birds. And it is from the Dinosauria that the birds are supposed to have arisen. (Plates XV and XVI.) The main line of the Reptilia, however, departs from both the mammalian and the avian type and loses in strength as compared with the former. In the latest orders, the Phythonomorpha and Ophidia, the pelvis is rudimental or absent.

As regards the limbs, the degeneracy is well marked. No reptilian order of later ages approaches so near to the Mammalia in these parts as do the Permian Theromorpha. This approximation is seen in the internal epicondylar foramen and well-developed condyles of the humerus, and in the well-differentiated seven bones of the tarsus. The epicondylar foramen is only retained in later reptiles in the Rhynchocephalian Hatteria (Dollo); and the condyles of the Dinosauria and all of the other orders, excepting the Ornithosauria and some Lacertilia, are greatly wanting in the strong characterization seen in the Theromorpha. The posterior foot seems to have stamped out the greater part of the tarsus in the huge Dinosauria, and it is reduced, though to a less degree, in all the other orders. In the paddled Sauropterygia, dwellers in the sea, the tarsus and carpus have lost all characterization, probably by a process of degeneracy, as in the mammalian whales. This is to be inferred from the comparatively late period of their appearance in time. The still more unspecialized feet and limbs of the Ichthyosaurus (Ichthyopterygia) can not yet be ascribed to degeneracy, for their history is too little known. At the end of the line, the snakes present us with another evidence of degeneracy. But few have a pelvic arch (Stenostomidæ Peters), while very few (Peropoda) have any trace of a posterior limb.

The vertebræ are not introduced into the definitions of the orders, since they are not so exclusively distinctive as many other parts of the skeleton. They nevertheless must not be overlooked. As in the Batrachia, the Permian orders show inferiority in the deficient ossification of the centrum. Many of the Theromorpha are notochordal, a character not found in any later order of reptiles excepting in a few Lacertilia (Gecconidæ). They thus differ from the Mammalia, whose characters are approached more nearly by some of the terrestrial Dinosauria in this respect.

Leaving this order, we soon reach the prevalent ball-and-socket type of the majority of Reptilia. This strong kind of articulation is a need which accompanies the more elongated column which itself results at first from the posterior direction of the ilium. In the order with the longest column, the Ophidia, a second articulation, the zygosphen, is introduced. The mechanical value of the later reptilian vertebral structure is obvious, and in this respect the class may be said to present a higher or more perfect condition than the Mammalia.

In review it may be said of the reptilian line, that it exhibits marked degeneracy in its skeletal structure since the Permian epoch; the exception to this statement being in the nature of the articulations of the vertebræ. And this specialization is an adaptation to one of the conditions of degeneracy, viz., the weakening and final loss of the limbs and the arches to which they are attached.

The history of the development of the brain in the Reptilia presents some interesting facts. In the Diadectid family of the Permian Theromorpha it is smaller than in a *Boa constrictor*, but larger than in some of the Jurassic Dinosauria. Marsh has shown that some of the latter possess brains of relatively very narrow hemispheres, so that in this organ those gigantic reptiles were degenerate, while the existing streptostylicate orders have advanced beyond their Permian ancestors.

There are many remarkable cases of what may now be safely called degradation to be seen in the contents of the orders of reptiles.* Among tortoises may be cited the loss of the rib-heads and of one or two series of phalanges in the especially terrestrial family of the Testudinidæ. The cases among the Lacertilia are the most remarkable. The entire families of the Pygopodidæ, the Aniellidæ, the Anelytropidæ, and the Dibamidæ are degraded from superior forms. In the Anguidæ, Teïdæ, and Scincidæ, we have series of forms whose steps are measured by the loss of a pair of limbs, or of from one to all the digits, and even to all the limbs. In some series the surangular bone is lost. In others the eye diminishes in size, loses its lids, loses the folds of the epidermis which distinguishes the cornea, and finally is entirely obscured by the closure of the ophthalmic orifice in the

* Such forms in the Lacertilia have been regarded as degradational by Lankester and Boulanger.

PLATE XV.

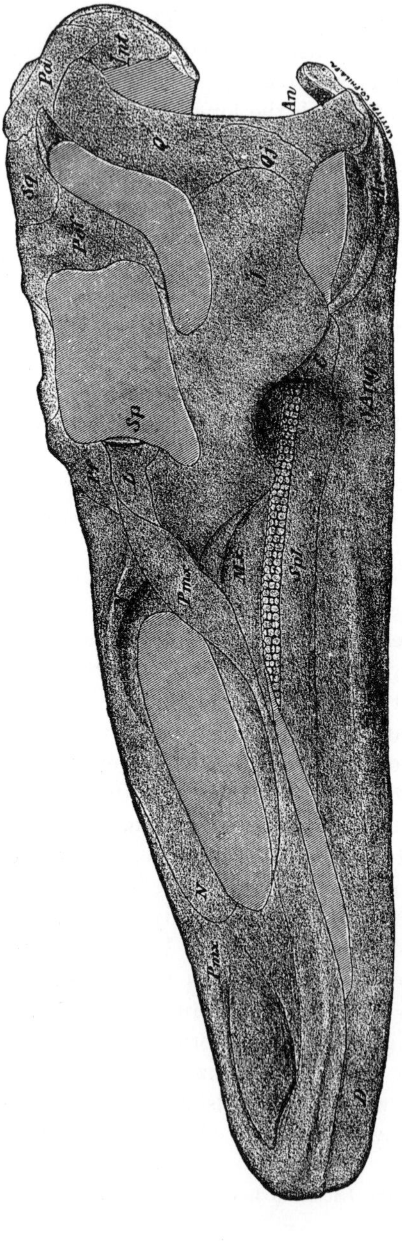

Diclonius mirabilis skull, one seventh natural size, profile. Dinosaurian from the Laramie bed of Dakota.

PLATE XVI.

Diclonius mirabilis skull, one seventh natural size, from below.

true skin. Among the snakes a similar degradation of the organs of sight has taken place in the order of the Scolecophidia, which live underground, and often in ants' nests. The Tortricidæ and Uropeltidæ are burrowing-snakes which display some of the earlier stages of this process. One genus of the true snakes even (according to Günther) has the eyes obscured as completely as those of the inferior types above named (genus Typhlogeophis).

VII. THE AVIAN LINE.

The paleontology of the birds not being well known, our conclusions respecting the character of their evolution must be very incomplete. A few lines of succession are, however, quite obvious, and some of them are clearly lines of progress, and others are lines of retrogression. The first bird we know at all completely, is the celebrated Archeopteryx of the Solenhofen slates of the Jurassic period. In its elongate series of caudal vertebræ and the persistent digits of the anterior limbs we have a clear indication of the process of change which has produced the true birds, and we can see that it involves a specialization of a very pronounced sort. The later forms described by Seeley and Marsh from the Cretaceous beds of England and North America, some of which have biconcave vertebræ, and all probably, the American forms certainly, possessed teeth. This latter character was evidently speedily lost, and others more characteristic of the subclass became the field of developmental change. The parts which subsequently attained especial development are the wings and their appendages; the feet and their envelopes, and the vocal organs. Taking all things into consideration, the greatest sum of progress has been made by the perching birds, whose feet have become effective organs for grasping, whose vocal organs are most perfect, and whose flight is generally good, and often very good. In these birds also the circulatory system is most modified, in the loss of one of the carotid arteries.

The power of flight, the especially avian character, has been developed most irregularly, as it appears in all the orders in especial cases. This is apparent so early as in the Cretaceous toothed birds already mentioned. According to Marsh the Hesperornithidæ have rudimental wings, while these organs are well developed in the Ichthyornithidæ. They are well developed among natatorial forms in the albatrosses and frigate pelicans, and in the skuas, gulls, and terns; among rasorial types the sand-grouse, and, among the adjacent forms, the pigeons. Then among the lower insessores, the humming-birds exceed all birds in their powers of flight, and the swifts and some of the Caprimulgidæ are highly developed in this respect. Among the higher or true song birds, the swallows form a notable example. With these high specializations occur some remarkable deficiencies. Such are the reduction of the feet in the Caprimulgidæ, swifts, and swallows, and the fœtal character of the bill in the same families. In the syndactyle families, represented by the kingfishers, the condition of the feet is evidently the result of a process of degeneration.

A great many significant points may be observed in the developmental history of the epidermic structures, especially in the feathers. The scale of change in this respect is in general a rising one, though various kinds of exceptions and variations occur. In the development of the rectrices (tail-feathers) there are genera of the wading and rasorial types, and even in the insessorial series, where those feathers are greatly reduced or absolutely wanting. These are cases of degeneracy.

There is no doubt that the avian series is in general an ascending one.

VIII. THE MAMMALIAN LINE.

Discoveries in paleontology have so far invalidated the accepted definitions of the orders of this class that it is difficult to give a clearly cut analysis, especially from the skeleton alone. The following scheme, therefore, while it expresses the natural groupings and affinities, is defective in that some of the definitions are not without exceptions :

I. A large coracoid bone articulating with the sternum.

Marsupial bones; fibula articulating with proximal end of astragalus 1. *Monotremata.*

II. Coracoid a small process co-ossified with the scapula.

α. Marsupial bones; palate with perforations (vagina double; placenta and corpus callosum rudimental or wanting; cerebral hemispheres small and generally smooth).

But one deciduous molar tooth.................................2. *Marsupialia.*

αα. No marsupial bones; palate entire (one vagina; placenta and corpus callosum well developed).

β. Anterior limb reduced to more or less inflexible paddles, posterior limbs wanting (Mutilata).

No elbow-joint; carpals discoid, and with the digits separated by cartilage; lower jaw without ascending ramus.................................3. *Cetacea.*

An elbow-joint; carpals and phalanges with normal articulations; lower jaw with ascending ramus ....4. *Sirenia.*

ββ. Anterior limbs with flexible joints and distinct digits; ungual phalanges not compressed and acute at apex * (Ungulata †).

γ. Tarsal bones in linear series; carpals generally in linear series.

Limbs ambulatory; teeth with enamel..........................5. *Taxeopoda.*‡

γγ. Tarsal series alternating; carpal series linear.

Cuboid bone partly supporting navicular, not in contact with astragalus. 6. *Proboscidia.*

γγγ. Both tarsal and carpal series more or less alternating.

Os magnum not supporting scaphoides; cuboid supporting astragalus; superior molars tritubercular..................................7. *Amblypoda.*

Os magnum supporting scaphoides; superior molars quadritubercular #
8. *Diplarthra.*‖

βββ. Anterior limbs with flexible joints. Ungual phalanges compressed and pointed △ (Unguiculata).

ε. Teeth without enamel; no incisors.

Limbs not volant; hemispheres small, smooth.......................9. *Edentata.*

εε. Teeth with enamel; incisors present.

No postglenoid process; mandibular condyle round; limbs not volant; hemispheres small, smooth ....10. *Rodentia.*

Limbs volant; hemispheres small, smooth.......................11. *Chiroptera.*

---

* Except the Hapalidæ. † Lamarck, "Zoölogie Philosophique," 1809.

‡ This order has the following suborders, whose association is now made for the first time:

Carpal series linear; no intermedium; fibula not interlocking with astragalus; no anapophyses; incisors rooted; hallux not opposable.. ....*Condylartha.*

Carpal series linear; an intermedium; fibula interlocking with astragalus; hallux not opposable ....*Hyracoidea.*

An intermedium; fibula not interlocking; anapophyses; hallux opposable; incisors growing from persistent pulps...........................*Daubentonoidea.*

An intermedium; fibula not interlocking; anapophyses; hallux opposable; incisors rooted; carpus generally linear.............. ....*Quadrumana.*

No intermedium; * nor anapophyses; carpal rows alternating; incisors rooted
Anthropoidea.†

The only difference between the Taxeopoda and the Bunotheria is in the unguliform terminal phalanges of the former as compared with the clawed or unguiculate form in the latter. The marmosets among the former division are, however, furnished with typical claws.

Some may prefer to use the term Primates in place of Taxeopoda, and such may be the better course.

\# Except Pantolestes.

‖ This order includes the suborders Perissodactyla and Artiodactyla. It is the Ungulata of some authors.

△ Except Mesonyx.

---

* Except in Simia and Hylobates. † Includes the Anthropoid apes and man.

A postglenoid process; mandibular condyle transverse; limbs not volant; no scapholunar bone;* hemispheres small, smooth............12. *Bunotheria*.†
A postglenoid process; limbs not volant, with a scapholunar bone; hemispheres larger, convoluted.......................................13. *Carnivora*.

Paleontology has cleared up the phylogeny of most of these orders, but some of them remain as yet unexplained. This is the case with the Cetacea, the Sirenia, and the Taxeopoda. The last-named order and the Marsupialia can be supposed with much probability to have come off from the Monotremata, but there is as yet no paleontological evidence to sustain the hypothesis. No progress has been made in unraveling the phylogeny of the Cetacea and Sirenia. The facts and hypotheses as to the phylogeny of the Mammalia may be represented in the following diagram:

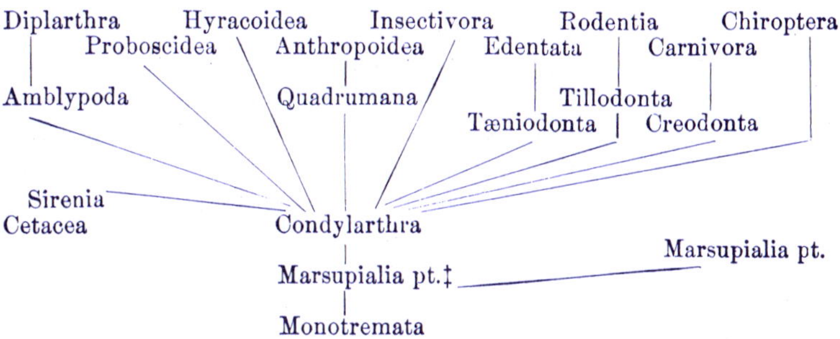

It will be readily seen from the above diagram that the discovery of the Condylarthra was an important event in the history of our knowledge of this subject. This suborder of the Lower Eocene epoch stands to the placental Mammalia in the same relation as the Theromorphous order does to the reptilian orders. It generalizes the characteristics of them all, and is apparently the parent stock of all, excepting, perhaps, the Cetacea. The discovery of the extinct Bunotherian suborders united together inseparably the clawed orders, excepting the bats; while the extinct order Amblypoda is the ancestor of the most specialized of the Ungulates, the odd- and even-toed Diplarthra.

The characters of the skeleton of the order Monotremata show that it is nearest of kin to the Reptilia, and many subordinate

* Except Erinaceus.

† With the suborders Insectivora, Creodonta, Tæniodonta, and Tillodonta.

‡ This was inadvertently omitted in the original. (Ed. 1886.)

characters point to the Theromorpha as its ancestral source.* In the general characters the Marsupialia naturally follow in a rising scale, as proved by the increasing perfection of the reproductive system. The Monodelphia follow with improvements in the reproductive system and the brain, as indicated in the table already given. The oldest Monodelphia were, in respect to the structure of the brain, much like the Marsupialia, and some of the existing orders resemble them in some parts of their brain-structure. Such are the Condylarthra and Amblypoda of extinct groups, and the Bunotheria, Edentata, Rodentia, and Chiroptera, recent and extinct. The characters of the brains of Amblypoda and some Creodonta are, in their superficial characters, even inferior to existing marsupials. The divided uterus of the recent forms named, also gives them the position next to the Marsupialia. In the Carnivora, Hyracoidea, and Proboscidia, a decided advance in both brain-structure and reproductive system is evident. The hemispheres increase in size, and they become convoluted. A uterus is formed, and the testes become external, etc. In the Quadrumana the culmination in these parts of the structure is reached, excepting only that, in the lack of separation of the genital and urinary efferent ducts, the males are inferior to those of many of the Artiodactyla. This history displays a rising scale for the Mammalia.

Looking at the skeleton, we observe the following successional modifications: †

First, as to the feet, and (A) the digits. The Condylarthra have five digits on both feet, and they are plantigrade. This character is retained in their descendants of the lines of Anthropoidea, Quadrumana, and Hyracoidea, also in the Bunotheria, Edentata, and most of the Rodentia. In the Amblypoda and Proboscidia the palm and heel are a little raised. In the Carnivora and Diplarthra the heel is raised, often very high, above the ground, and the number of toes is diminished, as is well known, to two in the Artiodactyla and one in the Perissodactyla. (B) The tarsus and carpus. In the Condylarthra the bones of the two series in the carpus and tarsus are opposite each other, so as to form continuous and separate longitudinal series of bones. This continues to be

* "Proceedings American Philosoph. Society," 1884, p. 43.

† See the evidence for evolution in the history of the extinct Mammalia. "Proceeds. Amer. Assoc. Adv. Science," 1883.

PLATE XVII.

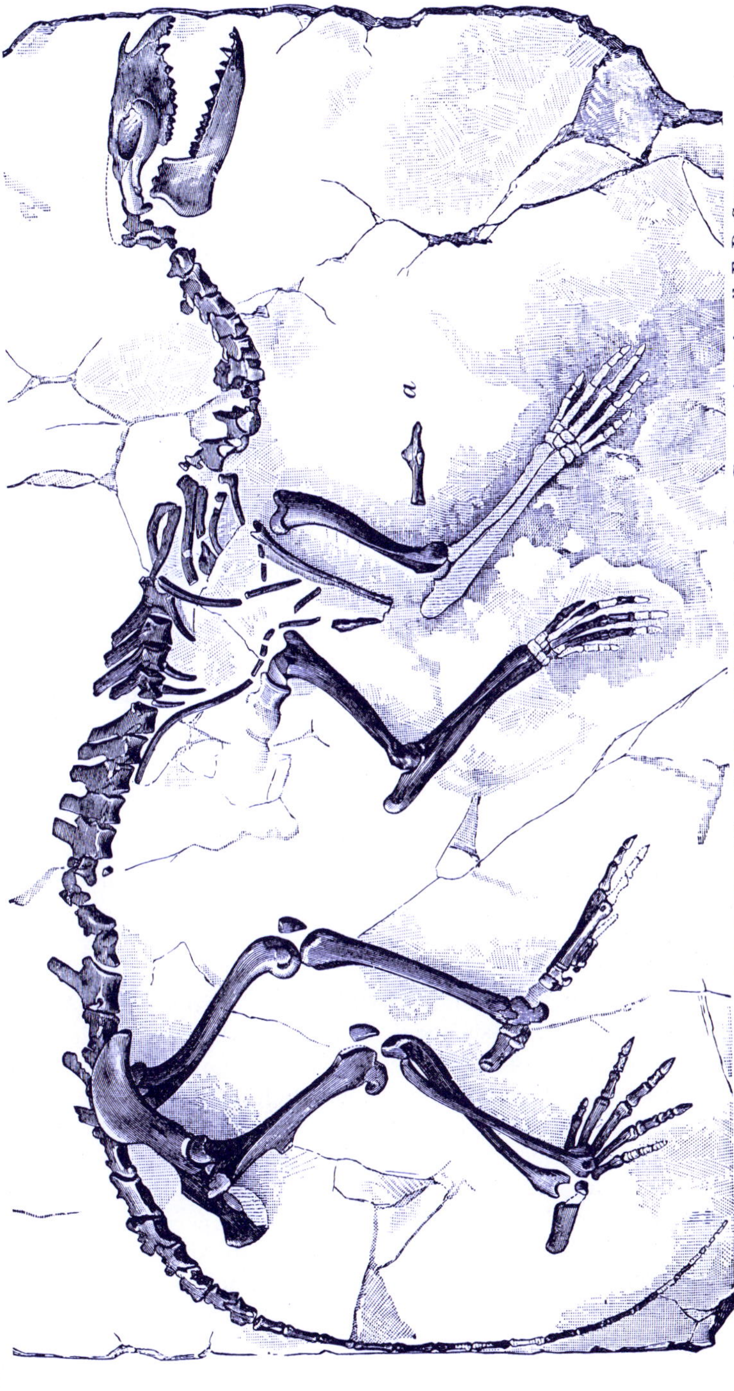

Phenacodus vortmani, representative of the order Condylarthra, 2/9 natural size. From specimen in coll. E. D. Cope.

the case in the Hyracoidea and many of the Quadrumana, but in the anthropoid apes and man the second row is displaced inwards so as to alternate with the first row, thus interrupting the series in the longitudinal direction, and forming a stronger structure than that of the Condylarthra. In the Bunotherian Rodent and Edentate series, the tarsus continues to be without alternation, as in the Condylarthra, and is generally identical in the Carnivora. In the hoofed series proper it undergoes change. In the Proboscidia the carpus continues linear, while the tarsus alternates. In the Amblypoda the tarsus alternates in another fashion, and the carpal bones are on the inner side linear, and on the outer side alternating. The complete interlocking by universal alternation of the two carpal series is only found in the Diplarthra. (C) As to the ankle-joint. In most of the Condylarthra it is a flat joint or not tongued or grooved. In most of the Carnivora, in a few Rodentia, and in all Diplarthra, it is deeply tongued and grooved, forming a more perfect and stronger joint than in the other orders, where the surfaces of the tibia and astragalus are flat. (D) In the highest forms of the Rodentia and Diplarthra the fibula and ulna become more or less co-ossified with the tibia and radius, and their middle portions become attenuated or disappear.

Secondly, as regards the vertebræ. The mutual articulations (zygapophyses) in the Condylarthra have flat and nearly horizontal surfaces. In higher forms, especially of the ungulate series, they become curved, the posterior turning upward and outward, and the anterior embracing them on the external side. In the higher Diplartha this curvature is followed by another curvature of the postzygapophysis upward and outward, so that the vertical section of the face of this process is an S. Thus is formed a very close and secure joint, such as is nowhere seen in any other Vertebrata.

Thirdly, as regards the dentition. Of the two types of Monotremata, the Tachyglossidæ and the Platypodidæ, the known genera of the former possess no teeth, and the known genus of the latter possesses only a single corneous epidermic grinder in each jaw. As the Theromorphous reptiles from which these are descended have well-developed teeth, their condition is evidently one of degeneration, and we can look for well-toothed forms of Monotremata in the beds of the Triassic and Jurassic periods. Perhaps some such are already known from jaws and teeth. In the marsupial order we have a great range of dental structure, which almost epitomizes that of the Monodelph orders. The dentition of the carnivorous forms is creodont; that of the kangaroos is perissodactyle, and that of the wombats is rodent. Other forms repeat the Insectivora. I therefore consider the placental series especially. I have already shown that the greater number of the types of this series have derived the characters of their molar teeth from the stages of the following succession. First, a simple cone or reptilian crown, alternating with that of the other jaw. Second, a cone with lateral denticles. Third, the denticles to the inner side of the crown forming a three-sided prism, with tritubercular apex, which alternates with that of the opposite jaw. Fourth, development of a heel projecting from the posterior base of the lower jaw, which meets the crown of the superior, forming a tubercular-sectorial inferior molar. From this stage the carnivorous and sectorial dentition is derived, the tritubercular type being retained. Fifth, the development of a posterior inner cusp in the superior molar, and the elevation of the heel in the inferior molar, with the loss of the anterior inner cusp. Thus the molars become quadritubercular, and opposite. This is the type of many of the Taxeopoda, including the Quadrumana and Insectivora as well as the inferior Diplarthra. The higher Taxeopoda (Hyracoidea) and Diplarthra add various complexities. Thus the tubercles become flattened and then concave, so as to form Vs in the section produced by wearing; or they are joined by cross-folds, forming various patterns. In the Proboscidia the latter become multiplied so as to produce numerous cross-crests.

The dentition of some of the Sirenia is like that of some of the Ungulata, especially of the suilline group, while in others the teeth consist of cylinders. In the Cetacea the molars of the oldest (Eocene and Miocene) types are but two-rooted and compressed, having much the form of the premolars of other Mammalia. In existing forms a few have simple conical teeth, while in a considerable number teeth are entirely wanting.

A review of the characters of the existing Mammalia as compared with those of their extinct ancestors displays a great deal of improvement in many ways, and but few instances of retrogression. The succession in time of the Monotremata, the Marsupialia, and the Monodelphia, is a succession of advance in all the characters of the soft parts and of the skeleton which define them (see table of classification). As to the monotremes themselves, it is more than probable that the order has degenerated in some respects in producing the existing types. The history of the Marsupialia is not made out, but the earliest forms of which we know the skeleton, Polymastodon (Cope) of the Lower Eocene, is as specialized as the most specialized recent forms. The dentition of the Jurassic forms, Plagiaulax, etc., is quite specialized also, but not more so than that of the kangaroos. The premolars are more specialized, the true molars less specialized than in those animals.

Coming to the Monodelphia, the increase in the size and complication of the brain, both of the cerebellum and the hemispheres, is a remarkable evidence of advance. But one retrogressive line in this respect is known, viz., that of the order Amblypoda,* where the brain has become relatively smaller with the passage of time. The successive changes in the structure of the feet are all in one direction, viz., in the reduction of the number of the toes, the elevation of the heel, and the creation of tongue and groove joints where plain surfaces had previously existed. The diminution in the number of toes might be regarded as a degeneracy, but the loss is accompanied by a proportional gain in the size of the toes that remain. In every respect the progressive change in the feet is an advance. In the carpus and tarsus we have a gradual rotation of the second row of bones on the first, to the inner side. In the highest and latest orders this process is most complete, and, as it results in a more perfect mechanical arrangement, the change is clearly an advance. The same progressive improvement is seen in the development of distinct facets in the cubito-carpal articulation, and of a tongue and groove ("intertrochlear crest") in the elbow-joint. In the vertebræ the development of the interlocking zygapophysial articulations is a clear advance.

Progress is generally noticeable in the dental structures; for, unlike the marsupial line, the earliest dentitions are the most simple, and the later the more complex. Some of the types retain the primitive tritubercular molars, as the Centetidæ, shrews and some lemurs, and many Carnivora, but the quadritubercular and its derivative forms are by far the most common type in the recent fauna. The forms that produced the complicated modifications in the Proboscidia and Diplarthra appeared latest in time, and the most complex genera, Bos and Equus, the latest of all. The extreme sectorial modifications of the tritubercular type, as

* See "Naturalist," Jan., 1885, p. 55.

seen in the Hyænidæ and the Felidæ, are the latest of their line also.

Some cases of degeneracy are, however, apparent in the monodelphous Mammalia. The loss of pelvis and posterior limbs in the two mutilate orders is clearly a degenerate character, since there can be no doubt that they have descended from forms with those parts of the skeleton present. The reduction of flexibility seen in the limbs of the Sirenia and the loss of this character in the fore limbs of the Cetacea are features of degeneracy for the same reason. The teeth in both orders have undergone degenerate evolution ; in the later and existing forms of the Cetacea even to extinction. The Edentata appear to have undergone degeneration. This is chiefly apparent in the teeth, which are deprived of enamel, and which are wanting from the premaxillary bone. A suborder of the Bunotheria, the Tæniodonta of the Lower Eocene period, display a great reduction of enamel on the molar teeth, so that in much-worn examples it appears to be wanting. Its place is taken by an extensive coat of cementum, as is seen in Edentata, and the roots of the teeth are often undivided as in that order. It is probable that the Edentata are the descendants of the Tæniodonta by a process of degeneracy.

Local or sporadic cases of degenerate loss of parts are seen in various parts of the mammalian series, such as toothless Mammalia wherever they occur. Such are cases where the teeth become extremely simple, as in the honey-eating marsupial Tarsipes, the carnivore Proteles, the Pteropod bats, and the aye-aye. Also where teeth are lost from the series, as in the canine genus Dysodus, and in man. The loss of the hallux and pollex without corresponding gain in various genera, may be regarded in the same light.

In conclusion, the progressive may be compared with the retrogressive evolution of the Vertebrata, as follows : In the earlier periods and with the lower forms, retrogressive evolution predominated. In the higher classes progressive evolution has predominated. When we consider the history of the first class of vertebrates, the Tunicata, in this respect, and compare it with that of the last class, the Mammalia, the contrast is very great.

PART III.

MECHANICAL EVOLUTION.

XII.

THE RELATION OF ANIMAL MOTION TO ANIMAL EVOLUTION.

To explain the origin of variation in animal structure is, *par excellence*, the object of the doctrine of evolution. There can be little doubt that the law of natural selection includes the cause of the preservation of certain modifications of pre-existent structure, in preference to others, after they have been brought into existence. In what manner or by what process the growing tissues of young animals have been so affected as to produce some organ or part of an organ which the parent did or does not possess, must be explained by a different set of laws. These have been termed *originative,* while those involved in natural selection are *restrictive* only.

I.

Of course we naturally look to something in the "surrounding circumstances" in which a plant or animal is placed, or its "environment," as the most probable stimulant of change of its character, because we know that such beings are totally dependent on cosmic and terrestrial forces for their sustenance and preservation. The difficulty has been to connect these forces with change of structure as *originative;* to show their operation as multiplying, restricting or destroying organisms already in existence is comparatively easy. This difficulty is partially due to the fact that such modifications must be realized during a limited portion of the life of an animal at least; that is, during the period of growth, when it is not at all or but little subject to the influence of external environment, but is usually protected or supported by the parent.

That the environment and changes in it affect the movements

of plants and animals is clear enough. The potency of such changes may be read in the physical history of the earth. A long series of modifications preceded the advent of life upon it, and change, both gradual and sudden, has been exhibited in the configuration and climate of all portions of the surface of the globe since that period. Animals have again and again been called upon to face new conditions, and myriads of species have fallen victims to the inflexibility of their organization which has prevented adaptation to new surroundings. But it is evident that if change of environment has had any influence in the progress of evolution, it has not been alone destructive. It has preceded life as well as death, and has furnished the stimulus to beings capable of change, while it has destroyed those which were incapable of it. It is a truism that change of physical conditions has preceded all great faunal changes, and that the necessity for new mechanism on the part of animals has always preceded the appearance of new structure in geologic times.

The embryology and paleontology of vertebrated animals show that the primary steps in the progress of this branch of the animal kingdom are marked by the successive changes in the structure of the circulatory system. First we have the various mechanical methods for the aëration of blood in a watery medium ; the result being a fluid whose metamorphosis in nutrition produces no heat. After the fishes followed *Batrachia*, the earliest air-breathers, whose long tarriance to-day in early aquatic stages is an epitome of the necessarily "amphibious" character of air-breathing vertebrate life, when land and fresh water, in constantly changing areas, were rising and separating from the universal ocean. The successive disappearances of the traces of the fish type of circulation in *Batrachia* and reptiles, are familiar facts ; and the exclusion of the unaërated blood from the systemic circulation in the birds and mammals marks the increase of general temperature which gives those classes one of their claims to superiority.

The appearance of land of course furnished the opportunity for aquatic animals to assume a terrestrial life. Marine animals which had acquired the habit of gulping air from the surface, which some of them now possess, perhaps because its richness in oxygen produced an agreeable exaltation or intoxication, would not find visits to the land difficult. And this would naturally follow the necessity of escape from aquatic enemies, or the search for new supplies of food.

In fine, it requires little argument to show that the environment has had in the past, as in the present, a primary influenc over the movements of animals.

II.

I will now endeavor to exhibit some reasons for believing tha the movements of animals affect their structure *directly*.

There are two alternative propositions expressive of the relations of the structures of animals to their uses. Either the us or attempt to use preceded the adaptive structure, or else th structure preceded and gave origin to the use. The third alternative, that use and structure came into being independently c each other, is too improbable for consideration in the presen article. Many facts render the first of these propositions muc the more probable of the two.

A general ground for suspecting that movement affects structure is the fact well known to systematic zoölogists, that adaptiv characters are the least reliable in systematic classification, i. e. are the most variable. What we call adaptive characters ar those whose teleological significance we can most easily perceive those whose uses are at the present time most obvious. Systematists habitually fall back on characters which are apparently th least related to the ordinary necessities of the life of the anima and this not from any theoretical considerations, but because suc characters are found to be the most constant. This is a very significant fact, showing as it does that it is the adaptive structur which are undergoing modification to-day. And this truth ca doubtless be discerned in all past ages, for many of the structur which are not now more related to the needs of an animal tha many others might be, were at one time most essential to its wel being, or necessarily related to its environment. Such are th structural characters of the heart and arteries already enumerate There seems to be no reason why all *Vertebrata* might not exi with equal comfort and success at the present if possessed of uniform organization in this respect. But the successive modifications which they present were, in past ages, most intimate connected with the progressive changes of the medium in whic they lived, as to the volume of oxygen supplied for respiratio as compared with that of the vapor of water, carbonic acid ga etc. But it must be here noted, in passing, that there are mar structures in animals which have never been adaptive, but whic re simply due to excess or defect of nutrition following a redistribution of force.*

The most direct evidence in support of the view that motion ffects structure directly is to be found in the well-known phe-omenon of the increase of the size and power of all organs by se. This increase is limited in the adult animal by the general ixity of all the organs, so that one of them cannot be developed eyond a certain point without injury to others, or without ex-austing the source of supply of nutritive material or special force erived from other organs. The syncope of the gymnast is an llustration of the natural limitation to the development of the nuscular system which proceeds at the expense of the digestive nd circulatory. But effort and exertion may become a habit of nind, which, even if limited in its executive means, is probably in-erited by offspring like all other mental traits. Such a quality ossessed by an infant or child doubtless tells on the growth of ts organs during their plastic stage, and produces structure by rowth which is impossible to the mature body.† And no one nows as yet how far mental bias may affect the nutrition of the arts of the infant in utero. Certain it is, that if use modifies utrition in the adult, it must have still greater influence in the oung; and it is in the young that the changes which constitute volution necessarily appear.

Change of structure during growth is accomplished either by ddition of parts ("acceleration") or by subtraction of parts "retardation").

Acceleration is produced either by multiplication of parts (as ells or segments) already present ("homotopy"), or by the ransfer of parts (cells) from one part of the organism to the ther ("heterotopy"). Homotopy or repetition is the usual and ormal mode of acceleration; it may proceed by an "exact repeti-on" of the parts already existing, as in the simplest animals and lants; or the new parts may differ from the old, as in higher nimals, where the process is called "modified repetition." Where ew forms traverse in their growth all the stages in which they reviously existed, they necessarily present at each stage the char-cters of those forms which have remained stationary in them, and ave not changed. This relation of "exact parallelism" is the

---

* "Method of Creation," 1871, p. 23.

† In man these changes are chiefly produced in the brain.

result of the simplest form of evolution or "palingenesis." When the history of growth of an advanced form does not show an identity between its stages and the various undeveloped or lower adult types, the relation is termed "inexact parallelism," and the type of development "cœnogenesis."

Change of structure is seen to take place in accordance with the mechanical effect of three forms of motion, viz.: by *friction*, *pressure* and *strain*. Under the first two, epidermal tissues become both dense and thick, as is seen on the palms and soles of the hands and feet, and in corns. There is no doubt that strength of the teeth is intimately connected with the hardness of the food. Density of osseous tissue and the coössification of parts of the skeleton, are directly associated with the force and duration of muscular contraction. Pathology abounds in illustrations of the determination of nutrition to new localities to meet the exigencies and demands arising from new stimuli. It is only necessary for a structure-producing supply of nutritive material to be habitually determined to a new locality by oft-recurring stimulus for the movement to become automatic and reflex ; and such a tendency would sooner or later be inherited, and produce structure in the growing organism of the young to a degree far exceeding anything that is possible in the adult.

In view of the above considerations, we can ascribe an extensive class of osseous projections at points of muscular insertion to the strength and duration of muscular contractions. To the same cause may be ascribed various anchyloses, such, for instance as is seen in the foot of the sloth. Transverse strains or their absence may be looked upon respectively as the cause of the hinge like or immovable articulations of the segments of the limbs of vertebrate animals. It is well known that in land animals, where easy flexibility of the limbs is essential to speed, these articulations are highly developed, while in marine animals, where the limbs are only used as paddles, they are almost or quite inflexible, and the extremities of the bones are truncate. In the most highly organized land Mammalia, the tibio-tarsal and humero-cubital articulations display an interlocking or tongue and-groove character. The same thing is seen in the ulno radial fixed articulation in the same types. These arrangements are especially adapted to prevent dislocation by side strains and if the preceding explanations be true, this structure is a corrugation due to the lateral pressure of a more or les convex surface, on a concave one which embraces it, and vice versa.

In the circulatory system, pressure has doubtless played an important part. Increased oxygenation of blood, the necessary consequence of the purification of the atmosphere, would stimulate the action of all the organs, including that of the heart. Greater pressure on its walls and septa would increase their size and strength, and ultimately close such foramina as were not in the course of the blood current, as the *foramen septi ventriculorum* of reptiles, and the *f. ovale*. Increased force of the current would, on the other hand, soon cause the enlargement of one or other of the four or five pairs of primary aorta bows, and develop it at the expense of the others, until finally the pre-eminence of one channel be secured and the aorta be the result. This part of the subject might be prolonged to an unlimited extent, but the above illustrations must suffice to indicate the meaning of my propositions.

III.

That movements change the environment of a plant or an animal, or parts of them, is obvious enough. If we consider only the reflex class, to which all the movements of plants and many of those of animals belong, we perceive that but for them the ordinary functions of assimilation, circulation, etc., could not be performed; there would be no change in the contents of their tubes and cells, and the environment of these would be unaltered. But when we view the movements of the higher animals, we perceive the immense importance of the powers and organs of movement as a factor in evolution. It may be safely assumed that, without powers of designed or adaptive movement, life would never have advanced beyond the stage presented by the vegetable kingdom.

The stimuli which are effective in animal consciousness are four, viz.: excessive temperature, hunger, danger from enemies, and the reproductive instinct. These prompt to the movements which we observe in animals in a wild state, and without which it is evident that the animals themselves would soon cease to exist.

It can not be denied that organisms which are incapable of moving from place to place in search of food, or of migration to escape vicissitudes of temperature, are much more completely subject to the influences of their environment than those that are capable of such movement. Hence animals are much more independent of the supply of food and of temperature than are plants. Hence also, other things being equal, the greater the powers of motion, the greater the independence.

Powers of movement then enable animals to avoid extremes of climate by migrations or by protective arts. They enable them to procure food by making journeys in search of it, and by all methods of capturing it. They furnish the agent of active defense against enemies, and of successfully reproducing their kind.

When, through changes of level of the earth's surface, drought has overtaken a region, animals capable of the necessary migrations have escaped. When an irruption of destructive animal enemies has threatened an animal population with death, those members of it whose strength or speed insured them safety, were the survivors. When land has been encroached upon by water to such a degree as to bring starvation on its animal inhabitants, those which could fly or swim have sought new localities.

Since all food-supply, as well as the ability to obtain food, is dependent on temperature, those portions of the organism which furnish means of resistance to climatic vicissitudes have the deepest significance in the life-history of any division of animals.

The organs of circulation and motion are generally recognized as primary in the classification of *Vertebrata*. All situations where animal life is permitted by climate, support vegetable life also ; so each of the primary divisions of animals presents types adapted to the use of all kinds of food ; herbivorous, omnivorous, and carnivorous. Accordingly, it has been found that dental and other structures connected with digestion define divisions of secondary value and minor extent. Paleontology shows that the origin of such divisions is of later date than that of the great classes first mentioned ; and each of the latter has in its day been modified in the subordinate directions indicated by the teeth and beak. But here also organs of movement are of great importance ; so that the herbivorous and carnivorous types at least, have ever in land animals (reptiles, birds and mammals) been characterized by the structure of their feet also.

IV.

It has been maintained above, that environment governs the movements of animals, and that the movements of animals then alter their environment. It has also been maintained that the movements of animals have modified their structure so as to render them more or less independent of their environment. The history of animal life is in fact that of a succession of conquests over the restraints imposed by physical surroundings. Man has attained to a wonderful degree of emancipation from the iron bonds that confine the lower organisms.

It becomes then all important to examine into the elements involved in animal movements.

These are of the two classes, reflex and conscious. To the former belongs the accelerated activity of muscular action and circulation, inferred to have accompanied increase in the percentage of oxygen in the atmosphere, during the periods of geological time. To the consciously performed acts belong all those due to states of pain or pleasure in animals; such as are excited by the four classes of stimuli already mentioned.

Doubtless physical changes in the surrounding medium have always produced new reflex movements in animals, and have been a first element in evolution. Such has been the immediate cause of change of structure in plants, and in animals so far as they are unconscious. But consciousness brings with it limitless possibilities, since it places an animal in contact with innumerable stimuli which leave unconscious beings unaffected. All the causes which provoke the movements of higher animals are appeals to consciousness, and the consequences due to movements of such beings have only been possible through consciousness.

It is evident then that sensibility to impressions has been the prime essential to the acquisition of new movements, and hence of new structure, other things being equal. Another essential, not less important, has been memory; because without this faculty, experience, and hence education and the acquisition of habits of movement, are not possible.

The ascending development of the bodily structure in higher animals has thus been, in all probability, a concomitant of the evolution of mind, and the progress of the one has been dependent in an alternating way on the progress of the other. The development of mind has secured to animals the greatest degree of independence of their environment of which they are capable. The first important acquisition leading to this end was aërial respiration; the second, rapid nutrition by hot blood. And as essential to the production and preservation of these, improvements in organs of movement have been superadded to every successive type of life.

Consciousness remains as the unresolvable factor in the process; as at once the measure of, and respondent to a large class of phenomena.

XIII.

ON THE TRITUBERCULATE TYPE OF MOLAR TOOTH IN THE MAMMALIA.

It is now apparent that the type of superior molar tooth which predominated during the Puerco epoch was triangular or tritubercular; that is, with two external and one internal tubercles.* Thus, of sixty-seven species of placental Mammalia of which the superior molars are known, all but four have three tubercles of the crown, and of the remaining sixty-five, all are triangular, excepting those of three species of Periptychus, and three allied forms, which have a small supplementary lobe on each side of the median principal inner tubercle.†

This fact is important as indicating the mode of development of the various types of superior molar teeth, on which we have not heretofore had clear light. In the first place, this type of molar exists to-day only in the insectivorous and carnivorous Marsupialia; in the Creodonta, and the tubercular molars of such Carnivora as possess them (excepting the plantigrades). In the Ungulates its persistence is to be found in the molars of the Coryphodontidæ of the Wasatch, and Dinocerata of the Bridger Eocenes. In later epochs it is occasionally seen only in the last superior molar.

It is also evident that the quadritubercular molar is derived from the tritubercular by the addition of a lobe of the inner part of a cingulum of the posterior base of the crown. Transitional states are seen in some of the Periptychidæ (*Anisonchus*), and in the sectorials of the Procyonidæ.

The tritubercular or triangular superior molar is associated with a corresponding form of the anterior part of the inferior molar. This kind of inferior molar ‡ I have called the tubercular

---

* See "American Naturalist," April, 1883, p. 407.

† This type is therefore only an extension of the tritubercular. (Ed. 1886.)

‡ See Report G. M. Wheeler, D. Chief of Engineers on Explor. Surv. W. 100th Mer., vol. iv, pt. ii; on the Creodonta.

sectorial, and is very variable as to the degree of development of the sectorial cutting edge. The anterior triangle is formed by the connection, by angle or crest, of the median and anterior internal crests with the anterior external. Its primitive form is seen in Didelphys, Pelycodus, Pantolambda, and the Amblypoda generally; in Centetes and Talpa; and in its sectorial form, in Stypolophus and Oxyæna, etc.

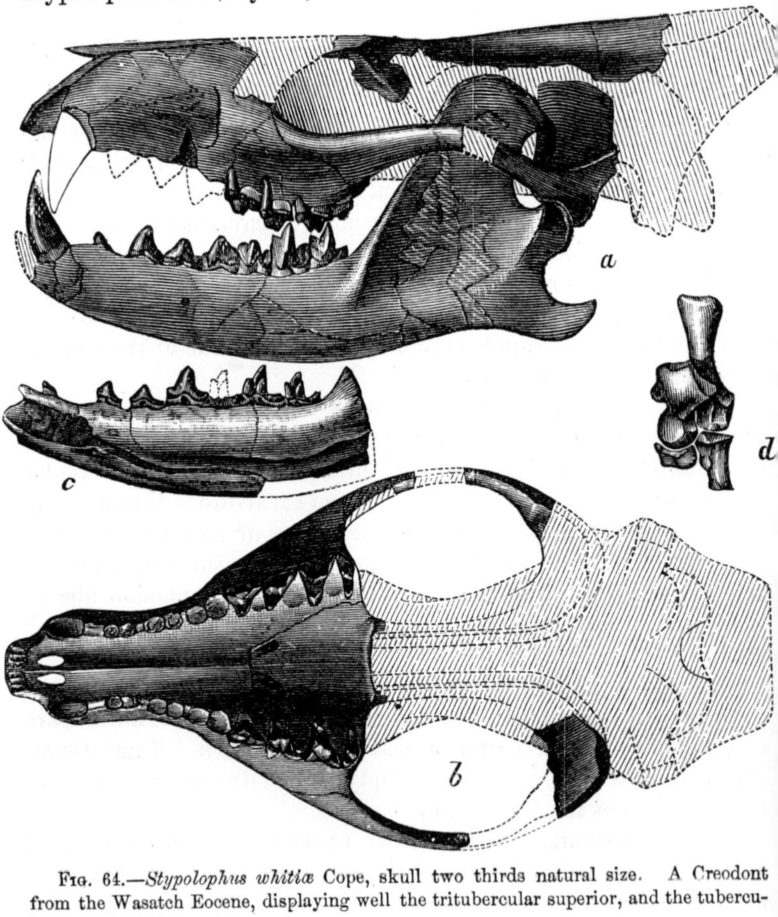

Fig. 64.—*Stypolophus whitiæ* Cope, skull two thirds natural size. A Creodont from the Wasatch Eocene, displaying well the tritubercular superior, and the tubercular-sectorial inferior molars.

The mechanical action of such teeth is as follows: Of course, it results from the form of the superior molars that the spaces between them are wedge-shaped, the apex external, the base opening to the palate. The base of the triangular section of the anterior part of the inferior molar is interior, and the apex exterior, and when the jaws are closed this triangular prism exactly fits the space between the superior molars. The lower heel of the inferior molar receives the impact of the crown of the superior molar. Thus the oblique edges of the inferior triangle shear on the edges of two adjacent superior molars. The anterior parts of the inferior molars, and the superior molars, form an alternate dental series as distinguished from the prevalent opposed dentition of most Mammalia. In so far it resembles the reptilian dentition.

This primitive dentition has been modified in two directions, viz., to form the grinding and the sectorial dentitions. As already remarked, the superior molars gradually acquire a posterior internal lobe, which produces the quadrituberculate type. This lobe, by opposing the anterior internal lobe of the next posterior inferior molar, precludes the entrance of the anterior triangle of the latter between the two superior molars. Hence we find in the types which possess quadritubercular superior molars, that the anterior triangle of the inferior molar is not elevated, if present, as, for instance, in Rhinocerus. It is, however, more frequently atrophied, and disappears, contributing to form the inferior quadritubercular molar so well known.

On the other hand, as I have pointed out,* the anterior internal cusp of the triangle of the inferior molar may be more developed antero-posteriorly, giving the antero-internal edge of the triangle much greater obliquity than the postero-internal. In correspondence with this modification, the superior triangular molar loses its equilateral character by the more anterior position of its internal angle, thus elongating the posterior internal side of the crown. The latter thus fits the corresponding form of the triangle of the inferior molar, forming with it the shear of the sectorial tooth.

In a former article, "On the Homologies of the Molar Teeth," etc., I traced the modifications of the superior and many of the inferior molars of the ungulate mammals to a parent quadrituberculate type. In a subsequent essay† I traced the origin of the

---

* On the origin of the sectorial tooth of the Carnivora, "American Naturalist," 1875.

† "Journal Academy Natural Sciences," Philadelphia, March, 1874.

inferior sectorial to a primitive five-tubercled, or "tubercular sectorial" type. Farther than this I did not go, and made no attempt to derive the few cases of triangular superior molars then known, nor the type of the superior sectorial. The revelations of the Puerco fauna show that the superior molars of both ungulate and unguiculate Mammalia have been derived from a tritubercular

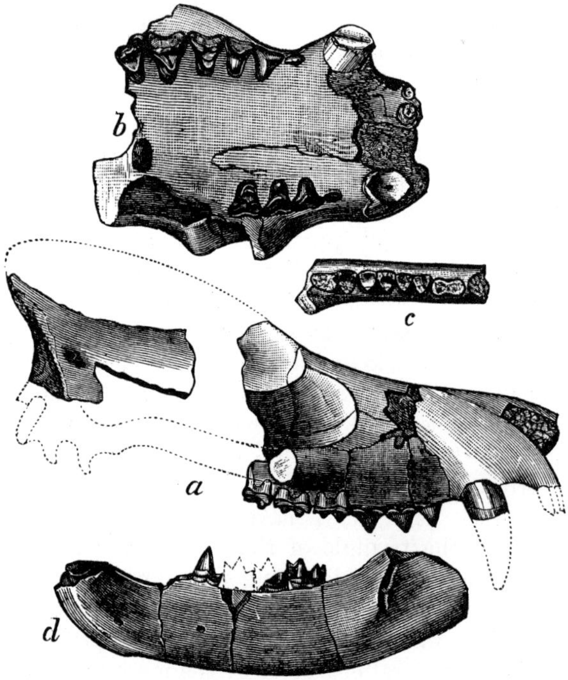

Fig. 65.—*Deltatherium fundaminis* Cope, skull and ramus mandibuli, two thirds natural size, from the Puerco beds of New Mexico. Figs. *a*, *b*, *c*, from one individual. Fig. *d*, from a second animal. Fig. *a*, right side of cranium; *b*, palate from below; *c*, mandible part, from above; *d*, left ramus, outer side. From the Report U. S. Geol. Surv. Terrs., vol. iii. Original.

type; and that the inferior true molars of both have been derived from a "tubercular sectorial" type. Shall we look for the origin of the latter in a tritubercular tooth also, i. e., tubercular sectorial without heel; and will the crowns of the true molars of the primitive mammals alternate with, instead of oppose each other? This is a probable result of future discovery.

XIV.

THE ORIGIN OF THE SPECIALIZED TEETH OF THE CARNIVORA.

THE specially developed teeth of the Carnivora are the canines and sectorials. The former are large in many orders of *Mammalia,* and their origin is probably to be sought among the Theromorphous reptilia,* as *Clepsydrops* and *Deuterosaurus,* if not in still lower types. The successive modifications of form which have resulted in the existing specialized single lower sectorial tooth of the *Felidæ* have been already pointed out.† They were shown to consist in the gradual obliteration of the internal and posterior tubercles, and the enlargement of the external anterior tubercle in connection with an additional anterior tubercle. The modification in the character of the dentition, taken as a whole, was shown to consist in the reduction in the number of teeth, until in *Felis,* etc., we have almost the entire function of the molar series confined to a single large sectorial in each jaw.

Observation on the movements of the jaws of Carnivora shows that they produce a shearing motion of the inferior on the superior teeth. This is quite distinct from the sub-horizontal movement of Ruminants, or the vertical motion of hogs and monkeys. Examination of the crowns of the sectorials shows that the inner side of the superior, and the external side of the inferior, are worn in the process of mastication. The attempt to cut the tough and stringy substances found in animal bodies is best accomplished by the shearing of the outer edge of the lower molar on the inner edge of the external tubercles of the superior molar, in an animal with simple tubercular teeth. The width of the mandible is too great to allow the inferior teeth to shear on the inner edge of the inner tubercles of the superior series. The

---

* "American Naturalist," 1878, p. 829.

† Cope, "Proceedings Academy Philada.," 1875, p. 22.

PLATE XVIII.

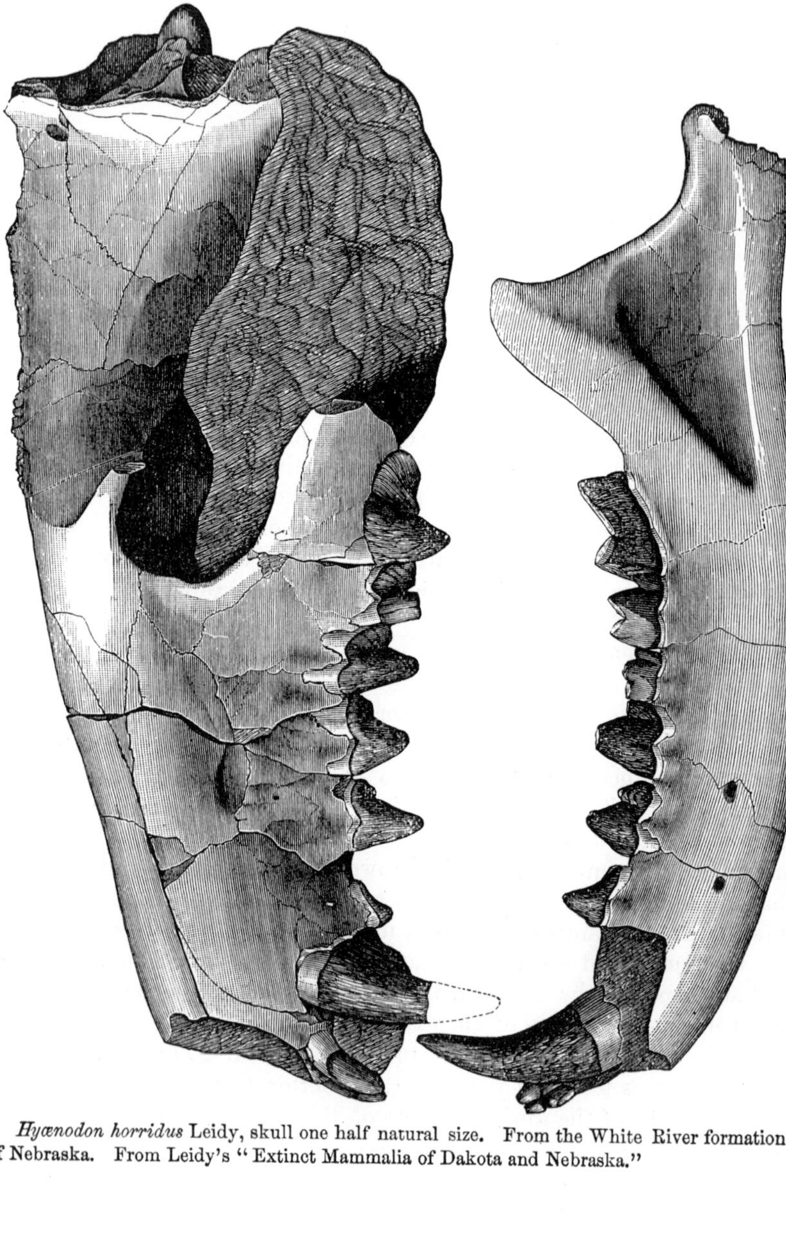

Hyænodon horridus Leidy, skull one half natural size. From the White River formation of Nebraska. From Leidy's "Extinct Mammalia of Dakota and Nebraska."

cusps of both superior and inferior teeth engaged in this process have developed in elevation, at the expense of those not engaged in it, viz.: the internal cusps of the same teeth. The atrophy of the latter can not have been due to friction, since the internal cusps of the inferior series, which have not been subjected to it, are reduced like those of the superior sectorial, which have. Indeed, it is possible that some of the *Creodonta,* the carnivores of

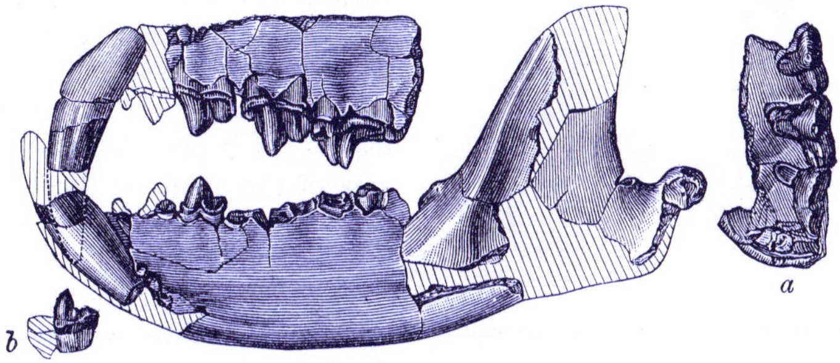

FIG. 66:—*Oxyæna lupina* Cope, jaws, one half natural size, from the Wasatch beds of New Mexico. Fig. *a*, maxillary bone with teeth, from below; *b*, last superior molar, from behind. Original; from the Report U. S. G. G. Survey W. of 100th Mer., G. M. Wheeler.

the Lower Eocene, may have been derived from ancestors without or with rudimental inner cusps. In any case the effect of use in lengthening the external cusps appears to have operated in the Carnivora, as it has done to a greater degree in the *Ungulata;* and the lateral vertical wear would appear to have resulted in the blade-form, as transverse wear in the Ungulates has resulted in the plane grinding surface.

The specialization of one tooth to the exclusion of others as a sectorial, appears to be due to the following causes. It is to be observed, in the first place, that when a carnivore devours a carcass, it cuts off masses with its sectorials, using them as shears. In so doing it brings the part to be divided to the angle or canthus of the soft walls of the mouth, which is at the front of the masseter muscle. At this point, the greatest amount of force is gained, since the weight is thus brought immediately to the power, which would not be the case were the sectorial situated much in front of the masseter. On the other hand the sectorial could not be situated farther back, since it would then be inaccessible to a carcass or mass too large to be taken into the mouth.

The position of the sectorial tooth being thus shown to be dependent on that of the masseter muscle, it remains to ascertain a probable cause for the relation of the latter to the dental series in modern Carnivora. Why, for instance, were not the last molars modified into sectorial teeth in these animals, as in the extinct *Hyænodon*, and various *Creodonta?* The answer obviously is to be found in the development of the prehensile character of the canine teeth. It is probable that the gape of the mouth in the Hyænodons was very wide, since the masseter was situated relatively far posteriorly. In such an animal the anterior parts of the jaws with the canines had little prehensile power, as their form and anterior direction also indicates. They doubtless snapped rather than lacerated their enemies. The same habit is seen in the existing dogs, whose long

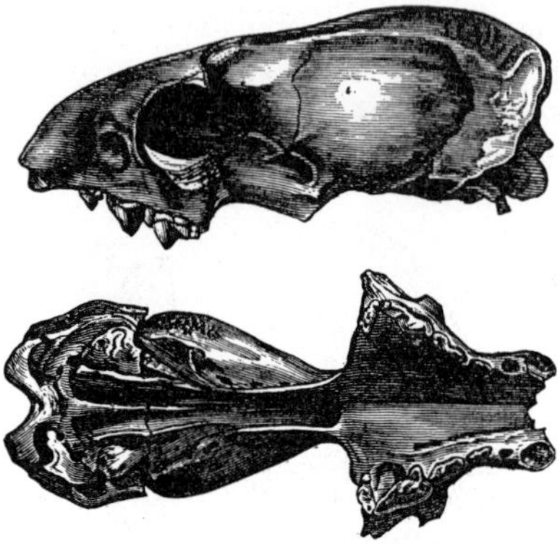

FIG. 67.—*Procælurus julieni* Filh., two thirds natural size. From the Phosphorites of France. From Filhol.

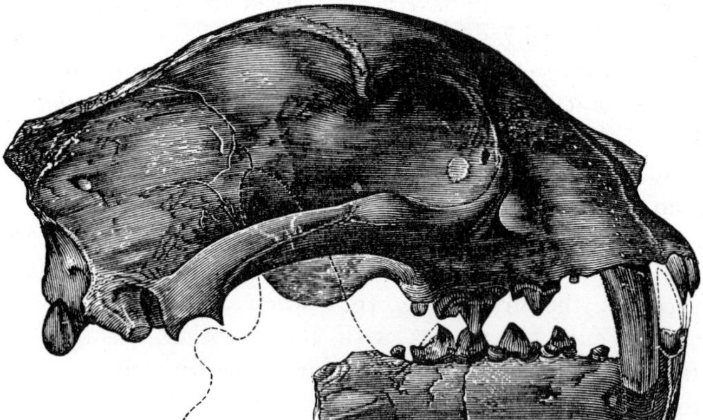

FIG. 68.—*Dinictis cyclops*, one half natural size. From John Day Miocene of Oregon. Mus. Cope. From vol. iii, U. S. Geol. Surv. Terrs.

jaws do not permit the lacerating power of the canines of the *Felidæ*, though more effective in this respect than those of the *Hyænodons*. The usefulness of a lever of the third kind depends on the approximation of the power to the weight; that is, in the present case, the more anterior the position of the masseter muscle, the more effective the canine teeth. Hence it appears that the relation of this muscle to the inferior dental series depended originally on the use of the canines as prehensile and lacerating organs, and that its insertion has advanced from behind forward in the history of carnivorous types. Thus it is that the only accessible molars, the fourth above and the fifth below, have become specialized as sectorials, while the fifth, sixth, and seventh have, firstly, remained tubercular as in the dogs, or, secondly, have been lost, as in hyenas and cats.

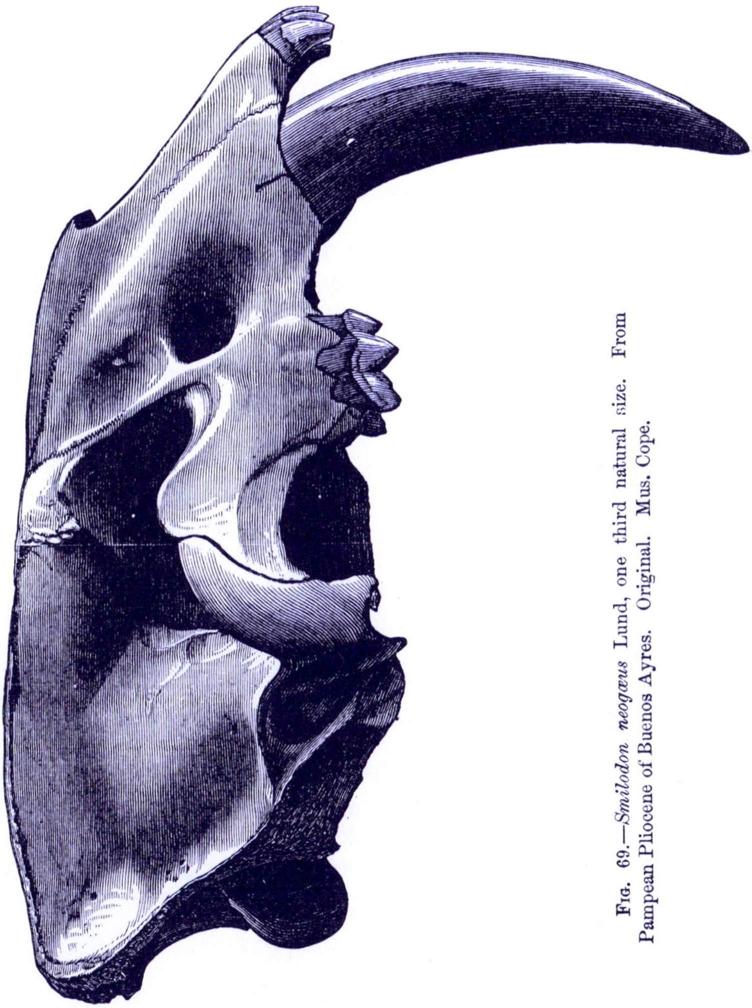

Fig. 69.—*Smilodon neogæus* Lund, one third natural size. From Pampean Pliocene of Buenos Ayres. Original. Mus. Cope.

XV.

ON THE ORIGIN OF THE FOOT STRUCTURES OF THE UNGULATES.

THE following considerations have been suggested by a study of the primitive types of the odd- and even-toed ungulates. I first, in 1874, recorded the opinion that the *Mammalia* with a reduced number of digits were derived from pentadactyle plantigrade types.* The ungulate order which fulfills this requirement is the *Amblypoda,* and from them, I doubt not, both the *Perissodactyla* and *Artiodactyla* have arisen, although not from any of the genera now known. Both of these great orders display a regular diminution in the number of the digits ; in the former, by reduction and extinction on both sides of the third digit ; in the latter, by reduction and extinction on each side of the third and fourth digits. Mr. John A. Ryder † has pointed out that reduction in digits is probably directly related to strains and impacts. He reminds us that the anterior digits are reduced in *Mammalia* of unusual scansorial or fossorial powers ; while in forms which display powers of running, the reduction is seen first in the posterior feet, which propel the body much more than the fore feet. This view is well illustrated in the Perissodactyle families, the majority of which have the digital formula 4—3.

No reason has ever been suggested, so far as I am aware, in explanation of the fact that one series of ungulates has retained two digits, and the other only one ; that is, why there should have been two kinds of digital reduction instead of one kind. In seeking for an explanation, we will remember that the tarsus in the odd or single-toed line is bound together by fixed articulations, while in the cloven-footed line it is interrupted by the hinge between the first (astragalus) and second rows of bones. The hinge-joint, being more liable to luxation than the fixed articula-

* "Journal Academy Philadelphia," March, 1874.

† "American Naturalist," October, 1877.

tion, requires a wider basis of support, such as would be furnished by two divergent digits, rather than by a single central one.

In the early types, where the median digits are slender, the mechanical advantage in favor of the bidigital over the undigital arrangement is much more obvious than in modern genera. Late in time, the horse developed the middle digit to such a width as to form almost as good a support as the bidigital structure. In the Eocene genera, the slender median digit could not have sustained the weight on a hinge, without great risk of dislocation.

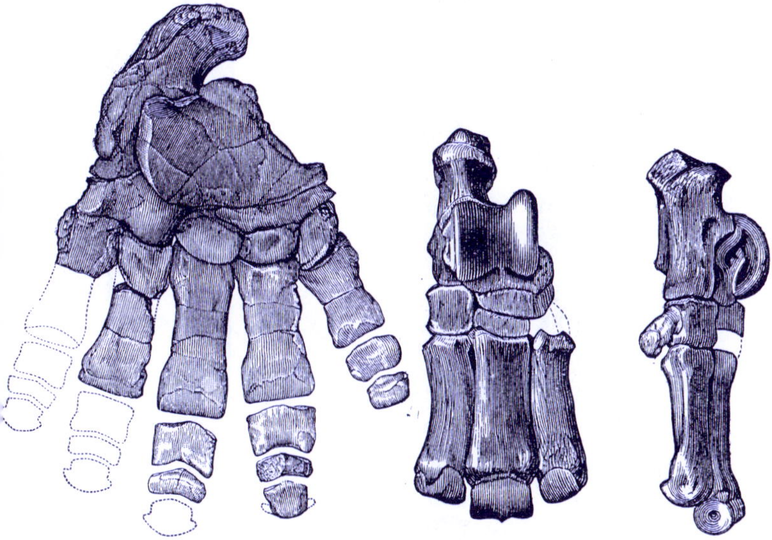

Fig. 70. Fig. 71.

Fig. 70.—Right posterior foot of a species of *Coryphodon* from New Mexico, one half natural size. From Report Expl. W. of 100th Mer., G. M. Wheeler, iv, Pl. lix.

Fig. 71.—Right posterior foot of the rhinoceros, *Aphelops megalodus* Cope, from Colorado, one half natural size. From Report U. S. Geol. Surv. Terrs. F. V. Hayden, iv, Pl. cxxx.

This explanation, it can be said, applies only to the posterior foot. The posterior foot has, however, led the way in the evolution of *Ungulata,* and the fore foot may have followed in accordance with the law of antero-posterior symmetry in growth. A curiously malformed deer from Mendocino County, Cal., throws some light on this subject. It has apparently a single functional digit on each foot. Examination shows that the posterior foot is bidigital, but that the phalanges are fused ; while the anterior foot is perissodactyle, all the digits but the third being rudimental! Similar evidence is furnished by the genus *Anoplotherium* * of the French Eocene. Its posterior foot is modified artiodactyle, while the anterior is modified perissodactyle. We may assume from

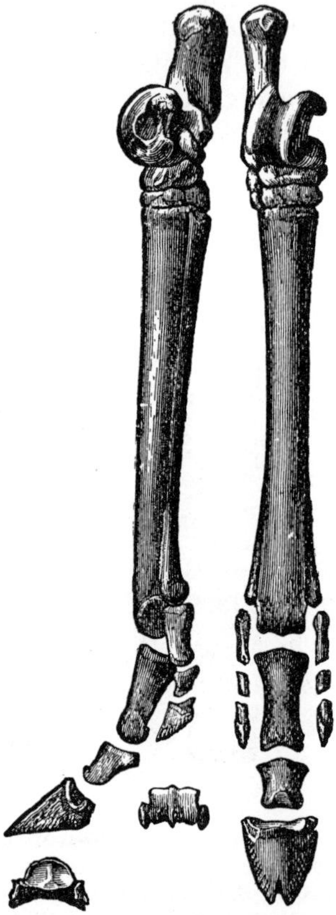

Fig. 72.

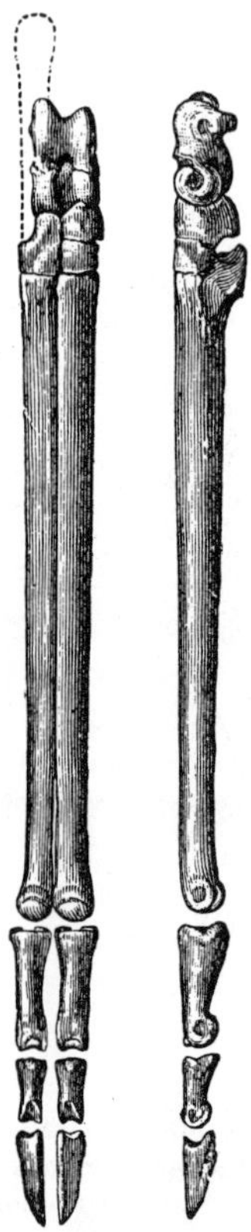

Fig. 73.

Fig. 72.—Right posterior foot of *Protohippus sejunctus* Cope, from Colorado, about one half natural size. From Report U. S. Geol. Surv. Terrs. F. V. Hayden, iv.

Fig. 73.—Right posterior foot of *Poëbrotherium labiatum* Cope, from Colorado, three fifths natural size. From Hayden's Report, iv, Pl. cxv.

these facts, that the posterior foot is more subject to the influences which tend to produce the bidigital structure than is the anterior limb.

---

* *Eurytherium* Gerv. in the original, a name shown by Schlosser to apply to Anoplotherium. (Ed. 1886.)

I suspect that the production of a ginglymus in the middle of the tarsus has been due to the use of the posterior limb in soft swampy ground. In the absence of this condition, as in a life on harder ground than swamp, no ginglymus would be formed. The action of an ungulate in walking through deep mud is very suggestive. The posterior foot is bent on the leg, and the antero-posterior strain of the weight or propulsive force is transverse to its long axis. In progression on dry land, the impact is in the direction of the length or axis of the foot. The obvious effect of a cross strain is to produce by degrees greater and greater mobility of some articulation. The one which has yielded is that between the two tarsal rows. Another effect of walking in swampy ground is to spread the digits apart. As the first digit of both feet is always of reduced size, there are practically but four digits to be considered. The weight falling nearly medially on these, would tend to spread them equally, two on each side. Thus the same cause may have been effective in producing both the artiodactyle structures. The perissodactyle structure, so soon as the lateral digits are much reduced, ceases to be favorable for progression in soft ground, owing to the liability of the lateral digits to injury, in following the principal one into the yielding material, filled with sticks and other hard *débris*.

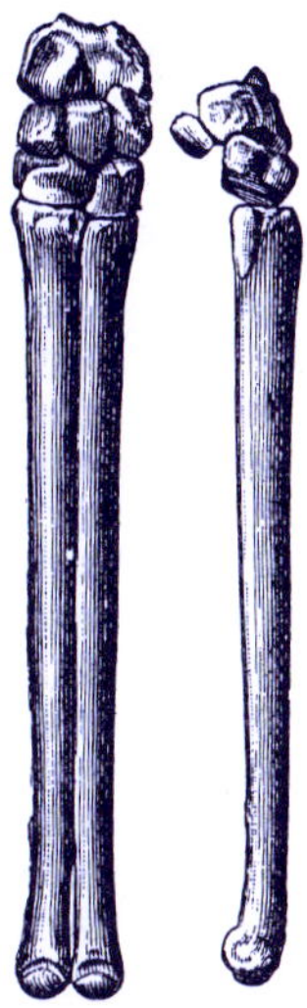

FIG. 74.—Left fore foot with part of radius of *Poëbrotherium vilsoni* Leidy, from Colorado, three fifths natural size. From Hayden's Report, iv, Pl. cxv.

The lowest existing forms of the *Artiodactyla*, the *Omnivora*, are universally swamp lovers and livers. So, we are told, are the lower existing *Perissodactyla*, the tapirs and rhinoceroses. The higher types of both orders are dwellers on plains and in forests. We do not know the habits of the Eocene *Perissodactyla*, but I doubt their having inhabited muddy ground to the same extent as the hogs and hippopotami, the lowest of the *Artiodactyla*. Now, in progression on dry land, any pre-existent inequality in the length of the digits would tend to become exaggerated. Such an inequality exists in the *Amblypoda*, the third digit being a little the longer. In rapid movement on hard ground the longest toe receives the greatest part of the impact, even if its excess of length is but little. The harder the ground, the larger the proportion of impact it will receive.

The fact that the *Perissodactyla* did not develop the solid ungulate or equine foot until a late geological period, or, in other words, that the orders so long retained the digital formula 4—3, would indicate that it did not adopt a habitat which required great speed as a condition of safety, so early as the *Artiodactyla*.

XVI.

ON THE EFFECT OF IMPACTS AND STRAINS ON THE FEET OF MAMMALIA.*

THE principal specializations in the structure of the feet of the Mammalia may be summarized as follows:

I. The reduction of the number of the toes to one in the *Perissodactyla* (horses, etc.), and two in the *Artiodactyla* (cloven feet).

II. The second hinge-joint in the tarsus of the *Artiodactyla.*

III. The trochlear ridges and keels at the various movable articulations of the limbs. These are as follows:

1. Looking downward—
   - α. Intertrochlear crests of the humerus.
   - β. On the carpal end of the radius.
   - γ. Metacarpals, distal ends.
   - δ. Tibia distally.
   - ε. Metatarsals distally.
2. Looking upward—
   - α. Radius distally.
   - β. Astragalus, edges.
   - γ. Astragalus distally (*Artiodactyla*).
   - δ. Phalanges (very weak).

The following observations may be made respecting the structures included under Division III: The trochlear keels which look downward are much the most prominent and important. Those enumerated as looking upward are weak and insignificant, or of a different character from the down-looking ones. The latter are all projections from the middles of the ends of the respective elements. The up-looking are generally projections of the edges of bones. Such are the lateral crests of the astragalus, and the adjacent edges of the cuboid and navicular bones which cause the distal emargination of the astragalus in the *Artiodactyla.* The

---

* Read before the National Academy of Sciences, April, 1881. Abstract. Some of the points of this paper have already been discussed in the preceding article, but the present abstract contains additional matter.

proximal ridges of the phalanges are very weak, and the concavities in the extremity of the radius can not be called trochlear, as they are adaptations to the carpal bones.

I. The reduction in the number of toes is supposed to be due to the elongation of those which slightly exceeded the others in length, in consequence of the greater number of strains and impacts received by them in rapid progression, and the complementary loss of material available for the growth of the smaller ones. This is rendered probable from the fact that the types with reduced digits are dwellers on dry land in both orders, and those that have more numerous digits are inhabitants of swamps and mud. In geological history it is supposed that the Perissodactyles (Figs. 72, 73) originated from the *Amblypoda*, or primitive *Ungulata* (Figs. 69, 70), which first assumed terrestrial habits, while the *Artiodactyla* (Fig. 74), originating from the same order, long continued as mud dwellers; as witness the hippopotami and hogs of to-day. The mechanical effect of walking in the mud is to spread the toes equally on opposite sides of the middle line. This would encourage the equal development of the digits on each side of the middle line, as in the cloven-footed types. In progression on hard ground, the longest toe (the third) will receive the greatest amount of shock from contact with the earth. There is every reason to believe that shocks, if not excessive, encourage growth in the direction of the force applied. This is strongly suggested by the relations between the length of the legs and the rate of speed of animals; and the lengths of the teeth and their long-continued use. Certain it is that the lengths of the bones of the feet of the Ungulate orders have a direct relation to the dryness of the ground they inhabit, and the possibility of speed which their habitat permits them, or necessarily imposes on them.

II. The hinge between the first and second series of tarsal bones in the *Artiodactyla* may be accounted for by reference to the habits which are supposed to have caused the cloven-footed character. Observation on an animal of this order walking in mud, shows that there is a great strain antero-posteriorly transverse to the long axis of the foot, which would readily cause a gradual loosening of an articulation like that connecting the two series of tarsals in the extinct *Amblypoda*. Any one who has examined this part of *Coryphodon* will see that a little additional mobility at this point would soon resemble the second tarsal joint of the hogs. In the case of animals which progress on hard ground, no such cross-strain would be experienced, and the effect would be to consolidate by flattening the fixed articulation.

III. The trochleæ. These prominences, which form the tongues of the tongue and groove articulations, exhibit various degrees of development in the different *Mammalia*. Those of different parts of the skeleton coincide in their condition in any one type of ambulatory *Mammalia*, and so may be all considered together. This fact suggests strongly that they are all due to a common cause.

They are all imperfect in the *Rodentia* and *Carnivora* (Figs. 75–77) (except the *Leporidæ*, which are especially characterized by their great speed). Among ungulates they are very imperfect in the *Proboscidea*. The orders mentioned all have elastic pads on the under sides of their feet or toes. The same is true of the lowest types of both the *Artiodactyla* and *Perissodactyla*, the hippopotami and rhinoceroses. In the *Ruminantia* the trochleæ are well developed (Fig. 80) with one exception, and that is the distal metacarpal and metatarsal keels of the *Camelidæ* (Figs. 79, 81). These animals confirm the probability of the keels being the effect of long-continued shocks, for they are the only Ruminants which have elastic pads on the inferior sides of their digits.

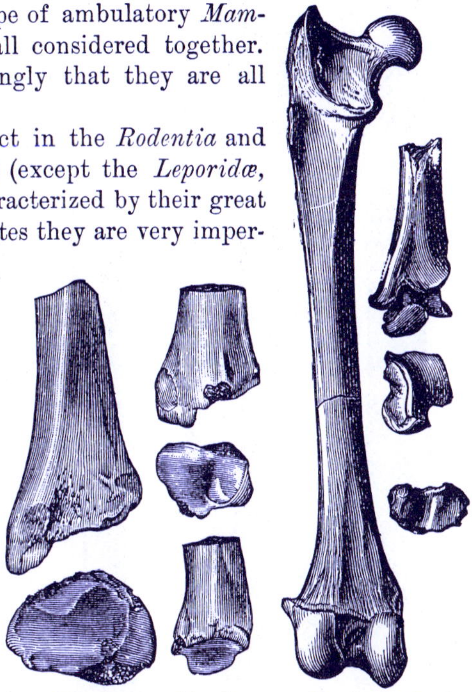

FIG. 75. FIG. 76. FIG. 78. FIG. 77.

Fig. 75, Distal extremity of tibia of *Amblyctonus sinosus* Cope. Fig. 76, Distal extremity of tibia of *Oxyæna morsitans* Cope. Both flesh-eaters, and two thirds natural size. From Report Expl. and Surv. W. of 100th Mer., G. M. Wheeler, iv, Pt. ii.

Fig. 77, End of tibia and astragalus of *Archælurus debilis*. Fig. 78, Femur of *Nimravus gomphodus*. Carnivora; one third natural size. Mus. Cope.

That these processes may be displacements due to shocks long continued, is rendered probable by the structure of the bones themselves. (1) They project mostly in the direction of gravity.

Constant jarring on the lower extremity of a hollow cylinder with soft (medullary) contents and flexible end walls, would tend to a decurvature of both inferior and superior adjacent end walls. If the side walls are wide and resistant, the projection will be median, and will be prolonged in the direction of the flexure of the joint. The groove of the astragalus deepens coincidently with the increase of digitigradism, showing that, while it is primarily produced by shocks, it is extended fore and aft by the repeated flexure of the ankle-joint. (2) They

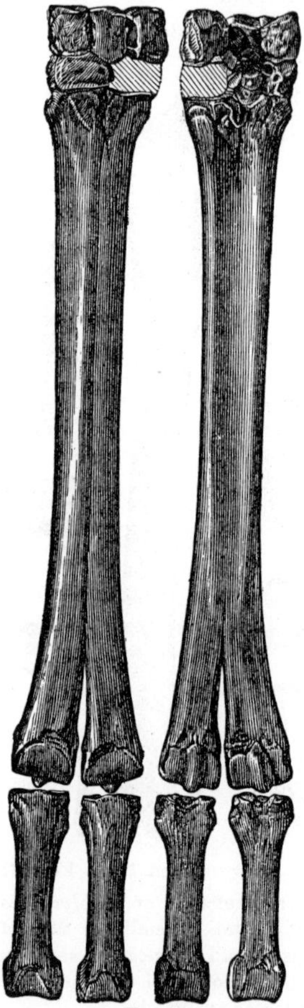

Fig. 79.

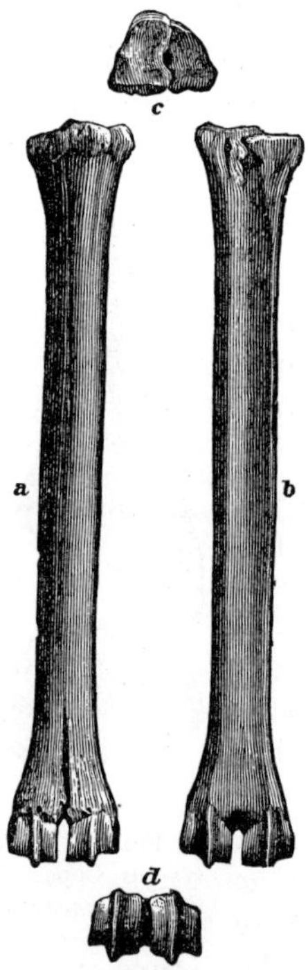

Fig. 80.

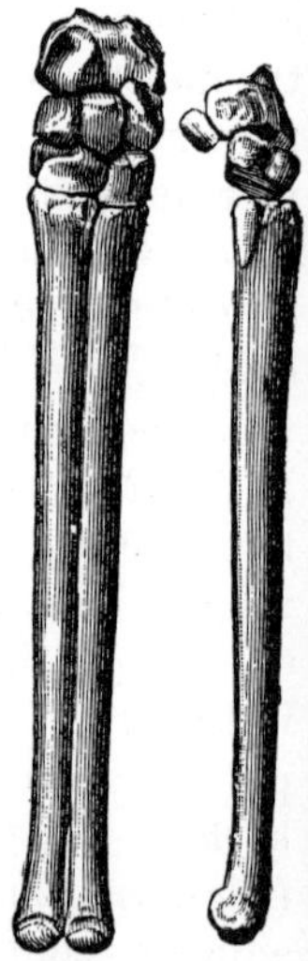

Fig. 81.

Fig. 79, Part of anterior foot of *Procamelus occidentalis* from New Mexico. Original from Report of Capt. G. M. Wheeler, vol. iv, Pt. ii.

Fig. 80, Metacarpals of *Cosoryx furcatus* from Nebraska, two thirds natural size. *a*, anterior face; *b*, posterior; *c*, proximal end; *d*, distal end.

Fig. 81, Left fore foot with part of radius of *Poëbrotherium vilsoni* Leidy, from Colorado, three fifths natural size. From Hayden's Report, iv.

fit entering grooves of the proximal ends of corresponding bones. These will be the result of the same application of force and displacement as the protrusion of the inferior, commencing with a concavity (*Elephas*); becoming more concave (Fig. 77), and becoming finally a groove. (3) When the dense edge of a bone, as in the case of the lateral walls of the astragalus, is presented upward, a groove is produced in the down-looking bone; e. g., the lateral grooves of the distal end of the tibia. (4) When the inferior bones are the denser, the superior articular face yields; e. g., the distal end of the radius to the first row of carpals (Fig. 81).

(5) The metapodial keels commence in the lower types on the posterior side of the distal extremity of the bone. This is partly due to the presence there of a pair of sesamoid bones which, with the tendons in which they are developed, sustain and press on the lateral parts of the extremities, and leave the middle line without support (Figs. 79–81).

PART IV.

METAPHYSICAL EVOLUTION.

XVII.

THE EVOLUTIONARY SIGNIFICANCE OF HUMAN CHARACTER.*

THE complicated constitution of the human mind is well impressed on the investigator as he seeks to understand the origin of any one of the many different types of character which come before him. The number of possible combinations of its numerous elements, each of which present developmental phases, is necessarily very great. The *species* of human minds, as one may properly term them, are probably as numerous as the species of animals, as defined by their physical structure. As in the case of anatomical species, however, analysis of the mind reduces its many details to a few leading departments. Although the classification of the elements of the mind is a classification of functions, it is, if correct, a sure index of the classification of structure also ; of the grosser and more minute structure of the brain, principally of the gray matter.

The division of mental activities into three primary divisions is generally admitted. These are : the emotions, the intellect, and the will. The emotions include the likes and dislikes, or the tastes, and their strongest forms, the emotions and the passions. The intellect includes those powers which rearrange the experiences in an order different from that in which they enter the mind. This new order may have sole reference to questions of liking and disliking, and is then a product of the imagination ;

---

* The present article is in continuation of the previous one on the Evolutionary Significance of Human Physiognomy, published in the "Naturalist" of June, 1883.

or it may be a result of experience of the laws of pure necessity, without regard to questions of taste ; then it is a process of reason. The will, properly so called, is the spontaneous power of the mind by which the other processes are originated, directed or restrained. The range of the will, and even its existence, are questions of dispute.

Below and behind these mental activities lies *sensibility* or consciousness, in its forms of general and special sensation ; that is, touch and hearing, taste, smell, sight, and the muscular sense, with many others, concomitants of both health and disease. It is well understood that these primitive mental qualities are more or less developed in animals in which the more purely mental functions are rudimental. The doctrine of evolution teaches that from this class the higher activities of the mind have been developed, during long ages, through the agency of memory. The nature of the present essay only permits a casual reference to the astonishing character of memory, and the remark that its phenomena demonstrate most clearly, of all others, that mind is an attribute of some kind of matter.

If we now consider these natural divisions of the mind as they present themselves in the combinations which we call human character, we shall observe a variety in the mode of their action which pervades all divisions alike. These variations fall under two heads, those of *quantity* and of *quality*.

Thus as to quantity ; one human mind may present a greater amount of intellectual than emotional activity ; of imaginative than rational intellection ; of affectionate than irascible emotion ; of gastronomic than musical taste, etc., etc. The quantity here indicated is probably an index of the proportion of brain tissue devoted to the functions mentioned. The intensity or force of the action is a matter of quality.

Of qualities the variety is much larger. They are often parallel to those of inorganic force, and suggest the same kind of modifications of the material bases as those which affect the latter. Two prominent qualities are *fineness* and *coarseness*. Fineness observes and uses detail in both rational and emotive acts, and is essential to the precision of finish. Coarseness neglects detail, but deals with the gross of things, and is sometimes accompanied by largeness of quantity. When it is not, the result is not good. Fineness is, on the other hand, often associated with smallness. It is a more feminine attribute, while coarseness is more masculine.

Another pair of antithetic qualities are *intensity* of action and the reverse. This probably means that a given bulk of brain tissue produces (i. e., converts) a greater amount of energy in a given time than an equal bulk of non-intense tissue.

The *speed* or rate of action in time, and its opposite, slowness, are related to the last-named qualities, but are not identical with them. Thus growth of the mind always witnesses a diminution in the rate of action, but an increase in intensity.

Tenacity of mental action is a very marked character, and of great importance. It signifies the persistence of mental action, or mental endurance, and may characterize the entire mind, or only a part of it. Its opposite, seen in changeability, desultoriness or fickleness, may also characterize all or a part only of the mind. According as it characterizes the intellectual or emotional departments are its exhibitions most varied, though they probably have a common histological basis.

Impressibility and *stolidity* express antitheses of character which are seen every day. The term impressibility is used as identical with irritability, and is preferred, because the latter has special physiological and popular meanings, some of which are only among its phases. These qualities are apt to pervade the entire mental organism, although, like others, they may characterize a part only. Impressibility is obviously a condition of tissue, since it varies greatly with physiological conditions in the same person. Its exhibitions in the department of the emotions may be confounded with strong development of the emotions themselves. A moment's thought, however, shows that easy excitation of emotion is a different thing from energy of emotion, and is often found apart from it. Impressibility of intellect shares with tenacity a leading position as an attribute of a first-class mind, and the combination of the two forms a partnership of superior excellence.

I may mention here a quality whose absence is pathological, and hence does not properly enter the field ; this is *tonicity*. In its normal condition, every organ should be supplied with sufficient nutriment or energy to insure the occupation of its entire mechanism. Anything short of this is followed by poor work. Debility of mental action in the emotional department is seen in abnormal irritability, such as peevishness or "spooning"; and in the intelligence, in absence of mind and blundering; and in both, in general frivolity.

Returning to the primary elements of mind, we may examine their divisions with reference to the question of growth. To begin with the perceptions, there are great diversities in the acuteness of the general and special senses, and greater and less susceptibilities to physical pleasure and pain. In the important representative faculty, memory, the differences between people are great. As perception as well as thinking involves a certain amount of structural change, it is evident that susceptibility or impressibility of the senses, which is the first stage of memory, signifies ready metamorphosis of tissue. Unimpressibility, which impedes memory, is a consequence of resistance on the part of tissue to the usual stimuli. Hence the effect of "sights, sounds, and sensations" is greatest in childhood, and the memory is most impressible, for at that time the nervous tissue is undergoing constant change, and nutrition, being in excess of waste, constantly presents new material to be organized. And I may here refer to the general truth, that consciousness of all kinds is the especial and distinguishing attribute of life as distinguished from death or no life.* Whatever other phenomena we may be accustomed to regard as "vital," are only distinguishable from inorganic motion or force, because they primitively took their form under the guidance of consciousness, and are hence, so to speak, its children. With the perfect working of most of the mechanism of the body, consciousness no longer concerns itself, although it may speedily do so in pathological conditions. This prerogative is now restricted to the nervous system, and to certain parts of it; the one which is, histologically speaking, the most generalized of the systems. And it is quite consistent with the "doctrine of the unspecialized," that nervous tissue in its unfinished state in childhood should be more impressible to stimuli than at later periods of life. But this statement requires this modification, that there is a stage of imperfection of mechanism which does not display high sensibility, as, for instance, in the earliest infancy. With age sensibility gradually diminishes.

Next in order of appearance in growth are the emotions. It is true that some of these are not fully developed until long after the appearance of many or all of the intellectual faculties; but it is also true that their full development precedes that of the intellect, in so far as they are developed at all. The primitive condi-

* "The Origin of the Will," "Penn Monthly," 1877, p. 440.

tion of the emotions is that of appetites. The first of these in the necessary physiological order, and hence in time, is the appetite of hunger. Second in order in the history of life, but not in the growth of individuals, is the instinct of reproduction, such as it is in animals that only multiply by fission. Very early in evolution the emotion of fear must have arisen, and it is probably the immediate successor of hunger in the young of most animals. Anger appears as early as the mind can appreciate resistance to its first desires, and no doubt followed as third or fourth in the history of evolution. The rudiments of parental feeling would follow the origin of reproduction at a considerable interval of time. One of the latest of the instincts to appear would be the love of power; while later still would be the emotions of relativity (Bain), because they are dependent on a degree of mental appreciation of objects. Such are admiration, surprise, and wonder. These, as well as all other consequences of inherited intellect, appear earlier in infancy than they did in evolution, as may be readily understood.

Of these instincts and emotions, it is to be supposed that hunger remains much as it has ever been. The reproductive instinct has, on the other hand, undergone the greatest modifications. Sex instinct could not have existed prior to the origin of the male sex, which must be regarded in evolution as a derivative from the female. Hence it is probable that the parental instinct preceded the sexual in time. These two instincts being the only ones which involve interest in individuals other than self, furnish the sources of sympathy in all its benevolent aspects. Hence it has developed in man into the powerful passion of love; into affection and charity in all their degrees and bearings. Fear being, as Bain shows, largely dependent on weakness, has varied in development in all times, but must be most pronounced in animals of high sensibility, other things being equal. Hence its power has, on the whole, increased until it probably reached its extreme in the monkeys or the lowest races of men. Increasing intelligence of the higher order diminishes the number of its occasions, so that it is the privilege of the highest types of men to possess but little of it. The earliest of the emotions of relativity to appear in time has probably been the love of beauty; how early it may have appeared it is difficult to imagine. Surprise and wonder, as distinct from fear, one can only conceive as following an advanced state of intelligence.

Thus in psychology as in physiognomy,* the paleontological order of development is somewhat different from the embryological. I might compare the two orders as follows :

PALEONTOLOGICAL.	EMBRYOLOGICAL.
Hunger.	*Hunger.*
Reproduction.	*Fear.*
Fear.	*Anger.*
Anger.	*Beauty.*
Parental instinct.	*Wonder.*
Sex.	*Power.*
Power.	*Admiration.*
Beauty.	*Pity.*
Wonder.	*Sex.*
	Parental instinct.

The qualities enumerated in the first column follow each other directly in order from the simple to the complex. In the second column this order is disturbed by the earlier appearance of the derivative emotions, beauty, wonder, admiration, and pity, or benevolence, and the later appearance of the simple emotion of sex. Thus in psychological as in other evolution, some of the products of development appear earlier and earlier in life in accordance with the law of *acceleration.*

The intelligence has already been considered under the two heads of the imagination and the reason. The action of the imagination, unmixed with the exercise of reason, is chiefly to be seen in the creative fine arts, as distinguished from the imitative, the mechanic, and other arts. The musician, the painter, the sculptor, the poet, the novelist and the playwright, so far as they are not imitators, present the best illustrations of the work of the imagination. It is a faculty which must be very little developed in the animals below man. They occasionally make mistakes in the nature of objects, and suppose them to be other than what they are. Thus the Antilocapra supposes the Indian disguised with a skin and horns to be one of his own species, and suffers the penalty. But this is a most rudimental act of imagination, if it be not mere curiosity.

The reason, properly so called, begins in its lowest grades with the simplest re-arrangement of the objects of sense and memory, in accordance with some principle of relation. As the principle or standard of relation varies, so does the intellectual process. If

* "Naturalist," 1883, p. 618.

the process be discovery, or the enlargement of knowledge, man experiences (or hypotheses) will be successively encountered an tested, and appropriate generalizations reached (inductions). I the process be to accomplish the practical ends of life by use o well-known means, the intellect uses the customary rules of actio as standards, be they moral or mechanical, financial or political and attains its deductions and applications. These two types o intellect are strikingly distinct, and produce the most diverse con sequences. The inductive type is the most generalized, and henc capable of the largest growth and adaptability, and the wides range of thought. The deductive is the more specialized, th more "practical," but less capable of growth or general thought Its most remarkable exhibitions are seen in the skill with whic some men conduct the game of chess, and corresponding enter prises in real life. Also the ingenuity of mechanical invention and the wonderful rapidity of calculation which some minds dis play. In intellectual as in many other vital phenomena, the facil ity once developed, the active process is often unaccompanied b consciousness in many or even all of its stages.

Rapid and exact control of the muscles in obeying the direc tions of the mind is essential to the practice of many arts, espe cially to that of the musician. This accomplishment is acquire through the medium of the conscious mind, and may be regarde simply as the reflex of impressions made on the senses directed b some simple rule which has been impressed on the memory. Th often surprising results involve the exercise of a very simple phas of intellection.

The appearance of the rational faculties in time, may be est mated by their relative development in the existing divisions c animals whose period of origin is known or inferred. The an mal mind is capable of simple forms of induction and deductior and sometimes acquires considerable artistic skill. Bees, ants an spiders display these in varying degrees, and their antiquity probably co-extensive with that of the known sedimentary rock The supposed Ascidian ancestors of the Vertebrata, and even th lowest vertebrate (Branchiostoma), display far less intelligence tha the articulates mentioned, which are really lower in the scale c organic types. From such unpromising sources did the noble vert brate line descend. It is probable that the inductive act precede by a little the deductive in time, as it does in logical order. B the elaboration of these powers was doubtless long delayed ; f ntold ages they involved nothing more than the discovery and pplication of general principles of the simplest kind ; such as the ustomary sequence of natural phenomena, and the anticipation f their operations, as, for instance, in the laying up of winter rovisions. Occasionally deductive application of an old rule to a ew case would arise, as in that of the Mygale spider which was bserved by Dr. McCook to substitute cotton for her own silk for ne lining of her nest. The development of the rational faculty as been rather in quantity and quality, than in the nature of its onstituent parts. I may remark, however, that the embryological rder is here again different from the paleontological. Inherited ptitudes, as for music, calculation, etc., precede, in children, any onsiderable powers of thought, while the order of development of ne race has been the reverse.

As regards the appearance of the qualities of mind already nentioned, which depend on character of tissue, it is difficult to resent an order which shall be generally true. Our ignorance f the subject is profound ; nevertheless observation of animals nd men leads to the following conclusions : First, the primitive nind is negative, unimpressible, and little sensitive. In evolution, ensibility has been developed under stimuli, and diminished by isuse and repose. The energy of high-strung sensibility has probbly ever won for its possessors success in the struggle for existence, nd more or less immunity from the pains which stimulate to ac on.* It is true that the non-aggressive and ever-harassed Herivora have developed the higher brain structure. The inferiority f brain type of the Carnivora is a well-known fact of present and ast time. The early ruminants were smaller than the contempoary flesh-eaters, and therefore subject to the greatest risks. The est-developed brains, those of the Quadrumana, have been develped in still more defenseless animals, who in their arboreal life ave been confronted by still more complex conditions.†

Impressibility or sensitiveness has evidently been the means of cquisition of some of the other qualities mentioned. Thus *intenity* may have resulted from active use accompanied by vigorous utrition, and the consequent construction of compact force-conerting tissue. *Rapidity* without intensity must also result from

* "The Relation of Man to the Tertiary Mammalia," "Penn Monthly," 1875.

† Mr. C. Morris very reasonably regards the social life of these animals as the ource of their development of intelligence. See "American Naturalist," June, 886. (Ed. 1886.)

exercise, with a less vigorous construction of tissue. *Fineness* and *tenacity*, on the other hand, can not be regarded as being so much produced by use as by very primitive conditions of tissue. Restraint under pressure might produce fineness. Long-continued freedom from sudden changes, under pressure, might account for the origin of tenacious tissue. As to quantity, deficiency or diversion of nutritive energy or material must produce smallness, and the reverse condition, largeness.

These qualities impress themselves on the external as well as the internal organization, and can be more or less successfully discerned by the observer. I reserve the question of physiognomy to a later article, and here consider only the evolutionary bearings of character itself. As in physiognomy, we may arrange the faculties and their qualities under the two heads of ancestral and embryonic, or that of the species and that of the individual. The order of succession is the same in both kinds of development.

SPECIES.	INDIVIDUAL.
Indifference.	*Indifference.*
Emotions.	*Emotions.*
Intellect.	*Intellect.*
a. *Imagination.*	a. *Imagination.*
b. *Reason.*	b. *Reason.*

It is not practicable to go farther than this into the order of evolution of characteristics. There is probably little uniformity of sequence other than that I have already pointed out under the head of the emotions.

As a complex outcome of the emotional and rational faculties must be now mentioned the moral sense, or the sense of justice. It consists of two elements, the emotion *benevolence*, and the rational power of *discrimination* or judgment. The former furnishes the desire to do what is right to a fellow-being. Without the aid of reason, it is benevolence, not justice, and may often fail of its object. The rational element has acquired from experience a generalization, the law of right. It perceives what is most conducive to the best interest of the object of benevolence in his relation to others or to society, or whether he be a proper object of benevolence at all. By itself, this quality is absolutely useless to mankind. When it guides the action of human sympathy, it displays itself as the most noble of human attributes. Many animals display sympathy and benevolence, but justice has not yet been observed in any of them. Hence it has been said that it can not be a derivative faculty, but is "intuitive" in man. The objection to this view is its great variability and occasional entire absence in man, individually and racially. It is the last to appear in individual growth, as it has doubtless been in the order of evolution, of mind.

I now devote a little space to the discussion of the distribution of these qualities in races and sexes.

As regards the relative preponderance in action of the emotive and intellectual faculties, it is an axiom that in the great majority of mankind, apart from the necessities imposed by the simple physical instincts, it is a taste or an affection or an emotion that lies at the basis of their activities. Perhaps the most universal is the affection of sex. Given two types of rational beings who are objects of admiration and of pleasure to each other, each of whom desires to possess the other, and who therefore employs many devices to please and attract the other, and we have an effective agent of general development. Then the parental, and especially the maternal, affections arouse and direct many labors. Fear of suffering and death is at the basis of many others. The love of power or of possession, including ambition, is a well-known stimulus. The love of beauty is a strong motive in many persons. The pleasure derived from the exercise of the intelligence is a sufficient motive for a life-work in a comparatively small number of persons. These are the artists and the scientists; but it is far from being an unmixed motive in many of them.

Intellectual motives, however, enter into association with the affectional in many instances, as, for example, in the profession of teaching. But it is as guide and agent in the accomplishment of the main ends of life that the intellect, especially the reason, has its great field, and displays itself in an endless variety of ways.

If we now survey men as we find them, it is a general truth that it is in the male sex that the greatest proportion of rational method is to be found, and in the female the greatest proportion of the affectional and emotional. As we descend the scale of humanity, the energy and amount of the rational element grows less and less, while the affectional elements change their proportions. The benevolent and sex elements diminish in force more rapidly than the other sentiments, but it is probable that all the emotions are less active in savages, excepting those of power and of fear. In the lowest races there is a general deficiency of the emotional qualities, excepting fear, a condition which resembles one of the stages of childhood of the most perfect humanity. To this must be added revenge, where hatred may be re-enforced by several other sentiments, with a feeble perception of equivalent suffering or punishment, which may or may not be just. The pleasure of muscular exercise is greatly developed in people of out-door habits.

The order of the appearance of the intelligence is nearly dependent on the development of the powers of observation. In most savages these are very acute, and vary according to the nature of the environment which impresses them. The character of most civilizations tends to diminish the power of the perceptive, while the higher departments of imagination and reason are enlarged. The imagination reached a high development before reason had attained much strength. With the exception of a few families, the intelligence of mankind has, up to within two or three centuries, expressed itself in works of the imagination. When exact knowledge first began to be cultivated, it was in the department of astronomy, where the least precision was attainable, and where the greatest scope for the imagination is to be found.* Next in time metaphysics was the throne of learning, a field in which much may be said with the least possible reference to the facts of observation. With the modern cultivation of the natural and physical sciences, the perceptive faculties will be restored, it is to be hoped, to their true place, and thus many avenues opened up for the higher thought-power of a developed race. Thus it is that in the order of human development there is to be a return to the primitive powers of observation, without loss of the later acquired and more noble capacities of the intellect.

The relation of the qualities of impressibility, fineness, intensity, speed and tenacity, to our development in time, may have been as follows: Impressibility of mind is no doubt an embryonic character of "retardation," parallel to, and probably a consequence of, the retardation which is also expressed in the human skull and face. The preponderance of the osseous and nutritive elements over the nervous is the usual accompaniment of non-impressibility, and *vice versa*. Hence this quality is of late origin in the

* The governments of antiquity required the knowledge of the Chaldean astronomers as important to the success of their undertakings, and the governments of Europe and America were, for a long period, more liberal in their support of astronomy than any other science. At present, however, geology shares in this aid and to a less degree botany and zoölogy.

history of the Vertebrata and of man, and is most developed in the young, and better developed in women than in men.

Tenacity has an opposite significance, being an especial characteristic of maturity in the human mind. Hence it may have been more general in early ages than at present, but could have little value so long as the mind remained small in quantity. Curiously, it is a quality which may co-exist with a good deal of impressibility.

Fineness can only be a quality of full development, and is totally independent of the other qualities. It is unknown among savages, and is developed apparently in a few animals. Of intensity it is difficult to say much definitely. The nervous operations of animals often display the highest degree of this quality, and it is not unlikely that its appearances differ as much in savages as in civilized people. Its importance in mental action depends of course on the kind and amount of mental function which exhibits it. The same may be said of speed. The faculties which exist are more or less affected by it. In the well-formed reason it is an important characteristic, and a special form of development.

Having gone as far into the origin and developmental relation of mental functions and qualities as the nature of this sketch permits, I refer briefly to the stimulus to their growth; always remembering that the percentage of inherited qualities is much larger in a given character than that of acquired ones. On this head one word expresses a good deal, and that word is *use*. No truth is better known than this one, that mental faculties develop with use more rapidly than those of any other organ of the human body. Brain and nerve are apparently the most plastic of all tissues; the one which retains the properties of the primitive protoplasm, multiplied and intensified a thousand fold. It has always been the seat of creation, throwing off its "formed matter" in useful directions. It is still doing so; and in the human brain, ever creating itself, is in addition the seat of a new creation, which it executes through its instruments, the other organs of the body. Hence the greatest sin against the brain is idleness, or disuse. The brain activity of to-day is an indication of health and happiness beyond what the world has seen hitherto.

The greatest stimulus to exercise of the brain is human society. Hence the greatest developments of mind have always been in the centers of population. Whatever may be the passive virtues of country life, it is the cities that furnish both the stimulus and the field for the triumphs of mind.

XVIII.

CONSCIOUSNESS IN EVOLUTION.*

I. PRELIMINARY.

THE evidence of what is termed "design" in the structure of beings exhibiting life, is often appealed to by one class of thinkers, as proving the intervention of a personal Deity in the creation of such; and the same feature exhibited in the movements of living creatures is regarded by metaphysicians of a similar class as an indication of their possession of a power of choice, or "free agency," at least in the case of man. The opposing school, of whom Professor Bain may be selected as an example, believes that designed acts are without an element of freedom, but are simply performed in obedience to stimuli of various kinds, motion following stimulus as inevitably as effect succeeds cause in the non living world. The evolutionists attempt to explain design in structure through the operation of the Darwinian law of the "survival of the fittest," showing that only those beings whose organization displays that adaptation to use in relation to its surroundings, which is termed "design," could possibly continue to exist. It is justly urged against this reasoning that it attempts no explanation of the *origin* of such structures. Another school of evolutionists have therefore maintained that such structures are due to the effect of effort, i. e., stimulus or use, exerted by the living being on its own body, and that the design thus displayed is an expression of the intelligence at some time possessed by itself.

So long as there is any probability of the last explanation proving valid, it will be important to examine into the question of metaphysics which it necessarily involves. The investigation is indeed but the necessary projection of those which have resulted in satisfying the great majority of biologists of the reality of evolution, or of the fact of the descent of existing living beings

---

* A lecture delivered before the Franklin Institute, Philadelphia, February, 1874.

pecies by species, order by order, and class by class, from others vhich have preceded them in time. Clearly, then, we enter the question by considering the nature of movements of plants and nimals in relation to the stimuli which are supposed to call hem forth.

II. THE UNCONSCIOUS.

A true study of metaphysics necessarily has for its objects lants, animals, idiots, and infants, as well as healthy men ; never-heless, necessity compels us, in discussing the question, to dwell n our own experiences as a *sine qua non*. Now experience, in a eneral sense, includes not only the memory of our conscious acts, ut a knowledge of our unconscious ones, and to the latter espe-ial attention must be directed, since they are most readily over-ooked. The marvelous character of memory can not be too uch considered. Of the millions of impressions which the mind as received and registered, in the course of a lifetime, but one an be clearly present in consciousness at one time. The remain-ng millions are not lost ; they are stored, each in its appropriate lace, to be sprung into consciousness when the appropriate uggestion presents. How much more vast, from this point of iew, is the unconscious mind than the conscious ! But the henomenon is not confined to memory. Who that has ever at-empted the digestion of a subject which includes a mass of de-ails, is not acquainted with the unconscious activity of the mind n classification ? How frequently a question involving many arts, is, on the first reception of the constituent facts, all con-usion ; but in time displays its symmetry clearly to the con-ciousness, every part in its proper place, and that with little or o further attention having been devoted to it. It is indeed proba-le that the every-day process of inductive reasoning is conducted n unconsciousness on the part of the subject. Induction con-ists in the generalization of some quality as common to a great umber of objects of memory ; a greater or smaller number of ther qualities being neglected in the process. When this act is erformed voluntarily, one or many qualities are successively assed in review before the mind—each one being in its turn im-ressed on the perceptive centers—so long as it is the object of nquiry, the others being excluded from consciousness for the ime being. It is simply a process of classification, and when per-ormed in consciousness, constitutes "experiment." But when

no generality is anticipated, and its existence is unknown, it ofte
happens that such generalization becomes known or rises into con
sciousness, without the bestowal of effort in classification of th
objects to which it refers. The impressions consciously receive
have been arranged out of consciousness, and when revived int
consciousness display an order which was not previously know
to exist. It is in the latter way that the "practical man" "find
out" the rules by which, as by an instinct, he regulates his inter
course with the world. He often can not explain the reasons o
their truth, nor does he know how he came by them, being gen
erally content to call them the results of "experience." In som
persons they are so feebly expressed in consciousness as to be calle
"feelings"; and many experiences or repetitions are sometime
necessary to impress on us the importance of these mental prod
ucts before we are willing to follow them in action. "Strengt
of mind" is an expression applied to a high degree of this uncon
scious reasoning; expressing the extent of ground the proces
covers continuously, as well as the exactitude of its results. Th
experimental investigator, on the other hand, performs this wor
deliberately, and is acquainted with the processes; he is, there
fore, at first more confident of his results. And we observe here
in passing, that a rule once discovered is as readily retained i
the cells of the unconscious as is the memory of a simple objec
or event.

Another form of unconscious cerebration is seen in deductiv
reasoning, which employs rules already discovered in applicatio
to new cases. Calculating prodigies are a case in point. It i
well known that those persons who have from time to time ap
peared possessed of the power of calculating with enormous num
bers with marvelous rapidity, have never been able to explain th
process by which they reach their conclusion, nor are they con
scious of going through the steps involved in the calculation the
perform; and it has been said that great calculators have rarel
been great mathematicians.

The explanation of these phenomena is not far to seek. I
simpler forms it is presented to us every day. Thus it is an eas
matter to read with but little consciousness of the process, and n
recollection of the subject-matter of what is read. Most manu
operations can be performed while the consciousness is occupie
with other objects.

If these be facts of human experience, how much more likel

are they to be true of animals? If man be unconscious of the process during the performance of some of his most complex acts, how much more probable is it that animals are so while pursuing the narrower circle of their simpler ones? Yet animals are not devoid of consciousness; indeed, it is scarcely credible that any one should deny to them consciousness, after experience in their education.

But let these automatic acts be ever so simple or complex, it is here claimed that they could not have *originated* out of consciousness. Whatever we call voluntary acts in ourselves undoubtedly have to be *learned.* The acquisition of the primary act of walking is accomplished by a slow and painful education; while knitting and other manual exercises necessarily require preliminary training, some of shorter, others of longer, duration. This is true of such voluntary acts as we perform most readily automatically, and such as might be supposed to be most probably acquired by hereditary transmission, as for instance speaking. The case is the same with animals. All those services which are useful to us, or tricks which amuse us, are acquired at the expense of training, which involves a system of stimuli, consisting of rewards and punishments, as in our own species. Is there any reason to suppose that those habits which we observe them to possess in a state of nature have had a different origin?

It is incontrovertible that a regular succession of muscular movements may be committed to memory as certainly as a color or a shape, and that a change of brain substance, such as causes the retention of the simple impression, is also involved in the retention of the complex. When this machinery is completed, through the repetition of conscious stimulus, it works thenceforth without necessary intervention of consciousness. The consciousness may then be engaged in fresh acquisitions, accomplishing new organizations, thus accumulating a store of powers. Once organized, these powers are at the disposal of their possessor, yet the organized machine will at some time undergo change, if not more or less frequently used. Without use it may indeed finally disappear, showing that the capacity for organization is identical with a facility of disorganization.

III. THE ORIGIN OF AUTOMATIC MOVEMENTS.

Is any habit originated in unconsciousness? Those who affirm this proposition point to the movements of plants in the extension of their tendrils, and the closing of some sensitive leaves; the timely expansion of the down of the *Asclepias* seed, and the insect-catching habits of *Drosera* and *Dionæa*. No one surely attributes consciousness to these. And there are many similar movements in animals which are as thoroughly unconsciously performed as are those of plants, from the first moment of the animal's birth; as, for instance, the involuntary activities of the circulatory and digestive systems, etc. Did these originate in consciousness or unconsciousness? The answer to this question constitutes the key to the mysteries of evolution, and around it the battle of the evolutionists of the coming years will be fought.

It may be asserted at the outset that those habits whose origin we have had the opportunity of observing in ourselves and in other animals were certainly acquired in consciousness, and that we do not believe that they could have originated out of it. The stimuli to action are divided into the two general classes of pleasures and pains, and each stimulus is potent in proportion to the intensity with which it is consciously apprehended. If many and complex acts may be performed automatically, through the organization of special machinery in the gray matter of the brain, it is altogether reasonable that similar powers should be found to be conferred on gray nervous tissues in parts of the body which are no longer seats of consciousness. It is well known that the spinal cord of the headless frog responds to stimuli, in the vigorous muscular contractions of the limbs which follow the application of acid to the skin. So the ganglionic centers of organic life respond to their appropriate excitants; the various glands of the digestive system discharging their contents into the ingesta at the proper moment, consciousness having no share in the proceeding. These phenomena are more readily explained on the theory of endowment than on that of physical movements; since by means of the former the evident design in the movements is accounted for, while the latter gives us no clew to this characteristic feature of these and all other vital processes.

The lowest form of consciousness is common sensibility; and, judging by the resemblance between our own experience and that of the higher animals, the lowest of animals also are not devoid of this quality. The structureless jelly of Rhizopods, such as Amœbas, Gromias, etc., evidently selects its food with regard to its nutritious qualities, in most instances preferring diatoms and desmids to sand and other innutritious substances. Its acquisitions in knowledge of articles of food can only be accounted for on the hypothesis of original, pleasurable or painful, consciousness of the effects of external and internal contact with these substances, and retention of the impression in unconsciousness. The impression reviving on the recurring of a similar contact, the substance is accepted or rejected as the former sensations were pleasurable or painful. And this is not incredible, if, as the researches indicate, the structure of the protoplasm of these creatures is of the same type as that of the bioplastic bodies of the gray tissue of the brain.

In accordance with this view, the automatic "involuntary" movements of the heart, intestines, reproductive systems, etc., were organized in successive states of consciousness, which conferred rhythmic movements, whose results varied with the machinery already existing and the material at hand for use. It is not inconceivable that circulation may have been established by the suffering produced by an overloaded stomach demanding distribution of its contents. The structure of the Cœlenterata offers the structural conditions of such a process. A want of propulsion in a stomach or body sack occupied with its own functions would lead to a painful clogging of the flow of its products, and the "voluntary" contractility of the body or tube-wall being thus stimulated, would at some point originate the pulsation necessary to relieve the tension. Thus might have originated the "contractile vesicle" of some protozoa, or contractile tube of some higher animals; its ultimate product being the mammalian heart. So with reproduction. Perhaps an excess of assimilation in well-fed individuals of the first animals led to the discovery that self-division constituted a relief from the oppression of too great bulk. With the increasing specialization of form, this process would become necessarily localized in the body, and growth would repeat such resulting structure in descent, as readily as any of the other structural peculiarities. No function bears the mark of conscious origin more than this one, as consciousness is still one of the conditions of its performance. While less completely "voluntary" than muscular action, it is more dependent on stimulus for its initial movements, and does not in these display the unconscious automatism characteristic of the muscular acts of many other functions.

Bearing in mind the property of protoplasm to organize machinery which shall work automatically in the absence of consciousness, we can glance at the succession of vegetable forms. The active movements of the primary stages of the Algæ are well known. After swimming actively through the water, they settle down, take root, and assume the role of plants. The *Aethalium*, swimming with the movements of a Rhizopod, has been known to take food before establishing itself on the damp piles of the tan-bark, where it speedily becomes a low form of fungus. The approximation of the lower forms of plants to animals is notorious. The fungi, it is said, are the only terrestrial plants which live like animals on organic matter, appropriating the humus of their rich nidus in a state of solution. Now the paleontology of animals has absolutely established the fact that the predecessors of all characteristic or specialized types have been unspecialized or generalized types, "neither one thing nor another." It may then be regarded as almost certain that the ancestors of the present higher types of plants were more animal-like than they; that the forms displaying automatic movements were more numerous, and the difficulty of deciding on the vegetable or animal nature of a living organism greater than it is now. Hence it may be concluded that "animal" consciousness has from time to time organized its machinery and then disappeared forever, leaving as result the permanent form of life which we call vegetable. But it is not to be supposed that all changes of structure cease with the departure of consciousness. Given spontaneous movement (i. e., growth), and surrounding conditions, and the resultant product must be structures adapted to their surroundings, just as the plastic clay is fitted to its mold. And this is essentially the distinguishing character of vegetable teleology as compared with animal. In the average plant we see adaptation to the conditions of unconscious nutrition; in the animal, adaptation to conditions of conscious contact with the world under a great variety of conditions.

IV. GROWTH-FORCE.

The active processes of living beings are examples of conversion of physical forces, only differing from the conversions observed to take place in inorganic bodies, in the nature of the machinery which exhibits them. The construction of this machinery, as in its use when finished, involves a conversion of force, the resultant consisting of the attraction of nutritious material in definite new directions. This determinate attraction has been regarded as a distinct force, to which the name of bathmic or growth force has been applied. It differs from all the physical forces in this, that while they are only exerted inversely as the square of the distance, this one is in addition most excessive where pleasure has been experienced, and weakest where pain has left its deepest traces. In other words, its movements express *design*, the essential condition of which is *consciousness*. It is thus evident that it differs utterly from all other forces, although a retrograde metamorphosis of matter is as necessary for its production as for that of any of the other forces. Now, although the evidences that stimulated consciousness, or if you choose, mind, can modify structure, are, as matter of observation, not very satisfactory ; yet, since the essential peculiarity of growth-force is its instant attendance on the needs of consciousness, it is a permissible hypothesis that its activity is immediately due to consciousness. This activity is located in bioplasts which do not exhibit consciousness ; whether it co-exists with consciousness in brain bioplasts is unknown. The successive exhibitions of this force, from the lowest to the highest of living beings, have ever been additions to the executive machinery of a more and more specialized consciousness. Thus it is that its results in structure have ever become more and more complex, that is, composed of an ever-increasing number of parts in some region of the organism. Hence another point of distinction from other forces exists, which has been pointed out in a previous paper. It is quite evident that the higher forms of life are the result of continued superaddition of one result of growth-force on another, some examples of subtraction or simplification of parts being generally accompanied by a great preponderance of additions. This is evidence of the accumulation of the property of producing this kind of force, since each successive addition imposes on the growing animals a great number of successive stages before the process reaches its termination, maturity. This involves the belief that the property of exhibiting frequent "repetitions" of growth-activity exists in a higher degree in the reproductive bioplasm of the more complex animal than in that of the lower ones. This is in accordance with the fact of the regular increase in relative complexity and bulk of the nervous system, which accompanies complexity of structure in other respects in the ascending scale of animals. Thus this force differs from all others, as remarked by Prof. Hartshorne, in that its expenditure ultimately increases the amount of its production, because it constructs machinery which feeds its especial organs more and more successfully. Although expended by becoming energetic, its energy produces the means of its own increase. Unlike the physical forces whose expenditure renders matter ever more inert, growth-force when expended adds material which, as a profitable addition, increases the power of the central machine from which the force emanates, by furnishing an increased supply of food.*

Thus it is evident that growth-force is not concentric nor polar in its activity, as are the physical forces, and that its determinations are antagonistic to these. Its existence in the earth has been a succession of conquests over polar force, and, if preceding assumptions be true, the gradual progress presented by animals in abandoning the symmetrical forms exhibited by the lower types has doubtless been due to the constantly increasing amount of consciousness.

V. THE DOCTRINE OF THE UNSPECIALIZED.

It is, however, evident that the directing power of consciousness is limited by the nature of the matter with which it has to deal. There are certain fundamental necessities to which it must conform. No one supposes that any degree of power can make twice two equal to six, cause two solid substances to occupy the same space at the same time, or make an absolutely solid substance out of incompressible atoms of different forms. These involve the absurdity that something can be made out of nothing, or nothing out of something. From the present conduct of the inorganic world, it would appear to possess properties which render consciousness impossible to it. This is doubtless due to the relations existing between the atoms or molecules of which its various species consist. The movements it displays are polar. The colloid molecular state is, so far as this planet is concerned, the only one which we know to be capable of consciousness, and then only while in a state of active transformation. As we have seen, when protoplasm is once organized and working automatically, consciousness need not be present; and when this is absent, the rate of transformation, that is, the amount of food consumed, is greatly lessened. The excess of expenditure during conscious

* It is incorrect to say that growth-force is "potential" in highly organized types, as it is undoubtedly expended in the movement of nutritive pabulum to a given locality. The maintenance of it in that locality is due to ordinary molecular cohesion, which can only be set free by greater molecular consolidation.

activity over that necessary to unconscious activity is well known. It is thus evident that organization renders consciousness unnecessary, so long as external conditions are unchanged, and most probably a degree of fixity may be attained which renders consciousness impossible. The history of the evolution of animal types is apparently an illustration of this truth. The relations of the divisions of the animal kingdom are those of the limbs, branches, and trunk of a tree. Although the termini of the branches are successively nearer the root or starting-point as we proceed from the apex downward or backward, yet the connection is not from end to end of these. To find this we pass down the limb to its junction with the trunk, and trace the branches from the axis outward. Thus with the branches of the animal kingdom. Although the divisions Vertebrata, Mollusca, Echinodermata, etc., stand in an undoubted relation of succession to each other, there is no connection between the highest representative of one and the lowest of another. It is the lower or less specialized forms of each which exhibit the relationship. Thus, among the articulates, the low group of the worms gives us connection with the Mollusca above by *Brachiopoda*, and the echinoderms connect themselves with the *Vermes* by the less specialized *Holothurida*. It seems highly probable also that the point of contact of the *Vertebrata* with these is by one of the lowest divisions, formerly regarded as molluscan, viz.: the *Ascidia*. The same principle holds good within the great divisions. The most specialized orders of Mammalia are the *Artiodactyla*, higher *Perissodactyla*, the *Carnivora*, *Quadrumana*, and perhaps *Cetacea*; but the higher of these have not been derived from the lower. Modern investigations show that several of them have been derived from a common type of mammals of the Eocene period, which is intimately connected with their lower forms, while wanting in the features which give them their special characters. These two illustrations serve to explain the universal law of zoölogical affinity, and therefore of evolution.

The conclusion derived from a survey of this field is, that structure, like habit, when once established, is closely adhered to, and that the movement of growth-force once determined or organized becomes automatic, i. e., independent of consciousness. Therefore a type which reproduces itself automatically becomes after a time so established as to be incapable of radical change, in consequence of a molecular fixity which precludes it. Nevertheless susceptibility to influences of conscious stimuli may remain in some portions of the organism, and thus subordinate modifications of structure have their origin. When conditions of life change, as they often have done during geologic time, those changes of structure which are possible take place under the stimulus of roused consciousness. But if the changes be radical, affecting the foundation processes of vital economy, the specialized forms must undoubtedly perish, and the life of the succeeding time be derived from forms of less pronounced character. The adaptability of generalized types, as to habits, and the absence of mechanical peculiarities in their structure, explain fully the cause of their standing in ancestral relation to all the typical faunæ of the earth.

Nowhere is this truth more remarkably illustrated than in the case of man, the predominant mammal of the present period. From the generalized mammalian fauna of the Eocene, the Carnivora developed a highly organized apparatus for the destruction of life and appropriation of living beings as food. The cloven-footed and odd-toed hoofed orders * are the result of constantly increasing growth of the mechanical appliances for rapid motion over the ground; the former superadding exceptional powers of assimilation of innutritious food. The proboscidians developed huge bulk and an extraordinary prehensile organ. The *Quadrumana* produced none of these things. In respect to speed of limb and powers of digestion, both in function and structure, they remain nearly in the generalized condition from which the other orders of mammals have risen. The limbs and teeth of man retain the characters of the primitive type. Yet but two species of proboscidians remain; the Perissodactyle multitudes are represented by but a few vanishing species. The day of the *Carnivora* has passed forever, and the remaining Artiodactyle herds exist but by the permission of their master, man. But past geologic time reveals no such abundance of true *Quadrumana* as the present period displays. These animals were evidently unable to compete with those of other types in seizing on the opportunities of living. They were excluded from the chase by the more sanguinary ancestors of the Carnivora, and from the field by the multiplying herds of the swifter or more resistant hoofed animals. They possessed neither bulk, nor speed, nor cruelty to serve them

* Represented by the ox and the horse.

in the struggle for existence. So they were doubtless compelled to assume an arboreal life, which required little or no modification of the limbs for its maintenance, although the ultimate production of the grasping thumb from their primitive squirrel-like feet may be traced to this mode of life. The acquisition of a hand must be regarded as the first step in that marvelous accession of experiences which is the condition of mental development. And this latter growth has taken the place of all other means of conquering a position in the world of life, so that man has even retrograded in the efficiency of bodily powers. He has lost the prehensile quality of the hind feet, and the special usefulness of his canine teeth. But the competition among men continues to be such as to render it in the highest degree improbable that he will, as a species, lose the position gained, or suffer any prolonged diminution of the power of intelligence.

Now it is obvious that the more restricted the conditions of the life of a given animal type, the more sensitive it will be to changes. Hence it is that the risks to the existence of *Carnivora, Artiodactyla, Proboscidia*, etc., are much greater than to the omnivorous, all-adaptive order of *Quadrumana*. The same is true of mind. The greater the proportion of unconscious automatism of habits, the less the power of adaptation; and this must be the condition of all animals whose structure is so specialized as to place them beyond reach of competition, or to cut them off from a wide range of experiences. The greater the degree of consciousness of stimulus, the greater will be the degree of adaptability to new relations, and to such constant rousing the unspecialized mind is always open. If, without strong natural weapons, vigilance is the price of existence; if not confined by organization to a peculiar kind of food, ceaseless investigation is stimulated. And these are the mental peculiarities which distinguish the monkeys among all the Mammalia.

The reverse of this picture may now be described, as has been done by Prof. Vogt. It is well known that the young of many parasitic animals are free and active, and discover during migration the localities to which they afterward attach themselves for life. During the early stages they present the characteristic marks of their order and class, and in some instances the males, remaining free, continue to do so. Such are the *Entoconcha mirabilis*, the *Sacculinæ* and the *Trematoda;* the first a mollusk, the second a cirrhiped crustacean, the third a worm. On their becoming attached to their host a successive obliteration of their distinctive characters takes place, so that they become so simplified as to be no longer referable to their proper class, but susceptible, as Prof. Vogt remarks, of being united in a single division. A similar process is observed in the structural degeneration of the Lernean parasites, which are at first free, but afterward become parasitic on fishes. There is in this instance a coincidence between degeneracy of structure and loss of compulsory activity : not only is every function of their sluggish lives automatically performed, but consciousness itself must experience little stimulus.

From what has preceded, it is evident that automatism is at once the product and the antagonist of evolution, and that it is represented in structure by specialization. It appears also that consciousness is the condition of the inauguration of new habits, and this is only possible to structures which are not already too far specialized. This is doubtless true, whether osseous and muscular tissue be concerned in evolution, or whether it be nervous and brain tissue. Hence in the highest form of development, that of brain mechanism, automatism is the enemy, and consciousness the condition of progress. As a product of development, automatism is the condition of stationary existence, and constitutes its effective machinery, but every additional step requires the presence of consciousness. This may be expressed in the every-day language of human affairs, by saying that routine and progress are the opposite poles of social economy.

VI. THE ORIGIN OF CONSCIOUSNESS.

This question has not yet been touched upon, nor is it necessary to give it prolonged attention at present. Consciousness is in itself inscrutable to us, and the contrast which it presents to physical and vital forces is the great fact of life. It is obvious enough that certain molecular conditions are essential to its appearance ; drugs intensify or obscure it ; concussions and lesions destroy it. It will doubtless become possible to exhibit a parallel scale of relations between stimuli on the one hand, and the degrees of consciousness on the other. Yet for all this it will be impossible to express self-knowledge in terms of force. The question as to whether the product of the force conversion involved is the consciousness itself, or only a condition of consciousness, may receive light from the following consideration.

Nowhere does "the doctrine of the unspecialized" receive

greater warrant than in the constitution of protoplasm. Modern chemistry refers compound substances to four classes, each of which is characterized by a special formula of combination. These are called the hydrochloric-acid type, the water-gas type, the ammonia type, and the marsh-gas type. These series are defined by the volumetric relations of their component simple substances : thus, in the first, a single volume unites with an equal volume of hydrogen ; in the second, two volumes of hydrogen unite with a single volume of another element ; in the third, three, and in the fourth, four volumes of hydrogen unite with the single volume of other elements. Hence the composition of these compounds is expressed by the following formulas—chlorine, oxygen, nitrogen, and carbon being selected as typical of their respective classes : HCl, H_2O, H_3N, and H_4C. Now it is an interesting fact that protoplasm is composed of definite proportions of four simple substances, each one representing one of the classes above named, or, in other words, the capacity for proportional molecular combination which characterizes them. The formula $C_{24}N_8OH_{17}$ expresses the constitution of this remarkable substance. Now, although the significance of these combining numbers is unknown, there is a conceivable connection between the characteristic peculiarities of protoplasm and the nature of the substances which compose it. It is probable that these, when in combination with each other, exert a mutually antagonistic control over each other's especial and powerful tendencies to form stable, and hence dead, compounds. It is therefore reasonable that the terms "unspecialized" or "undecided" should be applicable to the molecular condition of protoplasm, and in so far it is a suitable nidus for higher molecular organization, and a capacity for higher forms of force-conversion than any other known substance. If also in inorganic types, as in the organic, the generalized have preceded the specialized in the order of evolution, we are directed to a primitive condition of matter which presented the essentially unspecialized condition of protoplasm, without some of its physical features. We are not necessarily bound to the hypothesis that protoplasm is the only substance capable of supporting consciousness, but to the opposite view, that the probabilities are in favor of other and unspecialized, at present unknown, forms of matter possessing this capacity.

Consciousness constitutes then the only apparently initial point of motion with which we are acquainted. If so, we are at liberty to search for the origin of the physical forces in consciousness, as well as the vital; their present unconscious condition being possibly due, as in the case of the vital, to automatism: the automatism being the expression of the atomic type of the substance exhibiting it. And, doubtless, the simple quantitative relations of the lowest types of forces are related to correspondingly simple geometrical conditions of matter, both representing the simplest grade of automatic action and machinery. We may also suppose that all of these primary conditions were necessary to the production of protoplasm, the only form of matter known to us in which consciousness can persist.

In conclusion, it is obvious that the metastatic condition of protoplasm necessary to the *persistence of consciousness* could not be supported without a constant source of supply by assimilation. Hence it would appear that the preliminary creation of dead and unconscious substances and organisms were a necessary antecedent to the accomplishment of this end; at least under circumstances of temperature under which living beings or protoplasm exist on this planet. Without the unconscious inorganic and organic products of nature, consciousness could not exist on the earth for a day. No animal can maintain consciousness without food; and that food must be, in the main, protoplasm. Protoplasm is manufactured from inorganic matter by the (supposed) unconscious protoplasm of the plant. What form of matter originally gave origin to protoplasm is yet unknown, but it is obvious that the ordinary physical forces must have existed as conditions of its creation, since now they are absolutely necessary to its persistence. Hence we may view the succession of automatic activities somewhat in the light of the fagots used by the elephant to lift itself from the well into which it had fallen. One placed upon another finally raised the footing to an elevation which enabled the animal to obtain its freedom.

Consciousness is the essential condition of personality; so that in this view of the case we are led to a primitive personality, although not to what we call life. And the reason why this personality is to us so obscure a conception is probably to be found in the fact that it, as well as ourselves, is conditioned in its relations to matter by necessary laws of "mathematical" truth.

XIX.

ON ARCHÆSTHETISM.

I. THE HYPOTHESIS OF USE AND EFFORT.

THE claims of the theory of Lamarck, that use modifies structure in the animal kingdom, are being more carefully considered than heretofore, and are being admitted in quarters where they have been hitherto neglected or ignored. Eleven years ago I restated the question as follows: *

"The influences and forces which have operated to produce the type-structures of the animal kingdom have been plainly of two kinds: 1. *Originative;* 2. *Directive.* The prime importance of the former is obvious; that the latter is only secondary in the order of time or succession, is evident from the fact that it controls the preservation or destruction of the results or creations of the first.

"Wallace and Darwin have propounded as the cause of modification in descent their law of natural selection. This law has been epitomized by Spencer as the 'survival of the fittest.' This neat expression no doubt covers the case, but it leaves the origin of the fittest entirely untouched. Darwin assumes a 'tendency to variation' in nature, and it is plainly necessary to do this, in order that materials for the exercise of a selection should exist. Darwin and Wallace's law is, then, only restrictive, directive, conservative, or destructive of something already created. I propose then to seek for the originative laws by which these subjects are furnished—in other words, for the causes of the origin of the fittest.

"It has seemed to the author so clear from the first as to require no demonstration, that natural selection includes no *actively* progressive principle whatever; that it must first wait for the de-

---

* "The Method of Creation," 1871, pp. 2 and 18, Walker Prize Essay, Proceeds. Amer. Philos. Soc., pp. 230–246.

velopment of variation, and then, after securing the survival of the best, wait again for the best to project its own variations for selection. In the question as to whether the latter are any better or worse than the characters of the parent, natural selection in no wise concerns itself."

In seeking for the causes of the origin of variation, the following hypothesis was proposed :

"What are the influences locating growth-force ? The only efficient ones with which we are acquainted are, first, physical and chemical causes ; second, use ; and I would add a third, viz. : effort. I leave the first as not especially prominent in the economy of type-growth among animals, and confine myself to the two following. The effects of use are well known. We can not use a muscle without increasing its bulk ; we can not long use the teeth in mastication without inducing a renewed deposit of dentine within the pulp-cavity to meet the encroachments of attrition. The hands of a race of laborers are always larger than those of men of other pursuits. Pathology furnishes us with a host of hypertrophies, exostoses, etc., produced by excessive use, or necessity for increased means of performing excessive work. The tendency, then, induced by use in the parent, is to add segments or cells to the organ used. Use thus determines the locality of new repetitions of parts already existing, and determines an increase of growth-force at the same time, by the increase of food always accompanying increase of work done, in every animal.

"But supposing there be no part or organ to use. Such must have been the condition of every animal prior to the appearance of an additional digit or limb or other useful element. It appears to me that the cause of the determination of growth-force is not merely the irritation of the part or organ used by contact with the objects of its use. This would seem to be the remote cause of the deposit of dentine in the used tooth ; in the thickening epidermis of the hand of the laborer ; in the wandering of the lymph-cells to the scarified cornea of the frog in Cohnheim's experiment. You can not rub the sclerotica of the eye without producing an expansion of the capillary arteries and corresponding increase in the amount of nutritive fluid. But the case may be different in the muscles and other organs (as the pigment cells of reptiles and fishes) which are under the control of the volition of the animal. Here, and in many other instances which might be cited, it can not be asserted that the nutrition of use is not under the direct control of the will through the mediation of nerve-force. Therefore I am disposed to believe that growth-force may be, through the motive force of the animal, as readily determined to a locality where an executive organ does not exist, as to the first segment or cell of such an organ already commenced, and that therefore effort is, in the order of time, the first factor in acceleration."

A difficulty in the way of this hypothesis is the frequently unyielding character of the structures of adult animals, and the difficulty of bringing sufficient pressure to bear on them without destroying life. But in fact the modifications must, in most instances, take place during the period of growth. It is well known that the mental characteristics of the father are transmitted through the spermatozoöid, and that therefore the molecular movements which produce the mechanism of such mental characters must exist in the spermatozoöid. But the material of the spermatozoöid is combined with that of the ovum, and the embryo is composed of the united contents of both bodies. In a wonderful way the embryo develops into a being which resembles one or both parents in minute details. This result is evidently determined by the molecular and dynamic character of the original reproductive cells, which necessarily communicate their properties to the embryo, which is produced by their subdivision. Rud. Hering has identified this property of the original cells with the faculty of memory. This is a brilliant thought, and, under restriction, probably correct. The sensations of persons who have suffered amputation shows that their sensorium retains a picture or map of the body so far as regards the location of all its sensitive regions. This simulacrum is invaded by consciousness whenever the proper stimulus is applied, and the locality of the stimulus is fixed by it. This picture probably resides in many of the cells, both sensory and motor, and it doubtless does so in the few cells of simple and low forms of life. The spermatozoöid is such a cell, and, how or why we know not, also contains such an arrangement of its contents, and contains and communicates such a type of force. It is probable that in the brain-cell this is the condition of memory of locality. If now an intense and long-continued pressure of stimulus produces an unconscious picture of some organ of the body in the mind, there is reason to suppose that the energies communicated to the embryo by the spermatozoöid and ovum will partake of the character of the memory thus created. The only reason why the oft-repeated stories of birth-marks are so often untrue is because the effect of temporary impressions on the mother is not strong enough to counterbalance the molecular structure established by impressions oftener repeated throughout much longer periods of time.

The demonstration of the truth or falsity of this position, so as to constitute it the true doctrine of evolution, could only be verified from the prosecution of the science of paleontology. It is only in this field that the consecutive series of structures can be obtained which show the directions in which modification has taken place, and thus furnish evidence as to the causes of change. The most complete result of these investigations, up to the present time, has been the obtaining of sufficiently full series of the Mammalia of the Tertiary period to show their lines of descent. In this way the series of modifications of their teeth and feet has been discovered, and the homologies of their parts been ascertained.* Perhaps the most important result of these investigations is the following: The variations from which natural selection has derived the persistent types of life have not been general or even very extensive. They have been in a limited number of directions,† and the most of these have been toward the increase in perfection of some machine. They bear the impress of the presence of an adequate originating cause, directed to a special end. Some of the lines struck out have been apparently inadequate to cope with their environment, and have been discontinued. Others have been more successful, and have remained, and attained further modification.

The reader can estimate the chance of the production of an especially adaptive mechanism in the absence of any pressure of force directing growth to that end. It appears to me that the probability of such variation appearing under such circumstances is very slight indeed, and its continuance through many geologic ages directed to the perfecting of one and the same machine still smaller. For this reason, attempts have been made to demonstrate a mechanical cause for the modifications of structure ob-

---

* "Homologies and Origin of the Molar Teeth of the Mammalia educabilia." Journal Academy Nat. Sciences, Philadelphia, March, 1874. Proceedings Academy Nat. Sci., 1865, p. 22.

† See Hyatt on this point, "Tertiary Planorbis of Steinheim." Anniv. Mem. Bost. Soc. Nat. Hist., 1880, p. 20.

served. For these I refer to papers by Messrs. Alpheus Hyatt, J. A. Ryder and myself: by Prof. Hyatt . . . "Upon the Effects of Gravity on the Forms of Shells and Animals;" * Mr. Ryder "On the Mechanical Genesis of Tooth Forms;" † and "On the Laws of Digital Reduction;" ‡ by myself "On the Origin of the Specialized Teeth of the Carnivora;" # "On the Origin of the Foot Structures of the Ungulates;" ‖ "On the Effect of Impacts and Strains on the Feet of Mammalia." ^ Now demonstration of the mechanical effects of the application of force to matter can only be obtained by observation of the process, and this can not be seen, of course, by the observation of fossils. The relation of the observed facts to the hypothesis is, however, shown by the above papers to be so precise that it only needs observation on the production of similar changes by similar causes in living types to give us a demonstration by induction, which will satisfy most minds. That such facts have been observed among the lower animals is well known. The change of form of animals without hard parts, in adaptation to their environment, is an every-day occurrence.

That these views are now shared by many naturalists is becoming every day more evident. Prof. E. Dubois-Raymond ◊ has recently delivered a lecture before the physicians of the German army, on exercise or use, in which he makes some important admissions. We give the following extract: "We should be, therefore, free to admit, with some appearance of reason, that the vigor of the muscles of wings and of digging feet; the thick epidermis of the palm of the hand and of the sole of the foot; the callosities of the tail and of the ischia of some monkeys; the processes of bones for the insertion of muscles; are the consequences of nutritive and formative excitation, transmitted by heredity." In this position Prof. Raymond is in strict accord with the Lamarckian school of evolutionists. But Prof. Raymond still clings to the obscurities of the Darwinians, though Darwin himself is not responsible for them, in the following sentences: "It is necessary to admit, along with development by use, develop-

---

* "Proceeds. Amer. Assoc. Adv. Science," 1880, p. 527.

† "Proceedings Academy Philadelphia," 1878, p. 45; 1879, 47.

‡ *Loc. cit.*, 1877, October.

"American Naturalist," March, 1879, p. 171.

‖ *Loc. cit.*, April, 1881, p. 269. ^ *Loc. cit.*, July, 1881, p. 542.

◊ "Revue Scientifique," Paris, Jan. 28, 1882.

ment by natural selection, and that for three reasons. First, there are innumerable adaptations—I cite only those known as mimetic coloration—which appear to be only explicable by natural selection and not by use. Second, plants which are, in their way, as well adapted to their environment as animals, are of course incapable of activity. Thirdly, we need the doctrine of natural selection to explain the origin of the capacity for exercise itself. Unless we admit that which it is impossible to do from a scientific stand-point, that designed structures have a mechanical origin, it is necessary to conclude that in the struggle for existence the victory has been secured by those living beings who in exercising their natural functions have increased by chance ('*par hasard*') their capacity for these functions more than others, and that the beings thus favored have transmitted their fortunate gifts to be still further developed by their descendants."

To take up first the second and third of these propositions, Prof. Raymond does not for the moment remember that movement (or use) is an attribute of all life in its simplest forms, and that the sessile types of life, both vegetable and animal, must, in view of the facts, be regarded as a condition of degeneration. It is scarcely to be doubted that the primordial types of vegetation were all free swimmers, and that their habit of building cellulose and starch is responsible for their early-assumed stationary condition. Their protoplasm is still in motion in the limited confines of their walls of cellulose. The movements of primitive plants have doubtless modified their structure to the extent of their duration and scope, and probably laid slightly varied foundations, on which automatic nutrition has built widely diverse results. We may attribute the *origin* of the forms of the vegetable kingdom to three kinds of motion which have acted in conjunction with the physical environment; first, their primordial free movements; second, the intracellular movements of protoplasm; third, the movements of insects, which have doubtless modified the structure of the floral organs. Of the forms thus produced, the fit have survived and the unfit have been lost, and that is what natural selection has had to do with it.

The *origin* of mimetic coloration, like many other things, is yet unknown. An orthodox Darwinian attributes it to "natural selection," which turns out, on analysis, to be "*hasard.*" The *survival* of useful coloration is no doubt the result of natural selection. But this can not be confounded with the question of origin. On this point the Darwinian is on the same footing as the old-time Creationist. The latter says God made the variations, and the Darwinian says that they came by chance. Between these positions science can perceive nothing to choose.

I have attempted to explain the relation which non-adaptive structures bear to the theory of use and effort, in the following language: *

"*The complementary diminution* of growth-nutrition follows the excess of the same in a new locality or organ, of necessity, if the whole amount of which an animal is capable be, as I believe [for the time being], fixed. In this way are explained the cases of retardation of character seen in most higher types. The discovery of truly complementary parts is a matter of nice observation and experiment. Perhaps the following cases may be correctly explained.

"A complementary loss of growth-force may be seen in the absence of superior incisor teeth and digits in ruminating Mammalia, where excessive force is evidently expended in the development of horns, and complication of stomach and digestive organs. The excess devoted to the latter region may account for the lack of teeth at its anterior orifice, the mouth; otherwise, there appears to be no reason why the ruminating animals should not have the superior incisors as well developed as in the odd-toed (Perissodactyl) Ungulates, many of which graze and browse. The loss to the osseous system in the subtraction of digits may be made up in the development of horns and horn-cores, the horn sheath being perhaps the complement of the lost hoofs. It is not proposed to assert that similar parts or organs are necessarily and in all groups complementary to each other. The horse has the bones of the feet still further reduced than the ox, and is nevertheless without horns. The expenditure of the complementary growth-force may be sought elsewhere in this animal. The lateral digits of the *Equidæ* are successively retarded in their growth, their reduction being marked in *Hippotherium,* the last of the three-toed horses; it is accompanied by an almost coincident acceleration in the growth-nutrition of the middle toe, which thus appears to be complementary to them."

* "Method of Creation," p. 23, 1871.

II. THE OFFICE OF CONSCIOUSNESS.

If the law of modification of structure by use and effort be true, it is evident that consciousness or sensibility must play an important part in evolution. This is because movements of animals are plainly in part controlled by their conscious states. The question as to how many of the actions of animals are due to conscious states at once arises. It is well known that most of the more strictly vital functions are unconsciously performed. Not only these, but many acts which have to be learned come to be performed in unconsciousness. Further, movements appropriate to needs which arise at the moment, and which are ordinarily termed voluntary, because they require the introduction of more or less of the rational faculty, are readily performed by vertebrated animals deprived of a brain, through the agency of the spinal cord alone.* The history of the origin of these movements must then be traced.

The movements of living beings generally possess the peculiarity of design, in which they differ from the movements of non-living bodies. That is, their actions have some definite reference to their well-being or pleasure, or their preservation from injury or pain, and are varied with circumstances as they arise. This is not the case with non-living bodies, which move regardless of their integrity or of that of objects near them. This characteristic at once suggests that some element enters into them which is wanting to the movements of non-living masses. It has been suggested that the attraction of animals for their food and their repulsion from pain are derivatives from the attractions and repulsions of inorganic bodies, supposed to be the exhibitions of the force called chemism. But this supposition does not explain the wide difference between the two classes of acts. The adaptation to the environment seen in organic acts is unknown to the inorganic world, while the invariable character of the motions of inorganic force is greatly modified in beings possessed of life. Whether consciously performed or not, the acts of organic beings resemble those of conscious beings actuated by instincts of hunger, reproduction, and defense.

An explanation of these facts seems to be offered by a well-

* Such expressions as "unconscious sensibility" and "unconscious will" are not used here, as being self-contradictory in terms and without meaning.

known phenomenon. We know that it is true of ourselves and of many other animals, that while all new movements have to be learned by repeated attempts, with each succeeding movement the act becomes easier, and that finally it can be performed without requiring any attention whatever. If continued, the movement becomes automatic, so that it may be or is performed in a state of unconsciousness. In the words of Spencer, nervous currents move most readily along accustomed channels. Thus the "habits" of animals may be looked on as movements acquired in consciousness, and become automatic through frequent repetition. Not only this, but the organization thus produced in the parent is transmitted to the succeeding generation, so that the movements of the latter are automatically and often unconsciously performed. This view may be even extended to the purely vital functions, with every probability of its being the true explanation of their origin and development. On a former occasion * I wrote :

"In accordance with this view, the automatic 'involuntary' movements of the heart, intestines, reproductive systems, etc., were organized in successive states of consciousness, which conferred rhythmic movements whose results varied with the machinery already existing and the material at hand for use. It is not inconceivable that circulation may have been established by the suffering produced by an overloaded stomach demanding distribution of its contents. The structure of the Cœlenterata offers the structural conditions of such a process. A want of propulsive power in a stomach or body sac occupied with its own functions, would lead to a painful clogging of the flow of its products, and the 'voluntary' contractility of the body or tube wall being thus stimulated, would at some point originate the pulsation necessary to relieve the tension, Thus might have originated the 'contractile vesicle' of some protozoa, or contractile tube of some higher animals ; its ultimate product being the mammalian heart. So with reproduction. Perhaps an excess of assimilation in well-fed individuals of the first animals led to the discovery that self-division constituted a relief from the oppression of too great bulk. With the increasing specialization of form, this process would become necessarily localized in the body, and growth would repeat such resulting structure in descent as readily as any of the other structural peculiarities. No function bears the mark

* "Consciousness in Evolution." "Penn Monthly," August, 1875, p. 565.

of conscious origin more than this one, as consciousness is still one of the conditions of its performance. While less completely 'voluntary' than muscular action, it is more dependent on stimulus for its initial movements, and does not in these display the unconscious automatism characteristic of the muscular acts of many other functions."

It was not proposed in the preceding paragraph that the contractility of living protoplasm should be regarded as due to consciousness, but that the location in a particular place of a contractility already existing might be due to that cause.

The preceding hypotheses bring us to a general theory of the evolution of organic structures or species. It is that they are the result of movements long continued and inherited, and that the character of these movements was originally determined by consciousness or sensibility. It remains then to consider the nature of consciousness.

It may be mentioned that it is here left open whether there be any form of force which may be especially designated as "vital." Many of the animal functions are known to be physical and chemical, and if there be any one which appears to be less explicable by reference to these forces than the others, it is that of nutrition. Probably in this instance force has been so metamorphosed, through the influence of the originative or conscious force in evolution, that it is a distinct species in the category of forces. Assuming it to be such, I have given it the name of Bathmism ("Method of Creation," 1871, p. 26). Perhaps the contractility generally regarded as an attribute of living protoplasm may be a mechanical phenomenon dependent of course on nutrition; or it may be the exhibition of a force peculiar to living beings; and hence one of the 'vital' group.

III. ARCHÆSTHETISM.

The doctrine of evolution derives the organs of special sense from those of simple sensibility or touch. In other words, their history has been that of other organs; the complex have been derived from the general and simple. There are then generalized consciousness and specialized consciousness. A number of forms of consciousness multiplies its vividness, the one kind reënforcing the other by a slightly different appreciation of the same thing. In the case of persons deprived of the sense of touch, the sense of sight is not sufficient to convince them of their own existence, as a matter of intellectual reflection. When there is no nervous system we must suppose sensibility to be generally distributed throughout the protoplasmic substance of the animal. The localization of consciousness must depend on a localization of the kind and condition of protoplasm which sustains it; while in other parts of the body the protoplasm is modified in other directions and for other purposes. If this be true, the nervous tissue of the higher animals should retain the characters of the lowest simple organisms. In point of fact this is the case, the nucleated cell being the essentially active element in the functions of brain and nerve, and being more numerous in that tissue than in any other.

The remarkable evanescence of consciousness is one of its most marked characteristics. It is this peculiarity which has led many thinkers to deny its existence in the lower animals, and to nduce others to believe that it can have had but little place among he causes of evolution. Partly for the same reason many biolo-ists attempt to derive it by metamorphosis from some form of orce.

But the nature of consciousness is such that it can not be de-ived from unconsciousness, any more than matter can be derived rom no matter, or force from no force. The "unthinkable dogma f creation" (Haeckel) can not be applied to consciousness more han to matter or force. It is a thing by itself, and with matter nd force forms a trio of primitive things which have to be ccepted as ultimate facts. This is perfectly consistent with the osition that consciousness is an attribute of matter, and neither nore nor less difficult to comprehend than the fact that force is n attribute of matter. This view is maintained in a fashion of is own by G. H. Lewes. Prof. Raymond* says in support of ne same position:

"More temperate heads betrayed the weakness of their dia-ctics in that they could not grasp the difference between the ew which I opposed, that consciousness can be explained upon a echanical basis, and the view which I did not question, but sup-orted with new arguments, that consciousness is bound to mate-al antecedents." This position has been maintained by various riters, among them Prof. Allman † and the writer. But Prof.

* Address on the celebration of the Birthday of Leibnitz. "Pop. Science Month-" Feb., 1882.

† Address delivered before the British Association for the Advancement of ience.

Raymond has not found it to be acceptable to his nearest contemporaries. He says: "The opposition which has been offered to my assertion of the incomprehensibility of consciousness on a mechanical theory, shows how mistaken is the idea of the later philosophy, that that incomprehensibility is self-evident. It appears, rather, that all philosophizing upon the mind must begin with the statement of this point." In stating this point some years ago we used the following language: * "It will doubtless become possible to exhibit a parallel scale of relations between stimuli on the one hand and the degrees of consciousness on the other. Yet for all this it will be impossible to express self-knowledge in terms of force." And again: † "An unprejudiced scrutiny of the nature of consciousness, no matter how limited that scrutiny necessarily is, shows that it is qualitatively comparable to nothing else. . . . From this stand-point it is looked upon as a state of matter which is coëternal with it, but not coëxtensive."

It is probable then that consciousness is a condition of matter in some peculiar state, and that wherever that condition of matter exists consciousness will be found, and that the absence of that state implies the absence of consciousness. What is that state?

It would be a monstrous assumption to suppose that consciousness and life are confined to the planet on which we dwell. I presume that no one would be willing to maintain such an hypothesis. Yet it is obvious that if there be beings possessed of these attributes in the planets Mercury and Saturn, they can not be composed of protoplasm, nor of any identical substance in the two. In the one planet protoplasm would be utterly disorganized and represented by its component gases; in the other it would be a solid, suitable for the manufacture of sharp-edged tools.‡ But as it is probable that protoplasm is adapted for the phenomena of consciousness by a certain peculiarity of its constitution, it seems evident that other substances having a similar peculiarity may also be able to sustain it. I have elsewhere attempted to discover what this is, in the following language: #

"Nowhere does 'the doctrine of the unspecialized' receive greater warrant than in the constitution of protoplasm. Modern chemistry refers compound substances to four classes, each of

---

* "Consciousness in Evolution." "Penn Monthly," July, 1875.

† "The Origin of the Will." "Penn Monthly," 1877, p, 439.

‡ Frazer in "American Naturalist," 1879, p. 420.

\# "Consciousness in Evolution," 1875, p. 573.

which is characterized by a special formula of combination. These are called the hydrochloric-acid type, the water-gas type, the ammonia type, and the marsh-gas type. These series are defined by the volumetric relations of their component simple substances: thus in the first, a single volume unites with an equal volume of hydrogen; in the second, two volumes of hydrogen unite with a single volume of another element; in the third, three, and in the fourth, four volumes of hydrogen unite with the single volume of other elements. Hence the composition of these compounds is expressed by the following formulas—chlorine, oxygen, nitrogen, and carbon being selected as typical of their respective classes: HCl, H_2O, H_3N, and H_4C. Now it is an interesting fact that protoplasm is composed of definite proportions of four simple substances, each one representing one of the classes above named, or, in other words, the capacity for proportional molecular combination which characterizes them. The formula $C_{24}N_8OH_{17}$ expresses the constitution of this remarkable substance. Now, although the significance of these combining numbers is unknown, there is a conceivable connection between the characteristic peculiarities of protoplasm and the nature of the substances which compose it. It is probable that these, when in combination with each other, exert a mutually antagonistic control over each other's special and powerful tendencies to form stable, and hence dead, compounds. It is therefore reasonable that the terms 'unspecialized' or 'undecided' should be applicable to the molecular condition of protoplasm, and in so far it is a suitable nidus for higher molecular organization, and a capacity for higher forms of force-conversion than any other known substance. If also in inorganic types, as in the organic, the generalized have preceded the specialized in the order of evolution, we are directed to a primitive condition of matter which presented the essentially unspecialized condition of protoplasm, without some of its physical features. We are not necessarily bound to the hypothesis that protoplasm is the only substance capable of supporting consciousness, but to the opposite view, that the probabilities are in favor of other and unspecialized, but unknown, forms of matter possessing this capacity."

The condition of living protoplasm was also referred to in the following language in a later publication: *

* "The Origin of the Will." "Penn Monthly," June, 1877, p. 439.

"The cause of the difference between conscious and unconscious force must be secondarily due to different conditions of matter as to its atomic constitution; consciousness being only possible, so far as we can ascertain, to matter which has not fallen into fixed and automatic relations of its atoms."

Protoplasm in the form of food is not conscious; and tissue formed of protoplasm is not conscious, excepting certain cells where the forming process is in action. Nor is consciousness present in all cells where nutrition is active. From the increased consumption of energy, and the increased expenditure of energy (heat, Lombard) which takes place during conscious processes, we may well believe that the decomposition of protoplasm is more considerable in such processes than in other forms of nervous activity. We can imagine simple nutrition to be a condition of the elements of this substance in which the chemical force is simultaneously combining and dissolving its combination, and that during the process there is a condition in which the chemism is fo the time being unsatisfied, though present. The direction whicl this nutrition or metastasis takes, is due to the arrangement of th molecules already existing in the tissue, the new molecules takin the form of the old ones in replacement, so long as no extraneou force interferes. That they are rearranged under the influence o consciousness is apparent in the origin of variations of structur in accordance with the views of evolution already entertained. I is the arrangement of the molecules which constitutes the automatic machinery of nutrition as well as of other activities, so tha consciousness necessarily only appears in that stage of nutritio while the matter is in a transition state, and unformed. Whethe chemism must be regarded as suspended, or only unsatisfied, a this stage, can only be imagined. As non-satisfaction is probabl the temporary condition in all nutrition, it is not unlikely that suspension may be the condition of consciousness.

Perhaps the character of the components of protoplasm is suc that the movements of their atoms, i. e., their chemism, mutuall interfere and destroy each other, as in the cases of the interferenc of the waves of light and sound.

The colloid form of protoplasm is especially favorable to internal movements which shall not destroy the integrity of the mas perhaps more so than a gaseous state in a compound of similar constitution. It is, moreover, more favorable to the preservation molarity than a gas could be, on account of the ease with whic it adheres to solid substances, and transports and locates them as part of its external and internal supports. But it is not inconceivable that under other conditions of temperature, etc., the gaseous condition of matter might answer the same purpose. It must be borne in mind, however, that this is a subordinate question, and that the real characteristic of the "physical basis of life" is to be found rather in its generalized *dynamic* condition.

We must then believe that wherever this generalized condition exists, consciousness will be present. As soon as mechanical or chemical force appears in the molecules of the sustaining substance, consciousness disappears. The organism has taken the first step toward death, but is not dead, but is *anæsthetized.* Constant nutrition is essential to the performance of all life functions, including consciousness, and it is evident that this is necessary to the maintenance of the unspecialized condition in which the latter appears.

Is the appearance of sensibility on the development of its sustaining condition, evidence that the latter stands to the former in the relation of cause and effect? If the view of the pre-existence of consciousness be true, there is no more relation of cause and effect than in the case of the opening of a door which admits a wind. The force expended in opening the door is not converted into the energy exerted by the wind as it enters the room. It simply releases it, or admits it to a new field. It is, however, true, that consciousness having once entered, a larger conversion of force is necessary to its persistence than is expended during its absence. Like combustion, which is only communicable under suitable conditions, consciousness having once been transmitted to a new *æsthetophore,** lives on it, and requires constant supplies of material for its sustenance.

The hypothesis of the primitive and creative function of consciousness may be called *Archæsthetism.*

IV. PANÆSTHETISM.

It has been the custom of men from the dawn of thought to attempt to construct for themselves cosmogonies and theologies. Science is yet far from supplying the facts necessary to the construction of a true system of the universe, and philosophy can only stretch out a little further into the unknown by the use of neces-

---

* Æsthetophore, a substance which sustains consciousness.

sary inference. In spite, however, of the insufficiency of the data, men still suggest new views or cling to old ones, and an occasional flight into this region of thought at least brings the thinker into sympathy with the thoughts of his fellow-men.

The admission of the possibility of the existence of consciousness in other forms of matter than protoplasm, and in other regions than the Earth, lends countenance to a rational belief in the so-called "supernatural" (better called the supersensuous) so prevalent among men in irrational forms. The question naturally arises, Is there any generalized form of matter distributed through the universe which could sustain consciousness? The presumption is that such a form of matter may well exist. Evolution or specialization has only worked up part of its raw material in the organic world. Wherever primitive conditions remain, there primitive organisms abound. *Protozoa* are yet numerous on land, and the *Protobathybius* inhabits the depths of the sea. Highly specialized forms of life are in fact numerically a minority of living beings. May not this be true also of inorganic beings? It is thought that various celestial bodies represent unfinished worlds. Is it not probable that the grand sources of matter not yet specialized into the sixty odd substances known to us, may still sustain the primitive force not yet modified into its species, and that this combination of states may be the condition of persistent consciousness from which all lesser lights derive their brilliancy? There is much to warrant such a view in the observed facts of life, taken in connection with the general course of evolution. Moreover, that some form of matter connects the interstellar spaces, is thought to be proved by the transmission of light in some cases, and light and heat in others. That such a form of matter pervades all spaces whatever, is the theory of some physicists. If it be so generalized as to be capable of sustaining consciousness, it becomes the source from which other substances derive it, so soon as they, through the energy of nutrition, which resists death, maintain the same primitive and unformed constitution capable of exhibiting it.

Of course there is no evidence in our own memory of the existence of our personality prior to our human experience. No one on awaking from unconsciousness remembers having been anywhere in particular during the interval. These facts may be harmonized with the theory here presented, on the supposition that memory is lost on a transfer of consciousness from one physical basis to another. The arguments in favor of a transfer of consciousness do not sustain the idea of a transfer of memory. Memory requires an arrangement of molecules or atoms which when finished no longer exhibits consciousness. With proper stimulus, when the proper kind of force conversion is set up in them, consciousness extends into them, and, taking their form, produces reminiscence or conscious memory. The molecular arrangement would be probably lost on a transfer of consciousness to a new material basis. It might then be supposed that with every such transfer a new personality is established. Though the correct definition of personality includes memory as well as consciousness, when viewed as an objective concept, it may be questioned whether memory is necessary to the subjective belief in one's own personality. Those insane persons who believe that they have lost their personality, and think that they are some one else, nevertheless recognize the fact that what they now are has a continuity of existence with what they once were. The material limitations of consciousness are the authors of the kind of personality it presents. A limitation or an expansion of its range would not destroy the idea of personality, but would simply restrict or extend it. The possible confluence of many personalities would not destroy them, but each one would regard the others as additions to himself, and himself, therefore, as so much the greater being.

As a summary of the preceding conclusions, the following analysis of metaphysical systems may be given. It defines the place of the doctrine of archæsthetism, above proposed, as distinguished from the opposing view of metæsthetism, which is held by many monists :

I. Consciousness (" spirit ") is independent of matter.................. DUALISM.
II. Consciousness is an attribute of matter.......................... MONISM.
α. Consciousness is primitive and a cause of evolution......... *Archæsthetism.*
β. Consciousness is a product of the evolution of matter and force. *Metæsthetism.*

XX.

ON CATAGENESIS.*

I. THE EVOLUTION OF ORGANISMS.

THE general proposition that life has preceded organization in the order of time, may be regarded as established. It follows necessarily from the fact which has been derived from paleontological investigation, that the simple forms have, with few sporadic exceptions, preceded the complex in the order of appearance on the earth. The history of the lowest and simplest animals will never be known, on account of their perishability ; but it is a safe inference from what is known, that the earliest forms of life were the rhizopods, whose organization is not even cellular, and includes no organs whatever. Yet these creatures are alive, and authors familiar with them agree that they display, among their vital qualities, evidences of some degree of sensibility.

The following propositions were laid down by Lamarck, as established by facts known to him, in 1809 :†

I. "In every animal which has not passed the term of its development, the frequent and sustained employment of an organ, gradually strengthening it, develops and enlarges it, and gives it power proportional to the duration of its use ; whilst the constant disuse of a like organ insensibly weakens it, deteriorates it, progressively reduces its functions, and finally causes it to disappear.

II. "All that nature acquires or loses in individuals, through the influence of circumstances to which the race has been exposed for a long time, either by the predominant use of an organ or by the disuse of such part, she preserves by generation among new

---

* An address delivered before the Biological Section of the American Association for the Advancement of Science, at Philadelphia, September 4, 1884, by E. D. Cope, vice-president.

† "Philosophie Zoölogique," Pt. I, p. 235 (Edit. 1830).

individuals which spring from it, provided the acquired changes be common to both sexes or to those which produce new individuals."

The same proposition was previously enunciated by Lamarck in the following condensed form ("Recherches sur les Corps vivans," p. 50):

"It is not the organ, that is, the nature and form of the parts of the body, which have given origin to its habits and peculiar functions, but it is, on the contrary, its habits, its manner of life, and the circumstances in which individuals from which it came found themselves, which have, after a time, constituted the form of the body, the number and character of its organs, and the functions which it possesses."

Several years ago, not having read Lamarck, I characterized the above hypothesis as the "law of use and effort," * and I have subsequently formulated the *modus operandi* of this law into two propositions. The first of these is, that animal structures have been produced, directly or indirectly, by animal movements, or the doctrine of *kinetogenesis;* the second is, that as animal movements are primitively determined by sensibility, or consciousness, consciousness has been and is one of the primary factors in the evolution of animal forms. This is the doctrine of *archæsthetism.* The doctrine of kinetogenesis is implied in the speculations of Lamarck in the following language ("Philosophie Zoölogique," ed. 1830, p. 239): "With regard to the circumstances which [Nature] uses every day to vary that which she produces, one can say that they are inexhaustible. The principal arise from the influence of climates; from diverse temperature of the atmosphere and of the environment generally; from diversity of location; from habits, the most ordinary movements, and most frequent actions," etc. The influence of motion on development is involved in Spencer's theory of the origin of vertebræ by strains; † and I have maintained the view that the various agencies in producing change mentioned by Lamarck are, in the case of animals, simply stimuli to motion. ‡ The immediate mechanical effect of motion on animal structure has been discussed in papers by Ryder,

* "Method of Creation," "Proceedings American Philosophical Society," 1871, p. 247.

† "Principles of Biology," II, p. 195.

‡ "On the Relation of Animal Motion to Animal Evolution," "American Naturalist," Jan., 1878.

Hyatt, Clevenger, and myself, and I have cited the evidence of vertebrate paleontology as conclusively proving such an effect." *

The object of the present paper is to pursue the question of the relation of sensibility to evolution, and to consider some of the consequences which it involves. It is scarcely necessary to observe that, in the early stage which the subject presents at the present time, I can only point out the logical conclusions derivable from facts well established rather than any experimental discoveries not already known. And I will say here to those who object to the introduction of metaphysics into biology, that they can not logically exclude the subject. As in one sense a function of nervous tissue, mind is one of the functions of the body. Its phenomena are everywhere present in the animal kingdom. Moreover, when studied in the inductive, *a posteriori* method, metaphysics is an exact science. As Bain observes, it is a good deal easier to foretell the actions of a man than those of the weather. It is only want of familiarity with the subject which can induce a biologist to exclude the science of mind from the field.

For the benefit of those who are not familiar with the doctrine of archæsthetism,† I give an outline of its implications. In the first place, the hypothesis that consciousness had played a leading part in evolution would seem to be negatived by the well-known facts of reflex action, automatism, etc., where acts are often unconsciously performed, and often performed in direct opposition to present stimuli. But while it is well understood that these phenomena are functions of organized structure, it is believed that the habits which they represent were inaugurated through the immediate agency of consciousness. It is not believed that a designed act can have been performed *for the first time* ‡ without consciousness on the part of the animal of the want which the act was designed to relieve or supply. This opinion accords with our knowledge of ourselves, and, by irresistible inference, with our belief regarding other animals. From such familiar observation we also know that so soon as a movement of body or mind has been acquired by repetition, consciousness need no longer accompany the act. The act is said to be automatic when performed

* "The Evidence for Evolution in the History of the Extinct Mammalia," Am. Ass. Adv. Sci., 1883, p. 32.

† "American Naturalist," 1882, p. 454.

‡ The same view is expressed by Ribot, "Diseases of the Will," p. 38 (Humboldt Libr.).

without exertion, either consciously or unconsciously, and in those functions now removed from the influence of the unconscious mind such acts are called reflex. The *origin* of the acts is, however, believed to have been in consciousness, not only for the reasons above stated, but also from facts of still wider application. The hypothesis of archæsthetism then maintains that consciousness as well as life preceded organism, and has been the *primum mobile* in the creation of organic structure. This conclusion also flows from a due consideration of the nature of life. I think it possible to show that the true definition of life is, *energy directed by sensibility, or by a mechanism which has originated under the direction of sensibility*. If this be true, the two statements that life has preceded organism, and that consciousness has preceded organism, are co-equal expressions.

II. CONSCIOUSNESS, ENERGY, AND MATTER.

Regarding for the time being the phenomena of life as energy primitively determined by consciousness, let us look more closely into the characteristics of this remarkable attribute. That consciousness, and therefore mind, is a property of matter, is a necessary truth which to some minds seems difficult of acceptance. That it is not an attribute of all kinds of matter is clear enough, but to say that it is not an attribute of any kind of matter is to utter an unthinkable proposition. To my mind the absence of tridimensional matter is synonymous with nothingness or absolute vacuity. To say that phenomena have a material basis, is for me only another way of saying that they exist. It being granted then that consciousness is an attribute of matter, or a certain behavior of matter, it remains to trace its relation to energy, which is here used in the sense of motion. Consciousness is clearly not one of the known so-called inorganic forces. Objects which are hot, or luminous, or sonorous, are not, as is well known, on that account conscious. Consciousness is not then a necessary condition of energy. On the other hand, in order to be conscious, bodies must possess a suitable temperature, and must be suitably nourished. So energy is a necessary condition of consciousness. For this reason some thinkers regard consciousness as a form or species of energy. For my own part, in classification, I prefer to keep very different things apart. To classify consciousness with heat, light, sound, etc., does violence to my sense of fitness and to all proper definitions. This is well shown by Prof. Clifford in the following passage: "It will be found excellent practice, in the mental operations required by this doctrine, to imagine a train the fore part of which is an engine and three carriages linked with iron couplings, and the hind part three other carriages linked with iron couplings; the bond between the two parts being made out of the sentiments of amity subsisting between the stoker and the guard."* This satire, whether intentionally or not on the part of its learned author, expresses at once the distinctive character of consciousness *in esse* and the impossibility of dissociating it from energy *in posse*. For it is sufficiently clear that while the conscious feelings of the stoker and the guard could *by themselves* do nothing for the train, such a state is essential to the energy displayed by them when they are at work for its benefit. We all understand the absurdity of such expressions as the equivalency of force and matter, or the conversion of matter into force. They are not, however, more absurd than the corresponding proposition more frequently heard, that consciousness can be converted into energy, and *vice versa*.

The energetic side of consciousness, however, may be readily perceived by a little attention to its operations. Acts performed in consciousness involve a greater expenditure of energy than the *same acts* unconsciously performed. The difficulty of a given piece of labor is in direct proportion to its novelty; that is, is in direct proportion to the amount of endeavor we use in its performance. This is another way of saying that the labor is directly as the consciousness involved. Another evidence of the dynamic character of consciousness is its exclusive and therefore complementary character. Two opposite emotions can not occupy the mind at the same moment of time. An emotion excludes all high intellectual work, and *vice versa*.

But there is no fact with which we are more familiar than that consciousness in some way determines the direction of the energy which it characterizes. The stimuli which affect the movements of animals at first only produce their results by transmission through the intermediation of consciousness. Without consciousness, education, habits, and designed movements would be impossible. So far as we know, the instinct of hunger, which is at the foundation of animal being, is a state of consciousness in all animals. This incontestable fact is overlooked by the materialists properly so called.

---

* "Scientific Basis of Morals," Humboldt Library Ed., p. 21.

On the other hand, as consciousness is an attribute of matter, it is of course subject to the laws of necessity to which matter and energy conform. For instance, it can not cause two solid bodies to occupy the same space at the same time, nor can it add one body to one body and thus make three bodies. No more can it make ten foot-pounds of energy out of five foot-pounds of energy, and it can not abolish time more than it can annihilate space. These are fundamental truths which are overlooked by a majority of mankind. Moreover, nothing is more common than to hear life or mind spoken of as though it of itself is a "substance," and not, as it should be, as an *attribute* or condition of substance or matter.

What is, then, the immediate action of consciousness in directing energy into one channel rather than another? To take an illustration: Why, from a purely mechanical point of view, is the adductor muscle of the right side of the horse's tail contracted to brush away the stinging fly from the right side of the horse's body rather than the left adductor muscle? Why was the contraction-provoking energy deflected into the right interspinal motor nerves rather than into those of the left side? Why is the ear of the horse turned forward to catch the sound in front of him and backward to gather the sound coming from behind? The first crude thought is, that consciousness supplies another energy which turns aside the course of the energy required to produce the muscular contraction; either as the man with the rein in his hand turns aside the horse's head, or as the shield he holds deflects a moving body. But consciousness *per se*, that is, regarding it in its proper and distinctive definition, is not itself a force (= energy). How, then, can it exercise energy? Certainly no more than the bare good-will of the train hands can pull the train. Such an explanation is to admit the possibility of making something out of nothing.

III. THE RETROGRADE METAMORPHOSIS OF ENERGY.

The key to many weighty and mysterious phenomena lies in the explanation of the so-called voluntary movements of animals. I say "so-called," because true will is not at all involved in the question. I mean the acts directed by consciousness, the acts which would not take place at all if the animal were unconscious. That there are many such acts you well know. The explanation can only be found in a simple acceptance of the fact as it is, in the thesis, that *energy can be conscious.* If true, this is an ultimate fact, neither more nor less difficult to comprehend than the nature of energy or matter in their ultimate analyses. But how is such a hypothesis to be reconciled with the facts of nature, where consciousness plays a part so infinitesimally small? The explanation lies close at hand, and has already been referred to. *Energy become automatic is no longer conscious,* or is about to become unconscious. That this is the case is matter of every-day observation on ourselves and on other animals. What the molecular conditions of consciousness are, is one of the problems of the future, and for us a very interesting one. One thing is certain, the organization of the mechanism of habits is its enemy. *It is clear that in animals, energy, on the loss of consciousness, undergoes a retrograde metamorphosis,* as it does later in the history of organized beings on their death. This loss of consciousness is first succeeded by the so-called involuntary and automatic functions of animals. According to the law of catagenesis, the vegetative and other vital functions of animals and plants are a later product of the retrograde metamorphosis of energy. With death, energy falls to the level of the polar tensions of chemism, and the regular and symmetrical movements of molecules in the crystallization of its inorganic products. Let us now trace in more detail the energies displayed by animals and plants.

It has been already advanced (see page 425) that the phenomena of growth-force, which are especially characteristic of living things, originated in the direction given to nutrition by consciousness and by the automatic movements derived from it. There remain, however, some other phenomena which do not yield so readily to this analysis. These are: first, the conversion by animals of dead into living protoplasm; second, the conversion of inorganic substances into protoplasm by plants; and third, the manufacture of the so-called organic compounds from the inorganic by plants. To these points we may return again. It is also well known that living animal organisms act as producers, by conversion, of various kinds of inorganic eneregy, as heat, light, sound, electricity, motion, etc. It is the uses to which these forces are put by the animal organism, the evident design in the occasion of their production, that gives them the stamp of organic life. We recognize the specific ultility of the secretions of the glands, the appropriate distribution of the products of digestion, and adaptation of muscular motion to many uses. The increase of heat to protect against depression of temperature; the light to direct the sexes to each other; the electricity as a defense against enemies—display unmistakably the same utility. We must not only believe that these functions of animals were originally used by them under stimulus, for their benefit, but, if life preceded organism, that the molar mechanism which does the work has developed as the result of the animal's exertions under stimuli. This will especially apply to the mechanism for the production of motion and sound. Heat, light, chemism, and electricity doubtless result from molecular aptitudes inherent in the constitution of protoplasm. But the first and last production of even these phenomena is dependent on the motions of the animal in obtaining and assimilating nutrition. For without nutrition all energy would speedily cease. Now the motion required for the obtaining of nutrition has its origin in the sensation of hunger. So, even for the first steps necessary to the production of inorganic forces in animals, we are brought back to a primitive consciousness.

To regard consciousness as the primitive condition of energy, contemplates an order of evolution in large degree the reverse of the one which is ordinarily entertained. The usual view is, that life is a derivative from inorganic energies as a result of high or complex molecular organization, and that consciousness (= sensibility) is the ultimate outcome of the nervous or equivalent energy possessed by living bodies. The failure of the attempts to demonstrate spontaneous generation will prove, if continued, fatal to this theory. Nevertheless the order can not be absolutely reversed. Such a proceeding is negatived by the facts of the necessary dependence of the animal kingdom on the vegetable, and the vegetable on the inorganic, for nutrition, and consequently for existence. So the animal organism could not have existed prior to the vegetable, nor the vegetable prior to the mineral. The explanation is found in the wide application of the "doctrine of the unspecialized," * so clearly demonstrated by paleontology. From this point of view creation consists in specialization, an expression which describes the specific action of the general principle described by Spencer as the conversion of the homogeneous into the heterogeneous. To be more explicit, it consists of the

* The term specialized, introduced into biology by Prof. Dana, has been used in connection with energy in creation by the author, "Penn Monthly," 1875, p. 569.

production of mechanism out of no mechanism, of different kinds of energy out of one kind of energy. The material basis of consciousness must then be a generalized substance which does not display the more automatic and the polar forms of energy. From a physical standpoint protoplasm is such a substance. Its instability indicates weakness of chemical energy also, which suggests that the complexity of its molecule may be due to some form of energy not properly chemical. The readiness with which it undergoes retrograde metamorphosis shows that it is not self-sustaining, and furnishes a good illustration of creation of specialized substances by a running down in the scale of being. Loew and Bokorny * suggest that "the cause of the living movements in protoplasm is to be sought for in the intense atomic movements, and therefore easy metamorphosis of its aldehyde groups of components"; the molecular movements becoming molar, to use the language of Lester Ward. The position which I now present requires the reversal of the relations of these phenomena. Generalized matter must be supposed to be capable of more varied molecular movements than specialized matter, and it is believed that the most intense of all such movements are those of brain tissue in mental action, which are furthest removed of all from molar movements. From this point of view, when molar movements are derived from molecular movements, it is by a process of running down of energy, not of elevation; by an increase of the distance from mental energy, not an approximation to it.

The fact that the physical basis of consciousness is composed of four substances, which are respectively a monad, a dyad, a triad, and a tetrad, doubtless has something to do, as I have suggested,† with its exhibition of this remarkable attribute. It might be supposed that the presence of carbon had the effect of restraining the chemical and physical molecular tendencies of the three other substances. From this standing-ground we may imagine that other substances besides protoplasm might support consciousness and life. In other parts of the universe, other substances they would have to be, if consciousness exist there.

The manner in which protoplasm is made at the present time is highly suggestive. It is manufactured by living plants out of

* "Die chemische Kraftquelle in lebenden Protoplasma," von O. Loew u. T. Bokorny, Munich, 1882, I.

† "Penn Monthly," 1875, p. 574.

inorganic matter, the hydrogen, carbon, nitrogen, and oxygen contained in the atmosphere and in the earth. As dead plants will not perform this function, this action is regarded as in some way due to the presence of life. The energy peculiar to living protoplasm, and derived primarily in part only from the sun's rays, directs energy so that the complex molecular aggregation protoplasm is the result. This is the only known method of manufacture from inorganic matter of this substance. The first piece of protoplasm had, however, no paternal protoplasm from which to derive its being. The protoplasm-producing energy must, therefore, have previously existed in some form of matter not protoplasm. This is also suggested by the fact that it really antagonizes the chemical forces, and might be called, from this fact, *antichemism*. The protoplasm-sustaining energy of animal protoplasm may be a less energetic derivative, or *vice versa*. In terms of the theory of catagenesis, the plant life is a derivative of the primitive life, and it has retained enough of the primitive quality of self-maintenance to prevent it from running down into forms of energy which are below the life level; that is, such as are of the inorganic chemical type, or the crystalline physical type. A part of the energy does so run down, as can be seen in the few automatic movements of plants, and the phosphorescence of some. Also symmetrical crystals are made by some. But M. Pasteur has shown * that whenever the crystals are of the organic type, i. e., contain carbon, they are not symmetrical but are unilateral, or, as he terms them, dissymmetrical. This indicates that the presence of carbon has restrained a little the absolute symmetrical automatism of the formative force.

IV. ORIGIN OF LIFE ON THE EARTH.

If, then, some form of matter other than protoplasm has been capable of sustaining the essential energy of life, it remains for future research to detect it, and to ascertain whether it has long existed as part of the earth's material substance or not. The heat of the earlier stages of our planet may have forbidden its presence, or it may not. If it were excluded from the earth in its first stages we may recognize the validity of Sir William Thomson's suggestion that the physical basis of life may have reached us from some other region of the cosmos by transporta-

* "Revue Scientifique," 1884, Jan., p. 2.

tion on a meteorite. If protoplasm in any form were essential to the introduction of life on our planet, this hypothesis becomes a necessary truth. Here let me refer to the fact that hydrocarbonaceous substances have been discovered in meteorites. Here also the remarkable discovery of Huggins claims attention.* This veteran spectroscopist has detected the lines of some hydrocarbon vapor in the spectra of interplanetary spaces. The significance of this discovery is at once perceived if we believe that hydrocarbons are only produced under the direction of life.†

Granting the existence of living protoplasm on the earth, there is little doubt that we have some of its earliest forms still with us. From these simplest of living beings both vegetable and animal kingdoms have been derived. But how was the distinction between the two lines of development, now so widely divergent, originally produced? The process is not difficult to imagine. The original plastid dissolved the salts of the earth and appropriated the gases of the atmosphere and built for itself more protoplasm. Its energy was sufficient to overcome the chemism that binds the molecules of nitrogen and hydrogen in ammonia, and of carbon and oxygen in carbonic dioxide. It apparently communicated to these molecules its own method of being, and raised the type of energy from the polar non-vital to the adaptive vital by the process. Thus it transformed the dead mineral world, perhaps by a process of invasion, as when a fire communicates itself from burning to not burning combustible material. Thus it has been doing ever since, but it has redeposited some of its gathered stores in various non-vital forms. Some of these are in organic forms, as cellulose; others are crystals imprisoned in its cells; while others are amorphous, as waxes, resins, and oils. But consciousness apparently early abandoned the vegetable line. Doubt-

---

* See address of C. W. Siemens, Prest. British Ass. Adv. Science, 1882; "Nature," 1882, p. 400.

† Says Mr. S. F. Peckham ("American Journal of Science and Arts," 1884, p. 105), on the origin of bitumens: "These chemical theories [of the origin of bitumen] are supported by great names, and are based upon very elaborate researches, but they require the assumption of operations nowhere witnessed in nature or known to technology. . . . In the chemical processes of nature complex organic compounds pass to simpler forms, of which operation marsh gas, like asphaltum, is a resultant, and never the crude material upon which decomposing forces act."

The fact that many organic compounds are now produced in the laboratory, does not prove that such substances can be produced without the exercise of a species of energy different from the inorganic types with which we are acquainted.

less all the energies of vegetable protoplasm soon became automatic. The plants in general, in the persons of their protist ancestors, soon left a free-swimming life and became sessile. Their lives thus became parasitic, more automatic, and in one sense degenerate.

The animal line may have originated in this wise. Some individual protists, perhaps accidentally, devoured some of their fellows. The easy nutrition which ensued was probably pleasurable, and once enjoyed was repeated, and soon became a habit. The excess of energy thus saved from the laborious process of making protoplasm was available as the vehicle of an extended consciousness. From that day to this, consciousness has abandoned few if any members of the animal kingdom. In many of them it has specialized into more or less mind. Organization to subserve its needs has achieved a multifarious development. There is abundant evidence to show that the permanent and the successful forms have ever been those in which motion and sensibility have been preserved, and most highly developed.

This review of the history of living organisms has been epitomized in the following language: "Evolution of living types is then a succession of elevation of platforms, on which succeeding ones have built. The history of one horizon of life is that its own completion but prepares the way for a higher one, furnishing the latter with conditions of a still further development. Thus the vegetable kingdom died, so to speak, that the animal kingdom might live, having descended from an animal stage to subserve the function of food for animals. The successive types of animals first stimulated the development of the most susceptible to the conflict, in the struggle for existence, and afterward furnished them with food."

V. CATAGENESIS OF INORGANIC ENERGY.

If the principles adopted in the preceding pages be true, it is highly probable that *all forms* of energy have originated in the process of running-down or specialization from the primitive energy.

In the department of physics I am not at home, and touch upon it merely to carry out to a necessary conclusion the hypothesis presented in the preceding pages. It may be that physicists and chemists may find value in the suggestions which come from the side of biology. A cursory perusal of the general hypotheses current in these departments shows that the door is wide open to receive light from this quarter. What can be offered here is of the vaguest, yet it may suggest thought and research in some minds.

In the first place, it is highly probable that one of the problems to be solved by the physicists of the present and future is that of a true genealogy of the different kinds of energy. In this connection a leading question will be the determination of the essential differences between the different forms of energy and the material conditions which cause the metamorphosis of one kind of energy into another.

In constructing a genealogy of energies, it must be observed that we will probably obtain not a single line of succession, but several lines of varying lengths. It must also be remembered that, as in the forms of the material world, which are their expression, a greater or less extensive exhibition of all the types remains to the present day.

That the tendency of purely inorganic energy is to "run down," in all except possibly some electric operations, is well known. Inorganic chemical activity constantly tends to make simpler compounds out of the more complex, and to end in a satisfaction of affinities which can not be further disturbed except by access of additional energy. In chemical reaction the preference of energy is to create solid precipitates. In the field of the physical forces we are met by the same phenomenon of running down. All inorganic energies or modes of motion tend to be ultimately converted into heat, and heat is being steadily dissipated into space. Therefore the result has been and will be the creation of the mineral kingdom ; of the rocks and fluids that constitute the masses of the worlds.

The process of creation by the retrograde metamorphosis of energy, or, what is the same thing, by the specialization of energy, may be called *catagenesis*. It may be denied, however, that this process results in a specialization of energy. The vital energies are often regarded as the most special, and the inorganic as the most simple. If we regard them, however, solely in the light of the essential nature of energy, i. e., power, we must see that the chemical and physical forces are most specialized. The range of each species is absolutely limited to one kind of effect, and their diversity from each other is total. How different this from the versatility of the vital energy ! It seems to dominate all forms of conversion of energy, by the mechanisms which it has, by evolution, constructed. Thus, if the inorganic forces are the products of a primitive condition of energy which had the essential characteristics of vital energy, it has been by a process of specialization. As we have seen, it is this specialization which is everywhere inconsistent with life.

With these preliminary remarks we may now consider very tentatively the relations of the different kinds of energy to each other and to consciousness. In practice it is sometimes difficult to draw the line between conscious and unconscious states of energy. One reason is that, although a given form of energy may be unconscious, consciousness may apprehend the action by perceiving its results. The distinction is rendered clearer by the reflection that we can perceive by sight or touch any action of the body of whatever character The energy of the conscious type is therefore altogether mental. The relations may be expressed as follows :

	Examples.
A. Designed (always molecular).	
I. Conscious.	
1. Involving effort	" Voluntary " acts.
2. Not involving effort	Passive perception. Conscious automatism.
II. Unconscious.	
3. Involving mental process	Unconscious automatic.
4. Not involving mental process	Reflex.
B. Not designed.	
I. Molecular.	
5. Electric,	
6. Chemical, 7. Physical,	Crystallific and non-crystallific.
II. Molar.	
8. Cosmic.	

The only strictly molar energies of the above list are the cosmical movements of the heavenly bodies. The others are molecular, although they give rise to molar movements, as those of the muscles, of magnetism, etc. Some molar movements of organic beings are not, *in their last phases*, designed ; as those produced by nervous diseases.

The transition between the organic and the inorganic energies may be possibly found in the electric group. Its influence on life, its production of contractions in protoplasm, and its resemblance to nerve force, are well known. It also compels chemical unions otherwise impracticable, thus resembling the energy of the protoplasm of plants, whose energy in actively resisting the disintegrating inorganic forces of nature is so well known. Perhaps this type of force is an early-born of the primitive energy, one which has not descended so far in the scale as the chemism which holds so large a part of nature in the embrace of death.

Vibration is inseparable from our ideas of motion or energy, not excluding conscious energy. There are reasons for supposing that in the latter type of activity the vibrations are the most rapid of all those characteristic of the forces. A center of such vibrations in generalized matter would radiate them in all directions. With radiant divergence the wave lengths would become longer, and their rate of movement slower. In the differing rates of vibrations we may trace not only the different forms of energy, but diverse results in material aggregations. Such may have been the origin of the specialization of energy and of matter which we behold in nature.

Such thoughts arise unbidden as a remote but still a legitimate induction from a study of the wonderful phenomenon of animal motion; a phenomenon everywhere present, yet one which retreats, as we pursue it, into the dimness of the origin of things. And when we follow it to its fountain head, we seem to have reached the origin of all energy, and it turns upon us, the king and master of the worlds.

XXI.

THE ORIGIN OF THE WILL.

I. THE DEFINITION OF THE WILL.

Definitions of the term "Will," as we find them in metaphysical writers, are not identical; and much apparent difference of opinion depends, as usual, on this diversity of statement. Locke regards the concepts "will" and "freedom" as entirely distinct in their nature, and not essentially related to each other. He says: "Freedom belongs as little to the will, as swiftness to sleep, or squareness to virtue. Freedom to do is one power, will to do is another: will, a power of the mind exerting dominion over some part of a man by employing it in or withholding it from any particular action; freedom, again, a power which a man has to do or to forbear doing any particular action." It appears that what Locke here denominates will is that common activity of the mind which expresses itself in action, which may be readily considered apart from the question of choice. This doubtless expressed something to the metaphysicians of that time, but merely signifies to the physiologist of the present day the movement derived from the metamorphosis of nutritive material in the arterioles of the brain, which when consciously performed are called thoughts and feelings, and are the necessary precursors of a class of muscular acts. The question of will properly so called is not yet entered on at this point. Dr. Willis * elaborates Locke's position in the following language: "But there is, in fact, no one particular primitive faculty that wills in the human mind; will is a general term, and belongs to and is expressive of the activity of each of the primitive faculties of our nature; the benevolent faculty being active, causes us to will to do good and charitable offices; the reverential faculty being active, to will to feel respectfully or reverently; the musical faculty active, to will to sing or hear music,

* "Benedict de Spinoza: his Life, Correspondence, and Ethics," 1870, p. 145.

etc. ; and the willing here is necessary ; but whether we yield to the impulse of the benevolent, reverential, or musical faculty, and indulge therein their various willings, is not so ; here we are free, and can yield or abstain as we list." This passage renders it the more clear that the latter part of Locke's statement, in which he defines freedom, is that in which he really refers to the will as generally understood ; and Dr. Willis's assertion of the existence of our ability "to yield or abstain as we list," grants all that the advocate of "the freedom of the will" could desire.

The modern automatic school only avoid discarding the term will altogether by using it in the sense of Locke's definition. They make it merely the conscious mental activity that precedes the act ; the direction of that activity being necessary in its character ; i. e., the result of impinging stimuli. In other words, on the automatic theory, the spontaneous activity of the body is directed or deflected by stimuli, whose ultimate form depends on the existing mental machinery through which they pass. There is avowedly no room for a self-determination in such a process, and its existence is therefore denied by this school. Inasmuch as a faculty of self-determination is what is here understood by the term will, and the question in the present article is whether there be or be not such a faculty, the inquiry to which we address ourselves is whether a human will exist or not. Says Dr. Carpenter :* "The psychologist may throw himself into the deepest waters of speculative inquiry in regard to the relation between his mind and its bodily instrument, provided that he trusts to the inherent buoyancy of that great fact of consciousness that *we have within us a self-determining power which we call will.*" The existence of such a faculty is in these words assumed by Dr. Carpenter, but I have looked in vain in his writings for a demonstration of the truth of this position. The same is true of the works of many other metaphysicians.

Will may be considered in two aspects : first, as a control over the origin of mental or bodily movements ; and second, as a control over the direction which those movements take. The latter case is the one chiefly considered here, as the one involved in customary definitions of human will.

It need scarcely be added that the concept will is an abstraction from supposed special exhibitions of it, and represents a supposed mental property.

* "Mental Physiology," p. 28.

II. THE NATURE OF ACTIONS.

The discussion between the advocates of the freedom of the will on the one hand, and those of the doctrine of necessity on the other, has often been obstructed by a *petitio principii,* which yields the case to the latter side at the outset. This is the dictum which has often passed unchallenged by both parties, that "human action is the product of the strongest inducement," or, otherwise stated, that "the will is the result of a balancing of opposing motives," or that "the will obeys the strongest motive." This is simply the statement, in reversed order, of what we might suppose without examination to be a general truth, viz.: that the motives which precede the acts which we observe are stronger than all others at the time. If this proposition be true without qualification, there is no further need of discussion, since it involves the negation of freedom, or of a power of choosing. But as such, it is an assumption in advance of a conclusion in the case under consideration; a begging of the question in a clear sense. Such a position can only be adopted as a result of the fullest investigation into the phenomena; it can not be accorded before examination into the facts.

But the statement may be admitted with this important qualification, by which the argument is transferred to another stage of the subject, viz.: that we do not thereby explain why the inducements to act thus and so, do, in many obvious cases, overbalance all others in a given human mind. This inquiry is not fruitless, so long as we have before us every day examples of men acting differently under identical circumstances. If there be any "liberty," it is exercised at the point of permitting inducements or motives of one kind to occupy the mind to the exclusion of those of another kind; and secondly, such occupation being granted, freedom might be exercised in removing restraint from the pressure of the present motive, so that the act can take place. If there be no inherent power of controlling the attention, and none of restraining the pressure of motive, then there is no will in any proper sense of the word, and man is an irresponsible automaton. The proof or disproof of this proposition must, however, be the end, not the beginning, of the discussion.

An inquiry into the origin of actions must be preceded by an examination into the nature of the acts themselves. The following classification is offered, as expressing as nearly as possible their relations to the general developmental position of active beings, without any pre-suppositions as to their automatic or voluntary character. It is necessarily assumed that all acts are performed with reference to the acquisition of pleasure or the avoidance of pain ; in other words, that all acts are due to motive, and are the expression of design on the part of the actor. This is as true of the simplest as of the most complex actions of animals, whether consciously or unconsciously performed. The movement of the *Amœba* in ingulfing a Diatom in its jelly, is as much designed as the diplomacy of the statesman, or the investigations of the student. And the motive may be the same in all three cases ; viz.: hunger. But as the unconscious acts have been probably derived from conscious ones by organization, a fundamental classification must first recognize their relations to the two necessary terms of consciousness, the subject and the object. All actions may then be divided into two classes ; those which are performed with the design of securing the pleasure of the subject, and those whose motive is to secure pleasure for the object as distinct from, i. e., opposed to, that of the subject. The tendencies thus defined have been named, in other connections, the *appetent* and the *altruistic*, and these names may be preserved as equally appropriate for the present purpose. Actions of the appetent class differ according to the developmental grade of the animal displaying them, or the grade of the organ of the body to which they are proper. In their simplest form they are mechanical movements, following a stimulus without the intervention of any rational process ; the end being attained by movements, whose directions are determined by mechanical or physical laws only. Such acts belong to the lowest type of animals, and are also seen in the organic functions of all animals ; they may be called the *anæsthetic* division. They may be performed consciously or unconscioasly. Acts of another order are those which, while due to stimuli, are directed by a process of ratiocination. They are higher than those of the previous order, because they successfully accomplish their object under changing circumstances, to which they adapt themselves as the others can not. Like them they may be performed in consciousness or in unconsciousness, or in a still higher state of the mind, that of self-consciousness. The last condition is only possible to animals of a high order of intelligence, since it not only demands an exercise of the rational faculty, with reference to objects, but also with reference to itself—the subject. These three groups form the *rational* order. The unconscious actions of both the anæsthetic and rational kinds are called "reflex;" and all of them are "automatic," in so far as they are performed without will; terms more fully defined in the following pages. The process of intellection in unconsciousness is called unconscious cerebration.

Actions of the second great class, the altruistic, demand for their performance the attributes necessary for the highest of the appetent class. They require intelligence enough for the perception of what is the pleasure of the object, and self-consciousness, to know that that pleasure is inconsistent with its own, or subjective pleasure.

The arrangement may be summarized as follows:

I. Appetent class.

1. Anæsthetic. { Unconscious (reflex). Conscious.

2. Rational. { Unconscious (reflex). Conscious. Self-conscious.

II. Altruistic class; rational and self-conscious.

Under the definition of will above given, it can not be present in unconscious or reflex actions, and the inquiry is limited to the conscious groups exclusively. It may then be well to add a few words on the nature of consciousness.

This faculty is here understood in its broadest sense, namely, subjective perception. The term consciousness expresses the knowledge by the subject of the effects of stimuli on itself, which ranges all the way from the mere sense of contact to the sense of an idea. An unprejudiced scrutiny of the nature of consciousness, no matter how limited that scrutiny necessarily is, shows that it is qualitatively comparable to nothing else. The attempts to correlate it with the physical forces have so far been utter failures, although the vital forces, to which it gives direction, are evidently not excluded from the laws of quality and quantity. The common hypothesis that consciousness is the product of evolution appears to the writer, in view of this primary fact, to be irrational; while the converse, that evolution is a product of consciousness, is far more likely to receive ultimate demonstration. From this stand-point it is looked upon as a state of matter which is co-eternal with it, but not co-extensive. Itself in its totality a reservoir of force, it is the source of all physical and vital forces, with which it has therefore an equivalency of quantities, but not of qualities. The cause of the difference between conscious and unconscious force must be secondarily due to different conditions of matter as to its atomic constitution; consciousness being only possible, so far as we can ascertain, to matter which has not fallen into fixed and automatic relations of its atoms. The condition appears to be one of tension, in which the automatic (crystalline) tendencies antagonize each other, the material being all the while in the metastatic condition of nutrition. This idea is parallel to that of Heraclitus, who held that the essence of all things lay in perpetual modification, a universal becoming, an eternal emergence and disappearance.

In accordance with the preceding views, the relations between consciousness and matter are thus depicted as of a mutually necessary character, the movements of conscious force involving consequences to itself from which the properties of matter necessarily preclude its escape.

If we trace the consciousness of animals to such an origin, it may be asked, Why have not such beings the powers and perfections of their source, in quality if not in quantity? The answer to this query, in view of the fact that they have not such qualities, is only to be found in an investigation of the nature of memory. The absence of memory of the past would be equivalent to ignorance; while a new material vehicle might render memory possible for the future, and thus education, under new surroundings, create diverse beings from a primal common source.

We must include in our estimation of the distributions of consciousness and forces not our planet alone, nor our system only, but the universe. Hence Sir William Thompson's idea, that consciousness ("life") was originally exotic to our earth, is an altogether permissible hypothesis.

If there be such a state of consciousness as will, it must have appeared in course of the evolution of animals, at some point in the series of the stages of progress through which their mind has passed. Yet it is maintained by some thinkers that the doctrine of evolution necessarily excludes the idea of freedom from the nature of the minds thus produced. The case is, however, involved in that of consciousness, and the investigation of it must proceed in the same manner. If it has been shown that will does exist in connection with evolution, we must proceed to discover, if possible, the relation between the two facts.

The proof of the existence of a freedom, power of choice, or will, is found in the origin of the altruistic class of acts, which are probably only possible to the human species. These have been above defined as those in which the pleasure of the object, as distinct from, and therefore opposed to, that of the subject, is the design of the act. This definition excludes acts for the benefit of others in which the actor is also gratified, since the motive may be in that case the pleasure of the subject. No doubt, many generous acts are of this character, but they were not such the first time they were performed, since experience of their pleasurable character had not then been acquired, and the evidence of all past experience was of a diametrically opposite character. In other words, the motives already organized in the mind of the subject were all in favor of the subject. The laws of evolution render the introduction of a new element of character at this point absolutely necessary. It is well known that the development of mind, and through it of all the acts of the first or appetent class, has been due to the pursuit of pleasure and avoidance of pain on the part of the subject. The pursuit of an opposite course, by animals whose pleasures and pains are those of the successful fulfillment of the necessary functions of life, or the reverse, would insure their speedy extinction. Their survival has been due to their prompt discrimination of favorable and unfavorable conditions through their sensibilities, and the human species, as the product of evolution, displays these sensibilities in their highest form. Under these circumstances it is obvious, since none but the inherited motives, with refinements due to more complex circumstances, can be found in his mind, that without the intervention of will, an altruistic act is impossible.

It has also been pointed out that such can only be performed by a being capable of the highest state of consciousness, i. e., self-consciousness; in other words, by a being capable of recognizing its own mental states. Under such circumstance only can it distinguish the mental states of a being apart from itself, toward whom the altruistic act is directed.

It being then conceded that will is exhibited in certain human actions, it becomes important to determine, if possible, the conditions under which it appeared in the course of the evolution of man.

III. THE ELEMENTS OF MIND.

Prior to considering the origin of states of mind, it is necessary to go over the well-trodden field of its original constitution.

There are three primary conditions of consciousness, which naturally grade into each other, viz.: pain, indifference, and pleasure. Consciousness is of one or the other of these types in all animals. The constant flow of activity, either in movements of the whole body or of particular parts of the body, has brought animals from their beginning into contact with other bodies, either at rest, or animated by active forces, as light and heat, which have varied their sensations, rendering them more positive in each of the three directions named. These sensations soon cease, leaving consciousness where it was, but not without marks of their former presence in the organism. They are recorded, and continue in unconsciousness so long as the organism remains unchanged. This is the first part of memory, i. e., retention. Under the influence of what is called cohesion, the impressions may be returned to consciousness in a less distinct form by the occurrence of new impressions which have some near relation with them as to time, place, or qualities of other kinds. This is the second part of memory, or reminiscence. The sum of the impressions which are necessary to memory constitutes experience. It is evident that reminiscence is pleasurable or painful, as the experiences recalled were pleasurable or painful. Another quality is rendered possible by the two faculties of retention and cohesion, viz.: classification. This consists of a re-arrangement of retained impressions in accordance with different kinds of cohesions, i. e., different kinds of likenesses. The products of classification may be brought into consciousness just as sensible impressions are revived ; but unlike these, they constitute in their totality a new experience of internal origin. When a cohesion between two circumstances is due to a repeated experience of the one as following the other, men entertain the idea that one is necessary to the other. From memory of the necessary results of our own activity, we have come to regard necessary sequences as the result of activity somewhere. If activity be discerned in the first of two coherent events, we regard it as a cause of the second : if the first be passive, the idea of cause does not arise in connection with it, but in some other active agent. Finally, all processes involving reminiscences are less distinct than the original impressions. Spencer calls the former *faint*, the latter *distinct ;* the faint order are the processes of reason ; the distinct, of perception.

Whether these processes are pleasurable, painful, or indifferent, depends on the characteristics of the reminiscences which are their subjects. As the reminiscence is less distinct than the original impression, so there comes to be, as pointed out by Spencer, a faint order of pleasures and pains, which, with the indifferent class, form the material of the processes of reason. These mental states of pleasurable and painful consciousness constitute that primary division of the mind, the feelings or affections, as distinguished from the intellect.

The feelings co-exist with intellectual operations of all grades of complication, since pleasures and pains are states which follow all kinds of activities, and therefore also reminiscences. To seek pleasure and to avoid pain constitutes the business of the lives of all conscious organisms; and hence the feelings, as derived from experiences, are the directive and often originative conditions of movements or actions. In animals with higher intellectual powers, the general classification of experiences of given objects or actions results in a higher order of the mental feelings, which are called likes and dislikes. When these forms of consciousness assume an intense condition due to stimuli, they become emotions or passions.

These details are entered into in order to show that the feelings in their various grades are the motives of action in all animals, from the *Amœba* to man. In the former they are mere reminiscences; in the latter they are so generalized as to become enduring principles of action, which put the intellect to every conceivable labor. And it is evident, from this foundation fact, how the intellect itself has been constructed. The activity stimulated by the feelings has resulted in new experiences, and the accumulation and elaboration of these into new combinations of the faint type of consciousness, has been the law of their development. This we can observe in the education of one generation of living animals, and it has doubtless been the law of the generations of the past as well. We may then review the probable method of development of mind through the ages of past time.

IV. THE DEVELOPMENT OF MIND.

In the first place, it is evident that the evolution of mind has been due to the activity of animal life. Although not asserted, it is sometimes implied that "circumstances," in which the animal is passive, have been the efficient cause of mental development. That this could have been the case is inherently impossible, and since animals of the lowest types possess powers of movement, their activity has necessarily been an immediate cause, while surrounding circumstances have exercised a controlling influence.

Animal activity must be traced to the generation of force or motion by a protoplasmic body which is constantly supplied with nutriment. This production of force is the basis of the acts of animals, up to and including man; often, as remarked by Prof. Bain, displaying itself in aimless discharges or playful movements, as in lambs and boys.

But we observe even in very simple organisms, as the *Amœba*, etc., that many movements are not aimless, nor without design. We observe that these mere specks of jelly devour nutritious substances, and reject the innutritious; that they even distinguish between the Diatom which contains protoplasm within its shell, and the empty shell. Many facts of this kind lead us to believe in the consciousness of these pioneers of life, and seem to show that they have experiences of the pleasures of nutrition, and of the pains of retaining insoluble substances in the seats of assimilation. It would also appear that they *remember* these sensations, so as to seek the pleasures and reject the pains, when in the course of their wanderings they again come into contact with the material objects which have caused them. Given these two terms, sensibility (consciousnsess) and memory, and we have the conditions by means of which the entire complex superstructure of the affectional and the intelligent acts has been elaborated out of mere movements.

This has been accomplished by the well-known laws of organization of habits, and heredity. The performance of an act under stimulus so modifies the structure of the bioplasts of the brain as to facilitate its repetition. With further repetition the organization is soon complete, and action follows the stimulus without direction, so long as no adverse influence affects the consciousness. Whether the act be one of the Rational or Anæsthetic orders, from this habitual stage it becomes automatic in the true sense, between which and the reflex act no line can be drawn. It is evident that the degree of consciousness present in animals will depend on the number of changes appearing in their surroundings, whether due to modifications of the external world, or movements of their own bodies. Inasmuch as the habits of the lower animals are few and simple, most of them must be automatically performed, the consciousness being only present at the commencements of the several

›rocesses. It is probable that the organization of mental functions vas at the first identical with the so-called organic functions,* so ar as they consist of mechanical movements ; and that the latter arly became reflex and devoid of consciousness, developing subequent forms through mechanical causes. The organization of nental functions, on the other hand, was by continued education, vhich requires the presence of consciousness at every step.

The well-known fact of the inheritance of mental qualities hows that the evolution of mind has advanced by a continued rocess of accumulation as the product of animal experiences. 'he new generation has inherited the organization of the old, and ll the reflex and automatic activities entailed by it, and has proeeded in proportion to its activity to acquire new experiences, abits, and organization.

V. THE BEGINNINGS OF DEVELOPMENT.

(*a*) *The Feelings.*

The intellectual faculties of every animal thus belong to two lasses : first, those which have been inherited ; and second, those 'hich it has acquired by its own experiences. Of course progress onsists in accessions to the latter class, since inheritance without ddition is mere repetition. If no acquisitions were made, or to e made, the mental, i. e., the cerebral organization inherited by nimals would continually repeat the form of their actions as unrringly as the nature of a machine gives the character to the novements propagated through its wheels and cranks. That uch the larger proportion of animal acts are of this class, that , *are automatic*, there can be no room to doubt. With an imetus to movement given, the strongest liking or disliking selects ıe object or direction, and the reason furnishes the mode of acuisition or avoidance. The known past teaches of the unknown ıture, and the established circle of the functions of life is fulled. But without acquisitions, development is impossible.

Acquisitions to experience are gained by movements of the ›dy, and hence by the mental activity to which the latter are ıe. But it is evident that the primary movement has precedice in the order of time over the feeling which deflects it, or e experience which directs later actions. In the lowest animal

* See "On Consciousness in Evolution," "Penn Monthly," 1875, Aug., where is view is held.

the first movement was doubtless a mere discharge of force; bu the first designed action, the appropriation of food, was due to sense of want or hunger, which is a form of pain. This was fol lowed by gratification, a pleasure, the memory of which consti tuted a motive for a more evidently designed act, viz.: pursuit.

These two mental states, the one painful, the other pleasurabl form the basis in the feelings of all appetent acts. The painfu sense of want is the motive to the performance of the primar class of actions, and the experience of pleasurable gratification fu nishes the motive for a class which must be regarded as secondar

The primary organized feelings of animals are not numerous. In man, the most highly developed, Prof. Bain enumerates only eleven types, and some of these he states may be further r solved. From the stand-point of the evolutionist this is evidentl necessary, and a corresponding reduction in number can be mad The development of the feelings has proceeded from the early b ginnings above described, in subsequent ages, *pari passu* with th of the intellect. It is necessary in the nature of things that should be so, since the finer and fuller the sensibility to pleasur and pains in all directions, the greater will be the complexity experience, and hence of intelligence. It is not practicable trace the history of the feelings here, but I allude briefly to o class of them—the social affections—as they have been treated b Herbert Spencer, whose contributions to this department knowledge have been very important.

This author maintains that the social affections are the produc in the department of mind, of the function of reproductio They are the organized products of experiences of pleasures d rived from fellow beings, just as other kinds of likes and dislik are derived from experiences of the qualities of various object It is sufficiently evident that this faculty must survive, and t social instincts become more and more refined or specialized. is a remarkable fact in the successional relations, and hence ev lution, of the *vertebrata*, that the only system that has accomp nied the nervous in its progress from generalization to speciali tion and perfection, is the reproductive. Man, standing at t head of the series by his developed brain, possesses also the mo specialized reproductive system. He is inferior to many oth *Mammalia* in his osseous and muscular type, and in his digesti

---

* "The Emotions and the Will," p. 36.

rgans, including dentition, etc., but the orders which are his uperiors in these respects yield to him the supremacy in the two ystems mentioned.

Functionally the two systems oppose each other, and that xercise of the one is at the expense of the other is a physiological ıw. Health of the individual, and persistence of the species, epend on the maintenance of the equilibrium between them. 'his is because success in obtaining food on the one hand depends n intelligence, and undue power can not be expended in other irections without starvation. Thus the law of evolution lends ıll support to the doctrine first formulated by Kant, of the dual ature of the human mind, in its division into the intellect and ıe affections.

(*b*) *In the Intelligence.*

The intellect includes a record of experiences of resemblances ıd differences, of causes and effects, arranged in orders of place, me, and of qualities of all kinds. The importance of an intellect epends on the number of experiences it contains; on the clear- ess with which qualities can be brought into consciousness; on ıe correctness with which the classification expresses the quali- es; on the relation which the qualities preferred bear to an ob- ct of pursuit; and on the rapidity with which any or all of these ınctions may be performed. The triumph of reason is foresight r predication, in which it brings into consciousness the unknown, y reproducing its experiences of the known. This is the serv- e rendered by education, by the acquisition either of experiences ıemselves, or of the experiences of others.

Acquisitions then do not imply a predication of the unknown om the known, but an actual addition to the stock of the known. he automatic life above described includes no such process, but a routine varying only in unimportant details, and changing in) great feature. Progress evidently depends on something be- des knowledge, for in proportion to the degree of progress is the parture from the known, and in proportion to the novelty of a tuation is experience worthless as a guide.

Designed actions which are performed without a basis of knowl- ge which is sufficient for predication are not automatic. That , while the activity may be physically spontaneous and compul- ry, the direction it takes and the mode of its execution can not automatic, unless the machinery which must give the direction, d which creates the mode, be already in existence.

The field of the known is very limited, as compared with tha of the unknown, in the experience of the *Amœba*. In its firs movements, it has absolutely no basis on which to establish a anticipation of the future. Such is also the situation of the youn of every animal. But the cases of the inferior and superior spe cies present the important difference that in the former ther exist few or no mental powers derived by inheritance, while i the latter such are present in proportion to the position of th species in the scale of intelligence.

The facts of evolution teach that the habits of animals hav been modified during past geological ages, under the influence c changes in their physical surroundings. While these changes may perhaps, have furnished the stimuli to the adoption of new habit the conditions have not often been so rigid as to define exactl what those habits should be, in some or all of their details. Th animal has necessarily proceeded blindly in many instances; i others, his mental darkness has been illumined by a low grade c imagination. This may be believed in view of the many attemp which animals often make before succeeding in attaining a desire end. Imagination plays an important part in the origin of m tives and of actions, and is related to predication. It is define as the presentation or construction of images or representatio from items of experience, which representations so far differ i the connection of their details from actual experience, or so f lack the qualities of experiences, as not to constitute a predic tion of future events. Predication may be defined as the certa knowledge of the unexperienced from the experienced; whi imagination includes the grades of probable, possible, and impo sible concepts, constructed from the same material as predicatio Whether this faculty exists in the animals which can not spea is not readily ascertained; but, inasmuch as many of them pred cate, it is probable that they possess some degree of imaginati also. But it is obviously a quality of the highest types of min since its development depends primarily on the furniture of me ory, derived from a long period of experience, whose amount d pends on receptivity and retentiveness.

VI. THE ORIGIN OF MOTIVES.

It has been said that the operation ordinarily called choosin in which the will is popularly supposed to be free, consists mere of a sum in addition and subtraction, where various inducemen are balanced, the resultant preponderance being expressed in the act. It will be easily seen that while this statement is true in regard to cases where the elements of the calculation are known, it is not true where any or all of them are unknown. The difference in the two cases is very great. All likes and dislikes are based upon experience or knowledge; and when there is no knowledge, likes and dislikes can not be said to exist. Since likes and dislikes constitute motives, where the former are wanting the latter are also wanting. Whatever inducements are presented from beyond the field of knowledge are derived from the imagination, and are in self-conscious minds relatively weak as motives, or absolutely without weight. They might be regarded as motives in embryo, ready to become such on the acquisition of a corresponding experience. The imagination can prefigure one alternative as well as another, in a direction where experience is wanting, and might indeed be said under such circumstances to have no existence, and the expression, "I can't imagine," be thought to have foundation in fact. The influence of such a guide is not imperative, and raises no obstacle to the origin of a new feature of consciousness by an act of choosing, when the pressure to act at all is sufficiently great.

There is, perhaps, but one situation of the mind where the pressure of feeling is strong enough, and predication and imagination sufficiently excluded, to develop a will which shall create motives rather than obey them. This is in the cases where self-interest is weighed in the balance against the interest or good of other people. Here the feelings are most severely pressed, and the future results to self most uncertain. Self-sacrifice may be beneficial to self, or it may not: one may be the gainer by the general prosperity, or he may be the loser. Morality may promise future good to the community, but why sacrifice self for the community? Gratitude for services rendered is an uncertain anticipation. Man's most limited knowledge and greatest inability in predication is in the field of human motives and actions, and chiefly in respect to those which belong to his moral feelings. As already remarked, the complication in this direction is so great as to produce the effect of novelty: so that man, come into possession of an intellect which is the product of ages of development, finds before him a new field of his own making, where his inherited powers fail.

This is the field where the most momentous decisions possible in human life are made. Since questions of right and wrong relate to the happiness of men in their relations to each other, the social affections are the stronghold of the motives that bear on this result. It is evident that a thousand subordinate motives take their direction from the primary decisions between these two original alternatives of feeling.

It is true that the predication of human actions, necessitated by bodily functions alone, is easy, even when they come to be of a highly complex character, as in the mercantile transaction of a populous business center. But so soon as the ethical element enters into the calculation the difficulty is greatly increased, and with the majority of men predication ceases, and faith begins. This is illustrated in the many credit transactions, without which it is well known that trade on any but the most limited scale is impossible. So it must be admitted that many men practice faith in many affairs, and that this faith is chiefly reposed in the moral excellence of other men. Under these circumstances, that state of the affections arises in most men which is termed faith, and which is only present in the highest form of progressive action, whether the results of that action be beneficial or not. It is a condition of the affections, as imagination is a condition of the intellect. The lowest animal, when attempting a novel act in obedience to imperative stimuli, doubtless moves blindly, and adopts one of two or more alternatives through pure accident. In animals of a higher grade of intelligence, new situations are known to be such, and fear or suspicion is the usual result. Generally, animals of the higher orders do not adopt new habits excepting under severe pressure, and the majority of them have perished, in past geologic ages, on account of their inability to assume new modes of life. Nevertheless, in so far as an animal or a man ventures into an unknown field of action, where he is without the guidance of a past experience, he or it performs an act of trust in the broad meaning of the word. So far as this state of mind is known to the subject, the act is one of true faith in the restricted or proper sense of the word. Imaginations may and do assume to men the importance of truths, and in so far they are such to them. But in proportion as this is the case, faith in its proper sense is wanting, and the action following is automatic. The highest form of intellect is necessary to the highest form of faith, since it is only by a knowledge of the absence of knowledge that an act of faith is possible. In proportion to this knowledge of self is faith enlarged; in proportion to certainty, or supposed certainty, in affairs, is faith diminished in its scope.

It is evident then that, abstractly speaking, occasions must arise in human experience where a decision between two alternatives is dependent on choice alone. That these occasions have arisen, and the choice been made, is shown by the existence of the altruistic class of actions. The number of these occasions may not be very great, but the consequences are very important. In whatever direction these decisions are made, long series of automatic actions are organized.

Although the existence of the altruistic class of acts affords the clearest proof of the origin of will, it is not denied that corresponding situations may not occur in other directions. It is also probable that will, once organized as a faculty of consciouness, can be exercised in many acts in opposition to habits, differing in accordance with the constitution of the individual; and that it can be inherited like any other quality of mind. But I will show later, that the organization of altruistic habits has narrower limits than that of those of the appetent class, because self-preservation depends on the latter, and not on the former, so that the appetent qualities are more certain to be inherited and survive.

The conclusion of this portion of the subject is, that that department of mind called the feelings * is the primary source of action; that they act automatically, with or without the aid of the reasoning powers, when dealing with the known; but when dealing with the unknown may develop, in self-conscious beings, the state of faith and acts of will; that this freedom is born of tension of the affections and of inability of the intellect.

Thus have the irregular and fortuitous decisions of animals been replaced by volition, as the highest quality of the mind, and therefore the crown of evolution. No new "physical" force is here called into requisition. The determination of the direction of such forces already existing in or passing through the brain in executive action need not add to nor substract from them. Will is, under these circumstances, looked upon as developed consciousness.

All this is, of course, opposed by the doctrine of the origin of moral excellence by development, on the basis of the utilitarian theory of morals. I therefore proceed to a brief examination of its claims in this direction.

---

* Which appears to be identical with what Schopenhauer calls the will.

Good is well defined as the greatest happiness of the greatest number; and by a natural transfer, the term is applied to whatever is conducive to that object. It therefore includes not only present pleasures, but also the influences which conduce to future pleasures, and which may be sown in the mind long before they bring forth fruit. As present pleasures are not always consistent with greater ones in the future, so present pleasure is not always good. Evil being the reverse or negation of good, or happiness, is avoided by all beings to whom it is consciously known; but what they regard as evil will of course depend on their intelligence in determining or predicating the future effects of actions. But no matter what the degree of intelligence, no responsibility, as usually understood, can be expected of beings which have no power of choosing, or will.

The utilitarian theory of the evolution of morals asserts that the development of goodness is simply due to the discovery and enforcement of the law of self-protection and preservation. The selfish interests require the protection of person and property, without which a community is an impossibility. Law being thus established and enforced, moral habits are imposed upon men, which become incorporated into character and transmitted to succeeding generations. This is all doubtless true, but whether it is a fundamental or secondary truth is the point requiring attention.

The fundamental objection to this hypothesis is, that the altruistic affections are not inherited or transmitted. This is because the pains and penalties of wrong-doing as inflicted by law, can not (and ought not to) overcome the inherent instinct of self-preservation in man. It is true that moral character is inherited, and that changes in this department for better or worse are transmitted to offspring. The mental organization of a race may be improved by the weakening of the emotional or the strengthening of the rational faculties. But since the affections are at the foundation of all activity whatsoever, of right-doing as well as of wrong-doing, it is obvious that no amount of legal restraint can render them innocuous. Their existence is necessary for self-preservation, and law only restrains their activities to certain directions. That intelligence tends to restrain wrong-doing is true; but although intellect is inherited, the manner in which its teachings are applied in practice is not. Each man must learn the merits of different courses of action in regard to morals for himself; his intelligence places before him the facts, and shows him how to execute his wishes, but the state of his affections determines the direction of his acts. Moral amelioration has attended the progress of intelligence on the one hand, and moral abasement on the other. Intelligence is the condition of the perception of moral truth; in other words, intelligence, as applied to moral questions, is the conscience. Consequences of acts are understood, and their relations to the pleasures and pains of men are weighed. Thus, no doubt, the world has advanced in the knowledge of good and evil, and of right and wrong. That it has improved in the practice of right has not been due to the inheritance of respect for law, but to the self-destructive nature of wrong. That continued wrong sooner or later ends in the destruction of the wrongdoer, either from within or without, must be generally admitted. Thus is the truth of the doctrine of "the survival of the fittest" vindicated in moral as in natural law. But it is also true that this law is *restrictive* only, and that the school of Hume and Bentham has overlooked the deeper *originative* law in moral philosophy, as the school of Darwin has done in biological philosophy.

It may still be urged that, if it be granted that experience of the pains of evil-doing be not transmitted as an intellectual acquisition from generation to generation, nevertheless such experience is sufficient to educate each separate generation as it passes, without any other than automatic action on their part. It may be replied to this that the results thus obtained are not due to will, but simply follow compulsion, the motive thus created only varying in strength with the characters of the individuals. Its success is restricted to circumstances where the penalties are sufficiently certain to constitute counter-inducements to effect the necessary restraint. This can only be the case with the weaker members of society. Wherever there is sufficient power to escape penalties, wrong-doing has no restraint. Under such a system might and right are identical; for the strongest needs no protection of law. It is true that society can combine against a single malefactor, but it is also true that malefactors can combine. In fact, it is one of the usual phenomena of human society to find men becoming malefactors as soon as they attain to power; or to find society governed by a few malefactors who have an army to enforce their pleasure.

While then inheritance does not secure the performance of altruistic acts, appetent affections may be so increased by accumulation in descent as to become uncontrollable, so that will either does not come into existence, or is extinguished, so far as regards those affections. In such a situation there is no such equivalency between opposing motives as gives opportunity for the will, the experience of appetent pleasure being too strong to allow of hesitancy in the face of vague representations of imaginary consequences on the other side. Even in highly intelligent men, to whom consequences are best known, knowledge may be thrust from consciousness, by strong feeling in favor of one alternative at the moment of action.

VII. CONSEQUENCES.

It is now well to consider how far an automatic mind has any claim to personality or individuality, as generally understood. From the usual stand-point, a being without "liberty," or will properly so called, is without character, and is in so far a nonentity. Even the character of the Deity can not escape this destructive analysis; for according to Spinoza, if He is good, but a single line of action, without alternatives, lies open to God, if He be at the same time omniscient. All this is changed if the element of spontaneity in character be presupposed. The existence of such a quality in man renders foresight of its decisions in some cases no more than a calculation of chances, and in other cases impossible; thus offering the only conceivable limit to omniscience, and hence to omnipotence. And as we regard the goodness of God as the anchor of the universe, if that goodness be in some respect inconsistent with omnipotence, we are strengthened if we discover that there is ground for correcting our traditional suppositions in regard to the latter. Can we not find this ground in a liberty or freedom which is the condition of what we suppose, in the absence of knowledge, to be the characteristic of the highest class of conscious existences?

NOTE.—Another explanation of this conclusion of Spinoza's, quite apart from the question of human freedom, is to be found in some of the necessary properties of matter. The old theologians expressed this point obscurely in their phrase, "the intractability of matter." (Ed. 1886.)

SECOND NOTE. (Ed. 1886.)—The preceding essay was written several years prior to that on Catagenesis. In the latter article a mental quality was referred to as present in all animals, which constitutes "the will" of various writers, although it is not free. It is described in the following passage: "Why, from a purely mechanical point of view, is the adductor muscle of the right side of the horse's tail contracted to brush away the stinging fly from the right side of the horse's body, rather than the left adductor muscle? Why was the contraction-provoking energy deflected into the right interspinal motor nerves rather than into those of the left side?" The conclusion was that this phenomenon can not be explained on purely mechanical principles; but that it required the assumption of the thesis that "energy can be conscious." (*Supra*, p. 427.)

Further thought on this subject shows clearly that this conscious state of energy is self-directive in accordance with the lessons of experience, but without being necessarily free in the sense referred to in the preceding essay. In Mammalia the current of energy derived from stimulus, after reaching the cortical cells of the anterior cerebral lobes, issues from them in a direction appropriate to the needs of the animal, having undergone a deflection or determination within them. This capacity for "design" is peculiar to beings which can be conscious, and constitutes the most important proof in evidence of a degree of control of mind over matter. The evidence strongly suggests that this control is greatest when the environment is simple or generalized, since motives are then relatively simple.

NOTE TO CATAGENESIS BY THE AUTHOR, 1886. On p. 434 reference is made to the fact that "in chemical reaction the preference of energy is to create solid precipitates." This fact is but one illustration of the law of thermo-chemistry that "in any reaction, those bodies, the formation of which gives rise to the greatest development of heat, are formed in preference to others." This is a retrograde metamorphosis of energy, or catagenesis.

INDEX.

30

THE END.

THE PRIMARY FACTORS OF ORGANIC EVOLUTION

THE PRIMARY FACTORS

OF

ORGANIC EVOLUTION

BY

E. D. COPE, PH. D.

MEMBER OF THE U. S. NATIONAL ACADEMY OF SCIENCES; PROFESSOR
OF ZOÖLOGY AND COMPARATIVE ANATOMY IN THE
UNIVERSITY OF PENNSYLVANIA

CHICAGO
THE OPEN COURT PUBLISHING COMPANY
(LONDON: 17 JOHNSON'S COURT, FLEET ST., E. C.)
1896

The Lakeside Press
R. R. DONNELLEY & SONS CO., CHICAGO.

PREFACE.

THE present book is an attempt to select from the mass of facts accumulated by biologists, those which, in the author's opinion, throw a clear light on the problem of organic evolution, and especially that of the animal kingdom. As the actual lines of descent can be finally demonstrated chiefly from paleontologic research, I have drawn a large part of my evidence from this source. Of course, the restriction imposed by limited space has compelled the omission of a great many facts which have an important bearing on the problem. I have preferred the paleontologic evidence for another reason. Darwin and the writers of his immediate school have drawn most of their evidence from facts which are embraced in the science of œcology. Weismann and writers of his type draw most of their evidence from the facts of embryology. The mass of facts recently brought to light in the field of paleontology, especially in the United States, remained to be presented, and the evidence they contain interwoven with that derived from the sources mentioned.

Many of the zoölogists of this country, in common with many of those of other nations, have found reason for believing that the factors of evolution which were first clearly formulated by Lamarck, are really such. This view is taken in the following pages, and the book may be regarded as containing a plea on their behalf. In other words, the argument is constructive and not destructive. The attempt is made to show what we know, rather than what we do not know. This is proper at this time, since, in my opinion, a certain amount of evidence has accumulated to demonstrate the doctrine here defended, and which I have defended as a working hypothesis for twenty-five years.

In the following pages I have cited many authors who have contributed to the result, but it has been impossible to cite all who have written on one part or another of the subject. If some very meritorious essays have not been cited, it has been generally because I have confined myself to those in which facts or doctrines were first presented, and have not had so much occasion to refer to those of later date.

Mr. Romanes, in his posthumous book, Volume II. of his *Darwin, and After Darwin* on *Post-Darwinian Questions*, expresses the following opinion of the position which has been taken by the Neo-Lamarckians of this country. He says that they do not distinguish between the "statement of facts in terms of a proposition, and an explanation of them in terms of causality." Had Mr. Romanes been acquainted with the literature of the subject published in America and elsewhere during the last three years, he would have had reason to change this view of the case. I think he would have found in it demonstration "in terms of causality."

At the outset it must be stated that a knowledge of the history of organic evolution rests primarily on the science of morphology, and secondarily on the kinematics of the growth of organic structures. The phenomenon to be explained is the genealogical succession, or phylogeny of organisms; and the access to this subject is through the sciences of paleontology and embryology. The phenomena of the functioning of the organism, or physiology, are only incidentally referred to, as not the real object of inquiry. Since organic species are much more numerous than the tissues of which they are composed, organogenesis must claim attention more largely than histogenesis. It is true that histogenesis is fundamental, but it is a science as yet in its early infancy, and little space can be given to it. The exact *how* of organic evolution will never be finally solved, however, until our knowledge of histogenesis is complete.

The research depicted in the following pages has proceeded on the assumption that every variation in the characteristics of organic beings, however slight, has a direct efficient cause. This assumption is sustained by all rational and philosophical considerations. Any theory of evolution which omits the explanation of the causes of variations is faulty at the basis. Hence the theory of selection cannot answer the question which we seek to solve, although it embraces an important factor in the production of the general result of evolution.

In the search for the factors of evolution, we must have first a knowledge of the course of evolution. This can only be obtained in a final and positive form by investigation of the succession of life. The record of this succession is contained in the sedimentary deposits of the earth's crust, and is necessarily imperfect. Advance in knowledge in this direction has, however, been very great of recent years, so that some parts of the genealogical tree are tolerably or quite complete. We hope reasonably for continued progress in this direction, and if the future is to be judged of by the past, the number of gaps in our knowledge will be greatly lessened. In the absence of the paleontologic record, we necessarily rely on the embryologic, which contains a recapitulation of it. The imperfections of the embryonic record are, however, great, and this record differs from the paleontologic in that no future discovery in embryology can correct its irregularities. On the contrary every paleontologic discovery is an addition to positive genealogy. If the present work has any merit, it is derived from the fact that the basis of the argument is the paleontologic record.

E. D. COPE.

PHILADELPHIA, November 1, 1895.

CONTENTS.

PART II. THE CAUSES OF VARIATION.

PART III. THE INHERITANCE OF VARIATION.

LIST OF ILLUSTRATIONS.

INTRODUCTION.

THE doctrine of evolution may be defined as the teaching which holds that creation has been and is accomplished by the agency of the energies which are intrinsic in the evolving matter, and without the interference of agencies which are external to it. It holds this to be true of the combinations and forms of inorganic nature, and of those of organic nature as well. Whether the intrinsic energies which accomplish evolution be forms of radiant or other energy only, acting inversely as the square of the distance, and without consciousness, or whether they be energies whose direction is affected by the presence of con sciousness, the energy is a property of the physical basis of tridimensional matter, and is not outside of it, according to the doctrine we are about to consider.

As a view of nature from an especial standpoint, evolution takes its place as a distinct science. The science of evolution is the science of creation, and is as such to be distinguished broadly from the sciences which consider the other operations of nature, or the functioning of nature, which are not processes of creation, but processes of destruction. This contrast is especially obvious in organic evolution, where the two processes go on side by side, and are often closely intermingled, as for instance in muscular action, where both destruction of proteids and growth of muscular tissue result from the same acts, or use. Physiology, or the science of functions, concerns itself chiefly with destruction, and hence physiologists are especially prone to be insensible to the phenomena and laws of progressive evolution. The building of the embryo, remains a sealed book to the physiologist unless he take into account the allied biological science of evolution, as resting on the facts of botany, zoölogy, and paleontology. In his reflections on the relations of mind to matter he is likely to see only the destructive functioning of tissue, and not the history of the building of the same during the ages of geological time.

J. B. P. A. Lamarck[1] thus contrasts the theories of direct creation, and creation by evolution. The former asserts: "That nature or its author in creating animals has foreseen all possible kinds of circumstances in which they may have to live, and has given to each species a permanent organization as well as a predetermined form, invariable in its parts; that it forces each species to live in the place and the climate where one finds them, and to preserve there the habits which it has." He then states his own, or the evolutionary, opinion to be: "That nature in producing successively all species of animals, commencing with the most imperfect or simple, and terminating its work with the most perfect, has gradually complicated their organization; and these animals spreading themselves gradually into all habitable regions of the globe,—each species has been subjected to the influence of the circumstances in which it is; and these have produced the habits which we observe, and the modifications of

[1] *Philosophie Zoologique*, Paris, 1809, Vol. I., Chap. VII.

its parts." On an earlier page of the same chapter, Lamarck thus formulates the laws of organic evolution, to which his name has been attached.

First law. "In every animal which has not passed the time of its development, the frequent and sustained employment of an organ gradually strengthens it, de velops and enlarges it, and gives it power proportional to the duration of its use; while the constant disuse of a like organ weakens it, insensibly deteriorates it, progressively reduces its functions, and finally causes it to disappear."

Second law. "All that nature acquires or loses in individuals by the influence of circumstances to which the race has been exposed for a long time, and in consequence of the influence of the predominate employment of such an organ, or of the influence of disuse of such part, she preserves by generation, in new individuals which spring from it, providing the acquired changes be common to both sexes, or to those which have produced new individuals."

We have here a theory of the origin of characters; viz., of the increased development or loss of parts as a result of use or disuse. We have also the theory that the peculiarities thus acquired are transmitted to the succeeding generation by inheritance.

The next formal statement of the efficient cause of organic evolution was presented by Messrs. Charles Darwin and Alfred R. Wallace in 1859.[1] The cause assigned is natural selection, and Mr. Darwin thus states what is meant by this expression in his work *The Origin of Species*.[2] "If under changing conditions of life organic beings present individual differences

[1] *Proceedings of the Linnean Society of London.*

[2] Ed. 1872, p. 102.

in almost any part of their structure, and this cannot be disputed; if there be, owing to their geometrical rate of increase, a severe struggle for life at some age, season, or year, and this certainly cannot be disputed; then considering the infinite complexity of the relations of all organic beings to each other and to their conditions of life, causing an infinite diversity of structure, constitution, and habits, to be advantageous to them, it would be a most extraordinary fact if no variations had ever occurred useful to each being's own welfare, in the same manner as so many variations have occurred useful to man. But if variations useful to any organic being ever do occur, assuredly individuals thus characterized will have the best chance of being preserved in the struggle for life; and from the strong principle of inheritance, these will tend to produce offspring similarly characterized. This principle of preservation, or the survival of the fittest, I have called natural selection. It leads to the improvement of each creature in relation to its organic and inorganic conditions of life; and consequently in most cases, to what must be regarded as an advance in organization. Nevertheless, low and simple forms will long endure if well fitted for their simple conditions of life."

It is readily perceived that this statement makes no attempt to account for the origin of variations, but that it simply formulates, as observed by Mr. Darwin, the doctrine of survival of such variations as are most useful to their possessors. This fact is more distinctly pointed out in the same work (p. 63) where the author remarks: "Several writers have misapprehended or objected to the term natural selection. Some have even imagined that natural selection induces variability, whereas it implies only the preservation of such variations as arise and are beneficial to the being under its conditions of life. No one objects to agriculturists speaking of the potent effects of man's selection, and in this case the individual differences given by nature, which man for some object selects, must of necessity first occur." It is evident then that Mr. Darwin did not attempt to account for the origin of variations, but that the service rendered by him and by Mr. Wallace to the doctrine of evolution consists in the demonstration of the reality of natural selection. Darwin also assumes in the statement first quoted above, the inheritance of acquired characters.

In 1865 the *Principles of Biology* of Herbert Spencer appeared. In this work the attempt is made to set forth the laws of organic evolution, in a way which represents an advance beyond the positions of his predecessors. He adopts and harmonizes both the Lamarckian and Darwinian doctrines, and is at times more specific in his application of Lamarck's doctrine of the stimulus of the environment, and of use, than was Lamarck himself. Very often, however, Spencer contents himself with generalities; or takes refuge in the "instability of the homogeneous," as an efficient cause. This phrase, however, like his other one, "the unknowable," is but a makeshift of temporary ignorance, and is neglected by Spencer himself, when he can see his way through it. He approaches the cause of the varied forms of leaves of plants in this language:[1] "And it will also be remembered that these equalities and inequalities of development correspond with the equalities and inequalities in the incidence of forces." Language of similar significant but rather indefinite import is frequently used throughout this volume.

[1] *The Principles of Biology*, by Herbert Spencer, Amer. Ed., 1873, II., p. 143.

But in some cases Spencer is more specific. With reference to the inequality in the basal lobes of the erect leaves of Tilia and other plants, he says:[1] "A considerable deviation from bilateral symmetry may be seen in a leaf which habitually so carries itself that the half on the one side of the midrib is more shaded than the other half. The drooping branches of the lime show us leaves so arranged and so modified. On examining their attitudes and their relations one to another, it will be found that each leaf is so inclined that the half of it next the shoot grows over the shoot and gets plenty of light; while the other half so hangs down that it comes a good deal into the shade of the preceding leaf. The result is that having learned which fall into these positions, the species profits by a large development of the exposed halves; and by survival of the fittest acting along with the direct effect of extra exposure, this modification becomes established." In his discussion of the origin of the characters of animals, Spencer is also sometimes specific. Respecting the development of muscular insertions he remarks:[2] "Anatomists easily discriminate between the bones of a strong man and those of a weak man by the greater development of those ridges and crests to which the muscles are attached; and naturalists on comparing the remains of domesticated animals with those of wild animals of the same species, find kindred differences. The first of these facts shows unmistakably the immediate effect of function on structure, and, by obvious alliance with it, the second may be held to do the same, both implying that the deposit of dense substance capable of great resistance habitually takes

1 *Op. cit.*, p. 143.

2 *Loc. cit.*, p. 200.

place at points where the tension is excessive." Quite as specific is his ascription of the forms of epithelial cells to definite causes, as follows:[1] "Just the equalities and inequalities of dimensions among aggregated cells, are here caused by the equalities and inequalities among their mutual pressures in different directions; so though less manifestly, the equalities and inequalities of dimensions among other aggregated cells, are caused by the equalities and inequalities of the osmatic, chemical, thermal, and other forces besides the mechanical, to which their different positions subject them."

In spite of this not infrequent definiteness, Mr. Spencer occasionally falls into the error of ascribing the origin of structures to natural selection, as in the case of the forms of flowers,[2] and the armor-plates of paleozoic fishes.[3] Spencer assumes the inheritance of acquired characters throughout.

In 1866 Haeckel's *Schöpfungsgeschichte* appeared. In this work the author presents a mass of evidence which sustains the doctrine of evolution, and he combines the views of Lamarck and Darwin into a general system. He says:[4] "We should, on account of the grand proofs just enumerated, have to adopt Lamarck's theory of descent for the explanation of biological phenomena, even if we did not possess Darwin's theory of selection. The one is so completely and directly proved by the other, and established by mechanical causes, that there remains nothing to be desired. The laws of inheritance and adaptation are universally acknowl-

1 *Op. cit.*, p. 260.

2 *Op. cit.*, p. 153.

3 *Op. cit.*, p. 288.

4 *The History of Creation*, Amer. Edition, II., p. 355.

edged physiological facts, the former traceable to propagation, the latter to the nutrition of organisms." Apart from the statement that adaptation is traceable "to the nutrition of organisms," we find nothing in Haeckel's earlier writings which attempts the explanation of the origin of variations, beyond the general position assumed by Lamarck. The distinctive merit of Haeckel is his formulation of phylogeny. Much of this was speculative at the time he wrote, but so far as the Vertebrata are concerned, it has been largely confirmed by subsequent discovery.

Up to this period, the form in which the doctrine of evolution had been presented, was general in its application; that is, without exact reference to the structural definitions of natural taxonomic groups. No attempt was made to show the modes of the origin of any particular class, order, or genus, and only in the most general way in the case of a few species, by Mr. Darwin. Phylogeny was untried, except by Haeckel; and this distinguished author did not attempt to account specifically for the origins of the divisions whose filiations he set forth.

In the year in which Haeckel's work above cited appeared, Professor Hyatt of Boston and myself took the first step towards the formulization of a rational theory of the origin of variation, which should accord with specific examples of taxonomy. Quite independently, we selected the simple series presented by the characters of genera in their natural relations, Hyatt in the cephalopodous Mollusca, and I in the Batrachia Salientia. It is probable that Hyatt's[1] article was published shortly before mine. He says of the genera of Cephalopoda: "In other words, there is an increasing

[1] *Memoirs Boston Society Natural History*, 1866, p. 193.

concentration of the adult characteristics in the young of higher species and a consequent displacement of other embryonic features which had themselves also previously belonged to the adult period of still lower forms." My own language is:[1] "That the presence, rudimental condition, or absence of a given generic character can be accounted for on the hypothesis of a greater rapidity of development in the individuals of the species of the extreme type, such stimulus being more and more vigorous in the individuals of the types as we advance towards the same, or by a reversed impulse[2] of development, where the extreme is characterized by absence or 'mutilation' of characters." The phenomena of the aggregation of characters in progressive evolution, and the loss of characters in retrogressive evolution, were termed by me acceleration and retardation in an essay published in 1869.[3] In these papers by Professor Hyatt and myself is found the first attempt to show by concrete examples of natural taxonomy, that the variations that result in evolution are not multifarious or promiscuous, but definite and direct, contrary to the method which seeks no origin for variations other than natural selection. In other words, these publications constitute the first essays in systematic evolution that appeared. By the discovery of the paleontologic succession of modifications of the articulations of the vertebrate, and especially mammalian skeleton, I first furnished an actual demonstration of the reality of the Lamarckian factor of use, or mo-

[1] *Transactions American Philosophical Society*, 1866, p. 398; reprinted in *The Origin of the Fittest*, p. 92.

[2] The expression "reversed" is unfortunate, *diminished* being the proper word to convey the meaning intended.

[3] *The Origin of Genera*, Philadelphia, 1869.

tion, as friction, impact, and strain, as an efficient cause of evolution.[1] This demonstration led me to the necessary inference that when the agency directive of motion is consciousness, this also has been an important factor of evolution, in demonstration of the supposition of Erasmus Darwin.[2] Hyatt has demonstrated first on paleontologic evidence, the inheritance of a mechanically acquired character. Important contributions to corresponding histories of the Mollusca have been made by Hyatt,[3] Dall,[4] Jackson,[5] and Beecher.[6] Many other contributions, into which the paleontologic evidence does not enter, have also been made by various authors in Europe and America.

The authors quoted up to this point had all assumed that the progress of evolution depends on the inheritance by the offspring of new characters acquired by the parent, and had believed that such is the fact in ordinary experience. In 1883, Weismann, in an essay on heredity, announced the opinion that characters acquired by the body could not be transmitted to the reproductive cells, and could not therefore be inherited. This doctrine rests on the relation of the germ-cells to those of the rest of the body, which is expressed in the following language of his predecessor Jaeger: "Through a great series of generations the

[1] "The Origin of the Hard Parts of Mammalia," *American Journal of Morphology*, 1889, p. 137.

[2] *Origin of the Fittest*, 1887, p. 357.

[3] "Phylogeny of an Acquired Characteristic," *Proc. Amer. Philosophical Society*, 1893, p. 349; "The Genesis of the Arietidæ," *Memoirs Mus. Compar. Zoölogy*, Cambridge, Mass., 1889, XVI., No. 3.

[4] Dall, W. H., "The Hinge of Pelecypods and Its Development, *Amer. Jour. Sci. Arts*, 1889, XXXVIII., p. 445.

[5] Jackson, R. T., "Phylogeny of the Pelecypoda, the Aviculidæ, and Their Allies," *Memoirs Boston Society Natural History*, 1890, IV., p. 277.

[6] *American Journal Sci. Arts*, 1893.

germinal protoplasm retains its specific properties, dividing in every reproduction into an ontogenetic portion and a phylogenetic portion, which is reserved tc form the reproductive material of the mature offspring. This reservation of the phylogenetic material I described as the continuity of the germ-protoplasm. . . . Encapsuled in the ontogenetic material the phylogenetic protoplasm is sheltered from external influences, and retains its specific and embryonic characters." In other words, the reproductive cells are removed from the influence of those stimuli which affect and effect growth in the cells of the other parts of the body, so that no character acquired by the rest of the body can be inherited. The bearing of this theory on evolution is thus stated by Weismann:[1] "The origin of hereditary individual variations cannot indeed be found in the higher organisms, the metazoa and metaphyta, but is to be sought for in the lowest, the unicellular." "The formation of new species, which among the lower protozoa could be achieved without amphigony (sexual union), could only be attained by means of this process in the metazoa and metaphyta. It was only in this way that hereditary individual differences could arise and persist." In other words, variation in organic beings above the unicellular forms, has been and is, introduced only by sexual reproduction.

The conclusions of Weismann were derived principally from embryologic research, and his disciples have been chiefly recruited from embryologists. These conclusions have been supported by extensive and exhaustive investigations, which have added greatly to our knowledge of the subject. In order to account for

[1] *Essays*, p. 296. For a complete account of Weismann's views, see *The Germ-Plasm*, 1893.

the appearance of characters in the embryonic succession, through influences confined to the germ-plasma, Weismann invented a theory which requires the presence of distinct molecular aggregates within it, which represent the potentialities or causes. To these he has given the names of ids, idants, determinants, etc. As Weismann's contribution to evolution has been confined to the department of heredity, I will consider it more particularly in the third part of this book, which is devoted to that subject.

Weismann has, however, subsequently modified his views to a considerable extent. He has always admitted the doctrine of Lamarck to be applicable to the evolution of the types of unicellular organisms. His experiments on the effect of temperature on the production of changes of color in butterflies, showed that such changes were not only effected, but were sometimes inherited. This he endeavors to explain as follows.[1] "Many climatic variations may be due wholly or in part, to the simultaneous variation of corresponding determinants in some parts of the soma and in the germ-plasm of the reproductive cells." This is an admission of the doctrine which in 1890 I called Diplogenesis,[2] and which is adopted in the present work. It appears to have been first propounded by Galton in 1875. In the chapter on Heredity I hope to offer some reasons for believing that the suggestion of Galton embraces the true doctrine of heredity.

From what has preceded, two distinct lines of thought explanatory of the fact of organic evolution may be discerned. In one of these the variations of organisms which constitute progressive and regressive

1 *The Germ Plasm*, Contemporary Science Series, 1893, p. 406.

2 *American Naturalist*, Dec., 1889; published in 1890.

evolution appear fortuitously, and those which are beneficial survive by natural selection, while those which are not so, disappear. Characters both beneficial and useless or harmless, which are acquired by the adult organism, are not transmitted to the young, so that no education in habit or structure acquired by the adult, has any influence in altering the course of evolution. This is the doctrine of Preformation. From this point of view the cause of the variations of organisms has yet to be discovered.

The other point of view sees in variation the direct result of stimuli from within or without the organism; and holds that evolution consists of the inheritance of such variations and the survival of the fit through natural selection. This is the doctrine of Epigenesis. To this I would add that in so far as sensations or states of consciousness are present, they constitute a factor in the process, since they enable an organism to modify or change its stimuli. The position of each of these schools on each of the questions to which reference has been made, may be placed in opposition as follows:

1. Variations appear in definite directions.	1. Variations are promiscuous or multifarious.
2. Variations are caused by the interaction of the organic being and its environment.	2. Variations are "congenital" or are caused by mingling of male and female germ-plasmas.
3. Acquired variations may be inherited.	3. Acquired variations cannot be inherited.
4. Variations survive directly as they are adapted to changing environments. (Natural selection.)	4. Variations survive directly as they are adapted to changing environments. (Natural selection.)

5. Movements of the organism are caused or directed by sensation and other conscious states.	5. Movements of organism are not caused by sensation or conscious states, but are a survival through natural selection from multifarious movements.
6. Habitual movements are derived from conscious experience.	6. Habitual movements are produced by natural selection.
7. The rational mind is developed by experience, through memory and classification.	7. The rational mind is developed by natural selection from multifarious mental activities.

It is not the object of the present book to present all the available evidence on both sides of each of the questions above enumerated. I propose merely to submit certain facts, in support of the doctrines contained in the left-hand column of the above table. My aim will be to show in the first place, that variations of character are the effect of physical causes; and second, that such variations are inherited. The facts adduced in support of these propositions will be necessarily principally drawn from my own studies in the anatomy, ontology, and paleontology of the Vertebrata. It will be my aim, moreover, to co-ordinate the facts of evolution with those of systematic biology, so that the result may be as clearly presented as possible. The failure to do this by the founders of evolutionary doctrine has given their work a lack of precision, which has been felt by systematic biologists. The detailed application of the principles of Lamarck and Darwin has been the work of their successors, and has necessarily thrown much new light on the principles themselves. In pursuing the object above stated, I shall be obliged to consider briefly in the following pages, the validity of the general propositions on which the doctrine of evolution rests. Less space will be given, however, to those which are less relevant, than to those which are more relevant to the doctrine of neo-Lamarckism.

PART I.

THE NATURE OF VARIATION.

PRELIMINARY.

THE structural relations of organisms may be expressed in the following canons :[1]

1. *Homology.*—This means that organic beings are composed of corresponding parts; that the variations of an original and fixed number of elements constitute their only differences. A part large in one animal may be small in another, or *vice versa*; or complex in one and simple in another. The analysis of animals with skeletons, or Vertebrata, has yielded several hundred original elements, out of which the twenty-eight thousand included species are constructed. The study of homologies is thus an extended one, and is far from complete at the present day.

2. *Successional Relation.*—This expresses the fact that species naturally arrange themselves into series in consequence of an order of excess and deficiency in some feature or features. Thus species with three toes naturally intervene between those with one and four toes. So with the number of chambers of the heart, of segments of the body, the skeleton, etc. There are greater series and lesser or included series, and mistakes are easily made by taking the one for the other.

[1] *Origin of the Fittest*, p. 6. The laws here stated are as expressive of the relations of plants as of animals.

3. *Parallelism.*—This states that all organisms in their embryonic and later growth pass through stages which recapitulate the successive permanent conditions of their ancestry. Hence those which traverse fewer stages resemble or are *parallel* with the young of those which traverse more numerous stages. This is the broad statement, and is qualified by the details.

4. *Teleology.*—This is the law of fitness of structures for their special uses, and it expresses broadly the general adaptations of an animal to its home and habits.

The first and fourth of the laws above enumerated are taken for granted as generally accepted, and are not especially considered in the following pages. The second law, that of successional relation, is discussed and illustrated under the two heads of Variation and Phylogeny; the first expressing contemporary relations, and the second, successive relations in time. The third law, or that of parallelism, is considered in a chapter devoted to that subject.

CHAPTER I—ON VARIATION.

PRELIMINARY.

ALL species are not equally variable. Some species vary little or not at all, even under domestication. Thus the varieties of the turkey (*Meleagris gallopavo*) and the guinea-fowl (*Numida meleagris*) are few, and are confined to albinistic or melanistic conditions. The barnyard fowl (*Gallus sp.*), on the other hand, varies enormously, as does also the pigeon (*Columba livia*). Among domesticated Mammalia the variations of cats (*Felis domestica*) are few as compared with those of dogs (*Canis sp.*). Variability is not peculiar to domesticated animals. A large proportion of animals and plants are, in a state of nature, variable, and some of these are much more so than others. The common garter-snake (*Eutænia sirtalis*) varies exceedingly, while the variations of the allied ribband snake (*Eutænia saurita*) are minute or none. But little variation has been observed in the polar bear (*Ursus maritimus*), while the common bear (*U. arctos*) presents many varieties. Similar conditions are found among fishes. Thus the larger species of pike, the muskallonge (*Lucius nobilior*), the pike (*L. estor*), and the pickerel (*L. vermiculatus*) are constant in their characters, while the small pickerel (*L. vermiculatus*) presents numerous varieties.

Many of the varieties of the animals referred to inhabit the same territory, although some are restricted to particular regions. Of geographical varieties or races much is known. As a rule, all widely distributed species present them. Examples are the brown bear of the Northern Hemisphere (*Ursus arctos*); the cobra di capello snake of the warmer parts of Asia (*Naja tripudians*), and that of Africa (*Naja haje*). In North America the king-snake (*Ophibolus getulus*) and the milk-snake (*Osceola doliata*) are represented by distinct races in different regions. On the other hand, the copperhead (*Ancistrodon contortrix*) and the Eastern rattlesnake (*Crotalus horridus*), which have a wide range, scarcely vary at all. The chub (*Hybopsis biguttatus*) is an example of a fish distributed everywhere east of the Rocky Mountains, which presents scarcely any variation.

Variations are not promiscuous or multifarious, but are of certain definite kinds or in certain directions. Thus amid all their varieties, dogs never present black cross-bands on the back like those of the dog-opossum (*Thylacinus cynocephalus*) of Tasmania, nor do they present ocellated spots like those of the leopard, nor longitudinal stripes like those of certain squirrels. The same is true of the many varieties of cattle (*Bos taurus*), and of numerous other mammalia. Domestic fowls never vary to blue or green, colors which are common to many other birds; nor are canaries known to produce blue or red natural sports. All variations are in the first place necessarily restricted by the existing characters of the ancestor; but beyond this it is evident that other conditions determine the nature of the variation. It is not supposable, for instance, that the pale tints of animals which live in dry regions originated by an accident or without a determining cause. The increased amount of dark pigment observed in animals which dwell in especially humid regions must have a corresponding cause, and it is naturally to be supposed, of a kind the opposite of that which produces the pale colors.

I shall adduce some illustrations which show that color variations in species, as well as structural variations in higher groups, have appeared in certain definite series, and observe a successional relation to each other, which may or may not coincide with geographical conditions. The same relation is observed in the order of appearance of variations on the body.

Eimer and Weismann have shown that the gradual modification of color markings has originated in lizards and in caterpillars at the posterior end of the body and has gradually extended forwards. This has been discovered both by comparisons of the variations of the adults, and by studies of the order of their appearance in ontogenetic growth. Eimer shows that longitudinal bands have been produced in some animals by the confluence of spots placed in transverse series, which themselves are the remains of interrupted transverse bands. Thus he believes that the spots of the leopard group of the large cats were derived from the breaking up of transverse bands of the character of those now possessed by the tiger. The uniform coloration of the lion is the result of the obliteration of the spots. Traces of these spots may be distinctly seen in lions' cubs.

In plants variation is said to be equally definite by Henslow. He says: "In 1847 Professor J. Buckman sowed the seed of the wild parsnip in the garden of the Agricultural College at Cirencester. The seeds began to vary, but in the same way, though in different degrees. By selecting seed from the best rooted plants the acquired 'somatic' characters of an enlarged root, glabrous leaves, etc., became fixed and hereditary; and the 'Student,' as he called it, having been 'improved' by Messrs. Sutton and Sons, is still regarded as 'the best in the trade.' This is *definite variation*, according to Darwin's definition, for those weeded out did not differ from the selected, morphologically, except in degree, the variations towards improvement not being quite fast enough to entitle them to survive."

Finally I wish especially to point out that variation in animals, and probably in plants, (with which I am not so familiar,) gives no ground for believing that "sports" have any considerable influence on the course of evolution. This is apparent whether we view the serial lines of variations of specific, generic, or higher characters; or whether we trace the phylogeny of the animal and vegetable types by means of the paleontological record. The method of evolution has apparently been one of successional increment or decrement of parts along definite lines. More or less abruptness in some of the steps of this succession there may have been; since a definite amount of energy expended in a given direction at a given point of history might produce a much greater effect than the same amount expended at some other period or point of evolution. This might be due to the release of stored energy, which could only be accomplished by a coincidence of circumstances. A simple illustration of the phenomenon of abrupt metamorphosis is to be found in the passage of matter from the gaseous to the liquid, and from the liquid to the solid state. I have stated the case in the following language:[1]

"As one or more periods in the life of every species is characterized by a greater rapidity of development" (ontogenetic) "than the remainder; so in proportion to the approximation of such a period to the epoch of maturity or reproduction is the offspring liable to variation. During the periods corresponding to those between the rapid metamorphoses, the characters of the genus would be preserved unaltered, though the period of change would be ever approaching. Hence the transformation of genera may have been rapid and abrupt, and the intervening periods of persistency very long. Thus, while change is really progressing, the external features remain unchanged at other than those points, which may be called *expression points*. Now the *expression point* of a new generic type is reached when its appearance in the adult falls so far prior to the period of reproduction as to transmit it to the offspring and their descendants."

I. VARIATIONS OF SPECIFIC CHARACTERS.

a. Variations in the Coleopterous Genus Cicindela.

Dr. George H. Horn has traced the variations in the color patterns of the elytra of the North American species of this abundant and well-known genus. He shows that they form series, in the following language:[2]

"Any one in glancing over this series will perceive that there is a great similarity of marking between many species. This similarity, which may be considered as the type of marking, and is illustrated by

[1] *Origin of the Fittest*, p. 79.

[2] *Entomological News*, Philadelphia, Feb., 1892, p. 25.

No. 1 of the accompanying plate (Fig. 1) is the underlying pattern from which all the forms observed in our Cicindela have been derived.

"Before going further it is well to present the following propositions that the argument and the illustrations may be understood.

"1. The *type* of marking is the same in all our species.

"2. Assuming a well-marked species as a central type the markings vary,

a, by a progressive spreading of the white,

b, by a gradual thinning or absorption of the white,

c, by a fragmentation of the markings,

d, by linear supplementary extension.

"3. Many species are practically invariable. These fall in two series,

a, those of the normal type, as *vulgaris*, *hirticollis*, and *tenuisignata*,

b, those in which some modification of the type has become permanent, probably through isolation, as *marginipennis*, *togata*, and *lemniscata*.

"4. Those species which vary, do so in one direction only. That is, supposing a species begins typically with markings similar to *vulgaris*, the variation may be either in the direction of thickening and increase of white, as in *hyperborea*, *generosa*, and others, or in the direction of thinning or fragmentation of the white with perhaps an entire loss of markings as in *hæmorrhagica*, *splendida*, or *obsoleta*.

"The first two propositions must be considered as applying to the species of the genus collectively, the last two to the species separately.

"The accompanying plate has been prepared to illustrate these propositions. It must, however, be

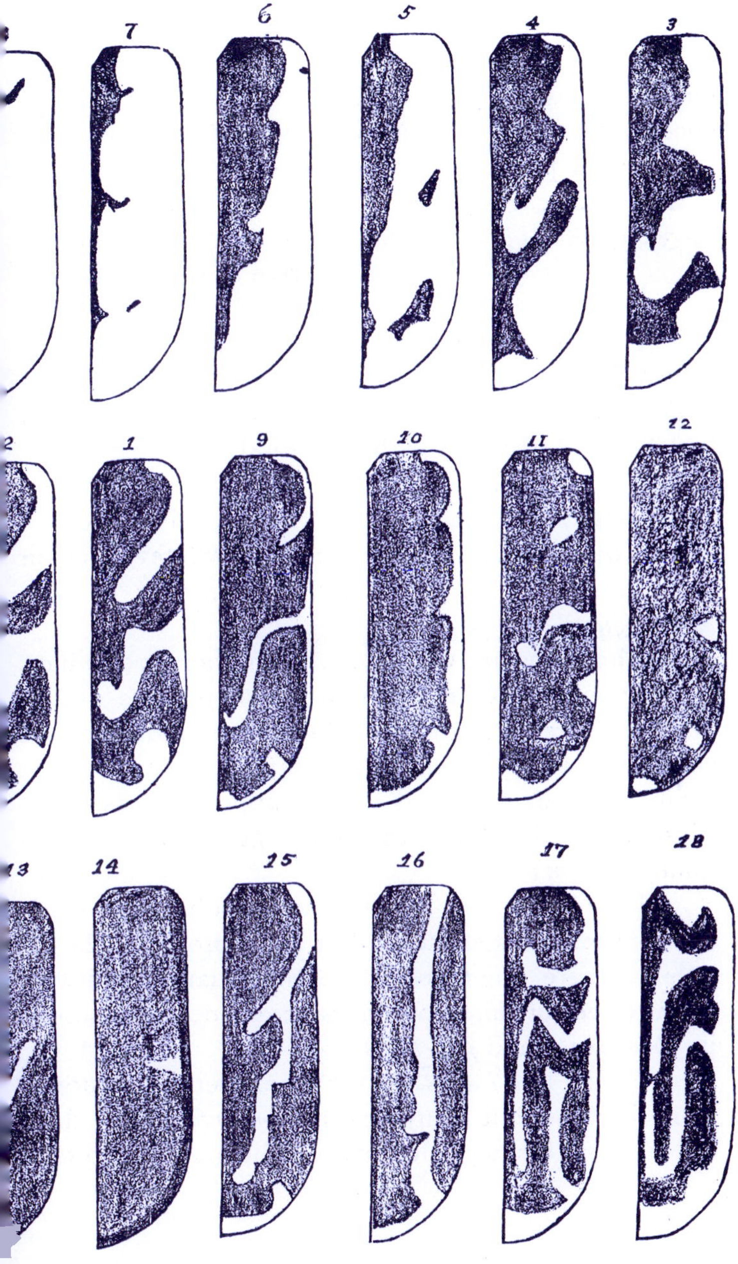

Fig. 1.—Horn on Cicindela.

understood that, in tracing the derivations from the typical, it is not possible to use one species, as these modifications go on gradually through a number of species, one sometimes beginning where another ends.

'In the plate, No. 1 represents *vulgaris*, which is a fairly typical species, following through *generosa* (2–3), *pamphila* (4), *hyperborea* var. (5), *togata* (6), *gratiosa* (7), *canosa* (8), we finally arrive at a perfect white elytron as seen in some varieties of *dorsalis*.

"Following in the other direction through *tenuisignata* (9), *marginipennis* (10), *hentzii* (11), *sexguttata* (12), *hæmorrhagica* (13), and *splendida* var. (14), it will be observed that through a gradual thinning or absorption of the markings, or by their fragmentation and obliteration, we arrive at the opposite result of elytra without any white markings whatever, as in many forms of *obsoleta*, *scutellaris*, *punctulata*, and *hæmorrhagica*.

"Those species which vary from the type in having the markings broken into spots, as in 12-*guttata* or *hentzii*, may lose the spots by a gradual decrease of size, so that they all seem to disappear nearly at the same time; or the spots may disappear successively, those on the disc being the first to go, while the marginal spots remain.

"From our series it would be difficult to say which spot is the most persistent, but it is probably the lunule, as there are more with entirely dark elytra with slight traces of this spot than with any other, as shown in *abdominalis* and *punctulata*.

"Forms like *lemniscata* (16) seem very far removed from the type, but many forms of *imperfecta* (15) show how the markings gradually leave the margin and tend by fusion to form a vitta at first somewhat oblique, but finally becoming nearly median.

"The last two figures on the plate represent the markings of *gabbii* (17) and *saulcyi* (18), in which the ends of the bands or lunules are greatly prolonged. The latter form, which represents *dorsalis* as well, is but rarely seen so perfectly marked, the tendency being toward a greater extension of the white. The other species is scarcely variable, although equally a coast form.

"Those species which retain a permanent divergence from the normal standard, such as *togata* (6) or *lemniscata* (16), are doubtless descendants from a normal type which has varied, and in which a variety has become isolated and perpetuated itself."

The accompanying plate is copied from the original drawing by Dr. Horn, and which accompanies the paper now cited.

b. Variations in the Osceola doliata.

The Milk-Snake, *Osceola doliata* Linn., ranges in North America over the Eastern, Central, and Austroriparian districts, and is absent from the Sonoran and Pacific districts. It is found also in the humid regions of Mexico and Central America, as far as the Isthmus of Darien. Beyond this point it does not occur, but a very similar snake (*Opheomorphus mimus*) is found in New Grenada.

I have called attention to the color variations of this species in a brief paragraph in the introduction to my check list of Batrachia and Reptilia in North America, 1875,[1] and have given the characters of the

[1] *Bulletin of the U. S. National Museum*, No. I, p. 4.

color types, or subspecies, in an analytical key, in a "Review of the Characters and Variations of the Snakes of North America," 1892.[1] I have also given a series of figures representing the North American color forms, for which I am indebted to the United States National Museum, which are here reproduced.

Before going further into the patterns of the *Osceola doliata*, I give a synoptic key of them.

I. No yellow band posteriorly from orbit (a yellow half-collar).
  a. Dorsal spots or saddles (red) open at the side, the borders of adjacent spots forming pairs of black rings.
    Interspaces between red saddles open below; scales not black-tipped; front more or less black; first black ring on nape only: *O. d. coccinea.*
    Interspaces between red saddles closed by black spot below; scales black tipped; front black; first black ring complete: *O. d. polyzona.*
    Interspaces not closed; rings, including first, complete on belly; first yellow band crossing occipital plates; front black; scales not black-tipped: *O. d. conjuncta.*
  aa. Dorsal spots closed at the sides below, forming saddles.
  b. Saddles closed by a single black tract on the middle of the belly; no spots between the saddles.
    Dorsal spots undivided medially; front black; first black ring complete: *O. d. annulata.*
    Dorsal spots divided longitudinally by a median black connection; front black: *O. d. gentilis.*
  bb. Inferior borders of saddles separate and not confluent with each other.
    Saddles completed on gastrosteges; no alternating spots; no black collar: *O. d. parallela.*
    Saddles completed on gastrosteges; spots opposite intervals forming a single series on the middle line of the belly: *O. d. syspila.*
    Saddles completed above the gastrosteges; alternating spots which do not meet on the middle line of the belly: *O. d. doliata.*

[1] *Proceedings of the U. S. National Museum*, XIV, p. 589-608.

II. A yellow band posteriorly from orbit, bounded below by a black or brown one.

a. Saddle spots closed laterally on gastrosteges; alternate spots entirely on gastrosteges.

A half collar behind parietal plates, no superciliary stripe: *O. d. temporalis.*

aa. Saddle spots closed above gastrosteges; alternate spots on scales.

A half collar nearly or quite touching occipital plates, no bands; alternate spots partly on gastrosteges: *O. d. collaris.*

Neck with longitudinal bands; alternate spots partly on gastrosteges: *O. d. clerica.*

Neck with bands; alternate spots entirely on scales: *O. d. triangula.*

In Fig. 2 are represented vertical, lateral, and inferior views of parts of the body of the subspecies *triangula*, taken from a specimen in my collection from Westchester County, New York, which I owe to the kindness of my friend, Mr. T. H. Mead.

The characters of this form are seen in (1) the presence of a light band extending from the posterior angle of the eye downward and backward, which is bounded by a black border above and below; (2) a black cross-band on the posterior border of the prefrontal plates; (3) chevron shaped mark with the apex on the posterior part of the frontal plate, whose limbs extend posteriorly as a band on each side of the neck, where they are fused together, and continue as a single, broad band for a short distance; (4) a series of lateral spots which do not extend beyond the scales on to the gastrosteges, and which alternate with the dorsal spots; (5) a series of spots on the ends of the gastrosteges which alternate with the last mentioned; (6) a series of spots on the centers of the gastrosteges which alternate with the spots mentioned under (5).

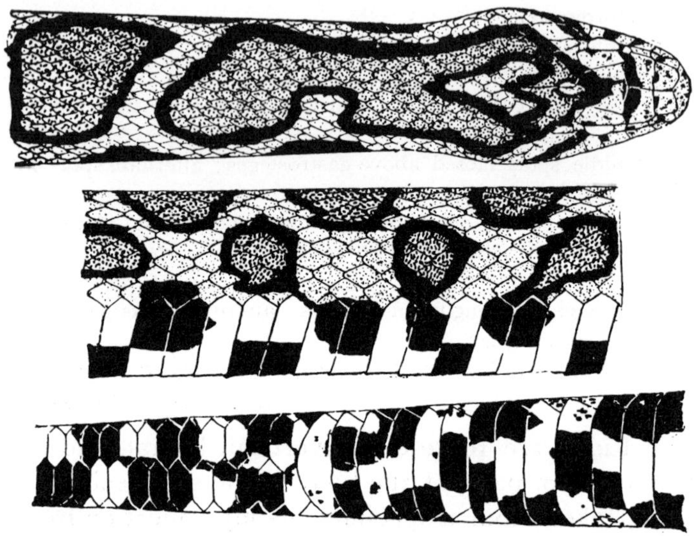

Fig. 2. *Osceola doliata triangula.*

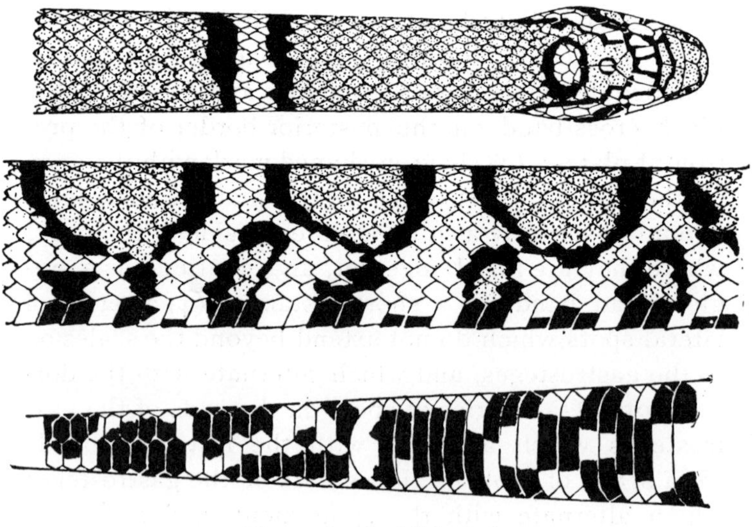

Fig. 3. *Osceola doliata clerica.*

The ground color in this form is gray, and the spots are a rich brown with black borders. The belly has a white ground color.

In Fig. 3 we have the subspecies *clerica,* where the following modifications appear. The fusion of the limbs of the chevron is more complete, and the dorsal spots are more expanded transversely. They extend to within two or three scales of the gastrosteges, while in the form *triangulus* they are five scales distant. The alternate spots touch the gastrosteges. This figure is taken from a specimen in the Museum of the Philadelphia Academy from southern Illinois.

In Fig. 4 we have an individual from Elmira, Illinois, which illustrates the characters of the form *collaris.* Here the chevrons are distinct from the first dorsal spot, whose anterior black border forms a half collar on the neck. This specimen is instructive, as it displays the last connection between the chevron and the first spot, in a black line on each side. This is wanting in the typical *collaris.*

The collar of ground color is complete in its anterior border, as well as the posterior in the form *temporalis* (Fig. 5), owing to the disappearance of the chevron. The transverse band on the prefrontals has also disappeared. The anterior extremity of the postorbital stripe is cut off, and consists of a spot of ground color. The dorsal saddle spots are wider, reaching the gastrosteges, while the intermediate spots are exclusively gastrostegal. The spots which alternate with them, have fused on the middle line. Fig. 5 is from a specimen from the State of Delaware.

In subspecies *doliata* the postocular stripe has disappeared, and the chevron is replaced by a black patch on the parietal and temporal plates. In other respects

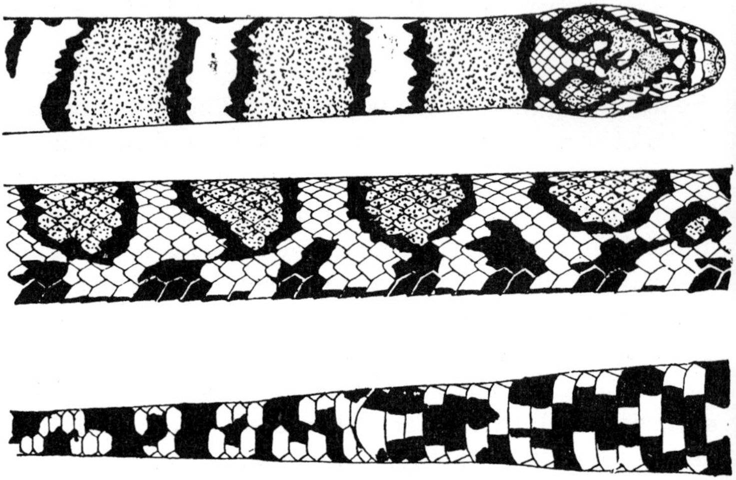

Fig. 4. *Osceola doliata collaris.*

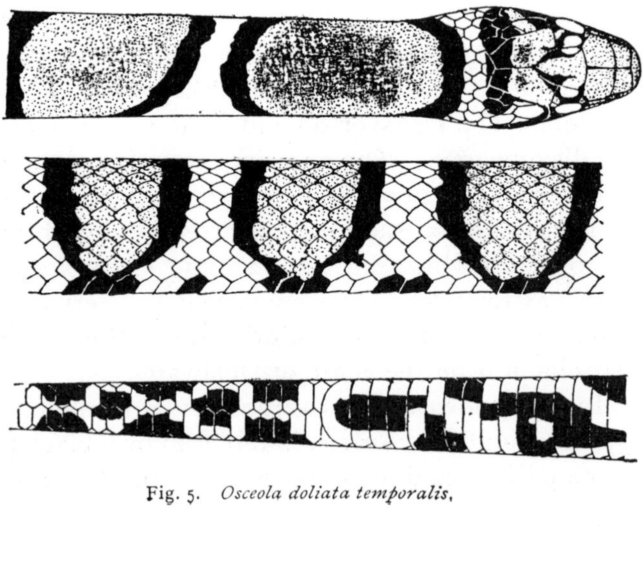

Fig. 5. *Osceola doliata temporalis.*

this form is more like the form *collaris*. The dorsal saddle spots are separated by a row or two of scales from the gastrosteges, and their alternating spots are partly on the scales. The ground color in this form, as in the *temporalis*, approaches red. This is the form of the tier of states between latitude 40° and the Gulf States.

The subspecies *syspila* is represented in Fig. 8. The head pattern is like that of *doliata* with the black patch more or less reduced—in the specimen figured being represented by a cross stripe. The dorsal saddle spots are more expanded than in any form yet encountered, their lateral borders being completed below the scales and entirely on the gastrosteges. The alternate spots now meet and fuse on the middle line of the abdomen, and the second series of alternating spots has disappeared. This is distinctively a southern form, extending west to central Oklahoma.

The dorsal saddles are so far extended in the next subspecies, *parallela*, as to form two parallel stripes with a narrow strip of ground color between, on the middle line of the abdomen. The alternating spots have disappeared. In the specimen figured, which is from Florida, and is in the United States National Museum, the supraocular spots seen in *temporalis*, are indicated. The ground color is red. Black begins to appear on the head.

From the form *syspila* two types of color modification may be traced. One of these brings the borders of the saddle spots together on the median line, forming a median black stripe; this is the subspecies *annulata*, which belongs to western Texas and the adjacent parts of Mexico. The top of the head is black (Fig. 10). In the other, the lateral borders of the saddle

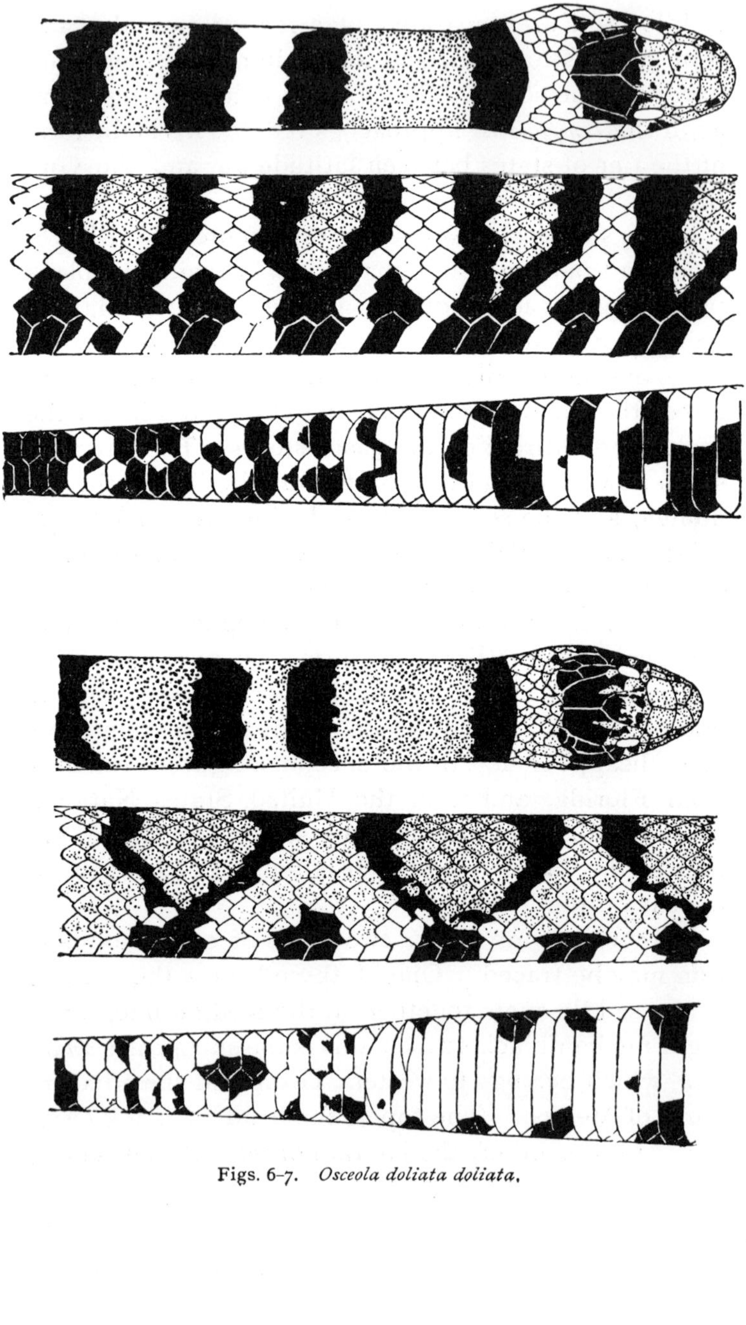

Figs. 6–7. *Osceola doliata doliata.*

spots have disappeared altogether, so that the body is more or less completely encircled by pairs of black rings, the alternating spots having disappeared. This might be supposed to have resulted from a continuation of the process by which the alternating spots have disappeared, and the edges of the saddles been brought closer and closer together. The continued transverse extension of the spot color would finally obliterate the lateral borders completely, as actually occurs in this last form, the *coccinea* of authors, which is the common type of the Gulf Coast. But the black has not covered the head and muzzle of this form as in the *annulata*. These regions are on the contrary red, as is the spot color generally, while the ground color is pale yellow.

A tendency to a development of black pigment in the saddle spots is seen in two other forms. The subspecies *gentilis* resembles *annulata*, but has a black longitudinal dorsal band which divides each saddle spot in two equal halves. This is a rare form, only known from the Indian Territory. The common Mexican form (*polyzona*) has the paired rings of *coccinea*, the black head of *annulata*, but each scale of the red intervals is tipped with black.

The relations of these forms may be expressed in a tabular form, given on page 39.

The main series corresponds with a distribution in latitude, commencing with the *triangula* of New England and New York, and passing gradually to the *coccinea* of the Gulf Coast regions, and *polyzona* of Mexico and Central America. The forms of the right-hand column are (except the *parallela*) from the central warmer parts of the continent.

This series of color-forms of the *Osceola doliata*

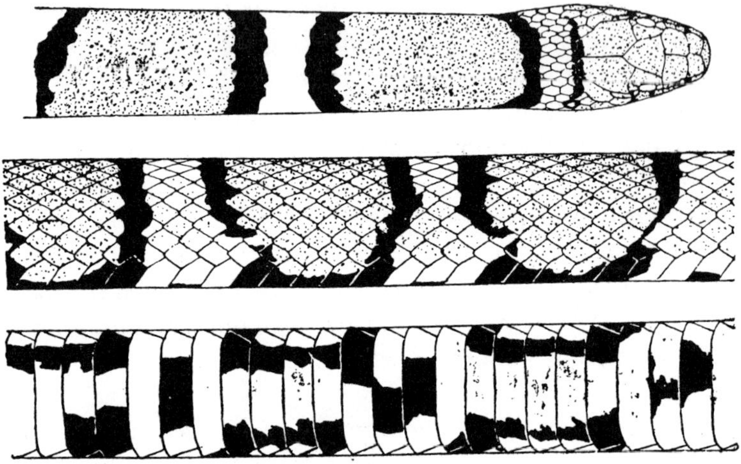

Fig. 8. *Osceola doliata syspila.*

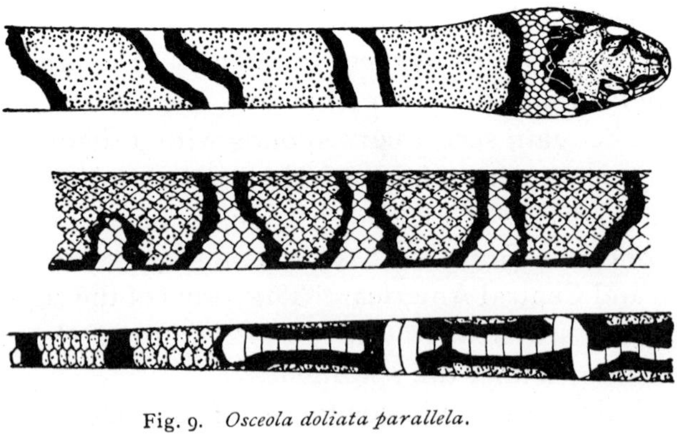

Fig. 9. *Osceola doliata parallela.*

demonstrates the following points. First: the color-variation is determinate and not indeterminate. It consists in, *a*, the successive enlargement of the dorsal spots toward, to, and across, the belly; *b*, the diminution and extinction of the longitudinal stripes on the head; *c*, do. of the spots of the inferior surface of the body; *d*, in the increase of red in the color of the dorsal spots, coincidentally with the changes mentioned. Second: these color-changes follow parallels

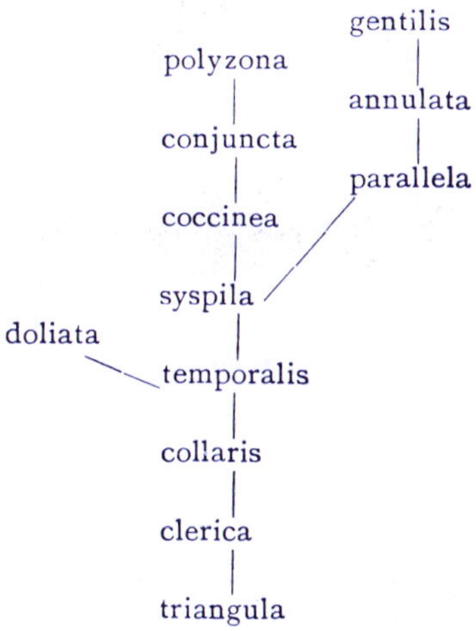

of latitude, the red color and accompanying changes developing from north to south. Third: so far as regards eastern North America, there is a diminution of size in passing from north to south; the *O. d. coccinea* being the smallest of the subspecies. In Mexico, the size is recovered, as the *O. d. polyzona* equals in dimensions the *O. d. triangula*.

The young of the northern *O. d. triangula* presents the colors of the dorsal spots nearly as brilliant as those of the southern *O. d. coccinea*, and they fade

Fig. 10. *Osceola doliata annulata.*

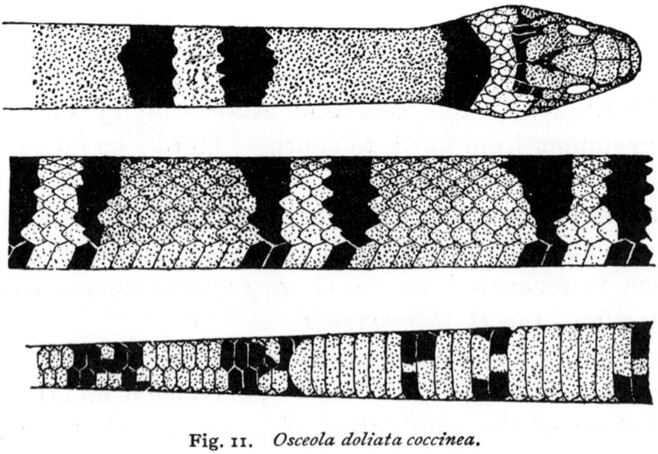

Fig. 11. *Osceola doliata coccinea.*

with age to the adult character. The pattern in the young at the period of hatching is the same as that of the adult.

c. Color-Variations in Cnemidophorus.

Another illustration of the nature of color-variation is to be found in certain species of the lacertilian genera Cnemidophorus in America, and Lacerta in Europe. In both genera the color-markings differ in the same individual at different ages, and the age at which the adult coloration is assumed, differs in different localities. Some of the species, e. g., *Cnemidophorus sexlineatus*, never abandon the coloration of the young of other species and subspecies. The same condition is characteristic of the *C. deppei* of Mexico, the *C. lemniscatus* of Brazil, and other species. The process of color-modification in the *C. tessellatus* and *C. gularis* of North America is, as I have pointed out,[1] as follows: The young are longitudinally striped with from two to four stripes on each side of the middle line. With increasing age, light spots appear between the stripes in the dark interspaces. In a later stage these spots increase in transverse diameter, breaking up the dark bands into spots. In some of the forms these dark spots extend themselves transversely and unite with each other, forming black cross-stripes of greater or less length. Thus we have before us the process by which a longitudinally striped coloration is transformed into a transversely striped one.

The large number of specimens of the *C. tessellatus* and *C. gularis* in the National Museum collection show that the breaking up of the striped coloration appears

[1] *Proceeds. Amer. Philos. Soc.*, 1885, p. 283. *Transac. Amer. Philos. Soc* 1892, p. 27.

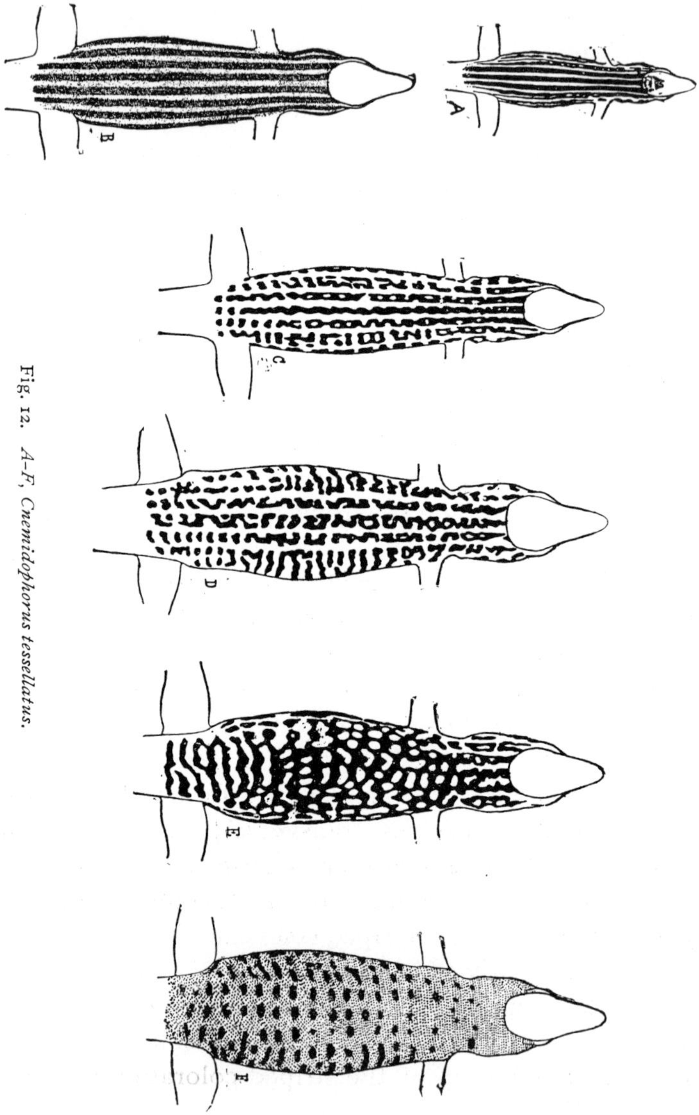

Fig. 12. *A–F, Cnemidophorus tessellatus.*

first at the posterior part of the dorsal region (i. e., the sacral and lumbar). The confluence of the spots ap-

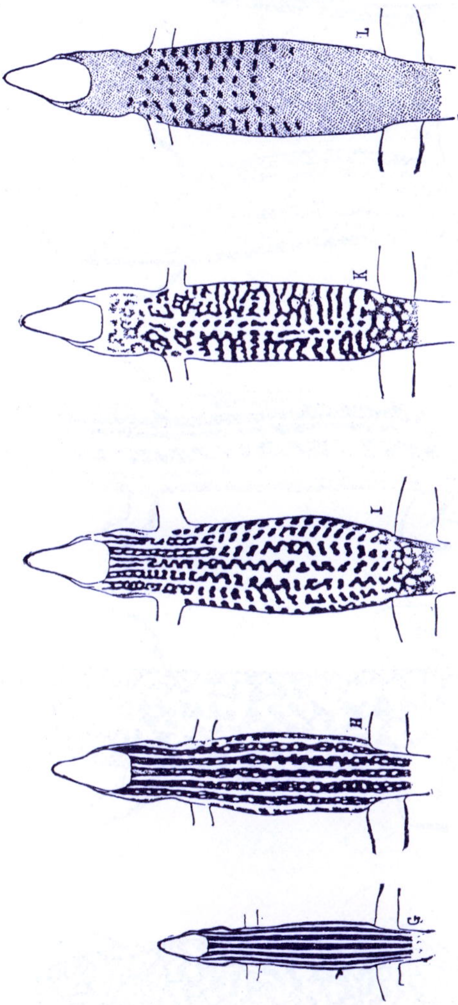

Fig. 13. *G–L, Cnemidophorus gularis.*

pears there first; and finally (*C. gularis semifasciatus*), where the color markings disappear, leaving a uniform hue, this also appears first at the posterior part of the

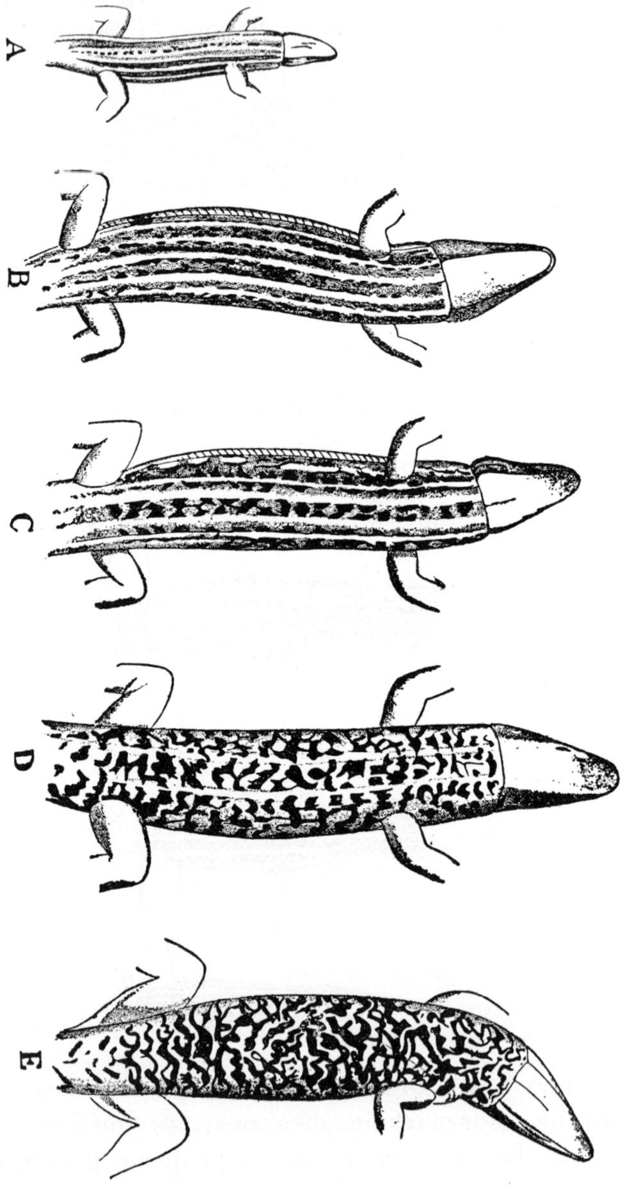

Fig. 14. *Lacerta muralis.*

body. In the *C. tessellatus rubidus* the dark spots disappear first on the anterior regions.

According to Eimer,[1] among many color-variations of the *Lacerta muralis* there exists a series of types closely similar to those observed by me to characterize the two species of Cnemidophorus mentioned. I give figures of these series in all three species. It will be observed that in the second and third forms (B and C) of the *L. muralis*, the pale portions of the dark stripes do not assume the very light hue of the ground color as they do in the corresponding phase of the *Cnemidophorus tessellatus* (C and D, Fig. 12), but this intermediate condition is exactly paralleled by the subspecies *mariarum* of the *Cnemidophorus gularis*. The correspondences are represented in the table on page 46.

There are some color forms in the *Lacerta muralis* which are not repeated in the North American Cnemidophori, particularly those which result in a strong contrast between the dorsal colors as a whole and the darker lateral colors, as a band. The color variety, No. 7, of the Cnemidophori is not reported by Eimer as occurring in the *Lacerta muralis*.

The variations from one to four form a direct series, and so do those represented by Nos. 1, 2, 3, and 5. Such variations cannot be regarded as promiscuous, especially when the same process of change is to be observed in three different species, one of which inhabits a continent remote from the other two.

d. Variations in North American Birds and Mammals in Relation to Locality.

The distinguished zoölogist, Dr. J. A. Allen of New York, has made a thorough study of this subject with

[1] *Archiv für Naturgeschichte*, 1881, p. 239.

	Cnem. tessellatus.	*Cnem. gularis.*	*Other Cnemidoph'i.*	*Lacerta muralis.*
1. Longitudinally striped.	*C. t. perplexus.*		*C. octolineatus.* *C. sexlineatus.*	*L. m. campestris.*
2. Dark interspaces pale spotted	*C. t. tessellatus α.*	*C g. gularis α.*	*C. labialis.* *C. septemvittatus.*	
3. Dark interspaces divided by light color	*C. t. tessellatus β.*	*C. g. scalaris α.*	*C. grahamii.*	*L. m. albiventris.* *L. m. striatomaculata.*
4. Dark spots confluent transversely, forming crossbars.	*C. t. tessellatus γ.*	*C. g. scalaris β.* *C. g. costatus.*		*L. m. reticulata.* *L. m. tigris.*
5. Light spots not confluent; light stripes broken up; pattern reticulated	*C. t. melanostethus.*		*C. variolosus.*	*L. m. punctulatofasciata.*
6. Dark stripes interrupted by darker color.		*C. t. mariarum.*		*L. m. maculostriata.*
7. Dark spots separate and on a brown ground...	*C. t. rubidus.*	*C. g. semifasciatus.*		

ample material at his disposal. Following lines already laid down by Prof. Spencer F. Baird, Dr. Allen has shown that variations in form, size, and color are directly related to latitude, and that they are not promiscuous or multifarious, but are definite and graded. I make the following extract from a summary of the subject published by him in *The Radical Review* (New Bedford, Mass.) for May, 1877:

"Geographical variation, as exhibited by the mammals and birds of North America, may be summarized under the following heads, namely: (1) variation in general size, (2) in the size of peripheral parts, and (3) in color; the latter being subdivisible into (*a*) variation in color with latitude, and (*b*) with longitude. As a rule, the mammals and birds of North America increase in size from the south northward. This is true, not only of the individual representatives of each species, but generally the largest species of each genus and family are northern. There are, however, some strongly marked exceptions, in which the increase in size is in the opposite direction, or southward. There is for this an obvious explanation, as will be presently shown; the increase being found to be almost invariably toward the region where the type or group to which the species belongs receives its greatest numerical development, and where the species are also most specialized. Hence the representatives of a given species increase in size toward its hypothetical center of distribution, which is in most cases doubtless also its original center of dispersal. Consequently there is frequently a double decadence in size within specific groups, and both in size, and numerically in the case of species, when the center of development of the group to which they belong is in the warm-temperate or tropical regions. This may be illustrated by reference to the distribution of the higher classes of vertebrates in North America. Among the species occurring north of Mexico there are very few that may not be supposed to have had a northern origin; and the fact that some are circumpolar in their distribution, while most of the others (especially among the mammals) have congeneric Old World allies further strengthens the theory of their northern origin. Not only do individuals of the same species increase in size toward the north, but the same is true of the species of different genera. Again, in the exceptional cases of increase in size southward, the species belong to southern types, or, more correctly, to types having their center of development within or near the intertropical regions, where occur, not only the greatest number of the specific representatives of the type, but also the largest.

"For more detailed illustration we may take three families of the North American Carnivora; namely, the Canidæ (wolves and foxes), the Felidæ (lynxes and wild cats), and the Procyonidæ (raccoons). The first two are to some extent cosmopolitan, while the third is strictly American. The Canidæ have their largest specific representatives, the world over, in the temperate or colder latitudes. In North America the family is represented by six species,[1] the smallest of which (speaking generally) are southern, and the largest northern. Four of them are among the most widely distributed of North American mammals, two (the gray wolf and the common fox) being circumpolar species; another (the Arctic fox) is also circumpolar, but

[1] The gray wolf (*Canis lupus*), the coyote (*C. latrans*), the Arctic fox (*Vulpes lagopus*), the common fox (*V. alopex*), the kit fox (*V. velox*), and the gray fox (*V. cinereoargentatus*).

is confined to high latitudes. The three widest-ranging species (the gray wolf, the common fox, and the gray fox) are those which present the most marked variation in size. Taking the skull as the basis of comparison, it is found that the common wolf is fully one-fifth larger in the northern parts of British America and Alaska than it is in Northern Mexico, where it finds the southern limit of its habitat. Between the largest northern skull and the largest southern skull there is a difference of *about thirty-five per cent. of the mean size!* Specimens from the intermediate region show a gradual intergradation between these extremes, although many of the examples from the upper Missouri country are nearly as large as those from the extreme North.

"The common fox, though occurring as far north as the wolf, is much more restricted in its southward range, especially along the Atlantic coast, and presents a correspondingly smaller amount of variation in size. The Alaskan animal, however, averages about one-tenth larger than the average size of specimens from New England. In the gray fox, whose habitat extends from Pennsylvania southward to Yucatan, the average length of the skull decreases from about five inches in Pennsylvania to considerably less than four in Central America—a difference equal to about thirty per cent. of the mean size for the species.

"The Felidæ, unlike the Canidæ, reach their greatest development, as respects both the number and the size of the species, in the intertropical regions. This family has but a single typical representative—the panther (*Felis concolor*)—north of Mexico, and this ranges only to about the northern boundary of the United States. The other North American representatives of the family are the lynxes, which, in some of their varieties, range from Alaska to Mexico. They form, however, the most northern, as well as the most specialized or 'aberrant,' type of the family. While they vary greatly in color, as well as in the length and texture of the pelage, at different localities, they afford a most remarkable exception to all laws of variation in size with locality; for a large series of skulls, representing localities as widely separated as Louisiana, Northern Mexico, and California, on the one hand, and Alaska and the Mackenzie River district on the other, as well as various intermediate localities, reveals no appreciable difference in size throughout this wide area. The true cats, however, as the panther and the ocelots, are found to greatly increase in size southward, or toward the metropolis of the family. The panther ranges from the Northern States southward over most of South America. Skulls from the Adirondack region of New York have an average length of about seven and a half inches, the length increasing to eight and three-quarters in Louisiana and Texas, from beyond which points there is lack of data. The ocelot (*Felis pardalis*) finds its northern limit near the Rio Grande of Texas, and ranges thence southward far into South America. The average size of Costa Rican examples is about one-fifth greater than that of specimens from the Rio Grande.

Instances of increase in size northward among the Carnivora of North America are so generally the rule that further space need not be taken in recounting examples in detail. It may suffice to state that the badger (*Taxidea americana*), the marten (*Mustela americana*), the fisher (*M. pennanti*), the wolverine (*Gulo luscus*), and the ermine (*Putorius ermineus*)—all northern types—afford examples of variations in size strictly parallel with that already noticed as occurring in the foxes and wolves.

"To refer briefly to other groups, it may be stated that the Cervidæ (deer family) are mainly rather northern in their distribution; that the largest species occur in the colder zones, and that individuals of the same species increase rapidly in size toward the north. Some of the species, in fact, afford some of the most striking instances of northward increase in size; among which are the Virginia deer and its several representatives in the interior of the continent and on the Pacific Slope. It is also noteworthy that the most obviously distinctive characteristic of the group—the large, annually deciduous antlers—reaches its greatest development at the northward. Thus all the northern species, as the moose, the elk, and the caribou, have branching antlers of immense size, while the antlers are relatively much smaller in the species inhabiting the middle region of the continent, and are reduced to a rudimentary condition—a simple, slender, sharp spike, or a small and singly forked one—in the tropical species; the antlers declining in size much more rapidly than the general size of the animal. This is true in individuals of the same species as well as of the species collectively.

"The Glires (the squirrels, marmots, spermophiles mice, and their affines) offer the same illustrations in respect to the law of increase in size as the species already mentioned, the size sometimes increasing to the southward, but more generally to the northward, since the greater number of the species belong decidedly to northern types. There is no more striking instance known among mammals of variation in size with locality than that afforded by the flying squirrels, in which the northern race is more than one-half larger than the southern; yet the two extremes are found to pass so gradually the one into the other, that it is hardly possible to define even a southern and a northern geographical race, except on the almost wholly arbitrary ground of difference in size. The species, moreover, is one of the most widely distributed, ranging from the Arctic regions (the northern limit of forests) to Central America.

"Among birds the local differences in size are almost as strongly marked as among mammals, and in the main, follow the same general law. A decided increase in size southward, however, or toward the warmer latitudes, occurs more rarely than in mammals, although several well-marked instances are known. The increase is generally northward, and is often very strongly marked. The greatest difference between northern and southern races occurs, as in mammals, in the species whose breeding-stations embrace a wide range of latitude. In species which breed from Northern New England to Florida, the southern forms are not only smaller, but are also quite different in color and in other features. This is eminently the case in the common quail (*Ortyx virginianus*), the meadow-lark (*Sturnella magna*), the purple grackle (*Quiscalus purpureus*), the red-winged blackbird (*Agelaeus phaeniceus*), the golden-winged woodpecker (*Colaptes auratus*), the towhee (*Pipilo erythrophthalmus*), the Carolina dove (*Zenædura macrura*), and in numerous other species; and is quite appreciable in the blue-jay (*Cyanurus cristatus*), the crow (*Corvus americanus*), in most of the woodpeckers, in the titmice, numerous sparrows, and several thrushes and warblers, the variation often amounting to from ten to fifteen per cent. of the average size of the species.

"As a general rule, certain parts of the organisms vary more than does general size, there being a marked tendency to enlargement of peripheral parts under high temperature, or toward the tropics,—hence southward in North America. This is more readily seen in birds than in mammals, in consequence, mainly, of their peculiar type of structure. In mammals it is manifested occasionally in the size of the ears and feet, and in the horns of bovines, but especially and more generally in the pelage. At the northward, in individuals of the same species, the hairs are longer and softer, the under fur more abundant, and the ears and the soles of the feet better clothed. This is not only true of individuals of the same species, but of northern species collectively as compared with their nearest southern allies. Southern individuals retain permanently, in many cases, the sparsely clothed ears and the naked soles that characterize northern individuals only in summer, as is notably the case among the different squirrels and sphermophiles.

"In mammals which have the external ear largely developed—as in the wolves, foxes, some of the deer, and especially the hares,—the larger size of this organ in southern as compared with northern individuals of the same species is often strikingly apparent. It is more especially marked, however, in species inhabiting extensive open plains and semi-desert regions. The little wood hare, or gray rabbit (*Lepus sylvaticus*), affords a case in point. This species is represented, in some of its varieties, across the whole breadth of the continent, and from the northern border of the United States southward to Central America, but in different regions by geographical races or subspecies. In addition to certain differences of color and general size, the ears vary still more strongly. In the form inhabiting the Great Plains, commonly known as the little sage-brush hare (*L. sylvaticus nuttalli*), the ears are considerably longer than in the eastern variety, and increase in size from the north southward, reaching their greatest development in Western Arizona and the desert region further westward and southward, where the variety is characterized mainly by the large size of its ears, which are in this race nearly twice the size they attain in the eastern variety. In the long-eared 'jackass' hares of the plains, the ear likewise increases in size to the southward. In *Lepus callotis*, for example, which ranges from Wyoming southward far into Mexico, the ear is about one-fourth to one-third larger in the southern examples than in the northern. The little brown hare of the Pacific Coast (*L. trowbridgei*) presents a similar increase in the size of the ear southward, as does, to a less extent, the prairie hare (*L. campestris*). Not only are all of the long-eared species of American hares confined to the open plains of the arid interior of the continent, but over this same region is the tendency to an enlargement of the ear southward stronger than elsewhere. It is also of interest in this connection that the largest-eared hares of the Old World occur over similar open, half-desert regions, as do also the largest-eared foxes. On our western plains, the deer are represented by a large-eared species. Among the domestic races of cattle, those of the warm temperate and intertropical regions have much larger and longer horns than those of northern countries; as is shown by a comparison of the Texan, Mexican, and South American breeds, with the northern stock, or those of the South of Europe with the more northern races. In the wild species of the Old World, the southern or sub-tropical are remarkable for the large size of their horns. The horns of the American prong-horn (*Antilocapra americana*) are also much larger at southern than at northern localities.[1] Naturalists have also recorded the existence of larger feet in many of the smaller North American Mammalia at the southward than at the northward, among individuals of the same species, especially among the wild mice, in some of the squirrels, the opossum and raccoon, as well as in other species.

"In birds, the enlargement of peripheral parts, especially of the bill, claws, and tail, is far more obvious and more general than in mammals. The bill is particularly susceptible to variation in this regard,—in many instances being very much larger, among individuals of unquestionably the same species, at the southward than at the northward. This accords with the general fact that all the ornithic types in which the bill is remarkably enlarged occur in the intertropical regions. The southward enlargement of the bill within specific groups may be illustrated by reference to almost any group of North American birds, or to those of any portion of the continent. As in other features of geographical variation, the greatest differences in the size of the bill are met with among species having the widest distribution in latitude. Among the species inhabiting eastern North America we find several strik-

[1] The deer tribe, in which the antlers increase in size toward the north, offer an apparent exception to the rule of increase in size of peripheral parts toward the tropics. The antlers of the deer, however, are merely seasonal appendages, being annually cast and renewed, and are thus entirely different physiologically from the horns of bovines, which retain a high degree of vitality throughout the life of the animal.

ing examples of this enlargement among the sparrows, black-birds, thrushes, crows, wrens, and warblers, in the quail, the meadow-lark, the golden-winged woodpecker, etc. Generally the bill, in the slender-billed forms, becomes longer, more attenuated, and more decurved (in individuals specifically the same) in passing from the New England States southward to Florida, while in those which have a short, thick, conical bill there is a general increase in its size so that the southern representatives of a species, as a rule, have thicker and longer bills than their northern relatives, though the birds themselves are smaller. There is thus not only generally a relative, but often an absolute, increase in the size of the bill in the southern races. The species of the Pacific Coast and of the interior afford similar illustrations, in some cases more marked even than in any of the eastern species. More rarely, but still quite frequently, is there a similar increase in the size of the feet and claws.

"The tail, also, affords an equally striking example of the enlargement of peripheral parts southward. Referring again to the birds of the Atlantic Coast, many of the above-named species have the tail absolutely longer at southern localities than at northern, and quite often relatively longer. Thus while the general size decreases, the length of the tail is wholly maintained, or decreases less than the general size; but, in some cases, while the general size is one-tenth or more smaller at the south, the tail is ten to fifteen per cent. longer than in the larger northern birds. Some western species are even more remarkable in this respect; and in consequence mainly of this fact the southern types have been varietally separated from the shorter-tailed northern forms of the same species.

"Variations in color with locality are still more obvious, particularly among birds, in which the colors are more positive, the contrasts of tints greater, and the markings consequently better defined than is usually the case in mammals. The soft finely-divided covering of the latter is poorly fitted for the display of the delicate pencilings and the lustrous, prismatic hues that so often characterize birds. Mammals, however, present many striking instances of geographical variation in color.

"As already stated, geographical variations in color may be conveniently considered under two heads. While the variation with latitude consists mainly in a nearly uniform increase in one direction, the variation observed in passing from the Atlantic Coast westward is more complex. In either case, however, the variation results primarily from nearly the same causes, which are obviously climatic, and depend mainly upon the relative humidity, or the hygrometric conditions of the different climatal areas of the continent. In respect to the first, or latitudinal variation, the tendency is always toward an increase in intensity of coloration southward. Not only do the primary colors become deepened in this direction, but dusky and blackish tints become stronger or more intense, iridescent hues become more lustrous, and dark markings, as spots and streaks or transverse bars, acquire greater area. Conversely, white or light markings become more restricted. In passing westward a general and gradual blanching of the colors is met with on leaving the wooded regions east of the Mississippi, the loss of color increasing with the increasing aridity of the climate and the absence of forests, the greatest pallor occurring over the almost rainless and semi-desert regions of the Great Basin and Colorado Desert. On the Pacific Slope, north of California, the color again increases, with a tendency to heavy, sombre tints over the rainy, heavily-wooded region of the Northwest Coast."[1]

2. VARIATION IN STRUCTURAL CHARACTERS.

Modifications of structural characters may appear quite independently of variation of specific ones. Indeed, generic characters have at times changed completely without the appearance of corresponding changes in the more superficial characters which define the species. Thus changes in the dentition of some of the Mammalia appear within the limits of species, which, should they become permanent, would entitle the two sets of individuals which display the different dentitions to be placed in different genera.

Some striking examples of how generic characters may undergo metamorphosis without corresponding changes in specific characters, have been brought to light by Dr. William H. Dall among the Brachiopodous Mollusca. Some of the species of different genera can scarcely be distinguished, except by comparison of their generic characters. I have cited the axolotls as illustrative of this phenomenon. Here the same species may reproduce as a permanent larva, or as an adult. Duméril has shown that the North American salamander (*Amblystoma tigrinum*) can lay and fertilize eggs before the metamorphosis is passed. I have since observed that the females of the allied species of Amblystomidæ, the *Chondrotus tenebrosus* B. and G., of California contain mature eggs ready for de-

[1] *The Radical Review*, May, 1877.

posit, and have supposed that this species has also the same power.[1] The difference between such larvæ and the adult which has passed the metamorphosis is great. It extends not merely to the branchial processes, but to the splenial teeth, which are shed, and to the palato-pterygoid arch, which is absorbed, and to the posterior ceratobranchial and epibranchial cartilages, which are absorbed. In the larva of the *C. tenebrosus* the palatopterygoid arches and epibranchials are ossified, so that the probability of its being able to maintain an independent existence as a larva is greater than in the case of the *A. tigrinum.* In this type, then, each species displays variations concomitant with reproductive maturity, which are not only of generic, but of family significance. In a third species, the *Siredon mexicanum*, no metamorphosis has yet been shown to take place, so that it is probable that it reproduces ordinarily while in the branchiferous stage. Yet it is only specifically different from the larva of the *Amblystoma tigrinum.*

Excellent illustrations of the serial appearance of generic characters may be seen in the family of the dogs (Canidae). In the true genus Canis, the dental formula is, I. $\frac{3}{3}$; C. $\frac{1}{1}$; P. m. $\frac{4}{4}$; M. $\frac{2}{3}$. The inferior sectorial (m. 1) has a metaconid, and the second inferior true molar has two roots. It not unfrequently happens, however, that the last inferior molar (m. 3) is wanting; and in some cases the inferior m. 2 has but one root. When in addition to this, as in some of the black-and-tans, in the Mexican naked dog, and in the pug, the inferior m. 1 loses its metaconid, we have the genus Synagodus. Occasionally the pug dog, and frequently the Mexican dog, loses one of its premolars

[1] *Batrachia of North America*, 1888, p. 113, Pl. xxii, xxiii.

from both jaws. The Japanese spaniel goes still further, and usually loses also its second superior true molar and frequently another premolar from each jaw; and we then have a dentition which indicates a third genus, which has been called Dysodus. Its dental formula is I. $\frac{3}{3}$; C. $\frac{1}{1}$; P. m. $\frac{2}{2}$–$\frac{3}{3}$; m. $\frac{1}{2}$. Transitions between this and the normal dentition of Canis, in all respects can be found in the smaller domesticated dogs. And these modifications are not pathological, but simply express a rapid metamorphosis of the dentition towards the reduced formula which is characteristic of the cats. And while the most characteristic dentitions belong to particular species (or races) of dogs, many of the single modifications are both absent and present in dogs of the same species or race. And these are the kind of characters which are observed to mark the slow progress during long geologic ages, of mammals of various other groups. These modifications are not promiscuous, but are in the direct line of change which has characterized all Mammalia during geologic time; i. e., the reduction of the numbers of the molar teeth. And in greater detail, the loss of metaconid of the inferior sectorial, and loss of posterior true molars, are the exact losses which the carnivorous type has undergone in the evolution of the cats.

A significant modification of the third superior premolar has been observed by Dr. Horace Jayne to be occasionally met with in the domestic cat. Sometimes an internal cusp (deuterocone), with a corresponding root is developed, giving rise to a tritubercular crown.

Similar observations have been made on the dentition of man, which presents two phenomena of variation of opposite phylogenetic significance. I have shown[1] that most of the Indo-European race, together with the Esquimaux, present a reversion to a lemurine form in the second and third superior molars, and sometimes, in the case of the Esquimaux, in the first superior molar also. I have also shown,[2] after a study of the dentition of the extinct Mammalia, that the more complex molars of later placental orders, have been derived from a tritubercular type, which prevailed throughout the earth just before the opening of the Eocene period. In the line of human and quadrumanous phylogeny, the lemurs of the Eocene period presented this type of molar in the upper jaw, and mostly continue to do so to the present time. The true monkeys, however, added the fourth tubercle or hypocone, in accordance with the developmental law in Mammalia generally, and the apes and men of the lower races present the same characteristic. Now, in the yellow race the hypocone of the last molar is generally wanting, while in the white race it is usual to find it absent from both the second and third molars. In this we have a case of reversion.

The reduction of the third (last) superior molar, and of the inferior as well, has gone further in the white race, since the tooth is frequently abnormally small, abortive, or totally wanting. The external superior incisor has a similar history, although its reduction and loss is not nearly so frequent as that of the last molars. These losses from the dental series are not of the nature of reversions, since the number of teeth is more and more numerous as we recede in time along the line of human ancestry. It is, on the contrary, the continuation of a process which has been,

1 *American Journal of Morphology*, 1888, p. 7.

2 *American Naturalist*, 1884, p. 350; *Origin of the Fittest*, 1887, p. 347.

as already remarked, common to all the Mammalia, of reduction in the number of teeth. Thus men with fewer teeth are more advanced than those with more numerous ones; while people with tritubercular superior molars have reverted to an ancient type; and both results are probably attained by the same physiologic process, i. e. defect of nutrition. It is to be remembered also, in connection with our argument, that these dental variations are modifications of generic characters, and that they are in definite directions, and are not promiscuous. With regard to the question as to whether dental variations in man are promiscuous or not, we have better opportunities of investigation than in the case of the lower animals generally. It may be safely asserted that the dental variations above cited are by far the most frequent in man, and that all others put together are relatively insignificant.

3. SUCCESSIONAL RELATION.

As the biologic types are variations become permanent, it is important to examine how the former stand related to each other. These relations express the direction which variation has taken, and throw a great deal of light on the nature of the process. That existing types of all grades are the result of the isolation of variations of species, is shown by the frequent examples of incomplete isolation, which follows inconstancy of the definitive characters. Groups of individuals which display this partial isolation are termed subspecies.

As an illustration of the mingling of isolated groups of individuals (species) with imperfectly isolated groups (subspecies), in a single genus, I refer to the American garter-snake (genus Eutaenia B. and G.). An examination of several hundred individuals of this genus yielded the following results: I found seventeen groups of individuals, which could be said to be completely isolated in characters, with very few exceptions. Eight other groups (species) are probably isolated, but they are not represented by a sufficiently large number of specimens to yield a satisfactory demonstration. Of the seventeen, four species embrace fifteen non-isolated geographical forms (subspecies), besides the typical forms (eight of which are included under the *E. sirtalis*); and two others include three color forms easily recognizable, besides the typical ones. Similar phenomena are presented in every part of the animal and vegetable kingdoms.

One of the most instructive natural divisions for the study of taxonomic relations as the result of variation, on account of the simplicity of the relations presented, is the Batrachia Salientia, or the order of Batrachia to which belong the toads, frogs, etc. Omitting the very restricted suborders of the Aglossa and Gastrechmia, the Batrachia Salientia fall into two divisions, which differ only in the structure of the lower portion of their scapular arch, or shoulder-girdle. In the one the opposite halves are capable of movements which contract or expand the capacity of the thorax; in the other the opposite halves abut against each other so as to be incapable of movement, thus preserving the size of the thoracic cavity. But during the early stages, the frogs of this division have the movable shoulder-girdle which characterizes those of the other division, the consolidation constituting a modification superadded in attaining maturity. Furthermore, young *Salientia* are toothless, and one section of the species with embryonic shoulder-girdle never acquire teeth. The suborder with embryonic shoulder-girdle is called the *Arcifera*, and that which is advanced in this respect is the *Firmisternia*. Now the frogs of each of these divisions present nearly similar scales of development of another part of the skeleton, viz., the bones of the top of the skull. We find some in which one of these bones (ethmoid) is represented

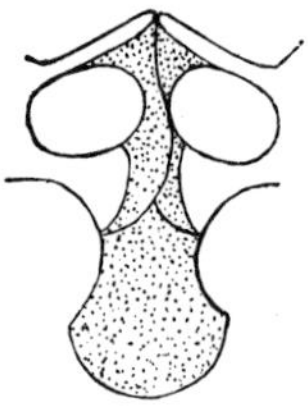

Fig. 15.

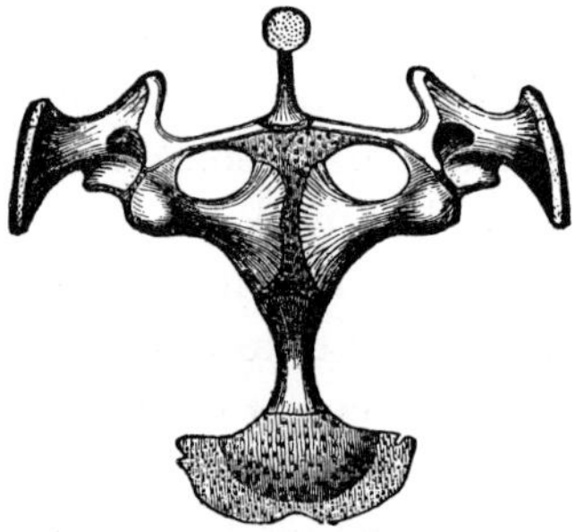

Fig. 17.

Fig. 16.

SHOULDER-GIRDLES OF "ANURA."

Fig. 15.—Of the Arciferous type (*Phyllomedusa bicolor*). Fig. 16, *Rana temporaria*, tadpole with budding limbs. Fig. 17, do., adult. Figs. 16 and 17 from Parker.

by cartilage only, and the frontoparietals and nasals are represented by only a narrow strip of bone each. In the next type the ethmoid is ossified; in the next, we have the frontoparietal completely ossified, and the nasals range from narrow strips to complete roofs; in the fourth station on the line, these bones are rough, with a hyperostosis of their surfaces; and in the next set of species this ossification fills the skin, which is thus no longer separable from the cranial bones; in the sixth form the ossification is extended so as to roof in the temporal muscles and inclose the orbits behind, while in the rare seventh and last stage, the tympanum is also inclosed behind by bone. Now all of these types are not found in all of the families of the *Salientia*, but the greater number of them are. Six principal families, four of which belong to the *Arcifera*, are named in the diagram below, and three or four others might have been added. I do not give the names of the genera which are defined as above described, referring to the explanation of the cuts for them, but indicate them by the numbers attached in the plate, which correspond to those of the definitions above given. A zero mark signifies the absence or non-discovery of a generic type.

	Sternum embryonic. Arcifera.				Sternum complete Firmisternia.
	Toothless.	Toothed.			
	Bufonidæ.	Scaphiopidæ.	Cystignathidæ.	Hylidæ.	Ranidæ.
1—	1	0	1	1	0
2—	2	2	2	2	0
3—	3	0	3	3	3
4—	4	4	4	4	4
5—	5	5	0	5	5
6—	6	6	6	6	6
7—	7	0	0	0	0

It is evident, from what has preceded, that a perfecting of the shoulder-girdle in any of the species of the arciferous columns would place it in the series of *Firmisternia*. An accession of teeth in a species of the division *Bufonidæ* would make it one of the *Scaphiopidæ*; while a small amount of change in the ossification of the bones of the skull would transfer a species from one to another of the generic stations represented by the numbers of the columns from one to seven.

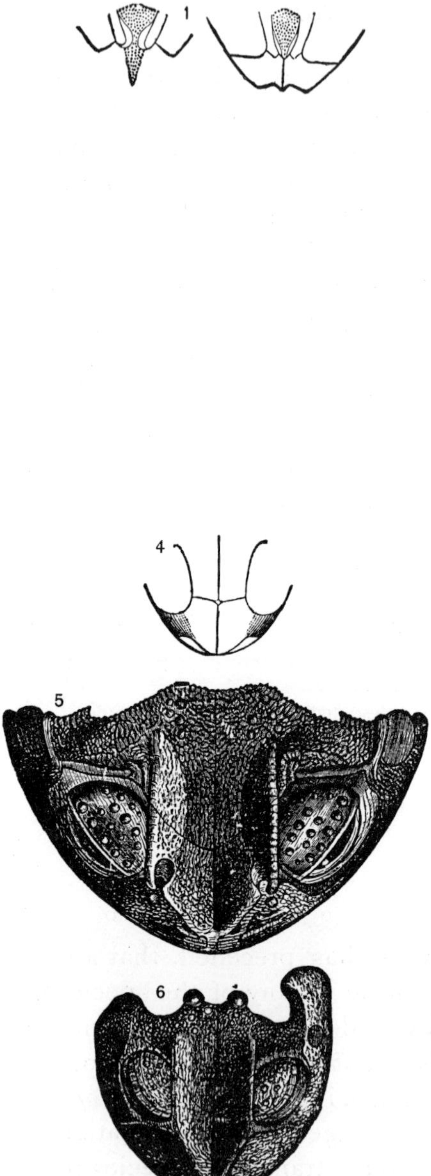

Fig. 18.
Bufonidæ.

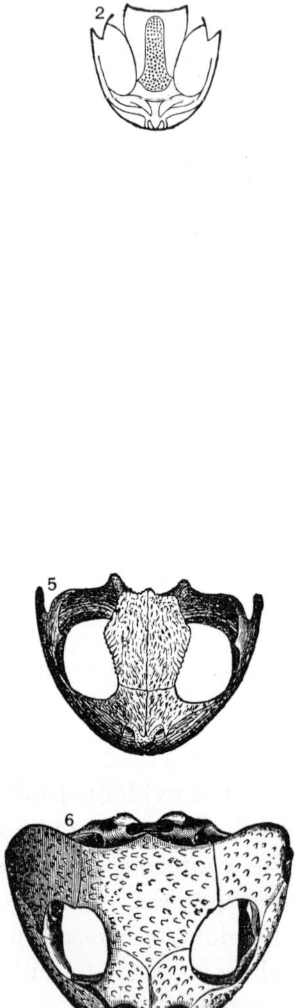

Fig. 19.
Scaphiopidæ and Pelobatidæ.

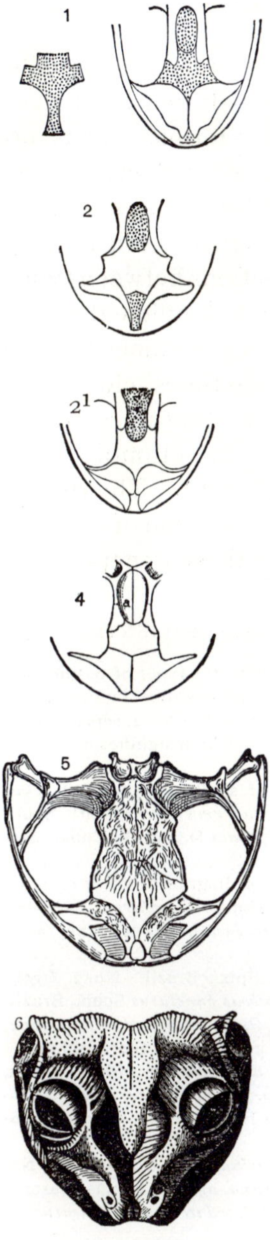

Fig. 20.
HYLIDÆ.

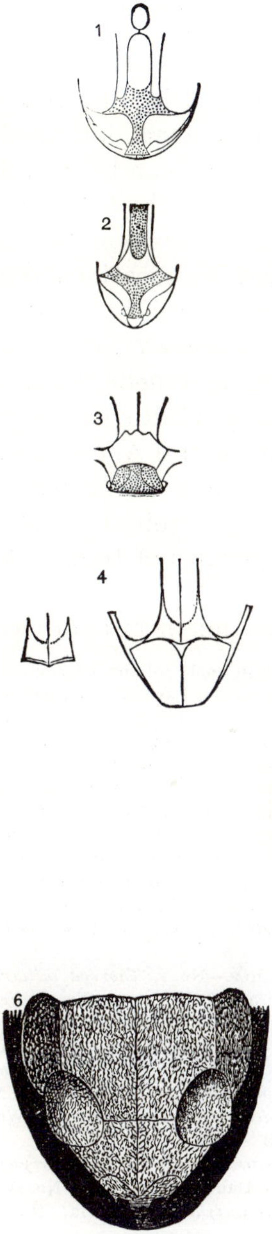

Fig. 21.
CYSTIGNATHIDÆ.

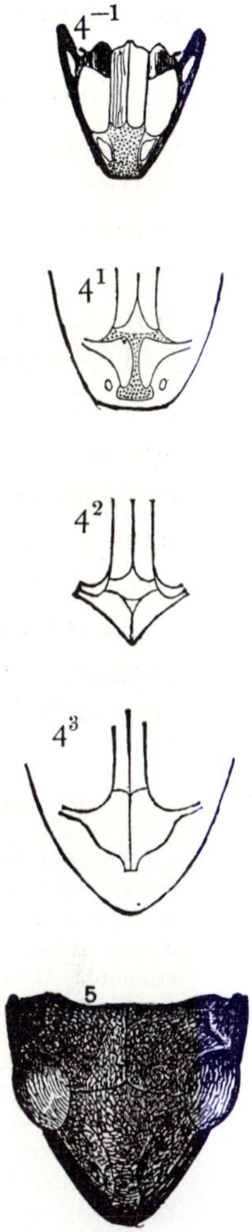

Fig. 22.
RANIDÆ.

That the above generic divisions have been actually developed from each other is demonstrated by the occurrence of occasional intermediate forms. Thus no generic distinction can be maintained between types third and fourth in the family of toads (Bufonidæ), so complete is the transition between them. In Hylidæ and Cystignathidæ occasional transitions between types second and third occur. In the Scaphiopidæ the subspecies *Spea hammondii intermontana* sometimes has the frontoparietal fontanelle open, sometimes closed. I have seen some adult specimens of *Rana virescens austricola* from Central America with the ethmoid bone unossified above, as in the genus Ranula. The rugose cranium is only acquired in old age of some of the species of Polypedates of India. Yet these genera are as

EXPLANATION OF CUTS OF CRANIA OF SALIENTIA.

The numbers in each column correspond with the types of ossification mentioned in the text, and are the same as those in the table of families given in the same connection. The power numbers attached to No. 4, represent the degree of ossification of the nasal bone, except -1, which signifies unossified ethmoid. Most of the cuts are original.

Fig. 18.—BUFONIDÆ.—No. 1, anterior part of skull of *Chelydobatrachus gouldii* Gray, from Australia. No. 4, do. of *Schismaderma carens* Smith, South Africa. No. 5, top of head of *Peltaphryne peltacephala* D. and B., Cuba. No. 7, top of head of *Otaspis empusa* Cope, Cuba.

Fig. 19.—SCAPHIOPIDÆ AND PELOBATIDÆ.—No. 2, diagram of top of cranium of *Didocus calcaratus* Micahelles, Spain. No. 5, skull of *Scaphiopus holbrookii* Harl., United States. No. 6, skull of *Cultripes provincialis*, from France, after Dugès.

Fig. 20.—HYLIDÆ.—No. 1, *Thoropa miliaris* Spix., Brazil. No. 2, *Hypsiboas doumercii* D. and B., Surinam. No. 2^1, *Hypsiboas punctatus* Schn., Brazil. No. 4^4, *Scytopis venulosus* Daudin, Brazil. No. 5, *Osteocephalus planiceps* Cope, E. Peru. No. 6, *Trachycephalus geographicus* D. and B., after Steindachner.

Fig. 21.—CYSTIGNATHIDÆ.—No. 1, *Eusophus nebulosus* Gir., Chili. No. 2, *Borborocœtes tasmaniensis* Gthr., Tasmania. No. 3, *Elosia nasus* Licht., Brazil. No. 4, *Hylodes oxyrhynchus* D. and B., West Indies. No. 6, *Calyptocephalus gayi* D. and B., Chili.

Fig. 22.—RANIDÆ.—No. 4^{-1}, *Ranula chrysoprasina* Cope, Costa Rica. No. 4^1, *Rana clamata* Daud., N. America. No. 4^2, *Rana agilis* Thomas, Europe. No. 4^3, *Rana hexadactyla* Less., India. No. 5, *Polypedates quadrilineatus* D. and B., Ceylon.

well defined as closely allied genera in most natural divisions.

It is seldom that so many stages of developmental series survive so as to be contemporaries, as in this case of the Batrachia Salientia. In order to obtain such series we usually have to explore the ages of the past. In the higher groups this is also the case, but here we have also occasional examples of the persistence of fairly complete series. Such a one is presented by the suborder Artiodactyla of the Diplarthrous Ungulate Mammalia. I give the definitions of the succession of the existing families.

I. Molars bunodont (tubercular); superior incisors generally present. No cannon or naviculocuboid bones.
  Lateral toes well developed; *Hippopotamidæ.*
  Lateral toes rudimental; *Suidæ.*

II. Molars selenodont (crescent-bearing). (Lateral toes rudimental or wanting).
 A. Premolars with one row of lobes.
  No naviculocuboid bone; one superior incisor; a cannon bone; *Camelidæ.*
  A naviculocuboid bone; no superior incisor; (cannon bone variable); *Tragulidæ.*
 AA. Premolars with two rows of tubercles; a naviculocuboid and cannon bones; no incisors above.
  Premolar iii with only one row of lobes; canine teeth, no horns; *Moschidæ.*
  Premolar iii with two rows of lobes; fixed horns; no canines above; *Bovidæ.*
  Premolar iii with two rows of lobes; horns deciduous; *Cervidæ.*

In this suborder we see a gradual complication of the structure of the molar teeth, and a loss of the incisors. In the limbs we observe the successive loss of the lateral digits, and the fusion of elements,—as the metapodials into cannon bones, and the elements of the tarsus, and, what is not stated in the above table, of the carpus also. Finally there is the remarkable development of horns on the head. When we come to examine the phylogeny of this order we will find how completely these characters are the result of the fixation of variations which have appeared in past geologic ages, and how various are the combinations and modifications presented by the extinct types.

Few natural groups permit of representation of their subdivisions in linear series. The only correct representation is in the form of a branching tree, and this cannot be well done in flat projection on the pages of a book. Each branch taken by itself, however, yields itself for a longer or shorter space to linear treatment.

For an example of such linear series in higher groups I turn again to the Batrachia Salientia. Here the two suborders of the Arcifera and Firmisternia present the following interesting parallels:

ARCIFERA.	FIRMISTERNIA
I. Without teeth.	
a. With sacral diapophyses dilated.	
Bufonidæ	Brevicipitidæ. Engystomidæ. Phryniscidæ.
aa. Sacral diapophyses cylindric.	
Dendrophryniscidæ.	Dendrobatidæ
II. With premaxillary and maxillary teeth only.	
a. With sacral diapophyses dilated.	
Pelodytidæ Pelobatidæ Hylidæ	Dyscophidæ. Cophylidæ.
aa. With sacral diapophyses cylindric.	
Cystignathidæ	Colostethidæ. Ranidæ.

III. Teeth in both jaws.

a. Sacral diapophyses not dilated.

Amphignathodontidæ } Hemiphractidæ ..... }	......................Ceratobatrachidæ.

In strict reference to the structure of the hind feet the following parallels may be drawn:

	Firmisternia. Ranidæ.	Arcifera.
External metatarsal free:		
Aquatic.	Rana.	Pseudis.
Subfossorial.	Hoplobatrachus.	Mixophyes.
External metatarsal attached:		
Feet webbed—		
Burrowing.	Pyxicephalus.	Ceratophrys.
Arboreal (vom. teeth).	Leptopelis.	Hyla.
Arboreal (no v. teeth).	Hyperolius.	Hylella.
Aquatic.	Heteroglossa.	Acris.
Feet not webbed—		
Terrestrial.	Cassina.	Cystignathus.
Terrestrial, spurred.	Hemimantis.	Paludicola.

Parallel series like those of the Arcifera and Firmisternia I have termed "homologous," and the corresponding terms of such series I have called "heterologous."[1] Such corresponding phylogenetic series are homologous to each other, while their terms or genera are heterologous in their relation to corresponding terms of other phyla. In such cases the genera or terms of a series owe their resemblances to each other to inheritance; but they owe their resemblances to their corresponding or heterologous genera, to identical evolutionary influences. Subsequently to my proposal to use the above terms, Prof. E. R. Lankester proposed the word "homogenous" to express what is conveyed by my term homologous, and "homoplastic" to express the sense of heterologous. For the two con-

[1] "Origin of Genera," *Proceedings Philadelphia Academy*, 1868, p. 281.

ditions he coined the words "homogeny" and "homoplassy." The terms introduced by Lankester differ from mine in that they convey implications as to the origin of the respective conditions.

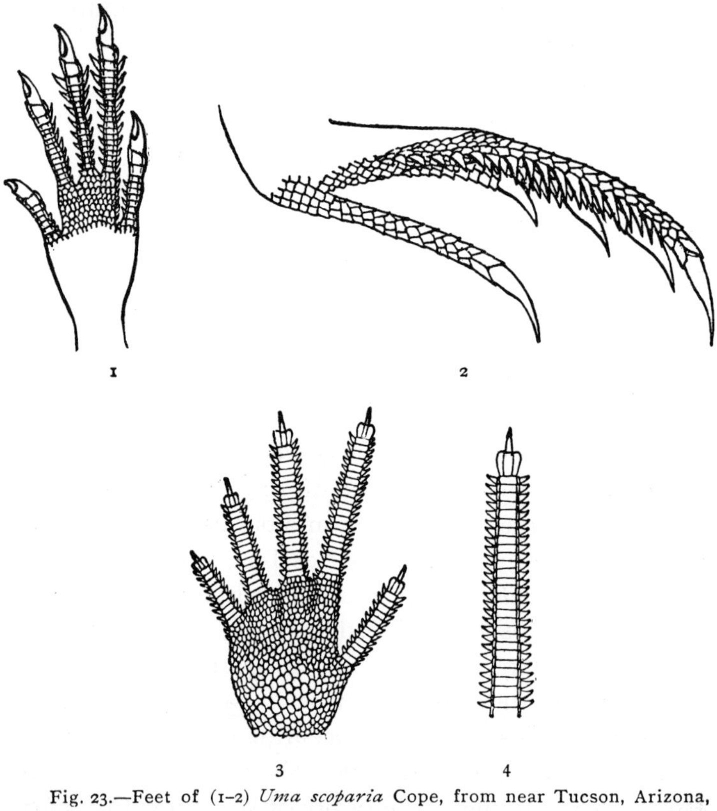

Fig. 23.—Feet of (1–2) *Uma scoparia* Cope, from near Tucson, Arizona, and (2–3) *Ptenopus garrulus* Smith, from South Africa. No. 1, manus; No. 2, pes; Nos. 3–4, pes. Nos. 1–2, original; Nos. 3–4, from Boulenger.

An illustration of homoplassy is to be seen in the spines on the head, tail, and feet of lizards which inhabit desert regions. The parallelism between Phrynosoma of the American dry regions, and the Moloch of the corresponding climates of Australia has been already noted. In the deserts of Asia, South Africa, and North America some of the lizards exhibit a great elongation of the lateral scales of the digits on one or both extremities. These become fringes of spines, freely articulated at their bases with the integument. By penetrating the sand, they increase the hold on its yielding surface, and greatly improve the speed of their movements. The genera in which this structure is conspicuous in the three localities in question, belong to as many distinct families. Thus in Asia it is the genus Phrynocephalus of the family Agamidæ; in South Africa it is Ptenopus of the Gecconidæ; while in North America it is Uma of the Iguanidæ. I give figures of the feet of Ptenopus and Uma for comparison. (Fig. 23.) Phrynocephalus is more like Uma than is Ptenopus.

In the succeeding chapters of this book many illustrations of the serial relation of characters will be given, so that it is not necessary to occupy more space with the subject here.

CHAPTER II.—PHYLOGENY.

1. GENERAL PHYLOGENY.

THE actual phylogeny or genealogy of organisms can only be positively determined by paleontologic research. We have been able in this way to obtain numerous lines of descent of animals and some general results as to the genealogic relations of the primary types of animals and plants. Many forms of both animals and plants are and have been without those hard parts which are susceptible of preservation in the formations of the earth's crust, so that no trace of their existence remains to us. In these cases our resort is embryologic investigation, since the embryonic history is a more or less complete recapitulation of the types of the past ages, from which the existing ones are descended. But since many representatives of the ancient and primitive forms of life still remain on the earth, we can trace, by the study of their structure, the larger features of general phylogeny. So far as we have compared the results derived from these three lines, it has been found that they coincide in their indications. We have in this a satisfactory proof that our conclusions are trustworthy contributions to the knowledge of the history of life.

The study of phylogeny shows that the evolution

of life-forms has been from the simple to the complex, and from the generalized to the specialized. These two forms of expression are not identical. In the phrase, "from the simple to the complex," is implied an ascending scale of evolution. In the phrase, "from the generalized to the specialized," we may include both progressive and retrogressive evolution. Retrogressive or degenerative evolution has been a frequent phenomenon in the past, and scarcely an organism exists which does not display degeneracy in some detail of its structure. Progressive evolution has, however, not been prevented by the frequent occurrence of an opposite process; and, indeed, degeneracy of parts, or of types of life, have been necessary to the advance of other and better organs or forms.

It is necessary to an understanding of the laws of evolution to get beforehand some idea of what that evolution has actually been. I will, therefore, give a general outline of the phylogeny of plants and animals, and will thus illustrate the subject in full detail in the case of the Vertebrata, where our facilities are especially good.

It is well known that the Protophyta and the Protozoa are not distinguishable by any sharp line of demarcation. Chlorophyll, which is so characteristic of plants, is absent from many of the lowest forms, including the entire class of Fungi, while it is present in a few of the lowest animals. The capacity for motion from place to place, so general in animals, at least in their earlier stages, is present in the earlier stages of some of the Algæ, and is universal, except at the period of reproduction, in the Myxomycetes. If it be denied that the latter are plants, then they are animals which do not reproduce by the ordinary process of fission, but by spores, which are borne together in a mass, or in a sporangium distinguished from the rest of the body. A distinction between the lower animals and plants is that the former introduce their food at a definite, though it may be a variable, point of the body; while plants absorb their nutriment in solution at all points. It is however demonstrated that some animals (many parasites) nourish themselves in the same manner as do plants; and in their early stages the Myxomycetes have the feeding habit of the lowest animals (Amœbæ). Both animals and plants from a morphological point of view have a common origin, in a nucleated, undivided, more or less globular piece of protoplasm or sarcode.

From freely moving Protophyta of this form, the vegetable kingdom took its rise. They first of all assumed a sessile position on the earth, and became what one may call earth-parasites. This abandonment of free mobility we cannot hesitate to regard as the efficient cause of a degenerate line of evolution. There can be no doubt about this, since fixity of habitat at once limits enormously the range of active influences which tend to modify an organism, whether they proceed from within it or from without it. The subsequent inclosure of the protoplasm in a structure of cellulose removes them from many of the stimuli which have so potent an influence in the life-history of animals, and the storage of other substances, as proteids, gums, resins, etc., in their cells, still further emphasizes the distinction.

From this beginning, progress in plants is seen chiefly in the modifications of their methods of reproduction. This function is the aim of the vegetable kingdom so far as their own condition is concerned.

Incidentally they are, however, essential to the existence of the animal kingdom, since they alone elaborate protoplasm and proteids from inorganic nature. In the simplest plants there is no sexuality, and reproduction is effected by spores which are mere fragments of the parental protoplasm (Protophyta). In the next stage sexual conjugation is necessary, but the sexes do not differ from each other in characters (Zygophyta). In the third stage (Oöphyta) the sexes are distinct, and the reproductive elements are distinguished as female germ-cell and male antheridium. In the remaining types of plants a distinct set of individuals, the prothallia, is produced by cell-division, whose function is sexual reproduction, thus constituting an alternation of generations. These plants may be entirely cellular (Carpophyta), or may be furnished with vascular canals. Of the latter the male and female prothallia may be naked and free (Pteridophyta or ferns, etc.), or may be enclosed in modified leaves, or flowers, the Phænogamia or flowering plants.

For the reasons already mentioned the order of "successional relation" above pointed out, is likely to prove to be the order of appearance of plants in time, and that such is the fact is demonstrated by their paleontology. In the earliest beds in which plants are positively known to occur, the Ordovician, we have only Algæ (Zygophyta and Oöphyta). In the Siluric we have a great predominance of the same classes, a very few species of which appear to have formed great erect stems. In the next period, the Devonic, probable Carpophyta are present, while the vascular Pteridophyta appear for the first time, and in considerable numbers. A few members of the gymnospermous Phænogamia (Coniferæ) appear. In the Carbonic pe-

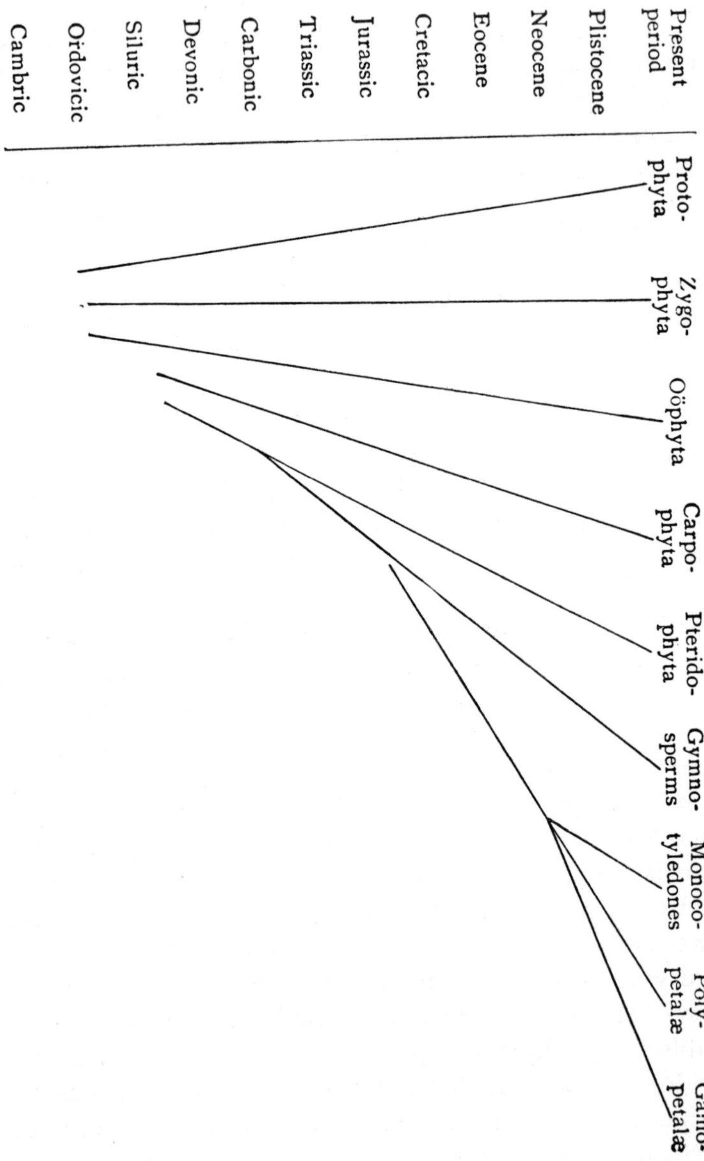
Present period
Plistocene
Neocene
Eocene
Cretacic
Jurassic
Triassic
Carbonic
Devonic
Siluric
Ordovicic
Cambric
Proto-phyta
Zygo-phyta
Oöphyta
Carpo-phyta
Pterido-phyta
Gymno-sperms
Monoco-tyledones
Poly-petalæ
Gamo-petalæ

riod, the greatest known development of the Pteridophyta (Lycopodia, ferns, Equiseta) took place, while the Gymnosperms were still represented by a few genera. Their period of development or acme arrived in the Mesozoic ages, and the Pteridophyta underwent a corresponding reduction of numbers and importance. Not until the upper Cretaceous epoch did the Angiospermous Phanerogams with their attractive flowers appear, and from that period to the present they have gained and maintained the ascendency. In accordance with the mode of origin of tubular flowers by the fusion of the separate petals of polypetalous forms, we find that the former succeeded the latter in time.

We may review this brief sketch of the paleontology of plants in the preceding phylogenetic table.

Turning now to the animal kingdom, its order of succession may also be perceived in existing species. As with plants we commence with unicellular asexual forms (Protozoa). Some of these increase by division only (Rhizopoda), while others must occasionally conjugate, since, according to Maupas, the reproductive energy is exhausted by continued self-division (Infusoria). Such structural specializations as the highest of the Protozoa possess, are merely vacuities or processes of their material, for the purpose of internal or external motion. In the second grade of organization animals are multicellular, or composed of more than one protoplasmic unit. The simplest of these, as Volvox, contains no specialized organs, but seems like a colony of Protozoa, although all its cells are not exactly alike (Ryder). An appreciable advance of structure defines the next class, the Cœlenterata. Here the multicellular mass contains a distinct digestive chamber, from which usually radiate tubes towards the periphery, which distribute the products of digestion. The first nervous threads appear. But there is as yet no body cavity which should form a sac in which the organs of nutrition and reproduction should be suspended. The Porifera (sponges) appear to be a much modified form of this type.

This grade of specialization belongs to the greater number of the five succeeding classes, the Echinodermata, Vermes, Mollusca, Arthropoda, and Vertebrata. Where it is absent from a few of the three lower classes, it is supposed to have disappeared by degeneracy. The Echinodermata come first in consideration. In these animals the form inclines to be radiated and the nervous system presents no single axis, but consists of branches which radiate from a ring surrounding the oral extremity of the digestive canal. In the second type, the general form is longitudinal and may be segmented, and a nervous axis may extend longitudinally from one or more points on the œsophageal ring. These are the Vermes (true worms). Embryology indicates clearly the common origin of the Echinodermata and the Vermes. No line of descent can be traced from the former, but from the latter we have traced the remaining branches of animals, three in number. Lowest of these is the Mollusca. The form of the body is sac-like, and the nervous system displays, typically, besides the œsophageal ring, a second ring, consisting of lateral commissures, making a single median ganglion of the foot. The body is not segmented, and there are no jointed limbs. In the second branch, that of the Arthropoda, the body is longitudinal and segmented, and segmented limbs are present. There is a median nervous axis, proceeding from the œsophageal ring along the inferior axis of the body, connecting several ganglia, (with some exceptions where the ganglia are fused or wanting). There is no internal skeleton. In the last and highest branch, that of the Vertebrata, the body is longitudinal and is segmented. It has a longitudinal nervous axis on the superior middle line, which is supported below by an axis of resistant material, usually bone, which forms the axis of an internal skeleton. Segmented limbs are present.

The lines of descent of these branches indicated by embryology are as follows :

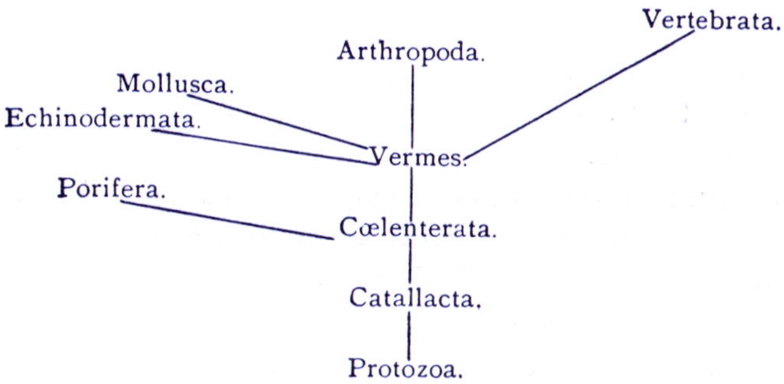

The above series present a history which is, on the whole, very different from that already described as characterizing the vegetable kingdom. Between the first and last terms of the series, there is exhibited a great progressive advance in all the higher features of life. These are mobility, and such control over the environment as it gives; and sensibility, through the development of a nervous system, which gives control over the movements. The highest development is that of mental characteristics, as emotions and intelligence, which are especially seen in the higher Vertebrata.

This progress has not been accomplished without much degeneracy by the way. All of the branches display divisions which have become sessile, and some of them are almost altogether so. Among Cœlenterata the Actinozoa are fixed, and often develop a calcareous skeleton. Many of the Hydroids are sessile. The great branch of the Echinodermata has its locomotive powers greatly curtailed, and many of them are permanently sessile. The same is true of the Mollusca. Both divisions are at one side of the line of progressive evolution as a consequence of this tendency. The Vermes display in their free representatives the conditions of progressive evolution. Being longitudinal and bilateral, one extremity becomes differentiated by first contact with the environment, as the seat of special senses, the basis having been secured by the location there of the nervous centers and ring. The Arthropoda present us with a great development of locomotive organs, and of special senses. As a whole, they have not made a considerable advance into the possession of the higher animal mental capacities, but display various degeneracies or degenerate tendencies among themselves. The moderately specialized as to structure are the most intelligent. These are the Hymenoptera, which display mental capacity superior to that of many Vertebrata. The latter branch, although presenting one sessile type, the Urochorda, has produced in its highest class, the Mammalia, the most general elevation in this, the highest of animal functions. This intelligence is in most of the types expended in preserving themselves from destruction against hostile environments, and the conquest of nature thus effected is remarkable from a physical point of view, but is an end of no great elevation of purpose from a mental standpoint. It is only in the most intelligent of the Mammalia, and in man, that we behold social and intellectual qualities which express something more than a mere routine of material existence.

Since Protozoa are very fragile, even when possessed of shells of mineral salts, we cannot expect to discover the actual date of their first appearance on the earth. Nevertheless they have been recently discovered in the later Archæan (=Huronian) beds of France. Some of the simplest Cœlenterata, however, (the Actinozoa), have deposited lime salts in the septa of their digestive chambers, and in some instances over their entire surface, so that their preservation has been assured. Thus we can prove that the simplest coral animals appear in the oldest rocks of sedimentary origin, the Cambric. Probably Vermes, and positively Echinodermata, Mollusca, and Arthropoda (Trilobita), also appear in the Cambric. Vertebrata appear definitely in the Siluric ; their supposed appearance in the Ordovicic being very doubtful.

The paleontologic history conforms to the systematic order in so far as it shows that the Cœlenterata appeared first, and the Vertebrata last, in time. A more complete correspondence between the two histories is found in the divisions of these branches, and I will take up the Vertebrata as the one of whose beginning we know the most, and are likely to know more.

2. THE PHYLOGENY OF THE VERTEBRATA.

a. Phylogeny of the Classes.

As the illustrations of evolution in the present work are mainly drawn from the Vertebrata, I go somewhat into detail in discussing the phylogeny of that branch.

They present the advantage, that, since they appeared last of the animal kingdom in time, we can obtain a clearer view of their beginnings than in the case of the other great branches.

Before going into the subject I wish to call attention to a prevalent source of error in the construction of phylogenies. This is the confusion of ideas general among naturalists who are not at the same time competent systematists, as to the subordination of characters. All correct phylogenetic inference depends on a correct appreciation of the value of characters. Failing this, error and confusion result. If, for instance, it is alleged that such a genus is ancestral to another genus, it is often forgotten that the descent of generic character, and not specific character, is meant. The usual type of critic attempts to contradict such hypothesis by showing some incongruity in *specific* characters, a matter which is quite irrelevant to the issue. Thus Madame Pavlov finds that *Hippotherium mediterraneum* is not the ancestor of *Equus caballus*, and comes promptly to the conclusion that the genus Hippotherium is not ancestral to the genus Equus. This is a *non sequitur*, for there are perhaps twenty species of Hippotherium, some of which are almost certain to have been ancestral to species of American Equus. In like manner, if it is alleged that the condylarthrous Mammalia are ancestral to the Diplarthra, if it should happen that no known genus of the former fits exactly the position of ancestor to any genus of the latter, in our present state of knowledge, the contention is not thereby vitiated, and it is implied that such genus will certainly be found. If it is also alleged that Condylarthra have been the ancestors of the anthropoid line, if some of the known genera of the former turn out to have no clavicle, a bone which is possessed by the latter, it is only to be concluded that the early lemuroids were derived from Condylarthra which possessed a clavicle. And in the discussion of the descent of one order from another, care must be taken that family, generic, and even specific characters are not imported into the discussion.

It is this confusion of ideas on the part of both phylogenists and their critics, that has brought phylogenetic schemes into a discredit in some quarters, which is sometimes deserved and sometimes undeserved. Embryologists are especially apt to construct impossible phylogenies, as they are generally not systematists, and frequently not anatomists. An excellent illustration of an impossible phylogeny is that of the fishes published a few years ago by the embryologist Dr. Beard. As an illustration of clean-cut phylogeny without confusion, I cite that of Haeckel; which I have shown to be, as regards the Vertebrata, mainly correct.

In attempting to ascertain the course of evolution of the Vertebrata, and to construct phylogenetic diagrams which shall express this history, among the difficulties arising from deficient information one is especially prominent. As is well known, there are many types in all the orders of the Vertebrata which present us with rudimentary organs, as rudimental digits, feet or limbs, rudimental fins, teeth, and wings. There is scarcely an organ or part which is not somewhere in a rudimental and more or less useless condition. The difficulty which these cases present is, simply, whether they be persistent primitive conditions, to be regarded as ancestral types which have survived to the present time, or whether, on the other hand, they be results of a process of degeneration, and therefore of comparatively modern origin. The question, in brief, is, whether these creatures presenting these features be primitive ancestors or degenerate descendants.

A great deal of light has been happily thrown on this question, as regards the Vertebrata, by the recent work done in North American paleontology. The lines of descent of many of the minor groups have been positively determined, and the phylogenetic connections of most of the primary divisions or classes have been made out. The result of these investigations has been to prove that the evolution of the Vertebrata has proceeded not only on lines of acceleration, but also on lines of retardation. That is, that evolution has been not only progressive, but at times retrogressive.

The Amphioxus (genus Branchiostoma) is generally regarded as the ancestral vertebrate. There are many reasons why this position must be accepted, although it possesses a few secondary modifications. Whether Branchiostoma derived its descent from an annelid worm, or from a tunicate, is a vexed question. Brooks[1] remarks as to this, "Up to this point I believe that the ancestral history of the tunicates was identical with that of the vertebrates; for the hepatic cœcum, the dilated pharynx, the pharyngeal clefts, the hypopharyngeal gland, and the peripharyngeal bands, have been inherited by all the Chordata (Vertebrata), and have impressed themselves so firmly in their organization that even the highest vertebrates still retain them, either as vestiges or as organs which have been fitted to new functions. I believe, however, that while they were acquired before the tunicates diverged from the chordate (vertebrate) stem, they were acquired by an

[1] *Studies from the Laboratory of the Johns Hopkins University*, 1893, p. 175.

organism whose environment and habits of life were essentially like those of the modern Appendicularia.' Appendicularia is well known as the tunicate which retains throughout life, the notochord and tail which characterize the larvæ of other Tunicata.

Omitting from consideration the two classes above mentioned (Acrania and Tunicata), whose remains have not yet been certainly found in a fossil state, there remain the following: the Pisces, Batrachia, Monocondylia, and Mammalia.[1]

I have traced the origin[2] of the Mammalia to the theromorous reptiles of the Permian epoch, and these to the Cotylosauria. The latter include the Pelycosauria, Procolophonina, Anomodontia, and perhaps other orders. In the Cotylosauria the temporal region is roofed over, which roof is reduced in the Pelycosauria to one postorbital arch of the skull, and this is the zygomatic of the Mammalia. In both Reptilia and Mammalia (excepting Prototheria and Procolophonina[3]) the coracoid element is of reduced size, and is co-ossified with the scapula. In both (except Cotylosauria) the capitular articulation of the ribs is intercentral. In both the humerus has distal condyles and epicondyles, and there is an entepicondylar foramen in the Pelycosauria as in the lower Mammalia. The posterior foot

1 See *The Evolution of the Vertebrata Progressive and Retrogressive; Amer. Naturalist*, 1885. Dohrn, *Der Ursprung der Wirbelthiere und das Princip des Functionwechsels*, Leipsic, 1875. "On the Phylogeny of the Vertebrata," Cope, *American Naturalist*, Dec., 1884. See also the following references: *American Naturalist*, 1884, p. 1136; *Proceedings of the Academy of Philadelphia*, 1867, p. 234; *Proceedings American Philosophical Society*, 1884, p. 585; *American Naturalist*, 1884, p. 27; *Proceedings American Association for the Advancement of Science*, XIX, 1871, p. 233; *Proceedings American Philosophical Society*, 1882, p. 447; *American Naturalist*, 1884, pp. 261 and 1121; *Report U. S. Geol. Survey W. of 100th Mer.*, G. M. Wheeler, 1877, IV, 2, p. 282.

2 *Proceedings American Philosophical Society*, 1884, p. 43.

3 Seeley, *Philos. Trans. Royal Society*, 1889, 269.

is constructed in the Pelycosauria almost exactly like that of the Prototheria. The single occipital condyle of the reptiles is not found in the Mammalia, but in some of the Lacertilia (Uroplates, Gecco) there are two condyles, the median (basioccipital) portion of the single condyle being rudimental, and Seeley has recently shown that it is deeply divided at the middle in the Permian Cynognathidae of South Africa.. The Pelycosauria could not, however, have given origin to the Prototheria, since in that subclass of mammals there is a well-developed coracoid. But in the Procolophonina this element is developed as in the Prototheria. Moreover, the Pelycosauria and the Procolophonina have the interclavicle, which is an element of membranous origin, while in the Prototheria we have the corresponding cartilage bone, the episternum. This element is present in the Permian order of the Cotylosauria, which is nearly related to the Pelycosauria. This order has, however, single-headed ribs, springing from the diapophyses, which is not usual in the Mammalia. But in some Cotylosauria the diapophyses are short, and in the Monotremata the postcervical ribs are single-headed, so this character is not an insurmountable one. It is evident that the Mammalia were derived from some type probably referable to a Permian reptilian order of the Theromorous series, although to which one is not yet known.

The Reptilia have been supposed by Haeckel to have taken their origin from the Batrachia. I have indicated that it is probable that the batrachian order, which stands in this relation to the Reptilia, is the Embolomeri of the Permian epoch. This conclusion rests on the following considerations. The reptilian order of the Cotylosauria approaches the Batrachia of the subclass Stegocephali in the overroofing of the posterior regions of the skull; in the presence of vomerine teeth, and in the absence of obturator foramen of the pelvis. In some Cotylosauria (Diadectidæ) the stegocephalian tabular bone of the skull is well developed. But in the Cotylosauria, the vertebral column consists mainly of centra, while in the Stegocephali it consists entirely or partly of intercentra. But in the Embolomeri the centra are well developed, and are larger than the intercentra anterior to the pelvis. Hence this is the only order of Stegocephali from which the Reptilia could have been derived.

Haeckel derived the Batrachia from the Dipnoi (Dipneusta), and I followed him in this belief, being strengthened in it by Huxley's ascription of an autostylic suspensorium of the mandible[1] to both divisions. This phylogeny is questioned by Pollard[2] and by Kingsley,[3] who would see the ancestry of the Batrachia in the crossopterygian fishes on embryological grounds derived from a study of Polypterus. In support of their view I would cite the absence of the maxillary arch in the Dipnoi, and its full development in the Stegocephali, which are the ancestral Batrachia. The large development of the dorsal and anal fins in the Dipnoi is not favorable to the Haeckelian view; nor do the paired fins approach as nearly to the limbs of Batrachia as do those of some other fishes. It has been shown by Huxley that the suspensorium of the Batrachia is hyostylic in its earliest stages, and that it becomes autostylic at a later period of development.

1 *Proceedings Zoölogical Society of London*, 1876, p. 59.

2 *Anatomischer Anzeiger*, VI, p. 338, 1891.

3 *American Naturalist*, 1892, p. 679. Kingsley would also derive the Dipnoi from Crossopterygia.

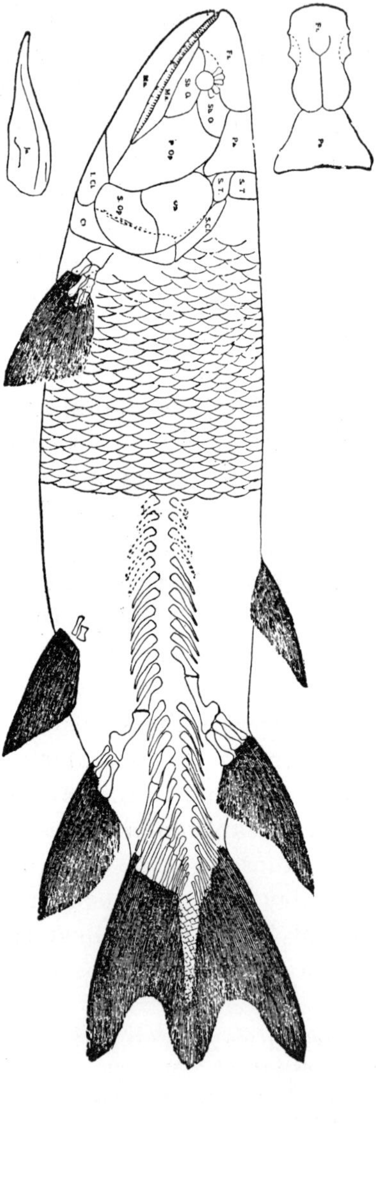

Fig. 24.—*Eusthenopterum foordii* Whiteaves; ⅓ natural size. Devonian of New Brunswick. From Whiteaves.

The Batrachia may then have originated from a hyostylic teleostomous (i. e., one with complete maxillary arch) fish. Among Teleostomata we naturally look for forms with limbs which approach nearest the batrachian type, and in which median fins are feeble or wanting. Such are the Rhipidopterygia (formerly included in the Crossopterygia), which include the families of Holoptychiidæ, Tristichopteridæ, Osteolepididæ, Cœlacanthidæ, and perhaps some others. These families, except the last, abounded in the waters of the Devonian period, at the time when the ancestors of the Batrachia also existed. All of them agree in possessing the median fins of greatly reduced proportions, and the mesodermal or internal elements of the paired fins more like the limbs of the Batrachia than are those of any known fishes. The constitution of the superior cranial wall is a good deal like that of the stegocephalous Batrachia. The characters of the fins can be learned from the accompanying figure of the *Eusthenopterum foordii* Whiteaves, one of the Tristichopteridæ. The pectoral fin well-nigh realizes Gegenbaur's theory of the derivation of the Chiropterygium from the Archipterygium.

The ancestral type of fishes is probably the acanthodean order of the subclass of sharks (Elasmobranchii).[1] Like other sharks, they are hyostylic and have no maxillary arch or cranial bones. They have the ptychopterygium, which is the primitive type of fin. In this fin the osseous elements which support the fin-rays are enclosed within the body-wall, the rays only being free. Such a fin sustains the hypothesis that the paired fins are parts of primitively continuous

[1] As represented by the Cladodontidae; see Dean, *Trans. N. Y. Acad. Sci.*, 1893. p. 124, and Cope, *American Naturalist*, 1893, p. 999.

longitudinal folds. This hypothesis is further sustained by the acanthodean genus Climatius, where a series of spines intervenes between the paired fins in

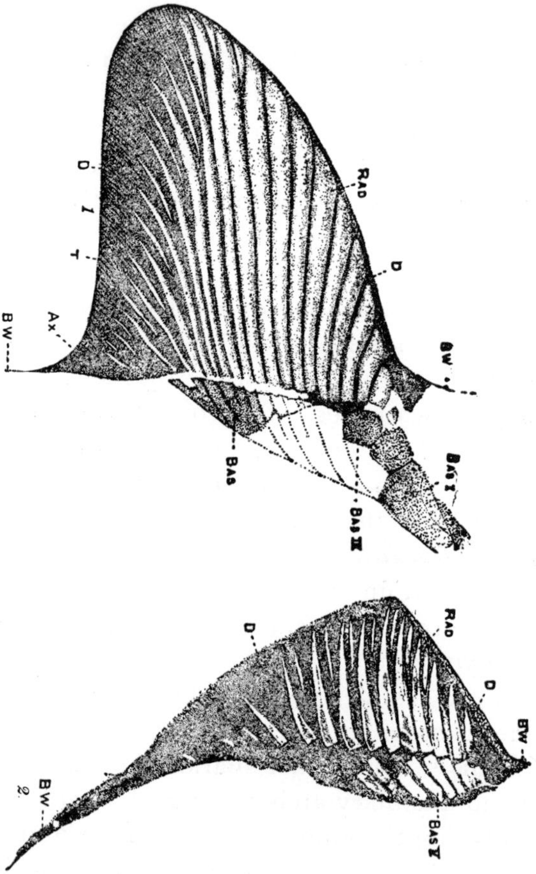

Fig. 25.—Paired fins of *Cladoselache Dean.* from the Carbonic of Ohio. No. 1, pectoral, No. 2, ventral fins. *BW*, body wall; *Ax*, axilla; *Bas*, basilar elements; *Rad*, "*radials*"; *T*, actinotrichia; *D*, dermal margin. From Dean.

line with them. From what the Acanthodeans can have been descended is at present conjectural. They trace their ancestry ultimately to Branchiostoma (Amphioxus) through forms not yet discovered. This genus represents the class Acrania, which is the ancestor of craniate Vertebrata.

The phylogeny of the Vertebrata may be represented diagrammatically as follows:

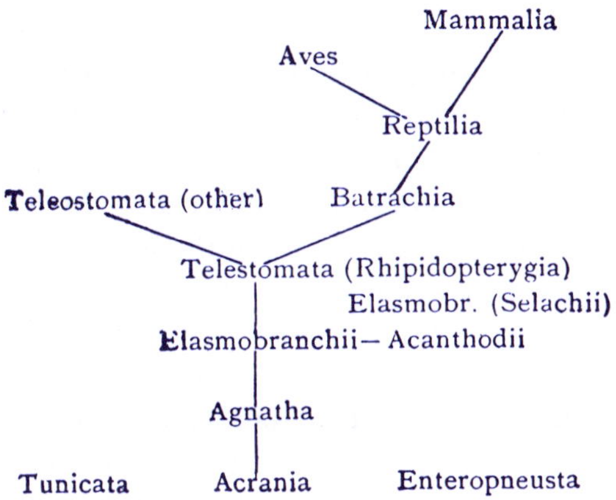

The Vertebrata exhibit the most unmistakable gradation in the characters of the circulatory system. It has long been the custom to define the classes by means of these characters, taken in connection with those of the skeleton. Commencing in the Leptocardii with the simple tube, we have two chambers in the Marsipobranchii and fishes; three in the Batrachia and Reptilia; and four in the Aves and Mammalia. The aorta-roots commence as numerous pairs of branchial arteries in the Leptocardii; we see seven in the Marsipobranchi, five in the fishes (with number reduced in some); four and three in Batrachia, where they generally cease to perform branchial functions; two and one on each side in Reptilia; the right-hand one in birds, and the left-hand one in Mammalia. This order is clearly an ascending one throughout. It consists of, first, a transition from adaptation to an aquatic, to an aërial respiration; and, second, an increase in the power to aërate and distribute a circulating fluid of increased quantity, and of increased calorific capacity. In other words, the circulation passes from the cold to the hot-blooded type coincidentally with the changes of structure above enumerated. The accession of a capacity to maintain a fixed temperature while that of the surrounding medium changes, is an important advance in animal economy.

The brain and nervous system also display a general progressive ascent. Leaving the brainless Acrania, the Marsipobranchs and fishes present us with small hemispheres with thin cortex, larger optic lobes, and well-developed cerebellum. The hemispheres are really larger than they appear to be, as Rabl Rückard has shown[1] that the supposed hemispheres are only corpora striata. But the superior walls are membranous, and support on their internal side only a layer of epithelial cells, as in the embryos of other Vertebrata, instead of the gray substance. So that, although we find that the cerebellum is really smaller in the Batrachia and most Reptilia than in the fishes, the better development of the hemispheres in the former gives them the preeminence. The Elasmobranchii show themselves superior to many of the fishes in the large size of their corpora restiformia and cerebellum. The Reptilia constitute an advance on the Batrachia. In the latter the optic thalami are, with some exceptions, of greater diameter than the hemispheres, while the reverse is generally true of the reptiles. The crocodiles display much superiority over the other reptiles in the larger cerebellum, with rudimental lateral lobes. The greater development of the hemispheres in birds is well known,

[1] *Biologisches Centralblatt*, 1884, p. 449.

while the general superiority of the brain of the living Mammalia over all other vertebrates is admitted.

The consideration of the successive relations of the skeleton in the classes of vertebrates embraces, of course, only the characters which distinguish those classes. These are not numerous. They embrace the structure of the axis of the skull; of the ear-bones; of the suspensors of the lower jaw; of the scapular arch and anterior limb, and of the pelvic arch and posterior limb. Other characters are numerous, but do not enter into consideration at this time.

The persistence of the primitive cartilage in any part of the skeleton is, embryologically speaking, a mark of inferiority. From a physiological or functional standpoint it has the same significance, since it is far less effective both for support and for movement than is the segmented osseous skeleton. That this is a prevalent condition of the lower Vertebrata is well known. The bony fishes and Batrachia have but little of the primitive cartilage remaining, and the quantity is still more reduced in the higher classes. Systematically, then, the vertebrate series is in this respect an ascending one. The Acrania are membranous; the Marsipobranchii and most of the Elasmobranchii cartilaginous; the other Pisces and the Batrachia have the basicranial axis cartilaginous, so that it is not until the Reptilia are reached that we have osseous sphenoid and presphenoid bones, such as characterize the birds and mammals. The vertebral column follows more or less inexactly the history of the base of the skull, but its characters do not define the classes.

As regards the suspensor of the lower jaw, the scale is in the main ascending. We witness a gradual change in the segmentation of the mandibular visceral arch of the skull, which clearly has for its object such a concentration of the parts as will produce the greatest effectiveness of the biting function. This is accom-

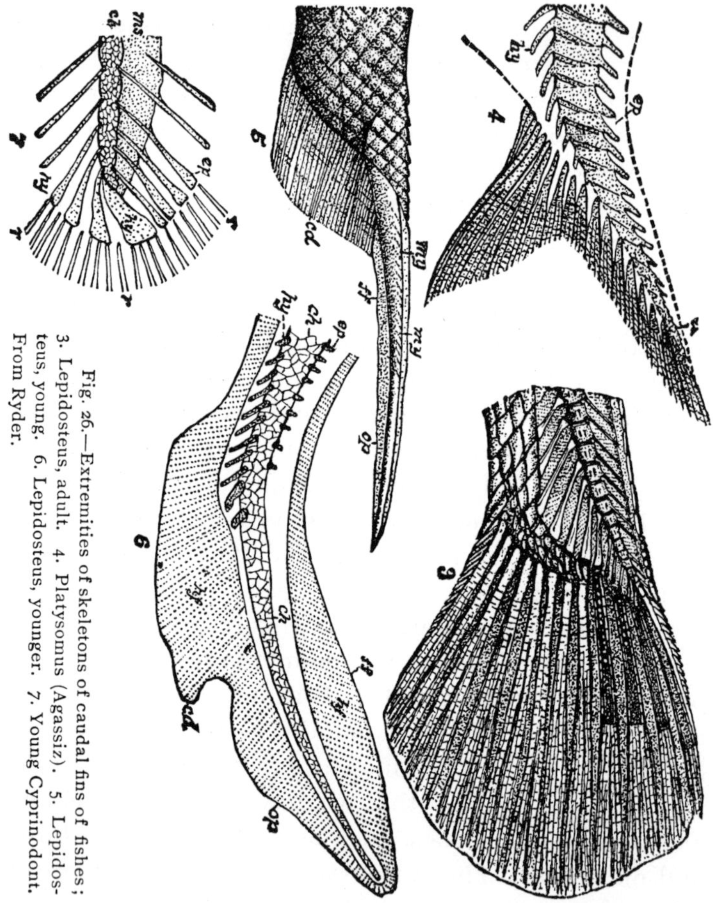

Fig. 26.—Extremities of skeletons of caudal fins of fishes; 3. Lepidosteus, adult. 4. Platysomus (Agassiz). 5. Lepidosteus, young. 6. Lepidosteus, younger. 7. Young Cyprinodont. From Ryder.

plished by reducing the number of the segments, so as to bring the resistance of the teeth nearer to the power, that is, the masseter and related muscles, and their base of attachment, the brain-case. This is seen in bony vertebrates in the reduction of the segments between the lower jaw proper and the skull, from four to none. In the fishes we have the hyomandibular, the symplectic, the inferior quadrate, and the articular. In the Batrachia, reptiles and birds, we have the quadrate and articular only, while in the Mammalia these elements also are wanting.

The examination of the pectoral and pelvic arches reveals a successive modification of the adaptation of the parts to the mechanical needs of the limbs. In this regard the air-breathing types display wide diversity from the gill-bearing types or fishes. In the latter, the lateral elements unite below without the intervention of a median element or sternum, while in the former the sternum, or parts of it, is generally present. Either arrangement is susceptible of much mechanical strength, as witness the siluroid fishes on the one hand, and the mole on the other. The numerous segments of the fishes' pectoral arch must, however, be an element of weakness, so that from a mechanical standpoint it must take the lowest place. The presence of sternal elements, with both clavicle, epicoracoid, and coracoid bones on each side, gives the Reptilia the highest place for mechanical strength. The loss of the bony coracoid seen in the tailed Batrachia, and loss of coracoid and epicoracoid in the Mammalia, constitute an element of weakness. The line is not then one of uniform ascent in this respect.

The absence of pelvis, or its extremely rudimental condition, in fishes, places them at the foot of the line in this respect. The forward extension of the ilium in some Batrachia and in the Mammalia, is to be compared with its backward direction in Reptilia, and its extension both ways in the birds. These conditions are all derived by descent from a strictly intermediate position in the Batrachia and Reptilia of the Permian epoch. The anterior direction must be regarded as having the mechanical advantage over the posterior direction, since it shortens the vertebral column and brings the grip of the posterior nearer to the anterior feet. The prevalence of the latter condition in the Mammalia enables them to stand clear of the ground, while the Reptilia move with the abdomen resting upon it, excepting the higher Dinosauria, where the arrangement is as in birds. As regards the inferior arches of the pelvis, the Mammalia have the advantage again, in the strong bony median symphysis connecting the ischium and pubis.[1] This character, universal among the land Vertebrata of the Permian epoch, has been lost by the modern Reptilia and birds, and is retained only by the Mammalia. So the lines, excepting the mammalian, have *degenerated* in every direction in the characters of the pelvis.

The limbs of the Pisces are as well adapted to their environment as are those of the land Vertebrata; but, from an embryological standpoint, their structure is inferior. The primitive rays are less modified in the fin than in the limb; and limbs themselves display a constantly increasing differentiation of parts, commencing with the Batrachia and ending with the Mammalia. The details of these modifications belong to the history of the contents of the classes, however, rather than to the succession of the Vertebrata as a whole.

In review, it may be said that a comparison of the characters which define the classes of the vertebrates shows that this branch of the animal kingdom has

[1] This is an advantage as a protection during gestation.

made with the ages successive steps of progress from lower to higher conditions. This progress has not been without exception, since, as regards the construction of the scapular arch, the Mammalia have retrograded from the reptilian standard as a whole.

In subsequent pages I shall take up the lines of the classes separately.

b. The Line of the Pisces.

The fishes form various series and subseries, and the tracing of all of them is not yet practicable, owing to the deficiency in our knowledge of the earliest or ancestral forms. Thus the origins of the three subclasses, Holocephali, Dipnoi, and Elasmobranchii, are lost in the obscurity of the early Paleozoic ages. The genus Paleospondylus of Traquair from the Carboniferous probably represents an Agnathous type from which all fishes may have sprung, although the genus, as now known, has not sufficient antiquity to claim this place. It may be a near descendant of the amphioxus.

A comparison of the four subclasses of fishes shows that they are related in pairs. The Holocephali and Dipnoi have no distinct suspensory segment for the lower jaw, while the Elasmobranchii and Teleostomata have such a separate element. The latter, therefore, present one step in the direction of complication beyond the former. It is, however, asserted by Huxley[1] that the absence of suspensorium is due to its appropriation by the hyoid arch in the Holocephali, and its rudimental condition in the Dipnoi. If this be the case, the Holocephali and Dipnoi are peculiar speciali-

[1] *Proceedings Zoölogical Society*, London, 1876, p. 45.

zations at one side of the main line of descent of the fishes. We look then for the ancestral type of the true fishes among the Elasmobranchii, and of these the Ichthyotomi display the greatest resemblances to the Teleostomata in all respects.[1]

Too little is known of the history of the subclasses, excepting the Teleostomata, for us to be able to say much of the direction of the descent of their contained orders. On the sharks much light is shed by the discovery of characters of the genus Cladodus Agass., in which the support of the paired fins consists of a metapterygium, which is enclosed in a lateral fold of the body wall, and which gives rise to simple external basilar rods only. Of the Teleostomata a much clearer history is accessible. It has four primary divisions or tribes which differ solely in the structure of the supports of the fins. In the first division, the Crossopterygia, the anterior limbs have numerous basilar bones which are supported on a peduncle of axial bones. The posterior limbs are similar. In the second division, or Podopterygia (the sturgeons, etc.), the posterior limbs remain the same, while the anterior limbs have undergone a great abbreviation in the loss of the axial bones and the reduction of the number and length of the basilar bones. In the third group, or Actinopterygia, both limbs have undergone reduction, the basilar bones in the posterior fin being almost all atrophied, while those of the fore limb are much reduced in number. In the fourth superorder, the Rhipidopterygia, the axial supports of the median fins are greatly reduced in number, presenting a marked contrast to the other superorders ; while the axial elements

[1] See *Proceedings American Philosophical Society*, 1884, p. 572, on the genus Didymodus.

of the paired fins are present and primitive, and resemble those of one of the suborders of sharks.

The phylogeny of the Teleostomata, as indicated by the fin-structure, will commence with the Crossopterygia. From this group the Podopterygia may be theoretically derived, and from these the Actinopterygia. The Rhipidopterygia appear to be a side group, not in the main piscine line. But the oldest known Crossopterygia are from the Carboniferous, while the Rhipidopterygia are abundant in the Devonian. Moreover, the superorder Actinopterygia, with its contracted fins, may have appeared in the Carboniferous, while the Podopterygia (Palæoniscidæ) certainly did so.

The descent of the fishes in general has witnessed, then, a contraction of the limbs to a very small compass, and their substitution by a system of accessory dermal radii. This has been an ever-widening divergence from the type of the higher Vertebrata, and from this standpoint, and also a view of the "loss of parts without complementary addition of other parts," may be regarded as a process of degeneration.

Taking up the great division of the Actinopterygia, which embraces most of the species of living fishes, we can trace the direction of descent largely by reference to their systematic relations when we have no fossils to guide us.

The three subtribes adopted by Jordan represent three series of the true fishes which indicate lines of descent. The Holostei include the remainder of the old ganoids after the subtraction of the Rhipidopterygia, the Crossopterygia, and the Podopterygia. They resemble these forms in the muscular bulbus arteriosus of the heart, in the chiasm of the optic nerves, and in the greater distinctness of the metapterygium. The two former characters are complexities which the two other divisions do not possess, and which, as descendants coming later in time, must be regarded as inferior, and therefore to that extent degenerate. Of these divisions the Malacopterygia approach nearest the Holostei, and are indeed not distinctly definable without exceptions. The third division, or Acanthopterygia, shows a marked advance beyond the others in : (1) the obliteration of the primitive trachea, or ductus pneumaticus, which connects the swim-bladder and œsophagus ; (2) the advance of the ventral fins from the abdomen forward to the throat ; (3) the separation of the parietal bones by the supraoccipital ; (4) the presence of numerous spinous rays in the fins ; and (5) the roughening of the edges of the scales, forming the ctenoid type. There are more or less numerous exceptions to all of these characters. The changes are all further divergencies from the other vertebrate classes, or away from the general line of ascent of the vertebrate series taken as a whole. The end gained is specialization ; but whether the series can be called either distinctively progressive or retrogressive, is not so clear. The development of osseous spines, rough scales, and other weapons of defense, together with the generally superior energy and tone which prevail among the Acanthopterygia, characterize them as superior to the Malacopterygia, but their departure from the ascending line of the Vertebrata has another appearance.

The descent of the Acanthopterygian fishes has probably been from Holostean ancestors, both with and without the intervention of Malacopterygian forms. This is indicated by increase in the number of basilar bones[1] in the fins of families which have pectoral ventral fins, and in the extinct genus Dorypterus.[2]

The Malacopterygia display three or four distinct lines of descent. The simplest type is represented by the order Isospondyli, and paleontology indicates clearly that this order is also the oldest, as it dates from the Trias at least. In one line the anterior dorsal vertebræ have become complicated, and form an interlocking mass which is intimately connected with the sense of equilibrium in the water. This series commences with the Characinidæ, passes through the Cyprinidæ, and ends with the Siluridæ. The arrangements for equilibration constitute a superadded complication, and to these are added in the Siluroids defensive spines and armor. Some of this order, however, are distinctly degenerate, as the soft purblind Ageniosus, and the parasitic Stegophilus and Vandellia, which are nearly blind, without weapons, and with greatly reduced fins.

The next line (the Haplomi, pike, etc.) loses the precoracoid arch and has the parietal bones separated, both characters of the Acanthopterygia. This group was apparently abundant during the Cretaceous period, and it may have given origin to many of the Acanthopterygia.

Another line also loses the precoracoid, but in other respects diverges totally from the Acanthopterygia and all other Malacopterygia. This is the line of the eels. They next lose the connection between the scapular arch and the skull, which is followed by the loss of the pectoral fin. The ventral fin disappeared sooner. The palatine bones and teeth disappear, and the suspensor

1 See Cope "On the Homologies of the Fins of Fishes"; *American Naturalist*, 1890, p. 401.

2 See *Proceedings of the American Association for the Advancement of Science*, 1878, p. 297.

of the lower jaw grows longer and loses its symplectic element. The opercular bones grow smaller, and some of them disappear. The ossification of most of the hyoid elements disappears, and some of their cartilaginous bases even vanish. These forms are the marine eels or Colocephali. The most extraordinary example of specialization and degeneracy is seen in the abyssal eels of the family Eurypharyngidæ. Here all the degenerate features above mentioned are present in excess, and others are added, as the loss of ossification of a part of the skull, almost total obliteration of the hyoid and scapular arches, and the semi-notochordal condition of the vertebral column, etc.

The Acanthopterygia nearest the Malacopterygia have abdominal ventral fins, and belong to several orders. It is such types as these that may be supposed to have been derived directly from Holostean ancestors. They appear in the Cretaceous period (Dercetidæ), along with the types that connect with the Malacopterygia (Haplomi). Intermediate forms between these and typical Acanthopterygii occur in the Eocene (Trichophanes, Erismatopterus), showing several lines of descent. The Dercetidæ belong apparently to the order Hemibranchi, while the Eocene genera named belong apparently to the Aphododiridæ, the immediate ancestor of the highest Physoclysti, the Percomorphi. The order Hemibranchi is a series of much interest. Its members lose the membrane of their dorsal spinous fin (Gasterosteidæ), and then the fin itself (Fistularia, Pegasus). The branchial apparatus has undergone, as in the eels, successive deossification (by retardation), and this in direct relation to the degree with which the body comes to be protected by bony shields, reaching the greatest defect in the Amphisilidæ. One more downward step is seen in the next succeeding order of the Lophobranchii. The branchial hyoid apparatus is reduced to a few cartilaginous pieces, and the branchial fringes are much reduced in size. In the Hippocampidæ the caudal fin disappears and the tail becomes a prehensile organ by the aid of which the species lead a sedentary life. The mouth is much contracted and becomes the anterior orifice of a suctorial tube. This is a second line of unmistakable degeneracy among true fishes.

The Acanthopterygia with pectoral ventral fins present us with perhaps ten important ordinal or subordinal divisions. Until the paleontology of this series is better known, we shall have difficulty in constructing phylogenies. Some of the lines may, however, be made out. The accompanying diagram will assist in understanding them.

The Anacanthini present a general weakening of the organization in the less firmness of the osseous tissue and the frequent reduction in the size and character of the fins. The caudal vertebræ are of the diphycercal type. As this group does not appear early in geological time, and as it is largely represented now in the abyssal ocean fauna, there is every reason to regard it as a degenerate type.[1] The Heterosomata (flounders) found it convenient to lie on one side, a habit which would appear to result from a want of motive energy. The fins are very inefficient organs of movement in them, and they are certainly no rivals for swift-swimming fishes in the struggle for existence, excepting as they conceal themselves. In order to see the better while unseen, the inferior eye has turned in-

[1] The general characters of the deep-sea fish-fauna are those of degeneracy.

ward, i. e., upward, and finally has penetrated to the superior surface, so that both eyes are on one side. This peculiarity would be incredible, if we did not know of

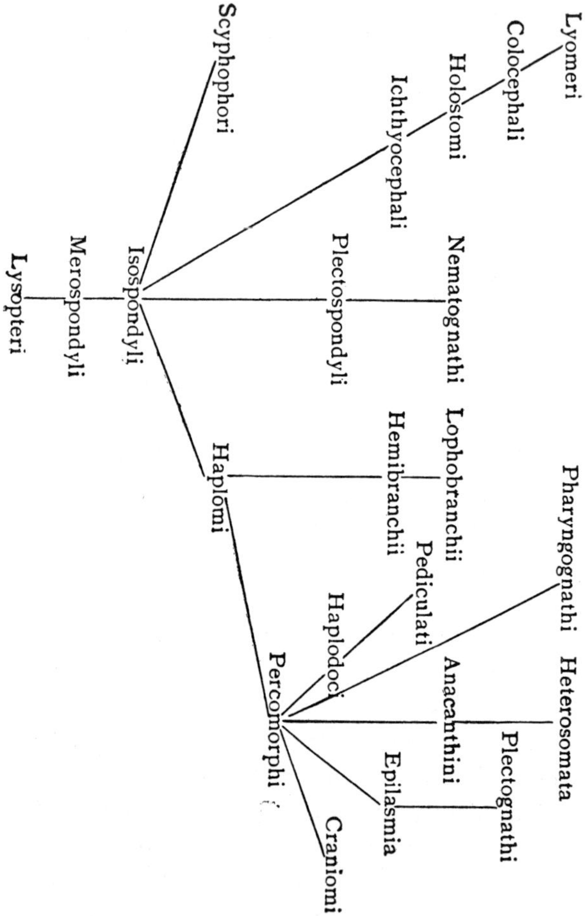

its existence, and is an illustration of the extraordinary powers of accommodation possessed by nature. The Heterosomata (flatfishes) can only be considered a degenerate group. The scyphobranch line presents a specialization of the superior pharyngeal bones, which is continued by the Haplodoci (Batrachidæ). This cannot be called a degenerate line, although the fin-rays are soft.

The double bony floor of the skull of the distegous percomorph fishes is a complication which places them at the summit of the line of true fishes. At the summit of this division must be placed the Pharyngognathi, which fill an important rôle in the economy of the tropical seas, and the fresh waters of the Southern Hemisphere. By means of their powerful grinding pharyngeal apparatus they can reduce vegetable and animal food inaccessible to other fishes. The result is seen in their multifarious species and innumerable individuals decked in gorgeous colors, and often reaching considerable size. This is the royal suborder of fishes, and there is no reason why they should not continue to increase in importance in the present fauna.

Very different is the line of the Plectognathi. The probable ancestors of this division, the Epelasmia (Chætodontidæ, etc.), are also abundant in the tropical seas, and are among the most brilliantly colored of fishes. One of their peculiarities is seen in a shortening of the brain-case and prolongation of the jaws downward and forward. The utility of this arrangement is probably to enable them to procure their food from the holes and cavities of the coral reefs, among which they dwell. In some of the genera the muzzle has become tubular (Chelmo), and is actually used as a blow-gun by which insects are secured by shooting them with drops of water. This shortening of the basicranial axis has produced a corresponding abbreviation of the hyoid apparatus. The superior pharyngeal bones are so crowded as to have become a series of vertical plates like the leaves of a book. These characters are further developed in the Plectognathi. The brain-case is very small, the face is very elongate, and the mouth is much contracted. The bones surrounding it in each jaw are coössified. The axial elements (pubes) of the posterior fins unite together, become very elongate, and lose the natatory portion. In one group (Orthagoriscidæ) the posterior part of the vertebral column is lost, and the caudal fin is a nearly useless rudiment. In the Ostraciontidæ (which may have had a different origin, as the pharyngeal bones are not contracted) the natatory powers are much reduced, and the body is inclosed in an osseous carapace so as to be capable of very little movement. The entire order is deficient in osseous tissue, the bones being thin and weak. It is a marked case of degeneracy.

There are several evident instances of sporadic degeneracy in other orders. One of these is the case of the family of the Icosteidæ, fishes from deep waters off the coast of California. Although members of the Percomorphi, the skeleton in the two genera Icosteus and Icichthys is unossified, and is perfectly flexible. Approximations to this state of things are seen in the parasitic genus Cyclopterus, and in the ribbon-fishes, Trachypteridæ.

Thus nearly all the main lines of the Acanthopterygii are degenerate; the exceptions are those that terminate in the Scombridæ (mackerel), Serranidæ, and Scaridæ (Pharyngognathi).

c. The Line of the Batrachia.

We know Batrachia first in the Coal Measures. They reach a great development in the Permian epoch, and are represented by large species in the Triassic period. From that time they diminish in numbers, and at the present day form an insignificant part of the vertebrate fauna of the earth. The history of their succession is told by a table of classification such as I give below:

I. Supraoccipital, tabular, and supramastoid bones present. Propodial bones distinct. STEGOCEPHALI.

Vertebral centra, including atlas, segmented, one set of segments together supporting one arch; *Rhachitomi.*

Vertebræ segmented, the superior and inferior segments each complete, forming two centra to each arch; *Embolomeri.*

Vertebral centra, including atlas, not segmented, one to each arch; *Microsauri.*

II. Supraoccipital and supramastoid bones wanting. Frontal and propodial bones distinct; URODELA.

a. An os tabulare.

A palatine arch and separate caudal vertebræ; *Proteida*

aa. No os tabulare.

A maxillary arch; palatine arch imperfect; nasals, premaxillaries and caudal vertebræ distinct; *Pseudosauria.*[1]

No maxillary or palatine arches; nasals and premaxilliary, also caudal vertebræ, distinct; *Trachystomata.*

III. Supraoccipital, tabular, and supramastoid bones wanting. Frontals and parietals connate; propodial bones and caudal vertebræ confluent; SALIENTIA.

Premaxillaries distinct from nasals; no palatine arch; astragalus and calcaneum elongate, forming a distinct segment of the limb; *Anura.*

The probable phylogeny of these orders as imperfectly indicated by paleontology is exhibited in the diagram on the following page.

An examination of the above tables shows that there has been in the history of the batrachian class a reduction in the number of the elements composing

[1] Includes the Gymnophiona.

the skull, both by loss and by fusion with each other. It also shows that the vertebræ have passed from a notochordal state with segmented centra, to biconcave centra, and finally to ball-and-socket centra, with a great reduction of numbers. It is also the fact that the earlier forms (those of the Permian epoch) show the most mammalian characters of the tarsus and of the pelvis. The later forms, the salamanders, show a more generalized form of carpus and tarsus and of pelvis also. In the latest forms, the Anura, the carpus and tarsus are reduced through loss of parts, except

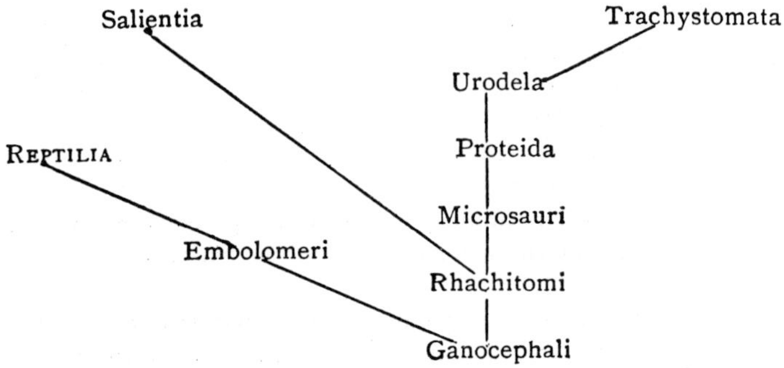

that the astragalus and calcaneum are phenomenally elongate. We have then, in the batrachian series, a somewhat mixed kind of change; but it principally consists of concentration and consolidation of parts. The question as to whether this process is one of progression or retrogression may be answered as follows: If degeneracy consists in "the loss of parts without complementary addition of other parts," then the batrachian line is a degenerate line. This is only partly true of the vertebral column, which presents the most primitive characters in the early, Permian, genera (Rhachitomi). If departure from the nearest approximation to the Mammalia is degeneracy, then the changes in this class come partly under that head. The scapular and pelvic arches of the Rachitomi are more mammalian than are those of any of their successors;

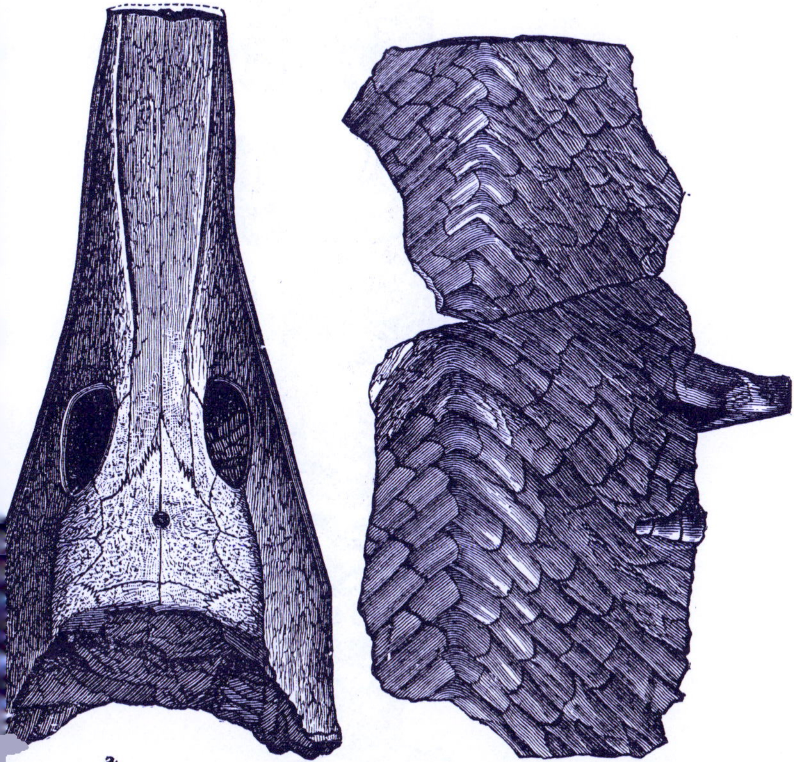

a b

Fig. 27.—*Cricotus crassidiscus* Cope, parts of individual represented in Fig. 28; one-third natural size. From Permian of Texas. *a*, head from above; *b*, part of belly from below. From Cope.

the carpus and tarsus are less so than that of the Anura.

There are several groups which show special marks of degeneracy. Such are the reduced maxillary bones

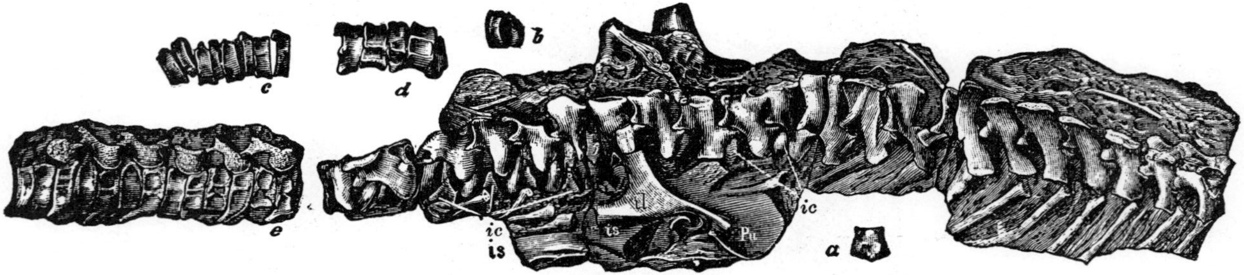

Fig. 28.—*Cricotus crassidiscus* Cope, vertebral column and pelvis; three-tenths natural size. Cope Coll. from the Permian formation of Texas. Fig. *a*, proatlas; *b*, *c*, cervical centra and intercentra; *d*, *e*, *f*, *g*, caudal centra; *is*, ischium; *il*, ilium; *Pu*, pubis; *is*, ischium.

and persistent gills of the Proteïda; the absence of the maxillary bones and the presence of gills in the Trachystomata; the loss of a pair of legs and feebleness of the remaining pair in the same; and the extreme reduction of the limbs in Amphiuma, and their total loss in the Cæciliidæ. Such I must also regard, with Lankester, the persistent branchiæ of the siredons. I may add that in the brain of the proteïd Necturus the hemispheres are relatively larger than in the Anura, which are at the end of the line.

It must be concluded, then, that in many respects the Batrachia have undergone degeneracy with the passage of time.

d. The Reptilian Line.

As in the case of the Batrachia, the easiest way of obtaining a general view of the history of this class is by throwing their principal structural characters into a tabular form. As in the case of that class, I commence with the oldest forms and end with the latest in the order of time, which, as usual, corresponds, with the order of structure. I except from this the first order, the Ichthyopterygia, which we do not know prior to the Triassic period:

I. The quadrate bone united with the adjacent elements by suture.
- A. Temporal region of skull with a bony roof; no postorbital bars.
  - Supramastoid bone present; an interclavicle; limbs ambulatory; *Cotylosauria.*
- AA. Cranium with one postorbital bar; no sternum. (Synaptosauria.)
- *a.* Paroccipital bone distinct.
  - A supramastoid bone; ribs two-headed on centrum; carpals and tarsals not distinct in form from metapodials; *Ichthyopterygia.*

No supramastoid; sub- and postpelvic ossifications; interclavicle and clavicles separated from and below scapular arch; ribs one-headed on centrum; coracoid large, free posteriorly; *Testudinata.*

aa. Paroccipital bone not distinct.

Ribs one or two-headed, capitulum intercentral; clavicles and interclavicles forming part of shoulder-girdle; scapula simple; pubis and ischium plate-like with small or no obturator foramen; no sub- or post-pelvic bones; no supramastoid; *Theromora.*

Supramastoid present; ribs one-headed; scapula triradiate; no sternum; pubis and ischium plate-like; no sub- or postpelvic bones; *Plesiosauria.*

AAA. Cranium with two postorbital bars; a sternum. (Archosauria.)

Paroccipital bone not distinct; no supramastoid.

Ribs two-headed; no interclavicle; external anterior digits greatly elongate to support a patagium; *Ornithosauria.*

Ribs two-headed; no interclavicle; acetabulum perforate; limbs ambulatory; *Dinosauria.*

Ribs two-headed; an interclavicle; acetabulum closed; feet ambulatory; *Crocodilia.*

Ribs one headed; an interclavicle; acetabulum closed; feet ambulatory; *Rhynchocephalia.*

II. The quadrate loosely articulated with the adjacent elements, and only proximally. (Streptostylica.)

One postorbital bar, when present; a paroccipital; supramastoid not distinct; ribs one-headed; *Squamata.*

An inspection of the characters of these ten orders, and their consideration in connection with their geological history, will give a definite idea as to the character of their evolution. The history of the class, and therefore the discussion of the question, is limited in time to the period which has elapsed since the Permian epoch inclusive, for it is then that the Reptilia enter the field of our knowledge. During this period two remarkable orders of reptiles inhabited the earth, those of the Cotylosauria and of the Theromora. The important character and rôle of these types may be inferred from the fact that the Cotylosauria are structurally nearer to the Batrachia and the Theromora to the Mammalia than any other, and the former presents characters which render it probable that all the other reptiles derived their being from them. The phylogeny may be thus expressed:

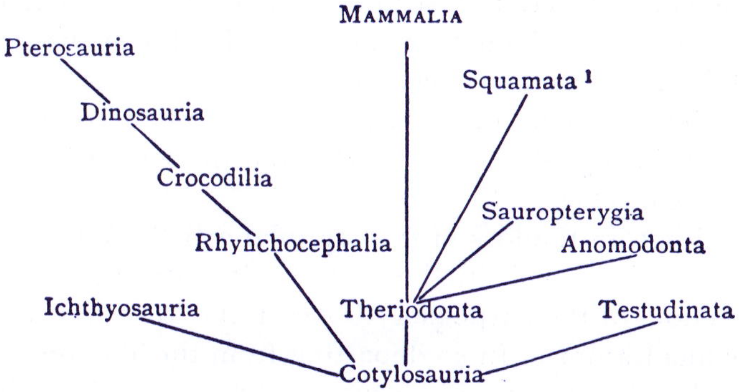

It is extremely probable that the characters of the posterior parts of the cranium of reptiles, as seen in the osseous bars posterior to the orbit, were derived by a kind of natural trephining of the cranial roof of the primitive order of the Cotylosauria. This order has left remains in the Permian beds of North America, South Africa, and Germany. This is the theory of Baur,[3] and I have rendered it probable by researches on the Permian genera of North America.[4]

[1] Some unknown type of Pythonomorpha will represent the ancestor of the Ophidia, while it is uncertain whether this order originated from the Theriodonta or the Rhynchocephalia.

[2] The Theromora include the Pelycosauria, Theriodonta, Anomodonta, and other suborders.

[3] *American Journal of Morphology*, 1889, p. 471.

[4] *Trans. Amer. Philos. Society*, 1892, p. 13; *American Naturalist*, 1892, p. 407.

The diagrams on pages 117–119 illustrate the successive changes in the structure of the posterior region of the skull in the types mentioned. The orders and suborders Pseudosuchia, Rhynchocephalia, Ichthyopterygia, Dinosauria, Crocodilia, Sauropterygia, and Testudinata commence at the beginning of Mesozoic time, after the Permian had closed. The Squamata (lizards and snakes) commence, so far as is certainly known, in the later Mesozoic, in the Cretaceous period.

The line which terminated in the Lacertilia and Ophidia (Squamata) may have originated directly from the Theriodonta, or it may have descended from the Rhynchocephalia. It departs from the former type in two respects:

First, in the loss of the capitular articulation of the ribs, and, second, in the gradual elongation and final freedom of the suspensory bone of the lower jaw (the os quadratum). In so departing from the Theromora, it also departs from the mammalian type. The ribs assume the less perfect kind of attachment which the mammals only exhibit in some of the whales, and the articulation of the lower jaw loses in strength, while it gains in extensibility, as is seen in the development of the line of the eels among fishes. The end of this series, the snakes, must therefore be said to be the result of a process of creation by degeneration, and their lack of scapular arch and fore limb and usual lack of pelvic arch and hind limb are confirmatory evidence of the truth of this view of the case.

Secondly, as regards the ossification of the anterior part of the brain-case. This is deficient in some of the Theromora, the ancestral series, which resemble in this, as in many other things, the contemporary Batrachia. The late orders mostly have the anterior walls

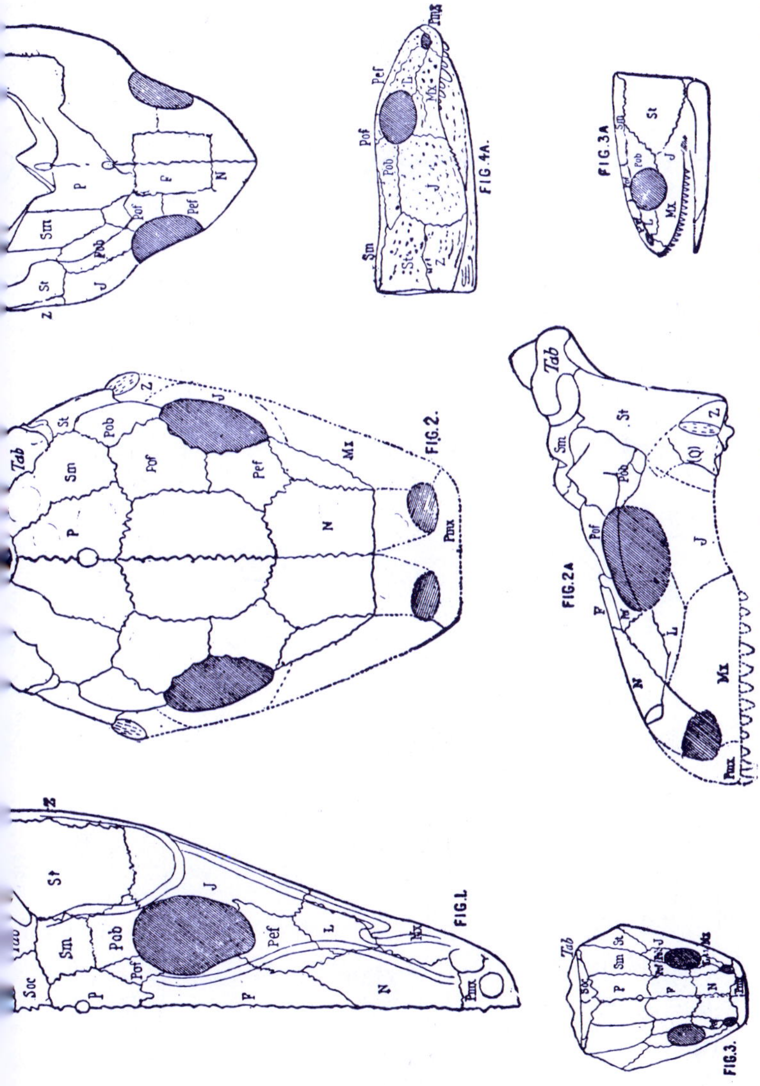

Fig. 29.—1. *Mastodonsaurus* Jaeger. 2. *Chilonyx* Cope. 3. *Pariotichus* Cope. 4. *Pantylus* Cope.—Fig. 1, Stegocephalian Figs. 2-4, Cotylosaurians.

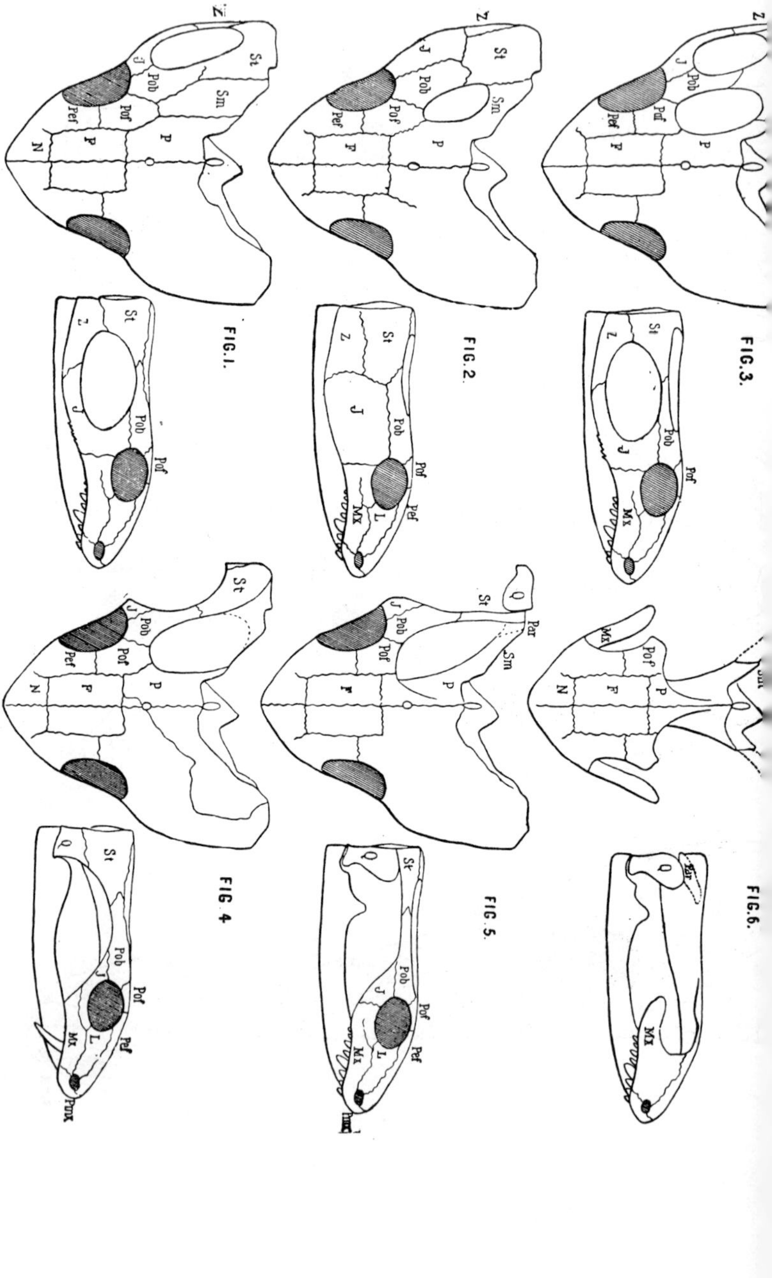

Fig. 30.—1. *Pelycosauria.* 2. *Pseudosuchia.* 3. *Rhynchocephalia.* 4. *Anomodontia.* 5. *Lacertilia.* 6. *Ophidia.*

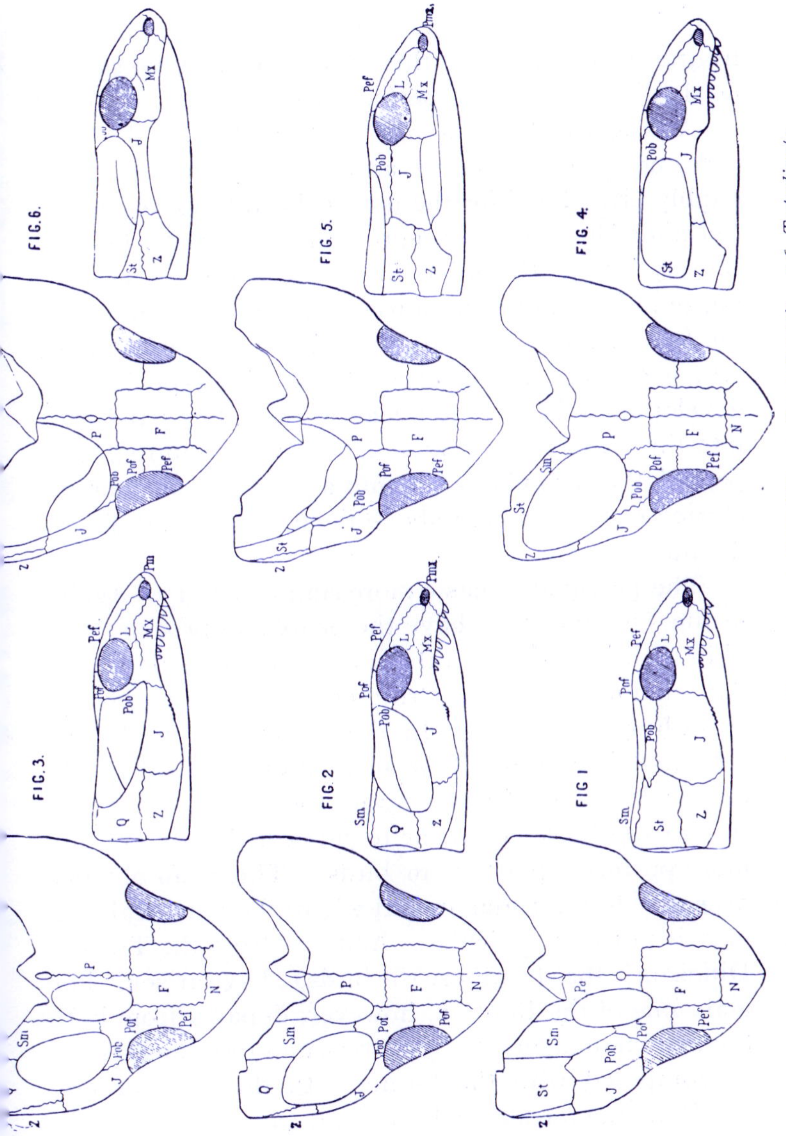

Fig. 31.—1. *Ichthyopterygia.* 2. *Dinosauria.* 3. *Crocodilia,* 4. *Sauropterygia.* 5-6 *Testudinata.*

membranous, but, in the streptostylicate series at the end, the skull in the snakes becomes entirely closed in front. In this respect, then, the latter may be said to be the highest or most perfect order.

As regards the scapular arch, including the sternum, no order possesses as many elements as thoroughly articulated for the use of the anterior leg as the Permian Theromora (excepting in the suborder Pelycosauria). In all the orders there is loss of parts, excepting only in the Ornithosauria and the Lacertilia. In the former the adaptation is to flying. The latter retain nearly the theromorous type. An especial side development is the modification of abdominal bones into three pairs of peculiar elements to be united with part of the scapular arch into a plastron, and the inclusion of the coracoid above them, seen in the Testudinata.

The pelvic arch has a more simple history. Again, in the Theromora we have the nearest approach to the Mammalia. The only other order which displays similar characters is the Ornithosauria (Dimorphodon, according to Seeley). In the Dinosauria we have a side modification which is an adaptation to the erect or bipedal mode of progression, the inferior bones being thrown backward so as to support the viscera in a more posterior position in birds. This is an obvious necessity to a bipedal animal where the vertebral column is not perpendicular. And it is from the Triassic Dinosauria that I suppose the birds to have arisen. The main line of the Reptilia, however, departs from both the mammalian and the avian type and loses in strength as compared with the former. In the latest orders, the Pythonomorpha and Ophidia, the pelvis is rudimental or absent.

As regards the limbs, the degeneracy is well marked. No reptilian order of later ages approaches so near to the Mammalia in these parts as do the Permian Theromora. This approximation is seen in the internal epicondylar foramen and well-developed condyles of the humerus, and in the well-differentiated seven bones of the tarsus. The epicondylar foramen is only retained in later reptiles in the rhynchocephalian Sphenodon (Dollo); and the condyles of the Dinosauria and all of the other orders, excepting the Ornithosauria and some Lacertilia, are greatly wanting in the strong characterization seen in the Theromora. The posterior foot seems to have stamped out the greater part of the tarsus in the huge Dinosauria, and it is reduced, though to a less degree, in all the other orders. In the paddled Plesiosauria, dwellers in the sea, the tarsus and carpus have lost all characterization, probably by a process of degeneracy, as in the mammalian whales. This is to be inferred from the comparatively late period of their appearance in time. The still more unspecialized feet and limbs of the Ichthyosaurus (Ichthyopterygia) can not yet be ascribed to degeneracy, for their history is too little known. At the end of the line, the snakes present us with another evidence of degeneracy. But few have a pelvic arch (Glauconiidæ Peters), while very few (Peropoda) have any trace of a posterior limb.

The vertebræ are not introduced into the definitions of the orders, since they are not so exclusively distinctive as many other parts of the skeleton. They nevertheless must not be overlooked. As in the Batrachia, the Permian orders show inferiority in the deficient ossification of the centrum. Many of the Theromora are notochordal, a character not found in any later order of reptiles excepting in a few Lacertilia (Gecconidæ). They thus differ from the Mammalia, whose characters are approached more nearly by some of the terrestrial Dinosauria in this respect. Leaving this order, we soon reach the prevalant ball-and-socket type of the majority of Reptilia. This strong kind of articulation is a need which accompanies the more elongated column which itself results at first from the posterior direction of the ilium. In the order with the longest column, the Ophidia, a second articulation, the zygosphen, is introduced. The mechanical value of the later reptilian vertebral structure is obvious, and in this respect the class may be said to present a higher or more perfect condition than the Mammalia.

In review it may be said of the reptilian line, that it exhibits marked degeneracy in its skeletal structure since the Permian epoch; the exception to this statement being in the nature of the articulations of the vertebræ. And this specialization is an adaptation to one of the conditions of degeneracy, viz., the weakening and final loss of the limbs and the arches to which they are attached.

The history of the development of the brain in the Reptilia presents some interesting facts. In the diadectid family of the Permian Cotylosauria it is smaller than in a *Boa constrictor*, but larger than in some of the Jurassic Dinosauria. Marsh has shown that some of the latter possess brains with relatively very narrow hemispheres, so that in this organ those gigantic reptiles were degenerate, while the existing streptostylicate orders have advanced beyond their Permian ancestors.

There are many remarkable cases of what may now be safely called degradation to be seen in the contents of the orders of reptiles.[1] Among tortoises may be cited the loss of one or two series of phalanges in several especially terrestrial families of the Testudinidæ. The cases among the Lacertilia are the most remarkable. The entire families of the Pygopodidæ, the Anniellidæ, the Anelytropidæ, and the Dibamidæ are degraded from superior forms. In the Anguidæ, Teidæ, and Scincidæ, we have series of forms whose steps are measured by the loss of a pair of limbs, or of from one to all the digits, and even to all the limbs. In some series the surangular bone is lost. In others the eye diminishes in size, loses its lids, loses the folds of the epidermis which distinguish the cornea, and finally is entirely obscured by the closure of the ophthalmic orifice in the true skin.[2] Among the snakes a similar degradation of the organs of sight has taken place in two suborders, which live underground, and often in ants' nests. The Tortricidæ and Uropeltidæ are burrowing-snakes which display some of the earlier stages of this process. One genus of the true colubrine snakes even (according to Günther) has the eyes obscured as completely as those of the inferior types above named (genus Typhlogeophis.)

e. The Avian Line.

The paleontology of the birds not being well known, our conclusions respecting the character of their evolution must be very incomplete. A few lines of succession are, however, quite obvious, and some of them are clearly lines of progress, and others are lines of re-

[1] Such forms in the Lacertilia have been regarded as degradational by Lankester and Boulanger.

[2] A table of the degenerate forms of Lacertilia is given in the chapter on Catagenesis.

trogression. The first bird we know at all completely, is the celebrated Archeopteryx of the Solenhofen slates of the Jurassic period. In its elongate series of caudal vertebræ and the persistent digits of the anterior limbs we have a clear indication of the process of change which has produced the true birds, and we can see that it involves a specialization of a very pronounced sort. The later forms described by Seeley and Marsh from the Cretaceous beds of England and North America, some of which have biconcave vertebræ, and all probably, the American forms certainly, possessed teeth. This latter character was evidently speedily lost, and others more characteristic of the subclass became the field of developmental change. The parts which subsequently attained especial development are the wings and their appendages; the feet and their envelopes, and the vocal organs. Taking all things into consideration, the greatest sum of progress has been made by the perching birds, whose feet have become effective organs for grasping, whose vocal organs are most perfect, and whose flight is generally good, and often very good. In these birds also the circulatory system is most modified, in the loss of one of the carotid arteries.

The power of flight, the especially avian character, has been developed most irregularly, as it appears in all the orders in especial cases. This is apparent so early as in the Cretaceous toothed birds already mentioned. According to Marsh, the Hesperornithidæ have rudimental wings, while these organs are well developed in the Ichthyornithidæ. They are well developed among natatorial forms in the albatrosses and frigate pelicans, and in the skuas, gulls, and terns, and are rudimental in their allies, the auks. They are developed among rasorial types in the sand-grouse, and, among the adjacent forms, the pigeons. Then

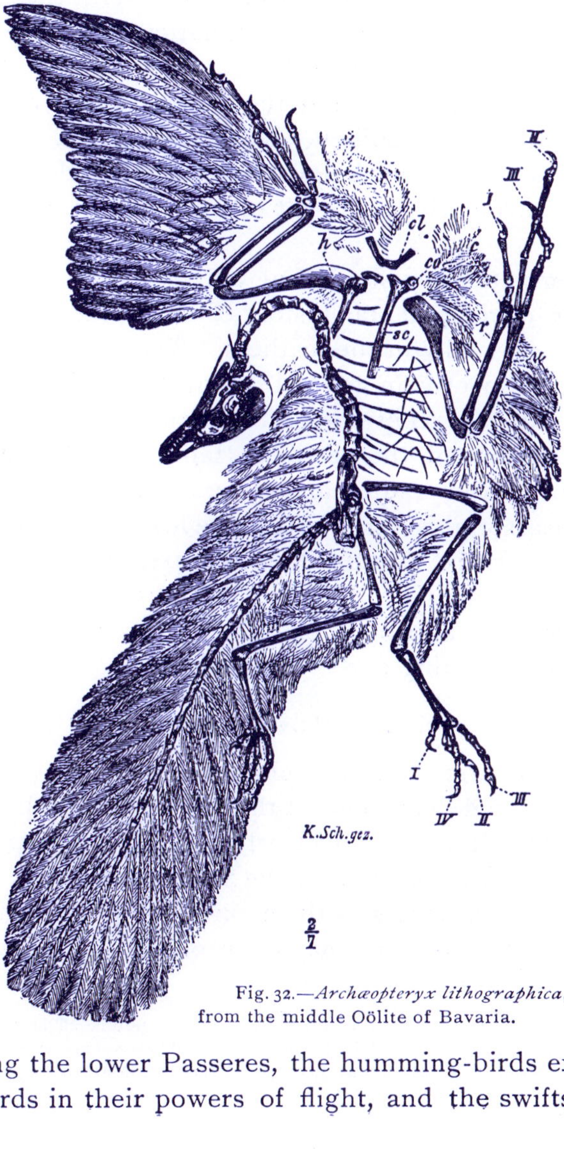

Fig. 32.—*Archæopteryx lithographica*, from the middle Oölite of Bavaria.

among the lower Passeres, the humming-birds exceed all birds in their powers of flight, and the swifts and some of the Caprimulgidæ are highly developed in this respect. Among the higher or true song birds, the swallows form a notable example. With these high specializations occur some remarkable deficiencies. Such are the reduction of the feet in the Caprimulgidæ, swifts, and swallows, and the fœtal character of the bill in the same families. In the syndactyle families, represented by the kingfishers, the condition of the feet is evidently the result of a process of degeneration.

A great many significant points may be observed in the developmental history of the epidermic structures, especially in the feathers. The scale of change in this respect is in general a rising one, though various kinds of exceptions and variations occur. In the development of the rectrices (tail-feathers) there are genera of the wading and rasorial types, and even in the insessorial series, where those feathers are of primitive structure (Menuridæ), are greatly reduced, or absolutely wanting. These are cases of degeneracy.

There is no doubt that the avian series is in general an ascending one.

f. The Mammalian Line.

Discoveries in paleontology have so far invalidated the accepted definitions of the orders of this class that it is difficult to give a clearly cut analysis, especially from the skeleton alone. The following scheme, therefore, while it expresses the natural groupings and affinities, is defective, in that some of the definitions are not without exceptions:[1]

[1] This classification of the Mammalia was first published by the writer in the *American Naturalist* for 1885; was improved in the same, 1889 (October); and appeared in a Syllabus of Lectures of the University of Pennsylvania, July, 1891.

I. A large coracoid bone articulating with the sternum. An interclavicle (Prototheria).

Epicoracoid and marsupial bones; fibula articulating with proximal end of astragalus: 1. *Monotremata.*

II. Coracoid a small process coössified with the scapula (Eutheria).

a. Marsupial bones; palate with perforations (uterus divided; placenta and corpus callosum rudimental or wanting; cerebral hemispheres small and generally smooth).

But one deciduous molar tooth: 2. *Marsupialia*

aa. No marsupial bones; palate generally entire (placenta and corpus callosum well developed).

β. Anterior limb reduced to more or less inflexible paddles posterior limbs wanting (Mutilata).

Elbow-joint fixed; carpals discoid, and with the digits separated by cartilage; lower jaw without ascending ramus: 3. *Cetacea.*

Elbow-joint flexible; carpals and phalanges with normal articulations; lower jaw with ascending ramus: 4. *Sirenia.*

ββ. Anterior limbs with flexible joints. Ungual phalanges compressed and pointed[1] (Unguiculata).

γ. Feet taxeopodous (with exceptions in the carpus).

δ. Teeth without enamel; generally no incisors.

Limbs not volant; hemispheres small, smooth: 5. *Edentata*

δδ. Teeth with enamel; incisors generally present.

No postglenoid process; mandibular condyle not transverse; limbs not volant; hemispheres small, smooth: 6. *Glires*

Anterior limbs volant; hemispheres small, smooth: 7. *Chiroptera*

A postglenoid process; mandibular condyle transverse; limbs not volant; no scapholunar bone[2]; hemispheres small, smooth: 8. *Bunotheria.*[3]

A postglenoid process; limbs not volant, with a scapholunar bone; hemispheres larger, convoluted: 9. *Carnivora.*

[1] Except Mesonychiidæ.

[2] Except Erinaceus and Talpa.

[3] With the suborders Pantotheria, Creodonta, Insectivora, and Tillodonta

$\gamma\gamma$. Feet diplarthrous.

Limbs ambulatory; a postglenoid process; molars quadritubercular: 10. *Ancylopoda.*

$\beta\beta\beta$. Anterior limbs with flexible joints and distinct digits; ungual phalanges not compressed and acute at apex[1] (Ungulata[2]).

ε. Tarsal bones in linear series; carpals generally in linear series.

Limbs ambulatory; teeth with enamel: 11. *Taxeopoda.*[3]

$\varepsilon\varepsilon$. Carpal series alternating; tarsal series linear.

Limbs ambulatory; median digits longest; teeth with enamel: 12. *Toxodontia.*

$\varepsilon\varepsilon\varepsilon$. Tarsal series alternating; carpals linear.

Cuboid bone partly supporting navicular, not in contact with astragalus: 13. *Proboscidia.*

$\varepsilon\varepsilon\varepsilon\varepsilon$. Both tarsal and carpal series more or less alternating.

Os magnum not supporting scaphoides; cuboid supporting astragalus; superior molars tritubercular: 14. *Amblypoda.*

Os magnum supporting scaphoides; superior molars quadritubercular:[4] 15. *Diplarthra.*[5]

[1] Except the Hapalidæ.

[2] Lamarck, *Zoologie Philosophique*, 1809.

[3] This order has the following suborders:

Carpal series linear; no intermedium; tibia not interlocking with astragalus; no anapophyses; incisors rooted; hallux not opposable: *Condylarthra.*

Carpal series linear; an intermedium; tibia interlocking with astragalus; hallux not opposable: *Hyracoidea.*

An intermedium; fibula not interlocking; anapophyses; hallux opposable; incisors growing from persistent pulps: *Daubentonioidea.*

An intermedium; fibula not interlocking; anapophyses; hallux opposable; incisors rooted; carpus generally linear: *Quadrumana.*

No intermedium;[6] nor anapophyses; carpal rows alternating; incisors rooted: *Anthropomorpha.*[7]

The only difference between the Taxeopoda and the Bunotheria is in the unguliform terminal phalanges of the former as compared with the clawed or unguiculate form in the latter. The marmosets among the former division are, however, furnished with typical claws.

[4] Except Trigonolestes.

[5] This order includes the suborders Perissodactyla and Artiodactyla. It is the Ungulata of some authors.

[6] Except in Simia and Hylobates.

[7] Includes the Anthropoid apes and man.

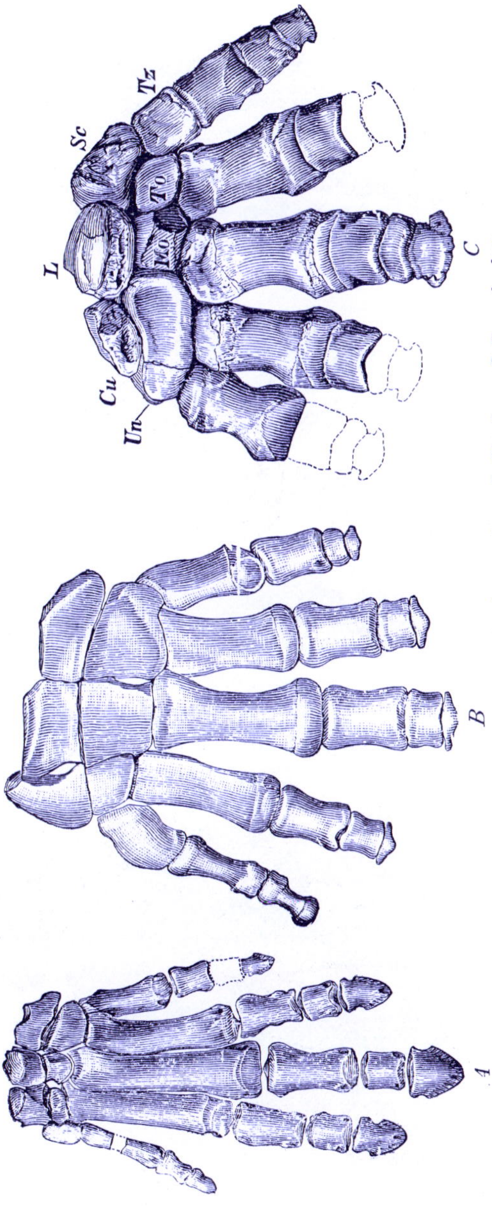

Fig. 33.—Ungulata, anterior feet. *A*, Phenacodus; *B*, Elephas; *C*, Coryphodon.

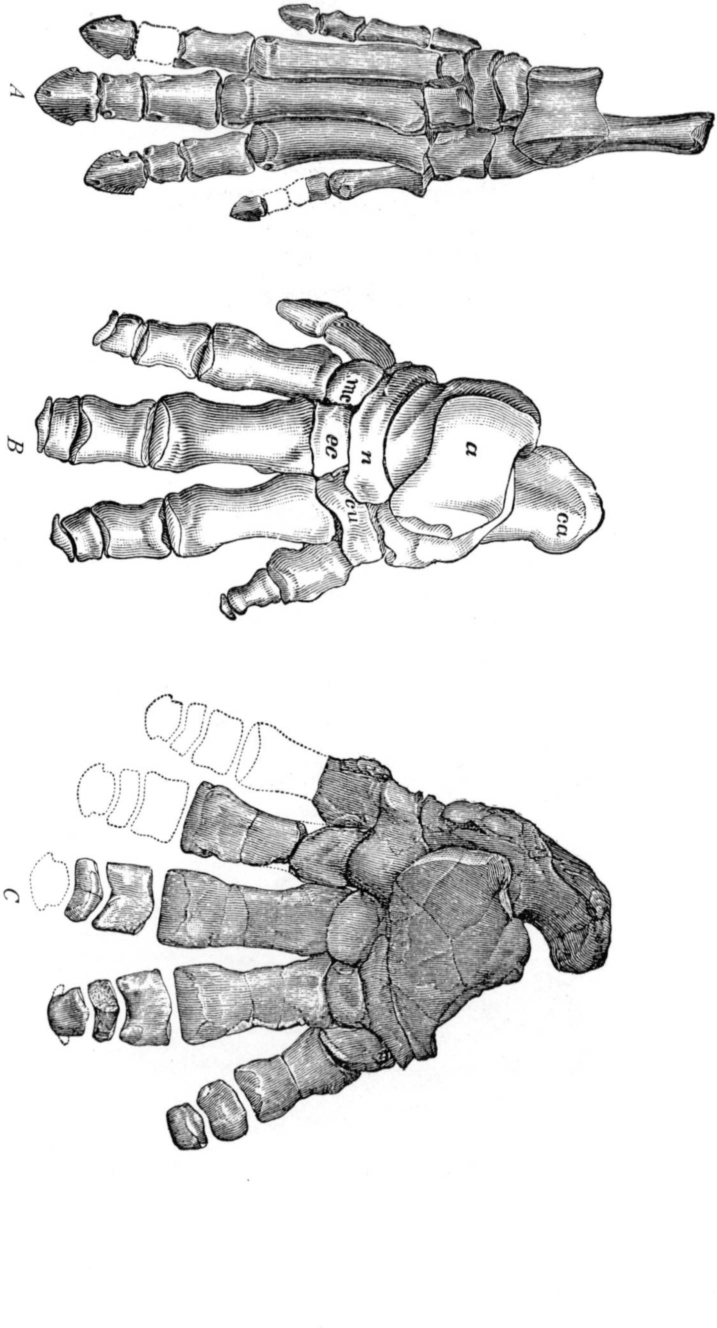

Fig. 34.—Ungulata, posterior feet. *A*, Phenacodus; *B*, Elephas; *C*, Coryphodon.

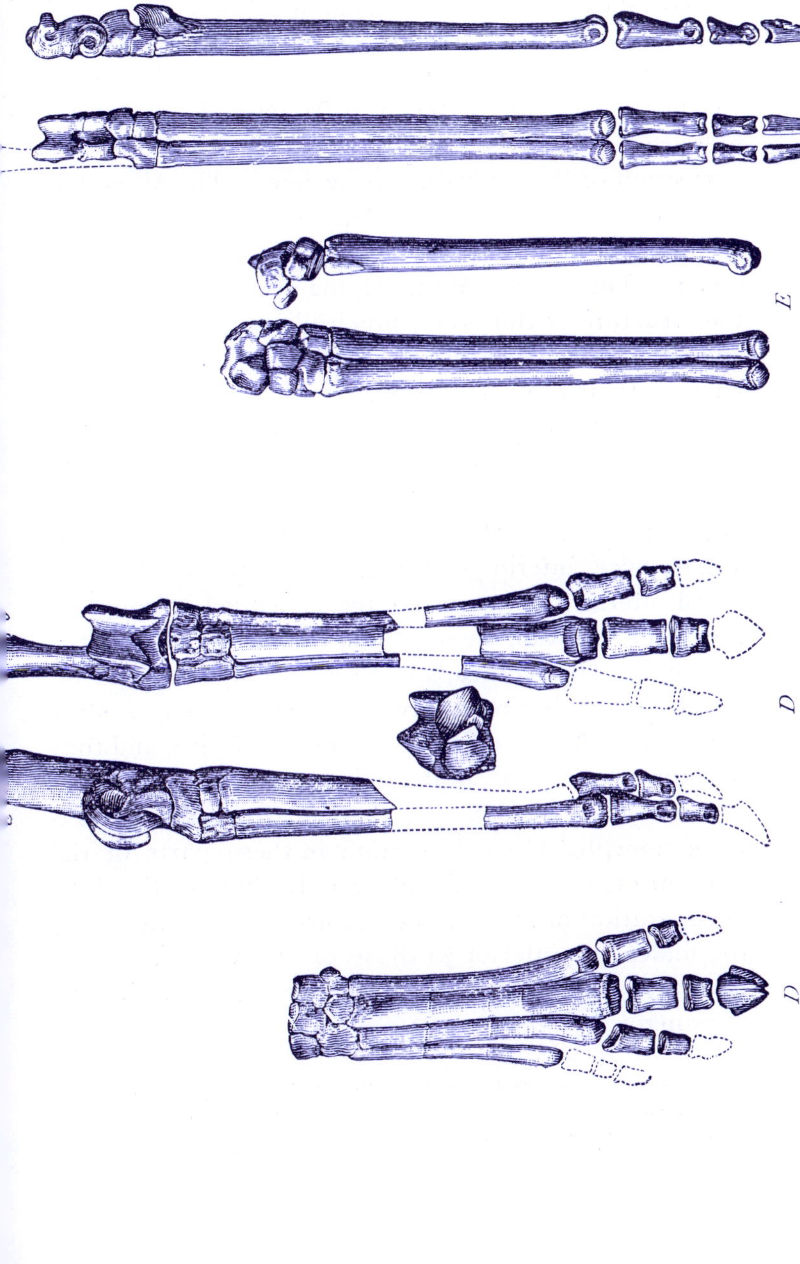

Fig. 35.—Ungulata, anterior and posterior feet. *D*, Hyracotherium; *E*, Poëbrotherium.

The characters of the skeleton of the order Monotremata show that it is nearest of kin to the Reptilia, and many subordinate characters, especially of the extremities, point to the Theromora as its ancestral source.[1] In the general characters the Marsupialia naturally follow in a rising scale, as proved by the increasing perfection of the reproductive system. The Monodelphia follow with improvements in the reproductive system and the brain, as indicated in the table above given. The oldest Monodelphia were, in respect to the structure of the brain, much like the Marsupialia, and some of the existing orders resemble them in some parts of their brain-structure. Such are the Condylarthra and Amblypoda of extinct groups, and the Bunotheria, Edentata, Glires, and Chiroptera, recent and extinct. The characters of the brains of Amblypoda and some Creodonta are, in their superficial characters, even inferior to existing marsupials. The divided uterus of the recent forms named, also gives them the position next to the Marsupialia. In the Carnivora, Hyracoidea, and Proboscidia, a decided advance in both brain-structure and reproductive system is evident. The hemispheres increase in size, and they become convoluted. A uterus is formed, and the testes become external, etc. In the Quadrumana and Anthropomorpha the culmination in these parts of the structure is reached, excepting only that, in the lack of separation of the genital and urinary efferent ducts, the males are inferior to those of many of the Artiodactyla. This history displays a rising scale for the Mammalia.[2]

[1] *Proceedings American Philosoph. Society*, 1884, p. 43. *Antea*, p. 87.

[2] See the evidence for evolution in the history of the extinct Mammalia. *Proceedings of the American Association for the Advancement of Science*, 1883.

Looking at the skeleton, we observe the following successional modifications:

First, as to the feet, and (A) the digits. The Condylarthra have five digits on both feet, and they are plantigrade. This character is retained in their descendants of the lines of Anthropomorpha, Quadrumana, and Hyracoidea, also in the Bunotheria, Edentata, and most of the Glires. In some of the Amblypoda and in the Proboscidia the palm and heel are a little raised. In the Carnivora and Diplarthra the heel is raised, often very high, above the ground, and the number of toes is diminished, as is well known, to two in the Artiodactyla and one in the Perissodactyla. (B) The tarsus and carpus. In the Condylarthra and most of the Creodonta the bones of the two series in the carpus and tarsus are opposite each other, so as to form continuous and separate longitudinal series of bones. This continues to be the case in the Hyracoidea and many of the Quadrumana, but in the anthropoid apes and man the second row is displaced inwards so as to alternate with the first row, thus interrupting the series in the longitudinal direction, and forming a stronger structure than that of the Condylarthra. In the bunotherian, rodent, and edentate series, the tarsus continues to be without alternation, as in the Condylarthra, and is generally identical in the Carnivora. In the hoofed series proper it undergoes change. In the Proboscidia the carpus continues linear, while the tarsus alternates. In the Amblypoda the tarsus alternates in another fashion, and the carpal bones are on the inner side linear, and on the outer side alternating. The complete interlocking by universal alternation of the two carpal series is only found in the Diplarthra. (C) As to the ankle-joint. In most of the Condylarthra

Fig. 36.—*Phenacodus vertmanii*, representative of the order Condylarthra, two-ninths natural size.

it is a flat joint or not tongued or grooved. In most of the Carnivora, in a few Glires, and in all Diplarthra, it is deeply tongued and grooved, forming a more perfect and stronger joint than in the other orders, where the surfaces of the tibia and astragalus are flat. (D) In the highest forms of the Rodentia and Diplarthra the fibula and ulna become more or less coössified with the tibia and radius, and their middle portions become attenuated or disappear.

Secondly, as regards the vertebræ. The mutual articulations (zygapophyses) in the Condylarthra have flat and nearly horizontal surfaces. In higher forms, especially of the ungulate series, they become curved, the posterior turning upward and outward, and the anterior embracing them on the external side. In the higher Diplarthra this curvature is followed by another curvature of the postzygapophysis upward and outward, so that the vertical section of the face of this process is an S. Thus is formed a very close and secure joint, such as is nowhere seen in any other Vertebrata.

Thirdly, as regards the dentition. Of the two types of Monotremata, the Tachyglossus, and the Platypus, the known genera of the former possess no teeth, and the known genus of the latter possesses only a single corneous epidermic grinder succeeding two deciduous molars, like those of certain extinct forms, in each jaw. As the theromorous reptiles from which these are descended have well-developed teeth, their condition is evidently one of degeneration. We probably have their ancestors in the Multituberculata, which range from Triassic to lower Eocene time in both hemispheres. In the marsupial order we have a great range of dental structure, which almost epitomizes that of the monodelph orders. The dentition of the carnivorous forms is creodont; that of the kangaroos is perissodactyle, and that of the wombats is rodent. Other forms repeat the Insectivora. I therefore consider the placental series especially. I have already shown that the greater number of the types of this series have derived the characters of their molar teeth from the stages of the following succession. First, a simple cone or reptilian crown, alternating with that of the other jaw. Second, a cone with anterior and posterior lateral denticles. Third, the denticles rotated to the inner side of the crown below, and outer side above forming with the principal (median) cone a three-sided prism, with tritubercular apex, which alternates with that of the opposite jaw. Fourth, development of a heel projecting from the posterior base of the lower jaw, which, in mastication, meets the crown of the superior, forming a tubercular-sectorial inferior molar. From this stage the carnivorous and sectorial dentition is derived, the tritubercular type being retained. Fifth, the development of a posterior inner cusp in the superior molar, and the elevation of the heel in the inferior molar, with the loss of the anterior inner cusp. Thus the molars become quadritubercular, and opposite. This is the type of many of the Taxeopoda, including the Quadrumana and Insectivora as well as the inferior Diplarthra. The higher Taxeopoda (Hyracoidea) and Diplarthra add various complexities. Thus the tubercles become flattened and then concave, so as to form V's in the section produced by wearing; or they are joined by cross-folds, forming various patterns, of which the most specialized is that of the horse. In the Proboscidia the latter

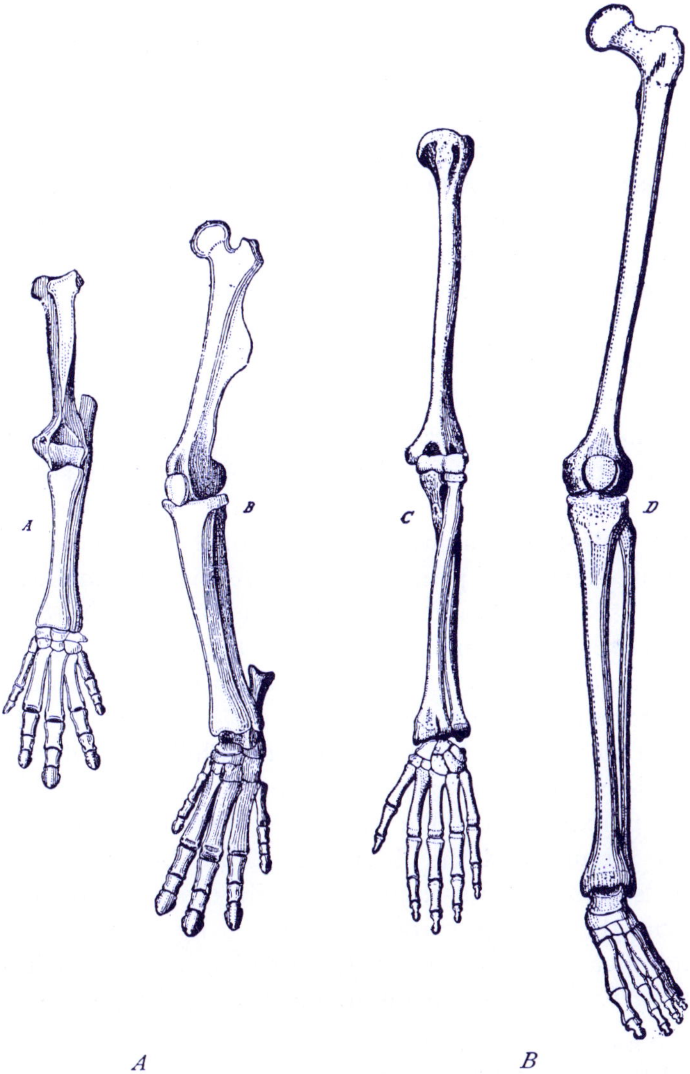

A *B*

Fig. 37.—*A*, *Phenacodus primævus*, fore and hind limbs; *B*, *Homo sapiens*, fore and hind limbs.

become multiplied so as to produce numerous cross-crests.

The molars of some of the Sirenia are like that of some of the Ungulata, especially of the tapirine group, while in others the teeth consist of cylinders. In the Cetacea the molars of the oldest (Eocene and Miocene) types are but two-rooted and compressed, having much the form of the premolars of other Mammalia. In existing forms a few have simple conical teeth, while in a considerable number teeth are entirely wanting.

g. General Review of the Phylogeny of Mammalia.

In the accompanying table some of the characters of the mammalian skeleton above described are thrown into a tabular form. They are exhibited in the order of their appearance in geological time, beginning with the oldest horizon at the bottom of the left-hand column. Continued primitive types are enclosed in brackets. These relations were pointed out by me in 1883,[1] and every discovery made since that date has confirmed their correctness. Some characters of the Mesozoic Mammalia are now added.

Paleontology has cleared up the phylogeny of most of the mammalian orders, but some of them remain as yet unexplained. This is the case with the Cetacea, the Sirenia, and the Edentata. The Marsupialia can be supposed with much probability to have come off from the Monotremata, but there is but little paleontological evidence to sustain the hypothesis. Little progress has been made in unravelling the phylogeny

[1] *Proceedings American Assoc. Adv. Science*, p. 40. The successional gradation in the limbs and teeth was announced by me in 1873 (*Proceeds. Academy Philadelphia*, p. 371, and *Journal of the Academy*, 1874, p. 20), and that in the size of the hemispheres of the brain by Marsh in 1874 (*American Journal Sci. Arts*, p. 66).

Formation.	No. Toes.	Feet.	Astragalus.	Carpus and Tarsus. Ungulata.	Radius. Ungulata.	Superior Molars.	Zygapophyses.	Brain.
Recent......	1–1	Digitigrade.	Grooved.	Interlocking.	Faceted.	4-tubercles, crested and cemented.	Doubly involute.	Hemispheres larger, convoluted.
Miocene.....	2–2	(Plantigrade.)	(Flat.)	(Opposite.)			Singly involute.	
Upper...........	3–3							
(Loup Fork.)	4–4							
	(5–5)							
Middle...........	2–2	Digitigrade.	Grooved.	Interlocking.	Faceted.	4-tubercles, and crested.	Singly involute.	Hemispheres larger, convoluted.
(John Day.)	3–3				Smooth.		Doubly involute.	
	4–4							
Lower	3–3	Digitigrade.	Grooved.	Interlocking.	Smooth.	4-tubercles, and crested.	? Singly involute.	Hemispheres small; and larger.
(White River.)	4–3	Plantigrade.			Faceted.			
	4–4							
Eocene......	3–3	(Digitigrade.)	Grooved.	Opposite.	Smooth.	4-tubercles.	Singly involute.	
Middle...........	4–3	Plantigrade.	(Flat.)	Interlocking.		3-tubercles, and crested.	Plane.	Hemispheres small.
(Bridger.)	4–5							
	5–5							
Lower	4–3	Plantigrade.	Flat.	Opposite.	Smooth.	4-tubercles.	Plane.	Hemispheres small; mesencephalon sometimes exposed.
(Wasatch.)	4–5	(Digitigrade.)	(Grooved.)	Interlocking.		3-tubercles, a few crested.	Singly involute.	
Cretaceous..	5–5							
Upper...........	5–5	Plantigrade.	Flat.	Opposite.	Smooth.	3-tubercles.	Plane	Mesencephalon exposed; hemispheres small and smoother.
(Puerco.)						(4-tubercles, none crested.)		
(Laramie.)			Flat.			Tritubercular.		
Jurassic.....						Tritubercular.		
Triassic.....						Triconodont. Protodont.		

of the Cetacea and Sirenia. The results attained by the study of the paleontology of the other orders may be summarized as follows:

First. It is probable that the common ancestors of the placental and implacental lines of Mammalia are known to us in some of the types of the Jurassic period. Whether they were marsupial in the sense of possessing an external pouch for the young or not, is immaterial. They were probably marsupial in brain characters, in the structure of their reproductive system, and in the absence of placenta. To this source the existing polyprotodont marsupials may be traced, through such forms as Myrmecobius. The multituberculate type has a contemporary history, and one distinct from that of the Polyprotodontia, and its ancestry has not yet been discovered. Their earliest forms (of the Jurassic and Triassic) are already highly specialized. They probably represent the Monotremata cf their time.

Second. The immediate didelphian ancestors of the monodelphous Mammalia have not yet been certainly discovered. In the oldest of the latter (of the Puerco epoch) numerous points of approach to the insectivorous Jurassic forms occur, especially in the prevalent trituberculy of the molars in both epochs.

Third. The phylogeny of the clawed group has been traced back to a common ordinal form which has been called the Bunotheria. Of these the most generalized are the Creodonta, from which we may trace the Carnivora, the Insectivora, and the Tillodonta, and probably all other Unguiculata. The Ancylopoda only have undergone the alternation of the carpal and tarsal bones, which obtains in the diplarthrous Ungulata.

Fourth. The phylogeny of the hoofed groups carries us back to the order Condylarthra, the hoofed cotemporary of the Bunotheria. The even and odd toed hoofed mammals are traceable back to the Amblypoda, whose oldest representatives are the Pantodonta of the Puerco. The Proboscidia and Hyracoidea come directly from the Condylarthra. Moreover, the phalanges of the lemurs are not distinguishable by any important characters from the hoofs of the Hyracoidea and Condylarthra. Not only this, but the structure of the foot in these three groups is identical in regard to the mode of articulation of the first and second rows of the tarsal and carpal bones.

Fifth. The characters of the feet of the Condylarthra agree with those of unguiculate placental Mammalia, and bind the two series together. The synthesis of the ungulate and unguiculate lines is accomplished by exceptions to the characters which define them. Thus the hoofs of Pantolambda (Amblypoda), Periptychus (Condylarthra), and Mesonyx (Creodonta) do not differ by any marked character. Claws occur in the Hapalidæ of the quadrumanous line, and the ungues of some Glires are absolutely intermediate between the hoofs and claws. Many Edentata have claws on the fore feet and hoofs on the hind feet. The Condylarthra with tritubercular molar teeth are then traceable to Bunotheria with tritubercular teeth, of which many are known from the Puerco beds; and the quadritubercular forms from corresponding quadritubercular or tritubercular Bunotheria, of which latter, some are known.

Sixth. The anthropoid line may be traced directly through the lemurs to the Condylarthra. The changes which have taken place in the skeleton are slight, and consist among other points in a rotation of the second row of carpal bones inwards on the first row, in the anthropoid apes and man, similar to that which has occurred among the Ungulata, but it has not become so pronounced.

As a result we get the general phylogenetic scheme as shown on the following page.

In this diagram, divisions of greater and lesser rank are mixed, so as to display better some of the relationships. Thus all the divisions whose names stand on the right side of the middle vertical line are unguiculates; and those on the left side of the line, excepting Sirenia and Cetacea, are ungulates. The three names in the middle vertical line are those of the suborders of the Taxeopoda.

A review of the characters of the existing Mammalia as compared with those of their extinct ancestors displays a great deal of improvement in many ways, and but few instances of retrogression. The succession in time of the Monotremata, the Marsupialia, and the Monodelphia, is a succession of advance in all the characters of the soft parts and of the skeleton which define them (see table of classification). As to the monotremes themselves, it is more than probable that the order has degenerated in some respects in producing the existing types. The history of the Monotremata is not made out, but the earliest forms of which we know the skeleton, Polymastodon (Cope) of the Lower Eocene, is as specialized as the most specialized recent forms. The dentition of the Jurassic forms, Plagiaulax, etc., is quite specialized also, but not more so than that of the kangaroos. The premolars are more specialized, the true molars less specialized than in those animals. The history of Marsu-

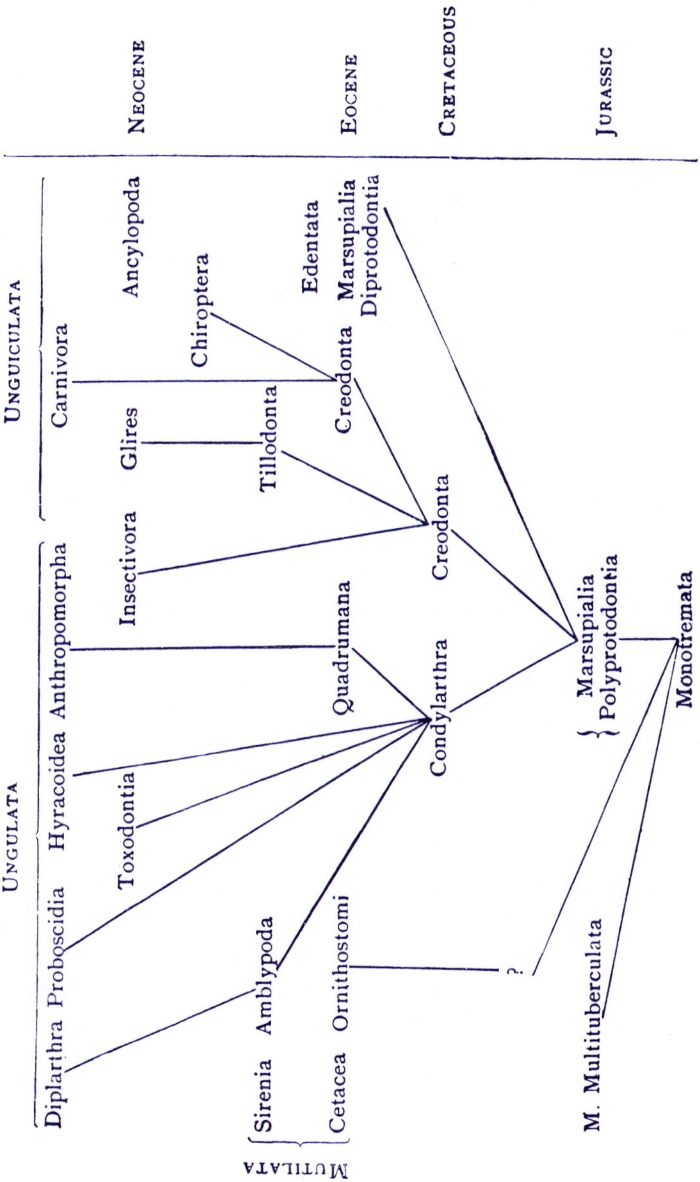
Neocene
Eocene
Cretaceous
Jurassic
Unguiculata
Ungulata
Mutilata
Carnivora
Ancylopoda
Chiroptera
Glires
Tillodonta
Creodonta
Edentata
Marsupialia Diprotodontia
Creodonta
Insectivora
Anthropomorpha
Quadrumana
Hyracoidea
Toxodontia
Condylarthra
Proboscidia
Amblypoda
Ornithostomi
Diplarthra
Sirenia
Cetacea
Marsupialia Polyprotodontia
Monotremata
M. Multituberculata
?

pialia indicates that the primitive types were all insectivorous, and possessed a larger number of molars than any of the recent forms. The latter have then followed the same course as the placentals in the reduction of the number of teeth and specialization of those that remain.

Coming to the Monodelphia, the increase in the size and complication of the brain, both of the cerebellum and the hemispheres, is a remarkable evidence of advance. But one retrogressive line in this respect is known, viz., that of the order Amblypoda,[1] where the brain has become relatively smaller with the passage of time. The successive changes in the structure of the feet are all in one direction, viz., in the reduction of the number of the toes, the elevation of the heel, and the creation of tongue and groove joints where plain surfaces had previously existed. The diminution in the number of toes might be regarded as a degeneracy, but the loss is accompanied by a proportional gain in the size of the toes that remain. In every respect the progressive change in the feet is an advance. In the carpus and tarsus we have a gradual extension of the second row of bones on the first, to the inner side. In the highest and latest orders this process is most complete, and, as it results in a more perfect mechanical arrangement, the change is clearly an advance. The same progressive improvement is seen in the development of distinct facets in the cubito-carpal articulation, and of a tongue and groove ("trochlear crest") in the elbow-joint. In the vertebræ the development of the interlocking zygapophysial articulations is a clear advance.

Progress is generally noticeable in the dental struc-

[1]See *Naturalist*, Jan., 1885, p. 55.

tures; for, the earliest dentitions are the most simple, and the later the more complex. Some of the types retain the primitive tritubercular molars, as the Centetidæ, shrews, and some lemurs, and most Carnivora (above), but the quadritubercular and its derivative forms are by far the most common type in the recent fauna. The forms that produced the complicated modifications in the Proboscidia and Diplarthra appeared latest in time, and the most complex genera, Elephas, Bos, and Equus, the latest of all. The extreme sectorial modifications of the tritubercular type, as seen in the Hyænidæ and the Felidæ, are the latest of their line also.

Some cases of degeneracy are, however, apparent in the monodelphous Mammalia. The loss of pelvis and posterior limbs in the two mutilate orders is clearly a degenerate character, since there can be no doubt that they have descended from forms with those parts of the skeleton present. The reduction of flexibility seen in the limbs of the Sirenia and the loss of this character in the fore limbs of the Cetacea are features of degeneracy for the same reason. The teeth in both orders have undergone degenerate evolution; in the later and existing forms of the Cetacea even to extinction.

The Edentata have undergone degeneration. This is chiefly apparent in the teeth, which are deprived of enamel, and which are wanting from the premaxillary bone. A suborder of the Bunotheria, the Tillodonta of the Lower Eocene period, display a great reduction of enamel on the molar teeth, so that in much-worn examples it appears to be wanting. Its place is taken by an extensive coat of cementum, as is seen in Edentata, and the roots of the teeth are often undivided as in that order.

Local or sporadic cases of degenerate loss of parts are seen in various parts of the mammalian series, such as toothless Mammalia wherever they occur. Such are cases where the teeth become extremely simple, as in the honey-eating masupial Tarsipes, the carnivore Proteles, the pteropod bats, and the aye-aye. Also where teeth are lost from the series, as in the canine genus Dysodus, and in man. The loss of the hallux and pollex without corresponding gain, in various genera, may be regarded in the same light.

In conclusion, the progressive may be compared with the retrogressive evolution of the Vertebrata, as follows: In the earlier periods and with the lower forms, retrogressive evolution predominated. In the higher classes progressive evolution has predominated. When we consider the history of the first class of vertebrates, the Tunicata, in this respect, and compare it with that of the last class, the Mammalia, the contrast is very great.

h. The Phylogeny of the Horse.

As an example of special phylogeny I select that of the horse, because it is the most completely represented by specimens in our museums.

I have already pointed out that the alternate type of carpus and tarsus of the Diplarthra has been derived from the linear of the Taxeopoda by a displacement inwards of the bones of their second rows. In the posterior foot this has changed the convex surface of the head of the astragalus into a bifacetted face. Thus was the condylarthrous astragalus modified into that of the Diplarthra. At the beginning of the line of the horses we find the condylarthrous genus of the Wasatch Eocene, Phenacodus Cope, to differ in this way from the perissodactylous genus Hyracotherium Owen. Phenacodontidæ are indicated as the ancestors of all the Ungulata by their character as "bunodont pentadactyle plantigrades," characters in which they agree with the ancestors of all placental mammals. That they are not the ancestors of all the latter is shown by the fact that their molar type is quadritubercular; but one has to go backwards but a short distance in time to the Puerco epoch, to find their tritubercular ancestors. Between these and Phenacodus, comes the quadritubercular genus Euprotogonia Cope, of the Puerco, which has simpler premolar teeth.

Between Phenacodus and Hyracotherium there is room for two or more genera with fully facetted carpals and tarsals, longer feet, and a rudimental first toe on the anterior foot, and first and fifth toes on the hind foot. In Hyracotherium these digits have disappeared. Further, in Hyracotherium the internal cusps of the molars are more or less connected with the external by low and indistinct ridges, which in the superior molars include the small intermediate tubercles or conules. Thus is the lophodont dentition foreshadowed. Hyracotherium was a contemporary of Phenacodus and continued later in Eocene time. Some of its forms developed an increased complexity of the last premolars in both jaws, forming the genus Pliolophus, and foreshadowing the development of molar-like premolars, which is so characteristic of the later members of the horse line. In the genus Epihippus Marsh, of one epoch later in time (the Uinta), two such premolars are developed in each jaw. We have seen very short interspaces next the canine teeth in Phenacodus, and these have become longer in Hyracotherium and Epihippus.

With the opening of the Neocene age, we have the descendant of Epihippus in Mesohippus Marsh, which differs from its predecessor as follows. There are but three toes on all the feet; three premolars resemble the true molars; the crests which connect the internal pair of cusps with the external in both jaws, are much more elevated, and soon form on wearing a part of the pattern of the crown. Hyracotherium already walked on the ends of its toes, and the feet of Mesohippus continue the character. The crowns of all the molars are short like those of its ancestors. In the Middle Neocene formations we have the genus Anchitherium Kaup, where the incisor teeth show the addition of a ridge or cingulum round the inner side, which bounds a cup; forming the cupped incisors so characteristic of the horses. The species have been all the while growing gradually larger.

Towards the end of Neocene time important progress was made. In the Loup Fork epoch the three toed horses were very numerous in species, but their lateral toes were all much shortened so that they did not reach the ground. The crests of the molar teeth were much stronger, and in the superior series the conules had assumed a greater importance, extending themselves posteriorly from the transverse crests, and showing crescentic sections, resembling those of the outer cusps, with which they are parallel. The anterior conule extended so far posteriorly as to join the posterior one, resembling in this respect also the anterior external cusp. So the crown came to have six modified cusps of which the two inner are the smallest and remain unconnected with each other. The crowns of the molars vary in length in these later three-toed horses. Some are short like the Anchitheriums, and others are longer, approaching the true horses. In the valleys between these high cusps cement is deposited, as in the true horses and other mammals with long-crowned molars. There are two types of these later three-toed horses. In one the posterior inner cusp is not joined to the conule by a transverse crest (genus Hippotherium Kaup), or it is so joined (genus Protohippus Leidy).

Plistocene times witnessed the perfection of the horse line. The lateral toes dwindled into splints concealed beneath the skin. The crowns of the molar teeth became very long, and in the upper jaw the inner posterior tubercle, now a column, joined the adjacent conule, and became extended very much in fore and aft diameter. The small anterior premolar disappeared, and the canines became the mark of the male sex only, in general. The lower molars acquire some additional complications, and the feet are longer than in any of its ancestors. The genus Equus L. is finished, and remains a permanent member of the human epoch, from which its only relatives, the rhinoceros and the tapir, are gradually disappearing.

This history may be duplicated in manner and mode, by the lines of the camels, the dogs and bears, the cats, the beaver, etc.

Examination of all these genealogical lines reveals a certain definiteness of end and directness of approach. We discover no accessions of characters which are afterwards lost, as would naturally occur as a result of undirected variation. Nor do we discover anything like the appearance of sports along the line, the word sport being used in the sense of a variation widely divergent from its immediate ancestor. On the contrary, the more thorough becomes our knowledge of the series, the more evident does it become that progressive evolution has advanced by minute increments along a definite line, and that variations off this line have not exerted an appreciable influence on the result.

i. The Phylogeny of Man.

In man the feet retain the pentadactyle plantigrade type with scarcely grooved astragalo-tibial articulation, which characterizes the Mammalia of the Puerco epoch, and most of those of the Lower Eocene.[1] His dentition is not lophodont, but is simply bunodont, like that of the Phenacodontidæ of the Lower Eocene. It is only in the structure of the brain and the reproductive system that man shows an advance over the Eocene type. In the former he greatly excels any mammal that has appeared since; a superiority already apparent in one of his early ancestors, the anaptamorphous lemur of the Lower Eocene. In the reproductive system he is about on a par with the higher Artiodactyla, although the male, in the persistent union of the genital and urinary efferent ducts is not so much specialized as some of the latter, where they are distinct. It is an interesting fact that man displays in his dentition strong tendencies to a greater specialization by simplification beyond the ordinary quadrumanous type, by reduction in the number of the true molars and incisors. Thus the $M.^{\underline{3}}$ is not unfrequently absent in the highest races, and some families display a rudimental condition and absence of the $I.^{\underline{2}}$.

Much importance attaches to the composition of the molar dentition. Many years ago, Owen[2] called

[1] This fact was first pointed out by myself in the *Penn Monthly Magazine*, 1875; see *Origin of the Fittest*, p. 268.

[2] *Odontography*, 1840–5, p. 454.

attention to the fact that in the dark races the roots of the last superior molar are distinct from each other, while in the Indo-Europeans they are known to be

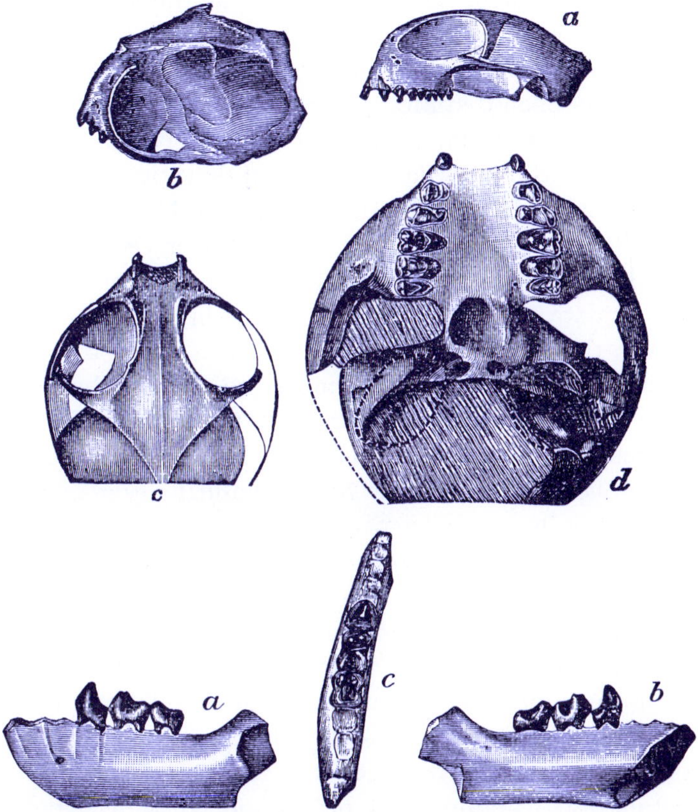

Fig. 38.—Fig. *a*, skull of *Anaptomorphus homunculus* Cope, natural size. Fig. *b*, same, oblique view, displaying the large cerebral hemispheres. Fig. *c*, superior view of skull, natural size. Fig. *d*, inferior view, three-halves natural size. Lower figs. *a*, *b*, and *c*, left branch of lower jaw of *Anaptomorphus æmulus* Cope, twice natural size; *a*, from left side; *b*, inner side; *c*, from above.

more or less fused together. These now well-known characteristics of human dentition constitute one of the examples of transition from a simian to a human type. I have pointed out a corresponding modification in the structure of the crown of the superior true molars, viz.: the transition from a quadritubercular to a tritubercular structure in passing from the lower to the higher races. As this point has some interesting implications in the earlier phylogeny of man, and as its value has been disputed, I give it a little attention.

The facts of the case are as follows: I have demonstrated[1] the fact that all forms of dentition exhibited by the eutherian mammals have been derived from a primitive tritubercular type. Professor Osborn says that he expects to be able to do the same for the multituberculate (? Prototherian) dentition. I have also shown that man exhibits a tendency to revert from his primitive quadritubercular molar to this tritubercular type.[2] As to the significance of these facts, I have expressed the view that this acquisition of a tritubercular molar is a reversion to the lemurine type. This conclusion is necessary because the lemurs are the last of the families in the line of the ancestry of man which present this dentition. The monkeys and anthropoid apes are all quadritubercular, except a few limited collateral branches of the former, which still retain the lemurine type. There are also a few collateral types of lemurs which have acquired one or more quadritubercular molars, but they are not typical. In many tritubercular mammals, a precocious form or two can be found, which has acquired the fourth tubercle. But the further back we go in time, the fewer they become, until, in the Puerco fauna, of eighty-two species of

[1] *Proceeds. Amer. Philos. Soc.*, Dec., 1883; *Origin of the Fittest*, 1887, pp 245, 347, 359.

[2] *American Journal of Morphology*, II., 1888, p. 7.

eutherian mammals, but four have true quadritubercular superior molars.

I take this opportunity of saying, however, that reversion is not necessarily a result of heredity. It may be simply a retrogression on a line of advance already laid down. What influence lemurine heredity may have had in the case of man, it is not easy to know. But it must be borne in mind that various forms of degeneracy of molar teeth are possible other than the resumption of the tritubercular type, yet the normal reduction generally presented is just this lemurine and primitive eutherian condition. The simplicity of the elements involved, has something, but not everything, to do with this reversion.

Dr. Paul Topinard has made an investigation [1] of the characters of the crowns of the molars in man, and has reached general conclusions identical with my own. He remarks (p. 665): "It is demonstrated, in conclusion, that the teeth of man are, at present, in process of transformation, and that in some future which is remote the inferior molars shall certainly be quadricuspid, and the superior molars tricuspid. It will be curious to have the statistics as to prehistoric man; unfortunately, their crania are rare, and their molars generally much worn." In the details of his examination, there are some divergencies from my results. Thus he finds the quadritubercular second and third superior molar relatively of more frequent occurrence in Europeans than I did. But the absence of Europeo-Americans from his tables reduces the percentage of trituberculars in the Indo-Europeans. He makes no report of Esquimaux. Had he observed this type, he would have found a higher per cent. of tritubercular

2 *L'Anthropologie*, 1892, p. 641 (Nov., Dec.).

upper molars than in any race that he has recorded. He confirms my conclusion as to the high percentage of quadritubercular superior molars in the Malays, Polynesians, and Melanesians.

The relation of this fact to phylogeny is to confirm Haeckel's hypothesis of the lemurine ancestry of man. I have advanced the further hypothesis that the Anthropomorpha (which include man and the anthropoid apes) have been derived directly from the lemurs, without passing through the monkeys proper. This close association of man with the apes, is based on various considerations. One of them is that the skeleton of the anthropoid apes more nearly resembles that of man in the most important respects than it does that of the monkeys. This is especially true of the vertebral column, where the anapophyses are wanting in the Anthropomorpha (insignificant rudiments remaining on one or two vertebræ, as pointed out by Mivart), while they are well developed in the monkeys and lemurs. The molar teeth of the apes and man resemble each other more than either do those of the monkeys, since they lack the crests which connect the cusps, which are general in the latter.

The frequent presence of the tritubercular molar in man suggests the superior claim of the lemurs over the monkeys to the position of ancestor. Another significant fact pointing in the same direction is the existence of large-brained lemurs with a very anthropoid dentition (Anaptomorphidæ) in our Eocene beds, which have the dental formula of man and the Old World monkeys and apes. This resemblance is very remarkable, much exceeding that lately observed by Ameghino in certain extinct forms of monkeys in Patagonia, which appear to be ancestors of the existing South American monkeys (Cebidæ), and possibly of the Old World monkeys also. The superior molars of the Anaptomorphus are tritubercular, while the premolars, canines, and incisors are essentially anthropomorphous, and rather human than simian. Anaptomorphus is probably at the same time the ancestor of the Malaysian lemurine genus Tarsius, and M. Topinard remarks that Tarsius has as good claims to be regarded as ancestral to Homo as Anaptomorphus. But M. Topinard must be aware that in the existing genus the character of the canine and incisive dentition is very unlike that of the Anaptomorphus and Homo. It is specialized in a different direction. The dentition of Anaptomorphus being so generalized as compared with Tarsius, I suspect that its skeleton will be found to present corresponding characters. Of course, if it be found hereafter to have the foot structure of Tarsius (which I do not anticipate), it cannot be included in the ancestry of the Anthropomorpha.

It must be further observed that the ancestral line of the Anthropomorpha cannot be traced through any existing type of Lemuridæ, but through the extinct forms of the Eocene period.[1] This is on account of the peculiar specialization of the inferior canines, which are incisor-like, and because of the peculiar character of the incisors themselves, in the modern lemurs in the restricted sense. But we have numerous lemurine types of the Eocene of both America and Europe which satisfy the conditions exactly, so far as the dentition is concerned. These are mostly referable to the family Adapidæ.

Unfortunately, we do not know the entire skeletons

[1] *On the Primitive Types of the Orders of the Mammalia Educabilia*, 1873, p. 8.

of these Eocene lemurs, but as far as we have them (genera Tomitherium and Adapis) they are monkey-like. But we have what is almost as useful, the skeleton of their Eocene and Puerco ancestors, the Condylarthra. I long since pointed out that the latter order (not the genus Phenacodus, as Lydekker has

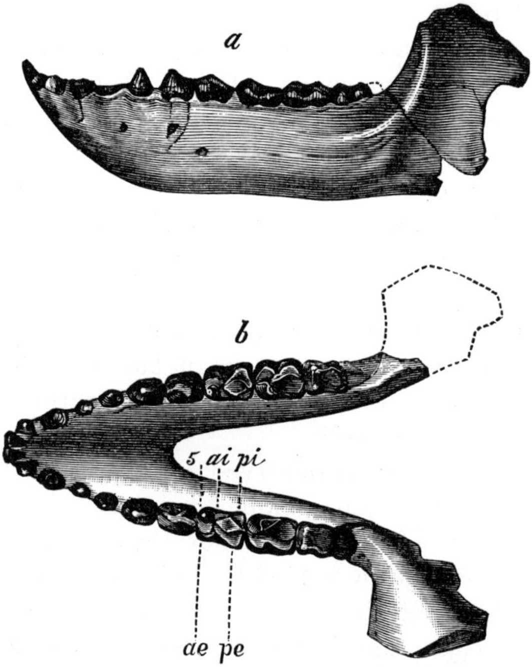

Fig. 39.—*Tomitherium rostratum* Cope, one of the Adapidæ, mandible, natural size; *a*, from left side; *b*, from above. Original, from Report U. S. Geol. Survey Terrs., Vol. III.

represented me as saying) must be the ancestors of the lemurs, basing my views expressly on the general structure of the Phenacodus, Periptychus, and Meniscotherium. The structure of the ungual phalanges of Periptychus is very significant, and even more so is that in Meniscotherium, as recently shown by Marsh, who adopts (without credit) my hypothesis of lemurine affinities of the Condylarthra (which he renames the Mesodactyla). From Condylarthra back to Creodonta is an easy transition, and I have always assumed that the Creodonta were derived from generalized polyprotodont Mursupialia. This view has been entirely confirmed by the recent discoveries of Ameghino in Patagonia, where he has found forms whose remains may be referred with equal propriety to the one group or the other. M. Topinard has been rather hasty in reaching the marsupial ancestry in supposing that Phenacodus belongs to that order. All the evidence shows that Phenacodus is a generalized ungulate placental.

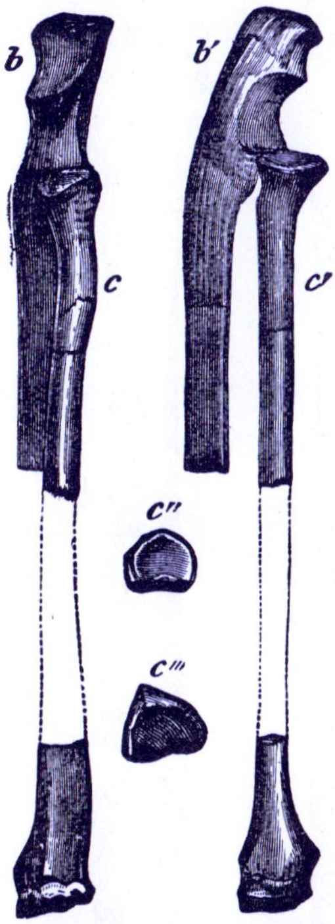

Fig. 40.—*Tomitherium rostratum* Cope, fore arm, five-sixths natural size. Original, *b*, ulna; *c*, radius.

To return to the more immediate ancestry of man. I have expressed,[1] and now maintain as a working hypothesis, that all the Anthropomorpha were descended from the Eocene lemuroids. In my system[2] the Anthropomorpha includes the two families Hominidæ and Simiidæ. The sole difference between these families is seen in the structure of the posterior foot; the Simiidæ having the

1 *American Naturalist*, 1885, p. 467.

2 *Origin of the Fittest*, 1887, p. 346, from *American Naturalist*, 1885, p. 345, where the classification of the Taxeopoda should be in a foot-note; *loc. cit.*, 1889, October.

hallux opposable, while in the Hominidæ the hallux is not opposable. This is not a strong character, since it depends on a slight difference in the form of the entocuneiform bone. In some vertebrates, as the tree-frogs, the same and similar characters (genus Phyllomedusa) are not regarded as of family value. It is then highly probable that Homo is descended from some form of the Anthropomorpha now extinct, and probably unknown at present, although we do not yet know all the characters of some extinct supposed Simiidæ, of which fragments only remain to us. It cannot now be determined whether man and the Simiidæ were both descended from a genus with opposable hallux, or without opposable hallux, or whether from a genus presenting an

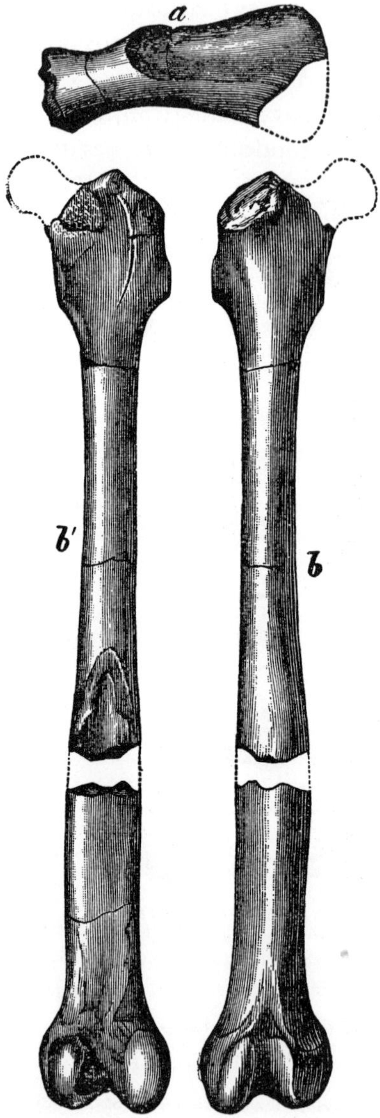

Fig. 41.—*Tomitherium rostratum* Cope, five-sixths natural size; *a*, ilium; *b*, femur. Original.

intermediate character in this respect. This genus was, in any case, distinct from either of the two existing genera of Simiidæ, Simia and Hylobates, since these present varied combinations of anthropoid resemblances and differences, of generic and specific value.

Professor Virchow in a late address[1] has thrown down the gage to the evolutionary anthropologists by asserting that "scientific anthropology begins with living races," adding "that the first step in the construction of the doctrine of transformism will be the explanation of the way the human races have been formed," etc. But the only way of solving the latter problem will be by the discovery of the ancestral races, which are extinct. The ad captandum remarks of the learned professor as to deriving an Aryan from a Negro, etc., remind one of the criticisms directed at the doctrine of evolution when it was first presented to the public, as to a horse never producing a cow, etc. It is well known to Professor Virchow that human races present greater or less approximations to the simian type in various respects. Such are the flat coössified nasal bones of the Bushmen; the quadritubercular molars of the Polynesians; the flat ilia and prognathous jaws of the Negro; the flat shinbones of various races; the divergent hallux of some aborigines of farther India, etc. Professor Virchow states that the Neanderthal man is a diseased subject, but the disease has evidently not destroyed his race characters; and in his address he ignores the important and well-authenticated discovery of the man and woman of Spy. These observations are reinforced by recent discovery of a similar man by DuBois at Trinil

[1] *Popular Science Monthly*, January, 1893, p. 373, translated.

in the island of Java. To these ancient people I will now devote some space.

What had been long suspected is now established,

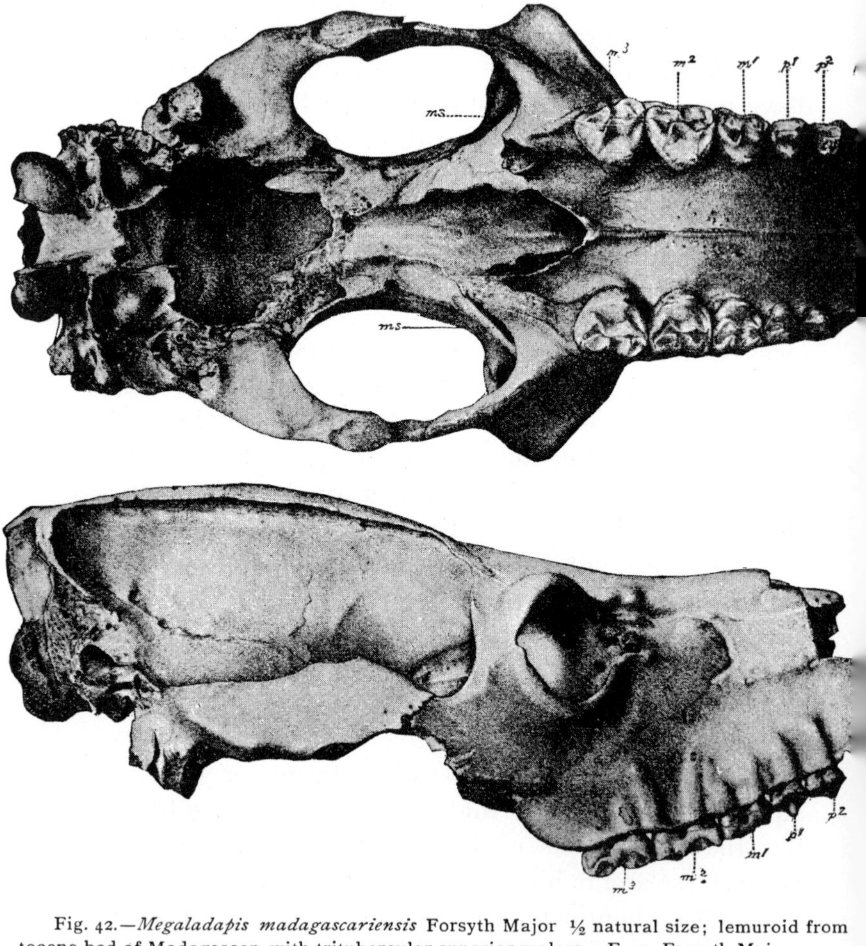

Fig. 42.—*Megaladapis madagascariensis* Forsyth Major ½ natural size; lemuroid from tocene bed of Madagascar, with tritubercular superior molars. From Forsyth Major.

through the discovery and descriptions of Messrs. Fraipont and Lohest of Liège; viz. that there dwelt in Europe during Paleolithic times a race of men which possessed a greater number of simioid characteristics than any which has been discovered elsewhere. The important discovery in the grotto of Spy of two skeletons, almost complete, served to unify knowledge of this race, which had previously rested on isolated fragments only. These skeletons proved what had been previously only surmised, that the lower jaws of Naulette, and of Shipka, and probably the skeleton of Neanderthal, belong to one and the same race. The

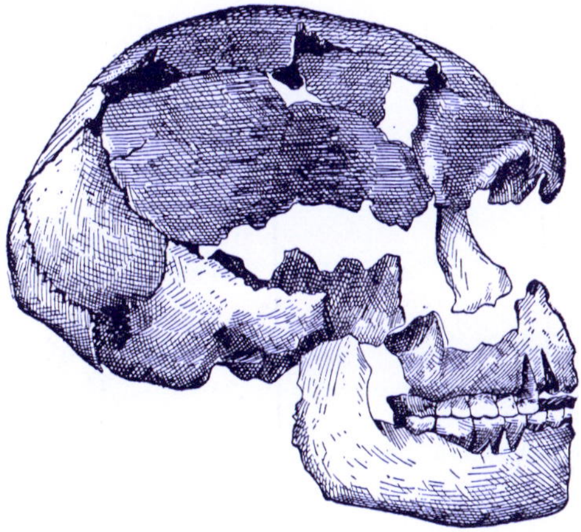

Fig. 43.—Skull of the man of Spy. From Prof. G. F. Wright's *Man and the Glacial Period*. From a photograph.

simian characters of these parts of the skeleton are well known. These are the enormous superciliary ridges and glabella; the retreating frontal region; the thickness of the cranial wall; the massive mandibular ramus with rudimentary chin, and the large size of the posterior molars. Messrs. Fraipont and Lohest have added other characters to these, viz.: the tibia shorter than in any other known human race; the sigmoid flexure of the femur; the divergent curvature of the bones of the fore-arm, and most important, a very peculiar form of the posterior face of the mandibular symphysis, already pointed out by Topinard in the jaw

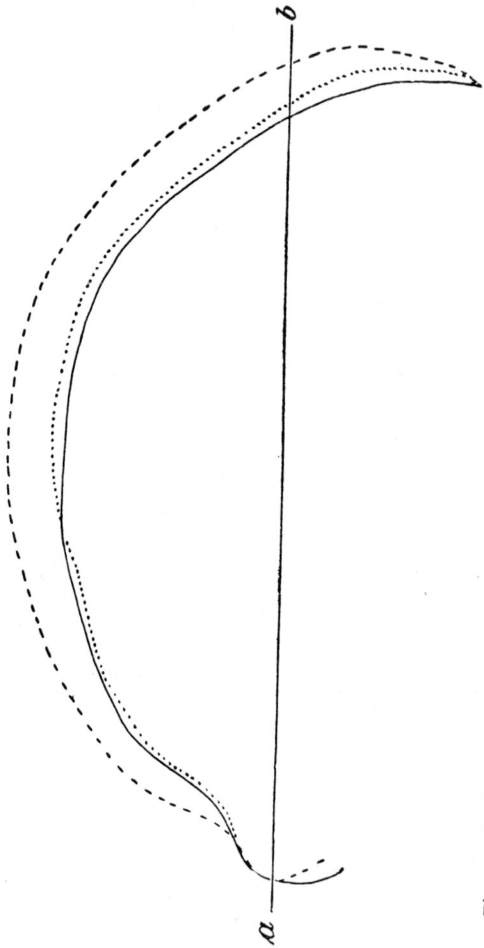

Fig. 44.—Outlines of calvaria; of the Neanderthal man in solid line; of the Spy man No. 1 in dots, and of the Spy man No. 2 in broken line. From Fraipont and Lohest.

of Naulette. On these characters the following remarks may be made.[1]

I. The prominent superciliary crests, which are

[1] *Archives Belges de Biologie*, VII., 1886, p. 731, Gand.

characteristic of the Neanderthal race. No existing race presents such a development, neither the Papuans, Australians, nor Negroes of any race. But we find the superciliary crests and underlying sinuses identical in adult female orangs and chimpanzees and young male gorillas. In the female chimpanzees the crests are almost inferior in size to those of the man of Spy.

II. The retreating forehead of the two crania of Spy is not found in any existing human race, while it is typical of that of Neanderthal. It is characteristic of female orangs and gorillas and of the young males of both species, and of adult males and female chimpanzees. It appears in existing men in rare and isolated cases; [probably as survivals].

III. The prominent transverse superior semicircular crest of the occipital bone is found in existing races among the Fellahs of Africa and the Nigritos. It is characteristic of the Neanderthal skulls, and presents exactly the same characters as the young male and female orang and gorilla and young male and adult female of the chimpanzee.

IV. No human race presents the characters of the lower jaw exhibited by those of Spy, Naulette, and Shipka. In this part of their osteology the anthropoids depart widely from man, the most conspicuous point in the latter being the presence of a chin. Accordingly, the angle formed by the anterior face of the symphysis with the inferior border of the horizontal ramus, is less than a right angle in man, and much more than a right angle in the anthropoids. According to Topinard, this angle in fifteen Parisians is 71.4°; in fifteen African Negroes, 82.2°; in fifteen Neocaledonians, 83.9°; in the jaw of Naulette, 94°. In the best preserved jaw of Spy the angle is 107°, if measured from the inferior symphysial border, or 90° if meas-

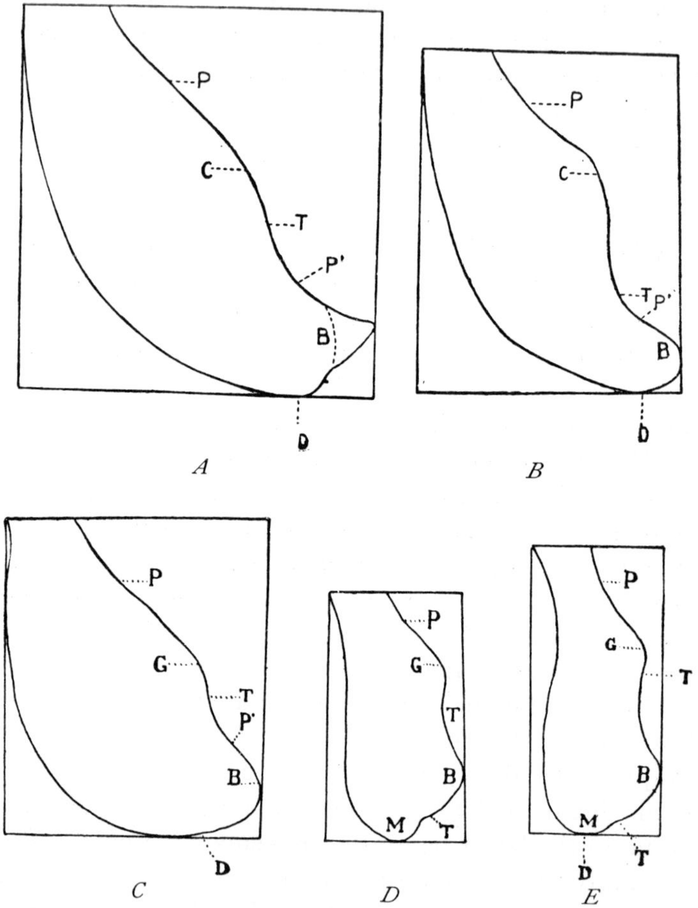

Fig. 45.—Vertical sections of symphysis mandibuli of gorilla (Fig. *A*), and orang (Fig. *B*), of chimpanzee (Fig. *C*), of Spy man No. 1 (Fig. *D*), and Spy man No. 2 (Fig. *E*). From Fraipont and Lohest.

ured from the inferior border of the ramus. There is no chin in the jaw of the Spy race, and the large angle approaches without nearly equaling that of the anthropoids. But the posterior face of the symphysis presents the most remarkable peculiarity. In the symphysis of the apes (Fig. 54, *A*, *B*, *C*) the posterior border is a continuous slope from the alveolar border to the inferior margin, interrupted by a slight concavity below the middle. In the human jaw this line slopes backward to near the middle, where are situated the small tuberosities for the insertion of the genioglossal muscles. (B in the accompanying figures.) The surface then slopes rapidly forward to pass into the narrow inferior border of the chin (Fig. 46, *F*, *G*). In the jaws Naulette and Spy the structure is exactly intermediate between the two, and quite different from both (Fig. 45, *D*, *E*). It commences above with a posterior slope similar to that of the apes, exhibiting what is called by Topinard "internal prognathism," as it appears in the lower human races. The surface then descends abruptly, forming a vertical concavity, which is bounded a considerable distance below by another protuberance, the insertion of the genioglossal muscles. This concavity is not present in the human symphysis, while it is less developed in the simian. The surface then slopes forward, as in the human symphysis, but this portion is shorter than in human jaws generally. It is represented by a convex face in the simian jaw. This character, taken in connection with the others cited, goes a long way toward justifying the separation of the Neanderthal race as a distant species, as has been done by some author under the name of *Homo neanderthalensis*. This name is objectionable but must be retained.

To these observations Messrs. Fraipont and Lohest add the following.

V. The curvature of the ulna and radius, which

produces a wide interosseous space, is not found in any human race, but is common to the apes. On the contrary, the shortness of these bones is entirely human.

VI. The anterior convexity of the femur, with its round section, is only found among living races among the Nigritos of the Philippine Islands. It is seen in a less degree in femora of Neolithic men, and occasional instances are seen among existing Europeans. It is the normal condition in the apes.

VII. The tibia is shorter in its relation to the femur than in any human race, and is more robust than in

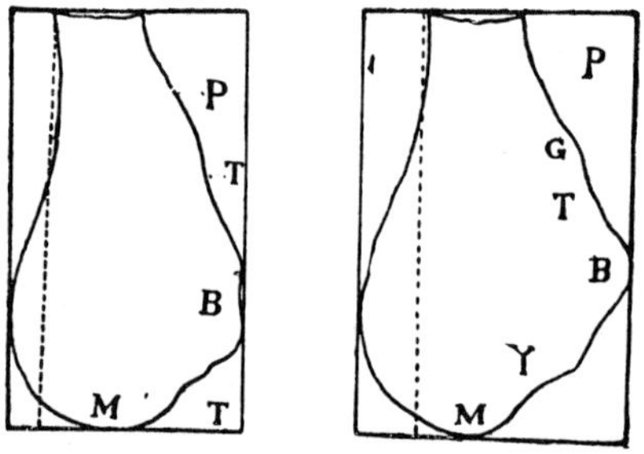

F *G*

Fig. 46.—Sections of symphysis mandibuli of modern Liègois (Fig. *F*) and of an ancient Parisian (Fig. *G*). From Fraipont and Lohest.

most of them. This character, with the oval section, while not identical with what is seen in the apes, forms an approximation to it.

Messrs. Fraipont and Lohest have pointed out the general characters of the dentition of the man of Spy. They show that the molars increase in size posteriorly to the same extent that they do in the apes, which is the reverse of what is usual in man, where they diminish posteriorly, or, in a few lower races (Australians, etc.), remain equal. They show that the superior molars are all quadritubercular, and that the internal root

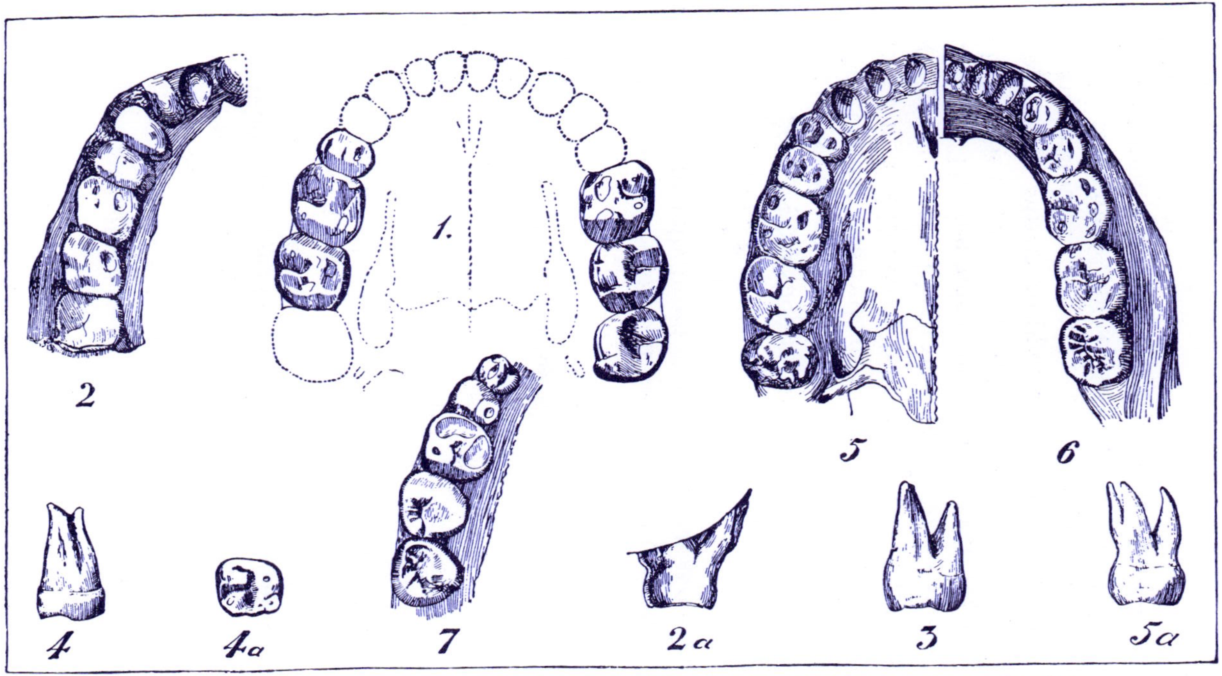

Fig. 47.—Molar teeth of man. 1–2, man and woman of Spy; 3, Maori; 4, Tahitian; 5–6, Fan, 7, Esquimau.

is distinct in all of them. Through the kindness of M. Lohest I received casts and photographs of these teeth, and I give here figures of the former (Fig. 47), which are more satisfactory than those in the memoir from which I have already quoted so fully, where, indeed, the grinding faces are not represented at all.

The figures accompanying[1] show the large size of the last superior molar, which exceeds in its proportions those of the corresponding tooth in the chimpanzee. The fourth tubercle, or hypocone, is especially large. In the male the crowns are more produced posteriorly than in man generally, and remind one of the character seen in the orang. The strong divergence of the internal root of the last molar is shown in No. 2 *a*, and the corresponding character in a Maori and a Fan from tropical Africa is shown in Nos. 3 and 5 *a*. The quadritubercular crown of the last superior molar of a Tahitian is shown in No. 4 *a*; and the roots, which are exceptionally fused nearly as much as in the typical Indo-European, are shown in No. 4.

Dr. Eugene Dubois of the Army of the Netherlands has recently published in Batavia, Java, in a brochure in quarto, an account of some bones of an interesting quadrumanous mammal allied to man, which were found in a sedimentary bed of material of volcanic origin of probably Plistocene age, near a village called Trinil. The remains consist of a calvarium which includes the supraorbital ridges and a part of the occiput; a last superior upper molar, and a femur. The tooth was found close to the skull and belongs probably to the same individual as the latter, while the reference of the femur is more uncertain, as it was found some fifty feet distant.

[1] From *The American Naturalist*, April, 1893.

The characters of the skull are closely similar to those of the men of Neanderthal and of Spy, but the walls are not so thick as those of the former, and more nearly resemble those of the latter. The frontal region is, therefore, much depressed, and it is also much constricted posterior to the postorbital borders. The sutures are obliterated. Dr. Dubois states that the cranial capacity is just double that of the gorilla, and two-thirds that of the lowest normal of man, bridging the gap which has long separated the latter from the apes. Thus the capacity of the former is 500 cubic centimetres, and the latter is 1500 cubic centimetres. In the Java man the capacity is 1000 cubic centimetres. The last upper molar has widely divergent roots, as in apes and inferior races of man, and the crown is large, with the cusps not clearly differentiated, showing a character commonly observed in the lower molars of the gorilla. The femur is long, straight, and entirely human. This discovery of Dr. Dubois adds to our knowledge of the physical characters of the Paleolithic man, and especially to his geographical range.

As regards the proper appellation of this being, Dr. Dubois is not happy. He proposes for him a new genus Pithecanthropus (after Haeckel), and even a new family, Pithecanthropidæ, without having shown that he is not a member of the genus Homo. It is not certain that he is not an individual of the species *Homo neanderthalensis*. His cranial capacity is less, it is true, than that of the man of Spy, but Virchow has pointed out that some of the Nigritos possess a remarkably small cranial capacity, as little as 950 cubic centimetres, and an inhabitant of New Britain only 860 cubic centimetres, a capacity even smaller than that of the man of Trinil. Until we learn the characters of the lower jaw of the latter we shall be in doubt as to whether this individual pertains to the *Homo sapiens* or the *Homo neanderthalensis.*

The characters of the dentition, cranium, and limbs which have been observed in the Paleolithic man, are not without parallel in existing races, though the characters do not generally occur together in the latter. The supposition that all the Paleolithic men so far found are all pathological subjects is not a probable solution of the question, although this type was no doubt subject to pathological conditions such as have been found in the leg-bones of the men of Neanderthal and Trinil. The characters of the symphysis of the lower jaw are quite sufficient to separate the Neanderthal man as a distinct species of the genus Homo.[1] This character is not pathological but it is zoölogical, and places that species between *Homo sapiens* and the apes.

In conclusion, it may be observed that we have in the *Homo neanderthalensis* a greater number of simian characteristics than exist in any of the known races of the *Homo sapiens*, although, so far as known, he belongs to the genus Homo. The posterior foot, so far as preserved, indicates this to be the case. The foot-character, which distinguishes the genera Homo and Simia still remains. There is still, to use the language of Fraipont and Lohest, "an abyss" between the man of Spy and the highest ape; though, from a zoölogical point of view, it is not a wide one.

The flints which were discovered in the stratum of cave deposit containing the human remains, are of the Paleolithic type known as Mousterien in France, which

[1] This view was first insisted on in an article on the Genealogy of Man in the *American Naturalist*, 1893, p. 331.

are of later origin than the Chelléen or older Paleolithic. The older Paleolithic man is not yet known. It is interesting to observe that these flints (Mousterien) are of the same form as the obsidian implements which I collected at Fossil Lake, in Oregon, with the bones of extinct llamas, horses, elephants, sloth, etc. The animals which accompanied the man of Spy are, *Cœlodonta antiquitatis* (wooly rhinoceros), *Equus caballus*, *Cervus elaphus*, *Cervus tarandus*, *Bos primigenius*, *Elephas primigenius*, *Ursus spelæus*, *Meles taxus*, *Hyæna spelæa* ; five extinct and four existing species.

As the evidence now stands, the most primitive and simian of human races inhabited the Old World. No trace of the *Homo neanderthalensis* has been found in America. As, however, Paleolithic implements are found in all continents, we may anticipate that this or some similar species of man will be discovered there also. The genealogy of man may be then represented as follows :

CLASS & BRANCH	ORDER AND FAMILY.	GEOL. SYSTEM
	Hominidæ	Plistocene
	Simiidæ	Neocene
	Adapidæ	Eocene
Mammalia	Condylarthra	Cretaceous
	Creodonta	
	Marsupialia polyprotodontia	Jurassic
	Monotremata	Triassic
Reptilia	Theromora	Carbonic
Batrachia	Batrachia Stegocephali	Carbonic
Pisces	Teleostomi Rhipidopterygia	
	Elasmobranchii Ichthyotomi	
Agnatha	[1]	
Cephalochorda	Leptocardii	
Vermes	[1]	
Cœlenterata	[1]	
Protozoa	[1]	

[1] Subordinate type not specified.

3. THE LAW OF THE UNSPECIALIZED.

The facts cited in the preceding parts of this chapter show that the phylogenetic lines have not been continuous, but that they may be represented by a system of dichotomy. In other words, the point of departure of the progressive lines of one period of time has not been from the terminal types of the lines of preceding ages, but from points farther back in the series. Thus it is not the highly developed or specialized plants which have given origin to the animal kingdom, but the lowest forms or Protophyta, which are not distinguishable from the Protozoa. Among animals it is not the specialized Arthropoda or Mollusca which present the closest affiliations with the Vertebrata, but the simple Vermes or Tunicata, from which the origin of the latter may be traced. In the Vertebrata it is not the higher fishes (Actinopterygia) which offer the closest points of affinity to the succeeding batrachian class, but that more generalized type of the Devonic period, the Rhipidopterygia, which probably occupies that position. The modern types of Batrachia (Urodela, Salientia) have plainly not furnished the starting-point for the reptiles, but the ancient order of the Stegocephali, which are also fish-like, is evidently their source. The Reptilia of the Permian present us with types with fish-like vertebræ (Cotylosauria, Pelycosauria), from which the class Mammalia may be distinctly traced. The later reptiles diverged farther and farther from the mammalian type with the advance of geologic time. The same principle has been found to be true in tracing the history of the subdivisions of the great classes, in the preceding section.

Agassiz and Dana pointed out this fact in taxonomy, and I expressed it as an evolutionary law under the name of the "Doctrine of the Unspecialized." This describes the fact that the highly developed, or specialized types of one geologic period have not been the parents of the types of succeeding periods, but that the descent has been derived from the less specialized of preceding ages. No better example of this law can be found than man himself, who preserves in his general structure the type that was prevalent during the Eocene period, adding thereto his superior brain-structure.

The validity of this law is due to the fact that the specialized types of all periods have been generally incapable of adaptation to the changed conditions which characterized the advent of new periods. Changes of climate and food consequent on disturbances of the earth's crust have rendered existence impossible to many plants and animals, and have rendered life precarious to others. Such changes have been often especially severe in their effects on species of large size, which required food in large quantities. The results have been degeneracy or extinction. On the other hand plants and animals of unspecialized habits have survived. For instance, plants not especially restricted to definite soils, temperatures, or degrees of humidity, would survive changes in these respects better than those that have been so restricted. Animals of omnivorous food-habits would survive where those which required special foods, would die. Species of small size would survive a scarcity of food, while large ones would perish. It is true, as observed by Marsh, that the lines of descent of Mammalia have originated or been continued through forms of small size. The same is true of all other Vertebrata.

It is not to be inferred from the reality of the law of "the unspecialized" that each period has been dependent on the simplest of preceding forms of life for its population. Definite progress has been made, and highly specialized characters have been gradually developed, and have passed successfully through the vicissitudes of geologic revolutions. But these have not been the *most* specialized of their respective ages. They have presented a combination of effective structure with plasticity, which has enabled them to adapt themselves to changed conditions.

In a large number of cases in each geologic age forms have been successful in the struggle for existence through the adoption of some mode of life parasitic on other living beings. Such habits reduce the struggle to a minimum, and the result has been always degeneracy. In other cases it is to be supposed that extremely favorable conditions of food, with absence of enemies, would have occurred, in which the struggle would have been almost nil. Degeneracy would follow this condition also. On the other hand, extreme severity of the struggle cannot have been favorable to propagation and survival, so that here also we have a probable cause of degeneracy. Degeneracy is a fact of evolution, as already remarked, and its character is that of an extreme specialization, which has been, like an overperfection of structure, unfavorable to survival.

In general, then, it has been the "golden mean" of character which has presented the most favorable condition of survival, in the long run.

CHAPTER III.—PARALLELISM.

IT IS now generally recognized that the successive types of organic beings present characters which are traversed in the embryonic life of those which attain the greatest complexity of development, and which occupy the highest places in the scale of life. This fact was observed by the early embryologists, as Von Baer and Agassiz, who did not admit its bearing on the doctrine of evolution. But Darwin and Spencer understood its significance, and Haeckel, Hyatt, and the writer applied it directly to the explanation of phylogeny. At the present time one of the chief aims of the science of embryology is to discover the record of the history of the past, recapitulated in the stages of embryonic life, and to unravel the phylogenesis of plants and animals by this method. The utility of these researches is attested by the results which they have attained, though for obvious reasons, these are not as definite and conclusive as those which are derived from paleontology. The general conclusion is however justified, i. e., that the records of embryology and paleontology are closely similar, and that any discordance between them may be explained on comprehensible principles.

A number of illustrations of the parallelism between taxonomy, ontogeny, and phylogeny may now be given.

1. PARALLELISM IN THE BRACHIOPODA.

For the following abstract I am indebted to Mr. C. E. Beecher of New Haven, whose excellent work in this field is well known.

The parallelism between the ontogeny and phylogeny in the Brachiopoda has been worked out in numerous instances.[1] To illustrate these, some more or less familiar genera may be taken as characteristic examples.

Lingula has been shown by Hall and Clarke (*Pal. New York*, Vol. VIII., 1892) to have had its inception in the Ordovician. In the ontogeny of both recent and fossil forms, the first shelled stage has a straight hinge line, nearly equal in length to the width of the shell. This stage may be correlated with the more ancient genus Paterina from the lowest Cambrian. Subsequent growth produces a form resembling Obolella, a Cambrian and Ordovician genus. Then the linguloid type of structure appears at an adolescent period, and is completed at maturity. Thus, Lingula has ontogenetic stages corresponding to (1) Paterina, (2) Obolella, and (3) Lingula, of which the first two occur as adult forms in geological formations older than any known Lingula.

Paterina represents the radicle of the brachiopods.

[1] C. E. Beecher, "Development of the Brachiopoda," Part I., Introduction, *Amer. Journ. Sci.*, Vol, XLI., April, 1891; "Development of the Brachiopoda," Part II., Classification of the Stages of Growth and Decline, *Amer. Jour. Sci.*, Vol. XLIV., August, 1892; "Development of Bilobites," *Amer. Jour. Sci.*, Vol. XLII., July, 1891; "Revision of the Families of Loop-bearing Brachiopoda," *Trans. Conn. Acad. Sci.*, Vol. IX., May, 1893.

It shows no separate stages of growth in the shell, is found in the oldest fossiliferous rocks, and corresponds to the embryonic shelled condition (protegulum) of other brachiopods.

The genus Orbiculoidea of the Discinidæ first appears in the Ordovician and continues through the Mesozoic. The early stages in the ontogeny of an individual are as in Lingula, first a paterina stage, followed by an obolella stage. Then from the mechanical conditions of growth a Schizocrania-like stage follows, and completed growth results in Orbiculoidea.

The elongate form of the shell in Lingula, as well as in many other genera, is determined by the length of the pedicle and freedom of motion. The discinoid or discoid of Orbiculoidea and Discinisca among the brachiopods, and Anomia among pelecypods, is determined by the horizontal position of the valves, which are attached to an object of support by a more or less flexible, very short organ, a pedicle or byssus, without calcareous cementation. This mode of growth is characteristic of all the discinoid genera, but, as already shown, the early stages of Paleozoic Orbiculoidea have straight hinge lines and marginal beaks, and in the adult stages of the shell the beaks are usually subcentral and the growth holoperipheral. This adult discinoid form, which originated and was acquired through the conditions of fixation of the animal, has been accelerated in the recent Discinisca, so that it appears in a free-swimming larval stage. Thus, a character acquired in adolescent and adult stages of Paleozoic species through the mechanical conditions of growth, appears by acceleration in larval stages of later forms before the assumption of the condition of fixation which first produced this character.

The two chief subfamilies of the Terebratellidæ undergo complicated series of metamorphoses in their brachial structure. Generic characters are based upon the form and disposition of the brachia and their supports. The highest genera in one subfamily, which is austral in distribution, pass through stages correlated with the adult structure in the genera Gwynia, Cistella, Bouchardia, Megerlina, Magas, Magasell, and Terebratella, and reach their final development in Magellania and Neothyris. The higher genera in another subfamily, boreal in distribution, pass through metamorphoses correlated with the adult structures of Gwynia, Cistella, Platidia, Ismenia, Mühlfeldtia, Terebratalia, and Dallina. The first two stages in both subfamilies are related in the same manner to Gwynia and Cistella. The subsequent stages are different except the last two, so that the Magellania structure is similar in all respects to the Dallina structure, and Terebratella is like Terebratalia. Therefore Magellania and Terebratella are respectively the exact morphological equivalent to, or are in exact parallelism with Dallina and Terebratalia.

The stages of growth of the genera belonging to the two subfamilies Dallininæ and Magellaniinæ are further correlated in the accompanying tables.

The simplest genus Gwynia, as far as known, passes through no brachial metamorphoses, and has the same structure throughout the adolescent period, up to and including the mature condition. In the ontogeny of Cistella the gwyniform stage, through acceleration, has become a larval condition. In Platidia, the cistelliform structure is accelerated to the immature period, and in Ismenia (representing an ismeniform type of structure in the higher genera), the gwyniform and cistelliform stages are larval, and the platidiform represents

MORPHOGENY FROM GWYNIA TO DALLINA.

Periods	Stages	Stages	Stages	Stages	Stages	Stages	Stages
Larval	gwyniform?	gwyniform	gwyniform	gwyniform cistelliform	gwyniform cistelliform	gwyniform cistelliform	gwyniform cistelliform
Adolescent	gwyniform	cistelliform	cistelliform	platidiform	platidiform ismeniform	platidiform ismeniform mühlfeldtiform	platidiform ismeniform mühlfeldtiform
Mature	*Gwynia*	*Cistella*	*Platidia*	*Ismenia*	*Mühlfeldtia*	*Terebratalia*	terebrataliform *Dallina*

MORPHOGENY FROM GWYNIA TO MAGELLANIA.

Periods	Stages	Stages	Stages	Stages	Stages	Stages	Stages	Stages
Larval	gwyniform	gwyniform	gwyniform	gwyniform cistelliform	gwyniform cistelliform	gwyniform cistelliform	gwyniform cistelliform	gwyniform cistelliform
Adolescent	gwyniform	cistelliform	cistelliform	bouchardiform	bouchardiform megerliniform	bouchardiform megerliniform magadiform	bouchardiform megerliniform magadiform magaselliform	bouchardiform megerliniform magadiform magaselliform
Mature	*Gwynia*	*Cistella*	*Bouchardia*	*Megerlina*	*Magas*	*Magasella*	*Terebratella*	terebratelliform *Magellania*

an adolescent condition. Similar comparisons may be made in the other genera. Progressively through each series, the adult structure of any genus forms the last immature stage of the next higher, until the highest member in its ontogeny represents serially, in its stages of growth, all the adult structures, with the larval and immature stages of the simpler genera. It is evident that in the identification of specimens belonging to the Terebratellidæ, whether recent or fossil, the strict specific characters must be given first consideration. Species, therefore, must be based upon surface ornaments, form, and color, within certain limits, and genera only upon structural features developed through a definite series of changes, the results of which are permanent in individuals evidently fully adult.

In each line of progression in the Terebratellidæ, the acceleration of the period of reproduction, by the influence of environment, threw off genera which did not go through the complete series of metamorphoses, but are otherwise fully adult, and even may show reversional tendencies due to old age; so that nearly every stage passed through by the higher genera has a fixed representative in a lower genus. Moreover, the lower genera are not merely equivalent to, or in exact parallelism with the early stages of the higher, but they express a permanent type of structure, as far as these genera are concerned, and after reaching maturity do not show a tendency to attain higher phases of development, but thicken the shell and cardinal process, absorb the deltidial plates, and exhibit all the evidences of senility.

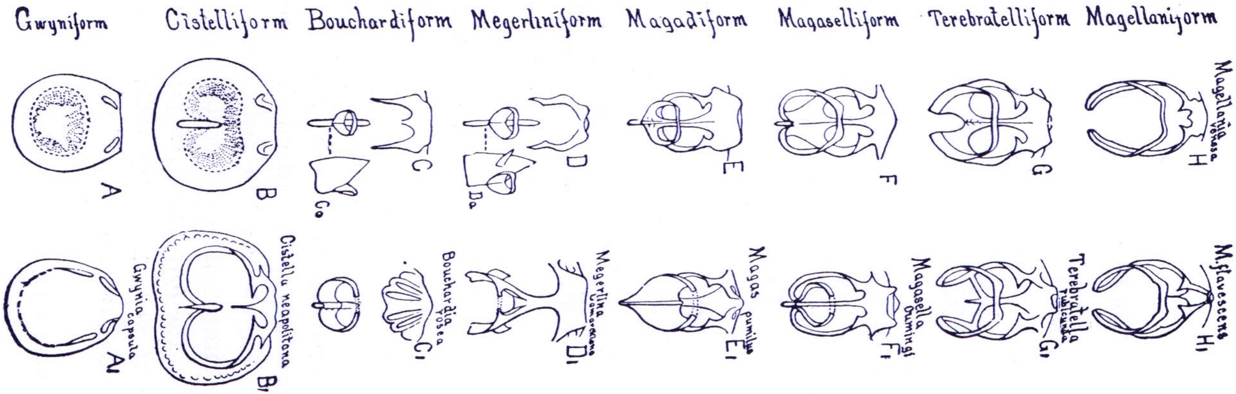

Fig. 48.—Parallelism in Brachiopoda; ontogeny and phylogeny of Magellania. (From Beecher.)

2. PARALLELISM IN THE CEPHALOPODA.

Among Mollusca it is well known that the Cephalopoda form a number of series of remarkable regularity, the advance being, in the first place, in the complication of the folds of the external margins of the septa, and, in the second place, in the degree of involution of one or both extremities of the shell to the spiral; third, in the position of the siphon.

Alpheus Hyatt, in an important essay on this subject,[1] points out that the less complex forms are in many cases identical with the undeveloped conditions of the more complex. He says: "There is a direct connection between the position of a shell, in the completed cycle of the life of this order, and its own development. Those shells occupying the extremes of the cycle" (in time), "the polar forms, being more embryonic than the intermediate forms. The first epoch of the order is especially the era of rounded, and, in the majority of the species, of unornamented shells with simple septa; the second is the era of ornamentation, and the septa are steadily complicating; in the third the complication of the septa, the ornamentation, and the number of species, about twice that of any other epoch, all combine to make it the zenith of development in the order; the fourth is distinguishable from all the preceding as the era of retrogression in the form, and partially in the septa.

"The four periods of the individual are similarly arranged, and have comparable characteristics. As

[1] *Memoirs of the Boston Society for Natural History*, 1866, p. 193. Hyatt was followed by Würtenberger in *Ausland*, 1873, who entirely confirmed his conclusions.

has been previously stated, the first is rounded and smooth, with simple septa; the second tuberculated, and the septa more complicated; the third was the only one in which the septa, form, and ornamentation simultaneously attained the climax of individual complication; the fourth, when amounting to anything more important than the loss of a few ornaments, was marked by a retrogression of the whorl to a more tabular aspect, and by the partial degradation of the septa."

I am indebted to Professor Hyatt for the following more detailed account of the results of his researches in this interesting field. The evidence as to the nature of evolution derived from the Cephalopoda is more complete than that obtained from any other source.

"Every group of nautiloids passes through, during its evolution in time, either a part or the whole of a certain series of changes. These modifications consist: first, of a straight or nearly straight cone, orthoceran; second, a curved cone, cyrtoceran; third, a coiled cone, gyroceran, which does not come in contact at any point; fourth, a coiled cone, nautilian, which does come in contact at the termination of the first volution and then during further growth remains in about the same condition, all of the internal whorls being visible as in a flat coil of rope; fifth, a coiled cone, involute-nautilian, which also comes in contact like the fourth but then the whorl growing with greater rapidity spreads internally, covering up more or less of the internal volutions sometimes to such an extent that even the centre is concealed from view. The examples which I have myself seen of the fifth kind range from the Silurian to the Nautili of the existing fauna, some being present in every period, and of other kinds, the first, second, and third kinds die out gradually, diminishing in the Devonian and Carboniferous and ultimately ceasing their existence altogether in the Trias. The fourth ceases in the Cretaceous, and the fifth alone survives in the Tertiaries and is still living in the *Nautilus umbilicatus* and *pompilus*, and two other species.

"Wherever found, the young of shells of the fifth kind are at first orthoceran or cyrtoceran like the first and second kind, then gyroceran in curvature like the third class, and then they become more or less rapidly nautilian like the fourth class in succeeding stages. In Silurian, Devonian, and Carboniferous forms this succession is so marked that about all of the young shells of the fifth class may be described as palingenetic, that is as cyrtoceran, gyroceran, nautilian, and then involute-nautilian in their individual or ontogenetic development. In the Trias, Jura, Cretaceous, Tertiary, and present, as the fifth class increases in numbers, there is a decided tendency to shorten and supersede the gyroceran or third stage and introduce the fourth kind or the tendency to spread by growth inwards, at earlier stages.

"The characteristics of the sutures are correlative with these stages of development, and it may be said in a general way, that all other characteristics correlate more or less when studied comparatively in different series with the differences in the curvature and coiling of the whorls. The curvature and amount of involution is therefore the most important single characteristic of the nautiloids, so far as the comparative study of change by evolution is concerned, whether the whole order be considered statistically as above, i. e. with reference to the existence or non-existence of certain forms orthoceran, cyrtoceran, etc., or genetically, i. e. with sole regard to the evolution of distinct series which may be traced from their origin to their termination in time.

"Of these last there are some in every period traceable with more or less completeness by gradations of adults back to orthoceran or cyrtoceran ancestors. Of these series of adults, some pass through only the orthoceran and cyrtoceran modifications, others have the orthoceran, cyrtoceran, gyroceran, and nautilian, but those having the latter and the nautilian-involute are of extreme rarity until the Carboniferous is reached. After this the nautilian shells begin to predominate in every series, ultimately becoming the sole representatives of genetic series.

"Such series are, of course, frequently so closely parallel that it is possible to follow them, and show they are distinct only by means of certain genetic characters, the apertures, the structure of the siphuncle, the sutures and septa, and sometimes, although very rarely, all of these internal characters may show differences peculiar to some one genetic series in which the regular gamut of forms is passed through in the usual succession. Neglect of the comparative study of the stages of development and decline, and of the obvious parallelisms between these and adults of ancestral forms, have caused naturalists, notably Barrande, to make artificial classifications in which about all straight forms, with the exception of some in which the siphuncles were notably distinct to be classed as Orthoceras, most of second kind as Cyrtoceras, most of the third kind as Gyroceras, most of the fourth and fifth kinds as Nautilus.

"To such authors the involute-nautilian forms of the Silurian and the existing fauna were considered to be only specifically distinct, although any prolonged study and comparison of the young would have shown that they were widely separated in development and really only morphic equivalents evolved from entirely distinct ancestors.

"A good example of this is the Eudoceratidæ[1] including the Silurian and Devonian Eudoceras and Tripteroceras, and probably gyroceran form Edaphoceras of the Carboniferous and the close-coiled nautilian shells of Endolobus of the Carboniferous. The peculiar forms of this series and their remarkable sutures enable the observer to follow the line both in the gradations of the adults and by means of the parallelisms of the development.

"Another good series easily distinguished by the remarkable sculpture of the shells is Zittelloceras of the Silurian with cyrtoceran forms, and the gyroceran and nautilian Halloceras of the Devonian.

"One of the best is Thoraceras, a rough spinous cone of the Silurian, Devonian, and Carboniferous, which has straight and cyrtoceran shells; the gyroceran Triboloceras of the Carboniferous, and the nautilian shells of Vestinautilus and its allies in the same period.

"There is no possible explanation of the parallelisms of development of these nautilian shells and the adult stages of others except heredity in the same genetic series. It is useless to waste time in discussion unless the facts are specifically denied after having been properly reëxamined.

"When the ammonoids are taken up, it is easy to demonstrate[2] by the study of the young of the Gonia-

[1] "Genera of Fossil Cephalopods," *Proc. Bost. Soc. Nat. Hist,*, p. 287.

[2] See "Genera of Fossil Cephalopods," *Proc. Bost. Soc. Nat. Hist.*, XXII., 1883, p. 303.

titinæ that they had straight forms among their ancestors and that these forms have a central siphuncle and suture as among nautiloids. The Devonian Goniatitinæ and some of the Carboniferous forms had also gyroceran forms and loosely coiled nautilian forms, indicating an ancestry with similar cones, but at these stages the siphuncle is invariably ventral as in the adults. The young of all of the Ammonitinæ, however involute the shell may afterwards become, have an invariably straight or curved cyrtoceran cone in the apical part, and when they come in contact by growth, the first whorl or whorls are equally invariably open coils like the coils of the fourth grade in nautiloids. The fifth kind of shell, the involute-nautilian, follows in precisely similar succession to what it does in the ontogeny of nautiloids. Farther than this the degree of involution increases according to the species, with age, and the amount of this involution is often an important part of the specific diagnosis.

"Among Ammonitinæ one finds at once that there are no orthoceran or cyrtoceran shells except among the large group designated by the author as Bactrites. This genus begins early in the Silurian with shells that are not distinguishable from true Orthoceras except by having the siphuncle in adults and later stages close to the venter. Some of these forms have no bulb or protoconch and have a large scar on the apex as in true Orthoceras, others have a calcareous bulb or protoconch on the apex as in true Ammonitinæ. There are also open or gyroceran shells in the adults of the genus Mimoceras which are repeated in the young of Anarcestes and other genera of Goniatitinæ figured in my 'Embryology of Fossil Cephalopods.'[1]

[1] *Bull. Mus. Comp. Zoöl.*, III.

The shells of the Ammonitinæ, however, are of the fourth and fifth kinds almost exclusively, and in following out the separate genetic series one has to distinguish the progressive gradations by means of the greater or less amount of involution even in the Goniatitinæ of the Devonian and Carboniferous.

"There are also some very remarkable facts showing that the coiling is closer in the Mesozoic than in the Paleozoic forms as a matter of hereditary derivation. The young of the Silurian and Devonian forms have the open, slowly coiling whorls figured by Sandberger and Barrande and repeated by myself as referred to above, but the young of all Mesozoic forms are close coiled so far as known. This is shown in the centre of the umbilicus by the perforation or central opening which is extremely large in most of the Paleozoic Gonititinæ but becomes almost obliterated in the true Ammonitinæ of the Jura.

"In tracing parallels between development of the individual and the series among Ammonitinæ it has been found by Branco and the author, that in ornamented shells the young are first like a nautilus in the sutures, then have a goniatitic stage like the first representatives of the order of ammonoids in the Paleozoic, and that during these stages it is invariably smooth and similar in general form to these same ancestors. After this nepionic stage is passed through the sutures and the characteristics alter with greater or less rapidity, but the stages show decisive parallelisms with the immediate ancestors of the same genetic series. Some of the best examples of palingenetic development of this kind, where the later stages of growth present parallels with proximate ancestors, are cited in my 'Genesis of the Ariëtidæ'[1] and others have been given by Buckman and Würtenburger.

"Some of the most remarkable occur in the least expected quarters. As usual, when one has a true law, it leads him into conclusions that are, perhaps, more surprising to himself than to his readers, or to any subsequent investigator. This was certainly my own case in being led to recognise the perfect examples of parallelisms in retrogressive series. Quenstedt and all students since his time agree that the so-called genera Crioceras, Hamites, Ancyloceras, Baculites, forms that are successively more and more uncoiled until in Baculites they are absolutely straight cones, were derived from normal, close-coiled, involute-nautilian shells of the Ammonitinæ. Their young have been repeatedly shown to be close-coiled and they grade into the normal progressive shells by all of their adult characters.

"The ultimate fact in this demonstration has been added by Dr. Amos Brown in the discovery of a close coiled nepionic stage in the straight Baculites of the Cretaceous, the only form whose development had not been ascertained and whose exact relations had not been determined.

"It is now admitted by all students of Ammonitinæ that these retrogressive groups are not true genera; but that as first demonstrated by Quenstedt, Baculites, Crioceras, etc., are retrogressive stages in the evolution of distinct genetic series and that they do not exist as natural groups of species. In other words, different genetic series of the Ammonitinæ die out by passing through a series of modifications which are parallel and which are just the reverse of the parallel

[1] *Smithson. Contrib.*, 673, p. 41 et seq.

series of the orthoceran, cyrtoceran, gyroceran shells with which each distinct complete genetic series of nautiloids arose in time.

"While the nautiloids coil up in their progressive evolution and the Ammonitinæ increase this coiling up tendency in the primitive and progressive forms of each genetic series, the latter in becoming retrogressive reverse the processes of progressive evolution. They become more and more uncoiled, each complete retrogressive series ending with a straight cone. All other characters correlate with this uncoiling and in a general way may be said to degenerate in greater or less proportion to the amount of the uncoiling. To make this extraordinary picture complete it is only necessary to add that these retrogressive series followed out to their ultimate development are distinctly parallel with changes or stages of modification taking place in the senile stages of individuals of the same genetic group.

"In old age the highly ornamented shell gradually parts with its spines and other ornaments, the whorls slowly diminish, the involution decreases and eventually in extreme age it becomes separated from the spiral and completely rounded and smooth. The aperture becomes correlatively modified, and also the sutures. If an old ammonite could have its life prolonged, it would become Baculites, and the full-grown part of the shell would, in some forms of Lytoceratinæ be very similar to the minute nepionic shell of the Baculites as described and figured by Dr. Brown. If now the coiled adult part of this imaginary shell were broken off and lost, the straight senile fragment would be referred to the old genus Baculites. The morphic characters of the gerontic or old-age stage of ontogeny are therefore parallel with the forms evolved in the paraplastic or retrogressive stage of evolution of the phylum. In other words, the morphic modifications which may occur as permanent, specific, and generic characters in the adults of retrogressive descendants of any progressive individuals may be predicted from the study of the similar changes that take place in the senile stages of the progressive individuals. As it has been stated by the writer on several occasions, the embryonic, nepionic, and later stages of development up to the adult repeat with greater or less clearness in proportion to their removal in time and organization from the point of the origin of the genetic group to which they belong the permanent characteristic of their ancestors; the adult gives the existing essential differentials acquired by its own species, genus, and group, being the index according to the time of its occurrence of the progression or retrogression of its group; the old, in its invariably retrogressive course, indicates the path that must be followed by degraded series after the acme of the group to which the individual belongs has been reached. This, of course, is a generalized statement of the correlations of the ontogenic cycle and the phylocycle when they occur as in the Ammonitinæ, but it will be found eventually that this law is true of all animals to some degree. It is obvious from all past experience that every law of correlation of structures cannot be true in any one group without being found more or less in all organisms. I have therefore ventured upon the basis of this and Beecher's, Clarke's, and Schuchert's researches among Brachiopoda, corals, and trilobites, Dr. Jackson's among pelecypods, and after the confirmations by the independent researches of Würtenburger and Buckman among Ammonitinæ, and those of Bather among crinoids, to designate the complete study of the correlations of the ontocycle and phylocycle as Bioplastology. Bioplastology is easily separable from the study of growth, and from that of heredity, for which last I have proposed the term Genesiology,[1] and from that of Ctetology or the study of the origin of acquired characteristics. By properly defining these different branches of research it is practicable to see that bioplastology includes the results of the action of growth, the laws of growth, as well as those of genesiology and ctetology, but has a field entirely distinct from all of these in so far as it deals essentially with the study of parallelism in all its phases."

The parallelism of the gyroceras with an early stage of all coiled Cephalopoda is represented in Fig. 119 page 410, as illustrative of the inheritance of an acquired character.

3. PARALLELISM IN THE VERTEBRATA.

Parallels between the ontogeny and phylogeny are well known in the Vertebrata. The primary relations of the Vertebrata are discernible in the successive types of structure of the nervous system, and of the skeleton, but most clearly in those presented by the circulatory system. It is well known that the central organ—the heart, is, in the amphioxus, a straight tube. In the next higher group, the Marsipobranchii (lampreys), it is a bent tube, with a constriction which divides it into two chambers. In the Pisces (fishes) the heart is composed of two chambers related to each other in a reversed longitudinal direction. In the Ba-

[1] See *Proceedings of the Boston Society of Natural History*, 1893, p. 59.

trachia and Reptilia the cephalad (auricular) division is divided into two chambers by a septum; while in the birds and Mammalia the caudad division (ventricle) is also so divided, making four chambers in all.

The sources of the great vessels which distribute the blood to the body and return it to the heart, display the same successional relation of types. In the Acrania (amphioxus), the Marsipobranchii, and most of the fishes, the vessel (truncus communis) which receives the blood from the central organ, gives off several branches on each side, which are distributed to skeletal bars or arches which are in immediate contact with water, which aerates the blood. They then return, and, first sending the carotids anteriorly, unite dorsad to the heart, and form the aorta posteriorly. In the Batrachia, where aerial respiration succeeds to an aquatic one during the life of the animal, the number of the vessels contributing to form the aorta is reduced from five to three in the sucessive types. One of the arches is aborted as an arch, and sends the circulating fluid to the modified swim-bladder of the fish, or lung, where it is aerated. This aerated blood is returned to the heart with non-aerated blood from other organs, and the mixture is sent throughout the body. In the reptiles we have essentially the same system, but the aorta-origins are reduced to two, and one, on each side. Next a division of the truncus communis ensues, which corresponds functionally with that in the ventricle, so that the impure blood from one auricle is sent into the ventricle (right) which communicates with the lung; and the aerated blood is then returned to the other auricle, which pours its contents into the left auricle, which drives it into the aorta, and thus throughout the body. Thus pure or

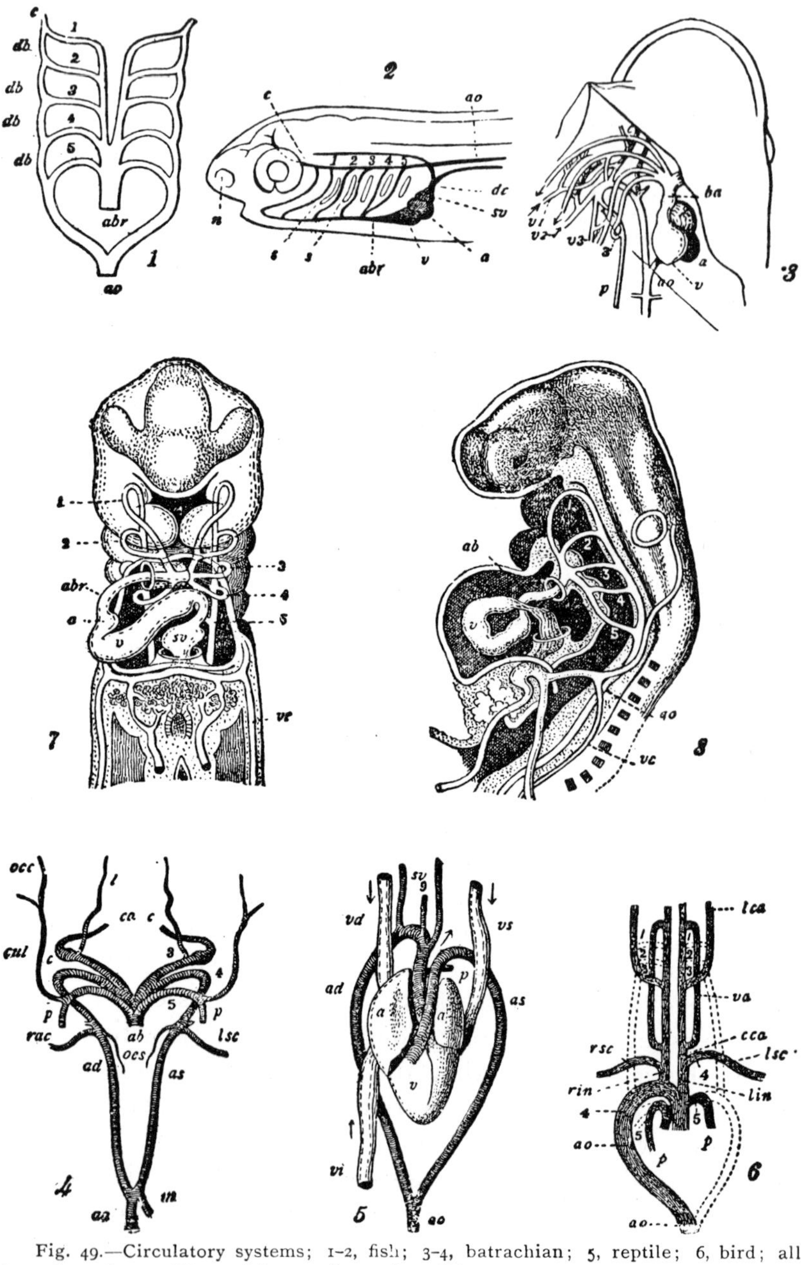

Fig. 49.—Circulatory systems; 1–2, fish; 3–4, batrachian; 5, reptile; 6, bird; all from Gegenbaur. Figs. 7–8, human fœtus, from His.

aerated blood is distributed to the organs, and all but one of the old roots of the aorta have ceased to function as such.

This evolutionary succession is preserved with much fidelity in the ontogeny of the respective classes of Vertebrata. The representatives of each class pass through the stages which are permanent in the classes below them in the series. The Mammalia, as the highest class, pass through all the stages. (Fig. 49.) This series coincides also with phylogenetic succession. The order of appearance in time of the Vertebrata is, first Agnatha, then Pisces, Batrachia, Reptilia, and Mammalia.

In all the details of structure the same relation may be observed. Referring to the illustrations of phylogeny and variation of character described in the preceding pages, many of the characters definitive of natural divisions have been observed to appear in the course of the embryonic life of those types which possess them. Those of greater systematic significance appear earlier, and those of less importance in a taxonomic sense, later. I select some illustrations of this principle.

I have shown that the primitive type of superior molar in the placental Mammalia is tritubercular, the fourth tubercle being added internally and posteriorly in the later forms. Dr. Taeker has recently observed that in the development of the superior molars in the horse, at an early stage the crown is tritubercular, and that the fourth cusp or hypocone is subsequently added, as in the phylogenetic history. As the horse presents the most complex molar among Mammalia, this survival of the record is interesting.

In the Artiodactyla and Edentata which lack superior incisor teeth, rudiments of them can be found in the early stages. We now know early extinct forms of both of these types where these teeth are permanent throughout life. In the toothless whalebone whales the same phenomenon has been observed.

It is well known that the highest deer (Cervidæ) add an antler to the simple spike horn in the third year, and an additional antler with each successive year for several years. Also they develop a basal snag of the antler (see Cuvier, *Ossem. Fossiles* ; Gray, *Catal. Brit. Mus.*) at the third year. Now a majority of those of the New World (genera Cariacus, Coassus) never develop it except in "abnormal" cases in the most vigorous maturity of the most northern Cariacus (*C. virginianus*); while the South American Coassus retains to adult age the simple horn of the second year of Cervus.

Among the higher Cervidæ, Rusa and Axis never assume characters beyond an equivalent of the fourth year of Cervus. In Dama the characters are on the other hand assumed more rapidly than in Cervus, its third year corresponding to the fourth of the latter, and the development in after years of a broad plate of bone, with points, being substituted for the addition of the corresponding snags, thus commencing another series.

Returning to the American deer, we have Blastocerus, whose antlers are identical with those of the fourth year of Cariacus.

The oldest known deer (Palæomeryx) have no horns, or they are undivided.

Among Batrachia excellent illustrations are furnished by the two series of Salientia, the Arcifera and the Firmisternia.

The firmisternial structure is a modification of the arciferous, which comes later in the history of growth, and probably also in geological time. During the early stages the Firmisternia have the movable shoul-

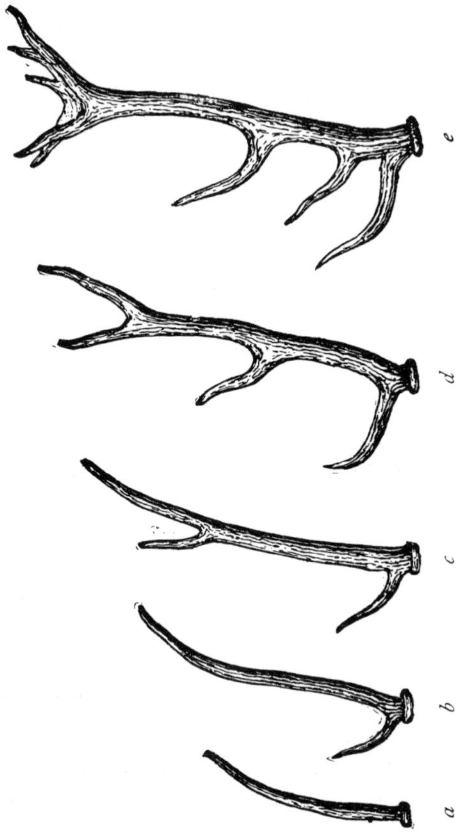

Fig. 50.—Succession of horns of *Cervus elaphus* L. from Gaudry. *a*, second year represented by permanent horn of Coassus; *b*, third year represented by permanent horn of Furcifer; *c*, fourth year represented by permanent horn of Rusa; *d*, fifth year; *e*, sixth year.

der girdle which characterizes those of the arciferous division, the consolidation constituting a modification superadded in attaining maturity. Furthermore, young Salientia are toothless, and one section of the species of Arcifera never acquire teeth. In these (the Bufonidæ) we have a group which is imperfect in two points instead of one.

The genera of these salientian suborders exhibited on a preceding page as forming indentical series in five different families (pp. 66–67) are related to each other as developmental stages in the history of the genera that attain the extreme development on each line. For example we select the family Hylidæ of which the terminal genus is Trachycephalus. Nearly allied

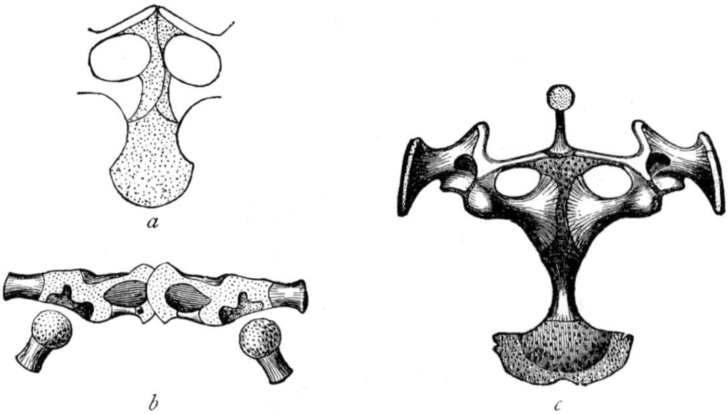

Fig. 51.—Shoulder girdles of Anura; *a*, of the arciferous type (*Phyllomedusa bicolor*); *b*, *Rana temporaria*, tadpole with budding limbs; *c*, do., adult, firmisternial type; *b* and *c* from Parker.

to it is the genus Osteocephalus, which differs in the normal exostosis of the cranium not involving the derm, as it does in the former. Close to this is Scytopis, where the fully ossified cranium is not covered by an exostosis. Next below Scytopis is Hyla, where the upper surface of the cranium is not ossified at all, but is a membranous roof over a great fontanelle. Still more imperfect is Hylella, which differs from Hyla in the absence of vomerine teeth. Now, the genus Trachycephalus, after losing its tail and branchiæ, possesses all the characters possessed by the genera Hylella and Hyla, either at or just before the mature state of the latter, as the ethmoid bone is not always ossified in advance of the parietals. It soon, however, becomes a Scytopis, next an Osteocephalus, and finally a Trachycephalus. It belongs successively to these genera, for an exhaustive anatomical examination has failed to reveal any characters by which, during these stages, it could be distinguished from these genera. The same succession in development of the genera of the other families is well known, the genus Otaspis of the Bufonidæ attaining a point beyond any of the others, in the enclosure of its membranum tympani posteriorly by dermoössification.

Finally reaching in our review the relations of specific characters, the readers will call to mind that the species of the lacertilian genus Cnemidophorus (page 41) are either striped, spotted, or cross-banded, and that the *Lacerta muralis* agrees with them in this respect. It was also shown that the young of all the species are striped, and that the cross-banded forms pass through not only a striped, but an intermediate spotted stage, before attaining the adult coloration.

The young of spotted salamanders are without spots (genera Amblystoma and Salamandra e g.); so that unspotted species resemble the young of the spotted. In many species of birds of more or less uniform patterns of coloration, the young are spotted. In some of these the females remain spotted throughout adult life. In some other species both sexes retain the spotted coloration of the young. The young of most deer are spotted. In the fallow-deer (Axis) the adults retain the spotted coloration, thus resembling the young of most of the species.

3. INEXACT PARALLELISM OR CÆNOGENY.

When the early or transitional stage of a higher form is exactly the same as a permanent lower form, the parallelism is said to be "exact." Such is the relation of a *Cnemidophorus gularis scalaris* to a *Cnemidophorus gularis gularis* as to color characters ; and of an *Amblystoma tigrinum* to a permanent breeding *Siredon lichenoides* in characters of higher structural value.

When the transitional stage of the higher only resembles the lower form in some one or more features, but not in all, the parallelism is said to be "inexact." It is evident that "exact parallelism" can only exist between ancestor and descendant in the same restricted line, and can be therefore only demonstrated in the case of the nearest relatives, between which a perfect phylogeny is known. So soon as new subordinate characters are assumed, or a change in the order of appearance of characters supervenes, the parallelism becomes "inexact," and such is the kind of parallelism usually observed. And it is more inexact the more widely removed in relationship are the forms compared. Thus the parallelism between the embryo man with five branchial slits, and the adult shark, is very inexact; but that between a true fish and a shark is much less inexact. That between a higher and a lower shark is still more exact, and so on. Exact parallelism in growth is called by Haeckel palingenesis or palingeny. The growth which has, through changes introduced subsequent to the origin of a line of descent, become inexact, or "falsified," is termed by the same author cænogenesis or cænogeny.

The superposition of characters which constitutes evolution, means that more numerous characters are possessed by the higher than the lower types. This involves a greater number of changes during the ontogenetic growth of each individual of the higher type. In other words, characters acquired during the phylogenetic history are continually assumed by the progressive form at earlier and earlier periods of life. This process has been metaphorically termed by Professor Alpheus Hyatt and myself "acceleration." All progressive organic evolution is by acceleration, as here described. Retrogressive evolution may be accomplished by a retardation in the rate of growth of the taxonomic characters, so that instead of adding, and accumulating them, those already possessed are gradually dropped; the adults repeating in a reversed order the progressive series, and approaching more and more the primitive embryonic stages. This process I have termed "retardation." Retardation is not however, always exact, even in retracing a true phylogenetic line, whence in such instances the process may not be correctly described as retardation. Professor Hyatt has applied to such types the term "senile," and gerontic; and to the resulting condition, the term "senility." His observations on this subject have been made on Mollusca, and principally on the Cephalopoda, and are of fundamental importance in this connection.

The history of a type which has passed through a full cycle of life, from its earliest appearance to its extinction, is divided by Haeckel into three stages, viz.: those of its rise; full vigor, as displayed by predominance of variations and numbers; and decadence. For these stages he uses the expressions Anaplasis,

Metaplasis, and Cataplasis. For the processes which bring about the first and last of these conditions, Professor Hyatt has used the terms Anagenesis and Catagenesis. Catagenesis is equivalent to degeneracy and has played an important part in organic evolution. I had used the term previously to Professor Hyatt for the same process, but with a wider application; extending its use to inorganic nature as well.[1] (See Chapter IV. of this book.)

Embryology has, however, revealed another series of phenomena which in many instances obscure the simplicity of the problem of ontogeny as presented in the preceding pages. It was the merit of Haeckel to generalize from the facts brought to light by this science, so as to present the relations which subsist between the primitive stages of all multicellular animals. This is known as the Gastræa theory. He showed that the primitive gastric cavity of all such animals is produced by an invagination of a portion of the surface of a primitive sphere or morula, which results from the segmentation of the oösperm. This hollow half-sphere he termed the gastrula, and the theoretical primitive animal which corresponds to it he called the Gastræa. Marine animals very similar to this Gastræa have been discovered. Haeckel showed, however, that gastrulas are not all alike, since they differ in the extent to which the segmentation of the oösperm may be carried, and the rate of segmentation of different parts of it. Thus early do inexact parallelisms arise. From this point onwards special peculiarities of the various developmental lines appear, some of which have especial reference to the necessities of embryonic life. Hence the trochosphere stage of so many

[1]*Origin of the Fittest*, p. 422.

invertebrate forms, and the nauplius and zoæa of the Crustacea.

Such are the statoblasts which are resting-stages for the embryos of fresh-water sponges and Polyzoa, and the glochidia of the Unionidæ, which are wanting in the marine forms of the same orders. Such are the amnion and allantois of certain Vertebrata and the placenta of certain Mammalia, which have no reference to any structures but their own residua, found in the adults of those animals.

A remarkable instance of this state of things appears in the history of the evolution of the insects. It is quite impossible to understand this history without believing that the larval and pupal states of the highest insects are the results of a process of degeneracy which has affected the middle periods of growth, but not the mature results. The earliest insects are the Orthoptera, which have active aggressive larvæ and pupæ, undergoing the least changes in their metamorphosis (Ametabola), and never getting beyond the primitive mandibulate condition at the end. The metamorphosis of the jawed Neuroptera is little more marked, and they are one of the oldest orders.

The highest orders with jaws undergo a marked metamorphosis (Coleoptera, Hymenoptera), the Hymenoptera even requiring artificial intervention in some instances to make it successful. Finally, the most specialized orders, the suctorial Diptera and Lepidoptera, especially the latter, present us with very unprotected more or less parasitic larval stages, both active and inactive. These animals have evidently degenerated, but not so as to prevent their completing a metamorphosis necessary for purposes of reproduction. As is well known, many imagines (Saturniidæ, Œstridæ) can perform no other function, and soon die, while in some Diptera the incomplete larvæ themselves reproduce, so that the metamorphosis is never completed.

This history is parallel to that proposed by Dohrn to account for the origin of the Ammocœtes larval stage of the Marsipobranchii. He supposes this form to be more degenerate than the corresponding stage of its probable ancestral type in the ancestral line of the Vertebrata. An inactive life in mud is supposed by Dohrn to have been the effective cause. An inactive life on the leaves of plants, or in dead carcases, has probably been the cause of the same phenomenon in the Lepidoptera and Diptera.

Thus we have developed an ontogeny within an ontogeny, and a phylogeny within a phylogeny. These facts do not, however, affect the general result in the least. They only show us that the persistent larvæ of those animals which possess them, have a history of their own, subject to the same laws of evolution as the adults. It results that in many cases the phylogeny can only be determined by the discovery and investigation of the ancestors themselves, as they are preserved in the crust of the earth. In all cases this discovery confirms and establishes such definite conclusions as may be derived from embryology. It is also clear that on the discovery of phylogenetic series it becomes at once possible to determine the nature of defective types. It becomes possible to ascertain whether their rudimental parts represent the beginnings of organs, or whether they are the result of a process of degeneration of organs once well developed.

An excellent illustration of inexact parallelism is to be found on comparison of man with the lower Vertebrata. I have pointed out[1] that in the structure of his extremities and dentition, he agrees with the type of Mammalia prevalent during the Eocene period (cfr. Phenacodus). Hence in these respects he resembles the immature stages of those mammals which have undergone special modifications of limbs and extremities, such as Ungulata in which cænogeny has not obliterated the early stages from the embryonic record. These forms are probably extinct. I have also shown[2] that in the shape of his head man resembles the embryos of all Vertebrata, in the protuberant forehead, and vertical face and jaws. In this part of the structure most Vertebrata have grown farther from the embryonic type than has man, so that the human face may be truly said to be the result of a process of retardation. Nevertheless, in the structure of his nervous, circulatory, and for the most part, of his reproductive system, man stands at the summit of the Vertebrata. It is in those parts of his structure that are necessary to supremacy by force of body only, that man is retarded and embryonic.

5. OBJECTIONS TO THE DOCTRINE OF PARALLELISM.

An objection to the theory of parallelism in its full sense has been recently put forward by Mr. C. Herbert Hurst.[3] He says, "My object now is to show that in neither case can a record of the variation at any one stage of evolution be preserved in the ontogeny, much less can the ontogeny come to be a series of

[1] "The Relation of Man to the Tertiary Mammalia," *Penn Monthly*, 1875; *Origin of the Fittest*, 268.

[2] "The Developmental Significance of Human Physiognomy," *American Naturalist*, June, 1883; *Origin of the Fittest*, 1887, p. 281.

[3] *Natural Science*, 1893, p. 195.

stages representing in proper chronological order some of the stages of adult structure which have been passed through in the course of evolution." Again: "The early stages of the fish embryo are very like those of the bird embryo. These two do correspond to each other. The statement that the embryonic structure of a bird follows a course which is from beginning to end roughly parallel with, but somewhat divergent from, the course followed by a fish, is borne out by the actual facts. A bird does not develop into a fish and then into a reptile, and then into a bird. There is no fish-stage, no reptile-stage, in its ontogeny. The adult resembles an adult fish only very remotely. Every earlier stage resembles the corresponding earlier stage of the fish more closely. There is a parallelism between the two ontogenies. *There is no parallelism between the ontogeny and the phylogeny of either a bird or any other animal whatever.* A seeming parallelism will fall through when closely examined." "The promise that this theory gave of serving as the guide to knowledge of past history without the labor involved in paleontological research, was indeed tempting: and where the royal road to learning has been shown by it, it is not surprising that some zoölogists should have entered for the race along this road. To what goal that road has led may be learned by a comparison of the numerous theories as to the ancestry of the 'Chordata' which have been put forward by those who have adopted the theory without enquiring as to its validity."

I have made this quotation as showing the point of view from which the doctrine of parallelism when incorrectly stated may be assailed. There is truth in the author's accusation that embryologists who have not used their results with proper caution, have been frequently led to incorrect and even absurd results. The errors of this class of biologists are mainly due to their ignorance of species in the adult state, and their ignorance of systematic biology or taxonomy. They profess to regard this branch of the science as only suitable for beginners, and as comparatively unimportant, as compared with their own; yet one might as well attempt the study of philology without a knowledge of alphabets, as to study phylogeny without the knowledge of natural taxonomy. The correct discrimination of species, genera, etc., imposes much greater burdens on the faculty of judgment, than does anything to be found in any science which includes observation and record only. But Mr. Hurst's statement is somewhat overdrawn, and he does not give embryologists the credit which is due to their theory of recapitulation. I think he will find the following, which I wrote in 1872[1] to be a correct statement of the facts, and a fair induction as to principles.

"The smaller the number of structural characters which separate two species when adult, the more nearly will the less complete of the series be identical with an incomplete stage of the higher species. As we compare species which are more and more different, the more necessarily must we confine the assertion of parallelism to single parts of the animals, and less to the whole animal. When we reach species as far removed as man and a shark, which are separated by the extent of the series of vertebrated animals, we can only say that the infant man is identical in its numerous origins of the arteries from the heart, and in the cartilaginous skeletal tissue, with the class of

[1] *Penn Monthly*, 1872. *Origin of the Fittest*, 1887, p. 8.

sharks, and in but few other respects. But the importance of this consideration must be seen from the fact that it is *on single characters of this kind that the divisions of the zoölogist depend.* Hence we can say truly that one order is identical with an incomplete stage of another order, though the species of the one may never at the present time bear the same relation in their entirety to the species of the other. Still more frequently can we say that such a genus is the same in character as a stage passed by the next higher genus; but when we can say this of species, it is because their distinction is almost gone. It will then depend on the opinion of the naturalist as to whether the repressed characters are permanent or not. Parallelism is then reduced to this definition: that each separate character of every kind, which we find in a species, represents a more or less complete stage of the fullest growth of which the character appears to be capable. In proportion as those characters in one species are contrasted with those of another by reason of their number, by so much must we confine our comparison to the characters alone, and the divisions they represent; but when the contrast is reduced by reason of the fewness of differing characters, so much the more truly can we say that the one species is really a suppressed or incomplete form of the other. The denial of this principle by the authorities cited has been in consequence of this relation having been assigned to orders and classes, when the statement should have been confined *to single characters*, and divisions characterized by them. There seems, however, to have been a want of exercise of the classifying quality or power of 'abstraction' of the mind on the part of the objectors."

It is nevertheless true that the records brought to light by embryologists are very imperfect, and have to be carefully interpreted in order to furnish reliable evidence as to the phylogeny of the species examined. An illustration of this is the fact that the species characters appear in many embryos before those which define the order or the family, although it is certain that the latter appeared first in the order of time. Most of the important conclusions as to the phylogeny of Vertebrata demonstrated by paleontology have never been observed by embryologists in the records of the species studied by them. Thus I have shown that it is certain that in the amniote vertebrates the intercentrum of the vertebral column has been replaced by the centrum; yet no evidence of this fact has been observed by an embryologist. If we could study the embryonic development of the vertebral column of the Permian and Triassic Reptilia, the transition would be observed, but in recent forms cænogeny has progressed so far that no trace of the stage where the intercentrum existed can be found.

Again I have demonstrated by paleontological evidence that the lines of the ungulate Mammalia originated from a bunodont pentadactyle plantigrade ancestor; but embryonic research has failed to discover the preservation of a record of this fact in the ungulates at present existing. The embryo of the horse is not pentadactyle, nor even tridactyle, although tridactyle horses persisted late in geologic time. Nor has embryonic research demonstrated a four-toed stage in the Bovidæ (oxen, etc.), although there is no doubt that they descended directly from an ancestor so characterized. Any number of similar cases might be cited to show the prevalence of inexact parallelism or cænogeny. If we could study the embryology of the many extinct forms of life, the missing stages would all be found, but as we have not the opportunity of pursuing this important research, we have to rely on paleontology for our phylogeny. Paleontology is and always will be imperfect, but all that we get is palingeny, or the phylogeny itself, and not an inverted and distorted record of it.

CHAPTER IV.—CATAGENESIS.

WE HAVE been principally occupied so far with progressive evolution or anagenesis. Reference has, however, been made to retrogressive evolution or degeneracy, in Chapter III., in describing the evolution of the Vertebrata, and will be in Chapter V., under the caption "Disuse in Mammalia." Degeneracy has, however, played a more important part in creation than would be suspected from these references, and I propose in the present chapter to go more fully into its phenomena, which, in the broadest sense, I have called collectively Catagenesis.

As evidence for degeneracy as a factor in evolution we naturally appeal first to examples in the life histories of plants and animals which are known to us; and then examine the records of the past, in the light thus gained, for evidence of degeneracy in vegetable and animal phylogeny. In both directions we are met by an *embarras de richesse*, and a few conspicuous cases will have to suffice.

The parasitic copepod Crustacea undergo a retrograde metamorphosis, which commences at different periods of the growth history of different genera. Says Claus: "Many parasitic Copepoda, however, pass over the series of nauplius forms [which are traversed by other copepods] and the larva, as soon as hatched, undergoes a moult, and appears at once in the youngest Cyclops form with antennæ adapted for adhering, and mouth-parts for piercing. From this stage they undergo a retrogressive metamorphosis, in which they become attached to a host, lose more or less completely the segmentation of the body, which grows irregular in shape, cast off their swimming feet, and even lose the eye, which was originally present (*Lernæapoda*). The males, however, in such cases often remain small and dwarfed, and adhere, frequently more than one, firmly to the body of the female in the region of the genital opening. In the *Lernæa* such pigmy males were for a long time vainly sought for upon the very peculiarly shaped body of the large female (Fig. 52), which carries egg-tubes.

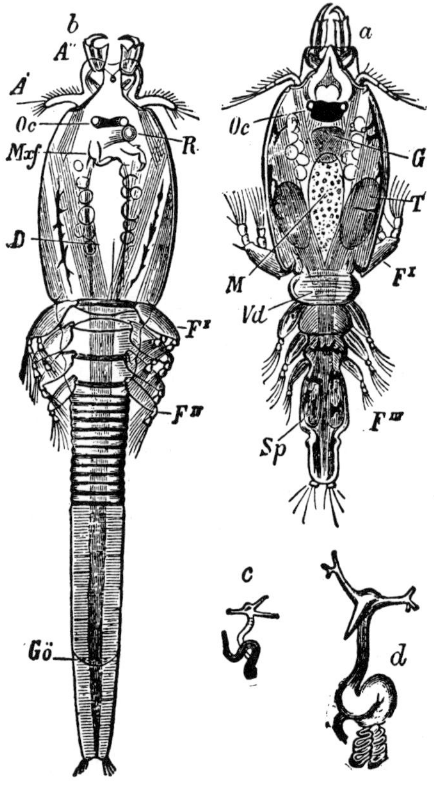

Fig. 52.—*Lernæa branchialis*; *a*, male; *b*, non-degenerate female; *c*, female after fertilization undergoing metamorphosis; *d*, *do.* with egg sacs, natural size. From Claus.

At last it was discovered that the small Cyclops-like males lead an independent life and swim about freely by means of their four pairs of swimming feet, and that the females in their copulatory stage resemble the males, and that it is only after copulation that they (the females) become parasitic and undergo the considerable increase in size and modification of form which characterizes the female with egg-tubes."

A degeneracy of the females of a remarkable character occurs in the insects of the order Strepsiptera. Here the female during the larval stage, bores its way into the body of a hymenopterous insect and soon undergoes a moult. At this time they shed their three pairs of well-developed legs, and become a parasitic maggot, which lives on the body of the host. The males do not undergo this degeneracy but retain the six legs and two pairs of wings common to the class Insecta.

A notorious example of degeneracy among the Mollusca is offered by the *Entoconcha mirabilis*. Says J. S. Kingsley: "So greatly has parasitism altered the form of the body, and all of the organs, that the proper position of this form among the gastropods is far from certain, some placing it near Natica. Indeed, were it not for the characters afforded by the young, its position among the Mollusca would not be suspected. Some thirty years ago [before 1885] Johannes Müller found in some specimens of *Synapta digitata* an internal worm-like parasite, attached by one extremity to the alimentary canal, while the other end hung free in the perivisceral cavity." "In one specimen of Synapta out of one or two hundred this strange form occurs. It is a sac, the upper part bearing the female, and the lower the male reproductive organs, while the centre of the body serves for a while as a broodpouch, the embryos later passing out from an opening at the free end of the body of the parent. The eggs undergo a tolerably regular development, producing a velum, shell, and operculum, the later stages being found free in the body-cavity of the host."

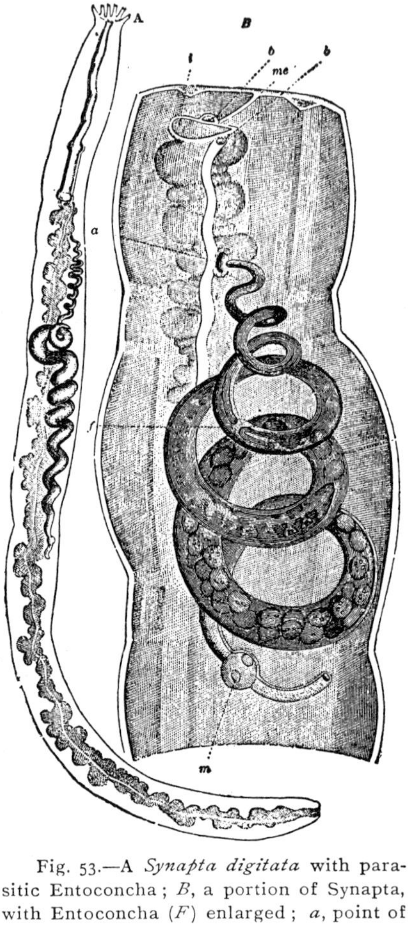

Fig. 53.—A *Synapta digitata* with parasitic Entoconcha; *B*, a portion of Synapta, with Entoconcha (*F*) enlarged; *a*, point of attachment; *b*, blood vessels; *f*, female portion; *i*, intestine; *m*, male portion; *me*, mesentery. From Kingsley.

The preceding examples illustrate the degenerating or catagenetic effect of a parasitic life. We will now observe the corresponding effect of a sedentary life, which may be called earth-parasitism. As an example of this I select the well-known case of the lowest of the Vertebrata, the Tunicata.

The embryo ascidian has the form of a tadpole-like larva which swims actively through the sea by vibrating its long tail. After a short free-swimming existence the fully developed, tailed larva fixes itself by its anterior adhering papillæ to some foreign object, and then undergoes a remarkable series of retrogressive changes, which convert it into the adult ascidian. The tail atrophies, until nothing is left but some fatty cells in the posterior part of the trunk. The adhering papillæ disappear and are replaced functionally by a growth of the test over neighboring objects. The nervous system with its sense-organs atrophies, until it is reduced to the single small ganglion placed on the dorsal edge of the pharynx, and a slight nerve-cord running for a short distance posteriorly. Slight changes in the shape of the body and a further growth and differentiation of the branchial sac, peribranchial cavity, and other organs now produce gradually the structure found in the adult ascidian (Herdman). It is, however, to be noted that in the order Larvacea, this retrograde metamorphosis does not take place. It embraces the single family Appendiculariidæ, which includes Tunicata which preserve the tail, notochord, and other larval features, and lead a free-swimming existence in the ocean.

On the Tunicata, Herdman makes the following general observations. "(1) In the ascidian embryo all the more important organs (e. g. notochord, neural canal, archenteron) are formed in essentially the same manner as they are in amphioxus and other Chordata. (2) The free-swimming tailed larva possesses the essential characters of the Chordata, inasmuch as it has a longitudinal skeletal axis (the notochord), separating a dorsally placed nervous system (the neural canal) from a ventral alimentary canal (archenteron); and therefore during this period of its life history the animal belongs to the Chordata. (2) The Chordata larva is more highly organized than the adult ascidian, and therefore the changes by which the latter is produced

	Diploglossa			Lepto	
	Pygopodidæ	Zonuridæ	Anguidæ	Teidæ	Gerrhosauridæ
I. Limbs, two pair *a.* Digits 5-4				Tejus	
b. Digits 4-5				Tretioscineus Micrablepharus Gymopthalmus	
c. Digits 4-4			Sauresia	Scolecosaurus	Saurophis
d. Digits 4-3					
e. Digits 3-4					
f. Digits 3-3				Microdactylus	
g. Digits 3-2				Herpetochalcis	
h. Digits 2-4					
i. Digits 2-3					
j. Digits 2-2					
k. One or both monodactyle		Chamæsaura	Panolopus	Cophias Ophiognomon	Cætia
II. Fore limbs only				Propus (digits o)	
III. Hind limbs only	Pygopus Cryptodelma Delma Pletholax Aprasia Lialis	Mancus	Pseudopus Opheodes Hyalosaurus		
IV. No limbs			Opheosaurus Dopasia Anguis		

LOSSA				Anniel- loidea	Annulati
Scincidæ	Acontiidæ	Dibamidæ	Anelytropsidæ	Anniellidæ	
Hagria					
Heteropus Ristella Menetia					
Gongyloseps Chiamela Rhinoseineus Tetradactylus Miculia Chalcidoseps Blepharactisis Sphenops					
Zygnopsis					
Allodactylus					
Tridentulus Chalcides Hemiergis Siaphus Phaneropis Sepomorphus Sphenoscineus Sepsina	Nessia				
Hemipodium					
Anisoterma					
Lerista Eumecia Heteromeles					
Dimeropus Chelomeles					
Brachystopus Oncopus Brachymeles Anomalopus Coloscincus Furcillus Dicloniscus	Evesia				
					Euchirotidæ (digits 3–5)
Ollochirus Dumerlia Scelotes Soridia Podoclonium		Dibamus			
Opheoscincus Herpetosaura Sepophis Herpetoseps Opheomorus	Acontias Typhlacontias		Anelytropsis Feylinia Typhlosaurus	Anniella	Amphisbæna Rhineura Lepidosternum Trogonophidæ

from the former may be regarded as a process of degeneration. The important conclusion drawn from all this is, that the Tunicata are the degenerate descendants of a group of the primitive Chordata" (=Vertebrata).

The degeneracy of the Tunicata follows immediately their assumption of the sessile condition. Some of the degenerate forms which are not sessile, are supposed to be the free descendants of sessile forms.

Among the craniate Vertebrata, most conspicuous examples of degeneracy are to be seen in the reduction and loss of limbs in certain Batrachia and in many Reptilia. In both classes successive loss of phalanges and digits form series in several groups of salamanders and lizards, and in both these orders there are forms with the limbs rudimental or altogether wanting. In Batrachia, the genus Amphiuma displays rudimental limbs with minute digits numbering two or three on each limb. In the Cæciliidæ, the limbs are wanting. Both types are subterranean in their habits. I give the annexed table of the Lacertilia with degenerate limbs, which it will be observed are found in eleven distinct families. (Pp. 216–217.)

Finally, in the snakes (Ophidia) the limbs have totally disappeared, rudiments only remaining in the boas and pythons and their allies.

Paleontology renders it clear that this reduction is a case of degeneracy, since both the Ophidia and Lacertilia can be traced to Reptilia of the Permian epoch, which have well-developed limbs. This degeneracy is allied to subterranean or terrestrial habits. It is probable that the primitive snakes sought concealment in cavities of the earth and beneath rocks and logs, and spent much of their time in narrow quarters, where limbs would be of no use to them. Some of them, the Angiostomata, are now subterranean in their habits, and most of them are blind, or nearly so. These forms present rudiments of limbs, which leads to the supposition that they are near to the ancestral types. From such forms they developed a type which has proved competent to compete successfully with other vertebrates on the ground, in the water, and in the trees of the forest.

From what has gone before it is now clear that while kinetogenesis is a factor in progressive evolution, the reverse process, or akinetogenesis, is as definite a factor in degeneracy. The evidence derived from parasitism and sedentary modes of life is conclusive in this direction.

I now cite another example of catagenesis which throws much light on the origin of the vegetable kingdom. I have advanced the hypothesis[1] that plants are the degenerate descendants of protozoan animal ancestors, and I will now produce some of the evidence on which the hypothesis rests. The Myxomycetes or Mycetozoa occupy debatable ground between the vegetable and animal kingdoms. They seem at one period of their history to pertain to the former and at another to the latter.

These organic beings are claimed by both botanists and zoölogists, the former placing them with the Fungi, the latter including them in the Protozoa. The fact is that in their mature form they enter the Fungi, while in their early stages they are Protozoa. They have distinct reproductive structures, which produce spores. From each spore issues a "flagellula," which is a simple cell with a flagellum, not apparently

[1] *Origin of the Fittest*, pp. 431-432.

different from a monad. The flagellum is early lost, and the cell is then termed an amœbula, since it does not differ materially from an amœba. Its movements are similar, and it puts forth short pseudopodia. When these amœbulæ come in contact with each other they

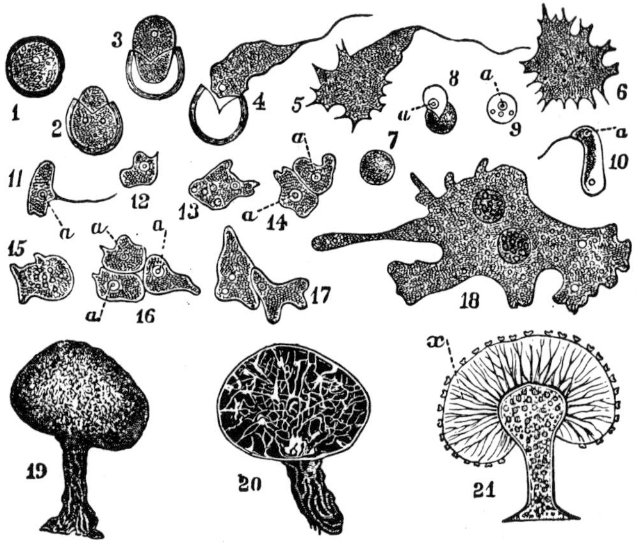

Fig. 54.—Mycetozoa (from Lankester after Du Bary). 1–6, Germination of spore (1) of *Trichea varia*, showing the emerging flagellula; (4–5) and its conversion into an amœbula (6). 7–18, Series leading from spore to plasmodium phase of *Chondrioderma difforma*; 7, spore; 10, flagellula; 12, amœbula; 14, apposition of two amœbulæ; 15–17, fusions; 18, plasmodium. 19–20, Spore-fruit (cyst) of *Physarum leucophæum* × 25; the former from the surface, the latter in section with the spores removed to show the sustentacular network or capillitium. 21, Section of the spore-cyst of *Dydymium squamulosum*, with the spores removed to show the radiating capillitium *x*, and the stalk.

fuse, often in large numbers, forming a continuous gelatinous sheet, the plasmodium (Fig. 54), which may have several square inches, and even feet of surface. At the proper time reproductive organs form on this surface in the form of capsules (sporangia), which may or may not be supported on peduncles, and which are filled with minute cyst-like masses of protoplasm, or spores. As already stated, these spores give issue to flagellula.

We have in the life of the Mycetozoa, if not the actual origin of the vegetable from the animal kingdom, a case closely similar to it in a collateral phylum. The process is one of degeneracy through the assumption of a sessile life, or earth-parasitism; an example of akinetogenesis. The paleontology of animals has absolutely established the fact that the predecessors of all characteristic or specialized types have been unspecialized or generalized types, "neither one thing nor another." It may then be regarded as almost certain that the ancestors of the present higher types of plants were more animal-like than they; that the forms displaying automatic movements were more numerous, and the difficulty of deciding on the vegetable or animal nature of a living organism greater than it is now. Hence it may be concluded that "animal" bathmism has from time to time undergone retrograde metamorphosis producing as a result the permanent form of life which we call vegetable. Given spontaneous movement (i. e. growth) and surrounding conditions, and the resultant product must be structures adapted to their surroundings, just as the plastic clay is fitted to its mould. And this is essentially the distinguishing character of vegetable teleology as compared with animal. In the average plant we see adaptation to conditions permitted by unconscious nutrition and reproduction; in the animal, adaptation to a greater variety of conditions, due to the presence of sensation or consciousness.

In closing Part I. of this book, I desire to point out the conclusion which has, I think, been reached.

It has been proved, as it appears to me, that the variation which has resulted in evolution has not been multifarious or promiscuous, but in definite directions. It has been shown that phylogeny exhibits a progressive advance along certain main lines, instead of having been indefinite and multifarious in direction.

It is not denied that many lines of variation have been at one geologic period and another discontinued. It is also true that certain divergences from the main lines have appeared, and that minor and secondary variations have occurred. Such variations do not seem to have had any material effect on the general course of evolution. In many cases such variations from main lines might be compared to the undulations in the course of a stream, which nevertheless seeks its lowest level in spite of all temporary obstacles. Professor Scott has termed these temporary variations "nutations," in an able article on the subject.[1] "Sports" seem to have been of no importance in evolution whatever.

[1] *American Journal Sci. Arts*, Vol. XLVIII., 1894, p. 355.

PART II.

THE CAUSES OF VARIATION.

PRELIMINARY.

IN Part II., which treats of the causes of variations, I propose to cite examples of the direct modifying effect of external influences on the characters of individual animals and plants. These influences fall naturally into two classes, viz., the physico-chemical (molecular), and the mechanical (molar). The modifications so presented are supposed to be the result of the action of the causes in question, continued throughout geologic time. To the two types of influence which thus express themselves in evolution, I have given the names Physiogenesis[1] and Kinetogenesis. The inheritance of character is assumed in this section, and the reason for so doing will be considered later, in the third section of this book.

In the animal kingdom we may reasonably suppose that kinetogenesis is more potent as an efficient cause of evolution than physiogenesis. In the vegetable kingdom it is quite evident that evolution is more usually physiogenetic than kinetogenetic. Atmospheric and terrestrial conditions play a major rôle in the

[1] "The Energy of Evolution." *American Naturalist*, March, 1894. "The Origin of Structural Variations" in *New Occasions*, Chicago, May, 1894. C. H. Kerr & Co.

determination of plant-structure, but motion has also had an important influence. The motion, however, has originated in small degree in the plant itself, but has been derived from without. Some importance must be ascribed to the effects of winds, but the principal source of the especial strains to which plants have been subjected, has been the insect world. Insects have been inhabitants of land-plants since their origin in early Paleozoic ages, and the mutual relations of plants and insects have ever been intimate. As has been insisted by Müller and Henslow, the uses to which the floral organs have been put by hymenopterous and other insects have been probably a principal cause of the forms assumed by the former. From this direction has been derived the kinetogenetic influence in plant evolution. The few independent movements displayed by plants may have had some influence on the evolution of their structure. We have no reason as yet to suppose that such movements have any other than purely physical factors.

mal Life, to which I refer my readers for a fuller exposition than can be given here.

a. Relation of the Size of Shells of Mollusca to the Environment.

It has been observed that both in natural conditions and in confinement, shells of fresh-water Mollusca grow to a larger size in larger bodies of water, and become reduced in size as the bulk of water in which they live is reduced. Varigny has shown that the reduced size follows a reduction of the air-surface of the water rather than a reduction of the actual bulk, though the two conditions may often coincide. He also shows that, other things being equal, the size of individuals is inversely as their numbers in a given enclosure.

b. The Conversion of Artemia Into Branchinecta.[1]

In 1871 the spring flood broke down the barriers separating the two different lakes of the salt works near Odessa, diluting the water in the lower portion to 8° Beaumé, and also introducing into it a large number of the brine shrimp, *Artemia salina.* After the restoration of the embankment, the water rapidly increased in density, until in September, 1874, it reached 25° of Beaumé's scale, and began to deposit salt. With this increase in density, a gradual change was noticed in the characters of the Artemiæ, until late in the summer of 1874 forms were produced which had all the characters of a supposed distinct species, *A. muelhausenii.* The reverse experiment was then tried. A small quantity of the water was then gradually diluted,

[1] Abstracted from an account by J. S. Kingsley, *Standard Natural History*, Vol. II.

by M. Vladimir Schmankewitsch, who conducted the experiments, and though continued for only a few weeks, a change in the direction of *A. salina* was very apparent. Led by these experiments, he tried still others. Taking *Artemia salina*, which lives in brine of moderate strength, he gradually diluted the water, and obtained as a result a form which is known as *Branchinecta schæfferii*, the last segment of the abdomen having become divided into two. Nor is this change produced by artificial means alone. The salt pools near Odessa, after a number of years of continued washing, became converted into fresh-water pools, and with the gradual change in character, *Artemia salina* produced first a species known as *Branchinecta spinosa*, and at a still lower density *Branchinecta ferox*, and another species described as *B. media*. Here not only new species were produced, but a new genus.

c. The Production of Colors in Lepidopterous Pupæ.

The following important contribution to this subject has been made by Poulton.[1] As an illustration of the direct effect of the environment in the production of color-changes, it is of the greatest value. Several lepidopterists, among others Weismann and Merrifield, had shown that by exposing the pupæ of butterflies to low temperatures material changes in the coloration of the mature insects can be produced. Says Poulton: "In 1867 Mr. T. W. Wood exhibited to the Entomological Society of London a number of chrysalides of the large and small garden white butterflies (*Pieris brassicæ* and *P. rapæ*), which corresponded in color to the surfaces to which they were attached. Dark pupæ

[1] *The Colors of Animals*, International Scientific Series, Vol. LXVIII, by E. B. Poulton, London, 1890.

had been found on tarred fences and in subdued light; light ones on light surfaces; while green leaves were shown to produce green chrysalides, at any rate in certain cases.

"During the following nineteen years, gradual confirmation of Mr. Wood's central position was afforded. In 1873 Professor Meldola supported the observations upon the chrysalides of the "garden whites." He compared large numbers of individuals and found that the pupæ upon black fences were darker than those upon walls.

"In 1874 a paper by Mrs. M. E. Barber, and communicated by Mr. Darwin to the Entomological Society of London, was printed in the transactions of that society. Mrs. Barber had experimented with a common South African swallow-tailed butterfly (*Papilia nireus*), and had found the chrysalis wonderfully sensitive to the colors of its environment. When the pupæ were attached among the deep green leaves of the food-plant, orange, they were of a similar color; when fixed to dead branches covered with withered, pale, yellowish-green leaves, they resembled the latter. One of the caterpillars affixed itself to the wood frame of the case, and then became a yellowish pupa of the same color as the wooden frame.

"Mr. Maurel Weale also showed that the color of certain other South African pupæ can be modified, and Mr. Roland Trimen made some experiments upon another African swallow-tail (*Papilio demoleus*) confirmatory of Mrs. Barber's observations. He covered the sides of the cage with bands of many colors, and found that green, yellow, and reddish-brown tints were resembled by the pupæ, while black made them rather darker. Bright red and blue had no effect. The larvæ did not exercise any choice, but fixed themselves indiscriminately to colors which their pupa could resemble and those which they could not. In the natural conditions the latter would not exist, for the pupæ can imitate all the colors of their normal environments.

"I began work with the common peacock butterfly (*Vanessa io*), of which the chrysalis appears in two forms, being commonly dark gray, but more rarely, bright yellowish-green; both forms are gilded, especially the latter. Only six caterpillars could be obtained, and these were placed in glass cylinders surrounded by yellowish-green tissue-paper. Five of them became chrysalides of the corresponding color; the sixth was removed immediately after the caterpillar skin had been thrown off, and was placed in a dark box lined with black paper, but it subsequently deepened into a green pupa exactly like the others. Obviously the surroundings had exercised their influence before the pupa was removed.

"Being unable to attain more larvæ of the peacock, I worked upon the allied small tortoise-shell butterfly (*Vanessa urticæ*), which can be obtained in immense numbers. In the experiments conducted in 1886, over seven hundred chrysalides of this species were obtained and their colors recorded. Green surroundings were first employed in the hope that a green form of pupa, unknown in the natural state, might be obtained. The results were, however, highly irregular, and there seemed to be no susceptibility to the color. The pupæ were, however, somewhat darker than usual, and this result suggested a trial of black surroundings, from which the strongest effects were at once witnessed: the pupæ were as a rule extremely dark, with only the smallest trace, and often no trace at all, of the golden spots which are so conspicuous in the lighter forms. These results suggested the use of white surroundings, which appeared likely to produce the most opposite effects. The colors of nearly one hundred and fifty chrysalides obtained under such conditions were very surprising. Not only was the black coloring matter as a rule absent, so that the pupæ were light-colored, but there was often an immense development of the golden spots, so that in many cases the whole surface of the pupæ glittered with an apparent metallic lustre. So remarkable was the appearance that a physicist, to whom I showed the chrysalides, suggested that I had played him a trick and had covered them with gold-leaf.

"These remarkable results led to the use of a gilt back-ground as even more likely to produce and intensify the glittering appearance. By this reasoning I was led to make the experiment which had been suggested by Mr. Wood nineteen years before. The results quite justified the reasoning, for a much higher percentage of gilded chrysalides, and still more remarkable individual instances, were obtained among the pupæ which were treated in this way.

"These observations and experiments had been made when I began to work at the subject in 1886: they appeared to prove that the power certainly exists, but nothing was really known as to the manner in which the adjustment is effected. Mr. S. W. Wood's original suggestion, that the 'skin of the pupa is photographically sensitive for a few hours only after the caterpillar's skin has been shed,' was accepted by most of those who had worked at the subject. And yet the suggestion rested upon no shadow of proof; it depended upon a tempting but overstrained analogy to the darkening of the sensitive photographic plate under the action of light. But the analogy was unreal, for, as Professor Meldola stated in the discussion which followed Mrs. Barber's paper, 'the action of light upon the sensitive skin of a pupa has no analogy with its action on any known photographic chemical. No known substance retains permanently the color reflected on it by adjacent objects.' The supposed 'photographic sensitiveness' of chrysalides was one of those deceptively feasible suggestions which are not tested because of their apparent probability. It would have been very easy to transfer a freshly formed pupa from one color to another which is known to produce an opposite effect upon it; and yet if this simple experiment had been made the theory would have collapsed, for the pupa would have been found to resemble the first color and not the second. Furthermore, Mr. Wood's suggestion raised the difficulty that chrysalides which had become exposed in the course of a dark night would have no opportunity of resembling the surrounding surfaces, for the pupal colors deepen very quickly into their permanent condition.

"Having thus defined the time of susceptibility. the next question was to ascertain the organ or part of the larva which is sensitive. At first it appeared likely that the larvæ might be influenced through their eyes (ocelli), of which they have six on each side of the head. Hence, in many experiments the eyes of some of the larvæ were covered with an innocuous black opaque varnish, and they, together with an equal number of normal larvæ from the same company, were placed in gilt or white surroundings. The pupæ from both sets of larvæ were, however, always equally light-colored. It then seemed possible, although highly improbable, that the varnish itself might act as a stimulus similar to that caused by gilt or white surroundings, and therefore the experiments were repeated with black surroundings in darkness, but the pupæ of the two sets were again almost identical, so that it appeared certain that the eyes can have nothing to do with the influence.

"It then seemed possible that the large branching bristles, with which the larvæ are covered, might contain some organ which was affected by surrounding colors, but experiments in which half of the larvæ were deprived of their bristles showed conclusively that the sensitive organs must have some other position, for the pupæ from both sets of larvæ were identical.

"I was thus driven to the conclusion that the general surface of the skin of the caterpillar is sensitive to color during stage ii, and part of stage iii. In order to test this conclusion, I wished to subject the body of the same larvæ to two conflicting colors, such as black and gold, producing the most opposite effects upon the pupa. Such an experiment, if successfully carried out, would decide some important points. If the part of the body containing the head was not more sensitive than the other part, a valuable confirmation of the blinding experiments would be afforded. Mrs. Barber's suggestion that particolored pupæ may be produced by the influence of two colors would be tested in a very complete manner; if particolored pupæ were obtained, it seemed probable that the light acts directly upon the skin, but if they could not be obtained, it seemed more probable that the light influences the termination of nerves in the skin, and that the pupal colors are produced through the medium of the nervous system. The experiments were conducted in two ways. In the first, the larvæ were induced to suspend themselves from sheets of clear glass, by placing them in wide shallow boxes, so that the ascent to the glass roof was easily accomplished. As soon as suspension (stage iii.) had taken place, each larva was covered with a cardboard tube, divided into two chambers by a horizontal partition, which was fixed rather below the middle. There was a central hole in the partition just large enough to admit the body of the larva. The tube was fixed to the glass sheet with glue; the upper chamber was lined with one color, e. g. gilt, and the lower chamber with the opposite color, e. g. black, with which the outside of the cylinder was also covered, in case the larva should stretch its head beyond the lower edge. The partition was fixed at such a height that the larval head and rather less than half of the total surface of skin were contained in the lower chamber, while rather more than half of the skin surface was contained in the upper chamber.

"The second method of conducting the conflicting color experiments was superior in the more equal illumination of the upper and lower colors. The bottom of a shallow wooden box was covered with alternate areas of black and gilt papers, and partitions were fixed along the lines where the two colors came in contact. Each partition was gilt toward the gilt surface, and black toward the black surface, and was perforated close to the bottom of the box with holes that would just admit the body of a larva. The box was then placed in a vertical position towards a strong light, so that the partitions became strong shelves, while the black and gilt surfaces were uppermost alternately. As soon as a larva was suspended to a glass sheet, the boss of silk was carefully scraped off and was pinned on the upper color, above one of the holes, so that the head and first five body-rings passed through the hole on to the color beneath, which tended to produce opposite effects. Other larvæ were similarly fixed between the shelves upon one color only, so as to afford a comparison with the results of the conflicting colors.

"A careful comparison of all the pupæ obtained in the conflicting color experiments showed that, when the illumination of the two surfaces was equal, the effective results were produced by that color to which the larger area of skin had been exposed, whether the head formed part of that area or not. Particolored pupæ were never obtained. It therefore appears to be certain that the skin of the larva is influenced by surrounding colors during the sensitive period, and it is also probable that the effects are wrought through the medium of the nervous system. The latter conclusion receives further confirmation from other observations."

Professor Poulton has since produced remarkable color-changes in the larvæ of Lepidoptera by confining them to the branches of plants of distinct colors. Thus geometrid larvæ confined to the stems of a black color, became correspondingly dark; while those restricted to white twigs became very pale. These larvæ, and, still more strikingly, those of *Cossus ligniperda*, when confined on branches which supported lichens, became of variegated colors, corresponding with those of the lichens, and affording an admirable means of concealment.

d. The Effect of Light on the Colors of Flatfishes.

It is well known that the side of the body which is uppermost in the normal position in the flatfishes (Pleuronectidæ) is colored generally with dark tints, and frequently with a distinct pattern, while the lower side is white. This is due to the absence from the lower side of the chromatophoræ or pigment-containing cells, which are abundant on the upper side. The young fish has chromatophoræ on both sides as it has its eyes also in the normal position, but as the fish turns the left side upwards and the right eye gradually rotates to the left side, the chromatophoræ disappear from the right side, which thus becomes colorless.

Prof. J. T. Cunningham[1] experimented with young flounders taken at the beginning or middle of their metamorphosis, by placing a mirror below the aquarium in which they were kept, at an angle of 45°, and cutting off the light from above by an opaque cover. In the great majority of the specimens treated in this way, after several months, although no effect was produced upon the eyes, more or less of the skin of the lower side was pigmented. He thus showed that the absence of pigment on that side in the normal fish is due to its position in shadow, where little light can reach it.

e. The Effect of Feeding on Color in Birds.

Mr. F. E. Beddard cites the following remarkable example of the direct effect of internal physical causes in producing change of coloration.[2]

1 *Zoologischer Anzeiger*, 1891.

2 *Animal Coloration, an Account of the Principal Facts and Theories Relating to the Colors and Markings of Animals.* By Frank E. Beddard, M. A. Oxon., F. R. S. E. With Four Colored Plates; and Wood-Cuts in the Text. London: Swan Sonnenschein & Co. New York: Macmillan & Co. 1892.

"That the yellow color of canaries can be altered to an orange red by mixing cayenne pepper with their food, has been known for a long time. This curious fact was first discovered in England, as was also the fact that the different races of canaries vary in their susceptibility to the action of the pepper; some kinds are more, others are less, affected, while one race is absolutely without any power of having its coloration altered by these means. The color-change is produced by feeding the newly hatched young with the pepper conveyed in their food or the old birds while sitting upon the nest are furnished with food containing the cayenne, which they in turn feed their offspring. The color change can, in fact, be only brought about in very young birds whose feathers are not completely matured; it is quite impossible to produce any alteration upon the full-grown canary. Clearly, therefore, here is an instance of the direct effect of food upon color. An interesting paper upon the subject, which has also furnished me with the facts already mentioned, has lately appeared,[1] and it will be of interest to give some account of the author's (Dr. Sauermann's) experiments for reasons that will appear. Cayenne pepper, of course, is a composite substance, from which a number of distinct chemical substances can be extracted: the red color is caused by a pigment termed capsicin, which can be separated from the pepper; and it might easily be supposed that the change from yellow to red in the feathers of the canary was simply caused by a transference of the pigment, as in the cases mentioned on page 127; but Dr. Sauermann shows that it is not so. Yellow-colored canaries were

[1] *Archiv für Anatomie und Physiologie.* 1889. Physiologische Abtheilung, 543.

not in the very slightest degree affected by the pigment alone; but, curiously enough, particolored birds did react,—the brown parts of the feathers became distinctly lighter in hue. It is a fatty substance (triolein) which appears to convey the pigment, and to produce thus a changing of the color from yellow to red; and further experiments were made with other birds, showing that it is not only canaries which are influenced by their food in this way. Some white fowls, belonging to a special breed, showed traces of yellow among the feathers after feeding with cayenne; but in this case there were not racial but individual differences in susceptibility, for all the specimens of birds experimented with did not react to the stimulus.

"A similar series of experiments was made with some other colors: it was found with carmine that the yellow color was destroyed and the birds became white. This unexpected effect is explained by the fact that a mixture of violet and yellow produces white. The fact that the fatty constituent, triolein, plays the chief part in the coloring of the feathers may perhaps help to explain the very singular fact that the Amazon parrots change from green to yellow when fed upon the fat of certain fishes.

"With regard to the white fowls referred to, the experiments made by Dr. Sauermann were particularly interesting. The interest lies in the fact that the pigment was not absorbed equally by all the feathers; only special tracts were affected; the breast feathers, for instance, became red, while the head remained white. It is therefore quite credible that in a state of nature partial alteration of color may be produced by a change of diet."

In a chapter of Dr. Beddard's book relating to protective resemblances will be found an account of several examples of animals which have apparently acquired a resemblance to their surroundings by the transference of pigment to their bodies in their food.

f. The Blindness of Cave-Animals.

Neo-Lamarckian writers have always ascribed the absence or rudimentary condition of the eyes characteristic of animals which dwell exclusively in caves, to disuse consequent on the absence of light. Lamarck ascribed the rudimentary condition of the eyes in the mole to this cause. As the removal of so important an organ as the eye is not accomplished in a single generation, the element of heredity enters the proposition. This subject is reserved for Part Third of this book; nevertheless, for the present suspending judgment as to this question, it has been rendered exceedingly probable by embryological investigations into the history of dwellers in darkness, that the Neo-Lamarckian view is the correct one.[1] Says Packard:

"In my essay on *The Cave Fauna of North America* (p. 139), I record the fact that in the young of the blind crayfish (*Orconectes pellucidus*), the eyes of the young are perceptibly larger in proportion to the rest of the body than in the adult, the young specimen observed being about half an inch in length. Previously to this, Dr. Tellkampf, in 1844, remarked that 'the eyes are rudimentary in the adults, but are larger in the young.' Mr. S. Garman states, regarding the blind Cambarus of the Missouri Cave: 'Very young specimens of *C. setosus* correspond better with the adults of *C. bartonii*; their eyes are more prominent in these

[1] I am indebted for a *résumé* of this subject to Packard, *American Naturalist*, 1884, p. 735.

stages, and appear to lack but the pigment.' In the blind cave-shrimp (Troglocaris) of Austria, Dr. Joseph discovered that the embryo in the egg is provided with eyes.

"In this connection should be recalled the observations of Semper in his *Animal Life* (pp. 80, 81) on *Pinnotheres holothuriæ*, which lives in the 'water-lungs' of holothurians, where, of course, there is an absence of light. The zoëa of this form has large 'well-developed eyes of the typical character. Even when they enter the animal they still preserve these eyes; but as they grow they gradually become blind or half-blind, the brow grows forward over the eyes, and finally covers them so completely that, in the oldest individuals, not the slightest trace of them, or of the pigment, is to be seen through the thick skin, while, at the same time, the eyes seem to undergo a more or less extensive retrogressive metamorphosis.'

"In this connection may be mentioned the case of the burrowing blind shrimp (*Callianassa stimpsonii*), which has been found by Prof. H. C. Bumpus, at Wood's Holl, Mass., living in holes at a depth of between one and two feet. He has kindly given me a specimen of the shrimp, which is blind, with reduced eyes, smaller in proportion to the body than those of the blind crayfish. He has also obtained the eggs, and has found that the embryos are provided with distinct black, pigmented eyes, which can be seen through the egg-shell.

"Recently, Zeller has studied the embryology of the Proteus of Adelsberg Cave, and has confirmed the statement of Michahelles, who, in 1831, discovered that the eyes of this animal are more distinct in the young and somewhat larger than in the adult. We quote and translate from Zeller's account:

"'The development of the eyes is very remarkable; they are immediately perceived and present themselves as small, but entirely black and clearly drawn circular points, with a slit which is very narrow, and yet, at the same time, well defined, and which penetrates from the lower circumference out to the middle.

"'Indeed, one can hardly doubt that this astonishing development of the eye has been accomplished by the influence of light, as has also the pigmentation of the skin, the reddish-white ground-color of which appears thickly studded with very small brownish-gray points mixed with detached white ones, over the upper surface of the head and over the back, down over the sides of the yellowish abdomen. Even on the edge of the fins (*Flossensaum*) the pigment is found. On the other hand, there is a whitish spot over the snout, as is likewise the case in the adult creatures which have been colored by the light. Both the under surface of the head and the entire abdomen are shown free from pigment like the limbs. . . .

"'I cannot specify very exactly as to when the pigmentation of the skin begins, but, in any case, it is very early, and often earlier that the first beginning of the eyes can be discovered. The latter occurs toward the end of the twelfth week, at which time a thin, light gray line, which still appears overgrown, may be perceived, forming a half-circle, open underneath. Then while this line subsequently becomes clearer and darker, and its ends grow further under and towards each other, there also takes place simultaneously a progression of the pigment larger towards the middle point, and the circle finally seems closed and filled up to the narrow slit mentioned above, which proceeds from the lower circumference and penetrates to the middle of the eye' (pp. 570, 571).

"But the most striking discovery bearing on this subject is that of the condition of the eyes in the embryo and young compared with the adult of the blind goby of San Diego.

"In his essay on *The Fishes of San Diego*, Professor Eigenmann briefly refers to and gives four figures (Fig. 55) of the embryo of Typhlogobius, Mr. C. L. Bragg having been fortunate enough to discover the egg in the summer of 1891. 'The eyes develop normally, and those of No. 4 differ in no way from the eyes of other fish embryos.' In this case, then, we have the simplest and clearest possible proof of the descent of this blind fish from individuals with eyes as perfect as those of its congeners.

"We have been permitted by the Director of the United States National Museum to reproduce Professor Eigenmann's excellent figures on the embryo, which tell the story of degeneration of the eye from simple disease of the organ, the species being exposed to conditions of life strikingly different from those of its family living in the same bay.

"Before the discovery of the eggs, the youngest individual ever seen is represented in Fig. 55, No. 1, its eyes being, though small, yet distinct, and 'apparently functional.'

"From these data it is obvious that future embryological study on cave-animals will farther demonstrate their origin from ancestors with normal eyes."

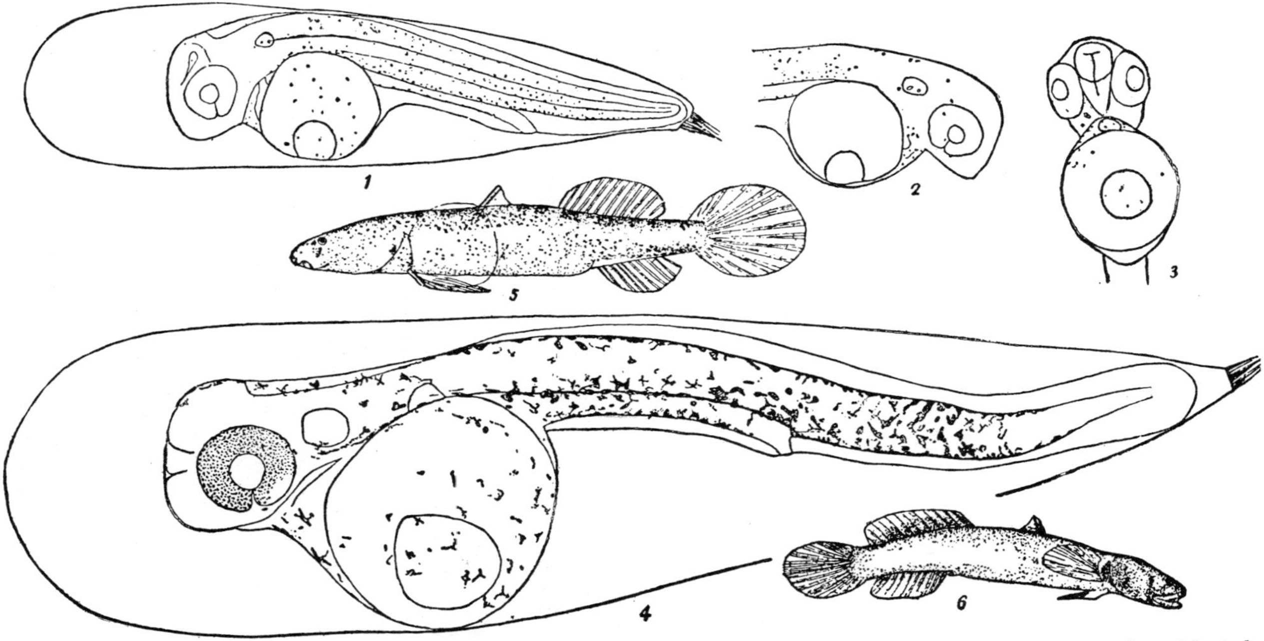

Fig. 55.—*Typhlogobius californiensis* Steind.; showing obliteration of eye in course of growth from the egg to maturity. Adapted from Eigenmann. No. 1. The youngest *Typhlogobius* seen; Zeiss A and 2, No. 2. The head of an older one from the side. No. 3. The head of one similar to No. 3, seen from below. No. 4. Another larva forty-eight hours older than No. 1; the oldest one seen; it is still inclosed in its club-shaped zona. No. 5. Young *Typhlogobius*; ×4 2–9, showing color-markings and the eye. No. 6. Adult *Typhlogobius*, natural size, in the position usually assumed in the aquarium.

CHAPTER VI.—KINETOGENESIS.

UNDER the head of kinetogenesis (development by motion) comes the consideration of the effect of use and disuse.* Use necessarily conditions the evolution of useful characters. These characters are such by reason of their adaptation to the life-functions of the beings which possess them. It is perfectly well known, however, that all plants and animals possess more or less numerous peculiarities which are not useful to their possessors. Such are the mammæ of male animals; the incisions forming palmate or pectinate leaves and petals of plants; rudimental organs of all kinds; great elongation of the vertebral column, especially of the caudal series in certain species; patches of color, or of hairs, at particular places, etc. These, and many others may be arranged in divisions according to their probable origin, as follows:

I. Excess of growth energy.

Examples: the recurved tusks of the mammoth, babirussa, etc.; the elongate feathers of some birds, etc.

II. Defect of growth-energy.

1. Atavism: examples; the tritubercular superior molar of certain races of man.

2. Degeneracy from disuse: examples; the rudimental legs and digits of numerous lizards.

3. Degeneracy from disuse and complementary excess elsewhere: examples; reduction in number of molars and incisors in man; reduced mammæ in male Mammalia; reduction of lateral digits in the true horses (Equus).

4. Physico-chemical causes: here must be probably included various color-patches and color-tints, for which no other explanation is accessible.

Darwin considers several of the above conditions, and endeavors to explain some of them as consequences of natural selection. Equivalent to the Section I. above, he enumerates extraordinary developments of particular parts.[1] He says, "A part developed in any species in an extraordinary degree or manner in comparison with the same part in an allied species, tends to be highly variable." He does not attempt any explanation of the origin of such characters, except through natural selection. Of the characters coming under Section II. above, he says, "Multiple, rudimentary and lowly organized structures are variable." Of these he remarks, "I presume that lowness here means that the several parts of the organism have been but little specialized for particular functions; and as long as the same part has to perform diversified work, we can perhaps see why it should remain variable, that is, why natural selection should not have preserved or rejected each little deviation of form so carefully as when the part has to serve for some one special purpose." Here Mr. Darwin well illustrates his unwillingness to look to disuse as the cause of the conditions he describes. Under "Compensation and Economy of Growth" he quotes from Goethe that "In order to

[1] *The Origin of Species*, Ed. 1872, p. 119.

spend on one side, nature is forced to economize on the other side." I have expressed the same view in the following language:[1]

"*The complementary diminution* of growth-nutrition follows the excess of the same in a new locality or organ, of necessity, if the whole amount of which an animal is capable be, as I believe, fixed. In this way are explained the cases of retardation of character seen in most higher types. The discovery of truly complementary parts is a matter of nice observation and experiment." An apparent illustration is that of the increase of the median digits in the diplarthrous Ungulata contemporaneously with the diminution and atrophy of the lateral digits. This is, however, an example of the relative effects of use and disuse, which proceed contemporaneously, and it is probable that most if not all cases of complementary growth-relations may be expressed in this way. Thus the orthognathism of the higher human races is accompanied by full frontal development, the two modifications constituting a retardation of the postembryonic growth of the face. But this change can be traced to use, increased brain action enlarging that organ, and expanding its osseous case, probably at the expense of lime salts which would otherwise go to the jaws. Reduction of teeth in Artiodactyla and in man cannot be regarded as a useful character in itself, but it is complementary to the development of other characters which are useful.

Under the same head may come perhaps, the facts included by Mr. Darwin under the head of "Correlated Variation." Of these he says, "I mean by this expression that the whole organization is so tied together

[1] "Method of Creation," *Proceedings of the American Philosophical Society*, 1871, p, 257; *Origin of the Fittest*, 1887, p. 201.

during its growth and development, that when slight variations in any one part occur, and are accumulated through natural selection, other parts become modified." After referring to various characters of composite and unbelliferous plants in illustration of such a law, he says (*l. c.*, p. 116), "Hence modifications of structure, viewed by systematists as of high value, may be wholly due to the laws of variation and correlation, without being, so far as we can judge, of the slightest service to the species." Here Mr. Darwin admits the insufficiency of natural selection as an explanation of the origin of such characters; for he says (p. 119), "Natural selection, it should never be forgotten, can act solely for the advantage of each being." He goes further, and admits (p. 114) that the Lamarckian doctrine has some claims to credence, where he says, "On the whole we may conclude that habit, or use and disuse, have in some cases, played a considerable part in the modification of the constitution and structure; but that the effects have often been largely combined with, and sometimes overmastered by the natural selection of innate variations."

1. KINETOGENESIS OF MUSCLE.

The fundamental condition of the molar movements of organic beings is the contractility of protoplasm. In the Amœba this contractility is a generally diffused characteristic of its body-substance, and this is the case with Rhizopoda generally. In higher Protozoa the contractility is already especially developed in certain regions where most needed for the movement of the body in definite directions; generally immediately beneath the denser sarcode of the external surface. In

Stentor this substance presents the appearance of longitudinal threads; in Gregarina they encircle the body. From these simple beginnings we can follow the muscular tissue to its various expressions in all classes of animals; to the concentric threads of the Cœlenterata; to the longitudinal bundles beneath the integument of worms; and the variously directed masses attached to the external skeleton of the Arthropoda, and the internal skeleton of the Vertebrata. The ease with which muscular tissue is grown in the higher animals under use, permits us to infer that its development in all animals has been due to the same cause. Muscular structure is directly related to the needs of the structures to which it is attached, in the performance of movements. In rudimental limbs muscles are reduced in both size and number, distinct slips or bodies becoming fused. In enlarged limbs the reverse process takes place. Muscular bellies increase in size, and in number by subdivision. The muscular system in the middle and higher Vertebrata is variable, and its plasticity to the stimuli to movements is well known. It is evident that definite muscular bands have been developed in the lines of resistance which it has been necessary to overcome in moving the body or parts of it. The movable segments which have become adapted for contact with the surrounding media, by development of hardness or extent of surface, as limbs (feet, wings), and jaws, have naturally become the points of origin of the most efficient muscular bands. No one can doubt the mutual stimuli which the muscular and skeletal systems have exchanged during the process of evolution, since they are necessary to each other from a mechanical point of view.

Hüter,[1] a distinguished specialist in the diseases of the joints, gives the following positive information as to the easy formation of new modifications of muscular structure:

"Muscular tissue everywhere possesses the capacity to shorten itself in consequence of continued approximation of its points of insertion; that is, to become shorter by the disappearance of tissue, in proportion to the duration of the approximation. This law is of the greatest importance for the muscular contraction of joints, that is, for such restriction of the freedom of movement of the articulations as has its origin in muscular movements. We have experimental opportunities for the observation of this fact, such as the effect of a stiff bandage on an articulation. When it is necessary, in consequence of a fracture of the fore-arm, to fix the elbow-joint for several weeks in a right-angled flexure, we find on the removal of the bandage that the power of extension of the fore-arm has been much restricted. That the cause is nutritive change is proven by the fact that considerable force of muscular contraction is necessary before the normal extension can be effected."

Similar phenomena are to be observed in consequence of a prolonged lying in bed, where no injury to the innervation exists.

"The muscles adapt themselves to the permanent positions of the articulations, as in joint-contractions which are due to muscular paralysis. Those muscles which are habitually stretched, increase in length, while those whose insertions are approximated, are shortened, producing joint-contraction." He then goes

[1] Hüter, "Studien an den Extremitätengelenken Neugeborener und Erwachsener." Virchow's *Archiv für pathologische Anatomie*, Bd. XXV., 6–8.

on to describe the effect of such fixation of joints on the bones themselves, to which I will refer on a later page, under the head of the origin of articular surfaces.

Professor Eimer of Tübingen has given us a synopsis of the nature of the evolution of the characters of the muscular tissue, which is highly instructive, and of which I present here an abstract.[1] The conclusions reached by Eimer are derived from a general study of the subject, both in the laboratory and in the literature. He says:

"(1) It is apparently continued contractions of the protoplasm in definite directions, which have produced muscular masses. Since plants do not display continuous and vigorous movements, they have not developed muscular bodies.

"(2) Undoubted facts indicate that from a primitively identical substance muscular tissue has developed in the direction of effective contractions, while connective tissue has developed where no contractions have been present.

"(3) Muscle-masses first appeared almost everywhere in the external layer of the contractile region:

"*a*, in Protozoa in the outer layer of the body;

"*b*, in Metazoa in the tegumentary sheath of the body.

"*c*, They consist first either of muscle-cells, or muscle-fibers, from which develop mantle muscle-cells and mantle muscle-fibers. Mantle muscle-fibers compose the otherwise highly developed striped muscles of

[1] "Die Entstehung und Ausbildung des Muskelgewebes, insbesondere der Querstreifung desselben als Wirkung der Thätigkeit betrachtet." *Zeitschrift für wissenschaftliche Zoologie*, LIII., Suppl., 1892, p. 67.

Arthropoda, and some Vertebrata (Batrachia). And when the entire muscle-fiber is divided into fibrillæ, there can appear an external layer of muscular threads.

"(4) That muscle cells and fibers first appear in the external stratum of the active body of animals, is due to the especially active movements necessary to this part of the body. This is a simple mechanical consequence of the fact, that in a more or less longitudinally extended body of protoplasm, whether it be Protozoön or worm, or muscle-cell or muscle-fiber, that its movements must be more vigorous on the external than the internal portion of it. Therefore, the former would first display muscular structure.

"(5) If the first stage is the development of masses of plasma, which display contraction in definite directions, the next step is the division of such masses into muscle-threads or fibrillæ. These threads must be regarded as a result of contractions, whose inequality produces subdivisions of the original mass. A compound structure is also more effective than a simple one in effecting contractions.

"(6) The next stage of evolution of muscular tissue consists of the appearance of the cross-striping. The mechanical effect of the cross-striping is to distribute the contractility equally throughout the length of the fiber. The contractions of the unstriped muscular fiber are less vigorous, and also less uniformly distributed than those of the striped fiber. *The cross-striping is a result of contractions.* It commences as simple undulations of the surface of the fiber. The contraction of the plasma is wave-like and is propagated rapidly through the fiber, and is not due to a flow and return of the contained protoplasm. The frequent repetition of these local contractions and enlargement of the fiber have resulted in a permanent difference in its intimate structure, the alternate waves becoming fixed as cross-bands."

As evidence of the truth of this proposition, (6), Eimer cites many facts. In the muscles of the Mollusca, striped fibers occur in those forms, as Pecten, where the closing of the shell is especially vigorous, this being their mode of progress through the water. In other forms, where the muscles have no such vigorous use, the fibers are smooth. In Arthropoda, the muscles of the legs of swiftly running forms are striped, while those of the alimentary canal are smooth. It is a general law that muscles which have energetic contractions are striped, while those in which the contractions are slow or feeble, are smooth. In the common house-fly, Eimer records some remarkable observations. Flies examined in winter, during the period of torpidity, were found to have the fibers of the thoracic muscles smooth. With advance of the spring the striping gradually made its appearance, and in the summer it was fully developed. An artificial imitation of winter, by refrigeration in an ice-cellar, caused the cross-striping to disappear. The striping in some other animals is shown by Eimer to be strongly influenced by physical conditions.

In fact, muscular tissue is highly plastic, and as it is directly under the control of nervous or equivalent stimuli, the effect of the latter in building structure is evident. That the motion communicated to the hard parts through the agency of the muscular system is effective in building the hard structures will be shown in a subsequent section.

2. KINETOGENESIS IN MOLLUSCA.

a. The Origin of the Plaits in the Columella of the Gastropoda.

Mr. W. H. Dall has developed the mechanics of evolution in the Gastropoda, and I quote extracts from one of his papers to show the harmony of his views with those of other Neo-Lamarckians.[1] "The question which first arises is as to the origin of the columellar plications and their function. In considering the dynamic relations of the animal to its shell we may obtain satisfaction on this point. In the fusiform Rhachiglossa an anatomical difference exists to which I believe attention has not hitherto been called. Indeed, unless the principles of dynamic evolution are granted, it is a difference which would appear to have little or no significance. These principles, however, afford a key which seems to unlock this and many other mysteries. In the recent forms of this sort the adductor muscle, which in all gastropods is attached to the columella at a certain distance within the aperture, is attached *deeper within the shell* than in non-plicate forms. The point of attachment may be an entire turn, or even more, behind the aperture, while in short globose few-whorled shells and in the non-plicate forms it is, as a general rule, little more than half a turn behind the aperture.

"Now let us consider the dynamics of the case. We have, reduced to its ultimate terms, a twisted, shelly, hollow cone, subangulate or even channelled at two ex-

[1] *Transactions of the Wagner Free Institute of Science*, Philadelphia, Aug., 1890, p. 58.

tremes corresponding to the canal and the posterior commissure of the body and outer lip. Inside of this we have a thin, loose epithelial cone, the mantle, of which the external surface, especially toward the margin, is shell-secreting; lastly, inside of the mantle-cone we have a more or less solid third cone, consisting of the foot and other external parts of the body of the animal, which can be extended beyond the mantle-cone outwardly, as the mantle-cone can be beyond the shell-cone. The body-cone and the mantle-cone are attached at one of the angles of the shell-cone some distance within the opening of the spiral of the latter. The two outer cones constitute a loose, flexible funnel within a rigid, inflexible funnel, while the body-cone forms a solid, elastic stopper inside of all.

"What will happen according to mechanical principles (which can be tested by anybody with the simplest apparatus) when the mantle-cone is withdrawn into a part of the shell-cone too small for the natural diameter of the contracted mantle-cone? It must wrinkle longitudinally. Where will the wrinkles come? They will come at the angles of the shell-cone first; they will be most numerous toward the aperture, since toward the aperture the mantle-cone enlarges disproportionately to the caliber of the shell, owing to its processes, the natural fold of the canal, etc., etc.; the deepest and strongest wrinkles will be on the pillar, owing to the fact that the attachment of the adductor prevents perfect freedom in wrinkling and the groove of the canal will mechanically induce the first fold in that vicinity. The most numerous small wrinkles will be near the aperture opposite the pillar, because of the mantle-edge this is the most expanded part, and there will be a tendency to a ridge near the angle of the posterior commissure. Repeated dragging of a shell-secreting surface, thus wrinkled, over a surface fitted to receive such secretion, will result in the elevated shelly ridges which on the pillar we call plications; and on the outer lip liræ, if long, or teeth, if short. The commonly existing subsutural internal ridge on the body of the shell near the posterior commissure will mark the special conditions in that part of the aperture.

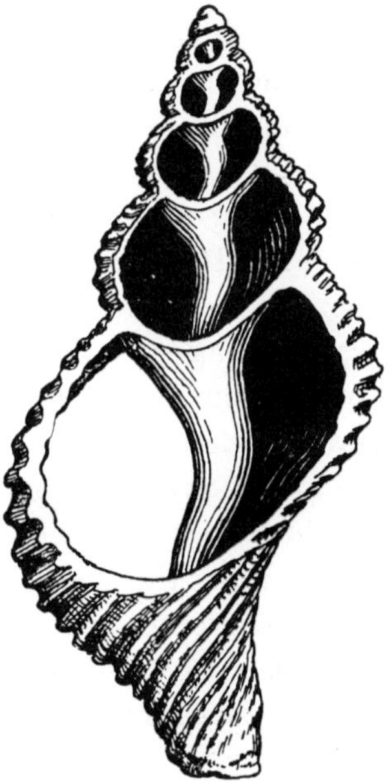

Fig. 56.—*Fusus parilis* Con. the body whorl opened, displaying non-plicate columella. From Dall.

"When the secreting surface is thus wrinkled or corrugated longitudinally, the wrinkles and the concave folds between them will be directed in the sense or direction in which the body moves in emerging from or withdrawing to the whorl. The summits of the convex wrinkles will be appressed more or less forcibly against the shell-wall exterior to them in which they are confined. The semi-fluid, limy secretion of which the shell-lining is built up, exuding from the whole surface of the mantle, will be rubbed away from the lines of the summits of the wrinkles and tend to accumulate in lines corresponding to the concave furrows between the wrinkles. This secretion hardens rapidly and these lines would become somewhat elevated ridges which would by their presence (when once initiated) tend to maintain the furrows and wrinkles in the same place with relation to the thus-initiated liræ, as these elevated lines are called when on the outer lip; or plaits, when situated on the pillar.

"The modification referred to generally takes place during resting stages of the animal's growth, since while the animal is rapidly extending its coil the secretions seem to be directed toward the extreme margin, and the general mantle-surface resumes its secretive function (or the latter becomes active) somewhat later, after the formation of a definite varix, or thickened margin; indicating a resting stage in the animal's career. It is probable also that during rapid growth there is less compression of the tissues than during the resting-stages. The external sculpture and some of the modifications of the aperture are connected with the functions of the extreme edge of the mantle; those we are at present considering relate more especially to the function of its general surface by which the layer which lines the whorls, the pillar, plaits, and liræ are solely secreted and deposited.

"In species with the abductor attached to the pillar near the aperture, the wrinkles would be fewer, and their action, if any, confined to the vicinity of the margin of the aperture. The deeper the attachment, the greater will be the compression of the secreting surface and the distance over which it is constantly dragged back and forth, and the consequent length of the ridges of shelly matter deposited. If the inner or mantle-cone had the whole cavity to itself, it is evident that it could and would infold itself in a manner which might not appress its folds against the inner surface of the rigid outer or shell-cone. But there the mass of the solid and elastic foot and external body comes into play, and by its withdrawal inward forces the wrinkled mantle-cone against the shell. The mantle is thus confined between a rigid outer and an elastic inner surface, with the result that it cannot recoil from the former and that a certain uniformity of size and direction is imposed upon the wrinkles, except where the recess of the canal allows them to become more emphatic, or to a less degree the posterior angle permits a slight expansion. The mechanical principles involved may be readily illustrated by the experiment of pulling a handkerchief through the neck of a bottle, or funnel, followed by a cork in the center. Of course, the more nearly the apparatus conforms to the form and twist of a spiral shell, the more nearly the results will approximate those of nature. It is difficult, however, to find any artificial tissue which will correspond in elasticity, or capacity for partial self-contraction, to the living tissues concerned in nature. Hence, an exact conformity is not to be expected, though the mechanical principles may be reasonably well illustrated.

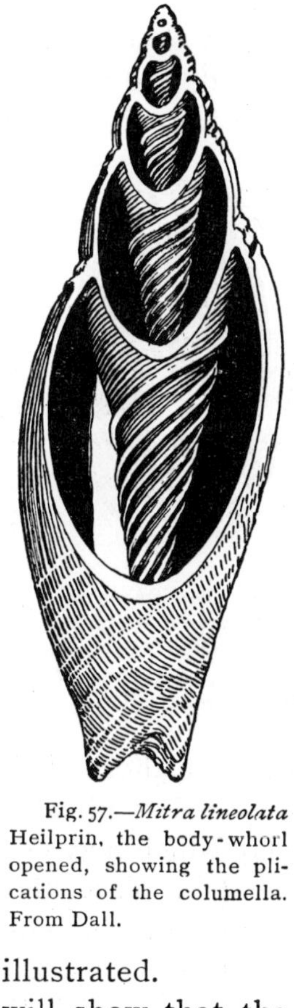

Fig. 57.—*Mitra lineolata* Heilprin, the body-whorl opened, showing the plications of the columella. From Dall.

"A comparison of specimens will show that the results exhibited agree with marvellous precision with the results called for by the preceding hypothesis, based on the dynamical status of the bodies concerned, their motions and secretions. The agreement is so complete as to amount to a demonstration, though in certain cases there may be complications which need additional explanation.

"A point which may be noted in regard to the Volutidæ, to which my attention was called by Mr. Pilsbry, is that in this group the mantle is greatly extended, and there would be more of it to be wrinkled than in such forms as Buccinum, etc. It may be added that the forms in which we note the beginning of plaits for this family, many of them, such as Liopeplum and Volutomorpha, had the mantle so extended as to deposit a coat of enamel over the whole shell, as in the modern Cypræa, so that here we have an additional reason why plication should be emphasized in this group.

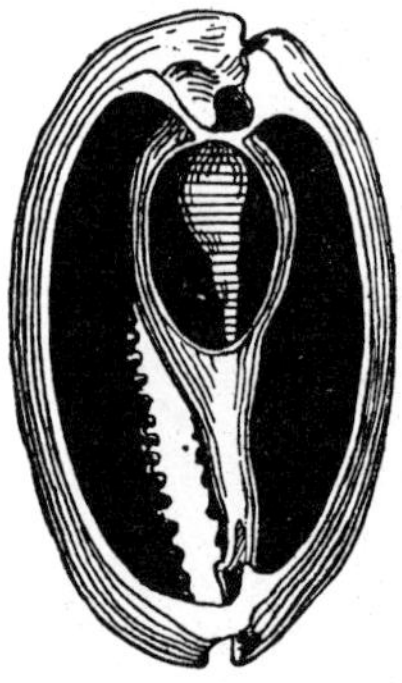

Fig. 58.—*Siphocypræa problematica* Heilprin; body-whorl opened, showing plications of lips. From Dall.

"Of course, as before noted, the mechanical principles are the same in any group of gastropods, but among those in which the wrinkling is confined to the region of the aperture, or those shells which are lirate or dentate, as opposed to plicate, several other principles come into play which may be briefly referred to in passing. In the first place, those species which have a very extended mantle, with hardly an exception have a lirate aperture (Oliva, Olivella, Cypræa, Trivia, etc.). With species in which there is a widely expanded mantle and yet no lirations, it will usually be found that the mantle is not entirely withdrawn into the shell in such forms, or is permanently external to the shell (many

Opisthobranchs, Marseniidæ, Sigaretus, Harpa, etc.). In a group like the Cypræidæ, where nearly all the species are lirate on both lips, there are a few which want these liræ, and these are species which have a wider aperture in the adult than most of the genus, and in which we should expect the wrinkles to be less emphatic."

b. Mechanical Origin of Characters in the Lamellibranchs (Pelecypoda).

Dr. Robert T. Jackson has pointed out[1] the history of the characters of the retractor muscle and some of those of the list, of bivalve Mollusca. I take the following abstract of his conclusions: "In the development of pelecypods we find in a late embryonic stage (the phylembryonic) that the shell has a straight hinge-line. This is characteristic of Ostrea (Fig. 59), Cardium, Anodonta, and so many widely separated genera that it apparently represents a primitive ancestral condition common to the whole class. Embryology shows that the bivalve shell doubtless arose from the splitting on the median line of a primitive univalvular ancestor. If that ancestor had a saddle-shaped[2] or a cup-shaped[3] shell, as is probable, the first result of the introduction of a hinge in the median line would have been to straighten the shell on the hinge-line. This is a simple problem in mechanics, for if one tries to break by flexion a piece of metal which is saddle-shaped or cup-shaped, it will tend to form a straight line on the axis of flexion. A parallel case is seen in the development of a bivalve shell in ancient crus-

[1] *Memoirs of the Boston Society of Natural History*, Vol. IV., No. 8, p. 277 July, 1890; *American Naturalist*, 1891, p. 11.

[2] Characteristic of young Dentalium.

[3] Characteristic of the extreme young of cephalous mollusks.

taceans. The ancient Ostracoda, Leperditia, Aristozoe, etc., have a straight hinge-line and subcircular valves, which are united dorsally by a ligament. The resulting form of the early condition of the bivalvular shell in these two distinct classes is so strikingly similar, that it lends weight to our supposition that the form is induced by the mechanical conditions of the case.

"I think that the adductor muscles which close the valves may also be demonstrated to be the necessary consequence of the bivalvular condition. In the phylembryo stage (Fig. 59) the valves are closed by a single adductor muscle, which is the simplest condition mechanically possible to effect the desired end.[1] This muscle does not seem homologous with any muscle in other classes of mollusks, and is probably developed from the mantle muscles as a consequence of the conditions of the case. In support of this view, bivalvular crustaceans may again be cited. They have an analogous adductor muscle, developed, of course, on an entirely different line of descent, but under closely similar mechanical conditions. At the completed prodissoconch stage in all pelecypods, as far as known, there are two adductor muscles, a second one having developed in the posterior portion of the body. In later life the anterior, the posterior, or both adductors

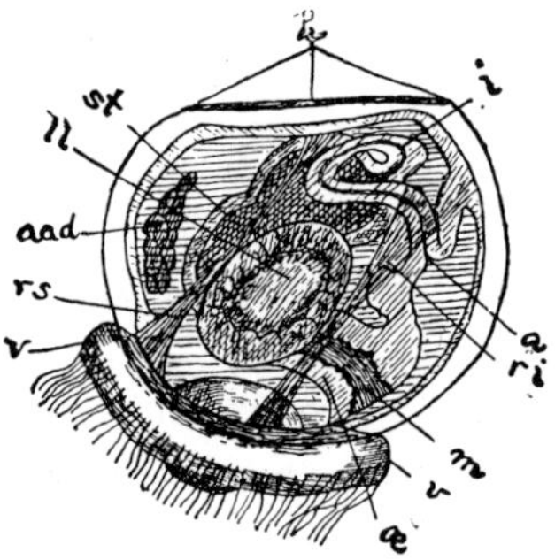

Fig. 59.—*Ostrea edulis*, embryo; *a ad*, anterior adductor muscle; *m*, mouth; *a*, anus; *v*, velum; *h*, hinge of shell. (After Huxley.)

[1] This early adductor appears in the same position in many genera, and is apparently characteristic of the class. It is the anterior of the two adductors found in the later stages; but it may be retained or lost in the adult.

may be retained, reduced, or lost, according as the persistence or changes in correlated anatomical features retain in use or bring into disuse the muscles in question.

"Let us look at examples of the retention or loss of the adductors. In typical dimyarian pelecypods, as in Mya (Fig. 60) or Venus, the adductors lie toward either end of the longer axis of the shell. As the hinge occupies a position on the borders of the shell about midway between the adductors, both muscles are nearly or quite in a position to be equally functional in clos-

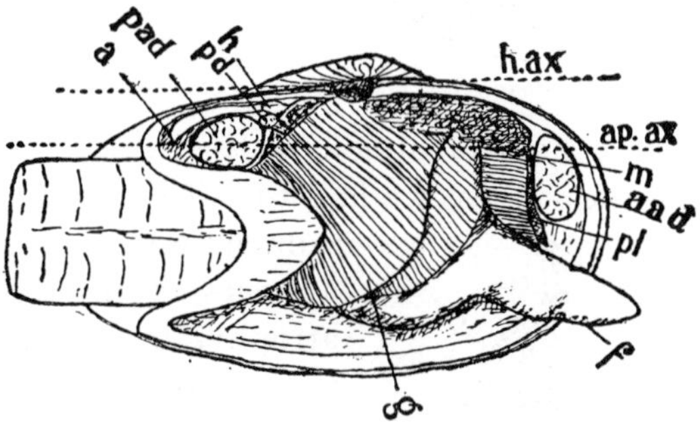

Fig. 60.—*Mya arenaria.* Lettering: *ap ax*, antero posterior axis; *h ax*, hinge axis; *a ad*, anterior, and *p ad*, posterior adductor muscle; *m.* mouth; *pl*, palps; *a*, anus; *g*, gills; *pd*, pedal muscle; *f*, foot; *b*, byssus; *h*, heart.

ing the valves. As a result, both muscles are of about the same size. The condition described is that existent in the completed prodissoconch stage in all pelecypods, as far as known. In later life, however, a revolution of the axes of the soft parts may take place, so that the antero-posterior axis (represented by a line drawn through the mouth and middle of the posterior adductor muscle), instead of being parallel to the hinge-axis (the axis of motion of the valves) as in dimyarians, may present a greater or less degree of divergence from the parallel. In progressive series, as in Modiola (Fig. 61), Perna, etc., as the adductor muscle is brought nearer and nearer to the hinge-line,

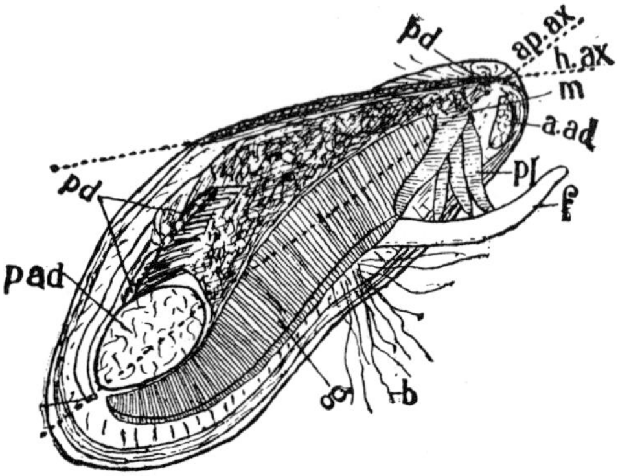

Fig. 61.—*Modiola plicatula.* Lettering same as in Fig. 60.

where its mechanical action is less and less effectual in closing the valves, we find that it is more and more reduced until it finally disappears from disuse and atrophy, as in Ostrea (Fig. 62), and Pecten. Conversely, the posterior adductor in the same series in the revolution of the axes is pushed farther and farther from the hinge-line and nearer to the central plane of the valves, where its mechanical action is most effectual in closing the valves. With its increase in functional activity the muscle increases in size. The revolved position of the axes, and the consequent reduction or loss of the anterior adductor and increase of the posterior adductor, is

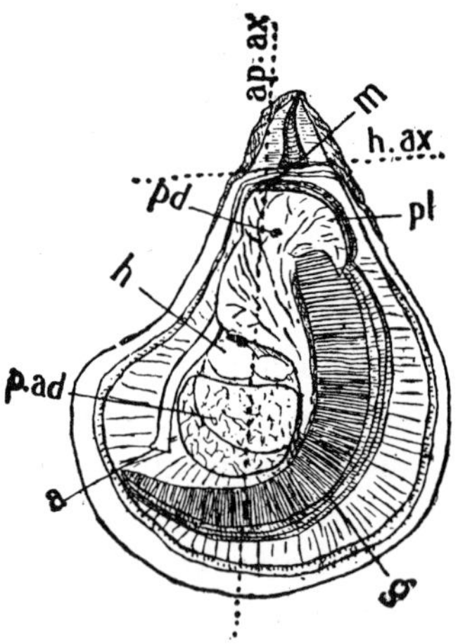

Fig. 62.—*Ostrea virginiana.* Lettering same as in Fig. 60.

found in many widely separated genera of pelecypods, as Ostrea, Mulleria, and Tridacna; thus proving the development of the same features on different lines of descent.[1] In Aspergillum the two valves have concresced so as to form a truly univalvular, tubular shell, so that the adductors would evidently be functionless if existent. The posterior adductor has disappeared and the anterior is reduced to a few disconnected shreds (Fisher), though evidently existent in the young, as attested by the form of the shell in the nepionic stage.

"Ordinarily there are two posterior retractor-muscles of the foot in pelecypods, one situated on either side. In adult Pecten either the left retractor alone exists, or both retractors are wanting (the left doubtless always exists in the young). In studies of young *Pecten irradians,* I found that the animal always crawled while lying on the right side, with the foot extended through the notch in the lower valve and pressed against the surface of support. It is evident that while crawling in this position the left retractor is in the plane of traction, and it is retained; on the other hand, the right retractor would not be in the plane of traction, and it has disappeared through disuse and atrophy.[2] A similar disappearance of the right retractors of the foot is seen in *Anomia glabra*, and is explained on similar bases of argument.

"In *Mya arenaria* we find a highly elongated siphon. In the young the siphon hardly extends beyond the borders of the valves, and then the animal lives at or

[1] Dr. B. Sharp and I published almost simultaneously closely similar views on the mechanical aspect of the relative size of the adductors. See *Proceeds. Phila. Acad.*, 1888, p. 122, and *Proceeds. Boston Soc. Nat. Hist.*, Vol. XXIII., 1888, p. 538.

[2] Both retractors doubtless exist in the prodissoconch stage of Pecten and allies.

close to the surface. In progressive growth, as the animal burrows deeper, the siphon elongates, until it attains a length many times the total length of the valves. The ontogeny of the individual and the paleontology of the family both show that Mya came from a form with a very abbreviated siphon, and it seems evident that the long siphon of this genus was brought about by the effort to reach the surface, induced by the habit of deep burial.

"The tendency to equalize the form of growth in a horizontal plane in relation to the force of gravity acting in a perpendicular plane, or the geomalic tendency of Professor Hyatt,[1] is seen markedly in pelecypods. In forms which crawl on the free borders of the valves the right and left growth in relation to the perpendicular is obvious, and agrees with the right and left sides of the animal. In Pecten the animal at rest lies on the right valve, and swims or flies with the right valve lowermost. Here equalization to the right and left of the perpendicular line passing through the center of gravity is very marked (especially in the Vola division of the group); but the induced right and left aspect corresponds to the dorsal and ventral sides of the animal, —not the right and left sides, as in the former case. Lima, a near ally of Pecten, swims with the edges of the valves perpendicular. In this case the geomalic growth corresponds to the right and left sides of the animal.

"The oyster has a deep or spoon-shaped attached valve and a flat or flatter free valve. This form, or a modification of it, we find to be characteristic of all

[1] "Transformations of Planorbis at Steinheim, with Remarks on the Effects of Gravity Upon the Forms of Shells and Animals." *Proceeds. A. A. A. S.*, Vol. XXIX., 1880.

pelecypods which are attached to a foreign object of support by the cementation of one valve. All are highly modified, and are strikingly different from the normal form seen in locomotive types of the group. The oyster may be taken as the type of the form adopted by attached pelecypods. The two valves are unequal, the attached valve being concave, the free valve flat; but they are not only unequal, they are often very dissimilar,—as different as if they belonged to a distinct species in what would be considered typical forms. This is remarkable as a case of acquired and inherited characteristics finding very different expression in the two valves of a group belonging to a class typically equivalvular. The attached valve is the most highly modified, and the free valve is least modified, retaining more fully ancestral characters. Therefore, it is to the free young before fixation takes place and to the free, least-modified valve that we must turn in tracing genetic relations of attached groups. Another characteristic of attached pelecypods is camerated structure, which is most frequent and extensive in the thick attached valve. The form as above described is characteristic of the Ostreidæ, Hinnites, Spondylus, and Plicatula, Dimya, Pernostrea, Aetheria, and Mulleria; and Chama and its near allies. These various genera, though ostreiform in the adult, are equivalvular and of totally distinct form in the free young. The several types cited are from widely separated families of pelecypods, yet all, under the same given conditions, adopt a closely similar form, which is strong proof that common forces acting on all alike have induced the resulting form. What the forces are that have induced this form it is not easy to see from the study of this form alone; but the ostrean form is the base of a series, from the summit of which we get a clearer view."

c. Mechanical Origin of the Impressed Zone in Cephalopoda.

Prof. Alpheus Hyatt has shown that the groove on the dorsum or inside of each coil of the Cephalopoda is due to the pressure exercised by contact with the ventral side of the coil within it. He has shown that this groove persists in cases where the shell in the course of evolution has become more or less unwound, and he regards this as an example of the inheritance of a mechanically acquired character. This subject is presented in greater detail in the part of this book devoted to heredity.

3. KINETOGENESIS IN VERMES AND ARTHROPODA.

It is believed with good reason that the Arthropoda have descended from some of the forms included in the branch Vermes, and perhaps Peripatus furnishes the nearest living approach to that type. The ancestor, whatever it may have been, developed limbs from processes of the body-wall, and used them to aid in progression. Peripatus has soft flexible limbs, and a non-chitinous integument generally. With the beginning of induration of the integument, segmentation would immediately appear, for the movements of the body and limbs would interrupt the deposit at such points as would experience the greatest flexure. The muscular system would initiate the process, since flexure depends on its contractions, and its presence in animals prior to the induration of the integuments in the order of phylogeny, furnishes the condition required. It is a matter of detail how the diverse segmentations of existing forms were produced. We can believe, however, that, as in Vertebrata, there has been a gradual elimination of less important segments of the limbs, until the highest mechanical efficiency was attained. We well know how the segments of the head and body have been modified by fusion, etc.

Prof. B. L. Sharp has shown the mechanical conditions of segmentation in Arthropoda as follows:[1]

"It occurred to me that if the theory [of kinetogenesis] had a general application, some additional proofs could be shown to exist among the invertebrates, where we have the action of muscular force upon hard and resisting parts of the skeleton. Those which present the best study for this purpose appear to be the crustaceans, where we find an immense variety of articulations in the body and in the limbs; highly complicated locked joints, others allowing motion in but one plane, as well as loose joints, where the hard parts scarcely come in contact with one another, and cases of degeneration of the hard parts, leading to total disappearance of a previously existing joint.

"In the Annelides, from which, there is no doubt, the arthropod branch sprang, we find no deposit of inorganic salts in the epidermis. The outer layer of the body is generally of a horn-like character, adhering closely to the secretive cells of the epidermis, very flexible, and thrown into folds by the vermicular motion of its possessor. In the leeches the body consists of a flexible cylinder, made up of two sets of muscles, an outer longitudinal cylinder and an inner cylinder of circular fibers, the contraction of which causes the animal to increase in length, while shortening is ef-

[1] *American Naturalist*, 1893, p. 89.

fected by the contraction of the longitudinal layer. The external surface of the medicinal leech, for example, is thrown into a regular series of very fine folds, extending across the longitudinal axis of the body. These folds do not correspond in numbers to the somites of the body, which are not indicated externally, five, six, or more of them belonging to one somite. When the animal shortens its length, these folds are deepened and the segments thrown closely together; when extension takes place, the folds are flattened, spread open, although not wholly disappearing, as they are a fixed quantity, so to speak. I believe these folds are due to mechanical action; by the disposition of the different fibers of the longitudinal series, in being inserted in a series of planes bounded by the valleys between the folds, this being aided by some of the circular fibers which pass through the longitudinal sheath, and find their attachment to the bases of the valleys.

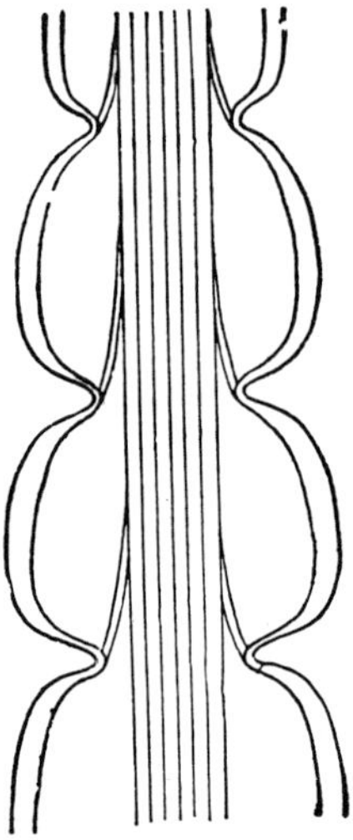

Fig. 63.—Diagrammatic representation of the segments of the leech, showing the folds, valleys, and muscular fibers.

"Starting from this point, and supposing the regularity of the folds to have become established from preëxisting folds by the regularity and stress of muscular action, we can conceive that when deposits of calcareous matter took place, rings similarly formed by a folding of a soft skin would receive that deposit at the most prominent portion of this fold, the convex face, and not in the protected valleys, as there would be more friction or pressure from external causes, and no deposits would take place in the valleys themselves, because they would not be subject to external friction, and their continual flexion would prevent any such deposits. Should such a deposit take place in the valleys, there would be a stiffening of the whole surface, which would defeat motion. In fact, in the leech the cuticle is already much thicker on the crests of the folds than in the valleys.

"In the more primitive Crustacea, we find the animal made up of rings extending over the whole length of the body, similar to the rings of the leech, save that there is but one ring to one somite, and instead of a perpendicular valley between the folds, this valley has an inward and a forward direction, allowing the anterior edge of a caudad ring to fit into the posterior edge of a cephalad ring.

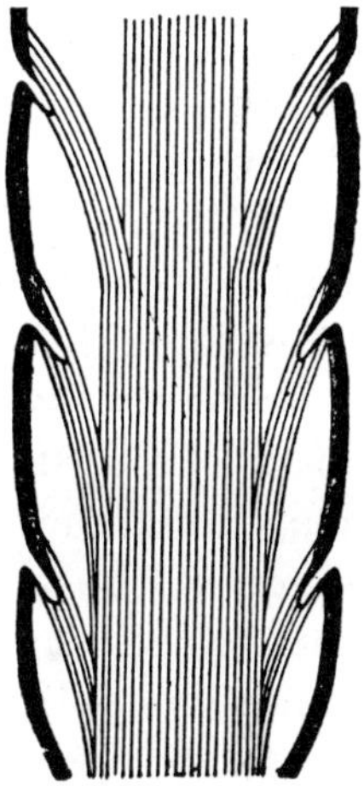

Fig. 64. — Diagrammatic representation of the rings of a primitive crustacean, showing the action of the muscles.

"In the higher Crustacea, several of the anterior rings have coalesced and form a solid shield which is known as the carapace. This has no doubt arisen by the lessening of the action between the anterior rings when the posterior portion of the body became the more active propelling organ. As the action ceased forward the valleys came to rest, and became exposed to friction and pressure, and consequently a deposit of calcareous matter took place producing the stiffening above hinted at.

"The formation of jointed appendages from parapodic paddles of the annelids can be followed out in the same manner, since the manner of mutual relation of the segments is the same as in the case of the body-segments.

"It has been stated that in the leech the folds do not correspond in number to the somites of the body, while they do in the Crustacea. All annelids do not move by means of a muscular system built upon the plan found in the leech. In many the circular layer has to a large extent disappeared, for the longitudino-circular plan is undoubtedly ante-annelidan. The movement of the free medusoid forms, and of the Ctenophora, is the result of a modified arrangement of this plan.

"With the disappearance of the circular layer, we find a peculiar modification of the longitudinal layer. This layer becomes broken up and the fibers act in moving the setæ, which answer to limbs. In a segment of a setiferous annelid, we may observe that the longitudinal muscles of the somite in section at the position of the seta are arranged like the letter 'V' in the fork of which the seta lies, the fibers to the left (anterior) pull the seta externally backward, those on the right (posterior) pull the seta forward. The introduction of the setæ, the origin of which I do not here attempt to explain, has no doubt been, together with the establishment of the external segmentation, a strong factor in causing the breaking up of the muscular tube into sections (myotomes), which by use and consequent increase have extended each arm of the 'V' into the segment on each side, while the insertion of the end of the seta has caused a break in the muscle by the formation of an aponurosis. This gives us the peculiar disposition of a myotome to extend across the union of two somites.

"If we examine the segments of the so-called abdomen of the macrurous Crustacea, as the lobster, we will find that the anterior face of one abdominal ring is pulled into the posterior orifice of the ring lying anterior to it, forming a kind of tubular ball and socket-joint, but with a flexible part of the integument with no calcareous deposit, folded upon itself, and acting physiologically as a tubular *ligamentum teres*. On examining the different joints, we will find that commencing at a fixed point, as at the base of the thorax, the movable ring of the first abdominal somite is *pulled into* the fixed part. Then the first abdominal somite becomes the fixed point for the movable ring posterior to it, and so on, so that we find that as the rings proceed away from the thorax, each is pulled into the opening of the one in advance. This is true of all those forms where the abdomen is well formed, strong, and an active organ in the economy of the animal; when this organ, the abdomen, ceases to be an active organ of motion, as in the burrowing forms, as in Callianassa, Gebia, some of the Squillidæ, etc., or where it is folded upon the sternum of the thoracic region, the muscles becoming weaker through disuse, the rings are not subject to the powerful muscular strain, and they as a rule overlap but little, if at all, but lie so that the edge of one ring rests upon the edge of another. In those forms where degeneration of the abdomen has proceeded so far as not to have even the usual deposit of calcareous matter, as in the hermit crabs, there are simply indications of rings on the abdomen, and this organ is but little more than a fleshy sac containing some of the viscera, and supplied with a few muscles which act together, with the form of the organ, to keep the abdomen curled so that it may hold as a hook, the animal within the molluscan shell which it habitually inhabits.

"When the limbs are examined, the same rule will be found to hold good, viz.: that the movable part is pulled into the fixed part. A modification of this is well illustrated in the evolution of the large chelæ. In some forms, take for example Ibacus, the first pair, and in fact all of the thoracic limbs end in a sharp-pointed segment, there being not the slightest suggestion of a chela. In Crangon, on the other hand, the terminal segment is pulled against the broad face of the penultimate one thus making a shift for a chela. In the Stomatopoda this step has been developed, for the last segment can be drawn against the whole length of the penultimate one (which is sometimes grooved to protect the points of the spines of the latter) and forms with it a very effective grasping organ. The continual use of the terminal segment, the increase of the muscular power will tend to draw this terminal segment backward (into) on the penultimate which enlarges with the increase of bulk of muscle, so that a well-developed chela, as in the lobster is found, where the ultimate segment is pulled backward to about the middle of the penultimate segment."

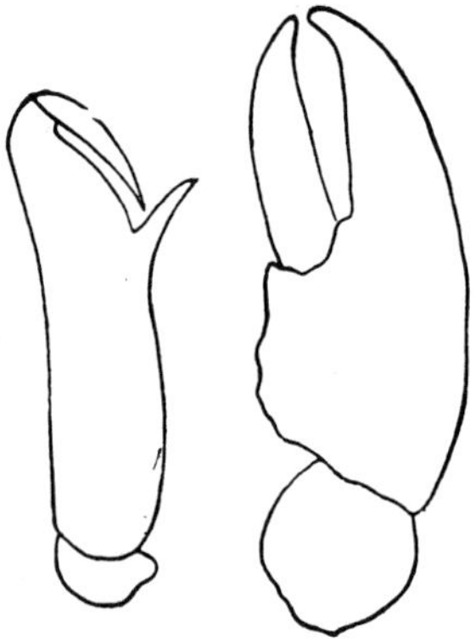

Fig. 65.—Diagrams of, *a*, hand of a form of Crangon; *b*, hand of a form of Astacus.

4. KINETOGENESIS IN THE VERTEBRATA.

I have already adduced the evidence in support of the doctrine that the structures of the hard parts of invertebrates have been produced by muscular movements. In turning to the Vertebrata we shall find that the evidence indicating that the details of their hard parts have had a similar origin, is quite convincing. This branch of the animal kingdom presents two distinct advantages for this study. First, we have a more complete paleontologic series than in any other. Second, we have the best opportunity for observation and experiment on their growth processes, since we ourselves, and our companions of the domesticated animals, belong to this branch of the animal kingdom.

I shall show first, the conditions under which abnormal articulations of the skeleton have been formed; and then the process involved in the formation of normal articulations. I shall then apply these facts to the phylogeny of the Mammalia as we know it, and then in a more general way to the Vertebrata as a whole.

i. KINETOGENESIS OF OSSEOUS TISSUE.

a. Abnormal Articulations.

Hütter, from whom I have quoted under the head of "Kinetogenesis of Muscle," thus describes the effect of abnormal conditions of joints on the articular surfaces of the bones which form them. He says: "We have abundant opportunity to investigate the change of condition which the joints undergo during a year of fixed muscular contraction.

"The ligaments and bursæ undergo similar changes

to those described for the muscles concerned. They elongate at points where the articular surfaces are spread apart, and correspondingly shorten where the flexure produces a folding. These changes proceed more slowly than those of the muscles and tendons. Very remarkable are the changes undergone by the articular cartilages. When a joint is permanently flexed, a part of the extremity of each bone is separated from contact with the other, and the articulation is finally destroyed at this point, because the cartilage begins to vanish. One must conclude that the existence of the articular cartilages is dependent on their mutual contact; for dislocated articular surfaces which remain in contact with soft tissues only, lose their cartilaginous covering. . . . Finally it is possible by a consideration of the etiology of the effects of joint contractions to reach some hitherto unnoticed conclusions regarding the changes of articular surfaces, and bone forms. The results of joint-contraction are most conspicuous when the latter occurs in childhood. During maturity, a dislocation which causes an articular border or prominence to rest abnormally on the opposed articular face in the act of walking, will be followed by the penetration of the former into the latter, and a deformation of the articulation; but the corresponding changes under like conditions in the growing skeleton are much more conspicuous."

Hütter thus describes the formation of new articular surfaces as a consequence of dislocation of joints. "If the head of the femur or humerus leaves its socket, and rests on the side of the ilium or the scapula, the periosteum of the bone which receives the new impact is excited to active bone-production, and the result is the deposit of new osseous tissue. The thin bones become thicker, not uniformly, but in correspondence with the periphery of the head of the humerus or femur, rather than with the point of contact of the latter. This point is irritated, but the contact of the ball restrains osseous deposit. So it occurs that gradually a new socket is developed, whose mechanical relations correspond exactly with those of the articulating bone. The head also acquires a strictly spherical shape, by such contractions and atrophies as are necessary to produce that result. Further, cartilage appears in the place of the periosteum of the socket, which functions like the primitive articular cartilage. It is characteristic of both connective and periosteal tissue to develop cartilage under the stimulus of continued friction of hard surfaces, such as occurs in dislocations and fractures."

These observations of Hütter have been confirmed by Henke, Reyher, Moll, and Lesshaft. Henke and Reyher state that the artificial prevention of flexure of articulations in young dogs renders them immobile, and their restraint of flexure to one direction, results in a curving of the articular faces in that direction.

I cite here two examples of modifications of structure under abnormal conditions which imposed new impacts and strains on the parts. I have described these cases, which are examples of false elbow-joints in man and in the horse, in the *Proceedings of the American Philosophical Society* for 1892.

In the first case, that of the human elbow, the cubitus was luxated posteriorly, so that the humeral condyles articulate with the ulna anterior to the coronoid process. The head of the radius is in contact with the external epicondyle on its posterior inferior face. The results are as follows. A new coronoid process was developed in front of the abnormal position of the humeral condyle to an elevation above the shaft of the ulna exceeding that of the normal coronoid. Between it and the normal coronoid was developed a perfectly functional cotylus which embraces the humeral condyle like the normal cotylus. The latter has its articular surface, buried under osseous deposit, so as to be no longer visible. The region of contact between the head of the radius and the external epicondyles, has developed in the latter a large articular cotylus which permits of both rotary and vertical movement of the former. The articular surface of the humeral condyles, except where in articulation with the ulna, is roughened, and partially overgrown with exostoses, so as to alter its form to a great extent. The opportunity of examining this specimen I owe to Provost Pepper of the University of Pennsylvania, in whose museum it is preserved.

In the case of the horse's elbow, the luxation of the cubitus is inward, so that the olecranon articulates with the external epicondylar surface, and the humeral condyles are not adapted to the head of the radius; their internal border falling considerably internal to the inner border of the radius. The horse from which this specimen was derived lived for two years after the luxation took place, and became able to use the limb in some degree. The effect on the articulation is as follows.

A large part of the inferior extremity of the posterior rib of the shaft of the humerus, which is the place of insertion of the external *flexor metacarpi* muscle, has been removed, so as to present a wedge-shaped outline with the apex downward. This removal permits the close articulation of the inner face of the olecranal process with the epicondyle, which has developed a considerable articular face, on which movement takes place in extension and flexion. The posterior border of this face has developed a ridge which borders the facet behind, and retains the olecranon in place. Two other facets are developed on the humeral condyles, and two on the head of the radius. The most important of the latter is a bevel of the external part of the surface to the border, due to the contact of the expanded internal humeral condyle. The articular face of the olecranon is much depressed in consequence of its articulation with the external epicondyle of the humerus. Besides these new and changed facets, the effect of the luxation is seen in the development of osseous crests at the points of insertion of the articular ligaments. One of these on the humerus has been already referred to. Another is concentric with and posterior to the internal humeral facet of the olecranar process, and serves as a guide to the humeral crest above described. A third is an extensive osseous deposit on the internal face of the head of the radius, which partially builds an extension of the head of the radius, which if completed would articulate with the overhanging portion of the internal humeral condyle. A third modification of normal structure is similar to that observed in the human elbow. It consists of osseous deposit beneath the synovial bursa at points where the luxation causes a gaping of the surfaces. This occurs at the trochlear groove of the head of the radius, which is partially filled up with exostosis.

The preceding observations lead to the following conclusions:

First. Continued excessive friction removes osseous tissue from the points of contact until complete

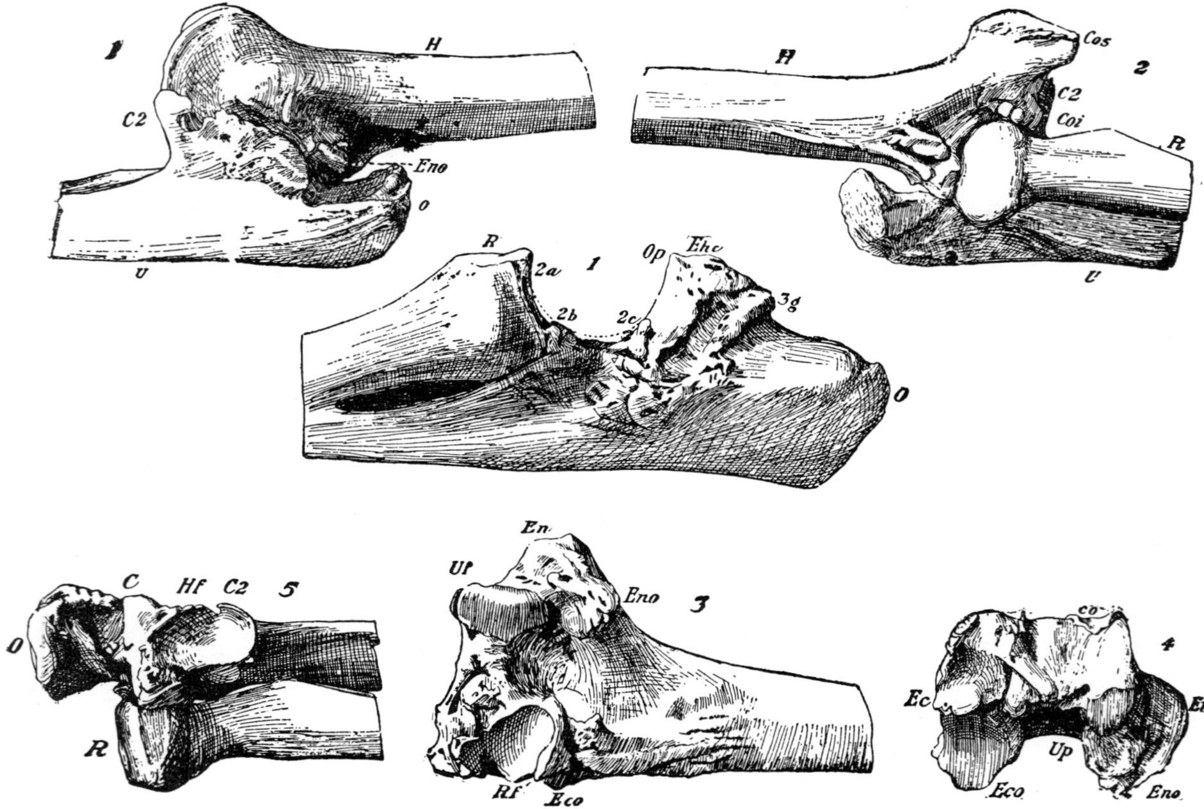

Fig. 66.—Elbows of man and horse.

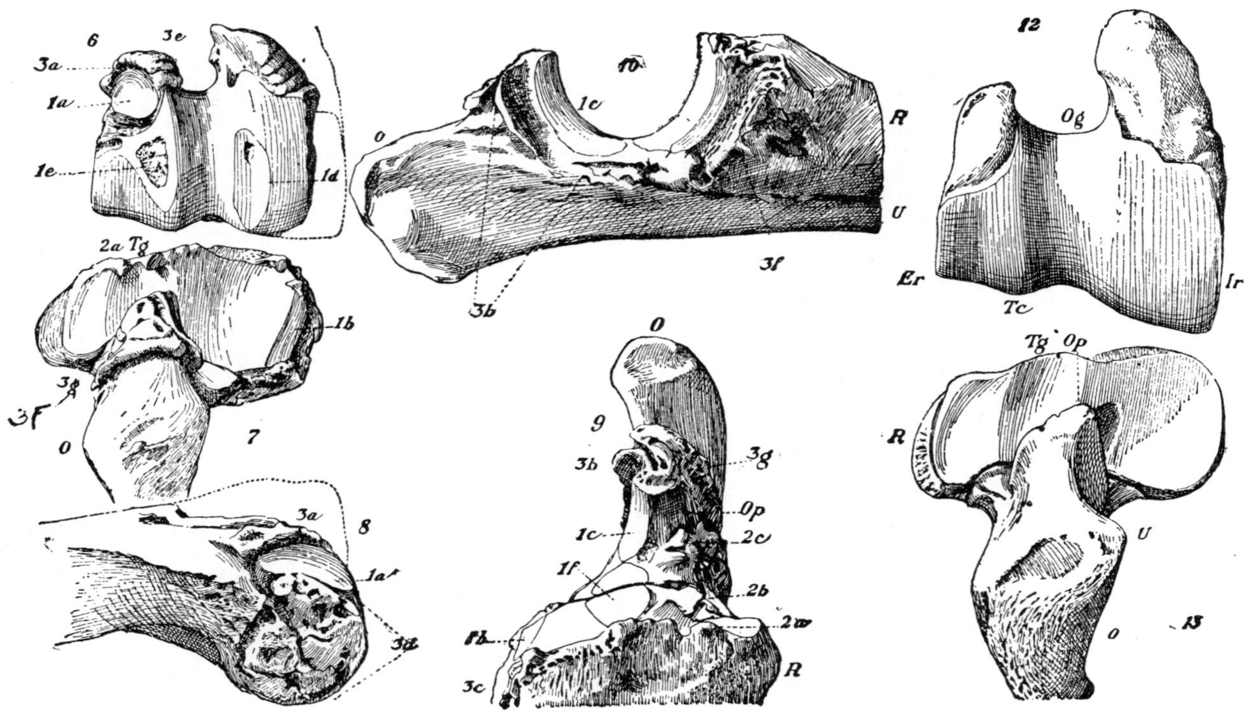

Fig. 67.—Elbow of horse.

adaptation is accomplished and the friction is reduced to a normal minimum. Then a normal articular surface is produced.

Second. Where the normal friction is wanting, and an inflammatory condition is maintained by a pulling stress on the investing synovial membrane, excess of osseous deposit is produced.

Third. Stress on the articular ligaments and tendons stimulates osseous deposit at their insertions, which deposit may be continued into their substance. This is a pulling stress.

These observations therefore show that osseous deposit is produced by different forms of mechanical stimulus.

EXPLANATION OF FIGURES 66 AND 67.

1–5, *Homo sapiens*, luxated elbow joint (one-half natural size); 1, luxated elbow joint, from within; 2, luxated elbow joint, from outer side; 3, humerus, posterior view of distal region; 4, humerus, distal view; 5, ulna and radius, anterior (superior) view; 6–11, bones of abnormal left elbow joint of horse (one-half natural size); 12, 13, normal bones of elbow joint of horse (one-half natural size); 6–12, humerus, distal views; 7–13, cubitus, proximal views; 8, humerus, external view of distal extremity; 9, humeral articulation of cubitus, from above; 10, cubitus, internal view; 11, cubitus, external view. Lettering.—*H*, humerus; *U*, ulna; *R*, radius; *C*, coronoid process; *C2*, second (abnormal) coronoid process; *O*, olecranon; *En*, entepicondyle; *Ec*, ectepicondyle; *Eno*, entepicondylar exostosis; *Eco*, ectepicondylar exostosis; *Co*, condylar exostosis; *Cos*, superior condylar exostosis; *Coi*, inferior condylar exostosis; *Hf*, humeral facet; *Rf*, radial facet; *Uf*, ulnar facet; *Op*, olecranar process of ulna; *Cp*, coronoid process of ulna; *Og*, olecranar groove of humerus; *Tc*, trochlear crest of humerus; *Tg*, trochlear groove of humerus; *Ehc*, external humeral facet of coronoid process; *Ihc*, internal humeral facet of coronoid process; *1a*, abnormal facet for coronoid process of ulna; *1b*, do. for internal roller of humerus; *1c*, do. for abnormal facet of humerus; *1d*, do. for internal border of radius; *1e*, do. for olecranar process of ulna; *1f*, do. for trochlear crest of humerus; *2a*, *2b*, *2c*, exostoses of radius and ulna to fill vacuity between humerus and radius and ulna, *3a*, abnormal crest which serves as a guide to the olecranar process of the humerus; *3b*, abnormal crest which serves as a guide to abnormal crest *3a*; *3c*, exostosis extending head of radius inwards to equalize its width with inward luxation of humerus; *3d*, exostoses of external epicondyle of humerus, to equalize its width with outward luxation of radius; *3e*, abnormal exostosis of insertion of external *flexor metacarpi* muscle; *3f*, *3g*, abnormal crest at insertion of external articular ligament on olecranar process of ulna.

b. Normal Articulations.

The origin of condyles and their corresponding cotyli has been made the subject of investigation by several German anatomists. L. Fick[1] expressed the opinion that the concavo-convex surfaces were produced by a wearing away of the surface which became concave, by the free action on it of the surface which became convex, the former being fixed, and the latter free. He found the conditions of muscular insertions to correspond with the conditions of fixity and freedom required ; for the insertions are always nearer to the concave surface than to the convex surface. He constructed plaster models of joints, and by moving one on the other obtained a convex surface on the moving, and a concave surface on the fixed extremities. These observations were confirmed by Henke,[2] but he very properly does not regard the result as due to wearing, but to the stimulation of metabolic action in the required directions. R. Fick[3] has confirmed these positions in an extended memoir, and recently Dr. E. Tornier has devoted a still more thorough research to the same subject.[4] R. Fick applied his observations to the question of the phylogeny of the articulations, but did not see in it proof of the operation of mechanical causes, but ascribed it to "inheritance and natural selection" in accordance with the meaningless formula usual at the time he wrote. W. Roux,[5] however, in reviewing Fick's article saw in the obser-

1 *Ueber die Ursachen der Knochenformen, Experimental-Untersuchung*, 1859, Göttingen, G. Wiegand.

2 *Anatomie und Mechanik der Gelenke*, Leipsic, 1863, p. 57.

3 *Archiv für Anatomie und Physiologie*, 1890, p. 391.

4 *Archiv für Entwickelungsmechanik*, I., 1894, p. 157.

5 *Biologisches Centralblatt*, 1891, p. 188.

vations of Fick proof of a direct mechanical cause of the structure. I have pointed out the phylogeny of the articulations in the Mammalia in various papers from 1877[1] to 1889,[2] and in 1881 I advanced the view that their successive evolution was due to impacts and strains (*American Naturalist*, July, 1881; *Origin of the Fittest*, p. 373). The opinion of Roux entirely supports my position, and it is further established by the elaborate memoir of Tornier just cited. This author adopts the view that bone-development is controlled by *Druck und Zug* or impact and strain, and he adds some important considerations to those previously advanced. He asserts that "in all existing Vertebrata true bones may appear as secondary structures, since all of these animals possess bands and threads of connective tissue which possess the latent capacity to be changed wholly or in part into cartilage." Thus is accounted for the development of sesamoid bones in tendons, in which category is included the patella. Tornier also shows that the concave articular face (cotylus) is that to which the flexor and extensor muscular insertions are nearest, while the convex face (condyle) is the one most remote from the muscular insertions.

It must be observed that Tornier adopts the language of the American Neo-Lamarckians in using the expression "impact and strain." Impact and strain are different modes of motion. "Impact" implies pressure, while "strain" implies a pulling stress, either direct or torsional. It is therefore alleged by Tornier, as it has been by myself, that opposite modes of mo-

[1] *Report U. S. Geol. Survey W. of 100th Meridian*, 1875, Vol. IV., p. 277-279. *Proceeds. Amer. Philosoph. Soc.*, 1884, p. 44.

[2] *Amer. Journal Morphology*, 1889, p. 163.

tion may produce metabolic changes in osseous tissue. For this reason it is possible to account for the lengthening of the limb-bones in heavy animals, as an effect of impact, while the astragalus of bats may have been elongated by a stretching strain.

c. The Physiology of Bone Moulding.

Dr. Koelliker has summarized the results of the observations made by himself and his predecessors on the processes of the growth and absorption of bone, which determine the forms of the elements of the skeleton.[1]

Bone is deposited through the agency of uninuclear cells, or osteoblasts, which may under peculiar conditions become enlarged and multicellular, when they are termed osteoclasts. These osteoclasts produce an absorption or destruction of the bone or dentine with which they are in contact, the bone or dentine being passive under the operation. How this is done is not known. Pieces of ivory which have been used to replace bone removed by surgical methods, have been found to be both corroded by osteoclasts, and overlaid by layers of living bone by osteoblasts.

In explanation of the causes which induce the formation and action of the osteoclasts, Koelliker remarks that: "the totality of changes of the jaws during the development of teeth appears to show that it is pressure by the soft parts which causes the absorption of bone. One can admit in the case of the jaw that the dental sacs in process of growth produce by their enlargement a state of irritation in the layer of osteoblasts which originally border the alveolar edge, and

[1] "The Normal and Typical Absorption of Bones and Teeth," *Verhandl. der Phys. Med. Ges. von Würzburg*, II., III., 1872.

that in consequence of this irritation the cellules transform themselves into osteoclasts, and acquire a new power, that of absorbing bone. The function will cease as soon as the teeth are formed, by the termination of pressure, and then the formative action of the cellules adjacent to the bone will repair it as a consequence of a retransformation of these elements into osteoblasts.

"I will not push further this first attempt at an explanation of the normal absorption of bone, but I content myself with observing, that in any case, pressure exercised by the soft parts counts for much in this phenomenon. Who does not remember in the face of these facts, numerous cases of pathological absorption of bone due to aneurisms, tumors, and hypertrophied organs? Who will not admit the great effect of the disappearance or arrest of development of organs on the size of their osseous surroundings; as Fick, formerly professor of anatomy at Marburg, has shown to take place in the orbit after the extirpation of the eye? It is possible to go a step further in the proposition, that external pressure has much to do with absorption. Thus the growth of the brain and spinal cord produce the resorption seen in the interior of the skull and of the spinal canal; that of the eye and of the nasal mucosa, and of the cranial vessels and nerves, have resulted in the enlargement of their cavities; and in the case of foramina, in their wider expansion. . . . The medullary cavities of bones are produced in the process of growth by the corrosive activity of osteoclasts."

It is then pressure which produces the excavations which form new cotyli in the construction of new articulations due to dislocations. By such excavations elevated portions remain adjacent to them. Other elevations, as already described, are due to deposit of bone stimulated by the absence of accustomed pressure, as in the filling up of the old ulnar cotylus in the human subject above described. Other elevations or osseous deposits such as occur at muscular and ligamentous insertions appear to follow a pulling stress.

Many other examples of the abnormal production of articulations might be cited, but the above are sufficient to show the plasticity of osseous tissue. It is also evident that if such results follow the stimulus of the parts during a short period of months or years, the continuance of the appropriate mechanical stresses through geologic ages must have been quite sufficient to produce all the characters which we observe in the articulations of the vertebrate skeleton.

I will now present the inferences which may be derived from consideration of the facts hitherto presented in this chapter. We have not been witnesses of the process of evolution, yet we believe that it has been in active operation. We have not been able to observe its *modus operandi*, but we may safely infer what it has been from the facts which are before us. Kinetogenesis having been observed in both the soft sarcode (muscle) and in the hard parts of animals, the law of uniformity obliges us to believe that similar changes have taken place in past ages whenever the necessity arose, and the energy at nature's disposal was sufficient.

ii. MOULDING OF THE ARTICULATIONS.

a. The Limb Articulations.

This part of the subject has the advantage of many facts of paleontology in our possession. We have now discovered the outlines of the phylogeny of many mammalian types, and many detailed histories of special lines of descent are known. Our knowledge is most complete in the unguiculate and ungulate placentals, while it is least as regards the Mutilata, and the implacentals. We have excellent series of skeletal parts, and I have given the successional modifications of some of them on page 139.

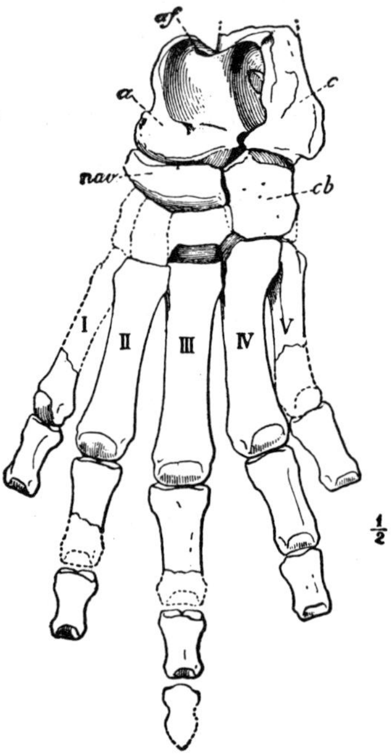

Fig. 68.—*Periptychus rhabdodon* Cope, a condylarthrous genus of the Puerco epoch of New Mexico; posterior foot, one-half natural size, showing pentadactyle plantigrade foot without groove of astragalus, as in the probable ancestor of the Diplarthra. From Scott and Osborn.

In the first place, I will select an illustration of the effects of use on the articulations of the limbs and feet of the Mammalia. I take first the ankle and wrist-joints. In the ruminating animals (ox, deer, camel, etc.) and in the horse, among other living species, the ankle-joint is a very strong one, and yet admits of an extensive bending of the foot on the leg. It is a treble tongue-and-groove joint; that is, two keels of the first bone of the foot, the astragalus, fit into two grooves of the lower bone of the leg, the tibia, while between these grooves a keel of the tibia descends to fill a corresponding groove of the astragalus. Such a joint as this can be broken by force, but it cannot be dislocated.

Now, in all bones the external walls are composed of dense material, while the centers are spongy and comparatively soft. The first bone of the foot (astragalus) is narrower, from side to side, than the tibia which rests upon it. Hence the edges of the dense side-walls of the astragalus fall within the edges of the dense side-walls of the tibia, and they have pressed into the more yielding material that forms the end of the bone, and causing bone absorption, pushed it upward, thus allowing the side-walls of the tibia to embrace the side-walls of the astragalus. Now, this is exactly what would happen if two pieces of plastic dead material, similarly placed, should be subjected to a continual pounding in the direction of their length. And in view of the facts already cited we cannot ascribe any other immediate origin to it in the living material.

The same active cause that produced the two grooves of the lower end of the leg produced the groove of the middle of the upper end of the astragalus. Here we have the yielding lower end of the tibia resting on the equally spongy material of the middle of the astragalus. There is here no question of the hard material cutting into soft, but simply the result of continuous concussion. The consequence of concussion would be to cause the yielding faces of the bones to bend downward in the direction of gravity, or to remain in their primitive position while the edges of the astragalus were pushed into the tibia. If they were flat at first they would begin to hollow downward, and a tongue above and groove below would be the result. And that is exactly what has happened. This inclusion of the astragalus in the tibia does not occur in the reptiles, but appears first in the Mammalia, which descended from them. See Figs. 68–69. I have shown that without exception, every line of Mammalia commenced with types with an astragalus which is flat in the transverse direction, or without median groove.

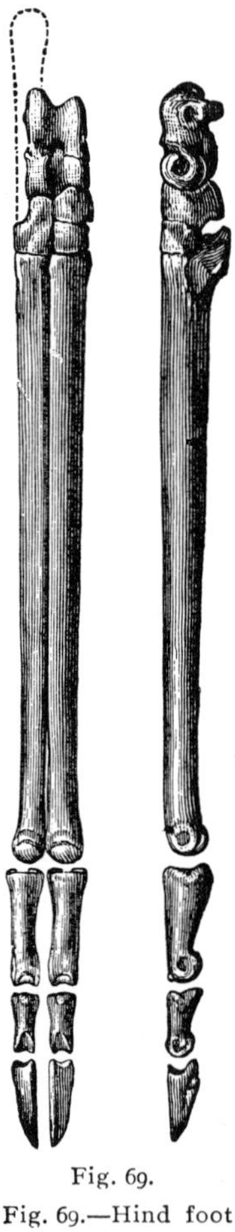

Fig. 69.

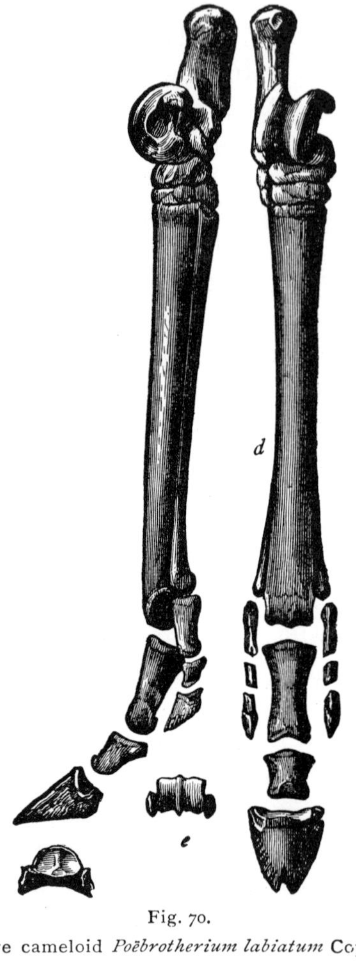

Fig. 70.

Fig. 69.—Hind foot of primitive cameloid *Poëbrotherium labiatum* Cope, showing grooved astragalus and first toe-bones without keel in front at lower end. (From Colorado.)

Fig. 70.—Hind foot of three-toed horse (*Prothippus sejunctus* Cope) (from Colorado), showing grooved astragalus, and trace of keel on front of lower end of first bone of middle toe.

From early Tertiary times to the present day, we can trace the gradual development of this groove in all the lines which have acquired it. The upper surface became first a little concave; the concavity gradually became deeper, and finally formed a well-marked groove.

The history of the wrist-joint is similar. The surface of the fore-arm bones which joins the fore-foot is in the early Tertiary Mammalia uniformly concave. In the ruminating mammals it is divided into three fossæ, which are separated by sharp keels. These fossæ correspond with the three bones which form the first row of the carpus or palm. The keels correspond to the free sutures between them. The process has been evidently similar to that which has been described above as producing the side-grooves in the end of the tibia. The dense walls of the sides of the three bones impinging endwise on the broad yielding surface of the fore-arm (radius) have gradually, under the influence of countless blows, impressed themselves into the latter. On the contrary, the surface above the weaker lines between the bones not having been subject to the impact of the blows, and influenced by

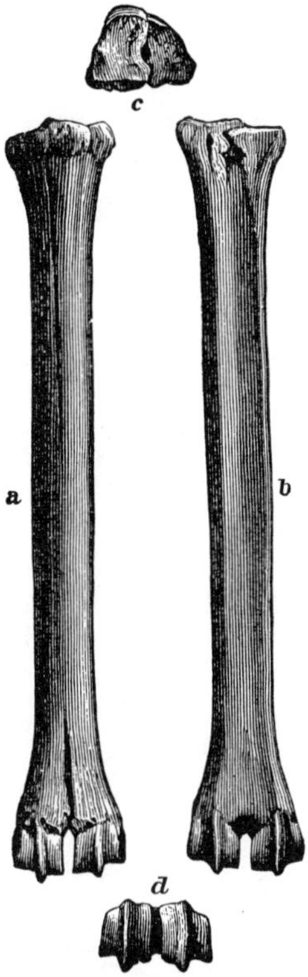

Fig. 71.—United first bones of two middle toes of deer-antelope (*Cosoryx furcatus* Leidy), showing extension of keel on front of lower end. (From Miocene of Nebraska.)

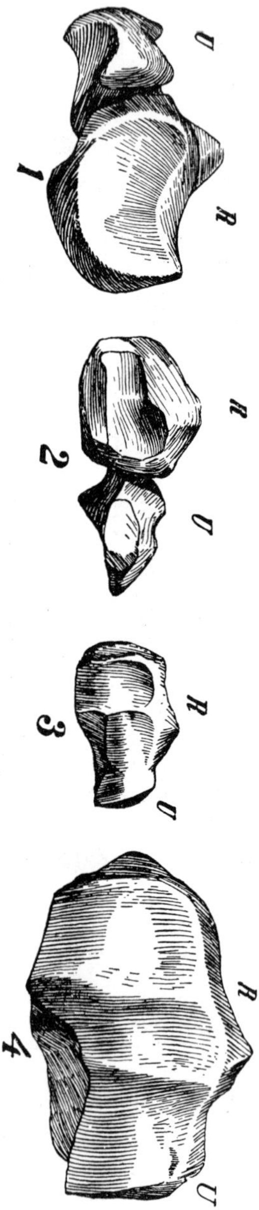

Fig. 72.—Wrist-joint at distal extremity of fore arm. 1, Pachyæna (Creodont Eocene). 2, Phenacodus (Eocene Condylarthron). 3, Prothippus (Upper Neocene three-toed horse). 4, Holomeniscus (Plistocene Camel).

gravity, remains to fill the grooves, and to form the keels which we observe. (See Fig. 72.)

There is another striking instance of the same kind in the feet of Mammalia; that is, in the development of the keels and grooves which appear at the articulation of the first set of bones of the toes (metapodials) with the bones of the second set (phalanges). These keels first appear on the posterior side of the end of the first set of bones, projecting from between two flexor tendons. These tendons, in many mammals, contain two small bones, one on each side, each of which acts like the knee-pan, and resembles it in miniature, which are called sesamoid bones. These tendons and bones exercise a constant pressure on each side of the middle line, when the animal is running or walking, and this pressure, together with the concussion with the ground, appears to have permitted the protrusion of the middle line in the form of a keel, while the lateral parts have been supported and even compressed. The reptilian ancestors of the mammals do not possess these keels.

Now, I have shown that the lines of mammalian descent displayed by paleontology are characterized, among other things, in most instances, by the gradual elevation of the heel above the ground, so that the animal walks on its toes. It is evident that in this case the concussion of running is applied more directly on the ends of the bones of the foot than is the case where the foot is horizontal. As a consequence we find the keel is developed farther forward in such animals. But in many of these, as the Carnivora, hippopotamus, and the camels, there is developed under the toes a soft cushion, which greatly reduces this concussion. In these species the keel makes no further progress. In other lines, as those of the horse, the pig, and of the ruminants, the ends of the toes are applied to the ground, and are covered with larger hoofs, which surround the toe, and the cushion is nearly or quite dispensed with. These animals are especially distinguished by the fact that their metapodial keels extend entirely round the end of the bone, dividing the front, as well as the end and back (Fig. 71) ; since the front of the metapodial is out of the reach of the sesamoid bones, its keel would seem to be a moulding to the groove of the first phalange, which is itself moulded by the middle and posterior part of the metapodial keel (Wortman.)

A third and similar example is furnished by the elbow-joint of the Quadrumana and Diplarthra. In the lower Mammalia, including the Carnivora (Fig. 73), the distal end of the humerus presents a submedian groove, which receives the ulna, and on the inner side of it, a more or less convex surface, which is applied to the head of the radius. The coronoid process of the ulna is narrow, and its dense bounding walls impinge on the broad face of the humeral condyle in flexion and extension, and transfers to it the force of impact when the foot strikes the ground. In either case strong pressure has been brought to bear on the humeral condyle, and it has yielded to the denser body of the ulna, thus forming the groove in question. In such Mammalia the effect of impact of the limb on the ground has been to impress the head of the radius on the humeral condyle upwards. The dense edges of the former have impressed themselves on the latter, while the unsupported middle portion has yielded in the direction of gravity, and the result is what we find, i. e., a cup-shaped surface of the head of the radius, and a convexity of the humeral condyle adapted to it.

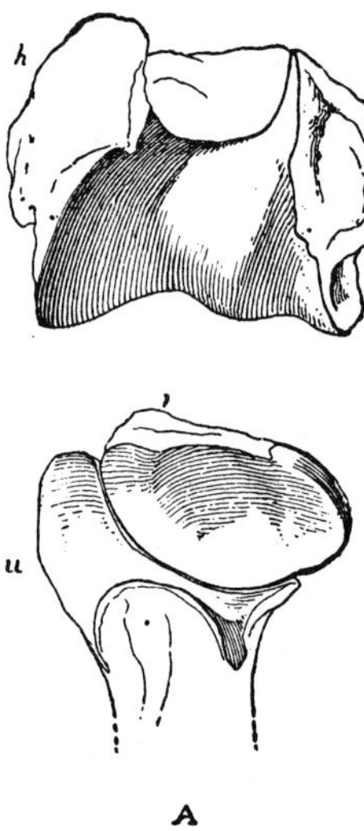

Fig. 73.—Elbow-joint of *Crocuta maculata* L. (spotted hyena) seen from behind; *h*, humerus; *r*, radius; *u*, ulna. Original.

Among specializations of the elbow-joint, I call attention to two. In Quadrumana the head of the radius, probably owing to continued supination of the manus, occupies a position at the external side of the coronoid process of the ulna, and impinges on the outer part of the condyle of the humerus. The concavity of its head, and the convexity of the humeral condyle, are visible as before, but a prominent tongue or keel, which has been called the intertrochlear crest, separates the ulnar and radial surfaces of the humerus (Fig. 74). This keel occupies the groove or interval which separates the head of the radius from the coronoid process of the ulna. It is plain that we have here another tongue and groove-joint, produced by the mutual adaptation of parts under strain, pressure, and impact. The other extreme of elbow-joint is found in that of the diplarthrous Ungulata (Fig. 75). Here the head of the radius, while retaining its normal position on the inner side of the fore-arm, is extended to the external side of the ulna and even beyond it, adapting itself to the entire width of the humeral condyles. The same structure is found in the specialized forms of both series of Diplarthra, the Perissodactyla and Artiodactyla. This expansion of the head of the radius appears to be in direct relation to the duration through long geologic ages of the impacts which have affected the limbs of these, the swiftest of the Mammalia. That the head of the radius should be spread so as to fit the entire surface of the humerus, under all circumstances, seems to be a mechanical necessity. But in addition to this we find a tongue-and-groove adaptation, in which the crest (which I have called the trochlear crest) articulates with a groove in the head of the radius. The internal articulation of the humerus with the radius has the usual form, convex and concave distad. The trochlear crest marks the external border of the olecranar groove of the humerus. But

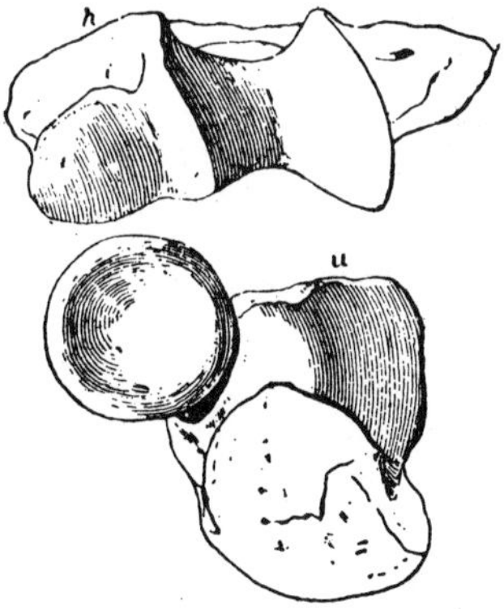

Fig. 74.—Elbow-joint of chimpanzee from behind.

the external part of the humeral condyles is converted into a roller which is set off from the trochlear crest by the abrupt contraction of its diameter; while the corresponding part of the head of the radius projects to fit it exactly.

A probable explanation of the form of this roller may be derived from a consideration of the almost identical structure of the metapodio-phalangeal articulation of the Artiodactyla. The internal and external sides of the distal metapodial condyles are not similar, the external being more strongly impressed than the internal (Fig. 77D). This is simply due to the unequal pressure exerted on the two extremities of the condyle by the phalanges, owing to the divergent direction of the digits when serving as a support. In the distal end of the humerus the same effect is seen, the external part of the condyle nearly resembling the corresponding part of the metapodial bones. This is traceable to the same cause, viz.: the divergent position assumed by the fore arm on the humerus, when the weight is supported on one fore leg only. This brings the line of pressure through the external part of both the head of the radius and the humeral condyle (Fig. 77A). That the higher ungulates are "knock-elbowed" may be readily observed by watching their gaits (Fig. 76).

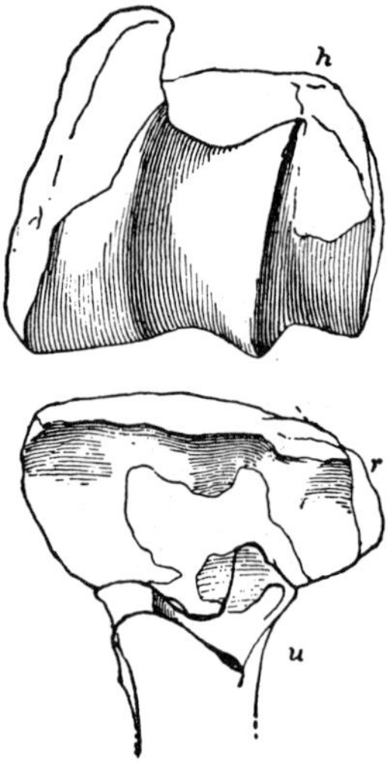

E

Fig. 75.—Elbow-joint of *Cervus elaphus* (red deer) from behind.

A distinct consequence of combined impact and strain is seen in the evolution of the carpus and tarsus of the Diplarthra. In primitive Mammalia, as in most Unguiculata, the bones of the carpus and tarsus succeed each other in such a way that the principal lines of separation between the elements coincide in the two rows, thus producing a linear relation between the former. In the Diplarthra, on the other hand, the elements of the two rows alternate with each other so as to produce a strong interlocking. I have shown that in the primitive Ungulata, the Taxeopoda, the linear arrangement is observed, while in three orders of ungulates, the Proboscidea, Toxodontia, and Amblypoda, there are various degrees of alternation intermediate between the linear type of the Taxeopoda and the completely interlocked condition of the Diplarthra. It has been already pointed out in the chapter on phylogeny that the taxeopodous type

Fig. 76.—*Cervus canadensis* in motion: from Muybridge's *Animal Motion*; showing the "knock-elbow" position of the fore leg, in both plantation and recover.

of foot preceded the diplarthrous in time. Besides the alternation mentioned, it is quite general in both types for the metapodial bones to possess a facet for contact with that element of the carpus and tarsus next exterior to the one to which they have their principal articulation. From these facts it is evident that the bones of the second carpal and tarsal rows have, in

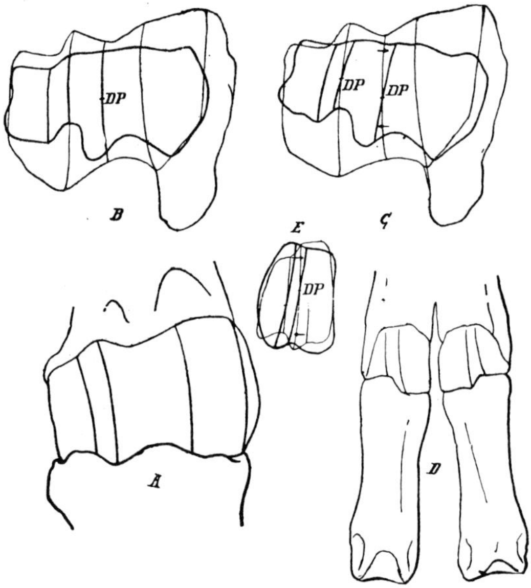

Fig. 77.—*Cervus elaphus; A, B, C,* humero-radial articulation; *A* and *B*, with the radius in position; *C*, with radius twisted; *D. E*, metatarsophalangeal articulation; *D*, front; *E*, distal view, twisted.

the process of evolution, assumed a position interior to their primitive position, with reference to the first row proximad to them, and the metapodials distad to them. The cause of this shifting of position is to be found in the movements of the limbs in progression, and especially in rapid progression (Fig. 78).

If we observe the movements of the limbs in a

diplarthrous ungulate, we shall see that as the foot is planted on the ground the prominent flexures of the limbs, the elbow and gambril joints, are turned in-

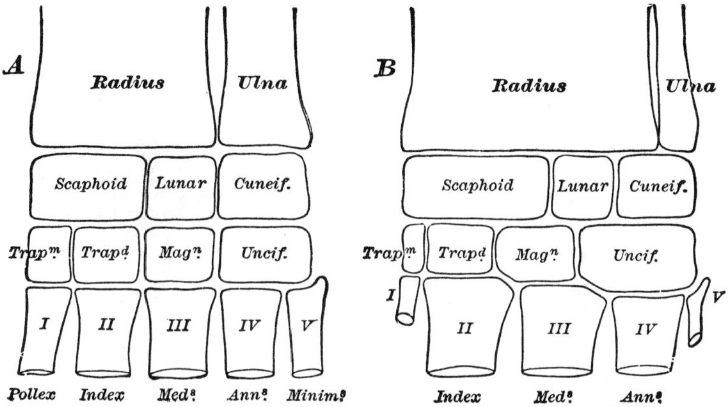

Fig. 78.—Diagram of carpus of a Taxeopod (*A*) and (*B*) of a diplarthrous ungulate. From Osborn.

wards, so that the limb, were it free from the ground, would be twisted or rotated on its long axis from within, forwards and outwards. As the foot rests on

Fig. 79.—Raccoon pacing, showing right fore foot just before recovery. From H. Allen.

the ground, the limb experiences a torsion strain in the directions mentioned. This throws the weight on the interior bones of the lower legs, the radius and the tibia. Thus these bones have acquired a great superiority in dimensions over the external elements (ulna and fibula) in all the Diplarthra. The bones of the inner side of the first carpal and tarsal rows have thus transmitted an ever-increasing share of the impact, as the radius and tibia have developed, and have grown

Fig. 80.

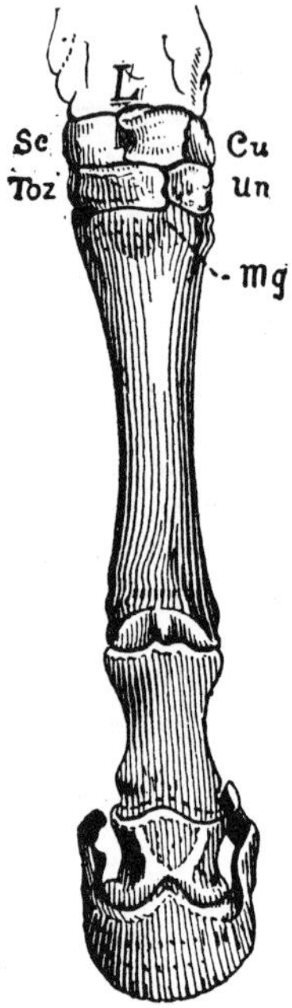

Fig. 81.

Fig. 80.—*Rhinocerus unicornis* carpus. Arrow ending in *P*, line of impact in plantation; *do.* ending in *R*, line of strain in recover.

Fig. 81.—*Equus caballus* fore foot. *Sc*, scaphoid; *L*, lunar; *Cu*, cuneiform; *Toz*, trapezium and trapezoideo; *Un*, unciform; *mg.*, magnum.

with their growth at the expense of the external elements, the cuneiform in the carpus, and the calcaneum in the tarsus, which have become very narrow elements in the higher Diplarthra. As the pressure has been obliquely from within outwards, the growth of the proximal elements, the scaphoid and lunar in front,

Fig. 82.—*Gazella dorcas*, gazelle. From Brehm.

and the astragalus behind, has been in the same direction. It has been shown by Dr. H. Allen that just before the recover of the foot, the latter is directed outwards from a line parallel with the axis of the body, so that the weight falls on the inner part of the sole of the former. This naturally causes the bones of the foot to press inwards on the heads of the metapodials, so that the latter tend to grow outwards on the second tarsal row. In this way were produced the facets on the external side of the heads of the metapodials. Thus is accounted for, on simple mechanical principles, the phenomenon of carpal and tarsal displacement exhibited in its highest development, by the Diplarthra.

It is significant that diplarthrism has not appeared in mammals which possess an elastic pad of connective tissue on the soles, as in Unguiculata generally, and especially in the Carnivora. Diplarthrism is present in the camels, which have a pad, but I have shown that this pad did not appear until a comparatively late geologic epoch, and long after diplarthrism had become established in the camels' ancestors, the Poëbrotheriïdæ.

The faceting of the head of the astragalus as the result of impacts, is seen on comparison of the astragali of Phenacodus and Hyracotherium in Ungulata (Figs. 33–35), and of Dissacus and Mesonyx among the Creodonta. In this last genus we have the only faceted astragalus among carnivorous mammals, but this genus is at the same time subungulate.

b. The Forms of Vertebral Centra.

The mutual articulations of the vertebral column are those of the centra and of the zygapophyses. Many important modifications in these articulations are to be seen in Vertebrata, the Reptilia presenting the greatest variety, excepting in the zygapophyses, which are tolerably uniform in that class. In the Mammalia, modifications of the central articulations are not more striking than are those of the zygapophyses.

The forms of central articulation are four, viz.: the amphicœlous, the ball-and-socket, the plane, and the saddle-shaped. The first type is only seen in a very imperfect degree in Mammalia and in but very few vertebræ, where it is indeed but a modification of the plane. The ball-and-socket is chiefly found in the neck of the long-necked Mammalia, as the higher Diplarthra, and to a less degree in their lumbar regions, while the dorsal vertebræ present an approach to the same type in the same groups. The saddle-shaped centrum is only found in Mammalia in the necks of certain genera of monkeys. The majority of Mammalia present the plane articulation of all the vertebral centra.

In Mammalia in which movement of the vertebræ on each other has become impossible, the centra coössify, as for instance in the sacrum. In this region the number of vertebræ coösified is directly as the length of the iliac bone, which supports and holds them immovable. Such is their condition throughout the dorsal region in the extinct Edentata of the family Glyptodontidæ, where the carapace is, as in the tortoises, inflexible, and which therefore limits the possibility of motion of the vertebral column. Another illustration is seen in the necks of the balænid Cetacea, and to some degree in the Delphinidæ and Physeteridæ. The lack of present mobility of this part of the column is due to its extreme abbreviation, a character which has been gradually developing during Cenozoic time; since the earliest Cetacea had considerably longer necks than the later ones, and had their vertebral centra distinct. It appears to me probable that the shortening was the result of disuse. This disuse would arise from gradually increasing powers of locomotion through the water, a progress, which, judging from the character of the limbs of the Zeuglodon, was evidently made after the time of the Eocene. The increase of speed would enable the animal to overtake and capture its prey, without the necessity of using a long prehensile neck in seizing it in the pursuit.

The ball-and-socket articulation of the vertebræ is well known to be the predominant condition in the Reptilia, and the fact that it is necessarily associated with the flexibility of the column is equally well understood. The flexibility is directly as the weakness of the limbs, for in the large and long-limbed terrestrial Reptilia of the order Dinosauria, the vertebral articulations of the dorsal region, at least, are plane. That it is chiefly confined to, and best developed in, the most flexible regions, i. e., the cervical and lumbar, of the column of the Mammalia, also shows this necessary connection. There can be no doubt but that the ball-and-socket vertebral articulation has been produced by constant flexures of the column in all directions, as has been suggested by Marsh.

iii. INCREASE OF SIZE THROUGH USE.

Under this head I enumerate examples where the mechanical causes in operation are less self-evident than those included under the preceding section. They are, however, probably due to the same process, viz., impact and strain.

a. The Proportions of the Limbs and of Their Segments.

The length of the legs of terrestrial Mammalia has increased with the passage of time. The inferior types of Mammalia now existing, as Marsupialia, Glires, Insectivora, Edentata, have short legs, with a few cases of extreme specialization as exceptions, such as kangaroos, rabbits, and jerboas (hind legs only), the *Dolichotis patachonica*, the Rhynchocyonidæ, and the sloths. In the orders which stand at the summit of the series, as the Diplarthra, Proboscidia, Carnivora, and Anthropomorpha, the legs are much increased in length, and this is especially marked in certain forms which stand in all respects at the summit of their respective orders. Thus in Diplarthra, the deer, antelope, and horse are distinguished for length of limb; in the Proboscidia, the elephant; in the Carnivora, the large cats and hyænas; in the Anthropomorpha, the fore limbs are long in all, the hind ones especially so in man.

The cause of this elongation is apparently use. It is the hind legs that are elongated in a straight line in animals that walk on them, as man; and both, in those that walk on both, as the elephant. In animals that leap with the hind legs these are still more elongated, and are folded when at rest, and rapidly extended when in motion. In animals that climb with the fore legs, these are elongated, as in the Anthropomorpha, except man. In those that climb with all fours, all are elongate, as in the sloths. It must be remembered that these elongations are the sum of increments added one to the other through long ages of use in geologic time. The mechanical character of that use has not been identical. It is of two principal kinds, viz.: impact and longitudinal strain. These two forms of energy move in directions opposite to each other; the one as compression in the direction of the length of the bone; the other, as a stretching in the direction of the length of the bone. Both processes alike appear to have stimulated growth in the direction of the length of the bone.

The increase in the length of the legs has not been always due to increase in length of the same segment. In a majority of the higher mammals, the increase has been principally in the foot, and especially in the metapodials and digits, producing digitigradism. In the forms which have remained plantigrade, the femur (Proboscidia), or femur and tibia (Quadrumana), or all three segments (Tarsius), have been the seat of the elongation. We can again trace these especial elongations to special uses. In animals which leap, the distal segments of the limbs are elongated; in those which do not leap, but which merely run or walk, it is the proximal segments of the limbs which are elongated.

Animals which run by leaping are divided into those which run and leap with all fours, as Diplarthra; and those which run and leap with the posterior limbs only, as the jerboas and kangaroos. In both types, the distal segments of the hind limb are elongated, and in the Diplarthra, those of the fore limb also.

Animals which do not leap in progression (elephants, Quadrumana, bears) are always plantigrade, and have very short feet, but elongate thighs, and, mostly, tibias.

These facts show that those elements which receive

the principal impact in progression are those which increase in length. In digitigrade animals it is the feet which receive the impact of the repeated blows on the

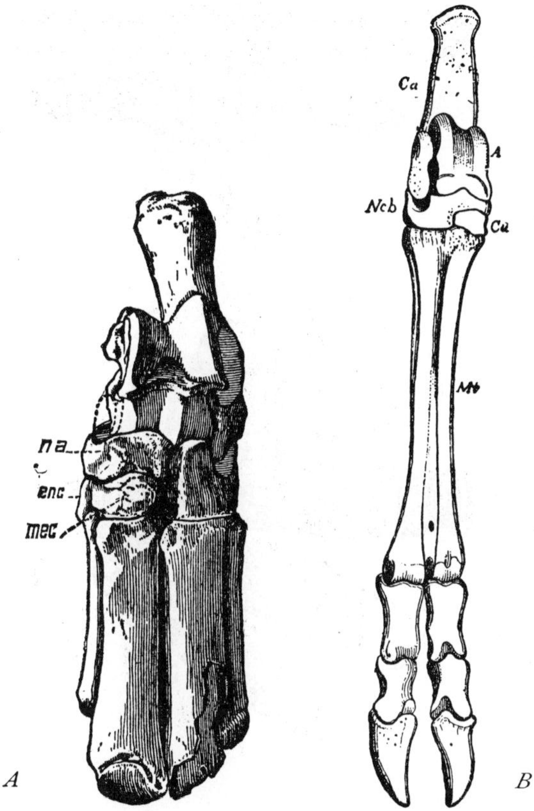

Fig. 83.—Pes of (*A*) *Merychochœrus montanus* from Scott; (*B*) *Bos taurus*, much reduced. *Ca*, Calcaneum; *As*, Astragalus; *Na*, Navicular; *Neb*, Naviculocuboid; *Cu*, *Mec*, Ecto-mesocuneiform; *Mt*, Metatarsals (cannon bone); *Enc*, Entocuneiform.

earth while in progression, while supporting the weight of the body at every stage of the process. In plantigrade animals it is the soles of the feet, and the bones of the leg in line with them, which receive the impact,

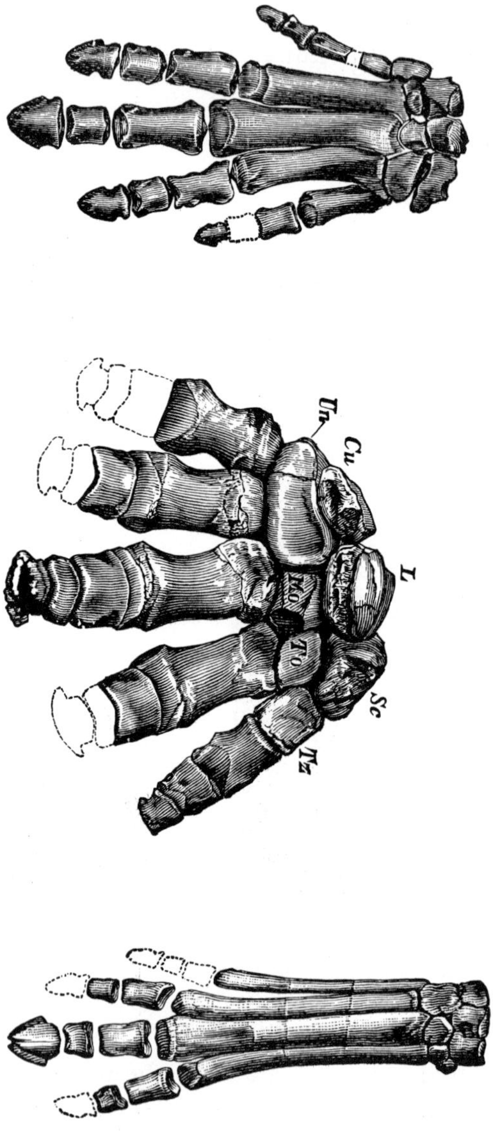

A *B* *C*

Fig. 84.—Anterior feet of primitive Ungulata, reduced. *A*, *Phenacodus primævus*. *B*, *Coryphodon elephantopus*. *C*, *Hyracotherium venticolum*.

while the feet beyond this point receive none, and do not support the body, except very partially at the moment of leaving the earth.

b. The Number of the Digits.

The reduction in the number of toes is supposed to be due to the elongation of those which receive the greater number of strains and impacts in rapid progression, and the complementary loss of material available for the growth of those not subject to this stimulus. This is rendered probable from the fact that the types with reduced digits are dwellers on dry land, and those that have more numerous digits are inhabitants of swamps and mud, or are more or less aquatic. That this inequality is due to these mechanical causes is still further indicated by the fact that in those forms where the soles are thickly padded (Carnivora, Proboscidia) the reduction has either not taken place, or has made little progress, amounting to the loss of only one digit. (An apparent exception in the case of the camels will be mentioned later.) A still more important body of evidence which shows that the inequality in size and number of digits is due to impacts and strains unequally distributed, has been brought forward by Ryder. He points out that definite results are to be observed in those limbs of a given type of animal which experience correspondingly definite influences; while in the limbs where the strains are equal, the modifications do not appear. Examples of this kind are to be found in the unguiculate Mammalia and in the Marsupialia. Thus in the jerboas which use the hind limbs in leaping, these only display reduced digits, the fore limbs remaining of primitive character. The same is true of the kangaroos. In digging genera, the fore limbs experience the modi-

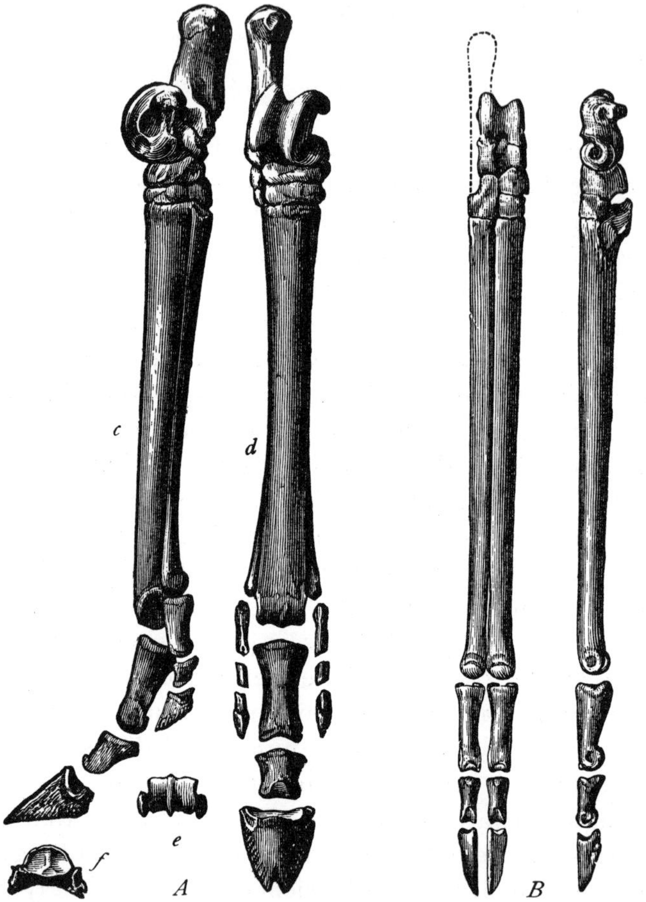

Fig. 85.—*A*, Right posterior foot of *Prothippus sejunctus* Cope, from Colorado, about one-half natural size. From *U. S. Geological Survey of Territories*, F. V. Hayden, IV. *B*, Right posterior foot of *Poëbrotherium labiatum* Cope, from Colorado, three-fifths natural size. From Hayden's Report, IV., Plate CXV.

fications, while the hind limbs are more normal, as in Chrysochloris and various Edentata.

Ryder sums up the evidence in two propositions, as follows: [1]

"I. The mechanical force used in locomotion during the struggle for existence has determined the digits which are now performing the pedal function in such groups as have undergone digital reduction.

"II. When the distribution of mechanical strains has been alike upon all the digits of the manus or of the pes, or both, they have remained in a state of approximate uniformity of development."

The application of the impact, or strain, or both, in progression, is easily understood. In recover (see p. 299), the leg is bent on the foot as it rests on the ground, and those digits which then leave the ground last, sustain greater strain than those which leave it sooner. In replacing the foot on the ground (plantation), those digits which strike it first experience greater force of impact than those which strike it later. Supposing the five primitive digits to have been of equal length, the distribution of the impact and of the strain will depend on the angle at which the foot is directed with reference to the direction of motion. If the feet are pointed forwards, the middle digits will experience strain and impact; if outwards, the inner digits bear the weight; if inwards, the external digits receive it.

Observation on five-toed plantigrade mammals shows that their feet are turned neither inwards nor outwards in progression, but straight forwards. It is probable that the primitive Mammalia moved in the same manner. This is also to be inferred from the fact that they were plantigrade, so that the leverage transversely in or out which results from the elevated heel of the digitigrade leg was very much less in them.

[1] *American Naturalist*, 1877, p. 607.

In progression of this type, the middle digits of course leave the ground last and strike it first. Thus the middle toes have been stimulated at the expense of the lateral ones, so that in the Diplarthra, either the mid-

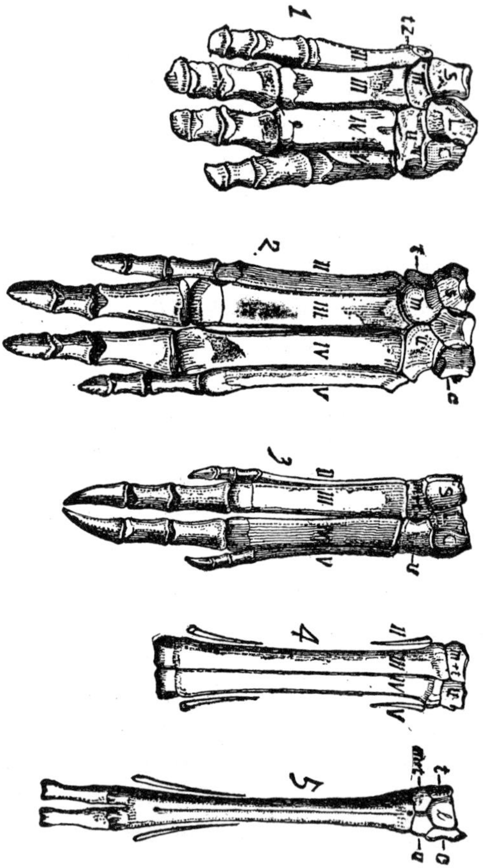

Fig. 86.—Manus of Artiodactyla, much reduced; from Kowalevsky. 1, *Hippopotamus*; 2, *Hyopotamus*; 3, *Sus*; 4, *Gelocus*; 5, *Cervus*.

dle one (Perissodactyla) alone remains, or the middle two (Artiodactyla). In the kangaroos the external toes have been chiefly used, so that the fourth and fifth digits have been principally developed. In man, who now turns his feet out when using them as bases of resistance to muscular labor, the inner digit has become most robust. The mechanical history of the human great toe is however yet unknown.

As regards the equal development of the third and fourth digits in the Artiodactyla, as distinguished from the development of the middle digit of the Perissodactyla, I have advanced the following hypothesis. I have supposed that the primitive members of this former division sprung from pentadactyl plantigrades who dwelt in swamps and walked on very soft ground. The effect of progression in mud is to spread the toes equally in all directions and on each side of the median line. Such feet remain in the mud-loving hippopotamus, and to a lesser degree in the true pigs. From such ancestry the cloven-footed Diplarthra derived their characters. The Hyracotheriinæ, the ancestors of all Perissodactyla, display on the other hand evidence of a life on harder ground, especially in the posterior foot, where articulations are already rigidly defined, and the third digit is longer than the others. Some of their descendants love swamps, as one or two species of tapirs and rhinoceroses, but others live on the dryest ground, as the Andean tapir and the African rhinoceros. As to the highest members of both even and odd toed groups, the Bovidæ and the Equidæ, their habitat is in the vast majority of cases the dry land (Figs. 80–81).

Continued and excessive prehensile strain with weight on the longest digits, must be assigned as the cause of the especial elongation; and disuse as the cause of the loss of the external and shorter digits, of the sloths; so that there remain but two and three (Cholœpus and Bradypus), and in the climbing ant-eater (Cycloturus) but one principal toe and two rudiments. The excessive strain and impact experienced by certain digits in leaping, accounts for the digital reduction in the hinder foot of the kangaroos and jerboas, precisely as in the perissodactyle ungulates.

c. The Horns.

Horns are developed in Mammalia and other Vertebrata on similar parts of the skull, principally on the posterior lateral angles, as in various Batrachia, Reptilia, and Mammalia, and on the nose, as in a few Mammalia and several reptiles, recent and extinct. These parts are the ones which are especially brought into contact with resisting bodies; the nose in pushing a path or way for the head and body; the lateral occipital region in defence and assault, when the sensitive nose and eyes are protected by being held near the ground. In the latter position the posterolateral angles, when present, receive more frequent collision with, and vigorous stimulation from, a body attacked or resisted, and in accordance with the observed results of irritation on dermal and osseous tissues, additional matter has been deposited. In Lacertilia and Batrachia Salientia there are distinct posteroexternal cranial angles; in Batrachia Urodela such angles are less prominent. In unguiculate Mammalia and in all others with a sagittal crest there are no such angles; hence this type of skull has never developed posterior horns. The rhinoceros has developed the dermal nasal horn, and the Elasmotherium, a median osseous horn, since posterolateral angles of the skull are wanting or close together. In the Dinocerata and the Artiodactyla, where the temporal crests are lateral, leaving a wide fronto-parietal plane with posterior lateral angles, horns are developed. In members of both groups horns have been developed over the orbits also (Fig. 87), and in the Dinocerata on the extremities of the nasal bones as well. These growths are all on parts which are subject to especial irritation by contact with other bodies, animate and inanimate.

Among Artiodactyla, the deer (Cervidæ) are especially distinguished by the periodical shedding of all but the bases of their horns. Extinct forms found in the Upper Miocene of the United States and France (the Loup Fork series) furnish the explanation of the origin of this remarkable peculiarity. In the genus Cosoryx we find that the horns may or may not possess a burr near the base of the beam, like that of the deer; the same species being indifferently with it or without it. This observation has been made on three species,—the *C. necatus*, *C. furcatus* and *C. ramosus*. The following explanation of these facts has been proposed by myself.[1] "From the facts of the case the following inference may be derived, premising that it is very probable that a genus allied to the present one has given origin to the family of the deer. It is obvious that the horns of (Dicrocerus) Cosoryx did not possess a horny sheath as in the Bovidæ, from the fact of their being branched. As the sheath grows by addition at the base, the presence of branches which necessarily obstruct its forward movement, would be fatal to the process. There is much to be said in favor of the view that the horns were covered with an integument, probably furred, as in the giraffe and young stage in the deer. Thus there are grooves in the surface of the beam for superficial blood-vessels, which

[1] *U. S. G. G. Survey West of the 100th Mer.*, G. M. Wheeler: IV., *Paleontology*, 1877, p. 348.

must have been protected by skin (I do not observe these grooves on the beam of *C. teres*). The retention of the broken extremity of an antler, so as to be re-united, as described (Fig. 87, *C*), could not have been accomplished without an integument. The presence of the burrs cannot be accounted for on any other supposition, as there are no foramina to give exit to nutrient vessels at the point where they exist; the irregularity of those positions also forbids the latter idea, and adds to the probability that the arteries which furnished the deposit of phosphate of lime were contained in a superficial dermal coating. The supposition is also strengthened by the fact that the only existing ruminants (the giraffes) with permanent horns without horny sheaths have them covered with hairy skin.

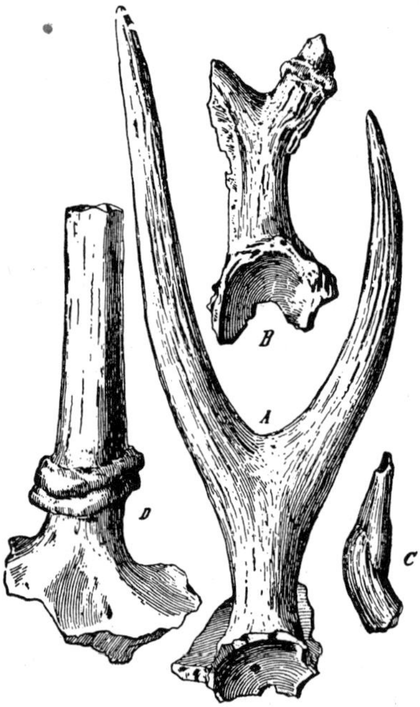

Fig. 87.—*A*, *B*, *Cosoryx necatus* Leidy; *A*, without, *B*, with, burr on antler; *C*, *D*, *Cosoryx ramosus* Cope; *C*, antler broken and reunited; *D*, beam with burr; one-half natural size; original; from *Report U. S. Gov. Geol. Expl. 100th Mer.*, G. M. Wheeler.

"It appears that in the antlers of the Cosoryx the deposit of a burr was immediately associated with the death of the portion of the horn beyond it, so that it disintegrated and disappeared. This was not the case with the beam in the specimens observed. Nevertheless it is probable that the death of the horn would be associated with the deposit of the burr in this case also, were the conditions the same. What those conditions were we can only surmise. It was very probably the death of the integument which invested and nourished the horn that produced that result; and this would more readily occur in the exposed antlers than in the more protected basal portion of the beam. It is very probable that this result would follow blows and laceration of the surface received during combat, or accidental contact with hard substances. The integument would be stripped up to near the junction of the antlers with each other, or of the beam with the cranium, and the arteries would be constricted or closed at those points. It is near these junctions that all of the burrs are found. But as such lesion would be necessarily less complete at the point where the horn has greatest circumference, so the entire death of the horn might be less usual than that of the branches. Should such lesions have occurred for a long period at the breeding season, nature's efforts to repair by redeposit of bony tissue might as readily become periodical as the increase in size and activity of the reproductive organs and other growths which characterize the breeding season in many animals. The subsequent death of the horn would be at some time followed by its shedding by the ordinary process of sloughing."

Cosoryx is not the true ancestor of the Cervidæ, as its teeth have already attained the prismatic type of the higher Bovidæ. But Blastomeryx is most probably the ancestor of the deer. The remains of this genus occur with those of Cosoryx, but the burr has not yet been observed on its horns.

IV. THE MECHANICAL ORIGIN OF DENTAL TYPES.

In investigating the origin of dental types it is necessary to become acquainted with the nature of the mutual movements of the series of the opposing jaws. I have classified them as follows :[1]

I. Inferior molars work within superior molars, but not between them. Psalidodect mastication.

1. The inferior molars shear on the interior side of the superior : *Triconodontidæ.*

II. Part or all of inferior molars work alternately to and between superior molars. Amœbodect mastication.

2. The inferior molar shears forwards on the superior molar. Proterotome mastication : *Creodonta; Carnivora.*
3. The inferior molars shear posteriorly against the superior molars. Opisthotome mastication : *Coryphodontidæ, Uintatheriidæ.*

III. Molar teeth of both jaws oppose each other. Antiodect mastication.

4. The movement of the lower jaw is vertical. Orthal mastication : *Suoïdea, Tapiridæ.*
5. The movement of the lower jaw is from without inwards. Ectal mastication : many *Perissodactyla.*
6. The movement of the lower jaw is from within outwards. Ental mastication : most *Artiodactyla;* some *Perrissodactyla.*
7. The movement of the lower jaw is from before backwards. Proal : some *Monotremata Multituberculata* and most *Glires.*
8. The movement of the lower jaw is from behind forwards. Palinal : *Proboscidea* (Ryder).

The distinction of teeth into incisors, canines, and molars appears independently at various points in the line of Vertebrata. Incisors and molars are distin-

[1] *Mechan. Origin Hard Parts of Mammalia*, 1889, p. 226.

guished in sparoid fishes, and in placodont and diadectid reptiles. Canine-like teeth, or pseudo-canines, appear in clepsydropid and crocodilian reptiles, and in saurodont fishes. Canine-like incisors appear in the Clepsydropidæ. The variety of character in these structures presented by the Mammalia to be considered is great, and the principles deduced from observation of them are applicable to the Vertebrata in general.

As mechanical causes of the origin of dental modifications, I have enumerated the following:

1. Increase of size of a tooth, or a part of a tooth, is due to increased use, within a certain maximum of capacity for increased nutrition.

2. The change of direction and use of a tooth take place away from the direction of greatest, and in the direction of least resistance.

3. It follows, from their greater flexibility, that crests of crowns of teeth yield to strains more readily than do the cusps.

4. The increase in the length of crests and cusps in all directions, and therefore the plications of the same, is directly as the irritation from use to which their apices and edges are subjected, to the limit set by the destructive effects of such use, or by the recuperative energy of nutrition.

5. The direction of growth of the branches of a V, or of the horns of a crescent, will be the direction of movement of the corresponding parts of the opposite jaw.

Before giving a review of the various dental types of Mammalia, I wish to describe some special examples where the effect of mechanical causes is most obvious. I therefore first repeat the observation of Ryder as to the origin of the selenodont molars of the Artiodactyla; and my own as to the origin of a similar structure in the molars of certain multituberculate Prototheria. In the former the mastication is ental; in the latter it is proal, as shown by Osborn.

In the accompanying figure from Ryder the movements of the lower jaw in mastication of lophodonts, are diagrammatically represented. *a* represents the movement in Carnivora, and in the orthal bunodonts, as the pigs. *b* shows a slight lateral movement believed by Ryder to exist in the wart hog (Phacochœ-

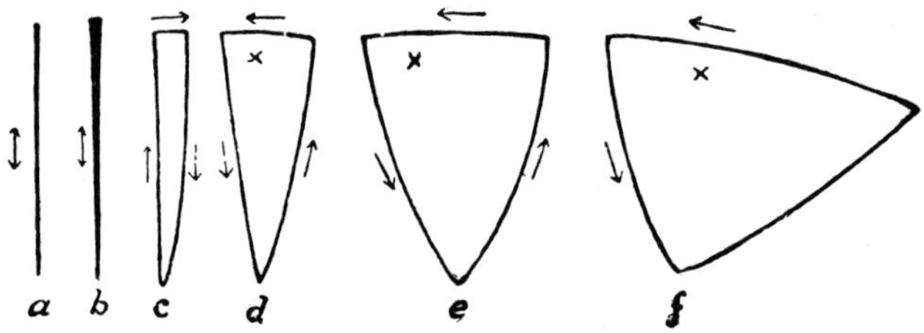

Fig. 88.—Diagram of excursion of lower jaw in mastication; from Ryder; *a–b*, orthal; *c–f*, ental.

rus). *c* represents the movement in kangaroos, phalangers, and tapirs. In *d* a theoretical intermediate movement is represented, such as Ryder supposed to have characterized the Anchitherium. In *e* the usual movement among ruminants is depicted, as is seen in the deer, etc. In *f* the wider excursion of the jaw is that seen in the giraffe, camel, and ox. In these movements from *b* to *f*, the lower jaw is moved transversely across the upper jaw from one side to the other. Some of the Diplarthra masticate on one side of the jaw when performing this movement, and some on the other. That is, in passing the lower jaw across the face of the upper, some masticate the food on the side where the external face of the lower jaw crosses the upper jaw from within outwards (ental); while in other types the food is masticated on the side where the lower jaw passes the external edge of the upper jaw from without inwards (ectal). While masticating with one side of the jaws, the opposing dental series of the other side are not in contact. All mutual effect of the teeth of one jaw on the other could therefore appear on the side temporarily used for mastication only. Among recent Ungulata the ruminants present the ental mastication; the rhinoceros and horses, the ectal; and rodents, the proal. Ryder is of the opinion that the mastication of the Proboscidia is palinal, but I have not been able to satisfy myself of this.

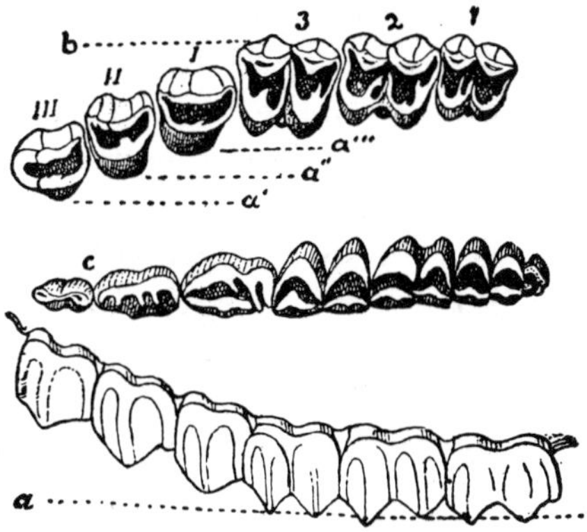

Fig. 89.—Cervus, molars: *a*, superior, external view; *b*, *do.* inferior view; *c*, inferior molars, superior view; from Ryder.

When the crests of the inferior molars were developed, their relation to the crests of the superior molars was always anterior in mastication. That is, the inferior crest, in the closing of the jaw, collides with the crest of the upper molar, with its posterior edge against the anterior edge of the latter. This is because: *first*, as to position, the two anterior cusps of the lower molar are the remains of the anterior triangle which fit originally between two superior molars, and because, in the closing of the jaw, these cusps continue to hold that position; and *second*, as to function, because the canine in the ungulate series diminishes in size, and does not, therefore, draw the inferior molars forwards by wedging on the superior molar, as in the Carnivora, but allows free scope to the posterior traction of the temporal muscle in its exercise on the lower jaw.

In those forms which masticate from the inside outwards (the ental type), the cusps of the inferior molars, passing between those of the superior molars, would tend to flatten the sides on which they exerted friction, and to extend those sides outwards beyond the median apex of the cusp. (Fig. 90.) The result would be, and taking into view the yielding of the tissue to

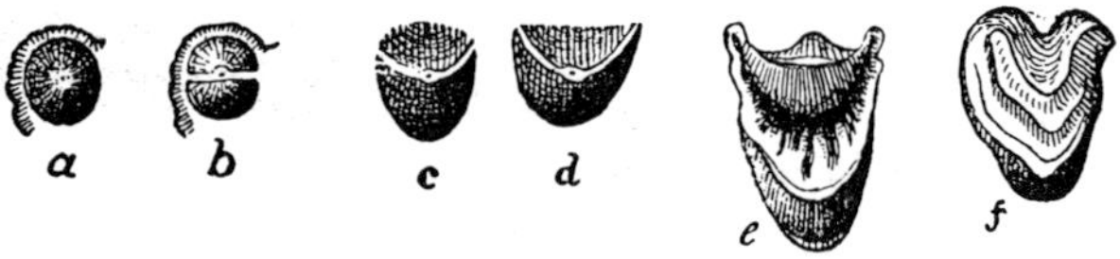

Fig. 90.—Cusps of superior premolars and molars: *a*, external cusps of molar of Sarcothraustes; *b*, of Phenacodus; *c*, of Anthracotherium; *d*, of Oreodon; *e*, half of inferior molar of Cervus; *f*, superior premolar of Coryphodon; from Ryder.

such strain, has been, to modify the shape of the cusp by pushing its side walls, so that a horizontal section of it would become successively more and more crescentic. The effect on the inferior teeth would be to produce the same result in their external cusps, but in the opposite direction. The sides of the cusps would be pushed inwards, past the apex, giving a crescentic section more or less perfect, as the operation of the cause had been of long or short duration. The result of the lateral movement in mastication may be understood by reference to the accompanying cut, Fig. 91. The external crescents of the inferior molars (*b*) are seen to pass between the internal crescents of the superior molars (a). The mutual interaction and effect on the form of the crescents may be readily understood. In Fig. 90 the successive stages of this effect

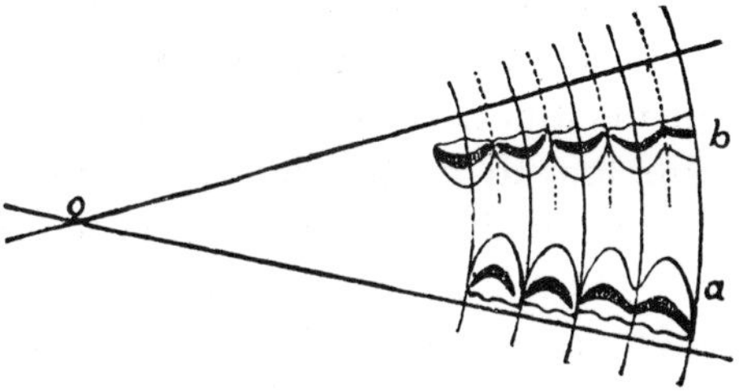

Fig. 91.—Two true molars of both jaws of a ruminant: *a*, superior molars, their inner crescents; *b*, inferior molars, their external crescents; the arcs show directions of motion of jaws in mastication; from Ryder.

on one or two cusps may be seen, beginning with a cone (a) and terminating with crescents (ef). Thus is the origin of the selenodont dentition of the highest

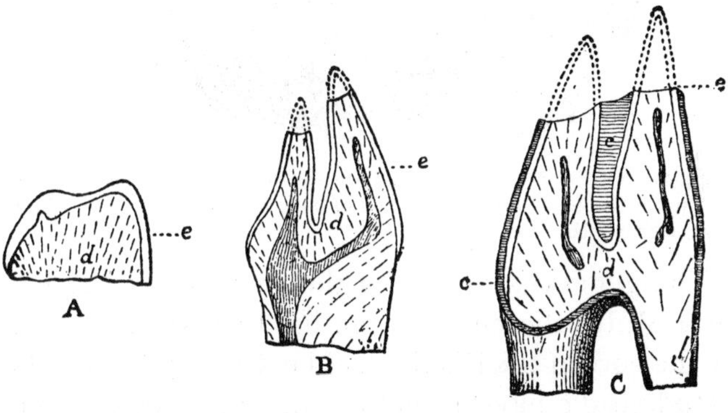

Fig. 92.—Transverse vertical sections of superior molar teeth, showing transition from bunodont (*A*) type to lophodonts (*B*, *C*). *A*, *Sus erymanthius*. *B*, *Ovis amaltheus*. *C*, *Bos taurus*. From Gaudry. Letters: *d*, dentine; *e*, enamel; *c*, cementum.

artiodactyle explained by Ryder, and, I believe, correctly.

Kowalevsky and I have shown that the types with selenodont (crescent-bearing) molars, have descended from tubercle-bearing (bunodont) ancestors. This descent has witnessed an increased depth of the infolding of the crown, as represented in the accompanying figure. Now, in the table of masticatory types above given it is shown that in the bunodont type of the Suoidea the mastication is orthal, and a gradual increase in the length of the lateral excursions of the lower jaw has been shown to have resulted in the ental mastication. Thus has structure kept pace with function in the evolution of the selenodont dentition.

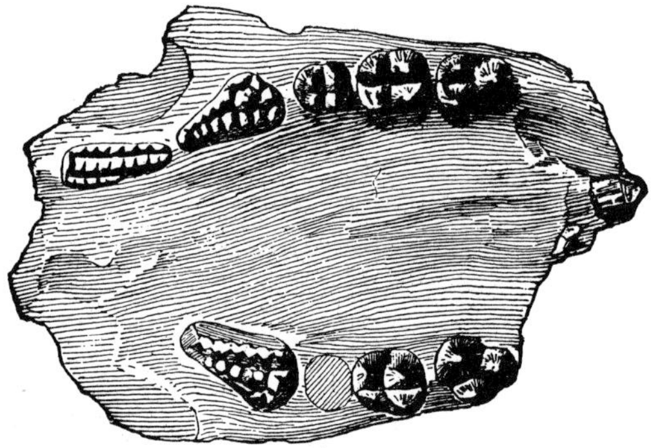

Fig. 93.– *Chirox plicatus* Cope, palate and molar teeth from below, three-halves natural size. From Puerco bed of New Mexico. From *American Naturalist*, 1887, p. 566.

The general structure of the dentition in the Prototheria Multituberculata is similar to that of the Glires. The incisors in the Plagiaulacidæ, Chirogidæ, and Polymastodontidæ have structure and functions generally similar to those of that order. The result in the form and function of the molar dentition has been similar to that observed in the Glires. The postglenoid process is probably absent in these animals; in any case the mandibular condyle is rounded, and is not transverse. Prof. H. F. Osborn has pointed out to me that mastication was performed by a fore-and-aft movement of the inferior molars on the superior in Plagiaulacidæ. This was no doubt the case in the other families named. The molar teeth of the lower types, as Tritylodon from the Trias, present conical tubercles in longitudinal series, two in the lower and three in the upper jaw. The two series of the lower jaw alternate with the three in the upper jaw, moving in the grooves between the latter, while the three series of the upper molars reciprocally embrace the two of the lower molars. This is demonstrated by the mutual wear of the tubercles seen in Ptilodus and Chirox (Fig. 93). The trituration was probably the same in Tritylodon, but in Polymastodon the increased thickening of the tubercles prevented their interlocking action in mastication. In this genus the tubercles slid over each other, and truncated the apices until in old specimens they were entirely worn away. In Meniscoëssus (Fig. 94) and Stereognathus we have an interesting illustration of the effect of the action of cusps on each other when under prolonged mutual lateral thrust. Their external sides have been drawn out into angles in the direction of thrust, converting their transverse sections from circles to crescents. As the thrust is in the Multituberculata longitudinal, the crescents are transverse to the axis of the jaw. In the selenodont Artiodactyla, where the thrust is transverse to the line of the jaw, the crescents are longitudinal. That similar effects should accompany similar move-

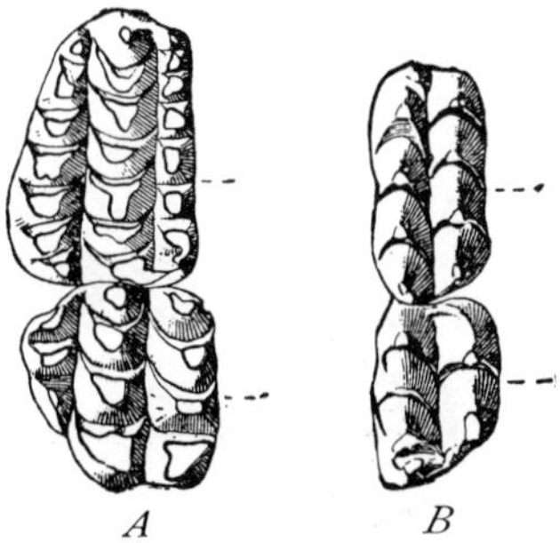

Fig. 94.—*A*, *Meniscoëssus conquistus* Cope, last two superior molars, from the Laramie of Wyoming, twice natural size. *B*, *Meniscoëssus*, second species, from Osborn.

ments in two groups of Mammalia so widely separated as these two is strong evidence in favor of the belief that the two facts stand in the relation of cause and effect.

I now present an example of the effect of strain, as shown by the direction of the inferior incisors of the lemurine Quadrumana. These teeth project horizontally from the extremity of the mandible, so as not to oppose the superior incisors, in consequence of which they are useless as organs of prehension. But they

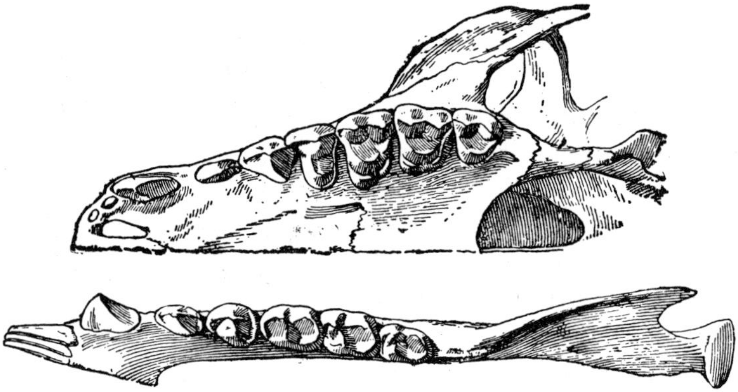

Fig. 95.—*Lemur collaris*, dentition from below and above; natural size; original.

are used by their possessors as a comb for the fur, drawing them from below upwards when thus employing them. The strain is always in one direction, and must have resulted in developing the procumbent position which they now display (Fig. 95). This is a direct deduction from the fact that the incisor teeth are similarly displaced by the pressure of the tongue in cases of the abnormal enlargement of that organ in man.

I now describe the general character of mammalian dentition, with the view of pointing out how strong, in the light of the facts already cited, is the evidence of their origin through mechanical strains and impacts.

a. The Origin of Canine Teeth.

The origin of canine, pseudo-canine, and canine-like incisor teeth is due to the strains sustained by them on account of their position in the jaws at points which are naturally utilized in the seizing of prey, or the fighting of enemies. In some reptiles (Dimetrodon) the end of the muzzle has been utilized ; in crocodiles, the side of the jaw ; while the intermediate position has been most used by Mammalia. The reason why the canine instead of the incisor teeth have been selected by carnivorous Mammalia for prehensile purposes is not at present clear to me. In accordance with Rule I., its increased size has been due to the especial and energetic strains to which it has been subjected while in use as a prehensile or offensive weapon, when buried in the body of its prey or enemy. The superior canine would acquire larger size earlier in time than the inferior canine, since it bears the greater part of such strain, as attached to the more fixed head and body of its possessor. The anterior teeth of the lower jaw would be less available for use, since they offer weaker and less fixed resistance to the opposing body. That the first tooth behind the canine was not generally enlarged is (under I.) due to the fact that its posterior position prevents it from having the same amount of use, and experiencing the strain that a tooth more anteriorly placed necessarily receives. It is excluded from considerable use by the projecting muzzle above and in front of it. That it was not drawn out into a horizontal position was due to the presence of teeth anterior to it.

That the increased size of canine teeth is due to strains is strongly indicated by the huge development of these teeth in the walrus. This animal uses its canines for the breaking of ice, and for lifting itself from the water on to the edge of strong ice. The fact that canines and not incisors have been thus developed is a necessary result of the fact that the walrus is a descendant of a line of animals which had already reduced incisors and larger canines.

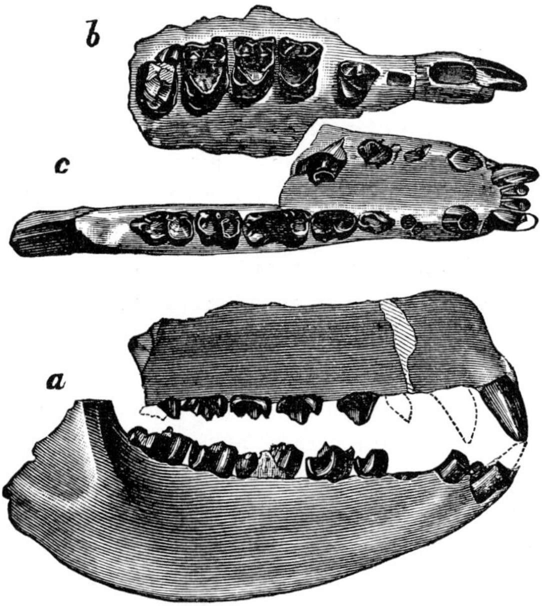

Fig. 96.—*Esthonyx burmeisterii* Cope, dentition: *a*, profile; *b*, superior; *c*, inferior dentition, grinding faces. Reduced.

b. Development of the Incisors.

The history of the incisor teeth of the Mammalia exhibits three processes, viz.: hypertrophy (e. g. Glires), specialization (e. g. Galeopithecus, Lemuridæ), and atrophy (e. g. Boöidea, Phacochœrus, Glossophaga, etc.).

Of hypertrophy we have two types: the first represented by the Glires, Multituberculata, Tillodonta and their ancestors; and second, by the Proboscidia, the narwhal and certain Sirenia. As the uses of the incisors present two types corresponding with their structure, we have ground for believing the uses in question to have been the efficient agent in producing the latter. Esthonyx furnishes us with an example (Fig. 96) where all the incisors are present in the lower jaw, and

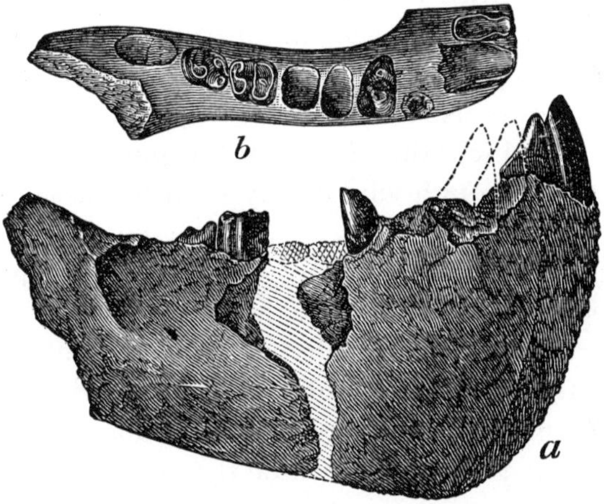

Fig. 97.—*Psittacotherium multifragum* Cope, mandibular ramus, one-half natural size; *a*, profile; *b*, from above.

where the function of one pair of them (the second) has evidently been partially rodent in character; that is, it has served as a scraper and gouger of food substances. Persistent use has apparently developed the size of this pair of teeth, until we find in Psittacotherium (Fig. 97) they have reached a greater efficiency, and that the external incisors of the lower jaw have disappeared. This disappearance can be accounted for on the ground of disuse, a retirement from service due to position, and the increased growth of incisor

No. 2. In Calamodon the first incisor has become rudimentary from the same cause, and in Anchippodus it has disappeared altogether, leaving a truly rodent incisor dentition, consisting of the second incisors only, in the lower jaw. Continued use as chisels has developed these teeth to the great proportions seen in such Glires as Castoroides, etc. (Fig. 107).

The use which the Proboscidia and Sirenia (Halicore) give their incisors, is, from a mechanical point of view, like that which the Carnivora give their canines; that is, it consists of impacts in the long axis, and strains transverse to the long axis of the tooth. The elephants use their tusks for prying up the vegetables on which they feed, or for pushing aside the vegetation through which they wish to pass. The ancestors of the Proboscidia are not certainly known, but they possessed incisors of enlarged proportions, such as we find in the Toxodontia and other late representatives of some of the primitive Ungulata. Use of such teeth in the manner referred to, without opposition from the inferior incisors, will account for the tremendous proportions which they ultimately reached in some of the species of Elephas.

The use made by the narwhal of its single huge superior incisor, that of an ice-breaker, indicates the origin of its large dimensions. So with the straight incisors of the hippopotamus; use as diggers has straightened them to a horizontal from their primitive vertical direction, a change which is also partially accomplished in the true pigs (Sus).

In the Sirenian genus Halicore the upper incisors have been used in excavating vegetable growths from the banks and bottom of shallow seas. The transition from three incisors (Prorastomus) to two (Dioplotherium), and to one (Halicore), is identical with what has taken place in the Proboscidia and Glires, and has resulted in the production of an effective digging-tool. In other genera it may be supposed that their habits of browsing on soft growing materials did not necessitate the use of digging incisors, hence these teeth became atrophied, as in the manatee and Rhytina.

c. Development of Molars.

In fishes and reptiles where teeth occasionally present very primitive conditions, the theory of the origin of particular types of molar teeth is more simple than in the case of Mammalia. The observations of Hüter on the action of osteoblasts under stimulus show that under moderate irritation osseous tissue is deposited, while under severe pressure osseous tissue is removed. Koelliker has shown that the action of these bodies is the same in dentine as in true bone. Hence modifications of dental structure must stand in close relation to the uses to which they are put. Thus severe pressure on a simple tooth crown would, if long continued, cause it to expand laterally, or in the direction of least resistance, and to grow but little in its vertical axis, i. e., in the direction of greatest resistance. The molar teeth have been subjected to much more severe direct irritation from use than any others in the jaws, and this will account for their increased diameters.

In the case of the eutherian Mammalia, molar teeth are not traceable back to ancestral types of reptilian molars, but to simple conic (haplodont) reptilian teeth. The process of the evolution of the complex mammalian molars from these, forms the subject of a paper in the *American Journal of Morphology* for 1889, from which I quote extensively in the present work.

I have there shown that the greater number of the types of this series have derived the characters of their molar teeth from the stages of the following succession. First a simple cone or reptilian crown, alternating with that of the other jaw (haplodont type). Second, a cone with lateral denticles (the triconodont type). Third, the denticles to the inner or outer side of the crown, forming a three-sided prism, with tritubercular apex, which alternates with that of the opposite jaw (tritubercular type). Fourth, development of a heel projecting from the posterior base of the lower jaw, which meets the crown of the superior, forming a tuberculosectorial inferior molar. From this stage the carnivorous and sectorial dentition is derived, the tritubercular type being retained. Fifth, the development of a posterior inner cusp of the superior molar and the elevation of the heel of the inferior molar, with the loss of the anterior inner cusp. Thus the molars become quadritubercular, and opposite. This is the type of many of the Taxeopoda, including the Quadrumana and Insectivora as well as the inferior Diplarthra. The higher Taxeopoda (Hyracoidea) and Diplarthra, add various complexities. Thus the tubercles become flattened and then concave, so as to form V's in the section produced by wearing; or they are joined by cross-folds, forming various patterns. In the Proboscidia the latter become multiplied so as to produce numerous cross-crests.

d. Origin of the Carnivorous Dentition.

The anterior cusplet of the triconodont crown is (Fig. 98 *A*), in the upper jaw, the paracone, and in the lower jaw the paraconid; and the posterior cusplet is the metacone or metaconid, respectively. As the principal cusps, or protocone and protoconid, alternate with each other, the cusplets stand opposite to them in the closing of the jaws, and a certain amount of interference results. As the lesser cusps are the less resistant to the wedging pressure of such contact, their position would change under its influence, rather than the large central cusps. The lower jaw fitting within the upper, the effect of the collision between the major cusps of the one jaw, and cusplets of the other, would be to emphasize the relation still more ; that is, the cusplets of the upper jaw would be wedged outwards, while those of the lower jaw would be pressed inwards, the major cusps retaining at first their original alternate position. With increase of the size of the teeth the cusps would soon assume in each jaw a position more or less transverse to that of the other jaw, producing, as a result of the crowding, a crown with a triangular section in both. The process may be rendered clear by the following diagram :

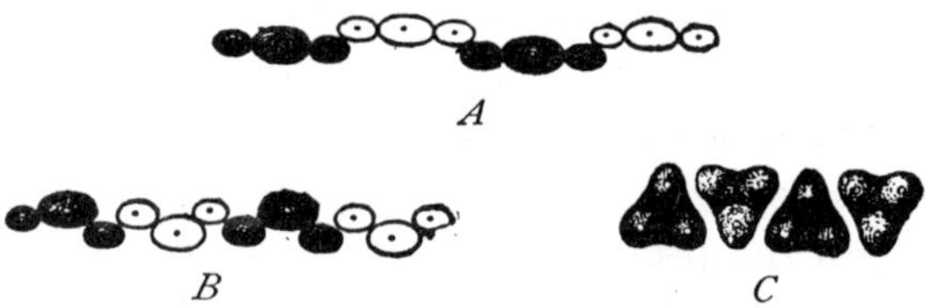

Fig. 98.—Diagrammatic representations of horizontal sections of tricuspidate molars of both jaws in mutual relation ; the shaded ones represent those of the upper jaw : *A*, Triconodon ; *B*, Menacodon ; *C*, ideal tritubercular molars, approached by Menacodon, *B*.

It is supposed on the contrary by Röse and Kükenthal that mammalian molars which support more than one cusp have been formed by the fusion of several simple reptilian cones. So far as regards the higher Mammalia this hypothesis is in opposition to all the facts of paleontology and is not worthy of discussion.

The only question that can arise is with reference to the origin of the multituberculate molar of the Prototheria.

It is further questioned by Forsyth-Major, whether the tritubercular molar has been derived from the triconodont. He believes, on the contrary, that it is derived from the multitubercular type by reduction. There are two objections to this view: (1) the cones of the tritubercular tooth or trigon should be subequal, were they derived from a multitubercular source. On the contrary, the two external cones of the upper, and the two inner cones of the lower series are in the earliest (Jurassic), as well as most of the tritubercular types, smaller than the single opposite cusp or protocone, precisely as are the anterior and posterior cones of the triconodont molar. (2) No paleontologic series from the multitubercular to the tritubercular types has been traced, while the series from the triconodont to the tritubercular is well known. Forsyth-Major's evidence that such a transition exists in the Glires, is better explained by tracing the moderate complexity he describes to a tritubercular origin.

It is also alleged by Allen and Scott that the internal cusps of the premolars, when present, originate by the development of internal cingula, and have no primitive tritubercular ancestry. The evidence at our disposal from paleontological sources is in favor of this view; hence it is reasoned that the history of the molar teeth must have been identical. This however does not follow, especially as the paleontologic evidence points the other way. The history of the two series has been different. In the first place the premolars have been subjected to much less use than the true molars; hence they retained the primitive reptilian simplicity for a much longer period, a simplicity which they retain in the Carnivora, except the $\underline{1}$, which became the sectorial. Secondly, the premolars, instead of increasing in size, have in many types decreased; the Diplarthra alone presenting an exception to this rule. That the internal cusps of the premolars may have arisen by growth of cingula in this order, is by no means improbable. We seem to have here an excellent illustration of the origin of two identical structures by different evolutionary routes.

The first modification of the tritubercular molar of the lower jaw is the addition of a low cingulum at the posterior base. This is seen in a rudimentary condition in various living species of the Centetidæ and Chrysochloridæ of the insectivorous order (Fig. 100); but in these existing forms the superior molar has added a posterior cingulum also, which widens internally, or towards the palate (Fig. 101). In the evolution of the dentition, the inferior posterior cingulum, or "heel," was developed first, as in the Deltatherium, Centetes, and Stypolophus (Figs. 99, 100, 102), where it is quite large; while the superior cingulum is wanting in Stypolophus and Didelphodus, but is present in a very rudimentary condition in *Deltatherium fundaminis*. In all of these genera the external cusps of the superior series have been pressed inwards, and more or less together, and are therefore removed in this respect from the primitive condition. The more primitive state of the superior cusps is seen in some species of Mioclænus, where, however, a posterior cingulum may be developed. The primitive type of tritubercular superior molar is that of Sarcothraustes, and in the same genus the inferior molar only differs from the primitive type in having a well-developed heel. Among recent Mammalia the carnivorous and insectivorous Marsupialia generally have the tritubercular lower molar with heel. In the Chiroptera and many Insectivora the heel is largely developed, and supports two cusps, as it does in some Creodonta.

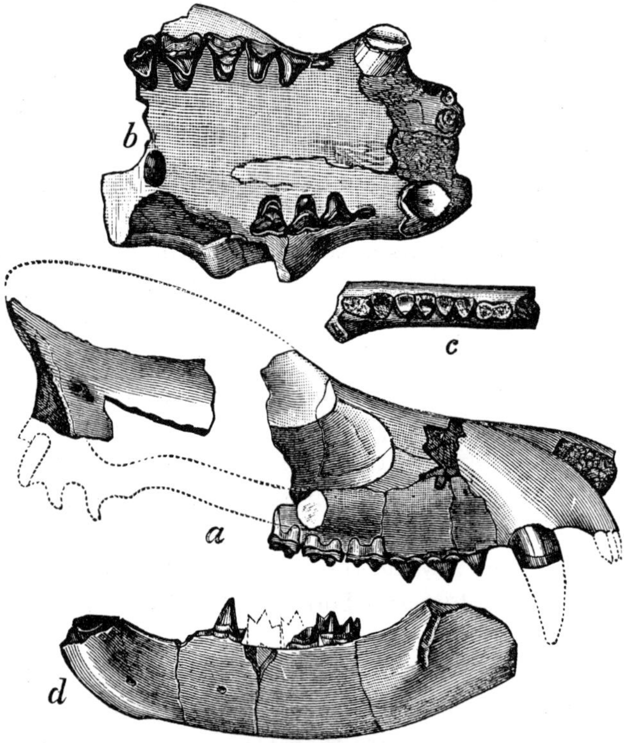

Fig. 99.—*Deltatherium fundaminis* Cope, fragmentary skull, two-thirds natural size; from the Puerco bed of New Mexico. *a*, *b*, *c*, from one individual; *d*, from a second animal; *a*, right side of cranium; *b*, palate from below; *c*, mandible, part from above; *d*, left ramus, outer side; from the *Report of the U. S. Geol. Surv. Terrs.*, Vol. III.

From this point the evolution of the tritubercular molar must be considered from two standpoints. The first is the mechanical cause of the changes of its form; and the second is the mechanical cause of its definite location in a particular part of the jaw. For it has been already stated that in the evolution of the sectorial dentition of the Carnivora, the number of molars and premolars has considerably diminished, while those that remain have become relatively much larger.

In the tritubercular dentition the crowns proper of one jaw alternate with those of the other (Fig. 100); but when heels are added in either jaw, they will oppose such part of the crowns of the teeth in the opposite jaw as comes in contact with them when in use.

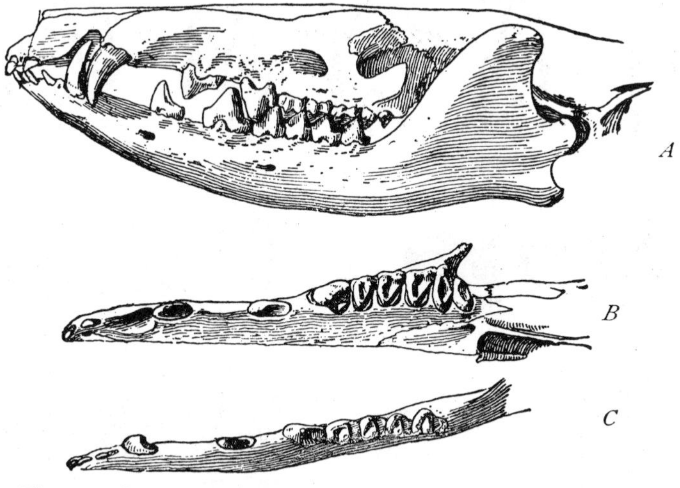

Fig. 100.—*Centetes ecaudatus*: *A*, skull, side seen obliquely from below; *B*, superior molars from below; *C*, inferior molars from above.

The development of the heel in the inferior molars produced a type which is known as the tuberculosectorial. This type characterizes the Creodonta and a few Carnivora. In the former there are generally three such teeth, in the latter but one.

In the tuberculosectorial type of inferior molar the primitive tritubercular part of the crown (trigonid of Osborn) stands principally anterior to the posterior root of the tooth. It appears that the posterior root has been extended backwards, so as to occupy a position below the middle of the superior molar, while the tritubercular crown has been confined to the space between the crowns of the superior molars. This would follow of necessity from the alternating action of the crowns of the opposite series, in connection with a general increase in size of the teeth. In the opening of

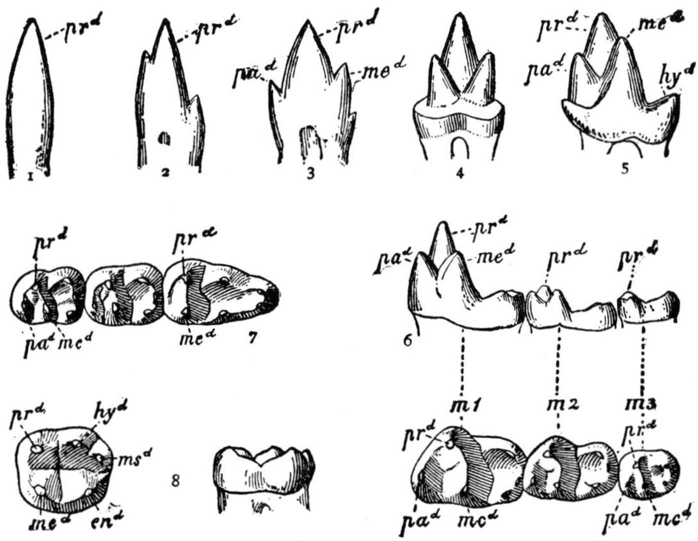

Fig. 101.—Series of inferior molar crowns representing the transition from the simple (haplodont) to the quadritubercular. From Osborn.

the jaws in a Creodont, the elevated portion of the inferior crown shears by its posterior face against the anterior face of the superior molar, thus restraining its extension posteriorly. The stimulus of use, however, develops a low extension posteriorly, or a heel, which covers the posterior root, and opposes in mastication the internal extremity or tubercle of the crown of the superior molar above it. Thus a molar element in mastication is added to the sectorial in some Creodonta, and in Canidæ and Ursidæ, etc., among Carnivora. This function predominates over that of the anterior triangle in the Lemuridæ. (Fig. 95.)

I have already pointed out the successive modifications of form which have resulted in the existing specialized single inferior sectorial tooth of the Felidæ. They consist in the gradual obliteration of the posterior-internal cusp, and of the heel, and the enlargement of the external and anterior internal tubercles of the primitive triangle. The modification in the character of the dentition taken as a whole was shown to consist

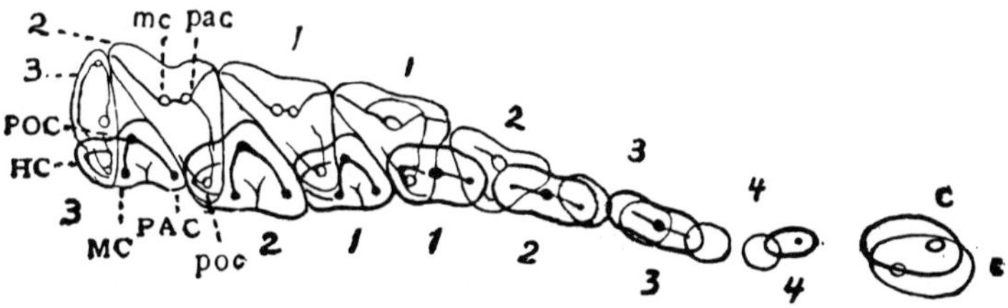

Fig. 102.—*Stypholophus whitiæ* Cope; diagram representing the apposition of the inferior and superior molars. The superior are in light, the inferior in heavy lines. The numbers represent the molars and premolars: *C*, canine; *poc*, protocone; *pac*, paracone; *mc*, metacone; *POC*, protoconid; *PAC*, paraconid; *MC*, metaconid; *hc*, hypocone; *HC*, hypoconid.

in the reduction of the number of the teeth, including the sectorials, until in Felis, etc., we have almost the entire function of the molar series confined to a single large sectorial in each jaw.

The genesis of the superior sectorial tooth has been explained as follows. In consequence of the fact that the lower canine tooth shuts anterior to the superior canine, the result of the enlargement of the diameters of those teeth will be to cause the crowns of the inferior teeth to be drawn from behind forwards against those of the superior teeth when the jaw is closed (Fig. 102). Thus a shearing motion would result between the anterior external edge of the lower triangle and the posterior internal edge of the superior triangle. Now the characters of the true sectorial teeth consist in the enormous extension of these same edges in a fore and aft direction, the inferior shutting inside of the superior. To account for the development of these blades we must understand that the oblique pressure of the front edge of the lower tooth, on the hind edge of the superior tooth, has been continued for a very long time. We must then observe that the internal tubercle of the superior triangle has been pushed continually forwards and been reduced to a very small size. Why should this occur? Why should not the corresponding tubercles of the inner side of the lower crown have been pushed backwards, since action and reaction are equal? The reason is clear: The superior tubercle is on the internal apex of the trigon, and is supported by but one root, while the resistant portion of the inferior crown is the base of the trigonid, and is supported by two, thus offering twice the resistance to the pressure that the superior does. But why should the anterior part of the inferior tooth move forwards? even if it be in the direction of least resistance? This is due to the regular increase in size of the teeth themselves, an increase which can be traced from the beginning to the end of the series. And this increase is the usual result of use (Fig. 102).

The mechanics of the above proposition I believe to be correct, but I have had occasion to modify the statement as to the initiatory cause of the process. In many primitive Ungulata the canines have been as well developed as in the Carnivora, yet the forward pressure of the inferior molars on the superiors has not resulted, or has not been sufficient to produce sectorial molars in those types. In the Amblypoda, the lower molars even shear backwards on the upper ones. It seems then that this growth of the canines is not in all instances sufficient to cause a proterotome mastication. I suspect that the more usual cause is to be found in the voluntary effort of the primitive flesh-eater, to masticate flesh by the manipulation of his lower jaw and the body to be divided. The presence of the inferior canine forbids a posterior shearing movement of the molars, so that the anterior shear is the only one possible to most of the Creodonta. The ab-

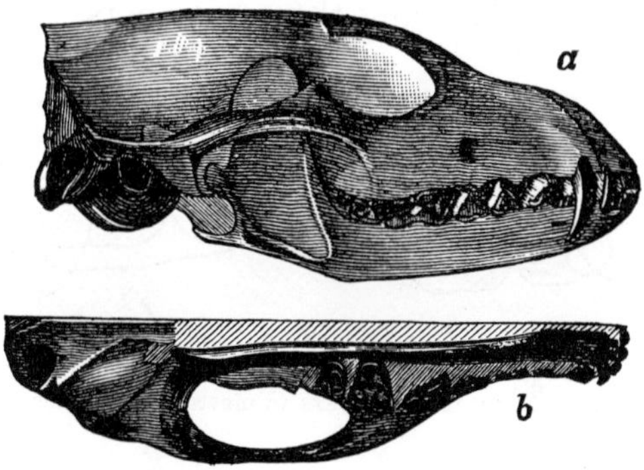

Fig. 103.—*Cynodictis geismarianus* Cope; skull one-half natural size: *a*, right side; *b*, left side from below.

sence of preglenoid crest in primitive Creodonta will permit a manipulation such as we observe in various ungulates to-day. The formation of a habit of a proterotome mastication would result, and the structural results would succeed as above pointed out.

The excess of the forwards pressure of the inferior teeth against the superior over any backwards pressure, has left the posterior internal cusp of the triangle of the inferior molar (metaconid) without contact or consequent functional use. It has, consequently, gradually disappeared, having become small in the highest

Canidæ, and wanting in some Mustelidæ, and all Felidæ. The heel of the same tooth has had a similar history. With the diminution in size of the first superior tubercular, with which it comes in opposition in mastication, its functional stimulus also diminished; and it disappeared sometimes a little sooner (Felidæ) and sometimes a little later (Hyænidæ) than that tooth.

The specialization of one tooth to the exclusion of others as a sectorial, appears to be due to the following causes. It is to be observed in the first place that when a carnivore devours a carcass, it cuts off masses with its sectorials, using them as shears. In so doing

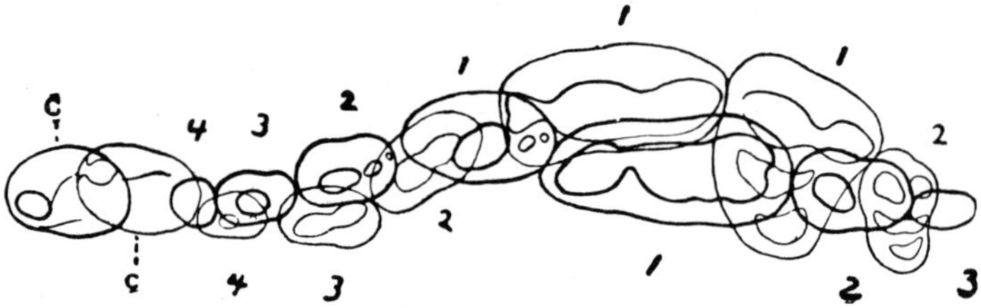

Fig. 104.—*Aelurodon sævus* Leidy; diagram representing coadaptation of crowns of superior and inferior molars in mastication; lines and lettering as in Fig. 102.

it brings the part to be divided to the angle or canthus of the soft walls of the mouth, which is at the front of the masseter muscle. At this point the greatest amount of force is gained, since the weight is thus brought immediately to the power, which would not be the case were the sectorial situated much in front of the masseter. On the other hand, the sectorial could not be situated farther back, since it would then be inaccessible to a carcass or mass too large to be taken into the mouth.

The position of the sectorial tooth being thus shown to be dependent on that of the masseter muscle, it remains to ascertain a probable cause for the relation of the latter to the dental series in modern Carnivora. Why, for instance, were not the last molars modified into sectorial teeth in these animals, as in the extinct Hyænodon, and various Creodonta. The answer obviously is to be found in the development of the prehensile character of the canine teeth. It is probable that the gape of the mouth in the Hyænodons was very wide, since the masseter was situated relatively far posteriorly. In such an animal the anterior parts of the jaws with the canines had little prehensile power, as their form and anterior direction also indicates. They doubtless snapped rather than lacerated their enemies. The same habit is seen in the existing dogs, whose long jaws do not permit the lacerating power of the canines of the Felidæ, though more effective in this respect than those of the Hyænodons. The usefulness of a lever of the third kind depends on the approximation of the power to the weight; that is, in the present case, the more anterior the position of the masseter muscle, the more effective the canine teeth. Hence it appears that the relation of this muscle to the inferior dental series depended originally on the use of the canines as prehensile and lacerating organs, and that its relative insertion has advanced from behind forwards in the history of carnivorous types. Thus it is that the only accessible molars, the fourth above and the fifth below, have become specialised as sectorials, while the fifth, sixth, and seventh have, firstly, remained tubercular as in the dogs, or, secondly, have been lost, as in hyænas and cats.

The reduction of the number of molars in relation to the increase in the size of the canines commenced as early as the Jurassic period. It is seen in the genera Triconodon (Owen) and Paurodon (Marsh), where

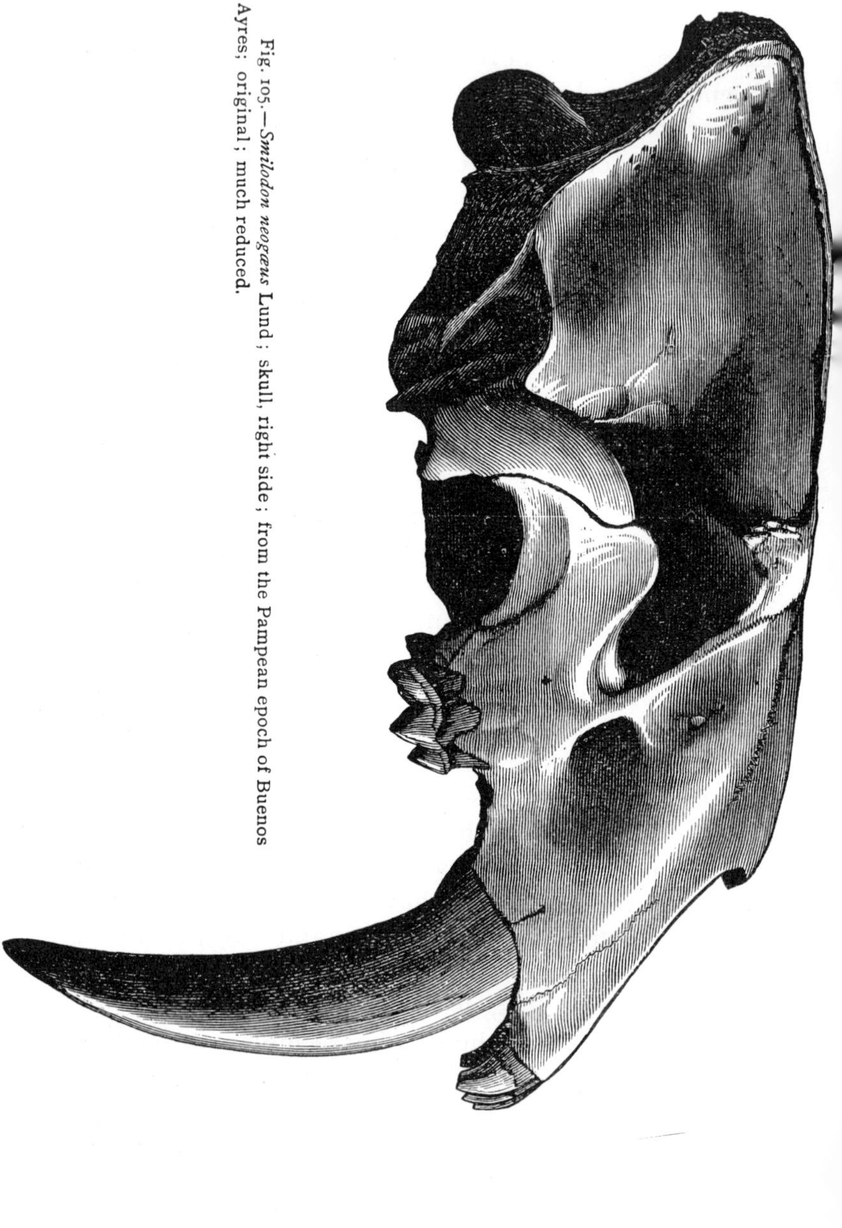

Fig. 105.—*Smilodon neogæus* Lund; skull, right side; from the Pampean epoch of Buenos Ayres; original; much reduced.

the canines are large and the molars few. In the Plagiaulacidæ a similar relation is seen between the development of the incisors and the reduction in number of the molars. This is the modification of relation

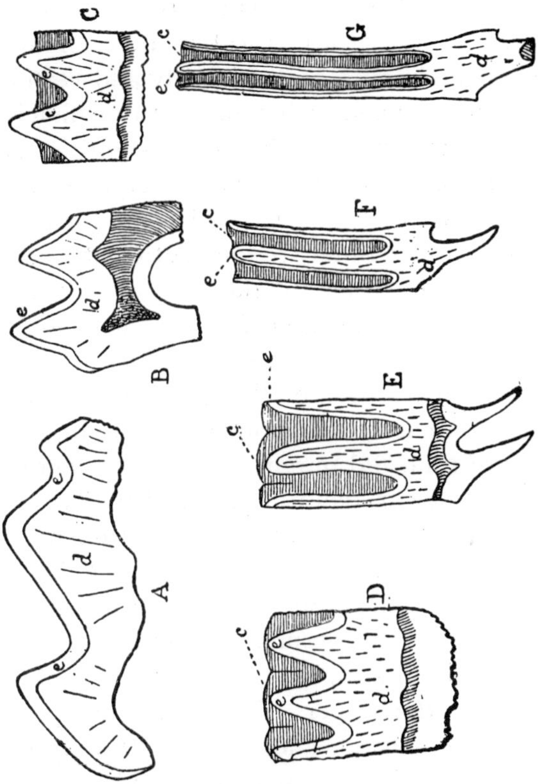

Fig. 106.—Sections of crowns of molars of Ungulata from Dinotherium to *Elephas primigenius*. *A, B,* Dinotherium; *C, D,* Mastodon; *E, F, G,* Elephas. From Gaudry.

observed in existing Mammalia of the orders Proboscidia and Glires, which will be mentioned later, under the head of proal dentition.

e. Origin of the Dental Type of the Glires.

The peculiarities of the rodent dentition consist, as is well known, in the great development of the incisors;

the loss of the canine and of all but one, or rarely of two, of the premolars, which leave a wide diastema; and the posterior position of the molar teeth, as relates to the rest of the skull. A peculiarity which belongs to the highest types of the order is the prismatic form of the molars, and the deep inflection of their always transverse enamel folds both laterally and vertically. A peculiarity of the masticating apparatus, which is the basis of distinction from the bunotherian order, is the lack of postglenoid process, and the consequent freedom of the lower jaw to slide backwards and forwards in mastication. Appropriately to this motion the condyle of the mandible is either subglobular, or is extended anteroposteriorly, and the glenoid cavity is a longitudinal instead of a transverse groove.

The mechanical action of the development of the rodent dentition has been as follows. The first factor in the order of time and importance was the increasing length of the incisor teeth. Those of the lower jaw closed behind those of the upper in the progenitors of the Glires (e. g. Psittacotherium) as in other Mammalia. Increase of length of these teeth in both jaws would tend to keep the mouth permanently open, were it not for the possibility of slipping the lower jaw backwards as it closed on the upper. This backward pressure had undoubtedly existed, and has operated from the earliest beginning of the growth of the rodent incisors. The process has been precisely the opposite of that which has occurred to the Carnivora, where the pressure has been ever forwards owing to the development of the canines. The progressive lengthening of the incisors through use has been dwelt on by Professor Ryder (*l. c.*). The posterior pressure on the lower jaw, produced by its closing on the upper, has been increased directly as the increase in the length of the incisors, especially those of the lower jaw.

The first effect of this posterior pressure will have been to slide the condyle of the mandible posteriorly

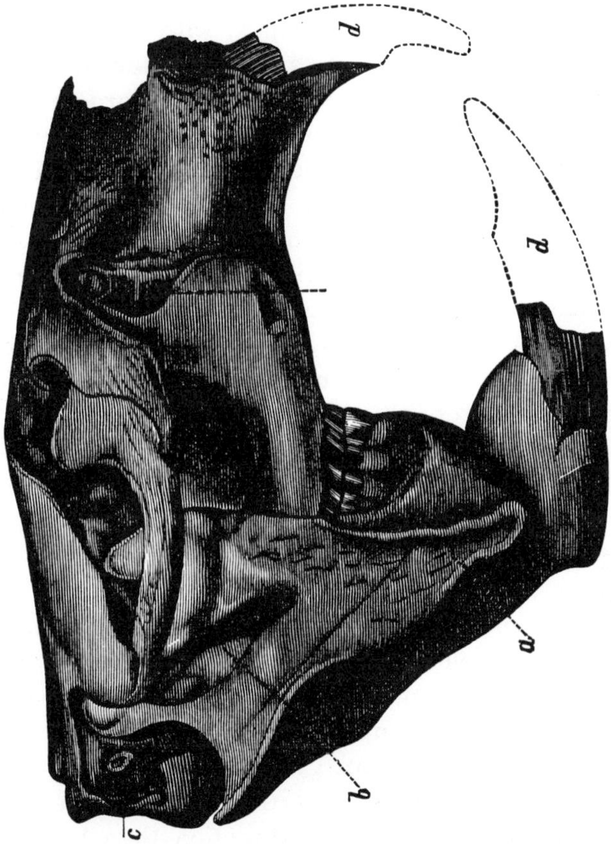

Fig. 107.—*Castoroïdes ohioensis* Foster, skull, right side; two-fifths natural size. *a*, inferior insertion of masseter muscle; *b*, fossa inside of ascending ramus: *c*, external auditory meatus; *d*, incisors; *e*, foramen infraorbitale; from Hall and Wyman.

over the postglenoid surface, if any were present, as is probable, in the bunotherian ancestor of the rodent. Continued repetition of the movement would probably push the process backwards so as to render it ineffective as a line of resistance, and ultimately to flatten it out against the otic bulla, and atrophy it. The lower jaw would thus come to occupy that peculiarly posterior position which it does in all rodents.

The anteroposterior (proal[1]) type of mastication becoming necessary, an appropriate development of the muscles moving the lower jaw, with their insertions, follows, *pari passu*. As a result we see that the insertion of the temporal muscle creeps forward on the ramus, until in the highest rodents (Cavia) it extends along the ramus to opposite the first true molars. The office of this muscle is to draw the ramus backwards and upwards, a movement which is commenced so soon as the inferior incisor strikes the apex of the superior incisor on the posterior side. By this muscle the inferior molars are drawn posteriorly and in close apposition to the superior molars. Connected with this movement, probably as an effect, we find the coronoid process of the mandible to have become gradually reduced in size to complete disappearance in some of the genera, e. g. of Leporidæ. In these genera the groove-like insertion of the temporal muscle develops as the coronoid process disappears.

As third and fourth effects of the posterior position of the lower jaw, we have the development of the internal pterygoid and masseter muscles and their insertions and origins. The angle of the ramus being forced backwards, these muscles are gradually stretched backwards at their insertions, and their contraction becomes more anteroposterior in direction than before. The internal pterygoid becomes especially developed, and its point of origin, the pterygoid fossa, becomes much enlarged. The border of the angle of the mandible becomes more or less inflected. In their effect

[1]Page 318.

on the movements of the ramus they oppose that of the temporal muscle, since they draw the ramus forwards. They are the effective muscles in the use of the incisor teeth; that is, in the opposition of the inferior incisors against the superior from below and posteriorly. Hence the great development of the internal pterygoid, and, in a less degree, of the masseter. Both muscles tend also to close the jaws, but at a different point in the act of mastication from that at which the temporal acts. If we suppose the mouth to be open, the action of the masseter and internal pterygoid muscles draws the mandible forwards and upwards until the incisors have performed their office, or the molars are in contact with each other or with the food. They then relax, and their temporal muscle continues the upward pressure, but draws the ramus backwards to the limit set by the adjacent parts, causing the act of mastication.

A fifth effect of the development of the incisors and of the proal mastication, is seen in the position of the molar teeth. The indefinitely repeated strain and pressure applied to the superior molars from forwards and below has evidently caused a gradual extension of the maxillary bone backwards, so that the last molars occupy a position much posterior to that which they do in other orders of mammals. This is especially the case in such forms as Bathyergus, Arvicola, and Castoroïdes (Figs. 107–108), where the last molars are below the temporal fossa, and posterior to the orbit.

A sixth effect of the causes mentioned has been referred to by Ryder.[1] This is the oblique direction of the axes of the molar teeth. These directions are opposite in the two jaws; upwards and forwards for the

[1] *Proceedings Philadelphia Academy*, 1877, p. 314.

lower, and downwards and backwards for the upper. The mechanics of this change of direction from vertical in the primitive forms (Sciuridæ) to oblique in the

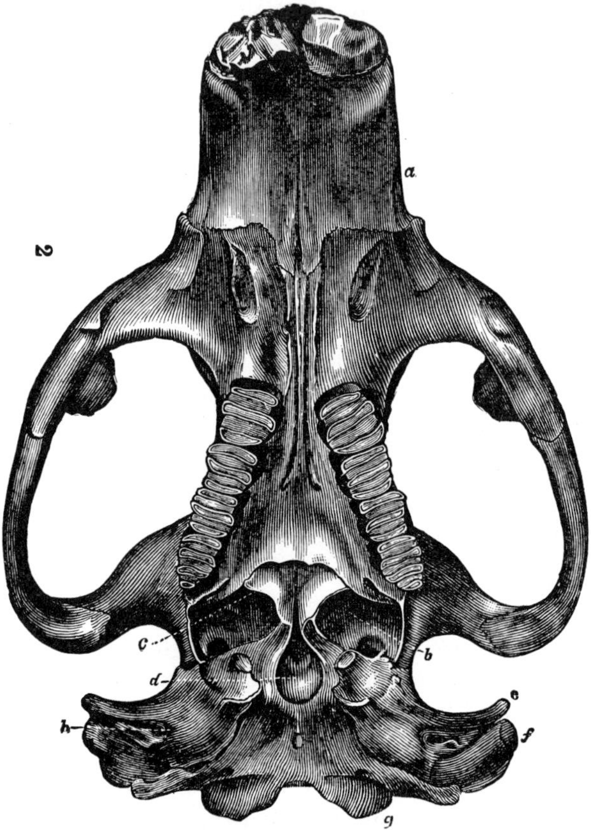

Fig. 108 —*Castoroïdes ohioensis* Foster; two-thirds natural size; skull from below. *a*, incisine foramen; *b*, pterygoid fossa; *c*, internal pterygoid plates; *d*, fossa in basioccipital bone; *e*, external auditory meatus; *f*, mastoid process; *g*, occipital condyles; *h*, tympania bulla, after Hall and Wyman.

genera with prismatic molars, is simple enough. The inferior crowns when closely appressed to the superior, and drawn posteriorly in the direction of the long axis of the jaw, press and strain the teeth in the two directions mentioned. The development of the long prismatic crowns which has proceeded under these circumstances, has been undoubtedly affected by the pressure and strain, and the direction we find has been the result.

The seventh effect is in the detailed structure of the teeth themselves. Beginning with short crowns with simple transverse crests (Psittacotherium and Sciuridæ, Figs. 106, 109), we reach through intermediate forms, crowns with vertical laminæ of enamel, which sometimes divide the crown entirely across (Chinchillidæ, Caviidæ, Castoroïdidæ) or appear only on the side of the crown, through the continued coalescence of the prisms of which each molar crown is composed (Arvicola). In many instances the crowns increase in transverse at the expense of their longitudinal diameter (Castor, Lepus). The vertically laminated structure is evidently due to the crowding together of transverse crests by the same pressure which has given the crowns their oblique direction. In many genera the lengthening of the crown has included the lengthening of the longitudinal connection between the transverse crests, as in Arvicola, Castor, and Hystricidæ generally. In others this connection has not been continued, so that the crown is

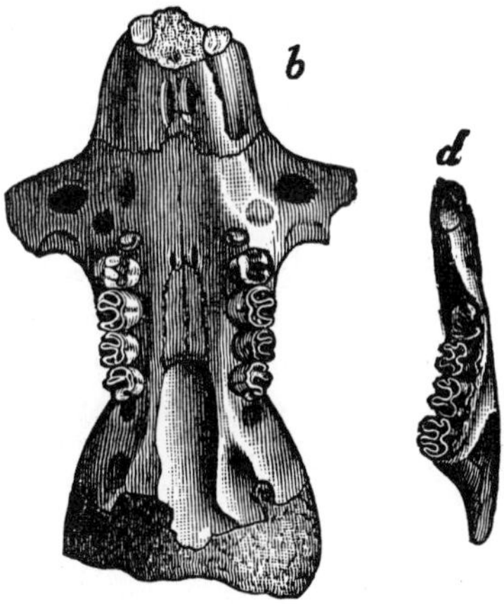

Fig. 109.—*Ischyromys typus* Leidy, from the White River beds of Colorado; original; from the *Report U. S. Geol. Surv Terrs.*: *b*, cranium from below; *d*, mandible from above.

composed of prisms which are separate to near the base, as in Amblyrhiza and Geomyidæ. In others, connection between the prisms has been lost by cænogeny, as in Chinchillidæ and Caviidæ generally. The latter families display also the greatest amount of crowding.

V. DISUSE IN MAMMALIA.

Modifications of structure of the mammalian skeleton accompany the disuse of parts, no less distinctly than in other divisions of animals. That these modifications are the direct consequences of this disuse may be reasonably inferred as the antithesis to the thesis of development of structure through use, maintained in the preceding pages. The evidence is more convincing from the fact that the same structures are observed to be related to similar dynamic conditions in groups of different taxonomic position. I select four illustrations from the Mammalia, from types in which the phylogeny is known, so that there is no question as to the degeneracy of the parts described.

a. Natatory Limbs.

The limbs have undergone great modifications of form in their gradual adaptation to aquatic habits. The stages of this process are to be observed first in the sea-otter (Enhydra), then in the seals, then in the sirenians, and last in the Cetacea. This succession of groups is not given here as a phylogeny, for paleontology does not warrant any such history, but the phylogeny of the limbs has been similar in the order of succession.

The use of a limb as an oar for propulsion in the water requires that it shall be, so far as the blade is concerned, inflexible. Such a structure has existed in all thoroughly aquatic Vertebrata. This implies the immobility of the articulations, which is due to the loss of their condylar surfaces. This may be traced to disuse of such articulations. This disuse would be at first voluntary, the limb being held stiffly while used as an oar in the act of swimming. Loss of power of extension and flexion is well known to result from disuse. It is well known that the flexors and extensors of the manus have become atrophied in the Cetacea. Not so, however, with the flexors and extensors of the humerus, which become those of the entire limb. In the whales the first segment of the fore limb is enclosed within the integument of the body, so that its motion being much restricted, the insertional crests are reduced in size. In the eared seals (Otariidæ) the hind limbs are somewhat free from the body integument, so that they can be turned forward when on land. They are further enclosed in the true seals (Phocidæ) so that their motion is very slight and they cannot be used for progression on land, and are available only for swimming.

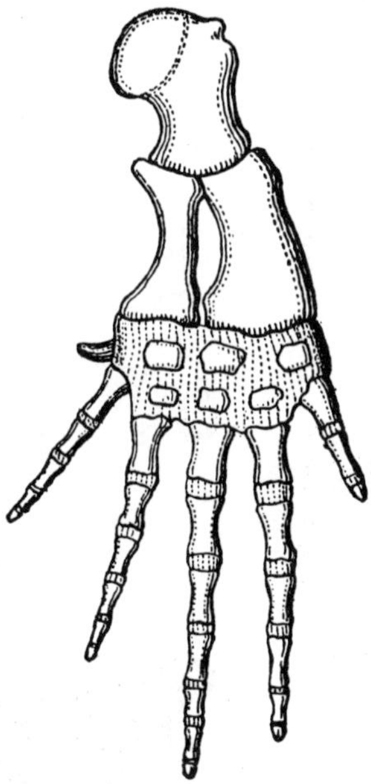

Fig. 110.—*Balæna mysticetus* fore limb: from Cuvier, *Oss. Fossiles.*

b. Abortion of Phalanges in Ungulata.

In the heavy Ungulata the longitudinal diameter of the phalanges is greatly reduced in relation to their transverse. The successive increase in depression in the bones of the feet with the advance of time is to be

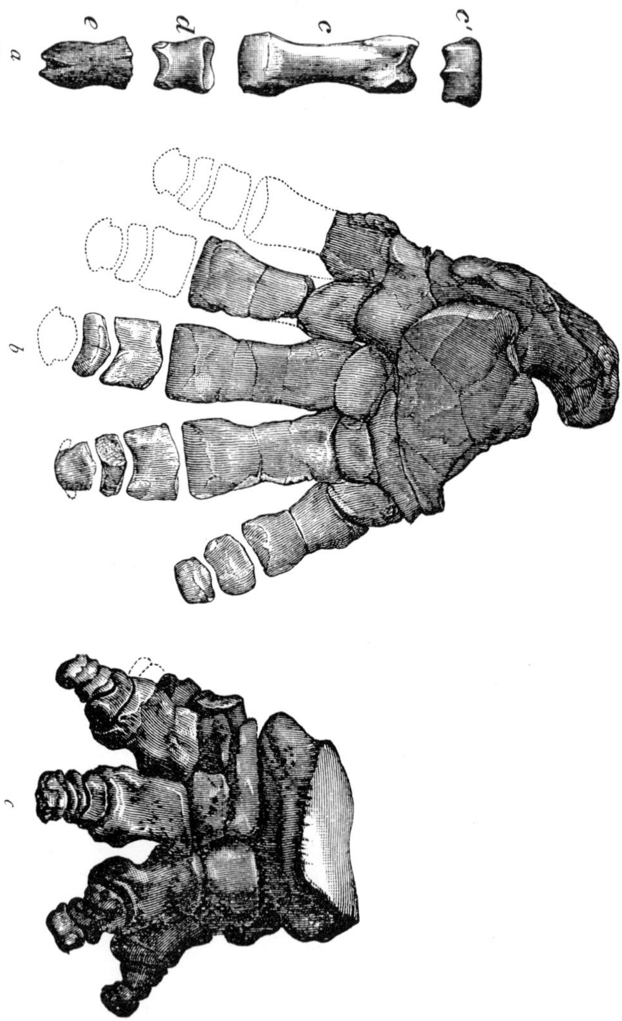

Fig. 111.—*a*, *Pantolambda bathmodon*, digit of posterior foot. *b*, Right posterior foot of a species of Coryphodon from New Mexico, one-half natural size. From Cope, *Report Expl. W. of 100th Mer.*, G. M. Wheeler, IV., Pl. LIX. *c*, *Eobasileus mirabilis*, right posterior foot; from Marsh, Dinocerata.

most readily seen in the order Amblypoda, where we pass from Pantolambda to Coryphodon and Uintatherium (Fig. 111.) A similar successive reduction is to be seen in the lines of the Perrisodactyla, as we pass from the smaller and lighter to the heavier and more bulky types. Such series are those which commence in the Lophiodontidæ, and terminate in the Menodontidæ on the one hand, and the rhinoceroses on the other. The elephants display the end of a similar line, which commences in the Condylarthra. In all of these bulky mammals the weight in progression is borne on the extremities of the metapodial bones, and the phalanges take but little share in it. They are turned forwards and are nearly useless. Their great reduction in dimensions in these forms appears to me to have followed disuse, and this is then the probable cause of it.

c. Atrophy of the Ulna and Fibula.

Successive atrophy of the ulna and fibula has been already referred to (p. 135). This is coextensive with reduction of the number of the digits in the ungulate Mammalia, and with the development of the digital patagium in the bats. This is in broad contrast to the subequal development of the ulna and radius in the Cetacea, where the fore limb functions as the blade of an oar. The cause of the reduction of the two elements in the Ungulata is the restriction of the functions of the fore and hind limb to the radius and tibia respectively. The distal extremities of the ulna and fibula are supported by the external bones of the carpal and tarsal series respectively. The reduction of the external digit deprives the external bones in question of their share in the support of the general weight, and consequently relieves them of the impact which passes through the longer median digits which remain. The median digits, on the other hand, support the radius and tibia through the medium of the carpus and tarsus, and it is these elements therefore which function in the use of the limb. We have here an evident illustration of the effect of disuse in effecting the atrophy of an element, and of use in increasing the size and complexity of an adjacent element, of the same organism. No other explanation seems possible, for the elements which are reduced and those which are enlarged, are subjected in every other respect to the same conditions.

d. Atrophy of Incisor Teeth.

This is complete in both jaws of existing Edentata; the upper jaw of Dinocerata and many Artiodactyla, and is partial in the upper jaw in various Chiroptera and Lemuridæ. We have already seen (p. 326) that the superior incisors of certain Lemuridæ are without utility, owing to the conversion of the inferior incisors into a horizontal comb. I have ascribed the reduction of the superior incisors of bats to disuse consequent on the adoption of a frugivorous diet.[1] Further reason, which is common to the living members of the orders mentioned, is to be found in the disuse which has followed the use of the tongue as an organ for the prehension of food. The fruit-eating bats with most reduced incisors (Glossophaginæ) carry the soft parts of fruits into the mouth with the tongue. The Edentata use the tongue for the collection of both insect and vegetable food, projecting it far exterior to the mouth. The Artiodactyla without superior incisors however, combine the prehensile use of the tongue with a use of the lower incisors, which bite off the grass thus seized,

[1] *Mechan. Origin, etc., Mammalia*, 1889, p. 224. See also Dr. H. Allen, *Proceeds. Academy, Philadelphia*, 1891, p. 451.

while it is pressed against the pad which replaces the superior incisors. Why the superior incisors should have disappeared in this group is not yet clear to my mind.

In this connection Dr. Allen (*l. c.*) reminds us that in hypertrophy of the tongue in man, the inferior incisors are thrown forward and are widely separated from each other. He considers it reasonable to infer that in lower animals where the tongue is used for prehension, the similar change which takes place in the teeth, from a vertical to a horizontal position, is induced by this cause.

vi. HOMOPLASSY IN MAMMALIA.

The direct evidence in favor of kinetogenesis above adduced is greatly strengthened by corroborative testimony presented by distinct phyla of animals. Restricting myself here to Mammalia, I will enumerate a number of cases where the same structures have appeared in distinct lines of descent under similar mechanical conditions, a phenomenon already referred to on page 72 under the name of Homoplassy.

Before reviewing the subject, I cite what is the most remarkable example of homoplassy in the Mammalia which has yet come to the knowledge of paleontologists. Ameghino has discovered in the Cenozoic formations of Argentina a group of Ungulata which he calls the Litopterna, and which I regard as a suborder of the Taxeopoda, allied to the Condylarthra (p. 128). Ameghino placed the group under the Perissodactyla, but the tarsus and carpus are of a totally different character, and indicate an origin from the Condylarthra quite independently of that division. The carpal and tarsal bones are in linear series, or if they overlap, it is in

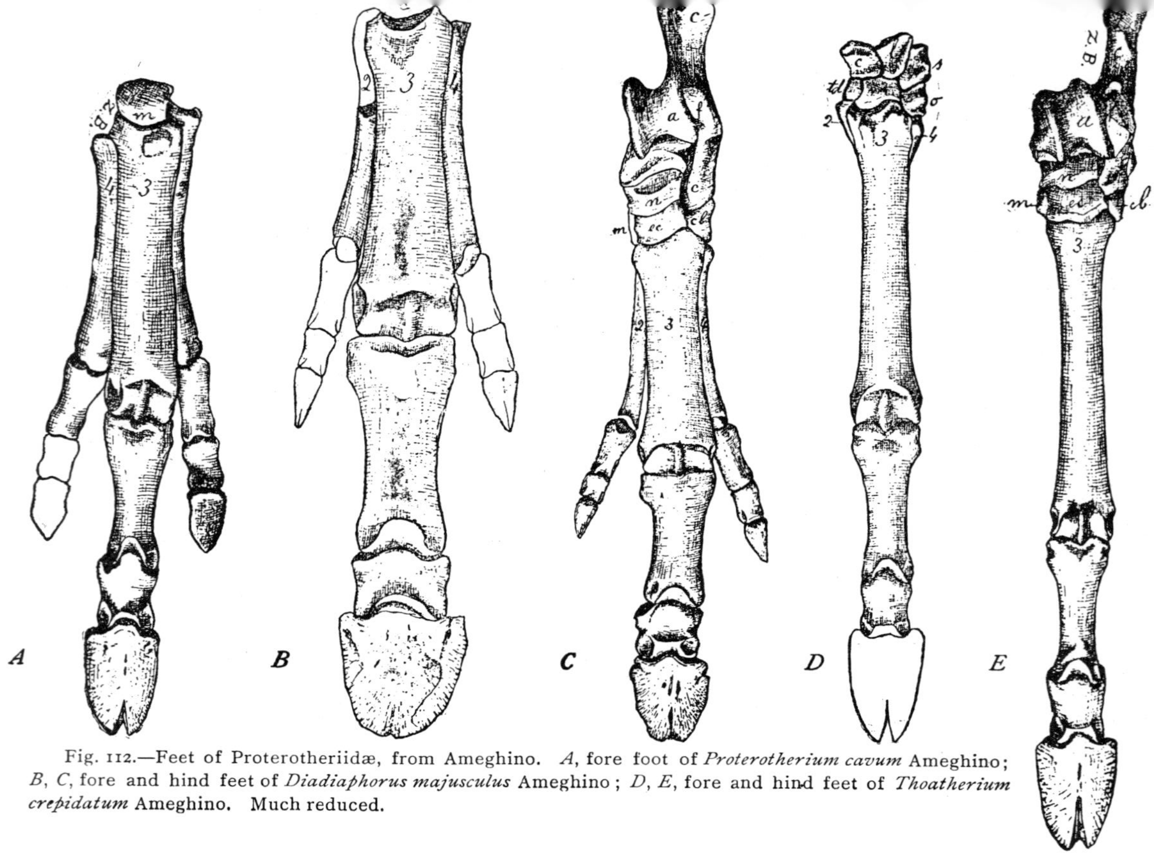

Fig. 112.—Feet of Proterotheriidæ, from Ameghino. *A*, fore foot of *Proterotherium cavum* Ameghino; *B*, *C*, fore and hind feet of *Diadiaphorus majusculus* Ameghino; *D*, *E*, fore and hind feet of *Thoatherium crepidatum* Ameghino. Much reduced.

a direction the opposite of that which characterizes the order Diplarthra (=Perissodactyla and Artiodactyla). But the Litopterna present a most remarkable parallelism to the Perissodactyla in the characters of both the feet and the dentition. No genus is known as yet which possesses more than three toes before and behind, and these are of equal length in Macrauchenia Owen. In this genus the teeth are not primitive but are much modified. The most primitive dentition is seen in the genus Proterotherium (Ameghino) where the superior molars are tritubercular as in many Condylarthra. In this genus (Fig. 112, *A*), there are three toes, but the lateral ones are reduced, about as in the Equine genus Anchitherium (p. 148). In the next genus, Diadiaphorus Ameghino, the superior molars are quadritubercular and crested, while the lateral toes are reduced still more, being quite rudimental (Fig. 112, *B*, *C*), as in the equine genera Hippotherium and Prothippus (p. 149; Fig. 70). The superior molars have not progressed so far as in these genera, but are not very different from those of Anchitherium. In the third and last type (Thoatherium Ameghino), the lateral digits have disappeared from both fore and hind feet (Fig. 112, *C*, *D*), so that the condition is that of the genus Equus (Fig. 81), but the splints in the *Thoatherium crepidatum* Ameghino are even more reduced than in the known species of horse. The superior molars have not assumed the pattern of the genus Equus, but resemble rather those of Macrauchenia, and could have been easily derived from those of Diadiaphorus.

Here we have a serial reduction of the lateral digits and their connections with the leg, and increase in the proportions of the middle digit and corresponding increase in the proximal connections, exactly similar to that which took place in the horse-line, in a different order of Mammalia.

In review I now cite as examples of homoplassy:

First, as regards the development of the tongue-and-groove ankle-joint. This has been developed independently along four distinct phyla, viz., in the leporid Glires, the Carnivora, and the even and odd toed Diplarthra.

Second, the wrist-joint. The faceting of the radial surface has appeared independently in the perissodactyle and artiodactyle lines, but is best developed in the latter. Also it appeared independently in the separate suoid and boöid lines in the latter suborder.

Third, the trochlear crest of the elbow-joint appeared independently in the perissodactyle and artiodactyle Diplarthra, and in the leporid Glires (the rabbit family).

Fourth, the round head of the radius appeared independently in the lines of the Edentata (ant-eater) and Quadrumana, under the stress of supination of the hand.

Fifth, the development of cusps with crescentic section out of cusps with round section has occurred in the widely different groups of the multituberculate Prototheria, and the selenodont Artiodactyla. In the former the crescents are transverse, since the thrust of the teeth in use is longitudinal; in the latter they are longitudinal, since the thrust of the jaws is transverse.

Sixth, the deep plication and hypsodonty of molars appeared independently in the Glires, Tillodonta, Proboscidia, Sirenia, Perissodactyla, and Artiodactyla; and probably in the Edentata and Toxodontia.

Seventh, increase in the length of the legs has ensued in the Marsupialia, Glires (Lepus, Dolichotis, Dipus), Carnivora, Ungulata, Quadrumana.

Eighth, reduction of digits has occurred under similar conditions in Marsupialia, Glires, Insectivora, Carnivora, Ungulata.

Ninth, the atrophy of the ulna and fibula occur in the distinct lines of the Perissodactyla and Artiodactyla, and the atrophy of the fibula in the leporid Glires; all in limbs which function in the most rapid progression.

Further confirmation of the law of kinetogenesis is to be found in those cases where different structures appear in different parts of the skeleton of the same individual animal, in direct correspondence with the different mechanical conditions to which these parts have been subjected. Examples: the diverse modifications of the articulations of the limbs in consequence of the uses to which they have been put, in mammals which excavate the earth with one pair of limbs only; as in the anterior limbs of the fossorial Edentata, Insectivora, and Glires. The reduction of the number of the digits in the posterior limb only when this is extensively used for rapid progression, as in leaping: this is seen in the kangaroo and jerboas, in the orders Marsupialia and Glires.

The development of a dental structure of premolars identical with that of the molars, from a different structural origin, in the Perissodactyla.

From the preceding facts I have inferred that in biologic evolution, as in ordinary mechanics, *identical causes produce identical results.*

VII. A HYPOTHESIS OF THE ORIGIN OF THE DIVISIONS OF THE VERTEBRATA.

In order to estimate the part which has been played by the movements of the Vertebrata in changing their environment in past geologic ages, we have to rely principally on inferences derived from the present physical characteristics of the earth. Formerly, as now, conditions of temperature, humidity, soil, shelter, food, etc., were avoided or appropriated by animals, through their capacity for moving from place to place. What concerns us chiefly here, is the effects on their structure produced by the movements of Vertebrata. In examining this question I will take it up in systematic order, so as to observe whether kinetogenesis has been the principal or only a subordinate agency in the evolution of this branch of the animal kingdom.

The most conspicuous index of the serial succession of the vertebrate classes, is, as has been already remarked, the circulatory system. The modifications of this system have been immediately connected with those of the method of respiration, which the exigencies of the environment induced in vertebrates. The existence of branchial arteries and veins dates from the earliest vertebrate, if not from prevertebrate life. They are already established in the Tunicata, and continued throughout the rising scale in diminished numbers, so long as Vertebrata were exclusively aquatic in their modes of life. When at the close of the Devonian system the land masses assumed great proportions in both the Eastern and Western Hemispheres, it is probable that many fishes were entangled in shallow water, which rapidly freshened, and ultimately were desiccated, and respiration by the swallowing of air into the alimentary canal began to take the place of respiration by gills. It is well known that respiration by this means may be carried on by fishes of various genera, e. g. Cobitis; and Professor Gage has shown that the same habit exists in Batrachia and in certain tortoises (Tronychidæ). In the middle Carboniferous shales tracks of land animals occur, and the bones of Batrachia abound in the coal measures. Already in the Permian these Batrachia are accompanied by numerous Reptilia, and air breathers of terrestrial habits had become numerous on the earth.

The habit of holding in the œsophagus large quantities of air while engaged in seeking food in foul water, or on land, on the part of vertebrates which normally oxygenated the blood by means of gills, was probably the mechanical cause of the development of a pouch, and afterwards of a diverticulum of the œsophagus, which became ultimately a swim-bladder or a lung. In vertebrates in which a return to aquatic life became necessary, it became the former; in those which remained for a shorter or longer period of time on land it became the latter.[1] It is noteworthy that among fresh-water fishes generally, the swim-bladder is more complex than among marine forms, showing that the varying conditions of shore and fresh-water life have been mainly responsible for its development.

The development of a lung at once produced a change in the uses to which the various branchial arches were put. The posterior, which supply the lung, would be subjected to greater pressure owing to the increased blood supply demanded by the lung,

[1] This view is adopted by C. Morris, *American Naturalist*, 1892, p. 975.

and a correspondingly diminished pressure would be experienced by the now unused branchial portions of the bows. The first would retain the importance of its basal portion, as the source of the carotids, while the middle arches would continue their existence as the bases of the central dorsal aorta. The loss of the right aorta-root in Mammalia was probably due to the fact that the great arteries which supply the digestive system are primitively branches of the left aorta-root, as they are to-day in the crocodiles and in many of the Batrachia. The right aorta-root disappeared through disuse. Probably in the immediate ancestors of the birds, as in the crocodiles, the right aorta-root gave off the carotides and the subclaviæ. As the birds demand an excessive blood-supply for the fore limbs, we have here probably the reason why the right root remained in this subclass.

The next index of successional development in Vertebrata is the brain. Our belief that use under stimulus has been the cause of its successive growth, can only be based on the analogy of our own experiences in the matter of education. No part of the human organism is so susceptible to stimuli as the nervous system, and the marvellous effects on faculty of continued exercise are well known to everybody. Since the changes of mental states are necessarily due to corresponding structural changes no one will find in ignorance of the mechanics of brain-evolution a serious obstacle to believing that it has taken place under the influence of the innumerable stimuli always present to animal life.

It is in the skeleton that we have the actual record and evidence of the effect of movement on structure. It must be remembered in this connection that skeletal and dental tissues exhibit the phenomena of nutrition and waste (metabolism), common to all living organic matter. Hence even the hardest osseous tissues are plastic and are subject to mechanical influences to a degree which is not possible to dead matter of equal density.

Fundamental differences between Vertebrata are displayed by their organs of movement, but before specially considering these I will refer briefly to certain other fundamental characters displayed by the skull. In advancing from the fishes to the Mammalia we observe a successive consolidation of the mandibular arch, and of its mode of connection with the cranium. The mandibular arch in its entirety displays in the fishes a segmented condition, generally comparable to that which characterizes the branchial arches. Among Batrachians and Reptilia various degrees of fixation of its suspensor (hyomandibular, quadrate) to the cranium exist, and in some of them it is closely united by immovable suture. The complete fusion with the squamosal seen in the Mammalia is its final status. The segmentation of the mandibular portion of the arch seems, from the discoveries of Ameghino, to have continued among some of the Lower Eocene mammals, but that finally disappeared, so that in the modern mammals the movable mandibular arch consists of a single element on each side. In this history we see an instance of the progressive coössification of parts, which results from the constant strain of use, of which many other normal and abnormal examples are known. This use is the act of mastication. Where there is no mastication, and the jaws are used only as prehensile organs, this coössification does not occur, as, for instance, in the snakes. In this most specialized and modern type of Reptilia, the segmentation is complete.

The segmentation of the limbs in the Vertebrata is a simple mechanical problem. Paleontology and embryology concur in proving that the limbs originated in primitive folds in the external integument, and that their connection with the internal skeleton was of later accomplishment, has been shown by Wiedersheim. At first free, they sought points of support on the skeleton, but did not lose their free mobility when this contact was attained. Appropriately to the mechanical conditions of rigidity and flexibility necessary to their use in a fluid medium, they were originally composed of slender rods which were segmented by interruptions at suitable points. The articulations of the fin-rays of fishes have been made the subject of an interesting research by Ryder, who finds them to be fractures, due to flexures during motion in the water medium.[1] The limb of land vertebrates (the chiropterygium) was derived from one of the forms of fins (rhipidopterygium) of water vertebrates. This is the simple type of primitive fin displayed by the Paleozoic Teleostomi of the superorder Rhipidopterygia. Whether the subdivisions of the chiropterygium, the propodial, metapodial, and phalangeal bones, etc., were divided from the primitive branches of the archipterygium, as held by Gegenbaur, or whether they have developed by sprouting from a simple axial series of segments, as held by Baur, or whether, as I have suggested, it is a derivation from the rhipidopterygian type of paired fin, is not yet decided. In either case, the limbs of the first land animals were segmented and flexible at the joints between the segments. The ne-

[1] *Proceedings of the American Philosophical Society*, 1889, p. 547.

cessities of such limbs are twofold: first, to serve as supports when at rest or in progression; second, to be applied to the body in protection from enemies, or in aiding the functions of feeding, reproduction, etc. The first function requires principally mobility at the point of connection with the body. The second, flexibility at some point on the shaft of the limb. The two kinds of movements in question would conserve two principal points of flexure, and these would be for the fore limb, just what we find, the shoulder and elbow joints; and for the hind limbs, the hip and knee joints. The two median joints are directed in opposite ways, the elbow backwards and the knee forwards. This diversity is clearly due to the diverse positions of the functioning regions. The opposite extremities of the alimentary canal, the posterior including the exits of the urogenital organs, requires that the fore limbs should bend forwards, and the posterior limbs backwards. And the constantly recurring necessity for the exercise of these flexures must necessarily have developed the appropriate articulations in preference to all others. The terminal flexure, that of the wrist or ankle, has been evidently due to a similar mechanical cause, viz., the flexure due to pressure of the weight of the body on the terminal segments when in contact with earth. The distal segments are the most slender in all types, and least able to maintain a linear direction under pressure, hence, they have flexed easily and thus the line of separation between leg and foot had its origin. The ankle and wrist in the Batrachia Urodela is still a mere flexure.

Mr. Herbert Spencer has endeavored to account for the origin of the segmentation of muscles into myotomes, and the division of the sheath of the notochord into vertebræ, by supposing it to be due to the lateral swimming movements of the fishes, which first exhibit these structures.[1] With this view various later authors have agreed, and I have offered some additional evidence of the soundness of this position with respect to the vertebral axis of Batrachia,[2] and the origin of limb articulations.[3] It is true that the origin of segmentation in the vertebral column of the true fishes and the Batrachia turns out to have been less simple in its process than was suggested by Mr. Spencer, but his general principle holds good, now that paleontology has cleared up the subject.

The Echinodermata, Mollusca, Arthropoda, and Vertebrata possess external or internal calcareous or chitinous skeletons for the most part. The lower forms of all these branches, however, are more or less deficient in this kind of protection, and embryology indicates that all of them are the descendants of the Vermes or worms, which are mostly without such hard supports and protections. Whether this be demonstrated or not, we have plenty of evidence to show that the primitive Vertebrata were without hard skeletons, and that their bodies were composed internally and externally of perfectly flexible tissues.

If we now imagine that either the integuments, or an axial rod, of a worm-like animal has become the seat of a calcareous or chitinous deposit, it is evident that the movements of the animal in swimming or creeping must have interrupted the deposit at definite points of its length. The lateral flexure of the body would be restricted to certain points, and the intervening spaces

1 *Principles of Biology*, 1873, pp. 198–204.

2 *Origin of the Fittest*, 1887, p. 305.

3 *Mechanical Causes of Origin of Hard Parts of Mammalia*, 1889, p. 163.

would become the seat of the deposit. At the lines of interruption joints would be formed, and if the movements were habitually symmetrical, these interruptions would be equidistant. In this way the well-known segmentation of the external skeletons of Arthropoda, and the internal skeletons of Vertebrata would be formed. We have more detailed evidence that this has been the case. Thus the segmentation of the osseous sheath of the chorda dorsalis in both primitive fishes and batrachians has been accomplished in wedge-shaped tracts precisely as may be observed in the folding of a tolerably stiff sleeve of a coat which ensheathes the arm, under the influence of lateral flexures. The wedge-shaped tracts are superior and inferior, the apices directed towards each other. Seen from the side they form two wedges with their apices together, and their bases one up and the other down. Now, if a person who wears a coat of rather thick material will examine the folds of his sleeve as they are produced on the inner side of his arm, he will see a figure nearly like that of the segments of the vertebral column described. The folds will correspond to the sutures, and the interspaces to the bony segments. He will find that the spaces are lens-shaped, or, when viewed in profile, wedge-shaped, with the apices together. This arrangement results from the necessary mechanics of flexure to one side. In flexure of a cylinder like the sleeve, or like a vertebral column, the shortest curve is along the line of the greatest convexity of the cylinder. Here is the closest folding of the sheath, and here, consequently, the lines of fold in soft material, or interruption in hard material, will converge and come together. That is just what they do in both the sleeve and the rhachitomous vertebral column, the only difference being that in the animal it is exhibited on both sides, and on the sleeve on only one side. This difference is, of course, due to the fact that the animal can bend himself in both directions, while the arm only bends in one direction.

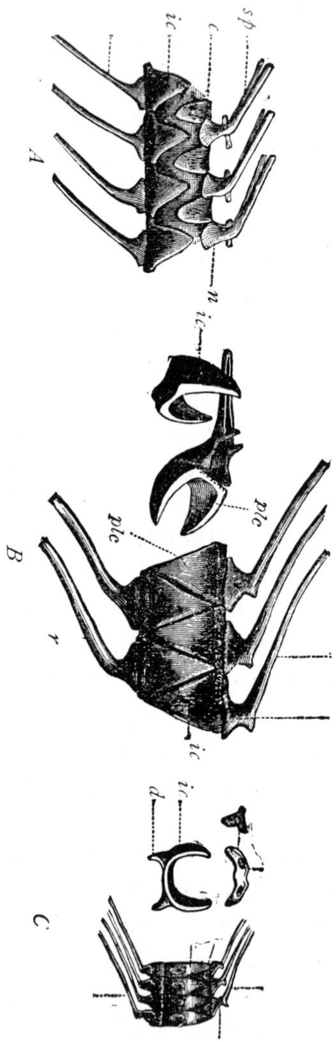

Fig. 113.—Portions of dorsal vertebræ of merospondylous fishes; from Zittel. *A. Caturus furcatus. B Callopterus agasizzi. C. Euthynotus micropodius. Sp*, neural spine; *n*, neurapophysis; *c*, *plc*, centrum; *ic*, intercentrum; *d*, diapophysis; *r*, rib.

It results from the above observations that the structure of the rhachitomous vertebral column *has been produced by the movements of the body from side to side, as in swimming, during the process of the deposit of mineral material* in and around the chorda dorsalis.[1]

See figures 113 to 114*a* where the coat-sleeve is compared with the "rhachitomous" vertebræ of primitive

[1] See *American Naturalist*, January, 1884.

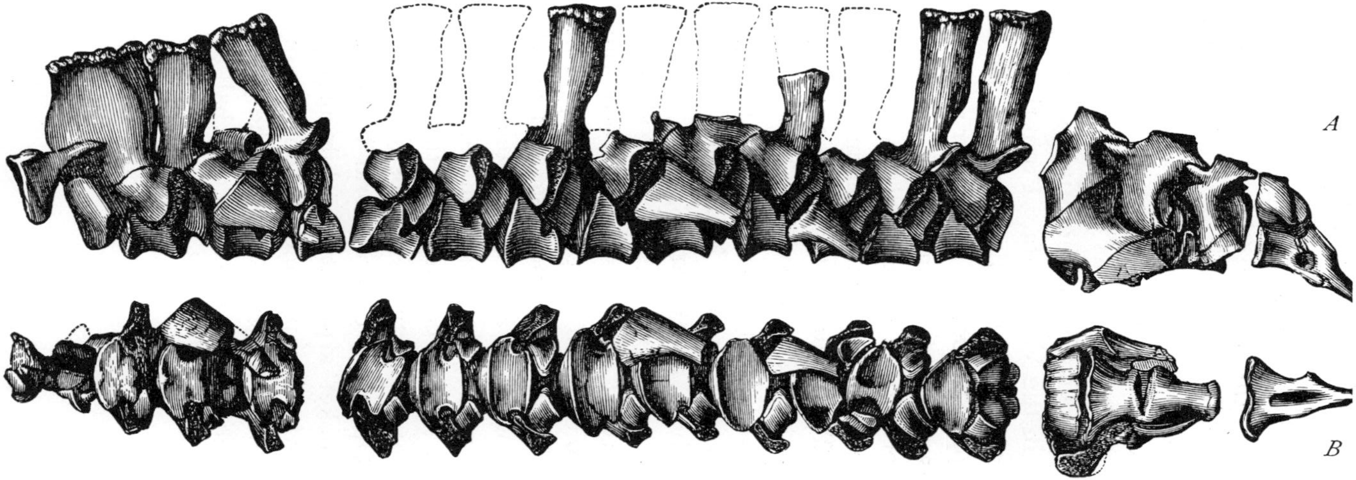

Fig. 114.—Rhachitomous vertebral column of *Eryops megacephalus Cope*, one-fourth natural size. *A*, profile; *B*, from below. From the Permian epoch of Texas.

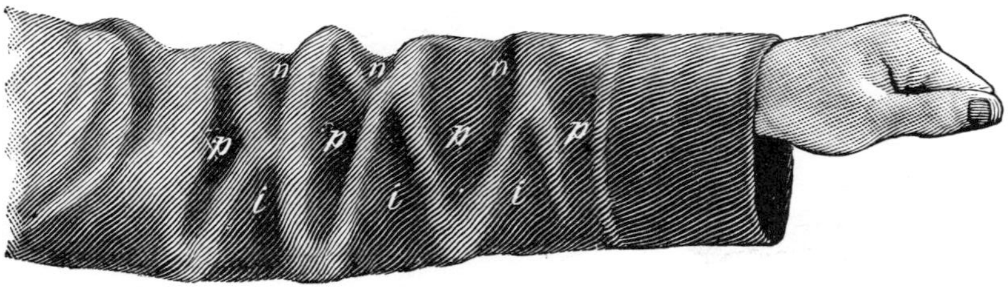

Fig. 114*a*.—Sleeve of a coat, showing folds produced by lateral flexure which leaves interspaces similar to the segments of a rhachitomous vertebra. Thus, *i* represents intercentrum; *p*, centrum or pleurocentrum; and *n*, neurapophysis.

fishes (Merospondyli[1]) and Batrachia (Rhachitomi). Such was the origin of the segmentation in the primitive sharks, Pleuracanthus, whose structure has been pointed out by Sauvage, and Hybodus, whose characters have been demonstrated by Smith Woodward. The segmented (or rhachitomous, as I have termed it,) condition may be then regarded as the primitive one of the osseous column in the Vertebrata. From the rhachitomous column two divergent lines have arisen as already remarked (pp. 89, 209). The inferior segment has been retained in the fish-batrachian line, whence I have termed their vertebræ "intercentral," while these bodies have disappeared or become rudimental in the higher Vertebrata. The pleurocentra (Figs. 113–114*a*, *c*, *pl.*) have, on the other hand, developed downwards, and, meeting below, have formed the effective centrum of the vertebra. Hence, in the Monocondylia and Mammalia the vertebræ are "central."

The Reptilia display a greater variety of vertebral articulation than any of the classes of Vertebrata. After the primitive biconcave (amphicœlous) type was abandoned, the two principal types assumed are the ball and socket (procœlous and opisthocœlous), and the plane (amphiplatyan). In those families in which the body is more or less in contact with the ground, owing to the absence, shortness, or position of the limbs (Lacertilia, Ophidia), the vertebral bodies exhibit the ball-and-socket articulation, while in types with longer limbs which supported the body in progression, so that the latter never reached the ground (Dinosauria), the articulations are plane. The ball-and-socket articulation may be inferred to have been pro-

[1] Zittel, *Handbuch der Palæontologie*, III., p. 138, 1887, where this character is first clearly pointed out in fishes.

duced by vermiform movements which utilize points of resistance on the earth as aids to progression, while the plane articulation has probably resulted from the persistence of the fixed relation which is appropriate to a body which should be relieved by the legs of all share in movements necessary to progression. That this position is correct is sustained by the fact that the cervical vertebræ of various reptiles and mammals which have plane dorsal vertebræ have the ball-and-socket structure. This is probably due to the constant flexures to which that part of the column has been subjected, as compared with the fixity of the dorsal region.

Owing to the comparatively advanced state of our knowledge of the phylogeny of the Mammalia (Chapter II.), this class furnishes especial opportunities for the study of kinetogenetic evolution. But our knowledge is not yet sufficiently complete to enable us to account on mechanical grounds for the origin of all the characters which distinguish all its subdivisions. This being the case, I have not presented the subject in taxonomic order, but have contented myself with offering it in the order of evidential value. I have first described certain cases where the action of kinetogenesis is self-evident. This has been followed by the presentation of cases where the evidence amounts to a high degree of probability.

Referring now to the table of definitions of the orders of Mammalia on pages 127–128, I will go over the characters *seriatim*, and show how far our knowledge warrants us in giving a kinetogenetic explanation of their origin. No mechanical cause can be at present assigned for the loss of the coracoid and episternal bones in the Eutheria. In the Eutheria the presence of the marsupial bones of the Didelphia is a survival, and it may be that their absence in the Monodelphia is due to disuse, on the withdrawal of strain on the abdominal walls which followed the abandonment of the oviparous habit of the Prototheria, and the young-bearing habit of the Marsupialia. The very considerable weight actually borne by existing forms and probable weight carried by extinct forms will probably account for the development of these bones through strain on the prepubic cartilage. The perforate palate is a reptilian survival, and the closure of the fontanelles may have been due to the increased strain due to the increased energy of mastication which early ensued, owing to the increased size and specialization of the molar teeth. Of the three divisions of the Monodelphia, the Mutilata, Unguiculata, and Ungulata, the Unguiculata possess modifications of a character of ungual phalanges inherited from the Reptilia, while the other two groups have experienced still greater modification. The limbs of the Mutilata display the result of disuse as to the posterior ones, and special use as to the anterior, as I have already pointed out. The Sirenia display a less degree of modification than the Cetacea. The hoofs of the Ungulata may well have assumed their laterally expanded and transverse forms by the extreme pressure and impact on the earth, incident to their function as supports.

The origins of most of the dental characters which characterize the orders of Mammalia have already been referred to mechanical causes, and have been already treated of. The same is true of tarsal and carpal characters, which are of so much importance among the Ungulata.

The characters enumerated on page 139 as indicating progressive modification in time have also been mainly accounted for on mechanical grounds.

5. OBJECTIONS TO THE DOCTRINE OF KINETOGENESIS

It has been objected that Neo-Lamarckians are self-contradictory and illogical in their defense of the doctrine of the development of structures by use, or by motion. It is asserted that they believe that stimuli of different kinds produce similar results, and that stimuli of the same kind may produce different results. The charge that Neo-Lamarckians hold those views is correct, but it is not correct to suppose that they are illogical or self-contradictory. This criticism is one of those generalities which will not bear examination, while the doctrine of kinetogenesis will bear examination.

Thus it has been experimentally shown that bone irritation will produce both bone deposit and bone absorption, according to the degree of irritation. Moderate irritation produces deposit, and greater irritation produces absorption. Hence it is that both impact and strain, or pressure and stretching, will elongate a bone, by stimulating growth, if not excessive. We have the illustrations in the elongation of ligaments and cartilages and their ossification under stretching, and the shortening of both in absence of use, from which we may infer their lengthening under use. The continued lengthening of the limbs and teeth of the higher Mammalia, in the course of geologic time, is an illustration of the effect of continued impact and transverse strain; while the lengthening of the limb bones of the sloth, and of the tarsal bones of many bats, is a consequence of longitudinal strain.

The differing directions of the elements entering into the articulations of the limbs of Mammalia may be cited in illustration of the supposed inconsistency of supposing them all to be the result of impact and strain. Thus most of the condyles are directed distad, but the heads of the humerus and femur, and the proximal surfaces of the carpus and tarsus are directed proximad (except the trochlear groove of the astragalus when present). So far as regards the distal ends of the radius and tibia (fore arm and leg) I have pointed out that here dense layers disposed longitudinally impinge on a similar layer disposed transversely, with the natural consequence the latter has yielded to the excess of impact so produced. The direct relation of the sculpture of the surfaces concerned, to the lengthening of the foot and increase of speed of the animal, and hence increase of force of impact, leads irresistibly to this conclusion. As regards the convexity of the heads of the humerus and femur, or rather the concavity of the corresponding surfaces of the scapula and acetabulum the explanation has been already given. Henke was apparently the first to call attention to the fact that the concavity and convexity of the articular surfaces is directly related to the positions of the insertions of the muscles which move them. He shows that a concave surface is developed at the extremity which is nearest to the muscular insertion, while the convex surface is developed on the extremity which is most remote from its muscular insertion. Thus is accounted for the apparently contradictory evidence of the limb articulations mentioned. In some cases at least, as those of the glenoid cavities of the scapula, ilium, and phalanges, the muscular insertions are so near to their borders, as to suggest that the growth of the latter is due to a pulling strain on them, as well as to the greater mobility of the element which becomes convex, as supposed by Fick.

It is claimed in the preceding pages, that impacts on the extremities of a bone or tooth, gradually increase its length. It may be hastily supposed that in this assumption I derive elongation of the shaft of a bone from the same stimulus which produces excavation and therefore abbreviation of its extremities. In the gross this charge is correct; but the position I have assumed is defensible, because in detail it is easy to perceive that effect of the use of a limb on an articular surface of a bone is quite distinct from that which it has on the shaft. At the articular faces we have discontinuity; and therefore *friction*; in the shaft we have only the concussions produced by impact, together with some torsion strain. That the former movement stimulates the development and activity of the osteoclasts has been shown by Koelliker; that the latter may stimulate the activity of osteoblasts is rendered highly probable from the facts of pathological anatomy. These show that a very slight modification of stimulus is sufficient to change the building cells into the absorbent cells and back again. For the same reason belief in the elongation of bones under stretching strain may not be inconsistent with belief in an elongation under impacts.

Cary makes specific objections against the kinetogenesis of the articulations of the mammalian skeleton.[1] After a study of the carpus of the Eocene perissodactyle genus Palæosyops he concludes that the trapezoid bone is too small to express properly the direct result of purely mechanical causes. He says that

[1] *American Journal of Morphology*, 1892, p. 305.

in reaching this result he has applied geometrical methods. "First, the volume of the bones was got at. Next the area of the bearing surfaces and their inclination to the digits were measured. Then giving to the thrust of each metacarpal a value proportional to its volume, the distribution of that thrust can by resolution and composition of forces, be traced through the foot, and the pressure on each surface and bone approximately obtained." Further than this the author does not explain how he reached the result that the trapezoid is too small. It is quite essential that this demonstration should be given if we are expected to accept his conclusion. An essential part of the problem is, however, unnoticed by Mr. Cary; and that is the condition of the trapezoid in the reptilian ancestors of the Mammalia. The phylogeny of an element must be known, since it furnishes the "physical basis" of the problem. The fact is that the trapezium, trapezoides, and the magnum owe their small size to their being the only carpal elements which have not been produced by the fusion of two or more primitive elements of the batrachian and reptilian carpus. The trapezoides moreover occupies a place in a longitudinal series of three elements in the primitive carpus, while the trapezium forms one of a series of only two elements. For similar reasons the cuneiforms are the smallest elements of the tarsus.

Mr. Cary then proceeds to criticize the explanations offered by Professor Osborn and myself, in accounting for the origin of certain structures. He finds our explanations to be self-contradictory, and that we also contradict each other. Osborn has supposed that the conules of the molars are produced by friction of the molars of opposite series on each other. I have expressed the opinion that the shear of the sectorial teeth of Carnivora was produced by lateral friction during vertical movement of the lower tooth on the upper. I have also asserted that the forms of facets of limb articulations are due to pressure. Mr. Cary sees here the attempt to explain the origin of totally different structures through identical mechanical processes, and believes that the attempt is a failure. Were the conditions of the problems alike, as Mr. Cary thinks them to be, he would have good reason for his opinion. But the conditions in the three cases are entirely different, and our author's conclusion is due to neglect of the elementary facts of the proposition.

The development of conules at the points indicated by Professor Osborn, has been supposed by him to be due to friction between existing ridges of enamel which cross each other when in action, at the points in question. In the case of the development of the sectorial shear, the faces between which the shearing motion takes place are smooth, and without ridges or crests. Hence the entire surface receives a homogeneous friction. In the third case, that of the foot articulations, there is no friction, but there is pressure which when abruptly applied in movement becomes impact. There is really no parity between the three cases.

The author of this paper also thinks that the explanation of the elongation of bones through use of different kinds is not a permissible hypothesis. He cites my attempt to account for the elongation of the leg bones of higher mammals through impact-stimulus; and of other limb bones of other mammals through stretching. But he does not prove that similar results may not flow from mechanical stresses applied in different ways. I suppose that any mechanical stress which determines nutritive processes to a part, will increase its size, *cæteris paribus* ; and the stretch as well as the impact has this effect. I have in fact shown, in the observations already cited (pp. 277–279), on the production of artificial elbow joints, that osseous deposit is stimulated by pulling strain as well as by pushing or impact.

In concluding, Mr. Cary admits one of the two contentions of the Neo-Lamarckians in his two closing propositions. He says "Plasticity of bone, using the word *plasticity* not in a physical sense merely, but to include absorption under pressure, will probably account for much structure in the foot and elsewhere, especially the connection with the joints, and in the fields of variation and correlation." In the second proposition he says that facts have been adduced by him which are inconsistent with the theory that the size of bones has been increased by the stimulus they receive, and with the theory that regions of growth are determined by regions of pressure and strain. "The testimony of the literature as to the latter point," he says, "is conflicting." I have shown that the supposed conflict is due to a misunderstanding on the part of the author of this paper. The proposition that pressure does not affect growth is in contradiction to the admission made by the author in his first proposition, where he admits that pressure determines structure; for in such change of structure there is always growth. Finally Mr. Cary remarks "That race changes follow those produced in the individual life, or that they are directly caused by their mechanical surroundings, I do not think it has been satisfactorily shown." The fact that the characters of bone structure admitted by Mr. Cary to have had a mechanical origin appear in the young before birth, is evidence that race characters are produced, in other words, that they are inherited.

Another objection proposed by Tomes, and quoted by Poulton and Wallace with approval, has reference to the kinetogenesis of teeth of Mammalia as described by Ryder and myself. Tomes asserts that it is quite impossible that the crowns of the teeth could have been altered by mechanical impacts and strains, since their form is determined in the recesses of the dental grooves, entirely removed from all the mechanical influences which affect the external surfaces of the jaws. But the observations of Koelliker and others show that osteoclasts are as active in dental as in ordinary osseous tissue. It is altogether probable that the modifications of dental structure have been produced by strain and friction under use in the adult, precisely as in the skeleton, and that the share that the unerupted crowns have in the process is that of inheritance only, as in the case of the skeleton. That teeth deposit dentine as process of repair in adult mammals is well known, and this repair is in direct relation to use. That the effects of dental wear are inherited is proven by the fact cited by Tomes and Wallace.

Another objection to the doctrine of kinetogenesis which has been made by some of the Neo-Darwinians is, that if growth under stimulus be true, how can it have limits, so long as the stimulus of use exists. In other words, what is to prevent, in the case of the vertebrate skeleton, of an indefinite increase in the length of the legs, of the teeth, and of their cusps, etc. The answer to this objection will vary more or less with the part of the structure considered. In general, however, it may be assumed that stimulus is stress due to a want of harmony between an organism and its environment, and that kinetogenesis is the result of the effort of the organism to adjust itself. So soon as equilibrium is attained, the stress of stimulus ceases or is much reduced, and evolution in this direction ceases. Such equilibrium is attained when the mechanism of an animal is sufficient for the satisfaction of its needs. When this is the case severe exertion is no longer necessary, and a period of easy use follows which is sufficient to maintain the mechanism in working condition. In the case where circumstances should become so favorable for the easy satisfaction of the necessities of life, as to call for little or no exertion, degeneracy of the organism is sure to follow. The well-known phenomena of degeneracy from disuse show that a large part, and in some cases all, of the stimulus of use, is only sufficient for the maintenance of the organism in working condition, and that there is no surplus to be expended in progressive evolution.

It is however true that some organs are stimulated to excessive growth by active use. Such are some of the teeth, which, if not worn at the crown by the opposition of those of the opposite jaw, soon grow to an inconvenient length. This occurs in the hypsodont molars of horses and artiodactyles, and in the prismatic molars and incisors of Glires. Hypsodonty in general is an illustration of continuous growth induced by long-continued stimulus in those orders of Mammalia, and in the Toxodontia and Proboscidia. The excessive growth of the canines in the South American saber-tooth tiger, and of the incisors of the mammoth, are cases where the energy of growth has not subsided in time to prevent excess.

Several years ago Prof. August Müller and I called attention nearly simultaneously to the probability that many of the forms of the reproductive organs of plants, especially the flowers, are due to the strains and other effects produced by their use by insects. Rev. George Henslow has written a book in which this subject is set forth in detail.[1] It is impossible to demonstrate this point with the same certainty as the kinetogenetic origin of the articulations of the vertebrate skeleton and their characters, owing to the absence of paleontologic evidence. Henslow, however, says: "When we find innumerable coincidences all tending in one direction, coupled with an indefinite capacity for varying in response to forces in all parts of plants, I still maintain that [this] theory does not utterly break down," as asserted by Mr. Wallace.[2] Wallace argues that since many regular flowers have been subject to the irritation of insects and have not become irregular, there is no reason to suppose that this is the cause of the irregularity in question. To this Henslow replies:[3] [Mr. Wallace] "will see that existing regular flowers being mostly terminal, have no lower petals at all, but are so situated as to offer access to insects from all points of the compass. Moreover, when a plant with normally irregular flowers (which are always situated close to the axis, so that insects can only enter them in one way) produces a blossom in a terminal position (as foxglove, larkspur, horse-chestnut, etc., often do), it at once becomes quite regular." This change may be brought about artificially, for, says

[1] *The Origin of the Floral Structures by Insect and Other Agencies.* International Science Series, Vol. LXIV.

[2] *Natural Science*, 1894, p. 178.

[3] *Natural Science*, 1894, p. 262.

Henslow, "flowers normally irregular in nature often revert to their ancestral regular forms under cultivation in the absence of insects, and then come true from seed, as the Gloxinias."

CHAPTER VII.–NATURAL SELECTION.

NATURAL SELECTION is the process of discrimination of variations, by which those which are most in harmony with the environment survive. It is, in short, as expressed by Spencer, "the survival of the fittest." Fitness is of various kinds, and is only determined by the nature of the environment. The success of a variety which appears in aquatic surroundings will depend on characters different from those which bring success in a forest. Variations which favor survival and increase among carnivorous animals differ from those useful to the life of herbivorous forms. So survival in human society depends on characters different from those which secure the same result among the lower animals, etc. It is thus evident that natural selection is of many kinds and that forms survive for various reasons; and it is hence of universal application. The reasons for survival may be divided into those which depend for survival on the relations of a type to the non-living environment, and those which depend on the living. The former may be divided into those which are passive and those which are active. The particular influences may be imperfectly enumerated as follows:

Non-living environment.

a. Passive.

Isolation of areas; continuity of areas; building material; food; place of concealment; temperature; humidity.

aa. Active.

Pressure of earth, water, and air.

II. Living environment.

Food; reproductive potency; sexual selection; digestive and other physiological power; muscular strength; superior weapons and other special mechanisms; intelligence.

I have already (page 4) quoted the language of Darwin, where he states that the supposition that natural selection is a cause of the origin of variations is a mistake. From the nature of selection, Darwin's position thus expressed, is self-evidently sound. The attempt has, however, been made to apply the term selection to the efficient cause of all variation, and to divide its exhibitions into two kinds, natural and artificial selection. If the primary assumption involved in this position is illogical, the dual division proposed is absurd. As may be readily seen in the table in the preceding paragraph, in which the factors of natural selection are enumerated, the conditions necessary to selection are mostly identical, whether imposed by nature or by the hand of man; i. e., whether natural or artificial. The physiological effects of food, temperature, exercise, etc., do not differ, whether due to natural conditions, or to the influence of man. The observation of man's influence is indeed especially instructive in increasing our knowledge of the effects of natural causes, since in the former case we have the process in action within our control, while in the latter case it is not.

The subject of natural selection has been ably

treated by Darwin, Wallace, and other writers, and it is one on which much further research may be profitably expended. It is the science of adaptations, and the name Chorology has been framed for it by Haeckel, but the earlier term Œcology is now generally used. It was not overlooked by biologists prior to Darwin and Wallace, and is stated in general terms by Lamarck in his *Philosophie Zoologique*, but it was reserved for the two authors just mentioned to create the science. I shall here only refer to a few aspects of the subject.

Isolation naturally tends to emphasize any peculiarities of structure which may harmonize with the conditions of the environment, by the barrier which it sets up against the entrance and mixture of forms from other localities where the environments are more or less different, and where the characters are correspondingly proportionately diverse. Isolation conversely prevents the emigration of forms, and the consequent mixture with the differing forms of other regions. Breeding in and in is produced on a large scale. Geographical isolation is a result of the formation and population of islands, whether this be accomplished by submergence below or by elevation above sea level. A noteworthy illustration of the former case is seen in the West Indian Islands, which represent the elevated regions of a former continent. Here the faunæ of the respective islands have been separated from each other since late Pliocene time. We find that while most of the genera of land Vertebrata are generally distributed, each island possesses peculiar species. This is even true of the birds, whose powers of migration are quite sufficient to enable them to pass from island to island. The restriction of land mollusca is still greater, several islands having genera peculiar to them.

Isolation is also produced by inequalities of land-surface, resulting from the elevation of mountain ranges, plateaus, etc. This is well seen in Mexico and Central America where the number of species of Vertebrata is large, owing to the fact that many of them are restricted to very narrow areas bounded by impassable barriers. The restriction of species of land Mollusca to each of the numerous valleys of the Hawaiian Islands has been made the subject of an especial study by Dr. Gulick, who treats of them with especial reference to the evolution of their forms.

The assimilation of inorganic matter necessarily preceded that of organic matter, so that this function characterized the first organic beings, whether animal or plant-like in other respects. Among animals we may regard the vegetable feeders as having by a little preceded the carnivorous forms. Omnivorous forms must have come into existence soon after, and from these, both classes of feeders have been from time to time recruited ever since. The primitive Vertebrata were probably carnivorous, and most of the fishes and Batrachia have always been such. Herbivorous forms have arisen from time to time among Reptilia, and of granivorous birds there are many. The early Mammalia were divided between omnivorous (Multituberculata) and insectivorous types (Protodonta, Pantotheria); while the higher Mammalia of all kinds were derived from more or less omnivorous forms (primitive Condylarthra and some Creodonta). We may account primitive insects to have been largely herbivorous, even more than they are at the present time, while Carnivora predominate in marine invertebrate life. It is not difficult to understand that circumstances of the environment would determine the food of animals, and would divert omnivorous forms into carnivorous or herbivorous habits as abundance of food and competition of rivals might dictate.

Sexual selection is of two general classes, that in which the male selects, and that in which the female determines the result. In the former case the most vigorous males, or those in which the mechanisms for seizing the females are most effective, propagate their kind most successfully. It is well known that in many animals, especially the Arthropoda, the males are furnished with especial organs of prehension. In vertebrates similar organs are especially conspicuous in some of the Batrachia Salientia. (See p. 65.)

The species of Arcifera exhibit peculiar structures during the breeding season ; either an extension of the natatory membrane, or the development of corneous plates or spurs, as aids to prehension. There is much variety and efficiency displayed in this point (except in Bufonidæ), in especial contrast to the apparent absence of all but the weakest modifications among the Ranidæ. This is in compensation for the structure of the sternum, whose lateral halves, being movable on each other, offer a slighter basis of resistance for the flexor and extensor muscles of the fore limbs of the male. In the Firmisternia the halves of the shoulder girdle do not overlap below on the middle line, but abut against each other, thus preventing compression (Fig. 51, page 198).

While no appendages of the season have been observed in some Cystignathidæ, in several genera two acute spurs appear on the superior aspect of the thumb and more rarely spur-like tubercles on the breast ; the body is sometimes shielded with hardened points on the rugosities, or the lip surrounded by an arched series of corneous rugæ. In the *Leptodactylus pentadactylus* Laur. a huge acute process of the metacarpal of the thumb, projects inwards. Its apex is covered by a horny cap, and it is a formidable grappling-hook to aid the male in retaining his hold. There is added to this in the same species a horny plate on each side of the thorax of the male, from which project three acute points. With these fixed in her back and the thumb spikes in her breast the females cannot escape. Structures like this do not appear in the Firmisternia. Here the inferior elements of the scapular arch abut against each other, so that the thoracic cavity does not contract on pressure, and the possibility of the male retaining a firm grip on the female is thereby greatly increased. In the *Cystignathus pachypus* the males exhibit a permanent enlargement of the brachium, dependent on largely developed anterior and posterior alæ of the humerus. (*Vide* Günther, *Ann. Mag. N. H.*, 1859.)

Another kind of male selection is accomplished by the combats of males for the possession of the females. This is usual in polygamous birds and Mammalia, and in some promiscuous species of both. Of the birds the Gallinæ form the best known example; and of the mammals, most Ungulata, and the eared seals (Otaridæ), are illustrations. In this way the weak males are eliminated either by death, or by exclusion from the opportunity of reproduction. The males in such species are armed with spurs, horns, or large teeth, except in some of the Perissodactyla, which have neither.

Female selection is seen in another direction. Here the male attracts by the superior brilliancy of his colors or peculiarity of physical appearance, as well as by his notions. The available growth-energy of the male being superior to that of the female in most animals, his structure is more liable to excess of development in useless directions. In many of the Arthropoda, especially the Insecta, the males possess processes of the head and thorax besides the especially useful prehensile peculiarities of the limbs. Among Vertebrata the male generally possesses the more brilliant colors. This is especially noteworthy in fishes and birds. It is also frequently the case in lizards, although in one genus (Liocephalus) the female has the brighter hues. The selection (taking no account of the origin of these characters) acts in the probable preference by the females for the most brilliant colors and most impressive forms, thus propagating both, and in the neglect of those males in which these characters are not so well developed. The plainness of the females aids in their concealment and enables them to perform their maternal functions in safety.

The desire of the males to attract the favor of the females leads to many peculiar performances among birds. The males display their plumage by spreading their wings, tail, tail-coverts, etc., and strut and go through many antics in the presence of the females. Familiar examples are seen in our barnyards in the turkeys, peafowls, and pigeons. In the paradise-bird the most remarkable exhibitions occur, according to Wallace. In song-birds the male is frequently the only or the best songster, and the development of the vocal powers resulting from the sexual impulse is most remarkable. Among Mammalia the female selection is less common than male selection. In the case of some of the old world monkeys (Macacus, e. g.) the female presents the greater physical indication of excitement in the extraordinary development of the nates at the season of heat. Mankind, appropriately to the high development of the mental powers, is selected with reference to qualities of mind as well as with regard to the physical attractions. Mental advantages being equal, beauty is preferred, but beauty is often neglected in favor of superior moral and intellectual characteristics. Both sexes take part in the selection; in the lower races chiefly the male, while in the higher races the choice rests ultimately with the female. It is probable that in future civilized mankind will exercise more care than now in the prevention of marriage of persons affected with serious physical and mental defects, such as chronic diseases, insanity, alcoholism, criminality, etc.; but beyond this, supervisory selection cannot go. The supposition which is sometimes entertained by some persons, that mankind will in a state of higher civilization prefer physical to mental perfection, is certainly ill founded. And among mental qualities, a high value will always be attached to those which render social life easiest and most pleasant; the standards of ease and pleasure being innumerable.

Among the most conspicuous examples of the action of natural selection are to be found those resemblances in color or form between animals and their environment which serve to conceal them from enemies. Animals possessing such protective appearances naturally escape the observation of enemies which prey on them, while those which do not possess them are more readily captured and eaten. Much is to be found of interest on this attractive subject in the writings of Wallace, Poulton, Beddard, and others. The two authors first named ascribe these color and form characters to natural selection *as a cause*. This is, however, impossible; yet natural selection has undoubtedly been the cause of their survival. Professor Poulton has demonstrated (p. 230) that the protective colors in lepidopterous pupæ are produced directly by the influences of light on the nerves of the animal and its reflex action on the pigment depository process.

The first objection to the belief that natural selection is the primary cause of organic evolution has been already stated as follows: "A selection cannot be the cause of those alternatives from which it selects. The alternatives must be presented before the selection can commence." But the supporters of the view that natural selection is the origin of variation allege that it produces this result by the continued survival of minute differences which are useful, thus accumulating variation. That minute advantageous differences will secure survival no one can doubt, but it must be remembered that the variations which constitute evolution have been in a vast number of cases too minute to be useful. But the general question is not affected by the supposition that advantageous variations may be sometimes minute. Minute or great, they have to be assumed in the argument for selection; and whether minute or great, they have a definite cause.

* * *

In conclusion of Part II. of this book, I trust that I have adduced evidence to show that the stimuli of chemical and physical forces, and also molar motion or use or its absence, are abundantly sufficient to produce variations of all kinds in organic beings. The variations may be in color, proportions, or details of structure, according to the conditions which are present.

PART III.

THE INHERITANCE OF VARIATION.

PRELIMINARY.

IN THE first section of this book I have endeavored to show that variation of character is not promiscuous or multifarious, but that it is limited to certain definite directions. That this rule applies to all kinds of characters, whether they are of the less fundamental kind which distinguish species, or of the more fundamental kind which distinguish the higher divisions.

In the second section I have endeavored to show that many characters, both those of the more superficial and those of the profounder kinds, are the direct result of chemical and physical stimuli, and of molar motion or use, or of the absence of the latter or disuse.

It now remains to ascertain whether the characters or variations so produced are inherited ; that is, whether characters so acquired are transmitted to succeeding generations. Unless this proposition is demonstrated, our knowledge of the method of evolution remains incomplete, and we must look for some new explanation of the progressive increments and decrements of structure which constitute the history of organic life. The present part of this book will be devoted to an examination of this question, and to the exposition of such laws as may be derived from such examination.

CHAPTER VIII.—HEREDITY.

I. THE QUESTION STATED.

IT IS the popular belief that characteristics of parents are transmitted to their offspring through the medium of the reproductive cells. This opinion is founded on an infinitude of observations easily made on plants, animals, and men, and in fact it is not denied as a general statement by anybody. It is a fact of ordinary observation that many and apparently most of the structural characteristics of one generation are inherited by its offspring. Not only is this the case, but the functionings of organs which depend on minute histological peculiarities are inherited. Such are points of mental and muscular idiosyncrasy; of weakness and strength of all or any of the viscera, and consequent tendencies to disease or vigor of special organs. Darwin has collected in his work on the Descent of Man numerous instances of the inheritance of various tricks of muscular movement of the face, hands, and other parts of the body. But it is a fact of equally ordinary observation that some peculiarities of parents are not, or may not be inherited, and among these may be enumerated mutilations and injuries, as well as characters which are normal. It is then a question of essential importance to ascertain what kinds of characters are inheritable, and what are not inheritable.

It has been insisted by Weismann and others that characters which are newly acquired by an organism are not inherited, whether they be products of normal or abnormal conditions. In support of this view, he points to the early isolation in embryonic life of the reproductive cells from the remainder of the organism, and their continued isolation during later life, so that they are protected from the stimulating influences which affect the remainder of the body. He also points to the permanence of this isolation of the germ plasma from generation to generation, which insures only the transmission of those characters which it contains, as distinct from those which are found in the remaining cells of the organism, which constitute the body or soma. The characters which are inherited, and which are present at birth are termed congenital, while those which appear in the body under the influence of external stimuli are termed acquired. The theory of Weismann then is, that the acquired characters are not inherited.

Besides the fact that sporadic injuries and mutilations of the soma are not inherited, there have been cited various cases of the non-inheritance of mutilations which have been often repeated and for long periods of time. Thus the rupture of the hymen in human females has not been followed by its abolition. The practice of circumcision by the Jews has not resulted in the disappearance from that race of the portion of the body thus artificially removed. The continued cutting of the hair of men of many races has not made it less abundant. The practice of distorting the feet of a class of their women by the Chinese has not modified the shape of that part of the structure in the race. I have myself cut off the tails of nine successive generations of mice without producing the slightest effect on the length of the tails of the tenth.

Nevertheless such negative evidence only demonstrates that such modifications of the structure *may not* be inherited. A single undoubted example of the inheritance of a mutilation would prove that no insurmountable barrier to such inheritance exists. And well authenticated examples of such cases are known and will be mentioned later on. But it is not with mutilations that the paleontologist has to do. The rupture of the hymen and circumcision, and most mutilations, can only occur once in the life of the individual, and generally they produce no appreciable direct effect on his or her metabolic physiology. Moreover, the mutilations above cited as not inherited are experienced by but one sex, except in the case of the tails of the mice. The question is widely different with regard to the parts of the structure in which we observe the real differences between organic types. The definitions of natural divisions rest to a great extent on the diversities displayed by the organs of motion and nutrition. Now these are in use in animals during most of the hours not spent in sleep. Their movements are perpetual, and their activities only cease with death. It is then quite unreasonable to cite the history of mutilations as evidence against the inheritance of natural characters produced by oft-repeated and long continued natural causes.

It has been shown in Part Second of this book that structural characters are produced by use and other stimuli to growth. It has also been shown in Part First that the characters so produced show a progressive increment or evolution from earlier or later geologic periods. There are two possible explanations of this phenomenon. The one is that the characters of one generation are inherited by the next, which adds to them by the activity of the same stimuli which gave them origin, thus producing progressive increase of growth. The alternative is, that these structural characters are produced by each generation for itself. It is obvious that the latter hypothesis provides for no additional development of a character in one generation above another. There are other objections to the latter view, but letting these pass for the present, it is only necessary to examine the embryonic history of animals to show that it is entirely untenable. For if some or all of these acquired characters can be found present in the early stages of growth, as in the egg, the pupa, the fœtus, etc., it becomes clear that such acquired characters have been inherited. That such is the fact is abundantly demonstrated by embryological researches. This fact alone is sufficient to set at rest by an affirmative answer the question as to the inheritance of acquired characters. And that this answer applies to all time and to all evolution is made evident by the fact, which is disclosed by paleontology, *that all characters now congenital have been at some period or another acquired.*

2. EVIDENCE FROM EMBRYOLOGY.

a. Vertebrata.

I have already (p. 292) pointed out the gradual evolution through mechanical causes of the tongue and groove-joints in the Mammalia as exhibited by the distal ends of the metapodial bones of the feet where they articulate with the phalanges. Mr. Carey having agreed with me that those have been produced by mechanical causes, he alleges that they are not inherited, but are produced by each generation for itself. To this Dr. J. L. Wortman remarks as follows:

"With reference to Mr. Carey's first proposition that the metapodial crests are produced during the life of each individual by the necessary interaction of parts, it appears to me to be a very simple one indeed. If they are produced, by pressure during the lifetime of each individual, and are not inherited, then clearly we should find the crests absent in new born animals that had never walked, and in which the metapodials had not been subjected to any impact or pressure whatever. I have taken the trouble to examine a number of such examples in which the distal ends of the bones were entirely cartilaginous, and I find that the keels and grooves are as well developed as they are in the adult animal. I will cite one case in particular in which I happen to know the history completely. During the past winter, a young hippopotamus was born in the Zoölogical Gardens in Central Park, New York, and it was stated to have been a premature birth; the animal lived but twenty-four hours, and I was informed by the keeper that it never stood upon its feet. An examination of the feet shows that the distal ends of the metapodials are entirely cartilaginous, and in them the keels are as well prefigured in cartilage as they are formed in bone in the adult animal. I have also found the same to be true of new-born rabbits and guinea-pigs. In another case of a young buffalo calf preserved in the American Museum Collection, the distal keels of the metapodials are complete notwithstanding the fact that the epiphyses of all the bones are very imperfectly ossified. This evidence, it appears to me, effectually disposes of the question of the production of these structures during the lifetime of the individual. They are as truly inherited as is the number of digits or any other important organ in the animal economy."

Such observations may be repeated indefinitely. Thus the astragali of the higher Mammalia are already grooved before birth, and are not flat up to that time as in their Puerco ancestry. The reduction of digits appears very early in fœtal life, and the ball and socket articulations of the cervical vertebræ of the Diplarthra are by no means introduced after birth.

The teeth possess the normal structure of their crowns while yet in the alveoli before eruption. In some cases the transition from a primitive to a modern type of tooth has been observed to take place in the embryo.

Dr. A. von Brunn[1] has shown that in the embryos of the rat, the enamel-producing epithelial layer of the molar teeth undergoes a remarkable change at the places where the transverse crests of the crowns are to appear. Before the enamel layer is deposited, the portion of the epithelial layer corresponding to the cross-crests undergoes degeneration, as a result of which it acquires the character of a stratified squamous epithelium. Thus no enamel is laid down on the summits of the cross-crests, which present the exposed dentine when erupted. Now it is a fact that the crowns of the molar teeth of the ancestors of the genus Mus, were covered with enamel at maturity, like all other

1 "Notiz über unvollkommene Schmelzentwickelung auf den Mahlzähnen der Ratte, *Mus decumanus*": *Archiv für Mikroskopische Anatomie*, 1880, XVII, pp. 241–243, Pl. XXVII. Ryder, *American Naturalist*, 1888, p. 547.

primitive Glires. The removal of the enamel from the apices of the tubercles and crests of their descendents is due to the abrasion consequent on ordinary use. On this Ryder (*l. c.*) remarks: "The great value which is to be attached to the fact that abrasions of the enamel of the adult, which have reacted upon the functional activity of the enamel organ of the embryo rat, so that such mechanically induced alterations could be inherited, does not consist so much in the proof it affords that mutilations can be inherited, as it does that mutilations incurred in the ordinary struggle for existence, may, under certain conditions in certain practically feral species, be transmitted."

Having shown by these examples that acquired characters can be inherited, I offer some other illustrations which are at hand.

b. Arthropoda.

It has been already rendered probable if not certain (p. 268) that the segments of the body and limbs of the Arthropoda were originally produced by the movements of definite tracts on each other, during the period that the external surfaces were becoming hardened by chitinous or calcareous deposits. It is well known that this segmentation is no longer produced by this mechanical cause during the adolescent or any other post-embryonic stage of the life of the individual, but that it appears during the various stages of embryonic life, and is therefore inherited. Thus segmentation of the body appears in the embryo while still attached to the yolk. During the larval life of many insects the process of segmentation is suspended, but during the repose of pupal life, it goes on with great rapidity. In this stage while protected from external mechanical stimuli, the limbs with their specialized segments are fully developed, so that the individual is mature as it issues from its prison. This illustration of inheritance derives its point in the present connection from the fact that it presents an example of the inheritance of characters which were plainly acquired by mechanical stimuli during post-embryonic life of the primitive ancestors of the Arthropoda.

3. EVIDENCE FROM PALEONTOLOGY.

a. The Impressed Zone of the Nautiloids.

I have already quoted Professor Hyatt on the parallelism which is characteristic of the various series of nautiloid Cephalopoda, as discovered by paleontologic research. (P. 182.) The impressed zone is a character which has been produced by mechanical causes (pressure), and Prof. Hyatt has observed cases where this acquired peculiarity has been inherited in instances where the mechanical cause which produced it no longer existed. He describes these instances as follows:[1]

"The characteristic dealt with in the paper of which this is an abstract, is of essential importance among nautiloids and ammonoids or all of the Cephalopoda having chambered shells and living within their shells. It consists mainly of an impression made on the inner side or dorsum of each outer whorl during the coiling up, as the whorl grows and is moulded over the venter or outer side of the next inner whorl.

"This matter will be better understood, if a short description is given of the following figures. Fig. 115 shows an almost complete fossil cast of a full grown

[1] *American Naturalist*, 1893. October, p. 865. Professor Hyatt has personally looked over and corrected these quotations.

Metatoceras cavatiformis Hyatt, and some of the lines or sutures made in the external surface of the cast by the intersections of the partitions or septa that cut up the coiled tube of the living shell into air chambers. Fig. 116 shows a broken specimen of the same species, but with the outer and older whorls in large part removed. The innermost septum near the center of the coil was built across the interior after the animal had constructed the hollow apex or point. It then moved along, adding to the external wall of the tube,

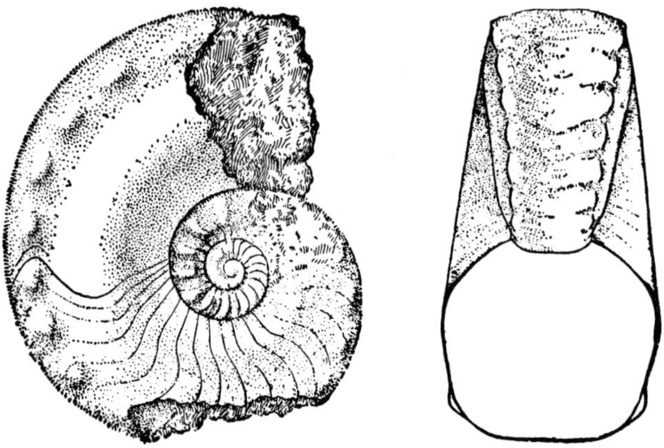

Fig. 115.

which has been destroyed and removed from this cast, and built the second septum, and so on until it reached the tenth septum. By some freak of fossilization a number of the septa beyond this have been destroyed, so that if we were to remove the fragment of the external whorl and take out the center which has just been described, this would have the exact aspect of a cast of a young shell with ten air chambers.[1] The

[1]The shaded area in the center, shaped like a large inverted comma, was an open space in the living shell. This is almost invariably filled by the

eleventh air space or chamber being open and without divisions would then appear to be the living chamber which the animal occupied when it built the tenth sep-

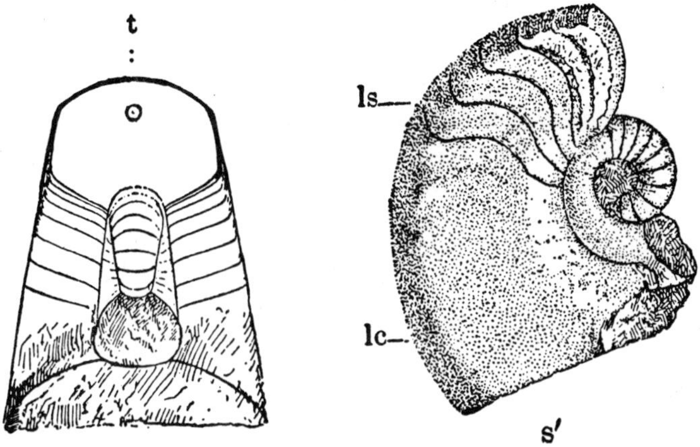

Fig. 116.

tum. Normally the shell really continued to progress from the tenth septum by additions to the outer wall and put in new septa behind it, together with the con-

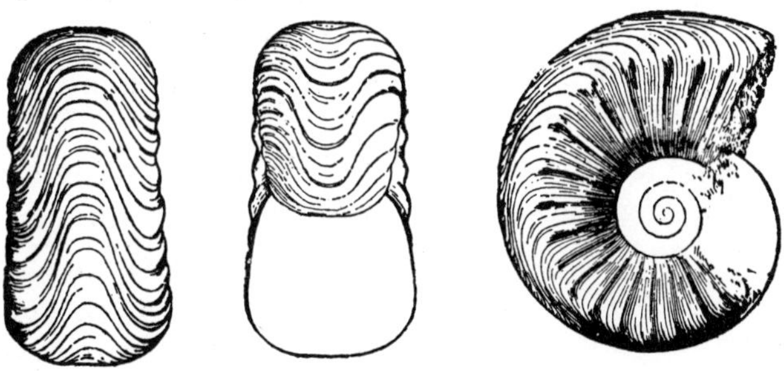

Fig. 117.

necting tube until it reached *s'*, and finally the last septum, *ls*. This one, *ls*, was really the last one built

rocky matrix in which the shells occur and is often, as in this specimen, allowed to remain. See also 4, 5, 6, of Fig. 119, which show the comma shaped umbilical perforations or openings left at the center through the crytoceran form of the young.

and it formed the floor of a true living chamber, *lc*, formerly occupied by the animal at the time of its death and burial in the sediment of the carboniferous period. Fig. 115 shows a similar fossil but with a longer, although still incomplete living chamber. If the external wall of shell had been preserved none of these structures could be seen. Fig. 117 shows a fossil *Temnochilus crassus*, a shell of the same family, with this external wall preserved, and all these internal structures covered up. The impressed zone is the reëntrant curve shown in all these figures and especially marked in the lower outline of an outer whorl of another carboniferous species, *Metacoceras dubium* Hyatt, Fig. 118, *im. z.*

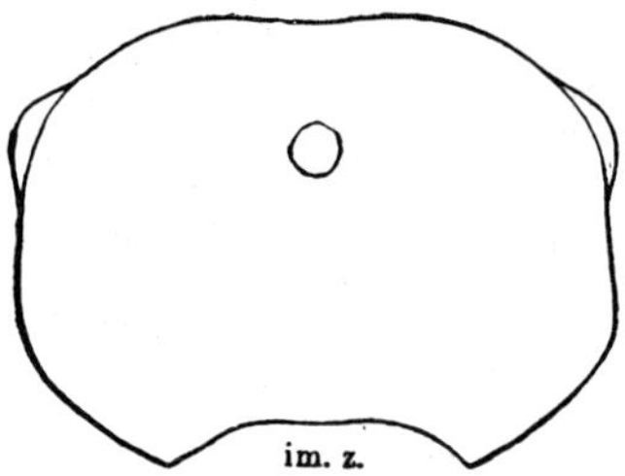

Fig. 118.

"It is not necessary to go into a discussion of the details of internal structures and their relations to the impressed zone in this abstract, but it is essential to give a general description of the morphogeny of the order of nautiloids.

"This group of chambered Cephalopods contains the following classes of forms: first, straight, conical shells, type Orthoceras, Fig. 119, No. 1; second, curved cones, Cyrtoceras, Fig. 119, No. 2; third loosely coiled, open whorled cones, do., No. 3; fourth, coiled cones with the whorls more or less enveloping, do., No. 5. The fourth and fifth forms are usually included in the old genus, Nautilus. Practically, it is better to designate the first class as orthoceran, the second as cyrtoceran, the third as gyroceran, and the fourth and fifth as nautilian. In tracing genetic series through time they are found to diverge in their evolution, starting with the orthoceran and passing through parallel lines of forms, many of the genetic series having in succession cyrtoceran, gyroceran, and even nautilian forms of the fourth and fifth classes. Others are not so perfectly parallel, stopping short with the cyrtoceran class of forms or the gyroceran. Many also begin with cyrtoceran shells, while others diverge from the gyroceran, and still other series have only nautilian shells of different grades of close coiling and involution.

"The application of the law of repetition in heredity to the chambered shell-covered cephalopods, shows that the straight orthoceran shells, Fig. 119, No. 1, were repeated in the young of the curved cyrtoceran forms, Fig. 119, No. 2, and these forms in their turn in the young of the gyroceran forms, Fig. 119, No. 3; and this may be seen by comparing the young or apical part of each shell represented in outline with the full-grown shells of the preceding figures. The apex of No. 2, with the whole of No. 1; the apex of No. 3, with the whole of No. 2. It will be understood, of course, that the figures in outline represent full-grown shells, except when otherwise explained, and that they were built like the shells of Nos. 1–2, by an animal living in their interiors and adding band after band of shelly matter to the exterior, but in these outlines the shell is supposed to be perfect and the internal structures concealed.[1] The young of Fig. 119, No. 4, which represents the fourth class of forms repeats the cyrtoceran form, then curves more closely, and just before it comes in contact there is a short time when

[1]Except in No. 9, in which a portion of the shell is broken away. showing the cast of the interior and the sutures.

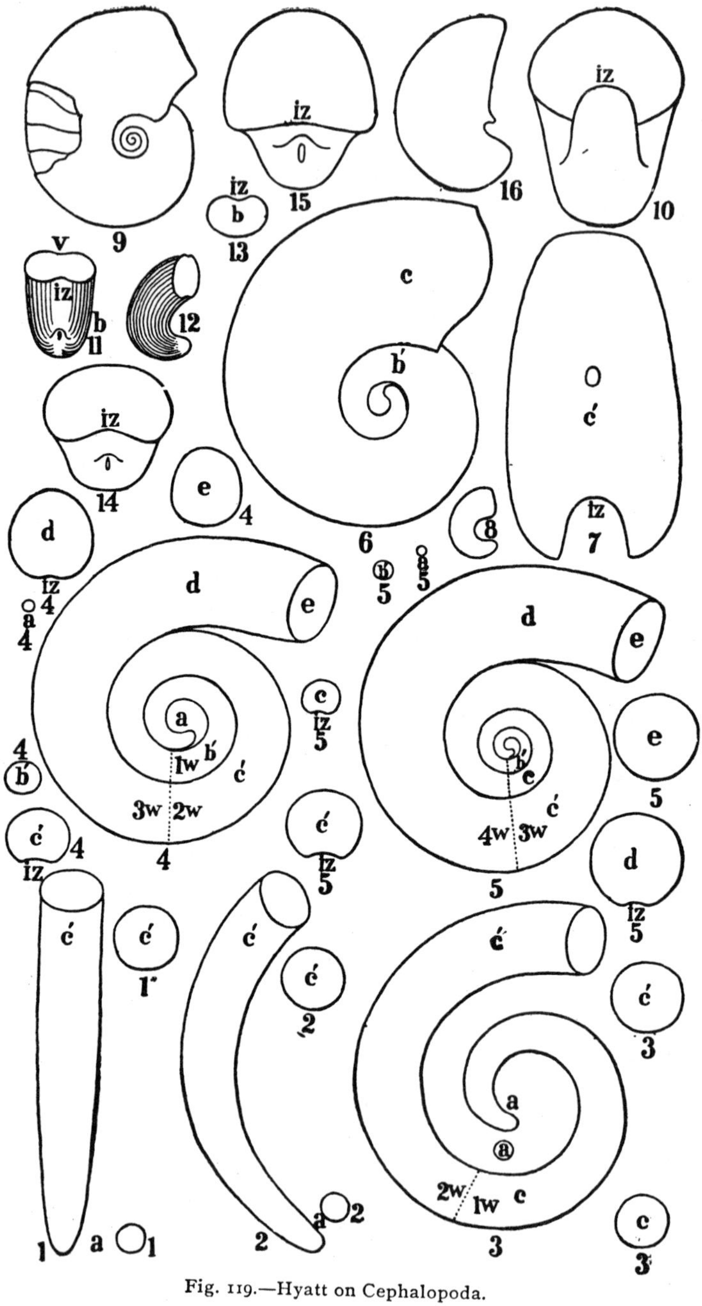

Fig. 119.—Hyatt on Cephalopoda.

EXPLANATION OF FIG. 119.

LETTERING.

a. Apex of shell. This usually bears a scar on the point, as shown in Nos. 14 and 15, but this has no bearing on the question discussed, and has not been described. This also represents the youngest (nepionic) or cyrtoceran stage in the growth of the shell, No. 8 being a young shell with complete living chamber. This letter also indicates the location of the sections correspondingly lettered in the figures.

b is used to indicate the section of the cyrtoceran stage in Nos. 11–13.

b' is used to indicate the place of the sections, Nos. 4–5*b'*, upon the whorls of Nos. 4–5. They were taken through the whorl in the gyroceran stage.

c is used for the adolescent (neanic) stage of growth in the whorl and the corresponding sections.

c' is used for the full-grown (ephebic) stage in the growth of the whorl and the corresponding sections.

d for the first part of the senile (gerontic) stage.

e for the final and most degenerative part of the senile stage.

i z for the impressed zone.

v venter or outer side of the shell, the dorsum being the inner side of the whorl.

w for the whorls, thus 1 *w* in Nos. 3 and 4 means the end of the first whorl, 2 *w* the beginning of the second whorl, 3 *w* that of the third whorl. These letters serve to show the progressive increase in numbers of the whorls in the different classes of forms.

FIGURES.

No. 1. Outline of an orthoceran shell.

No. 2. Outline of cyrtoceran shell.

No. 3. Outline of gyroceran shell.

No. 4. Outline of nautilian shell, having a larger umbilical perforation at (*a*) and fewer whorls at the same age, than in No. 5; in other words, it is less tightly and completely coiled up than the class of shells represented by that figure.

No. 5. A nautilian shell with tighter coils than in No. 4 and the whorls coming in contact and the impressed zone beginning at an earlier stage.

No. 6. *Barrandeoceras bohemicum* (sp. Barrande) Hyatt, showing the most involute of the Silurian shells so far as known; No. 6 is reduced in size, but the section No. 7 is natural size.

No. 8. A young shell of the same, natural size, with complete living chamber.

Nos. 9–10. *Coloceras globatum* (sp. De Koninck) Hyatt, adult. No. 9 has a part of the outer shell broken off, showing the edges of the septal partitions (sutures) as lines on the strong cast of the interior.

Nos. 11–13. Same to show the cyrtoceran stage and section, with its impressed zone.

No. 14. *Cenoceras clausum*, Hyatt.

Nos. 15–16. *Nautilus pompilius*, to show the cyrtoceran stage with its impressed zone.

it overlaps the apex without touching it. At this time it is plainly gyroceran, like the whole of No. 3. After it touches the first whorl just beyond the apex it remains in contact, and the inner side or dorsum of the second or overlapping whorl begins to show a flattening as a result of this collision of the whorls. The sections of the orthoceran, cyrtoceran, and gyroceran whorls show no such flattening in any of the specimens examined, although hundreds of different kinds have been studied. The sections are designated on the plate by the same letters as the supposed lines of the sections made through the tube, and although diagrammatic figures, they give a sufficiently clear general explanation of the facts observed. More specific figures could have been given in abundance and will be given in the paper now in course of preparation.[1]

"Fig. 119, No. 5, shows the same phenomena as No. 4. The young is at first cyrtoceran like the adult whorl of No. 2, and open, then becomes gyroceran in curvature and finally overlaps the apex when it has arrived at the end of the first volution, but does not at first touch it. Then coming into contact it acquires a flattened area or faint impressed zone on the dorsum or inner side of the second volution, as is shown in the section No. 5*c*. This is similar to the section of No. 4 shown in No. 4*c′*, which represents a cut through an adult whorl of the fourth class of forms. It differs only in being smaller, on account of the younger stage of growth at which it occurs.

"The entire series of forms from orthoceran to nautilian is more or less represented, even in the earliest period at which the nautiloids appear, namely, in the

[1]See "Phylogeny of an Acquired Characteristic," Hyatt, *Proceedings American Philosophical Society*, Philadelphia, XXXII., No. 143.

rocks of the Quebec group. There is, however, this qualification: the fifth class of forms, or the involute nautilian, are relatively rare and become more abundant in successive periods. The young of nautilian shells of the earlier periods are also apt to be less closely coiled, or, in other words, remain open and similar to cyrtoceras for a longer time during their growth. This is shown by the large size of the central hole, or umbilical perforation, left in the center of full-grown shells. This perforation is much larger, as a rule, in Paleozoic than in the Mesozoic forms.

"In each period the genetic series or groups of nautilian forms have peculiarities of structure in the sutures, ornaments, apertures, etc., by which they can be separated from each other, and these peculiarities are the same as those possessed by gyroceran, cyrtoceran, and often orthoceran shells which occurred often earlier in time, so that one can trace each group of nautilian shells back to its ancestors through the parallel stages of evolution above described. The groups, in other words, are parallel in their morphogenesis, like two individuals of the same parents in their development from youth to old age.

"As a general rule the impressed zone originates, as described above, after the whorls come in contact, rarely before this time in the growth of any individuals. Barrandeoceras is one of the most involute shells known in the Silurian, and Fig. 119, No. 6, gives a true sketch of this species; No. 7, shows a section of a full-grown shell with a decided impressed zone, and No. 8 is the young. This last is a purely cyrtoceran form with a compressed elliptical section like that of No. 7, but no impressed zone, the inner side being rounded like the diagram of Cyrtoceras, No. 2. The impressed zone is not present in the young of Ophidioceras, the closest coiled of all these forms, nor in the young of most species of the Silurian before the whorls touch, and all of the species likely to present this peculiarity have been investigated.

"The impressed zone is also, as a rule, lost in the oldest stage of the whorl of every individual when the whorls cease to continue to grow in contact. This condition is represented in the last part of the outermost whorl of Nos. 4 and 5 in sections, Nos. 4*e*, 5*e*, and in the outlines of their apertures, which are elliptical. The sections represent cuts through the whorls when, as is the case in extreme age, these cease to increase in size. As soon as this senile contraction begins to occur, the sides shrink, becoming narrower, the amount of involution becomes less, and the impressed zone shrinks in breadth, as shown in the sections. When the whorl finally parts company in consequence of continued contraction, the already shrunken impressed zone, Nos. 4*d*, 5*d*, rapidly disappears, and the apertures of such shells are frequently as round and free from indentations on the inner as on the outer side, as is shown at the free end of Nos. 4 and 5.

"In normally uncoiled forms, usually named Lituites, when the adult or young is coiled, and the succeeding stages, whether representing adults or old shells, are uncoiled, the phenomena are similar. The impressed zone is lost after the growth ceases to bring the whorls of the shell into contact.

"The young and the adults of many of these forms have now been observed in the earliest periods, and it is, therefore, obvious that during these early times the impressed zone must have been a modification of the whorl which took place in consequence of the mechanical effects produced by close coiling. This characteristic is slight when the coiling is slight and is developed in precise proportion to the increase of coiling or involution of the whorls, and, on the other hand, when through degeneration due to age, or to other causes, the whorls cease growing in contact, the impressed zone gradually disappears.

"Thus it generally appears preceded and accompanied by an observable tendency in the mode of growth toward closer coiling and that this tendency is quite capable of producing the impressed zone can hardly be denied with any show of reason, since the characteristic tends to disappear in proportion as the pressure is relieved through the degeneration of the powers of growth-force to continue the normal rate of progressive increase of bulk in old or young or prematurely degenerate shells and in uncoiled whorls of all kinds and all ages.

"The shells of Devonian series of nautiloids have also been extensively examined, especially in the more involute nautilian forms of the genus Nephriticeras, and so far not one has been found with the slightest indication of the presence of an impressed zone in the cyrtoceran or gyroceran stages of development. In several examples also, the disappearance of this characteristic has been observed in the last stages of old whorls. There is, therefore, every reason for regarding the impressed zone as a ctetic characteristic acquired in the later stages of growth and not hereditary so far as is known in any shells of the earlier Paleozoic periods.[1]

[1] Certain exceptions have been found since this was written, but their evidence is purely negative, it is impossible to say of them at present whether

"The same statement may also be made with regard to the majority of Carboniferous shells. There is, however, a notable exception in *Coloceras globatum* (sp. De Kon.) Hyatt, and very likely some other species of closely coiled nautilian shells. In *C. globatum* of Visé, Belgium, I found in seven specimens that the impressed zone appeared while the whorl was still in the cyrtoceran (or open) stage. Fig. 119, Nos. 9–10, give outlines of the adult of this species, and Nos. 11–12, of the young and the zone, showing that the impression appeared long before the whorls touched each other and began to assume nautilian characters. Section, No. 13*b*, shows the impressed zone occurring in the cyrtoceran stage while the venter or outer side of the whorl was rounded. Such facts admit of but one explanation, namely, that in this species the impressed zone had become hereditary and was in consequence repeated at an early age, previous to the occurrence of close coiling which usually produced it in the ancestral forms of the same group.

"There are certain correlative characters which lead me to think that this is only a partial statement and perhaps a more complete and better one would be as follows: that the impressed zone, together with a peculiar broadening out of the dorsum and helmet-shaped section of the whorl, and perhaps also certain forms of sutures occurred in the early stages of some Carboniferous species before the nautilian stage, and consequently they must have been introduced by heredity into the development of this species before the tendency to close coiling had completed the first whorl.

the impressed zone appeared as a genetic character or as a mechanical necessity. Either view can be taken, but the positive evidence is that they are very rare, and the impressed zone appears in them as a parallel character of distinct diverging series of forms.—*A. H.*

Thus these characters, although purely ctetic in origin, were repeated before the usual conditions recurred in the ontogeny of this species which had obviously and repeatedly produced them in the nautilian forms of the earlier Paleozoic and the more generalized genetic series of the Carboniferous. That this species, *Col. globatum*, is a highly specialized species is shown by other characteristics, especially the early inheritance of a furrowed abdomen, shown at *v* in Fig. 119, No. 11, and a peculiar aperture.

"The Triassic period is unimportant in this connection since it has but few nautilian species that are deeply involute and also sufficiently well known to throw any light upon this problem. All of the true orthoceran, cyrtoceran, and gyroceran forms diminish in the Carboniferous and disappear with the Trias.

"The Jura contains a considerable number of nautilian shells of different genera of which the cyrtoceran stages are sufficiently well known. *Cenoceros aratum*, of which several specimens have been studied, shows the impressed zone and correlative characters in this stage; *C. lineatum* is the same; *C. clausum*, same; *C. intermedium*, same. Fig. 119, No. 14, shows the cyrtoceran stage in a shell of *C. clausum*, with a well developed impressed zone, *i. z.* Endolobus is a characteristic Paleozoic type and there is a single survivor of this series in the Jura, *End.* (*Naut.*) *excavatum* sp. D'Orb. It is, therefore, very interesting and instructive to note that this species has the impressed zone, according to D'Orbigny's figure, during the cyrtoceran stage. This species has a large umbilical perforation and is slower in coiling up than other Jurassic species. The evidence that the impressed zone and its correlative characteristics are inherited in most species of the

Jura before the habit of close coiling could have acted upon the whorls so as to produce this modification is, therefore very general and convincing.

"The leading characteristic of parallelism in all genetic series of nautiloids is, as may be inferred from the facts cited, a tendency toward closer coiling and greater involution in the more specialized forms of each separate series and a correlative increase in the profundity of the impressed zone. When the impressed zone becomes inheritable in some closely coiled and involute specialized shells of the Carboniferous and in similar shells in all of the genetic series of the Jura this result is also directly connected with the observed fact of the quicker development of the coiling up tendency in the young of these Jurassic shells. This is shown by the small diameter of the umbilical perforation in the centers of the shells of the Carboniferous. It is also connected with the fact that the primitive uncoiled forms, orthoceran, cyrtoceran, and gyroceran shells begin to die out in the Carboniferous and cease with the Trias as mentioned above.

"This demonstration of the characters that accompany progress in close coiling, enables me to fill a gap which occurs in the evidence during the Cretaceous. In this period the existence of the impressed zone during the cyrtoceran stage of individuals has not been clearly established by observation except in two species, a form allied to *Cymatoceras pseudoelegans* D'Orbigny, from Faxoe, and *Cymatoceras elegans* from Rouen. In other shells, although a considerable number have been broken down, the state of preservation has been invariably imperfect. The coiling, however, in the young of all the shells examined is notably more accelerated than in the similar shells of the Jura, and the whorls broader and having more specialized characteristics correlative with closer coiling and the early existence of an impressed zone. It is, therefore, fair to infer that the evidence when accessible will confirm the facts observed in previous periods.[1]

"The same arguments apply also to the tertiary forms as far as known.

"The terminal members of the nautiloids are, of course, the existing species. *Nautilus pompilius* and *umbilicatus* have been examined in a considerable number of specimens, and in all of these the impressed zone and correlative helmet-shaped whorl and broad flattened dorsal side appears during the cyrtoceran stage. Fig. 119, Nos. 15–16, are outlines of the shell of *pompilius* during the cyrtoceran stage exhibiting the helmet-shaped whorl, broad dorsum, or inner side, and its impressed zone, *iz*. Thus, when the whorls touch, as in all the nautilian shells of the Carboniferous, Jura, and Cretaceous, in which the same acceleration of development also occurs, the whorl is already prepared to become involute and to mould itself more readily and rapidly over the surfaces of the apex and the side of the succeeding whorls. In other words, heredity has begun the work before the whorls touch, and before the deepening and enlargement of the impressed zone through the pressure of close coiling is begun. There are quite a number of characteristics of the species of existing Nautili which lead to the inference that they are survivors of Jurassic and generalized Cretaceous and Cenozoic forms; the size of the umbilical perforations, the smoothness of the shells, the simplicity of the sutures, and so on. These facts are

[1] This inference has been fully sustained by subsequent investigations.—*A. Hyatt.*

of importance only in so far as they show that the existing Nautilus does not represent the acme of progress of its order, but is a descendant of shells with less complicated structures than many of the genera of the Carboniferous, Jura, and Cretaceous."

In these cases it seems that the mechanically acquired impressed zone is inherited from the greater part of the soma where it existed to a part of the soma of the young where it could not be produced by mechanical causes, by reason of the non-contact of the parts. This acquisition appears in a few Carboniferous species, and then it is present in the cyrtoceran or mesozoid stage of all the Jurassic, Cretaceous, and Cenozoic species. Professor Hyatt, in his "Phylogeny of an Acquired Characteristic," thus summarizes his conclusions:

"The facts and arguments brought forward seem to justify the following conclusions:

"1. The impressed zone is primitively a contact furrow, an acquired characteristic of the dorsum of the whorls of nautilian shells having large umbilical perforations, which appeared either in the ananeanic or metaneanic (maturing) substages, and rarely later in their ontogeny. There is abundant positive evidence that in these primitive forms this furrow is a purely mechanical result of the nautilian mode of growth, not appearing in the ontogeny before contact, and either partially or entirely disappearing on the free gerontic (senile) volution.

"2. The impressed zone does occur independently of contact on the free dorsum of the paranepionic (adolescent) substage as a dorsal furrow in some close-coiled, highly tachygenic (accelerated) nautilian shells in the Quebec group and in the Devonian.

"3. While there is no positive proof that the dorsal furrow originated through heredity in the paranepionic substages of these nautiloids of pre-Carboniferous age, there is also no satisfactory evidence that it originated in the young of such species as have this character, through purely mechanical agencies.

"4. There is positive evidence that the similar dorsal furrow which also appears at the same age in the young shells of *Coloceras globatum* and perhaps *Cœlogasteroceras canaliculatum* among Carboniferous nautiloids can be explained only when it is considered as a transmitted, tachygenetic (accelerated) characteristic.

"5. This fourth conclusion is supported by the presence of a similar dorsal furrow in the paranepionic (adolescent) substage of the young shells of all the nautiloids of the Jura, so far observed.

"6. The fourth and fifth conclusions are rendered still more probable by the presence of the dorsal furrow at an earlier age, the metanepionic substage, in all of the nautiloids so far observed, from the beginning of the Cretaceous, through the Tertiaries, to and including the living species of the genus Nautilus. Its presence on this cyrtoceran volution in Cretacic shells can be explained only when it is considered as a transmitted, tachygenetic (accelerated) characteristic derived from ancestral nautilian shells of the Jura, which have the same characteristic at a later age, i. e., in the paranepionic substage.

"7. The first conclusion is also sustained by the parallel phylogeny of the impressed zone in the ancestral forms of the Ammonoidea, the Nautilinidæ, and especially in the Mimoceras, the radical genus of this family.

"8. The fourth, fifth, and sixth conclusions are

also supported by the presence of a contact furrow on the dorsum of the earliest age of the conch in the specialized and highly tachygenic forms of the Goniatitinæ of the Devonian and of all the remaining ammonoids to the end of the Cretaceous.

"9. These cumulative results favor the theory of tachygenesis (acceleration) and diplogenesis, and are opposed to the Weismannian hypothesis of the subdivision of the body into two essentially distinct kinds of plasm, the germ-plasm, which receives and transmits acquired characteristics, and the somatoplasm, which, while it is capable of acquiring modifications, either does not or cannot transmit them to descendants." (*Proceedings American Philosophical Society*, Vol. XXXII., p. 615).

4. EVIDENCE FROM BREEDING.

Under this head I cite the results of experience of breeding of the domesticated Vertebrata. It is here that we have had the best opportunity of testing the possibility of the inheritance of acquired characters, since the species in question have been the objects of observation and experiment for a long period of time. I especially avail myself of the writings in this connection of Prof. Wm. H. Brewer, of Yale University, President of the Agricultural Society of Connecticut. The result of his long-continued observations is contained in a series of papers in the journal *Agricultural Science* of the years 1892–1893. He considers the subject under the following heads, viz.: The inheritance of characters which are due to nutrition; of those due to the exercise of function; of those due to disease; of those due to mutilation and injuries; of those due to habit, training, and education; of those due to regional influences and to a combination of causes ; and of those of acquired plasticity and adaptation. I commence with an example of the

a. Inheritance of Characters Due to Nutrition.

"'One class of 'acquired characters,' the transmission of which by heredity is especially denied by Weismann, includes all 'those which are directly due to nutrition.'

"This denial strikes at the very foundation of what has heretofore been considered an essential factor in the practical improvement of breeds as to size. The size attained by adult, healthy domestic animals depends practically upon two causes—heredity and nutrition. Heredity is of course the chief one, for no amount of feeding will make the Shetland pony equal the Norman horse in size ; but whatever the heredity, the size of the adult individual as compared with the average of others of the same breed depends usually upon its food. The ultimate weight of the mature animal varies of course with the amount of fat assimilated, which may occur long after maturity ; but the size as dependent upon the frame, such as the weight, length, and general proportions, is modified by the quantity and quality of food available during the growing period of early life. This fact no one questions ; and if these acquired characters are in no degree whatever transmitted, then certain practices of breeders, which are founded upon the contrary belief are delusive and expensive mistakes.

"Practical breeders have hitherto believed that these characters are to some degree transmitted, and practice accordingly. I have searched extensively the writings of practical breeders to see if I could find a single one who questions it, and I fail to find even so much as an intimation of any such belief. All the recorded observations founded upon actual practice appear to point the other way, and consequently the fact of partial transmission is assumed.

"The practical value given to this factor is now much smaller than formerly given, but that it is a factor of some value is universally assumed.

"During the last century, and early part of this, many graziers had a maxim that in the profitable production of animals for slaughter, 'feed is more than breed,' but now both breeders and graziers know that heredity or 'breed,' is the more important. But no breeder claims that a breed is or can be kept up to extra size by selection alone. This belief is so universal, and is apparently so grounded upon long and extensive experience, that I cannot find there has ever been an attempt to increase the size of any breed without special attention to this factor, and consequently conclusive and direct experiment is entirely wanting. Positive proof either way cannot be deduced from the actual experiments of breeders; their belief that feed as well as selection is necessary, is a deduction from the observation of many facts which bear upon the question.

"In this connection it must be borne in mind that all the best breeders recognize the rule laid down by Darwin, that those characters are transmitted with most persistency which have been handed down through the longest line of ancestry. Breeders do not believe that the characters acquired through the feeding of a single ancestor, or generation of ancestors, can oppose more than a slight resistance to that force of heredity which has been accumulated through many preceding generations, and is concentrated from many lines of ancestry. Yet the belief is universal that the acquired characters due to food during the growing period has *some* force, and that this force is cumulative in successive generations. All the observed facts in the experience with herds and flocks point in this direction. It is the same whether the observations relate to the increase in the size of breeds, which has been brought about by systematic selection and feeding directed with this special aim, or to the local development of breeds under the combined influence of the food supply and unsystematic selection.

"Where both large and small breeds have been in process of improvement in the same region at the same time and with the same kinds of food, liberal feeding along with systematic selection is always practised where an increase of size is aimed at, and underfeeding during growth is practised when it is desired to reduce the size. We have examples of these going on together contemporaneously. Breeding for increase of size is more common than that for reducing, but the latter occurs not only in the small fancy breeds of poultry and dogs, but even of cattle. When small and delicate Alderney cows were a fashionable ornament for parks and lawns, some of the most successful breeders practised starving systematically, and at least one eminently successful breeder of these animals so underfed the growing calves that it led to legal interference by a local Society for the Prevention of Cruelty to Animals.

"So far as I know, all the breeds of especially large horses, cattle, and sheep have originated in districts of abundant food, usually in fertile valleys or on plains, and excepting fancy breeds of poultry and pets, all the smaller breeds have originated in districts of scantier forage. This can hardly be due to accident, for it is as true of local varieties of wild animals under natural selection as of domestic animals under artificial selection."

b. Inheritance of Characters Due to the Exercise of Function.

This class of cases has an especial bearing on the doctrine of kinetogenesis. We have a very conspicuous example of such inheritance in the case of the evolution of the trotting horse, which is described by Professor Brewer as follows:

"We have a copious literature relating to the development of this breed, and the 'records' of speed provide the data for a mathematical history of the rate of progress, and also the measure of amount of cumulative variation that has occurred up to this time. These data give to this breed a special interest for scientific study.

"The facts briefly stated are as follows: Trotters had their uses for ages, but fast trotters were not wanted until the improvement in roads and in wheeled vehicles during the last quarter of the last century caused an increasing demand for faster roadsters for light draft. Trotting is the gait of traction, as running is for riding, and trotting as a sport sprang up in nearly all the countries of Europe as well as in America so soon as faster trotters were needed for the road. Then trotting-horses began to be bred, and long before the close of the century there were trotting-stallions of considerable fame. There were also recorded statements as to the speed attained.

"Lawrence, a lover of trotters, in his *Treatise on*

Horses (London, 1796), considered that 'the utmost speed of the English trotter' (which he believes to excel all others), to be a mile in two minutes and fifty-seven seconds. During the next twenty years there were very many recorded trials of speed, and a few of the best animals, both here and in Europe, trotted a mile in three minutes, but none in less time than that given by Lawrence.

"Considering the number of animals that were tested, the widespread interest in the matter, and that these records were the best of both Europe and America, it is fair to assume that this was the utmost speed actually attained by the best trotting-horses until after 1820, although some specific selection in breeding trotters had been going on for half a century, and possibly much longer.

"By 1810, the taste for trotting as a sport had died out in western Europe, but it increased here, and in 1818 it became a recognized sport under specific rules. This is practically the beginning of technical 'trotting records' as we now know them. It soon became fashionable to drive a single horse for pleasure, a social factor in breeding that was lacking in the Old World. This created a demand for trotters, as well as increased the taste of trotting as a sport. Associations were chartered for the promoting of trotting, and special tracks built for the exercising and training of trotters.

"At the end of 1824, six years after the first accepted three-minute record, the record had fallen to 2:34, a reduction of twenty-six seconds. This great reduction so rapidly effected was, doubtless, due chiefly to better training, but also in part to special exercise of function, in part to heredity, and in part to the larger number of animals trained. It is not probable that mere exercise of training could materially further increase this speed, for the next ten years lowered the record only two and a half seconds, and twenty-one years more passed before the first 2:30 record in harness was made.

"By 1848, the record was lowered to 2:29½, and we have now a 2:30 class, with two or three horses technically in it, and perhaps half a dozen that had actually trotted at that speed. Now we began to have distinctively trotting blood, and heredity began to tell.

"The next decade lowered the record five seconds; and the next (ending in 1868), lowered it seven and a fourth seconds more; there were several horses in the 2:20 class, and nearly one hundred and fifty in the 2:30 list.

"The next decade lowered the record four seconds; and the next (ending in 1888), four and a half seconds, and the number of 2:30 horses had increased to 3,255 animals. At the close of last year, the record had been further lowered half a second, to 2:08¼; there were 5,908 in the 2:30 list, 507 in the 2:20 list, and seven in the 2:10 list. This is the history for seventy-three years of 'records.'

"Parallel with the evolution of this breed has been the development of a breed of pacers. The fast animals are not so numerous, but the speed is greater, and the gait, as a fast gait, is more distinctly artificial. The instincts involved will be discussed in a later paper; here I will notice only the development of speed, because that is the direct and obvious result of functional development, and because we have mathematical data as to the rate and amount of actual evolution.

"That the gain in speed has been cumulative, and

that for three-fourths of a century, that it has gone on along with systematic exercise of special function in successive generations of the present fast trotters, is indisputable and very evident. Selection has doubtless determined the proper correlation of the various organs involved in the special evolution, but the increase in speed has only come along with the functional development, which was enhanced by special exercise in the individuals of successive generations. The cumulative value of this, if transmitted, would be vastly more than enough to account for all the increase that has actually occurred, great as that is. Viewed as phenomena, there is every appearance and indication that the changes acquired by individuals through the exercise of function have been to some degree transmitted, and have been cumulative, and that this has been one factor in the evolution of speed. The cumulative variation has been most marked since we have had a 2:30 class, that is, since we have produced animals that are swift by heredity, and whose ancestors, as well as themselves, have been exercised and trained to trot. Studied as phenomena, there is not a particle of evidence that these special changes acquired by the individuals were totally lost to each successive generation, and that all that was 'transmitted by heredity,' was a something that did not exist in either parent or in any ancestor. There is nothing whatever in the actual phenomena observed anywhere along the line of this development of speed that would lead us to even suspect that the changes due to exercise of function had *not* been a factor in the evolution, and there is not a particle of evidence, other than metaphysical deductions, much less proof, that it would or could have gone on just the same by mere selection and adventitious variation."

c. Inheritance of Characters Due to Disease.

Under this head Brewer cites a well-known case. He says: "The most extensive and complete set of experiments yet published on the artificial production of disease by mechanical injuries are those of Dr. Brown-Sequard on the artificial production of epilepsy. This is a disease which is certainly sometimes hereditary and which may also be produced by art in previously sound animals. He experimented with guinea pigs and produced many artificial epileptics, and by breeding these he produced many congenital epileptics. The disease artificially produced in the parents was transmitted to the offspring in numerous cases. The acquired characters in those cases were certainly transmitted to the offspring and became hereditary. These experiments were continued and repeated by his assistant and pupil Depuy, and the results abundantly confirmed. It was shown, moreover, that in many cases it was the *tendency* to become epileptic that was transmitted rather than the disease itself. Just as in a great majority of cases of strictly hereditary disease it is the constitutional tendency rather than the disease itself that is commonly transmitted.

"These experiments have now been before the world some years, during which time ideas have greatly changed as to the causes of disease, and the nature of hereditary tendencies, but as yet there are no published accounts of experiments indicating that those of Brown-Sequard and of Depuy were not carefully performed, or that the conclusions were illusive. Medical literature abounds with alleged instances where nervous diseases acquired by parents through environment have been transmitted in some shape to children, but this evidence is not nearly so conclusive as the experimental proof cited.

"In conclusion we may say that the drift of all the collated observations on both man and brute seem to indicate that certain of the changes produced in the animal body by disease are often to some degree transmitted, that these may be cumulative and lead to degeneration if not indeed to the extinction of families. The experience of breeders as well as the observations of medical men practically establishes the fact that acquired weakness and defects occurring in successive generations may result in truly hereditary unsoundness."

d. Inheritance of Characters Due to Mutilation and Injuries.

While characters of this kind are relatively rarely inherited, there is little doubt that they can be. Dr. Brewer cites a few cases for illustration; "some of them have been already published, others have not. They are not the most striking, but are chosen because they are representative.

"*a.* A mare in foal had an eye seriously injured by burdocks entangled in the forelock. She suffered with violent ophthalmia, and in due time dropped a foal (a filly) which had the corresponding eye aborted. She afterwards bore several normal foals.

"(This case came under the observation of the eminent veterinarian, Professor Law of Cornell University. *Papers American Public Health Association*, 2, p. 254.)

"*b.* A game-cock, in his second year, lost an eye

in a fight. Soon after, and while the wound was very malignant (it never entirely healed), he was turned into a flock of game hens of another strain. He was otherwise healthy and vigorous. A very large proportion of his progeny had the corresponding eye defective. The chicks were not blind when hatched, but became so before attaining their full growth; some at the time of acquiring the pin-feathers, others later and before reaching maturity. The hens afterwards produced normal chickens with another cock. Both strains had been purely bred for ten or more years, and none of the fowls had been blind unless from fights.

"(This case was reported to me by an educated and reliable breeder of game-fowls.)

"*c.* A hunting mare had a split pastern and was then used for breeding. Her first, third, and fourth foals were sound, the second one had 'almost an exact reproduction of the mare's unsoundness.'

"(This is on the authority of the celebrated veterinary surgeon, Clement Stevenson, as occurring under his own observation, 'not hearsay.' *Live Stock Journal*, London, November 23, 1888, p. 508.)

"*d.* A female (and very prolific) cat, when about half grown met with an accident. 'Her fine, long tail was trodden on and had a compound fracture, two vertebræ being so displaced that they ever after formed a short offset between the near and far end of the tail, leaving the two out of line. At first I noted that out of every litter of kittens some had a tail with a querl in it.' With successive litters the deformity increased, until 'not a kitten of the old cat had a straight tail, and it grew worse in her progeny until now we have not a cat with a normal tail on the premises,' (in a cat-population of six or eight, exclusive of young kittens). 'The tails are now in part mere stumps, some have a semicircular sweep sideways, and some have the original querl. Perhaps the deformity was somewhat aggravated by in-and-in breeding and by artificial selection practised by my Chinaman, who, with the perversity of his race, preferred the crooked tails, and thus preserved them in preference to the normal kittens. There are no other abnormally-tailed cats in the neighborhood.'

"This is the essential part of an unpublished letter from that keen observer and eminent scientist, Prof. Eugene W. Hilgard of the University of California.

"Numerous cases have been recorded as occurring with mankind. I will give but two, both of which have not before been published.

"'*e.* A person, when a boy of ten years, cut the terminal phalange of the little finger of his left hand with a sickle. The joint was not injured, nor was the function of the finger seriously impaired. There was, however, an obvious deformity. The finger was ill-shaped and crooked, and the nail abnormal. He married and had two children, the first a son, with normal fingers, the second a daughter, who had the little finger of the corresponding (the left) hand deformed from birth in the same manner. The function of the finger was not seriously injured, but the deformity was precisely the same in shape, even to the malformation of the finger-nail. She died at thirty, without children, consequently no observation on a succeeding generation could be noted. None of his other kindred had malformed fingers, nor had any ancestor of the child for at least three generations, and there was no knowledge of any such in the more remote ancestry.

"(This case was related to me in full detail by the father with the deformed finger, and with whom I was personally acquainted. He was an eminent physician, the president of a large and reputable medical college, and his name is well known to the profession.)

"*f.* A woman thirty-five years of age had both kneepans broken. Erysipelas and other complications prevented the use of the usual surgical appliances for keeping the severed parts together while healing, so they never united by bony union, but became joined by intervening cartilage. The hurt was peculiarly painful and slow of healing, because of the complications alluded to, but the general health was fully restored. For some years after healing there was a very pronounced groove or furrow along the line of fracture over the connecting cartilage, especially in the right knee. The outer edges of the fractured bone were sharp at first, but ultimately became rounded by absorption. Both fractures were V-shaped. The right knee had the parts wider separated at the time of the accident, and was again partially torn asunder three and a half months later, and the furrow consequently remained very much broader and deeper than in the other knee. About four months (124 days) after the first accident, and while still unable to walk, she gave birth to a son. No abnormal appearance was noticed at the time, and later was not looked for until the child was ten or more years old, when he called attention to the matter himself. There was then a deep and well-defined groove across the surface of the right kneepan, very plainly perceivable through the skin. It corresponded precisely in shape and position with the fracture and the later furrow in the corresponding bone in the mother. It was most pronounced before the age of sixteen. After that the edges became modified by growth or absorption, becoming less sharp, following in this respect the changes that gradually occurred in the shape of the bone in the mother. The son is otherwise normal. Three other children of the same parents, one born before and two after the birth of the one described, are entirely normal. The ancestors of both parents are known for several generations (from three to eight in the several lines) and all were normal, so far as is known."

"(This case has been under my own observation during the whole period.)

"It will be noticed that in the cases *a*, *b*, and *f*, the injury to the parent occurred shortly before or during gestation, and that the healing had not taken place until after the birth of the offspring. Also, that the function of the organ involved, an important organ in the animal economy, was at the time suspended. Also that in all these cases, later offspring were normal."

e. Inheritance of Characters Due to Regional Influences.

Characters of this kind mostly come under the head of Physiogenesis. A case of inheritance is thus recorded by Brewer.

"The texture and certain other characters of wool which are of practical importance to manufacturers, depend in part on the breed and health of the animals, in part on the kind of food and on its uniformity of supply, and in part on local conditions of climate, soil, and forage. The wool grown in some regions is harsher than that grown in others, and this is certainly an acquired character in that it takes place in flocks taken from one region to another. I have specimens of wool alleged to have been taken from the same flock, the same individual animals, when pastured at two stations. The first were shorn when the sheep were pastured in southeastern Ohio, where the sheep were bred, a region noted for certain excellencies of its wool. Taken to a certain portion of Texas, and pastured on an alkaline soil, the wool of those sheep took another character, affecting both its texture and also its behavior with dyes. Treated in the same vats, as to dye, lac and mordant, the difference is very obvious.

"A certain harshness of the wools produced in some regions where the soil is alkaline or salt, the climate dry, and the forage plants characteristic of such regions, is widely known and is considered a defect by manufacturers. It is stated that when a flock is taken from a favorable region to such a less favorable one the change in the character of the wool begins immediately, but is more marked in the succeeding fleeces than in the first. It is also alleged that the harshness increases with succeeding generations, and that the flocks which have inhabited such regions several generations produce naturally a harsher wool than did their ancestors, or do the new-comers.

"Now, in this case, the deterioration in successive generations cannot possibly be due to panmixia, the withdrawal of selection; on the contrary, selection goes on under the new conditions just as carefully as under the old, and often more so, for this is the means used to lessen the evil.

"If this increase in the harshness of the wool of succeeding generations is due in part to the inheritance of an acquired character, it is very understandable. That it is a congenital adventitious variation coincident in all the individuals of immense flocks, is a mathematical absurdity.

"We have an analogous regional character in the hoofs of horses. From early times it has been a known fact that the feet of horses produced in mountainous and hilly regions stand travel on hard roads and on city pavements better than those bred on softer low lands, however rich and fertile the latter may be. European writers of previous centuries are very specific on this point. Jacquet, over two centuries ago, cites it as a fact true alike in Spain, Italy, and other countries of Europe. I have interviewed the livery-stable men in various eastern cities as to the relative character in that particular of the horses bred in the hilly regions of New England, New York, and Pennsylvania, compared with those produced on the prairies, and the testimony is almost unanimous to the same effect. The old Vermont bred horses are still famous.

"This regional character cannot be a matter of selection and adventitious variation. It must be related to the environment alone, and environment can only act on the living individual. If this fact is due to the inheritance of acquired characters, it is very easily understood. The different effects of exercise of the feet of the growing animal in the one case on the hard, stony soil of the hills, in the other, on the softer and fine soil of the prairies, makes a difference in the acquired characters, a difference of the very kind spoken of, and which becomes congenital."

At the close of this series of papers Brewer remarks: "The art of breeding has become in a measure an applied science; the enormous economic interests involved stimulate observation and study, and what is the practical result? This ten years of active promulgation of the new theory has not resulted in the conversion of a single known breeder to the extent of inducing him to conform his methods and practice to the theory. My conclusion is that they are essentially right in their deductions founded on their experience and observations, namely, that acquired characters may be, and sometimes are, transmitted, and that the speculations of the Weismann school of naturalists are unfounded."

5. THE CONDITIONS OF INHERITANCE.

Since the evidence adduced must be regarded as proving that characters acquired by an organism may be transmitted by inheritance, we next endeavor to ascertain what information is within our reach which can throw light on this mysterious process. Although Weismann has demonstrated the isolation and stability of the germ-plasma to exceed that of other tissues, he has not proven that it is entirely inaccessible to external influences. He admits that its continual subdivision by the development from it of the embryonic soma, would have speedily reduced it to an infinitesimal quantity, were it not that it grows by accession of nutritive material like other tissues, which nutritive material is furnished by the soma. The accessibility of the germ-plasma to stimuli which affect the soma is then clearly possible.

The effect of the specialization of tissues on their nutrition and repair after injury, is well known. Nutrition of each tissue produces only that tissue. Repair or restoration of parts is confined to the reproduction of a tissue similar to the part lost, or similar to some unfinished or embryonic stage of it. The lower we descend in the scale of life, the more complete is the reproduction of a lost part. The specialization of the higher organisms deprives the tissue of the capacity for exact reproduction. As an example of the reduction of this capacity, I cite the reproduction of the tail of lizards, where no vertebræ are reproduced, but in its place a notochord; while the squamation presents a simpler character than that of the normal tail. The possibility of reproducing the entire organism is restricted, in the multicellular animals, to the germ-plasma, which therefore may be regarded as retaining the characteristic of the protozoön, which reproduces itself by division. But in the multicellular plants the power of reproduction of the entire organism from any of its parts, is retained to a much greater degree than in multicellular animals. The reproduction of plants by cuttings, buds, tubers, and even by single leaves, is well known; a characteristic which is due to the general distribution of unspecialized protoplasm throughout the organism. Inheritance of characters is in these cases known to be very exact, and there can be here no isolation of the germ-plasma. This isolation is progressively more pronounced as we rise in the scale of specialization of structure, but that it ever becomes absolute, the facts before us forbid us to believe.

Having thus seen that the plasma of the germ-cells is open to the influence of stimuli, let us see how it is possible that such stimuli can be transmitted to it, and how they could affect growth of the embryo.

It has been shown that impressions experienced by an animal during one stage of development may be effective in causing the appearance of new structure in a later stage. I have already quoted (Chap. V.) from Poulton the results of experiments on the colors of

Lepidoptera by several English entomologists. By exposure to different colors, of larvæ which were approaching the period of pupation, corresponding colors were produced in the pupæ. Thus black, dark, green, and yellow larvæ, and larvæ with gilt spots or entirely gilded, were produced at will. In this instance the dynamic effect produced by the exposure was stored for the period which elapsed between the exposure of the larva and the full development of the pupa. In another experiment, larvæ which were in the act of weaving cocoons, on exposure to certain colors, were induced to weave cocoons of corresponding color. This experiment demonstrates that a stimulus may be transmitted to a gland so as to modify the character of its secretion in a new direction. From both experiments we learn the transmissibility of energy from the point of stimulus to a remote region of the body, and its conversion into growth energy (in this case by Physiogenesis). This prepares us to look upon heredity as an allied phenomenon, i. e., the transmission of a special energy from a point of stimulus to the germ-cells, and its composition there with the emphytogenic (inherited) energy into bathmism (or evolutionary energy).

The relation of inherited and acquired characters in a series of generations may be graphically represented as follows: Let S represent the aggregate of character of the body (*soma*) of a given species in process of progressive evolution or acceleration. Let g represent the aggregate of characters potential (or dynamically present) in the germ cells of the same individual. For the sake of simplification of the problem I consider here only one sex, and imagine the reproduction to be parthenogenetic. Let A represent the new character acquired by the *soma* under the appropriate stimulus, and let *a* represent the same characteristic as it is impressed on the germ-plasma of the same individual at the same time, and in consequence of the same stimulus. The history of the acquisition and incorporation of newly acquired characters by the line of descent originating with the species $S+g$, may be represented as follows, for successive generations, which are numbered 1, 2, 3, etc.

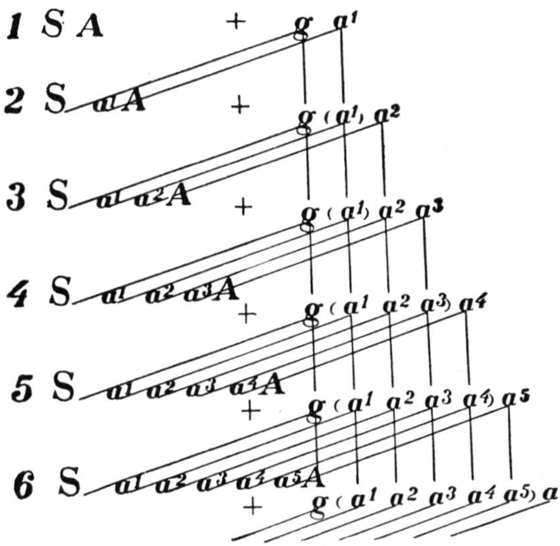

Fig. 120.—Diagram explanatory of Diplogenesis.

Under the appropriate stimulus the soma *S* acquires *A*, and the germ plasma *g* the identical a^1 as the first stage. The character $A\,a^1$ being only inheritable *via* the germ-plasma, it is represented by a^1 in the second stage or generation, where it appears as an addition to the characters of *S* and *g*, so that the soma of the second generation is represented by the expression $S\,a^1$, and the germ-plasma by $g(a^1)$; (on the suppositition that $SA + ga^1$ represents the first of a line in which a given character appears). A new character or an additional increment of the same character, appears in the second stage of acceleration "2," and is represented as before, by Aa^2, the A appearing in the soma, and the a^2 being added to the character of the germ-plasma. In the third stage, the new character a^2 appears in the soma, which now becomes Sa^1a^2. The a^2 acquired by the germ-plasma of the second stage, is inherited by that of the third, which is therefore represented by $g(a^1a^2)$. To the third stage is now added the acquisition Aa^3. The a^3 is again incorporated into the soma of the succeeding or fourth stage, which is therefore represented by the expression $Sa^1a^2a^3$; while the germ-plasma of the same (fourth, "4,") stage, is represented by $g(a^1a^2a^3)$, and so on. The lines of immediate inheritance are represented by straight lines. The vertical lines represent the descent of characters from one type of the germ-plasma to a succeeding one; and the oblique lines represent the transmission of the same characters to the soma into which it grows, as the succeeding generation or stage.

The letters a^1, a^2, etc., expressive of characters acquired by the germ-plasma, are numbered for identification only. Should the influences derived from the ancestry of the other sex be added to the diagram its complexity would become inconvenient, and they are therefore omitted. It is to be also observed, that the enumeration of generations as immediately successive, as 1–2–3 etc., is to be understood as indicating succession only, and not any exact number of generations.

In the hypothesis of heredity above outlined, it is insisted that the effects of use and disuse are two-fold; viz.: the effect on the soma, and the effect on the germ-plasma. Those who sustain the view that acquired characters are inherited, must, I believe, understand it as thus stated. The character must be potentially acquired by the germ-plasma as well as actually by the soma. Those who insist that acquired characters are not inherited, forget that the character acquired by the soma is identical with that acquired by the germ-plasma, so that the character acquired by the former is inherited, but not directly. It is acquired contemporaneously by the germ-plasma, and inherited from it. There is then truth in the two apparently opposed positions, and they appear to me to be harmonized by the doctrine above laid down, which I have called the Theory of Diplogenesis, in allusion to the double destination of the effects of use and disuse in inheritance.

From the preceding considerations we learn that a new character is not inherited unless it is acquired by germ-plasma, as well as by the soma. Should it fail of the former it will not be inherited, although it may appear in the soma. It is also evident that the same character appears in the soma of later generations by virtue of its inheritance by their germ-plasma. Hence should it fail to appear in the adult soma of one generation, it might arise in a later one; and hence the possibility of atavism, and an alternation of generations. Intermittent stimulus might be followed by intermittent activity of growth energy. This would be especially apt to occur during the assumption of sexuality by animals and plants whose reproduction had been performed by cell-division or budding only. And such is the character of most types of alternate generations; a sexual type alternates with a non-sexual type. The advantages being on the side of sexual reproduction on account of its increased opportunity of variation, it has replaced the more primitive method by the process of natural selection.

The time when the impressions of physical habits are conveyed to the reproductive elements has an important bearing on the question of inheritance. Professor Osborn[1] has thus classified the agencies which lie at the basis of organic evolution. Opposite to each he states the theories which have been proposed to account for them :

A. Ontogenic Variations.

a. Gonagenic, i. e., those arising in the germ-cells, including "blastogenic" in part of Weismann, the "primary variations" of Emery.	Theoretically connected with pathological, nutritive chemico-physical, nervous influences, including the doubtful phenomena of Xenia and Telegony.
b. Gamogenic, i. e., those arising from maturation and fertilization, including the "blastogenic" in part of Weismann, and secondary or Weismannian variations of Emery.	Theoretically connected with influences named above, also with the combination of diverse ancestral characters, Amphimixis of Weismann.
c. Embryogenic, i.e., those occurring during early cell-division, including the blastogenic and somatogenic of Weismann.	Theoretically connected with extensive anomalies due to abnormal segmentation, and other causes observed in the mechanical embryology of Roux, Wilson, Driesch, and others.
d. Somatogenic, i.e., those occurring during larval and later development after the formation of the germ-cells.	Connected with reactions between the hereditary development forces of the individual and the environment.

B. Phylogenic Variations.

Variations from types originating in any of the above stages which become hereditary.

[1] *American Naturalist*, 1895, p. 426: "On the Hereditary Mechanism and the Search for Unknown Factors of Evolution."

Osborn points out that Buffon appealed to the "direct action of the environment" as a cause of evolution, in so general a way, as to embrace all the conditions above enumerated. St. Hilaire dwelt on the embryogenic influences, while Lamarck laid stress on the somatogenic. Darwin only discussed variation after it came into being.

The distinctions pointed out by Osborn relate to the period of life at which modifying influences are experienced; that is, they are time distinctions. They must all, however, be included under two heads when the sources of influence are considered. That is, they must proceed from the organism itself, or from the environment directly. Those proceeding from the organism may also be divided into two classes, viz., those which are inherent in the physical and chemical characters of protoplasm, and those which have been acquired by generations prior to any given one under consideration. In this work I attend first to the probably efficient or phylogenetic causes, and these may be regarded as having been at some time or another during the history of the phylum as somatogenic. On this view, I have regarded the life of an animal as divided into three periods; those of embryonic life, of adolescence, and of maturity. During embryonic life impressions are exclusively somatic, and can be only obtained through or from parental stimulus and parental environment. Such will reach the embryo through nutrition, and through the direct mechanical contacts and strains of the environment. The environment of unprotected embryos is external to the parent; that of long protected embryos is the walls of the oviduct, uterus, etc., within the parent. Ryder has alleged with much reason that the nature of the contact of the chorion with the walls of the oviducts or uterus has determined the shape of the placenta ; and that the invagination of the embryo which resulted in the development of the amnion is a result of gravitation. While these facts have an important bearing on the study of inheritance, they have but a collateral relation to evolution ; since the embryo, whether in utero or in ovo, has little opportunity of experiencing the external influences which are only possible at later periods of life. It is during adolescence that the normal activities of maturity, except reproduction, are first practised, whether inherited or learned for the first time. The superior capacity of the adolescent stage for acquisition in all directions is well known, and it is reasonable to suppose that since growth is not completed, changes in its details can be most readily introduced. It is to this period of life then that we must look for the effective influence of the factors of evolution in the acquisition of new characters of the soma. And if the nervous, muscular and other tissues react at this period most readily to external stimuli, it is to be supposed that the developing reproductive cells possess the same characteristic, and record in their molecular movements the influences which are experienced by the entire body. Such influences on the reproductive cells, repeated millions of times from generation to generation, must produce a definite effect on them, in spite of the conservatism which their comparative isolation imposes on them.[1]

The transmission of acquired characters is evidently accomplished during the adult period. While the influence on the soma is greatest during adolescence, the influence on the germ plasma is probably important during maturity, because habits formed

[1] *American Naturalist*, December, 1889, "On Inheritance in Evolution."

during adolescence are now practised with especial energy and frequency. The influence on the constantly renewed germ-plasma is correspondingly greater, and transmission is of course more certain. Some characters seem to have been mainly acquired during maturity. Such is the permanent dentition of the higher Mammalia, which does not appear until or after maturity. In this case the influence of use on the germ-plasma must be more energetic than that on the soma. It is, however, not unlikely that the fundamental characters of mammalian dentition were laid during adolescence by direct influence on the temporary dentition. The tritubercular molar was established at that time and owes its present existence to inheritance. Only the sectorial and lophodont types have been added since the extensive development of the milk dentition in geologic time.

The chief source from which acquired characters are introduced into the germ-plasma, and hence into the soma of the next generation, is probably the spermatozoöid, since it is endowed with a greater kinetic energy than the ovum. The latter furnishes nutritive material for the supply of the needs of growth. That the male is the chief source of variation is also indicated in the numerous cases when he is more active than the female, and hence more capable of supplying the stimulus of use.

The manner in which influences which have affected the general structure are introduced into the germ-cells remains the most difficult problem of biology. For its explanation we have nothing as yet but hypotheses. The one which has seemed to me to be the most reasonable belongs to the field of molecular physics, and it must be long before it is either proved or disproved. I have termed it a "dynamic theory," and it is in some respects similar to that subsequently proposed by Haeckel under the name of the "perigenesis of the plastidule." I have already referred to the phenomena of the building or growth of the added characters which constitute progressive evolution as evidence of the existence of a peculiar species of energy, which I termed bathmism. This is to be explained as a mode of motion of the molecules of living protoplasm, by which the latter build tissue at particular points, and do not do so at other points. This action is most easily observed in the beginnings of growth, as in the segmentation of the oösperm, the formation of the blastodermic layers, of the gastrula, of the primitive groove, etc. In the meroblastic embryo the energy is evidently in excess at one part of the oösperm, and in defect at another. This is a simple example of the "location of growth force or bathmism." In all folding or invagination there is excess of growth at the region which becomes the convex face of the fold; i. e., a location or especial activity of bathmism at that point. All modifications of form can be thus traced to activity of this energy at particular points. A basis is thus laid for a more or less complex organism, and the channels of nutritive pabulum being once established, the location or distribution of the energy is assured in the directions in which they lead. Thus with the establishment of circulating channels nutrition is definitely guided to particular points. It is evident that on this hypothesis the bases of evolutionary change are laid in the embryonic tissues, where bathmism displays its activity in producing the base forms on which all subsequent structure is moulded.

The building energy being thus understood to be a mode of molecular motion, we are not at liberty to suppose that its existence is dependent on the dimensions of the organic body which exhibits it. It is as characteristic of the organic unit or plastidule as the mode of motion which builds the crystal is of the simplest molecular aggregate from which the crystal arises. Bathmism has, however, no other resemblance to crystalloid cohesion. The latter is a simple energy which acts within geometrically related spaces, without regard to anything else but the present compulsion of superior weight-energy. In bathmism we see the resultant of innumerable antecedent influences, which builds an organism constructed for adaptations to the varied and irregularly occurring contingencies which characterize the life of living beings. As this resultant is distinctive for every species, bathmism must be regarded as a generic term, and the characteristic growth-energy of each species as distinct species of energy, which presents also diversities expressive of the peculiarities of individuals.

The preceding statements do not, of course, constitute an explanation of the exact manner in which a stimulus which effects say the contraction of a muscle, effects molecular movements of the nuclei of the reproductive cells. This is a question of organic molecular physics, a science which has made scarcely a beginning. That the transmission of such influence is through nutritive channels, by the intermediation of a nervous structure where one exists, may be supposed. Poulton's experiments on Lepidoptera, already cited, led him to believe that the effect of color-environment was transmitted to the pigment-cells through the medium of the nervous system. That the *modus operandi* is similar to that which produces reflexes may be also reasonably supposed. How the record of these movements become reflexes, is concentrated in a reproductive cell, is a question to be solved only in a more advanced stage of knowledge of organic physics than we now possess.

Speculation in this direction takes the following forms. According to one view the energy or molecular movement must be transmitted to the germ-plasma through a material or molecular basis. This basis, it may be supposed, must be that which receives the mechanical impression which is to produce a corresponding modification of growth-energy in the ovum or spermatozoöid; that is, in the case of a modified bone-articulation, particles of matter must pass from the latter through the medium of the circulation to the reproductive cells. The alternative hypothesis is, that the energy which causes the active region to make or omit to make a given movement, the result of which is to be structural modification in the young, is impressed through protoplasmic channels on the germ-cells of either sex. In this case the transmission of particles of matter is not necessary, as material connection through the cells, nervous or other, already exists.

To the first of these points of view belong the pangenesis theory of Darwin, and the modified pangenesis of Weismann. These hypotheses present the difficulty that we must conceive of each particle or "gemmule" derived from a given part of the organism finding its way through the circulation to its exact place in the growing embryo; or otherwise, of transmitting its peculiar mode of motion to the correct molecules of the embryo, without error as to locality. The difficulties to be encountered in accomplishing such a feat seem to be insuperable. Hyatt well expresses these in the following language: "Every purely corpuscular throng . . . must not only account for a difficulty as great as that of the camel and the needle's eye, but must also account for putting the numberless characters derived from the entire caravan of its immediate progenitors and remote wild ancestors and their progenitors back to the origin of their phylum, through the same narrow tunnel. This physical difficulty is still further enhanced by the fact that the ova and spermatozoa do not increase in size in proportion to the increasing number of characters transmitted." (*Proc. Boston Soc. N. H.*, 1893, p. 70.)

The transmission of a mode of motion organized in a central nervous system, is less inconceivable. This central system is the seat of a composition of incoming stimuli and of outgoing energies, the resultant of both combined constituting the active agency in the production of automatic adaptive or intelligent adaptive movements of any and all of the organs. It appears to me that we can more readily conceive of the transmission of a resultant form of energy of this kind to the germ-plasma than of material particles or gemmules. Such a theory is sustained by the known cases of the influence of maternal impressions on the growing fœtus. Going into greater detail, we may compare the building of the embryo to the unfolding of a record or memory, which is stored in the central nervous organism of the parent, and impressed in greater or less part on the germ-plasma during its construction, in the order in which it was stored. This record may be supposed to be woven into the texture of every organic cell, and to be destroyed by specialization in modified cells in proportion as they are incapable of reproducing anything but themselves. The basis of memory is reasonably supposed to be a molecular (or atomic) arrangement from which can issue only a definite corresponding mode of motion. That such an arrangement exists in the central nervous organism is demonstrated by automatic and reflex movements. It is also demonstrated by the fact that the memory of the position and parts of amputated limbs is retained by the sensory center, so that irritation of the stump is referred to the lost limb. That the entire record is not repeated in automatic and reflex acts, but only that part of it which was last acquired, may be regarded as due to the muscular and other systems concerned in it having performed it most recently, and having for a longer or shorter period omitted to perform the older movement, because the latest structures of the organs would render the performance of the old movements impossible. In other words, the physiological division of labor extends to memory at the basis. In the case of the germ-plasma no other specialization exists, so that the entire record may be repeated stage after stage, thus producing the succession of type-structures which embryology has made familiar to us. In the process of embryonic growth, one mode of motion would generate its successor in obedience to the molecular structural record first laid down in the ovum and spermatozoöid, and then combined and recomposed on the union of the two in the oöspore, or fertilized ovum.

If the doctrine of kinetogenesis be true, this energy has been moulded by the interaction of the living being and its environment. It is the recorded expression of the habitual movements of the organism which have become impressed on, and recorded in, the reproductive elements. It is evident that these and the other organic units of which the organism is composed possess a memory-structure which determines their destiny in the building of the embryo. This is indicated by the recapitulation of the phylogenetic history of its ancestors displayed in embryonic growth. This memory has perhaps the same molecular basis as the conscious memory, but for reasons unknown to us, consciousness does not preside over its activities. The energy which follows its guidance has become automatic, and it builds what it builds with the same regardlessness of immediate surroundings as that which is displayed by the crystallific growth-energy. It is incapable of a new design, except as an addition to its record.

Were all cells identical in characters, every one would retain the structural record, or memory of its past physical history, as do the unicellular organisms. Evolution has, however, so modified most of the structural units of the organic body that none but the nervous and reproductive cells retain this record, in greater or less perfection. The nervous cells have been specialized as the recipients of new impressions, and the excitors of definite corresponding movements in the cells of the remainder of the organism. The somatic cells retain only the record or memory of their special function. On the other hand, the reproductive cells, which most nearly resemble the independent unicellular organisms, retain first the impressions received during their primitive unicellular ancestral condition; and second, those which they have acquired through the organism of which they have been and are only a part. The medium through which they can receive such impression is continuous protoplasm. Whether, in the higher animals, it is effected through that system of cells called the nervous system, which has been specialized through use and natural selection to receive impressions from without, and to transmit them to such parts of the organism as are capable of receiving them, or whether it is transmitted through other media, as in lower animals and in plants which possess no such system, is unknown. The only cells which can retain the entire record in the higher animals are the reproductive cells. In the lower animals and plants it is well known that germ-plasma is not confined to reproductive organs, but is widely disseminated throughout the organism. In some forms it seems that all of the sarcode is capable of reproduction.

This is the logical result of the considerations which have occupied the preceding pages, and is the carrying out of the bathmism theory of heredity, of which I have given hitherto only the bare outline.

Since Darwin, successive contributions have been made to the theory of heredity in its relation to evolution. In 1868 and 1871 the present writer advanced the dynamic hypothesis, but made no attempt to explain the mode of conveyance of dynamic impressions and modifications to the germ-cells. The theory of "perigenesis," proposed by Haeckel in 1873, is of the same character, and is deficient in the same way. The modified pangenesis theory of Brooks, published in 1883,[1] attempts to supply the defect found in the previous conceptions, but does so by assuming with Darwin the intermediation of gemmules, a hypothesis to which sufficient objection has been made by Galton and others. Brooks's theory also fails to admit the origin of variations through mechanical stresses, although he seeks for the origin of gemmules through

[1] *The Law of Heredity*, Baltimore, 1883, p. 80.

the lack of equilibrium between the organization and its environment, which embraces that proposition without definite specification. To Weismann we are indebted for the exposition of the separate origin and relative isolation of the germ-plasma, but no sufficient explanation of the origin and inheritance of new characters is offered. Ryder[1] has especially dwelt on the physiological division of labor seen in the tissues of the organism, and on the special function of the germ-plasma as the recipient of impressions through the processes of metabolism; but he does not go into greater detail.

What is true of the somatic cells is also true of those which follow immediately the segmentation of the oösperm. Each division contains the entire record, until a point is reached in which specialization of its growth-capacities begins.

Dr. Chalmers Mitchell thus discusses the question as to the location of specialized growth in the oösperm:[2]

"Loeb uses the term heteromorphosis to denote the power of organisms, under the stimulus of outer conditions, to produce organs on parts of the organism where they do not occur normally, or the power to replace lost parts by parts unsimilar to them. Regeneration is the reproduction of like parts. Heteromorphosis is the reproduction of unlike parts.

"If one cuts off part of the stem of almost any plant, on placing the stem in suitable soil, roots will grow out, although roots are not natural to that part of the stem. The prothallus of fern produces the male and female organs on the lower side turned away from the light. If the prothallus be darkened on the upper

1 *American Naturalist*, 1890, p. 85.

2 *Natural Science*, 1894, p. 187.

surface, and illumined by reflected light on the lower surface, then the antheridia and archegonia will be produced on the upper surface. Galls are produced under the stimulus of the insect almost anywhere on the surface of the plant. Yet in most cases these galls, in a sense grown at random on the surface of a plant, when placed in damp earth will give rise to a young plant. In the hydroid, *Tubularia mesembryanthemum*, when the polyp-heads are cut off, new heads arise. But if both head and root be cut off, and the upper end be inserted in the mud, then from the original upper end not head-polyps, but root-filaments, will arise, while from the original lower end, not root-filaments, but head-polyps will grow. In *Ciona intestinalis*, round a slit cut into the body-wall, a tubular process grew out, forming a new mouth, while around the base of this, a series of eye-spots, corresponding to the eye-spots round the real mouth, appeared. In all these cases, it is plain that there were present in parts affected, the determinants, to use Weismann's term, not only of the normal parts, but also of parts, which, under normal conditions, would never have appeared there; and these new parts growing in the unwonted places bore the normal species-stamp as characteristically as similar parts grown in their normal places. It can hardly be supposed that the architecture of the germ-plasm contains special determinants to be ready for occurrences so casual, especially as these are called into existence by circumstances quite foreign to the normal environment of the organisms. On the other hand, the facts are consonant with Hertwig's belief that, as all division is heirs-equal division, all the species-characters that depend upon cells are latent in every cell.

"The experiments of Driesch, Wilson, and Hertwig upon the early stages of developing ova show that heteromorphosis begins with the very earliest divisions of the egg. Thus Driesch, working upon echinoderm embryos, was able to flatten out the stage where there was a sphere of sixteen cells into a flat plate where all the cells were in the same plane. In such a plate, the nuclei of the cells occupied relative positions very different from the normal conditions. Yet from these Driesch obtained normal plutei larvæ. It was, in fact, as if the cells could be pushed about like billiard balls without destroying the future shape and characters of the embryo. Did each cell contain only the determinants that would correspond to the structures that would arise from it under normal conditions then change of its normal position would have arrested development. Each cell must, on the other hand, have contained the determinants for all the animal, and have allowed those to come into operation that were required by the new positions into which the cells were forced. Driesch, by separating the first two and the first four segmentation-spheres of an Echinus ovum, obtained two or four normal plutei, respectively one-half and a quarter of the normal size. Here again each sphere must have contained all the determinants for the whole organism. Heirs-equal division must have occurred. So, also, in the case of Amphioxus, Wilson obtained a normal, but proportionately diminished, embryo with complete nervous system from a separated sphere of a two or four or eight celled stage.

"Hertwig himself, some years ago, published the results of experiments he made upon the development of frogs' eggs under abnormal conditions. He showed that there could be no question of imperative divisions separating the germ-plasm into right and left halves, and so forth, but that the method of division was determined by pressures and relative gravities. Alteration of these made the ova divide into novel but symmetrical forms. Chabry obtained normal embryos in cases where some of the segmentation-spheres had been artificially destroyed.

"These cases all show that in its possibilities each segmentation-sphere is identical; that as a result of heirs-equal division, each cell contains all the material necessary to cause the development of a complete embryo. Weismann would have to suppose that in all these cases, in addition to its half of the nuclear matter resulting from heirs-equal division, it had also a stock of unaltered germ-plasm ready to be called into activity by unwonted stimuli. But even this hypothesis would not account for cells distorted by compression responding with the production of unwonted symmetries."

6. OBJECTIONS TO THE DOCTRINE OF INHERITANCE OF ACQUIRED CHARACTERS.

I will now mention some objections to the theory of epigenesis, or the inheritance of acquired characters. Some of them appear at first to have considerable force, but the explanations which have been offered seem to me to be sufficient.

Weismann's merit consists in having directed attention to the isolation and continuity of the germ-plasma, factors which must be taken account of in any theory of inheritance. The continuity of reproductive function which this substance displays is a fact of great interest, and one which has given rise to the statement that it is under normal conditions, immortal.

Isolation of the germ-plasma is however doubtfully complete anywhere, and in the vegetable kingdom it scarcely exists. Most plants may be propagated either by roots, cuttings, bulbs, buds, or even by leaves. The germ-plasma is evidently as widely distributed in these multicellular organisms, as it is in a Protozoön. The greater degree of isolation exhibited by the higher animals is one of their many specializations, but that it is not complete is shown by the facts already cited in the preceding pages. The continuity of protoplasm in the organism is likely to be true of the germ-cells as of other cells ; and they are not deprived of nutrition, so that they are evidently accessible to influences from or through the soma. As regards the immortality of the Protozoön there is reason to believe, that like its descendent the germ-cell, it requires renewal from another cell to escape death. According to Maupas, the Protozoa after reproducing by self-division for many generations, require conjugation, or they dwindle and die.

The old formula that variation is due to "natural selection and heredity" has derived new life from the fact that sexual conjugation is necessary for the renewal of the vitality of the ovarian cell. It is supposed by Weismann that variation as well as reproductive energy is introduced in this way, the process being termed by him Amphimixis. But like the old formula this explains nothing, for if the parents are the sources of variation, the question as to the source of the variation is simply relegated to the parents for answer. Moreover, Brooks, who made this suggestion prior to Weismann, points out that it has less force than appears at first sight to belong to it. He shows[1] that the ancestors of the individuals of a given species are in greater or less degree identical persons, and that they are on this account less numerous than has been sometimes assumed. Thus, if the population of a given district had for ten generations married first cousins, the total ancestry of each person for that period would number only thirty-eight persons. If, on the contrary, all the ancestors of each person had been distinct individuals, the total number of ancestors in ten generations would be two thousand and forty-six persons. An investigation into the ancestry of three persons, not nearly related, living on an island on the Atlantic coast where the records are complete for seven and eight generations, shows that the ancestry of each of the three averages only three hundred and eighty-two persons. That this consideration is of even greater importance in estimating the ancestry of the lower animals than in man, is evident from the fact that no consideration of kinship modifies their reproductive habits.

The fact that mutilations are not generally inherited is cited as evidence against the inheritance of acquired characters. A particular mutilation, however, as already remarked, rarely happens more than once or twice in the lifetime of a single individual; in fact its occurrence more than once is, in many cases, as in amputations, impossible. Such sporadic events must necessarily have little influence as stimuli to the organism, in comparison with the habitual movements of animals, or the continued exposure to especial physical conditions, as is experienced by both plants and animals, and are not worth considering in this connection.

[1] *Science*, 1895, February, p. 121.

One of the cases which is cited in opposition to the view here sustained, is the alleged fact that the artificial contraction of the feet undergone by high-caste Chinese female children, resulting in deformity of the feet of the women, is not inherited. That this abnormality has never been transmitted has not yet been satisfactorily shown ; but in any case there are some reasons why it should not be inherited. One of these is, that the deformity is confined to one sex. The male, who is without it, has the advantage of an ancestry possessing normal feet extending backwards indefinitely, while the modification of the female is a very modern interference with the law of the species. Moreover, a positive stimulus to ontogenetic growth, such as is in this instance furnished by the male, is always likely to be prepotent as compared with the negative part played by the female.

Professor Poulton, whose interesting experiments in the production of color changes in lepidopterous larvæ and pupæ have been previously cited, states that none of the color varieties which he has obtained, have been inherited. I cannot regard this result as conclusive until the experiments have been continued for a longer period than has yet been possible to devote to them.

Perhaps the strongest case that can be made out against the theory of use-inheritance as defended in the present book, is that of the variety of structure displayed by the neuter members of the colonies of ants and termites. Mr. W. P. Ball describes these briefly as follows : [1]

"But there happens to be a tolerably clear proof that such changes as the evolution of complicated

[1] *The Effects of Use and Disuse*, Nature Series, 1890, p. 24.

structures and habits and social instincts *can* take place independently of use-inheritance. The wonderful instincts of the working-bees have apparently been evolved (at least in all their later social complications and developments) without the aid of use-inheritance—nay, in spite of its utmost opposition. Working-bees, being infertile 'neuters,' cannot, as a rule, transmit their own modifications and habits. They are descended from countless generations of queen-bees and drones, whose habits have been widely different from those of the workers, and whose structures are dissimilar in various respects. In many species of ants there are two, and in the leaf-cutting ants of Brazil there are *three*, kinds of neuters which differ from each other and from their male and female ancestors 'to an almost incredible degree.'[1] The soldier caste is distinguished from the workers by enormously large heads, very powerful mandibles, and extraordinarily different instincts. In the driver ant of West Africa one kind of neuter is three times the size of the other, and has jaws nearly five times as long. In another case, 'the workers of one caste alone carry a wonderful sort of shield on their heads.' One of the three neuter classes in the leaf-cutting ants has a single eye in the midst of its forehead. In certain Mexican and Australian ants

1 *Origin of Species*, pp. 230–232; Bates's *Naturalist on the Amazons.* Darwin is surprised that no one has hitherto advanced the demonstrative case of neuter insects against the well-known doctrine of inherited habit as advanced by Lamarck. As he justly remarks, "it proves that with animals, as with plants, any amount of modification may be effected by the accumulation of numerous slight, spontaneous variations, which are in any way profitable, without exercise or habit having been brought into play. For peculiar habits confined to workers, however long they might be followed, could not possibly affect the males and fertile females, which alone leave any descendants." Some slight modification of these remarks, however, may possibly be needed to meet the case of "factitious queens," who (probably through eating particles of the royal food) become capable of producing a few male eggs.

some of the neuters have high spherical abdomens, which serve as living reservoirs of honey for the use of the community. In the equally wonderful case of the termites, or so-called 'white ants' (which belong, however, to an entirely different order of insects from the ants and bees), the neuters are blind and wingless, and are divided into soldiers and workers, each class possessing the requisite instincts and structures adapting it for its tasks. Seeing that natural selection can form and maintain the various structures and the exceedingly complicated instincts of ants and bees and wasps and termites in direct defiance of the alleged tendency to use-inheritance, surely we may believe that natural selection, unsupported by use-inheritance, is equally competent for the work of complex or social or mental evolution in the many cases where the strong presumptive evidence cannot be rendered almost indisputable by the exceptional exclusion of the modified animal from the work of reproduction.

"Ants and bees seem to be capable of altering their habits and methods of action much as men do. Bees taken to Australia cease to store honey after a few years' experience of the mild winters. Whole communities of bees sometimes take to theft, and live by plundering hives, first killing the queen to create dismay among the workers. Slave ants attend devotedly to their captors and fight against their own species. Forel reared an artificial ant-colony made up of five different and more or less hostile species. Why cannot a much more intelligent animal modify his habits far more rapidly and comprehensively without the aid of a factor which is clearly unnecessary in the case of the more intelligent of the social insects."

The explanation of this phenomenon will be probably some day found by paleontological discovery. We may suppose, on the basis of discoveries already made in other animals, that the primitive ants and termites presented homogeneous colonies, and that the varied structures which they present to-day have been primarily due to the usual process of specialization through use-inheritance. It is necessary to suppose that the varied functions of the different members of the community have modified the structures of the parts essential to their performance. It is probable that the earliest ants in an early geologic period became soldiers under the usual exigencies of their struggle for existence, and having thus secured a place in the economy of nature, certain members of the communities underwent degenerative changes, appropriate to their respective functions, of a less exacting character. In a second stage of evolution the community would present the character of a family of varied forms all of whose members would produce any or all of the types of form to be found in it, under slight diversities of conditions, just as now, all species produce young of two sexes. The differences between the members of an ant community are considerable in appearance, but not so great essentially as that between sexes.

Finally, in a third stage of the history, the functions of reproduction come to be the special office of a few members of the community. This may have been due to starvation, or to excessive labor on the part of certain individuals aborting the reproductive powers; but whatever may have been the cause, a majority of individuals became sterile. The reproducing members of the community, however, have continued to produce all the forms of the community.

They produce sterile workers and soldiers, sometimes of several forms, although themselves unlike most of their progeny. "This," says Mr. Ball, "is evidence that inheritance can have no share in the process." He believes that each one of the structural types of the community is produced by the treatment accorded to the young by the workers, *each generation for itself.*

As we have seen that the embryonic and paleontologic histories distinctly negative the idea that each generation has been produced by itself without inheritance, let us endeavor to read the riddle in the light of the knowledge we have gained from paleontology. I assume that the most specialized types, the soldiers, represent the type of the species in Mesozoic and possibly earlier time. They are already known from early Cenozoic formations (Scudder). The process of change into workers and breeders has been degenerative. I suppose, however, that in ants, as in the case of many other animals, slight differences in the supply of nutritive energy will prevent or produce these degenerative processes, as it appears to do in the case of the production of the sexes. (Experiments on lepidopterous larvæ have shown that excessive food supply produces females, and deficient supply produces males). In bees the larvæ of the female (queen) receives the largest food supply; those of the males less; and those of the neuters the least of all. How the food supply came to be varied so as to produce the several types in accordance with the exigencies of the community, is a question to be solved by future research. Perhaps it was due to variations in the supplies available at particular times of the year; and perhaps the ants ultimately learned the secret, and now practice it intelligently. It is enough for my present purpose to have shown that the basis of the entire community, the most specialized form, the original fertile soldier, acquired his characters in the usual way, by use, and that all other forms have been derived from him by inheritance modified by disuse, or degeneracy, under the influence of variations in the food supply.

This reply to Mr. Ball's argument was made by me at a meeting of the Philadelphia Academy of Natural Sciences on May 23, 1893. In the latter part of the same year an almost identical answer was published by Herbert Spencer. My remarks were not published until the end of the year.

Mr. A. R. Wallace[1] presents the fact of change of character under external stimulus as evidence of the non-inheritance of acquired characters. Thus he cites the cases of change of species of Artemia, in consequence of increased salinity of water (antea, p. 229); and of the change of color of a Texan Saturnia, when its normal food-plant *Juglans nigra* was replaced by *J. regia*. Under the new conditions the old characters were not continued. In the same way the appearance of all new characters might be assumed to prove non-inheritance of the old ones. The obvious interpretation of these facts is the one generally given them; that is, they demonstrate the superior potency of certain new stimuli over the inherited type of growth-energy. They demonstrate that the energy of inheritance is not unchangeable in its type, which is the condition of the possibility of evolution. They do not demonstrate that acquired characters cannot be inherited.

Objections have been made to the supposition that the simian characters of the lower human races are due to inheritance because it has been shown that

[1] *Nature*, 1893, p. 267.

some of them are due to mechanical causes acting after birth.

The demonstration of the mechanical origin of a given peculiarity, however, by no means precludes that such peculiarity may not be an inheritance from or reversion to pithecoid ancestors. It has been already pointed out that all of the form characters of the vertebrate skeleton, and for that matter, of the hard parts of all animals, have been produced by muscular pressures and contractions, and the friction, strains, and impacts, due to these. The demonstrations by Virchow and others that such is the origin of the platycnemic human tibia, is directly in the line of Neo-Lamarckian evolutionary doctrine, and shows us that atavistic and reversionary characters are found in the muscular system as well as in the skeleton. Such characters are inheritable as well as those of the skeleton. But the characters of the skeleton can generally be shown to be inherited, because they appear before birth, and are found at some stage or another of fœtal life. The later appearance of the muscular structures in the ontogeny, is simply a case of cænogeny, where the record has been falsified by retardation of the parts in question.

The variations in the characters of the human skeleton are of very various significance and value, and the zoölogist and paleontologist can perceive that they are sometimes misinterpreted by archeologists. Thus the presence of wormian (Inca) bones, and of a perforation of the olecranar fossa, have no zoölogical value; while the prognathous jaws, tritubercular molar, and platycnemic tibia have such a value. The tufted hair of the negro has a human value only, as it does not occur in any of the Quadrumana. But archeologists who are not zoölogists are not careful to point out these distinctions.

If the platycnemic tibia has been produced by muscular pressure in man, it has been probably so produced in the apes, where it is a universal character. If the early fusion of the sagittal suture is produced by the vigorous contractions of the temporal muscle as suggested by Brinton, in the black race, due to prognathous jaws, this is probably why it is a universal character of the apes, where the jaws are still more prognathous. What may be the cause of prognathism is not explained by archeologists, but has been discussed in my book on the *Origin of the Fittest*, and by Dr. C. S. Minot. That the prognathous jaws and platycnemic tibia are not found in the fœtus by no means proves that they are not inherited characters. Besides the fact already mentioned, that we are by this only thrown back on an inherited muscular structure, it is further to be remarked that characters which indicate the evanescence or degeneracy of parts, do not usually appear in the fœtus, but are disclosed at later stages. The prognathous jaws are disappearing from the higher races, and the process of disappearance is in this point accomplished by a retention of the fœtal face, which is excessively orthognathous. Prognathism is characteristic of most of the lower Mammalia, and whenever man displays it, if he be, as evolutionists believe, descended from some other mammal, he is simply continuing to develop the old character in the old manner. The same reasoning applies to the platycnemic tibia and the tritubercular molar.

As regards the lemurine character of the tritubercular molar, the term is a good one, as indicating the nearest of kin to man which present such molars. But this type can with equal propriety be called, as I have shown, the primitive placental molar. The lemur is the highest form next to man which displays it, but it was universal among the placentals at one geological epoch. It is possible that Topinard's suggestion as to the cause of its appearance in man is the correct one, as I made the same many years before, but that does not affect its value as an evidence of reversion, as in the cases already cited. There are various other ways in which molar teeth may degenerate, besides reversion to trituberculy, with which dentists are familiar, and which may be explained as Topinard and I have done ; i. e., by change of food ; but why the regular and normal mode should be trituberculy, and not one of those other modes, requires additional explanation. This explanation is that a regular or normal retrogressive modification of a structure is likely to be a return on the line by which it advanced. This is atavism or reversion.

That the anthropoids have been directly derived by descent from the false lemurs rather than from the Old World monkeys (Cercopithecidæ) is probable for various reasons which I have pointed out on page 157. I mention now that this view is somewhat confirmed by the recent discovery by Forsyth-Major, in Madagascar, in beds of Plistocene age, of a skull of a new genus of Lemuridæ with tritubercular molars, whose single species is nearly as large as a chimpanzee.

In closing these remarks, I call attention to the frequent muscular and occasional cerebral anomalies found in the negro, which are of simian character, and which indicate simian descent. An excellent synopsis of these has been given by Dr. Frank Baker in his address at Cleveland in 1888 as Vice-President of the

American Association for the Advancement of Science, and by Prof. H. F. Osborn in 1891 in the Cartwright lecture before the New York College of Physicians.

It is evident that evolutionists are reaching greater harmony of opinion on the question of inheritance. In fact, the discussion is sometimes a logomachy dependent on the significance which one attaches to the term "acquired characters." Thus Von Rath[1] remarks: "There is nothing in the way of the opinion that by the continued working of such external influences and stimuli the molecular structure of the germ-plasma also experiences a change which can lead to a transmission of transformations. Above all, it ought not to be forgotten in this case that the somatic cells are in no way the first to be modified by the stimulus, and that then by some sort of unexplained process (pangenesis or intracellular pangenesis), this stimulus is transmitted gradually by these cells to the plasma of the germ-cells. The influence on the germ-plasm is rather a direct one, and if by continued influence a transformation of the structure of this plasm takes place and transmission occurs, we have then simply a transmission of blastogenic, and by no means of somatogenic characters, and therein is not the slightest admission of the transmission of acquired characters."

This paragraph contains an admission of the doctrine of diplogenesis, and does not regard the phenomena as including a transmission of acquired characters. Nevertheless the stimuli traverse the soma in order to reach the germ-plasma. Such an energy is evidently then not of blastogenic origin, although it is

[1] *Berichte der naturforschenden Gesellschaft zu Freiburg in Baden*, Bd. VI., Heft 3.

such in its effects. Moreover, Von Rath omits to mention the fact that in traversing the soma, the stimulus, frequently, if not always, produces effects on the latter similar to those which it produces on the germ-plasma. I should call this process the inheritance of an acquired character, even in the case where no corresponding modification appears in the soma, since the causative energy is acquired by the soma and is not derived from the existing germ-plasma.

Romanes[1] says, in reviewing the opinions of Weismann: "(1) Germ-plasm ceases to be continuous in the sense of having borne a perpetual record of congenital variations from the first origin of sexual propagation. (2) On the contrary, *as all such variations have been originated by the direct action of external conditions* [italics mine], the continuity of the germ-plasm in this sense has been interrupted at the commencement of every inherited change during the phylogeny of all plants and animals, unicellular as well as multicellular. (3) But germ-plasm remains continuous in the restricted though highly important sense of being the sole repository of hereditary characters of each successive generation, so that acquired characters can never have been transmitted to progeny 'representatively,' even though they have frequently caused those 'specialized' changes in the structure of germ-plasm, which as we have seen, must certainly have been of considerable importance in the history of organic evolution."

Here the inheritance of characters acquired by the soma is admitted, and the process is after the method of diplogenesis. According to Romanes, Galton origin-

[1] *An Examination of Weismannism*, Chicago, 1893, p. 169.

ally propounded this doctrine. Galton's language[1] is as follows:

"It is said that the structure of an animal changes when he is placed under changed conditions; that his offspring inherit some of his change; and that they vary still further on their own account, in the same direction, and so on through successive generations until a notable change in the congenital characteristics of the race has been effected. Hence, it is concluded that a change in the personal structure has reacted on the sexual elements. For my part, I object to so general a conclusion, for the following reasons. It is universally admitted that the primary agents in the processes of growth, nutrition, and reproduction, are the same, and that a true theory of heredity must so regard them. In other words, they are all due to the development of some germinal matter, variously located. Consequently, when similar germinal matter is everywhere affected by the same conditions, we should expect that it would be everywhere affected in the same way. The particular kind of germ whence the hair sprang, that was induced to throw out a new variety in the cells nearest to the surface of the body under certain changed conditions of climate and food, might be expected to throw out a similar variety in the sexual elements at the same time. The changes in the germs would everywhere be collateral, although the movements where any of the changed germs happen to receive their development might be different."

This is the first statement of the doctrine of diplogenesis with which I have met, and it appears to furnish the most rational basis for the investigation into the dynamics of the process.

[1] *Contemporary Review*, 1875, pp. 343-344; *Proc. Royal Society*, 1872, No. 136.

CHAPTER IX.—THE ENERGY OF EVOLUTION.

IF we view the phenomena of organic life from the standpoint of the physicist, the first question that naturally arises in the mind is as to the kind of energy of which it is an exhibition. Ordinary observation shows that organic bodies perform molar movements, and that many of them give out heat. A smaller number exhibit emanations of light and electricity. Very little consideration is sufficient to show that they include among their functions chemical reactions, a conviction which is abundantly sustained by researches into the physiology of both animals and plants. The phenomena of growth are also evidently exhibitions of energy. The term energy is used to express the motion of matter, and the building of an embryo to maturity is evidently accomplished by the movement of matter in certain definite directions. The energy which accomplishes this feat is, however, none of those which characterize inorganic matter, some of which have just been mentioned, but, judging from its phenomena, is of a widely different character. If we further take a broad view of the general process of progressive evolution, which is accomplished by successive modifications of this growth-energy, we see further reason for distinguishing it widely from the inorganic energies.

In considering the dynamics of organic evolution, it will be convenient to commence by considering the claims of natural selection to include the energy which underlies the process. That natural selection cannot be the cause of the origin of new characters, or variation, was asserted by Darwin;[1] and this opinion is supported by the following weighty considerations:

1. A selection cannot be the cause of those alternatives from which it selects. The alternatives must be presented before the selection can commence.

2. Since the number of variations possible to organisms is very great, the probability of the admirably adaptive structures which characterize the latter having arisen by chance, is extremely small.

3. In order that a variation of structure shall survive, it is necessary that it shall appear simultaneously in two individuals of opposite sex. But if the chance of its appearing in one individual is very small, the chance of its appearing in two individuals is very much smaller. But even this concurrence of chances would not be sufficient to secure its survival, since it would be immediately bred out by the immensely preponderant number of individuals which should not possess the variation.

4. Finally, the characters which define the organic types, so far as they are disclosed by paleontology, have commenced as minute buds or rudiments, of no value whatever in the struggle for existence. Natural selection can only effect the survival of characters when they have attained some functional value.

In order to secure the survival of a new character, that is, of a new type of organism, it is necessary that

[1] *Origin of Species*, Ed. 1872, p. 65.

the variation should appear in a large number of individuals coincidentally and successively. It is exceedingly probable that that is what has occurred in past geologic ages. We are thus led to look for a cause which affects equally many individuals at the same time, and continuously. Such causes are found in the changing physical conditions that have succeeded each other in the past history of our planet, and the changes of organic function necessarily produced thereby.

1. ANAGENESIS.

It is customary to distinguish broadly between inorganic and organic energies, as those which are displayed by non-living and living bodies. This classification is inexact, since, as already remarked, nearly all of the inorganic energies are exhibited by living beings. A division which appears to be, with our present knowledge, much more fundamental, is into the energies which tend away from, and those which tend toward, the phenomena of life. In other words, those which are not necessarily phenomena of life, and those which are necessarily such. And the phenomena of life here referred to are the phenomena of growth and evolution, as distinguished from all others. I have termed[1] these classes the Anagenetic, which are exclusively vital, and the Catagenetic, which are physical and chemical. The anagenetic class tends to upward progress in the organic sense; that is, toward the increasing control of its environment by the organism, and toward the progressive development of consciousness and mind. The catagenetic energies tend to the creation of a stable equilibrium of matter, in

[1] *The Monist*, Chicago, 1893, p. 630.

which molar motion is not produced from within, and sensation is impossible. In popular language the one class of energies tends to life; the other to death.

Herbert Spencer has defined evolution as a process of "integration of matter and dissipation of motion";[1] "the absorption of motion and the diffusion of matter" he terms dissolution. If by evolution Mr. Spencer referred only to that of inorganic bodies and masses, his definition must be accepted; but the evolution of organic bodies, since it has proceeded in a direction the opposite of the inorganic, cannot be so characterized. Organic evolution has passed beyond the domain of the inorganic, and the terms applicable to the latter process cannot be correctly applied to the former. In organic anagenesis there is absorption of energy; dissipation of energy is only known in the functioning of organic structures, which is catagenetic; not in their progressive evolution, which is anagenetic.

Huxley, in a lecture delivered in 1854,[2] remarks: "Tendency to equilibrium of force and permanency of form then are the characters of that portion of the universe which does not live, the domain of the chemist and the physicist. Tendency to disturb existing equilibriums, to take on forms which succeed one another in definite cycles, is the character of the living world." In the letter to Professor Tyndall, prefatory to the volume *Lay Sermons and Addresses*, in which this essay appeared, Huxley says: "The oldest essay of the whole, that on 'The Educational Value of the Natural History Sciences,' contains a view of the differences between living and not-living bodies, which I have long since outgrown." Whatever might have

1 *First Principles*, ed. II., 1873, p. 542.

2 *Lay Sermons and Addresses*, 1880, p. 75.

been the cause of this change of opinion in Huxley's mind, the cause which has produced a similar change in the minds of many men, has been the discovery of means of producing in the laboratory numerous organic compounds, which, it had been previously supposed, could not be produced excepting through the action of living things, vegetable and animal. But it has been shown that all of these substances are the result of the running down of protoplasm, and are, hence, catagenetic, and not anagenetic.

That the catagenetic energies, whether physical or chemical, tend away from life is clear enough. Thus molar motion, unless continuously supplied, or directed by a living source, speedily ceases, being converted by friction into heat, which is dissipated. The same is true of molecular movements, under the same conditions. Chemical reactions, which are fundamental in world-building, result in the production of solids and the radiation of heat. This is the general result, although in the process, as it occurs in nature, irregularities occur, owing to local and temporary elevation of temperature. This arises from the decomposition of organic substances, which liberates heat; the oxydization of carbon, which owes its position as a terrestrial element to vegetable and animal organisms; and the access of heat from the interior of the earth, or from the sun's rays. Finally cosmic creation involves the perpetual radiation of heat into space, and the gradual reduction of all forms of matter to the solid state.

The endothermic chemical reaction, where inorganic matter undergoes a change of molecular aggregation the reverse of that just mentioned, with the absorption of heat, as in the case of several nitrogen compounds, is rare in nature, where free from organic complications, and is necessarily soon reversed by further reactions.

In the anagenetic energies, on the other hand, we have a process of building machines, which not only resist the action of catagenesis, but which press the catagenetic energies into their service. In the assimilation of inorganic substances they elevate them into higher, that is more complex compounds, and raise the types of energy to their own level. In the development of molar movements they enable their organisms to escape many of the destructive effects of catagenetic energy; by enabling them to change their environment; and this is especially true in so far as sensation or consciousness is present to them. The anagenetic energy transforms the face of nature by its power of assimilating and recompounding inorganic matter, and by its capacity for multiplying its individuals. In spite of the mechanical destructibility of its physical basis (protoplasm), and the ease with which its mechanisms are destroyed, it successfully resists, controls, and remodels the catagenetic energies for its purposes.

The anagenetic power of assimilation of the inorganic substances is chiefly seen in the vegetable kingdom. Atmospheric air, water, and inorganic salts furnish it with the materials of its physical basis. Then from its own protoplasm it elaborates by a catagenetic retrograde metamorphosis, the non-nitrogenous substances, as wood (cellulose), waxes, and oils, and the nitrogenous alkaloids, and it may take up inorganic substances and deposit them without alteration in its cells. Many of the compounds elaborated by plants and animals have been manufactured of latter time by chemists. The discovery that the living organism is not necessary for the production of these substances has led to the hasty conclusion that the supposed distinction between "organic" and "inorganic" energy does not exist. But the elaboration of these substances is not accomplished by anagenetic or "vital" energy, but by a process of running down of the higher compound protoplasm, which is catagenesis. No truly anagenetic process has yet been imitated by man.

All forms of functioning of organs, except assimilation, reproduction, and growth, are catagenetic. That is, functioning consists in the retrograde metamorphosis of a nitrogenous organic substance or proteid with the setting free of energy. The proteid is decomposed in the functioning tissue into carbon dioxide, water, urea, etc., and energy appears in the muscle as contraction, in the glands as secretion, and in all parts of the body as heat. The general result of physiologic research is, that the decomposition of the blood is the source of energy, while the tissue of each organ determines the character of that energy. That the tissue itself suffers from wear, and requires repair, is also true, but to a less extent than was once supposed.

In the anagenetic process of the growth of the embryo the case is different. Here the processes of functioning of organs are in complete abeyance, the plasma of the oösperm is not sensibly broken down in chemical decomposition, but it is in great part elaborated into tissues and organs. All the mechanisms necessary to the mature life of the individual are constructed by the activity of the special form of energy known as growth-energy or Bathmism. It is the modifications of this energy which constitute evolution, and it is these to which we will hereafter direct our attention. Its simplest exhibition is the subdivision of a unicellular protoplasmic body into two or more individuals or structural units of a multicellular organism. Further division of the latter does not abolish the individual, but extends it, and we now observe the elaboration of different structural types to become the conspicuous function of this form of energy. In other words, a once simple energy becomes specialized into specific energies, each of which, once established, pursues its mode of motion in opposition to all other modes not more potent than itself. Besides the evident truth of the proposition that a mode of building is a mode of motion, we have another very good reason for believing in the existence of a class of bathmic or growth-energies. This is found in the phenomena of heredity. The most rational conception of this inheritance of structural characters is the transmission of a mode of motion from the soma to the germ-cells. This is a far more conceivable method than that of the transmission of particles of matter, other than the ordinary material of nutrition. The bathmic theory of heredity bears about the same relation to a theory of transmission of the pangenes of Darwin, or the ids of Weismann, as the undulatory theory of light and other forms of radiant energy does to the molecular theory of Newton. I have therefore assumed as a working hypothesis the existence of the bathmic energy, and have inquired how far the facts in our possession sustain it. In doing so it has been necessary to elaborate the theory so as to render clearer its application to specific cases. The fact to be accounted for is its specialization into so many diverse specific forms.

A further indication of the existence of the bathmic energy is the quantitative limitation to which growth is obedient. Thus the successive stages of embryonic growth are limited in number in each species. The dimensions of most species are limited within a definite range. The duration of life, or of the functioning organic machine, has a definite limit in time. All this means that a certain limited quantity of energy is at the disposal of each individual organism.

In the preceding pages I have endeavored to show what causes have been and are efficient in the production of different types of organic life, through the modifications of the bathmic energy. We will now briefly consider the question of the origin of the living substance, protoplasm or sarcode, which exhibits bathmism.

On this subject Professor Manly Miles remarks :[1] "Omitting subordinate details, which represent the separate links in the chain of events, the processes of nutrition may be summarized in general terms as follows : In plants the chemical elements and binary compound on which they feed, are built up by successive steps of increasing complexity and instability into protoplasm, with a storing of the energy made use of in the constructive process, which is derived from the heat and light of the sun. The constructive processes are expressed by the term anabolism, and the products of the different upward steps are called anastatic. Protoplasm, the most complex and unstable of organic substances, is the summit of the ascending steps of anabolism ; and katabolism, which represents the succeeding downward steps of metabolism, then follows, and its products or katastates are starch, cellulose, proteids, etc., or what we recognize as the proximate

[1] *Proceeds. Amer Assoc. Adv. Sci.*, 1892, p. 203.

constituents and tissues of plants." I interject here the remark, that from a chemical point of view, protoplasm is, under certain conditions, not unstable.

If the tendency of the catagenetic energies is away from vital phenomena, it is impossible that they, or any of them, should be the cause of the origin of living matter. This logical inference is confirmed by the failure of all attempts to demonstrate spontaneous generation of living organisms from inorganic matter. Further, the principle of continuity leads us to infer that the energy which produced organic matter must be identical with or allied to that which is the efficient agent in progressive evolution of organisms, and is, therefore, anagenetic. Such a conclusion may seem to lead to a dualism which is itself opposed to the principle of continuity or uniformity, and which is opposed to experience of the phenomena of energy in general. How is uniformity to be harmonized with the hypothesis of two types of energy acting in different directions, apparently in opposition to each other? Since facts and logic do not support the derivation of the anagenetic from the inorganic energies, can the reverse process, the derivation of the catagenetic from the anagenetic be and have been the order of nature? In support of this hypothesis, we have the universal prevalence of the retrograde metamorphosis of energy in both the inorganic and organic kingdoms. Phenomena of structural degeneracy are well known in the organic world, and purely chemical phenomena in both organic and inorganic processes are all degenerate. It appears, then, much more probable that catagenesis succeeds anagenesis as a consequence, and does not precede it as a cause. In other words, it is more probable that death is a consequence of life, rather than that the living is a product of the non-living. I have therefore given to that energy which is displayed by the plant in the elaboration of living from non-living matter the name of antichemism.[1] Thus, while the heat of the sun is necessary to the building of protoplasm, within a certain range of temperature this form of energy has its opportunity.

The actual demonstration of this hypothesis can only come from researches into the thermochemistry of proteids and protoplasm. As these substances have not been made in the laboratory, these researches are not yet possible. We may, however, consider the problem as follows. In the process of making protoplasm, three gases, oxygen, hydrogen, and nitrogen are converted into a semisolid. In this case heat should be dissipated, to an amount reduced by the molecular dissolution of carbon. This is however not the case, for heat is absorbed with an integration of atomic bonds. In other words, it would seem that the manufacture of protoplasm by plants is an endothermic process. This view is strengthened by the discovery by Berthollet[2] that the production of numerous solid organic substances, in which organic bases are used, is also endothermic. These facts confirm the inference above recited, that the phenomena of organic growth involve the absorption of energy and not its dissipation.

Referring to the composition of protoplasm *C O H N*, I have called attention to the fact that each of its elements represents one of the great divisions defined

[1] *American Naturalist*, 1884, p. 979; *Origin of the Fittest*, 1887, p, 431.

[2] *Annales de Chimie et de Physique*, VI, 1895, p. 232.

by their valency, into which the ultimate substances of nature naturally fall. This combination I have suggested might account for the chemical inertness of protoplasm, through the mutual inhibition which each of these elements might be supposed to exercise over the other, owing to the diversity of their modes of chemical action.

In order to present more clearly the views enunciated in the preceding pages, I give a synoptic table of energies.

I. Anagenetic	Organic	Antichemism Bathmism
II. Catagenetic	Exclusively organic	Neurism Myism
	Inorganic and organic	Radiant Energy Chemism Cohesion Gravitation

2. BATHMOGENESIS.

The innumerable structures which are due to the activity of bathmisms may be supposed to result from the composition of the inherited form with energies which are derived from sources external to the germ-plasma, whether within the soma or external to it. These interferences produce new and specific types of energy. The inherited bathmism I have termed "simple growth force," and the modified forms I have termed "grade growth force."[1] It appears that these types of energy should be distinguished by special names. Hence I have proposed to restrict the term bathmism

[1] *Proceedings American Philosophical Society*, 1871, p. 253.

to the modified or "grade" growth force, and to term the inherited or "simple" type of growth force, emphytism.[1] As a matter of fact, pure emphytism can only be observed in the embryos of sexless or parthenogenetic origin, and in the repair of tissues.

Ryder has called the exhibition of growth-energy ergogenesis, and he calls attention to the fact that it appears under two aspects. In the first, ergogenesis is due to mechanical causes resident in the organism exclusively, and consists of the physical tensions inherent in protoplasm under all the conditions of growth. With these the growth-energies have to reckon, as they are the conditions which underlie them. They are not, however, strictly speaking, growth-energies, but would be exhibited by any similar colloid under similar conditions. To the movements due to physical causes under these circumstances, Ryder gives the name of Statogenesis.[2] The second aspect of the energies necessary to growth is present under the two forms already referred to, as emphytism and bathmism. The latter class, or interference energies, are naturally differentiated into those which are due to physical (or chemical) external agencies (molecular movements), and those that are due to molar movements as expressed in tissues, as impact, strain, etc. To the former I have given the name of physiobathmism, to the latter, kinetobathmism.[3]

The relations of these forms of energy may be represented in tabular form as follows:

[1] I have supposed in a late paper (*American Naturalist*, 1894, p. 212) that this is the statogenic energy of Ryder. This mistake has been corrected.

[2] *Proceedings American Philosophical Society*, 1893, p. 194.

[3] *American Naturalist*, 1894, p. 214. The two types of growth are then physiogenesis and kinetogenesis. (*Origin of the Fittest*, 1887, p. 423.)

		Ergogenesis			
Catagenetic	*Statogenesis*				
Anagenetic	Inherited	*Emphytogenesis*			
	With interference	*Bathmogenesis*	Molecular	*Physiogenesis*	
			Molar	*Kinetogenesis*	

Emphytogenesis I shall hereafter endeavor to show is an automatic (catagenetic) product of bathmogenesis, and a stationary factor in evolution.

The above table is designed to be a classification and formulation of the innumerable well-known facts of organic growth and evolution. It does not pretend to be an explanation of the processes involved, but it is the first step to be taken in attempting the explanation, i. e. a discrimination and classification of its factors.

Ryder thus expressed the relation between statogenesis and kinetogenesis.[1]

"So universal is this interference of the statical conditions of the plasma of segmenting ova with the ontogenetic processes, that not a single metazoan organism can be named, the development of which is not thus marred in some way or other. It is often a long time relatively after development has begun that there is any obvious delineation of the embryo. In fact, this cannot take place until the statical energies of surface-tension which have kept the egg globular are overridden. In so far as the ontogeny of any organism is marred by statical conditions of energy-display, its embryonic form is also modified. In so far as such statical interference affects the figure of the organism they are morphogenetic or form-determining. In so far the figure of a developing being is disturbed or

[1] *Proceeds. Amer. Philos. Society*, 1893, pp. 197–201.

modified by statical agencies its figure may be said to be subject to statogenetic influences. No existing larval form has escaped the influence upon its own shape of a constantly active statical equilibrium of its own substance. There is, therefore, a constant struggle going on during development between the phylogenetic and ontogenetic forces, determining the sequence and relations of the successive cleavages of the egg and the statical equilibria that obtain amongst its several parts. Statogenetic processes are, therefore, as constant and universal as the phylogenetic and ontogenetic. One may even go so far as to say that possibly the relations thus tending to be established by statical conditions may tend to become transmissible as hereditary tendencies. Such indeed is the view upheld by Prof. E. B. Wilson in his remarkable paper on 'The Cell-Lineage of Nereis.' I have myself seen no less than three consecutive recurrences of the same statical conditions in a fish egg, none of which can, for this reason, be definitely proved to be purely ontogenetic.

"While such phenomena as those of the genesis of the heterocercal or upwardly deflected condition of the axis in the tails of fishes, or the downwardly deflected condition of the axis in Ichthyosauri are almost purely kinetogenetic, the multiplicity of factors concerned, statogenetic as well as ontogenetic and phylogenetic, must always be considered and each given its due weight and importance in achieving the morphogenetic result. That there is an absolute conflict between statogeny and kinetogeny on the one hand, and of phylogeny and ontogeny on the other, in the case of the development of the ova of multicellular forms admits of no doubt. All metazoa pass through larval stages in which the statical conditions of equilibrium of the plasma of the egg is gradually, in a great measure, overriden by the hereditary energies represented by phylogeny and ontogeny. That there still remain traces of the effects of kinetogeny and statogeny in the adult organism cannot be denied in view of the facts to be derived from the shapes of tissue elements, and even of organs, as the foregoing paragraphs show."

The first appearance of bathmogenetic action is the first modification of the statogenetic and emphytogenetic energies from whatever source. Changes may be effected in the weight, color, and in functional capacity by temperature, humidity, food, etc., thus exhibiting physiogenesis. Or changes in the size and forms of parts of the body may be produced by movements of the organism, or of its environment, so displaying kinetogenesis. So long as these modifications of structure should be confined to the individuals thus modified, there would be no evolution. A second generation, if not subjected to the same stimuli, would not possess the modifications; and their possession of them would depend entirely on the amount of stimulus. In other words, there would be no accumulation of modification. It has, however, been generally believed that these modifications are inherited, and I think it has been shown that this belief rests on a solid basis. Meanwhile I have called the bathmogenesis which does not extend beyond the generation in which it appears, autobathmogeny.

The quantitative relation which necessarily exists between bathmism and its sources may be expressed as follows, with due recognition of the fact that such expression does not rest upon any experimental tests. Emphytogenesis is work done in the construction of tissues like those of the parent and without interference. Here we have the molecular energy of the parent converted into the molar movements observed to be concomitants of segmentation ; to be represented in the completed tissue by the mutual tensions by virtue of which each structural element maintains its integrity. It is evidently a process of metamorphosis of energy in which there is less waste than in any other known to us. Embryonic growth is accompanied by a very slight dissipation of heat, since a slight rise of temperature is noticeable in the eggs of cold-blooded animals and in flowers, when reproduction is active. The products of breaking down are equally rare in embryonic growth, and both this and the dissipation of heat are perhaps largely due to the changes wrought in non-cleavable nutritive substances with which the yolks are sometimes charged. It is probably to accomplish this process that the oxygen necessary for the embryonic growth is used. How much loss is due to cell-division itself is not known, but it must be very little if any. We have probably here a nearly perfect conversion of energy. Theoretically we have anagenesis wherever the up-building exceeds the down-breaking.

The attempt to realize in the imagination the *modus operandi* of bathmic energy in embryo-building takes the following form. It is to be supposed that movement which has been most frequently repeated, and for the longest period, is prepotent, and takes precedence of all others. This is clearly simple cell-division, which follows the nutrition supplied by the spermatozoön, and which represents the first act of animal life. Hence, segmentation of the oösperm is the first movement of bathmism. Each subsequent movement appears in the order of potency, which is, other things being equal, a time order, or the order of record. The cause of the localization of tissues and structures is much more difficult to understand than the cause of the order of their appearance. The more energetic part of the process naturally requires the greater space for its products. The ectoderm, which becomes the seat of the nervous axis and its muscular adjuncts, occupies the superficial portions of the yolk. Hence, we may regard this expression of the structural record of these functions as more energetic than that of the record-structure of the nutritive functions, which displays itself below the ectoderm. In meroblastic and amphiblastic embryos, the segmentation which develops the nutritive tissues is evidently more sluggish, for the cells are larger and fewer in number than those of the ectoderm.

In evolution external stimuli modify the course of emphytogeny above described, and by producing new structural records, cause a new form of energy, due to composition of the new with the old, and the process of growth then becomes bathmogeny. The external stimuli are molecular or molar, determining physiobathmism or kinetobathmism.

The effect of motion or use on the soma may be conveniently termed autokinetogenesis. Moderate use of a muscle is known to increase its size. Irritation of the periosteum is known to cause deposit of bone. Friction and pressure of the epithelium increases its quantity or changes its form. Increased activity of the functions of nervous tissues increases their relative proportions, as in the enlargement of nerves which replace others which are interrupted by mutilations, etc.

On the other hand, it is equally well known that disuse produces diminution of muscular tissues, and through it, a reduction in the quantity of the harder tissue (bone, chitin, etc.) to which it is attached (as muscular insertions, etc.). It was the observation of such well-known phenomena as these that led Lamarck to advance his doctrine of evolution under use and disuse, and which has led many others to give their adherence to such a view.

Thus much for cell-growth. Another class of modifications of a similar kind may be found in the parts of an organism which consist of a complex of cells, or tissues. Thus the lumen of a small artery is enlarged under the influence of pressure when it is compelled to assume the function of a larger vessel through the interruption of the latter. A part of an internal or external skeleton which is fractured will form an artificial joint at the point of fracture, if the adjacent surfaces are kept in motion. Marey (*Animal Mechanism*, pp. 88–89) says, "After dislocations the old articular cavities will be filled up and disappear, while at the new point where the head of the bone is actually placed, a fresh articulation is formed, to which nothing will be wanting in the course of a few months, neither articular cartilages, synovial fluid, nor the ligaments to retain the bone in place." I have given some illustrations of this fact,[1] which have come under my observation, and which have an important bearing on the origin of the articulations of the vertebrate skeleton as I have traced them throughout geological time. I have, as I think, conclusively shown that these varied structures have been produced by impacts and strains,

[1]Page 275 and *Proceeds. Amer. Philos. Soc.*, 1892, p. 285.

which are concomitants of the movements of the animals, acting through long periods of time.[1]

The term mnemogenesis is employed by Professor Hyatt[2] to characterize the manner in which kinetogenesis is supposed to produce results in inheritance. I have suggested that the phenomena of recapitulation, characteristic of ontogeny (*American Naturalist*, Dec., 1889), are due to the presence of a record in the germ cells, having a molecular basis similar to that of memory. This view is adopted by Professor Hyatt. I have already referred to it in the preceding pages.

A general statement of this doctrine was made by Mr. Sedgwick in *The British and Foreign Medico-Chirurgical Review* for July 1863 in the following language: "For atavism in disease appears to be but an instance of memory in reproduction, as imitation is expressed in direct descent; and in the same way that memory never, as it were, dies out, but in some state always exists, so the previous existence of some peculiarity in organization may likewise be regarded as never absolutely lost in succeeding generations, except by extinction of race." The next formulation of mnemogenesis is by Hering in 1870.[3]

It is concentrated in the following paragraph:

"The appearance of properties of the parental organism in the full-grown filial organism can be nothing else but the reproduction of such processes of organized matter as the germ when still in the germinal vesicles had taken part in; the filial organism remembers, so to speak, those processes, and as soon as

[1] "Mechanical Origin of the Hard Parts of the Mammalia," *Amer. Journal of Morphology*, 1889. *Origin of the Fittest*, 1887, pp. 305–373.

[2] *Proceeds. Boston Soc. Nat. History*, 1893, p. 73.

[3] Address before the Imperial Academy of Sciences of Vienna, May 30, 1870, by Ewald Hering; English translation, Chicago, 1895.

an occasion of the same or similar irritations is offered, a reaction takes place as formerly in the parental organism, of which it was then a part and whose destinies influenced it." In explanation of this theory, Hering says: "We notice, further on, that the process of development of the germs which are destined to attain an independent existence, exercises a powerful reaction both on the conscious and unconscious life of the whole organism. And this is a hint that the organ of germination is in closer and more momentous relation to the other parts, especially to the nervous system, than any other organ. In an inverse ratio, the conscious destinies of the whole organism, it is most probable, find a stronger echo in the germinal vesicles than elsewhere."

If heredity is a form of memory, its laws may resemble those of the psychic memory. In the latter, everything depends on what we call the strength of the impression. A single impression is often easily forgotten, and the certainty of recollection is largely dependent on the frequency of repetition of the stimulus. This is the essence of mental education, and it is probably the law of education of the germ-plasma as well. Thus may be understood how stimuli endlessly repeated through long geologic ages, must produce results far more profound and lasting than sporadic impressions of modern and artificial origin.

It must be here remarked for the benefit of the reader who may be unfamiliar with the explanation of the psychic memory, that it is the conscious part of memory which gives it its psychic character. This side is due to a fundamental molecular arrangement caused by stimuli, which may be retained for long periods without expression in consciousness. Thus

Bain regards memory as consisting of an unconscious and a conscious stage, and the latter he terms reminiscence. Other psychologists in common with mankind generally, use the word memory for the conscious expression. I have termed the unconscious condition of the molecular basis of mind, cryptopnoÿ.

CHAPTER X.—THE FUNCTION OF CONSCIOUSNESS.

1. CONSCIOUSNESS AND AUTOMATISM.

CONSCIOUSNESS is a general term, which embraces all forms of self-knowledge. Sentiency is sometimes used with an identical meaning. Consciousness must be distinguished from self-consciousness, which implies introspection. Consciousness may or may not be characterised by attention. There are two widely different types of consciousness, viz., the presentative and the representative. The former includes sense-perception only; the latter includes all the combinations of sense-perception which characterize mental action, from simple memory to the most comprehensive classification and conception. Most, if not all, animals appear to possess sense-perception, and all such possess the representative faculty of memory; but the higher grades of representative mental function are not so general among animals, and the extent of their occurrence is yet in dispute.

In the preceding pages I have endeavored to show that the factors of evolution are bathmogenesis corrected by natural selection. Bathmogenesis embraces the two factors physiogenesis and kinetogenesis, or the products of molecular and molar motion, respectively.

These two forms of motion have been coëxtensive with the existence of life, neither one preceding the other in time. Statogenesis is the expression of a form of energy which characterizes inorganic matter, while bathmogenesis is entirely peculiar to living things. Kinetogenesis is the fundamental principle in organic evolution, since it determines the amount and kind of physiogenesis, because it creates the environment which furnishes the conditions of physiogenesis. Progressive organic evolution may, then, be described as due to kinetogenesis corrected by natural selection. At the basis, however, molar organic motion, i. e., contraction of protoplasm, is probably molecular, but it is distinguished from other forms of molecular motion in the vast aggregate of molecules which move simultaneously in one direction, as in an amœba or a muscle, thus effecting a change in the position of all or a part of an organism. Hence the distinction is a real one.

Molar motion being, then, of such fundamental importance as a factor in evolution, the cause of such motion is also a capital question. Contraction of protoplasm is caused by stimuli, such as currents of electricity and chemical reagents; but such stimuli are not those which ordinarily produce the contractions to which the molar movements of living animals are due. In those animals which possess a nervous system it has been shown that contractions only follow stimuli which are conveyed to the contractile elements by nervous threads, and the internal energy which represents the external stimulus, is called nervous energy or neurism. In animals without a nervous system, and in plants, external stimuli must be justly supposed to be converted into the same form of energy, which in such organisms has a general circulation throughout the contractile protoplasm. The important point about these movements in most animals is, that their direction directly subserves the attainment of some position which is favorable for the procurement of relief from some unpleasant sensation, or the acquisition of some agreeable one, or both. We have the best reasons for believing this to be true of the vast majority of animals, because their structure is fundamentally like our own, and the inference that the same is true of the lowest forms of life is justifiable until it is proven to be mistaken.

Lamarck has attributed the movements of animals to the necessity of satisfying their instincts, without entering into the metaphysical questions which this involves. I have regarded the question as a metaphysical one by asserting that the necessary preliminary to movement is "effort," referring to what are called "voluntary" as distinguished from automatic motions.[1]

Without special organs of movement, a great part of the phenomena of kinetogenesis would have no existence, precisely as natural selection cannot act unless the materials for selection (i. e. variations) are already in existence. In explanation of the origin of organs of movement we have the general ability of the primitive animal, or protozoön, to project portions of its body-substance as pseudopodia, which, in more specialized forms, become persistent and more or less rigid, as flagella, cilia, etc.; which are the first distinct organs which subserve the transportation of the body from place to place. The causes which lead to these changes are as yet obscure, but that the use of these

[1] *Proceeds. Am. Philos. Soc.*, 1871, p. 18. *Origin of the Fittest*, 1887, p. 194,

organs when once called into existence is due to stimuli similar to those which affect the motions of the limbs of the higher animals, is altogether probable. Whatever be its nature, the preliminary to any animal movement which is not automatic, is an effort. And as no adaptive movement is automatic the first time it is performed, we may regard effort as the immediate source of all movement. Now, effort is a conscious state, and is a sense of resistance to be overcome. When an act is performed without effort, resistance has been overcome, and the mechanism necessary for the performance of the act has been completed. The stage of automatism has been reached. At the inception of a new movement resistance is necessarily experienced. It is generally believed that a mental state, as a sensation or a desire, which may or may not stimulate a rational process as an intervening element in the circuit, is concerned in overcoming this resistance.

A different view is held by certain physiologists and metaphysicians, as e. g. Wundt and Höffding. Huxley thus states his opinion in his Belfast address of 1874,[1] *à propos* of Descartes's doctrine that all animals below man are automata. "The consciousness of brutes would appear to be related to the mechanism of their body simply as a collateral product of its working, and to be as completely without any power of modifying that working as the steam-whistle which accompanies the work of a locomotive-engine is without influence on its machinery. *Their volition, if they have any, is an emotion indicative of physical changes, not a cause of such changes.*" (Italics mine.) In other words, stimulus excites conscious states, but the state thus produced has no influence on the resulting act.

[1] *Scientific Culture and Other Essays*, p. 243.

That sense-perceptions are stimuli to the immediate appearance of structural changes or movements is admitted. This is shown by the production of color changes in animals through changes in the condition of the organs of sight. Pouchet showed that the extirpation of both eyes of the turbot, and of a Gobius, paralyzed the chromatophorous cells, so that the usual color-adaptations to the color of the surface of the bottom on which they rested, could not be made. He also produced the same effect on one side of a trout by removing the eye of the opposite side. The chromatophoræ were permanently expanded, and the colors dark.

Some experiments which I tried with tree frogs of the species *Hyla gratiosa*, and *Hyla carolinensis*, are as follows: The color is usually green in both species, but it changes to dark brown where the frog rests on a brown surface, as of bark, etc. It appears that the maintenance of the brown color requires a more vigorous nervous stimulus than the green. The frogs are green at death; and limbs which I ligated remained green when the remainder of the surface became brown. Now in individuals with extirpated eyes, the color was always green, no matter what the surface on which they rested. The power of assuming the brown color was lost.

In these experiments it is difficult to connect the expansion of the chromatophorous cells as any effect of consciousness of color, by direct proof. If, however, muscular cells can be contracted under the influence of mental states, a similar mechanism may be supposed to exist in the case of chromatophoræ. The fact that the process may be now reflex does not exclude the other fact of the influence of consciousness at the inception, and its necessity for the continuance of the habit.

If we examine muscular movements, the evidence of control by consciousness becomes more distinct. New conditions bring forth new acts in animals too frequently to permit us to believe otherwise. Thus Mr. Belt tells of a procession of ants which crossed a railroad in Panama. Many were killed on the rail, so the column excavated a passage beneath the rail, and thus escaped further injury. No one can reasonably deny the intervention of a conscious state of a high order, as directly controlling the muscular movements of those ants. According to Beauchamp, termites in the same region display similar intelligence. A number of them were confined in a deep glass vessel with smooth sides which they could not scale. They thereupon deposited drops of their building secretion, which hardens on drying, on the glass, ascending backwards, and so made a stair out of their prison, by which they escaped.

A *Cebus capucinus* in my possession imitated some carpenters who were working in the room with a drawing-knife. He used a triangular piece of tin, and, holding the corners, drew the edge towards him over the surface of a piece of squared wood on which he sat. He did this rapidly and repeatedly, with many grimaces. A *Cebus apella* in the Philadelphia Zoölogical Garden lights matches whenever he can get them. He always selects the proper end, and holds it at a proper distance, so that the stick is not broken and his fingers are not burned. He strikes them on the rough outside of his drinking-kettle. My *Cebus* came direct from the forests of Venezuela, and he had not been educated among carpenters. The history of the *apella*

I do not know, but he had not probably been brought up among matches, and his act was in any case not reflex.

As an illustration of the simplest of movements, and their physical conditions, I cite those of the Myxomycetes, from Stahl.[1]

"The movement of Myxomycetes is influenced by:

"1. Moisture (Hydrotropism): In their young stages they wander from the parts of the substratum (i. e. of the deposit on which they are creeping), which are gradually drying up, toward those which continue moist longer; 'it is even possible, by bringing moist bodies into the proximity of any ramifications, to cause the production of pseudopodia, which elevate themselves from the substratum, and soon come into contact with the moist object, so as to enable the whole mass of the plasmodium to migrate on to it.' On the entrance of the plasmodia into the fructifying condition, positive hydrotropism gives place to negative; the myxomycete quits the moist substratum and creeps upwards on to the surface of dry objects.

"2. Unequal distribution of warmth in the substratum, and

"3. Unequal supplies of oxygen also cause locomotion in the myxomycete.

"4. Chemical substances soluble in water have a similar action. Contact of the plasmodia on one side with solutions of common salt, saltpetre, or carbonate of potash, cause them to withdraw from the dangerous spot, while infusion of tan, or a dilute solution of sugar, produces a flow of the protoplasm and ulti-

[1] E. Stahl, "Zur Biologie der Myxomyceten," *Botan. Zeitung*, 1884, No. 10–12. Abstract in *Sitzungsbericht der Jenaischen Gesellschaft für Medizin und Naturwissenschaft*, 1883, Sitzung vom 16. November.

mately translocation of the whole plasmodial mass towards the source of nourishment. Some solutions have an attractive or repulsive effect, according to their degree af concentration.

"5. Finally, they withdraw from light (negative heliotropism).

"With regard to the acceleration or definite direction of movement produced entirely by stimuli, compare the following:

"The knowledge of the remarkably delicate reaction of the plasmodia under external influences enables us to comprehend how these tender structures, destitute of every kind of external protection, are able to carry on their existence. The plasmodia which are not yet ripe for reproduction are kept in the moist substratum by positive hydrotropism, which is assisted by negative heliotropism.

"But within the darkness and moisture of the substratum the plasmodia by no means remain in one place, because the differences in the chemical composition of the substratum cause continual migrations. The plasmodia have the faculty in a wonderful way of avoiding harmful substances, and, traversing their substratum in all directions, of taking up the materials they require.

"When the internal changes have proceeded so far that the plasmodia approach the fructifying condition, they are brought by the negative hydrotropism which now sets in, from the moist parts of the ground in the forest or wood to the surface, where they creep up various upright objects, often only forming rigid reproductive capsules at some height from the ground.

"When in autumn the substratum becomes gradually colder a change which takes place from the surface downwards, the plasmodia migrate into deeper regions still having a higher temperature. When the cooling proceeds very gradually, which especially happens in large tan-heaps, the plasmodia may, in their migration reach somewhat considerable depths, where they then change into sclerotia. To find the sclerotia of Æthalium in winter it is, therefore, not seldom necessary to search through the mass of tan to a depth of several feet. When the temperature again begins to rise, the sclerotia again germinate, and movement in the opposite direction takes place from the deeper and cooler parts to the upper portions already warmed.

"In the locomotion of the Myxomycetes, then, we see extremely interesting cases of movements due to stimulation. Heliotropism, geotropism, hydrotropism, trophotropism, in general, are stimulus-movements, and ultimately all growth depends on stimulus-movement. It is the most primitive kind of protoplasmic movement. Stimuli in fixed directions and constantly repeated, produced, but only secondarily, fixed paths of conduction, and responses of a quite definite kind (reflexes). Thus arose nerves, and finally apparatus for stirring up stimuli, arose sensation and will—as acquired and inherited faculties."

In this lowest type of organic movement it is difficult to discern any cause for it different from those which actuate higher organisms. What form of inorganic energy can be cited as sufficient to cause the organism to change its position with regard to stimuli to self-preservation, and without regard to gravitation, or any known form of attraction or repulsion? In the Fuligo (tan flowers) a most pronounced example of designed movement has been observed. This form, in the amœbula stage, will, according to H. J. Carter, "confine itself to the water of the watch-glass in which it may be placed, when away from the sawdust and chips of wood among which it has been living; but if the watch-glass be placed upon the sawdust it will very soon make its way over the side of the watch-glass and get to it." This act probably depends on a sense-perception of the presence and position of the tan-bark, and of a feeling of desire to reach it. This may have been due to a sense of discomfort due to the surrounding water, or to a recollection of superior comfort associated with the tan-bark.

Ordinary observation of most animals leads to the belief that their movements are provoked by sensations, as of hunger, thirst, temperature, etc.; also of sight, hearing, smell, etc., when they possess those senses. There are physiologists who deny that such is the case, but I must insist on the importance of a psychological rather than a physiological study of animals as a most important source of information in this direction. The students of dead or mutilated animals miss important evidence as to the phenomena of consciousness. The attempt has been made to identify hunger, for instance, with chemical energy, a proposition which is simply irrational. It may be none the less true, however, that hunger is a necessary concomitant of a molecular condition. Observation on living animals shows in the most conclusive manner that by far the greater number of species are capable of the performance of acts in response to new situations and circumstances for the performance of which no automatic mechanism exists. Memory is clearly present in them, and, as a consequence, judgments are formed which determine the succeeding acts. The process, be it ever so simple, is "representative," and thus a mental act, at least one stage beyond sense-impression. The doctrine that conscious states have preceded organisms in time and evolution I have called archæsthetism. It seems to have been first clearly formulated by Erasmus Darwin, who believed that growth has been stimulated by "irritations" (of hunger, thirst, etc.) and by the pleasurable sensations attending those irritations, and by exertions in consequence of painful sensations, similar to those of hunger and suffocation,"[1] etc.

The weakness of the doctrine of archæsthetism consists in our ignorance of the characters of the Protozoa, with respect to the presence or absence of consciousness or sensation. While many of the acts of these low organisms need not be explained by supposing its presence, others seem to be impossible without it. We are, however, led to infer its presence rather on uniformitarian grounds, than by any certainty of explanation of the phenomena. We can trace sensation so far down in the scale of animal life, that it seems unreasonable to deny its presence when the same phenomena are exhibited by the Protozoa. We are confirmed in our belief in the presence of sensation in these low forms, by the knowledge that reflex acts are the product of conscious acts, whereas we have no evidence that conscious acts are the product of the reflex.

Although it is frequently alleged or assumed that designed conscious acts are the product of reflexes, no one has yet shown how this is possible. On the other hand, the development of automatic acts out of conscious ones is of ordinary occurrence, and is known under the name of education.

[1] *Zoonomia*, XXXIX., 3; Osborn, *From the Greeks to Darwin*, p. 148

The relation of consciousness to the physical basis is as yet a profound mystery, but that they exercise over each other a definite mutual control is unquestionable. The processes which produce thought, as conception, judgment, etc., are, however, not qualitatively related to the amount of nutritious proteids consumed in the central nervous system, but only quantitatively; yet it is the outcome of these processes that directs animal movements, when they are not automatic. In other words, the forms of thought, which have no weight, direct the movements of muscles, which have weight. This is not in accord with the doctrine of the correlation of energy. But what has the formation of a concept, or the development of a judgment, to do, *per se*, with the correlation of energy? I may give this idea a more definite expression by the following diagram :[1]

1

AFFIRMATIVE.	NEGATIVE.
1	5
2	6
3	
4	

2

AFFIRMATIVE.	NEGATIVE.
1	3
2	4
	5
	6

Let each square represent the grammes of energy necessary for the maintenance in consciousness of six propositions. Judgment issues from the side of the predominating number of propositions. They arrange themselves in consciousness in accordance with their qualities, in two aggregations represented by columns in the squares. Now if they arrange themselves in four affirmative and two negative, as in square 1, the

[1] This is in explanation of the same proposition as stated by me in the *Proceedings of the American Philosophical Society*, 1889, p. 504.

judgment is affirmative. If, on the other hand, they arrange themselves in two affirmative and four negative, as in square 2, the judgment is negative. The energy expended in the two cases is the same. So also in forming different concepts from the same set of particular sense-impressions or memories. Is there any difference in the energy expended in forming from them the concept of bigness as compared with that of redness? While, therefore, every mental process is expensive as a whole, the mental content is obedient to the forms of thought rather than the correlation of energy. This is what mind is. While it is doubtful whether any animal below man can form a concept (with a very few possible exceptions), the formation of simple judgments is general. Any decision based on experience is a judgment.

In order to render this proposition clearer, I have formulated it in the following language, although it is possible that the definition of energy will not bear the strain of the statement.

"The formal statement of this phenomenon may be found in the thesis, that *energy can be conscious*. If true, this is an ultimate fact, neither more nor less difficult to comprehend than the nature of energy or matter in their ultimate analyses. But how is such a hypothesis to be reconciled with the facts of nature, where consciousness plays a part so infinitesimally small? The explanation lies close at hand, and has been already referred to. *Energy become automatic is no longer conscious*, or is about to become unconscious. That this is the case is matter of every-day observation on ourselves and on other animals. What the molecular conditions of consciousness are, is one of the problems of the future, and for us a very interesting one. One thing is certain, the organization of the mechanism of habits is its enemy. *It is clear that in animals, energy, on the loss of consciousness, undergoes a retrograde metamorphosis.*

"To regard consciousness as the primitive condition of energy, contemplates an order of evolution in large degree the reverse of the one which is ordinarily entertained. The usual view is, that life is a derivative from inorganic energies as a result of high or complex molecular organization, and that consciousness (=sensibility) is the ultimate outcome of the nervous or equivalent energy possessed by living bodies. The failure of the attempts to demonstrate spontaneous generation will prove, if continued, fatal to this theory. With our present evidence it may be affirmed that not only 'has life preceded organization,' but that *consciousness was coincident with the dawn of life.*"

The facts cited, and the doctrines defended in the preceding pages lead to one inference as to the relation of consciousness to its physical basis. The condition of matter necessary to the maintenance of consciousness is, in the language of morphology, *generalized*; in the language of chemistry, *neutral*; in the language of physics, *non-equilibrated.* The materialist and the animist can alike agree as to this generalization. The difference between the two positions is a difference of view as to the mutual relations of the two members of the partnership. Is the permanent absence of equilibrium of living protoplasm due to control by consciousness? or is consciousness a product of an absence of equilibrium, which is due to chemical and physical action? The latter proposition is untenable, because the inevitable tendency of chemical and physical energies is to an equilibrium. Is, therefore, the other alternative true? There seems to be no escape from it, and it accords also with our personal human experience of the agency of conscious states in our various activities, physical and mental.

2. THE EFFECTS OF CONSCIOUSNESS.

From the facts cited it is evident that sensation (consciousness) has preceded in time and in history, the evolution of the greater part of plants and animals, both unicellular and multicellular. It appears also that if kinetogenesis be true, consciousness has been essential to a rising scale of organic evolution.

Animals who do not perform simple acts of self-preservation must necessarily perish sooner or later. In fact it is impossible to understand how the lowest forms of life, utterly dependent as they are on physical conditions of many kinds, should not have been all destroyed, were they not possessed of some degree of consciousness under stimuli at least. And the case is even plainer with the higher forms. We have only to picture to ourselves the condition of a vertebrate without general or special sensation, to perceive how essential consciousness is to its existence. If now, as maintained in Chapter IV., use has modified structure, and so, in coöperation with the environment, has directed evolution, we can understand the origin and development of useful organs. And we can understand how by parasitism or other mode of gaining a livelihood without exertion, the adoption of new and skilful movements would become unnecessary, and consciousness itself would be seldom aroused. Contiued repose would be followed by subconsciousness, and later by unconsciousness. Such appears to be the history of degeneracy everywhere, and such is, perhaps, the history of the entire vegetable kingdom. From their ability to manufacture protoplasm from inorganic substances, plants do not need to move about in search of food, so that they require no consciousness of conditions to guide their movements. They become fixed, and their entire organization becomes monopolized by the functions of nutrition and reproduction. Movements rarely occur, and when present are confined to those of one part of the structure or another. They are mostly rhythmic or rotary, and very seldom exhibit the quality of impromptu design. The satisfactory explanation of those that exhibit general design, as the adaptation for transportation often seen in seeds, may be chance coincidence and natural selection.

The ascending scale of development of intelligence observed among animals is strong evidence in support of the hypothesis here outlined. There can be no doubt that in the long run the most intelligent have survived. They have survived because they were capable of the most successful designed acts, thus directing their movements to the most useful ends. These movements ultimately modified their structure usefully, to the perfecting of mechanisms in every way important to their possessors. This much having been established as to the cause of anagenesis, let us look more closely into the history of catagenesis.

Movements of organic beings on frequent repetition become automatic, reflex, and finally, as it is termed, organic. This means the running down of energy through various grades, beginning with the highest or conscious stage, and ending with the purely reflex, which is as unconscious of changes in the environment as is any one of the physical energies. The conscious stage is evidently the most susceptible to the stimuli of the environment, and the process of catagenesis is one of degeneracy to less and less sensitive and to more and more mechanical conditions. The resemblance of the lowest grade of organic activities to physical mechanical energy is so great that it is almost universally supposed by evolutionists to be of purely mechanical origin, but I have endeavored to show that they are of totally different origin, and that the only explanation of their characteristics is the hypothesis of catagenesis.

In accordance with this view, the automatic "involuntary" movements of the heart, intestines, reproductive systems, etc., were organized in primitive and simple animals in successive states of consciousness, which stimulated "voluntary" movements, which ultimately became rhythmic; whose results varied with the machinery already existing and the material at hand for use. It is not inconceivable that circulation may have been established by the suffering produced by an overloaded stomach demanding distribution of its contents. The structure of the Infusoria offers the structural conditions of such a process. A want of propulsion in a stomach or body-sack occupied with its own functions would lead to a painful clogging of the flow of its products, and the "voluntary" contractility of the body or tube-wall being thus stimulated, would at some point originate the pulsation necessary to relieve the tension. Thus might have originated the "contractile vesicle" or contractile tube of some protozoa; its ultimate product being the mammalian heart. So with reproduction. Perhaps an excess of assimilation in well-fed individuals of the first animals led to the discovery that self-division constituted a relief from the oppression of too great bulk. With the increasing specialization of form, this process would become necessarily localized in the body, and growth would repeat such resulting structure in descent, as readily as any of the other structural peculiarities. No function of the higher animals bears the mark of conscious origin more than this one, as consciousness is still one of the conditions of its performance. While less completely "voluntary" than muscular action, it is more dependent on stimulus for its initial movements, and does not in these display the unconscious automatism characteristic of many other functions.

There remain, however, some phenomena which do not yield so readily to this analysis. First, we have the conversion of inorganic substances into protoplasm by plants. It is also well known that living animal organisms act as producers, by conversion, of various kinds of inorganic energy, as heat, light, sound, electricity, motion, etc. It is the uses to which these forces are put by the animal organism, the evident design in the occasion of their production, that gives them the stamp of organic life. We recognize the specific utility of the secretions of the glands, the appropriate distribution of the products of digestion and adaptation of muscular motion to many uses. The increase of heat to protect against depression of temperature; the light to direct the sexes to each other; the electricity as a defence against enemies—display unmistakably the same utility. We must not only believe that these functions of animals were originally used by them under stimulus, for their benefit, but, if life preceded organism, that the mechanism which does the work has developed as the result of the animal's exertions under stimuli. This will especially apply to the mechanism for the production of motion and sound. The production of heat, light, chemism, and electricity doubtless result from molecular aptitudes inherent in the constitution of protoplasm. But the first and last production of even these phenomena is dependent on the motions of the animal in obtaining and assimilating nutrition. For without nutrition all energy would speedily cease. Now the motion required for the obtaining of nutrition has its origin in the sensation of hunger. So, even for the first steps necessary to the production of inorganic forces in animals, we are brought back to a primitive consciousness. This hypothesis I have termed Archæsthetism.

"It maintains that consciousness as well as life preceded organism, and has been the *primum mobile* in the creation of organic structure. This conclusion also flows from a due consideration of the nature of life. I think it possible to show that the true definition of life is, *energy directed by sensibility, or by a mechanism which has originated under the direction of sensibility.* If this be true, the two statements that life has preceded organism, and that consciousness has preceded organism are coëqual expressions.

"Granting the existence of living protoplasm on the earth, there is little doubt that we have some of its earliest forms still with us. From these simplest of living beings both vegetable and animal kingdoms have been derived. But how was the distinction between the two lines of development, now so widely divergent, originally produced? The process is not difficult to imagine. The original plastid dissolved the salts of the earth and appropriated the gases of the atmosphere and built for itself more protoplasm. Its energy was sufficient to overcome the chemism that binds the molecules of nitrogen and hydrogen in ammonia, and of carbon and oxygen in carbonic dioxide. It apparently communicated to these molecules its own method of being, and raised the type of energy from the polar non-vital to the adaptive vital by the process. Thus it transformed the dead inorganic world, perhaps by a process of invasion, as when a fire communicates itself from burning to not burning combustible material. Thus it has been doing ever since, but it has redeposited some of its gathered stores in various non-vital forms. Some of these are in organic forms, as cellulose; others are crystals imprisoned in its cells; while others are amorphous, as waxes, resins, and oils. But consciousness apparently early abandoned the vegetable line. Doubtless all the energies of vegetable protoplasm soon became automatic. The plants in general, in the persons of their protist ancestors, soon left a free-swimming life and became sessile. Their lives thus became parasitic, more automatic, and in one sense degenerate.

"The animal line may have originated in this wise. Some individual protists, perhaps accidentally, devoured some of their fellows. The easy nutrition which ensued was probably pleasurable, and once enjoyed was repeated, and soon became a habit. The excess of energy thus saved from the laborious process of making protoplasm was available as the vehicle of consciousness and motion. From that day to this, consciousness has abandoned few if any members of the animal kingdom. In many of them it has specialized into more or less mind. Organization to subserve its needs has achieved a multifarious development. There is abundant evidence to show that the permanent and the successful forms have ever been those in which motion and sensibility have been preserved, and most highly developed.[1]

We must remember, however, that in the matter of the evolution of plant-types we have an especial factor in the influence which insects have exerted on the conditions of almost all of their organs. Insects originated early in geological time, and have closely accompanied plants in their evolution. As the source of the food, and as the dwelling-places of great numbers of insects, they have been subjected to a class of stimuli and strains similar to those which animals have experienced. It is believed that the forms of the organs of fructification and especially of the flowers, have been greatly modified by the influence of insects.[2] The general evolution of plants, however, presents us with a greater predominance of physiogenetic or simple dynamical conditions over the bathmic, than in the case of animals. Thus many peculiarities of the inflorescence directly result from the shortening of the axial growth in complementary relation to the increase of peripheral growth. Such is evidently the origin of flowers themselves; secondly of the umbel as compared with the spike or panicle and finally of the composite head as compared with the other modes of inflorescence. To the cohesion of the elements of a whorl, possible only in the case of an abbreviated axis, can we ascribe the formation of a seed vessel from discrete carpels, and a gamopetalous from a polypetalous corolla. Degeneracy or specialization is to be seen everywhere, as in the abortion of ovules, carpels, and perianth.

"Catagenesis of living organisms has been epito-

[1] *Origin of the Fittest*, p. 428, 432.

[2] Henslow.

mized in the following language: 'Evolution of living types is then a succession of elevation of platforms, on which succeeding ones have built. The history of one horizon of life is that its own completion but prepares the way for a higher one, furnishing the latter with conditions of a still further development. Thus the vegetable kingdom died, so to speak, that the animal kingdom might live, having descended from an animal stage to subserve the function of food for animals. The successive types of animals first stimulated the development of the most susceptible to the conflict, in the struggle for existence, and afterwards furnished them with food.' In the occupation of the world's fields, the easiest and nearest at hand have been first occupied, and successively those which were more difficult. The digging animals are generally those which first abandoned the open field to more courageous or stronger rivals; and they remain to this day generally of low type compared with others of their classes (e. g. *Monotremata*, *Glires*, *Insectivora*). All occupations have been filled before that one which requires the greatest expenditure of energy, i. e. mental activity. But all other modes of life have fallen short of this one in giving the supremacy over nature."

We now approach an explanation of the phenomenon of anagenesis. Why should evolution be progressive in the face of universal catagenesis? No other ground seems discoverable but the presence of sensation or consciousness, which is, metaphysically speaking, the protoplasm of mind. The two sensations of hunger and sex, have furnished the stimuli to internal and external activity, and memory, or experience with natural selection, have been the guides. Mind and body have thus developed contemporaneously and have reacted mutually. Without the coöperation of all these factors, anagenesis seems impossible.

From this point of view the study of the evolution of mind and its relation to the organic world assumes a new importance. Circumstances have forbidden my entering on this subject in the present volume, but I hope to be able to devote especial attention to it at a future time. One fundamental postulate of mental evolution may, however, be mentioned here. That is, as Spencer has pointed out, the instinct of hunger is at the basis of the activity which has developed the intelligence, while that of sex is at the basis of the development of the altruistic or social instincts and affections. With this proposition I leave this interesting part of the doctrine of evolution.

CHAPTER XI.—THE OPINIONS OF NEOLAMARCKIANS.

LAMARCK ascribed some of the evolutionary changes of structure to changes in the environment, some to the motions of organic beings, and others to both combined.[1] Spencer in 1865[2] devoted a short chapter to the effect of motion in producing variations, and specified the mechanical effect of flexure in producing segmentation of the vertebral column. The present writer in 1871[3] insisted on the importance of motion as a factor in determining growth, and in 1872[4] I approached the subject more definitely in the following language: "The first physical law is that growth force . . . must develop extent in the direction of least resistance, and density on the side of greatest resistance." In 1877 Ryder further applied the principle of motion to the origin of structural changes, chiefly reduction of digits, in the feet of Mammalia in language[5] which I have quoted on page 311.

1 *Philosophie Zoologique*, Chap. VII., 1809; translation in *American Naturalist* for 1888.

2 *Principles of Biology*, II., pp. 167 and 195.

3 *Proceeds. Amer. Philosoph. Soc.*, 1871, p. 259. *Origin of the Fittest*, 1887, p. 210.

4 *Penn Monthly Magazine*, July, 1872. *Origin of the Fittest*, 1887, p. 30.

5 *American Naturalist*, 1877, p. 607.

In the same year, in discussing the origin of the great development of the incisor teeth in the Rodentia,[1] Professor Ryder, in summing up, ventured "the reflection that the more severe strains to which they were subjected by enforced or intelligently assumed changes of habit, were the initiatory agents in causing them to assume their present forms, such forms as were best adapted to resist the greatest strains without breaking." In 1878 the writer[2] advanced the following proposition: "Change of structure is seen to take place in accordance with the mechanical effect of three kinds of motion, viz., by *friction*, *pressure*, and *strain.*" In the same year Professor Ryder went into a discussion of the specific application of strains in the evolution of the dental types of the diplarthrous Ungulata, and prepared the field for work in the Rodentia and Proboscidia.[3] In 1879 the writer gave mechanical reasons for the reduction of the sectorial teeth of Carnivora to one, and for their present position in the jaws.[4] In 1881 the writer[5] described the specific action of impacts and strains in the production of the existing characters of the articulations of the limbs in the higher Mammalia. In 1887 the same subject, together with that of the mechanical origin of the characters of the molar teeth, was more fully investigated in a paper on the Perissodactyla.[6] In 1888 the writer published a paper on the mechanical origin of the sectorial teeth of the Carnivora,[7] one on the mechanical

1 *Proceeds. Phila. Acad.*, 1877, p. 318.

2 *American Naturalist*, January, 1878. *Origin of the Fittest*, p. 354.

3 *Proceeds. Phila. Acad.*, 1878, p. 45.

4 *American Naturalist*, March, 1879.

5 *American Naturalist*, April and June, 1881.

6 *American Naturalist*, 1887, pp. 985, 1060.

7 Read before the American Association for the Advancement of Science, New York, 1887, p. 254.

origin of the peculiar dentition of the Glires,[1] and a third on the mechanical origin of the dentition of the Amblypoda.[2] In 1889 I discussed the mechanical causes of the structures of the elbow and other joints in the Artiodactyla and the origin of the peculiar intervertebral articulations in that order.[3] In the same year I published a *résumé* of the work done on this subject with reference to the Mammalia.[4] Since that time important contributions to the subject have been made by Ryder, Osborn, Wortman, Dall, Jackson, and others, to which reference will now be made.

Hyatt says as a result of his exhaustive studies of the phylogeny of the Cephalopoda[5]; "The action of physical changes takes effect upon an irritable, plastic organism which necessarily responds to external stimulants by an internal reaction or effort. This action from within upon the parts of organisms modifies their hereditary forms by the production of new growths or changes, which are therefore adapted or suitable to the conditions of the habitat, and are therefore physiologically and organically equivalent to the physical agents and forces from which they directly or indirectly originated. In so far then, as causes and habits are similar, they probably produce representation or morphological equivalence in different series of the same type in similar habitats: and in so far as they are different, they probably produce the differen-

1 *American Naturalist*, January, 1888, p. 3.

2 *Proceeds. Amer. Philosoph. Soc.*, 1888, p. 80.

3 *American Naturalist*, March, 1889.

4 *The American Journal of Morphology*, September, 1889, pp. 137–277, "On the Mechanical Causes of the Development of the Hard Parts of the Mammalia."

5 *The Genesis of the Arietidæ*, by Alpheus Hyatt; *Smithsonian Contributions to Knowledge*, and *Memoirs of the Museum of Comparative Zoölogy*, Vol. XVI., No. 3, 1889.

tials which distinguish series and groups from each other."

Packard[1] in discussing the causes of the blindness of cave animals, says: "Such a phrase as 'natural selection,' we repeat does not to our mind definitely bring before us the actual working-causes of the evolution of these cave organisms, and no one cause can apparently account for the result. The causes are 'change in the environment,' disuse of certain organs; 'adaptation,' 'isolation,' and 'heredity' operating to secure for the future the permanence of the newly originated forms as long as the physical conditions remain the same."

In 1889, Ryder described the ontogenetic origin of the articulations of the cartilaginous fin-rays of the *Salmo fontinalis*, and used the facts observed as evidence that these articulations are due to the mechanical strain experienced by the rays in use as motors of the body of the fish in the water.[2]

Prof. H. F. Osborn in 1890[3] discussed thoroughly the mechanical causes for the displacement of the elements of the feet of the ungulate Mammalia from linear to alternating relations. (See *antea*, p. 299.) In an article in *Nature*,[4] the same distinguished naturalist remarks: "The following views are adopted from those held by Cope and others, so far as they conform to my own observations and apply to the class of variations which come within the range of paleontological evidence. In the life of the individual, adap-

[1] "On the Cave Fauna of North America," *Memoirs of the U. S. National Academy of Sciences*, IV, pt., I., p. 137.

[2] *Proceedings of the American Philosophical Society*, 1889, p. 546.

[3] *Transactions of the American Philosophical Society*, XVI., February, 1890, p. 531.

[4] January 9, 1890, p. 277.

tation is increased by local and general metatrophic changes of necessity correlated, which take place most rapidly in the regions of least perfect adaptation, since here the reactions are greatest; the main term of variation is determined by the slow transmission, not of the full increase of adaptation, but of the disposition to adaptive atrophy or hypertrophy at certain points; the variations thus transmitted are accumulated by the selection of the individuals in which they are most marked, and by the extinction of inadaptive varieties or species; selection is thus of the ensemble of new and modified characters. Finally, there is sufficient paleontological and morphological evidence that acquired characters in the above limited sense, are transmitted. . . . Excepting in two or three side-lines the teeth of all the Mammalia have passed through closely parallel early stages of evolution, enabling us to formulate a law: *The new main elements of the crown make their appearance at the first points of contact, and chief points of wear of the teeth in preceding periods.* Whatever may be true of spontaneous variations in other parts of the organism, these new cusps arise in the perfectly definite lines of growth. . . . Now, after observing these principles operating in the teeth, look at the question enlarged by the evolution of parallel species of the horse series in America and Europe, and add to the development of the teeth what is observed in progress in the feet. Here is the problem of correlation in a stronger form even than that presented by Spencer and Romanes. To vary the mode of statement: what must be assumed in the strict application of the selection-theory? (*a*) that variations in the lower molars correlated with coincident variations of reversed patterns in the upper molars, these with metamorphosis in the premolars and pocketing of the incisor-enamel; (*b*) all new elements and forms at first so minute as to be barely visible, immediately selected and accumulated; (*c*) in the same individuals favorable variations in the proportions of the digits involving readjustments in the entire limbs and skeleton, all coincident with those in the teeth; (*d*) finally, all the above new variations, correlations, and readjustments not found in the hereditary germ-plasm of one period, but arising fortuitously by the union of different strains, observed to occur simultaneously and to be selected at the same rate in the species of the Rocky Mountains, the Thames Valley, and Switzerland! These assumptions, if anything, are understated."

I have already referred to the contribution by Dall to the doctrine of kinetogenesis which has resulted from his investigations of the origin of the characters of the lamellibranchiate (or pelecypod) Mollusca.[1] He observes: "In reflecting upon the origin of the complicated mechanical arrangements in bivalves which we call the hinge, I have come to the conclusion that here, as in the cases of the mammalian foot and tooth elaborated so clearly by Cope and Ryder, we have the result of influences of a mechanical nature operating upon an organ or apparatus in the process of development. . . . The shell is in one sense the product of secretion from the mantle, as the mammalian tooth is derived from the ectoderm of the jaw, or the skeleton from the periosteum and cartilages. Both are that and much more. It would be as reasonable to say that a steam boiler in process of construction is the product of the boy inside who holds the rivet-heads, as to claim that the shell has no more significance

[3] *American Journal of Science and Arts*, December, 1889, p. 447.

than is implied in the term 'secretion of the mantle.' The original theoretic protoconch may have been so, but as soon as it came into being, its development was governed by the physical forces impinging upon it from all sides, and through it influencing the growth and structure of the soft parts beneath. The gastropod shell is the result of the action and reaction between the physical forces of the environment, and the evolutionary tendencies of the organic individual. In the pelecypod we have the mechanical stresses and reactions of one valve upon the other, added to the category of influences. To some extent it is doubtless as true that the animal is moulded by its shell, as it is that the shell is shaped by the soft parts of the animal."

Dr. Dall in an able paper on the "Dynamic Influences in Evolution," read before the Biological Society of Washington, March 8, 1890, writes thus forcibly:

"It is often assumed, in discussing variation, that the possibility of variation is equal in every direction. A consideration of the dynamic conditions of life show that this is not the case, or, at least, if we grant its theoretic truth, in practice it can never be true. Under any conditions which would permit it, the resulting organic forms would have to pass their existence in constant rotation. The moment that any one of them came to rest, it would begin to be subjected to unequal stresses relatively to its different parts. Light, gravity, friction, opportunities for nutrition, would be unequally distributed, with the result of forcing an unequal growth, development, and specialization of its regions. Inequality of form once established, if it were a moving organism, friction and resistance of the circumambient medium, would confirm the inequality, and put individuals of its kind at a disadvantage when they varied towards the original shape."

This is an excellent statement of kinetogenesis in concentrated form.

Prof. A. S. Packard in an important essay "On the Evolution of the Bristles, Spines, and Tubercles of Certain Caterpillars," etc., remarks as follows:[1] "The Lamarckian factors (1) of change (both direct and indirect) in the *milieu*, (2) need, (3) habit, and the now generally adopted principle that a change of function induces change in organs, and in some or many cases actually induces the hypertrophy and specialisation of what otherwise would be indifferent parts or organs. These factors are all important in the evolution in the colors, ornaments, and outgrowths from the cuticle of caterpillars. . . .

"So far as I am aware no one has suggested why these horns and high tubercles and often pencils of hairs are restricted to these particular segments. As a partial explanation of the reason, it may be stated that the presence of these high tubercles, etc., is correlated with the absence of abdominal legs on the segments bearing the former. It will also be noticed that in walking the apodous segments of the caterpillar are more elevated and prominent than those to which the legs are appended. They tend to bend or hump up, particularly the first and the eighth abdominal, the ninth segment being reduced to a minimum, and the tenth simply represented by the suranal and paranal plates, together with the last pair of legs.

"As is well known, the loopers or geometrid worms, while walking elevate or bend up the part of the body

[1] *Proceedings of the Boston Society of Natural History*, 1890, p. 493, 510-513.

situated between the last thoracic and first pair of abdominal legs, which are appended to the seventh euomere. Now in the larva of *Nematocarpa filamentaria* which bears two pairs of remarkable filamental tubercles rolled up at the end, it is certainly very suggestive that these are situated on top of the loop made by the caterpillar's body during progression, the first pair arising from the second, and the hinder pair from the fourth abdominal segment.

"It seems, therefore, that the humps or horns arise from the most prominent portions of the body, at the point where the body is most exposed to external stimuli; and the force of this is especially seen in the conspicuous position of those tubercles which are voluntarily made to nod or so move as to frighten away other creatures. Perhaps the tendency of these segments to loop or hump up, has had a relation of cause and effect in inducing the hypertrophy of the dermal tissues entering into the composition of the tubercles or horns."

Prof. W. B. Scott[1] says: "To sum up the results of our examination of certain series of fossil mammals, one sees clearly that transformation, whether in the way of the addition of new parts or the reduction of those already present, acts just *as if* the direct action of the environment and the habits of the animal were the efficient cause of the change, and any explanation which excludes the direct action of such agencies is confronted by the difficulty of an immense number of the most striking coincidences. . . . So far as I can see the theory of determinate variations and of use-inheritance, is not antagonistic, but supplementary to natural selection, the latter theory attempting no explana-

[1] *American Journal of Morphology*, 1891, pp. 395, 398.

tion of the *causes* of variation. Nor is it pretended for a moment that use and disuse are the sole or even the chief factors in variation."

European authors, partly from their less favorable situation for the obtaining of paleontologic evidence, have not contributed as much as Americans to the doctrine of bathmogenesis. Nevertheless, in England Spencer, Cunningham, Henslow, and others have sustained this doctrine, and in France Giard and Edmond Perrier, and in Germany, Semper, Eimer, and others, who lean principally in their writings to its physiogenetic aspect. Says Eimer:[1]

"According to my conception, the physical and chemical changes which organisms experience during life, through light or want of light, air, warmth, cold, moisture, food, etc., and which they transmit by heredity, are the primary elements in the production of the manifold variety of the organic world, and in the origin of species. From the materials thus supplied, the struggle for existence makes its selection. These changes, however, express themselves simply as growth."

Nägeli discards completely the agency of natural selection, and sees an internal "Principle of Improvement" as the active agent in evolution. He apparently includes both statogenesis and bathmogenesis in his conception.[2] He says:

"Such internal causes must necessarily be supposed merely on the ground that modifications or variations of the phyla do actually take place in definite directions, are not irregular. The internal causes effect

1 *Organic Evolution*, English Translation, 1890, p. 22.

2 *Mechanische physiologische Abstammungslehre*, C. V. Nägeli, Munich, 1884.

a constant alteration of the phyla in definite directions, towards greater perfection, that is, towards greater complexity." Accordingly Nägeli describes his theory of internal causes, as the "Principle of Improvement." "Superficial reasoners," he says, "have pretended to discover mysticism in this. But the principle is one of a mechanical nature and constitutes the law of persistence of motion in the field of organic evolution. Once the motion of evolution is started it cannot cease, but must persist in its original direction."

For convenience of reference I give a list of papers by American authors on this subject. European authors, beginning with Lamarck, and including Spencer, have implicitly included in their theses the effects of mechanical strains and impacts in developing the structures of animals. Fick, Henke, Tornier, and others have attempted an exact demonstration of the manner in which these forms of mechanical energy have produced the results. These attempts have great merit as physiological studies, but they have not been used by their authors as illustrations of specific evolution, owing to the fact that they have not carried their researches into the field of paleontology.

LIST OF PAPERS BY AMERICAN AUTHORS WHO HAVE CONTRIBUTED TO THE EVIDENCE USED IN THIS BOOK.

1866. Hyatt, A. Memoirs of the Boston Society of Natural History, p. 203. On the Parallelism Between Stages in the Individual and those in the Groups of Tetrabranchiata.

1866. Cope, E. D. Transactions of the American Philosophical Society, p. 398. On the Cyprinidæ of Pennsylvania.

1871. Cope, E. D. Proceedings of the American Philosophical Society, p. 259. On the Method of Creation of Organic Types; reprinted in Origin of the Fittest, 1887, pp. 210, 195, 199.

1872. Cope, E. D. Penn Monthly Magazine, May–July. On Evolution and Its Consequences; reprinted in Origin of the Fittest, 1887, p. 30.

1877. Ryder, J. A. American Naturalist, p. 607. On the Laws of Digital Reduction.

1877. Ryder, J. A. Proceedings of the Academy of Natural Sciences of Philadelphia, p. 314. On the Significance of the Diameter of the Incisors of the Rodents.

1878. Cope, E. D. American Naturalist, January. The Relation of Animal Motion to Animal Evolution; reprinted in Origin of the Fittest, 1887, p. 350.

1878. Ryder, J. A. Proceedings Academy Natural Sciences, Philadelphia, p. 45. On the Mechanical Genesis of Tooth-Forms.

1879. Cope, E. D. American Naturalist, March. On the Origin of the Specialized Teeth of the Carnivora; Origin of the Fittest, 1887, p. 363.

1880. Hyatt, A. Memoirs Boston Society Natural History, Fiftieth Anniversary. Genesis of the Tertiary Species of Planorbis at Steinheim.

1881. Cope, E. D. American Naturalist, April and June. On the Origin of the Foot-Structures of the Ungulates. On the Effect of Impacts and Strains on the Feet of Mammalia; Origin of the Fittest, 1887, pp. 368, 373.

1883. Ryder, J. A. Popular Science Monthly, XIII., pp. 139–145, 4 figs. The Gigantic Extinct Armadillos and Their Peculiarities. [Discusses the mechanical genesis, degeneration, and coalescence of vertebral centra.]

1885. Ryder, J. A. American Naturalist, pp. 411–415. On the Position of the Yolk-Blastopore as Determined by the Size of the Vitellus.

1885. Ryder, J. A. American Naturalist, pp. 815–819 and 903–907. On the Availability of Embryological Characters in the Classification of the Chordata.

1885. Ryder, J. A. American Naturalist, pp. 1013–1016. On the Genesis of the Extra Terminal Phalanges in the Cetacea.

1885. Ryder, J. A. American Naturalist, pp. 90–97. An Outline of a Theory of the Development of the Unpaired Fins of Fishes.

1886. Ryder, J. A. American Naturalist, pp. 179–185. The Origin of the Amnion.

1886. Ryder, J. A. Annual Report U. S. Commissioner of Fish and Fisheries for 1884, pp. 981–1085, Pl. IX. On the Origin of Heterocercy, etc.

1887. Ryder, J. A. American Naturalist, pp. 780–784. A Theory of the Origin of Placental Types, etc.

1887. Cope, E. D. American Naturalist, pp. 985, 1060. The Perissodactyla.

1888. Cope, E. D. Proceedings of the American Association for the Advancement of Science, p. 254. On the Mechanical Origin of the Sectorial Teeth of the Carnivora.

1888. Cope, E. D. American Naturalist, p. 3. The Mechanical Causes of the Origin of the Dentition of the Rodentia.

1888. Cope, E. D. Proceedings of the American Philosophical Society, p. 80. The Mechanical Origin of the Dentition of the Amblypoda.

1888. Ryder, J. A. American Naturalist, p. 547. The Several Functions of the Enamel Organ in the Development of the Teeth of Mammals, and on the Inheritance of Mutilations.

1889. Cope, E. D. American Naturalist, March. The Artiodactyla. (The elbow joint; the zygapophyses.)

1889. Cope, E. D. American Journal of Morphology, III., p. 137. The Mechanical Causes of the Development of the Hard Parts of the Mammalia.

1889. Hyatt, A. Smithsonian Contributions to Knowledge, and Memoirs of the Museum of Comparative Zoölogy, Cambridge, XVI, No. 3. The Genesis of the Arietidæ.

1889. Dall, W. H. American Journal Science and Arts, XXXVIII., p. 445. Hinge of Pelecypoda and Its Development.

1889. Osborn, H. F. Transactions American Philosophical Society, XVI., p. 531. (Published February, 1890.) The Evolution of the Ungulate Foot; The Mammalia of the Uinta Formation.

1889. Ryder, J. A. American Naturalist, pp. 218–221. The Polar Differentiation of Volvox, etc.

1889. Ryder, J A. American Naturalist, pp. 271–274. The Quadrate Placenta of *Sciurus hudsonius*.

1890. Ryder, J. A. Proceedings American Philosophical Society, XXVIII., pp. 109–159. The Origin of Sex, etc.

1890. Ryder, J. A. American Naturalist, pp. 376–378. The Placentation of the Hedgehog and Phylogeny of the Placenta.

1890. Packard, A. S. Proceedings Boston Society Natural History, XXIV., p. 493. Hints on the Evolution of Certain Bristles, Spines, and Tubercles of Certain Caterpillars.

1890. Dall, W. H. Proceedings of the Biological Society of Washington, May. On Dynamic Influence in Evolution.

1890 Jackson, R. T. Memoirs Boston Society Natural History, IV., p. 277 (July); American Naturalist, 1891, p. 11. Phylogeny of the Pelecypoda; The Aviculidæ and Their Allies.

1890. Dall, W. H. Transactions of the Wagner Free Institute of Science, Philadelphia, III., p. 58 (September). American Naturalist, 1894, p. 909. Origin of the Plaits on Columella of Gastropoda.

1891. Scott, W. B. American Journal of Morphology, p. 378. On the Osteology of Mesohippus and Leptomeryx; On Some of the Factors in the Evolution of the Mammalia.

1891. Ward, Lester F. Proceedings Biological Society of Washington, Annual Address. Neo-Darwinism and Neo-Lamarckism.

1892. Elliot, D. G. The Auk, IX., January. The Inheritance of Acquired Characters.

1892. Ryder, J. A. American Naturalist, pp. 923-929. A Geometrical Representation of the Relative Intensity of the Conflict Between Organisms.

1892. Ryder, J. A. Proceedings Academy Natural Sciences, Philadelphia, pp. 219-224. On the Mechanical Genesis of the Scales of Fishes.

1893. Sharp, Benj. American Naturalist (February), p. 89. Joint Formation Among the Invertebrata.

1893. Ryder, J. A. Proceedings American Philosophical Society, p. 192. Energy as a Factor in Organic Evolution.

1893. Hyatt, Alpheus. Proceedings Boston Society Natural History, p. 59. Bioplastology and the Related Branches of Biologic Research.

1893. Riley, C. V. Proceedings Entomological Society Washington, II., June, No. 4. Parasitism in Insects.

1893. Orr, Henry B. A Theory of Development and Heredity. New York and London: Macmillan & Co. 8vo. pp. 255.

1893. Hyatt Alpheus. Proceedings of the American Philosophical Society, p. 349. The Phylogeny of an Acquired Characteristic.

1894. Ryder, J. A. American Naturalist, p. 154. Orr's Theory of Development and Heredity.

1894. Cope, E. D. American Naturalist, p. 205. The Energy of Evolution.

1894. Cope, E. D. New Occasions, No. 6 (May), Chicago. The Origin of Structural Variations.

1894. Scott, W. B. American Journal Science and Arts, XLVIII. p. 355. Mutation and Variation.

INDEX.

ERRATA

Page 193, next to last line, for "auricle" read ventricle.
Page 222, third line from end, for "nutations" read mutations.
Page 244, third paragraph, fifth line, for "disease" read disuse.
Page 285, fourteenth line, for "multicellular" read multinuclear.
Page 297, third line, for "Mammalia" read Ungulata.
Page 307, second line, for "feet" read digits.
Page 351, ninth line, for fig. "106" read fig. 97.
Page 366, eighth line from bottom, for "divided" read derived.
Page 391, first line, for "notions" read motions,

NATURAL SCIENCES IN AMERICA

An Arno Press Collection

Allen, J[oel] A[saph]. **The American Bisons,** Living and Extinct. 1876

Allen, Joel Asaph. **History of the North American Pinnipeds:** A Monograph of the Walruses, Sea-Lions, Sea-Bears and Seals of North America. 1880

American Natural History Studies: The Bairdian Period. 1974

American Ornithological Bibliography. 1974

Anker, Jean. **Bird Books and Bird Art.** 1938

Audubon, John James and John Bachman. **The Quadrupeds of North America.** Three vols. 1854

Baird, Spencer F[ullerton]. **Mammals of North America.** 1859

Baird, S[pencer] F[ullerton], T[homas] M. Brewer and R[obert] Ridgway. **A History of North American Birds:** Land Birds. Three vols., 1874

Baird, Spencer F[ullerton], John Cassin and George N. Lawrence. **The Birds of North America.** 1860. Two vols. in one.

Baird, S[pencer] F[ullerton], T[homas] M. Brewer, and R[obert] Ridgway. **The Water Birds of North America.** 1884. Two vols. in one.

Barton, Benjamin Smith. **Notes on the Animals of North America.** Edited, with an Introduction by Keir B. Sterling. 1792

Bendire, Charles [Emil]. **Life Histories of North American Birds** With Special Reference to Their Breeding Habits and Eggs. 1892/1895. Two vols. in one.

Bonaparte, Charles Lucian [Jules Laurent]. **American Ornithology:** Or The Natural History of Birds Inhabiting the United States, Not Given by Wilson. 1825/1828/1833. Four vols. in one.

Cameron, Jenks. **The Bureau of Biological Survey:** Its History, Activities, and Organization. 1929

Caton, John Dean. **The Antelope and Deer of America:** A Comprehensive Scientific Treatise Upon the Natural History, Including the Characteristics, Habits, Affinities, and Capacity for Domestication of the Antilocapra and Cervidae of North America. 1877

Contributions to American Systematics. 1974

Contributions to the Bibliographical Literature of American Mammals. 1974

Contributions to the History of American Natural History. 1974

Contributions to the History of American Ornithology. 1974

Cooper, J[ames] G[raham]. **Ornithology.** Volume I, Land Birds. 1870

Cope, E[dward] D[rinker]. **The Origin of the Fittest:** Essays on Evolution and **The Primary Factors of Organic Evolution.** 1887/1896. Two vols. in one.

Coues, Elliott. **Birds of the Colorado Valley.** 1878

Coues, Elliott. **Birds of the Northwest.** 1874

Coues, Elliott. **Key To North American Birds.** Two vols. 1903

Early Nineteenth-Century Studies and Surveys. 1974

Emmons, Ebenezer. **American Geology:** Containing a Statement of the Principles of the Science. 1855. Two vols. in one.

Fauna Americana. 1825-1826

Fisher, A[lbert] K[enrick]. **The Hawks and Owls of the United States in Their Relation to Agriculture.** 1893

Godman, John D. **American Natural History:** Part I — Mastology and **Rambles of a Naturalist.** 1826-28/1833. Three vols. in one.

Gregory, William King. **Evolution Emerging:** A Survey of Changing Patterns from Primeval Life to Man. Two vols. 1951

Hay, Oliver Perry. **Bibliography and Catalogue of the Fossil Vertebrata of North America.** 1902

Heilprin, Angelo. **The Geographical and Geological Distribution of Animals.** 1887

Hitchcock, Edward. **A Report on the Sandstone of the Connecticut Valley,** Especially Its Fossil Footmarks. 1858

Hubbs, Carl L., editor. **Zoogeography.** 1958

[Kessel, Edward L., editor]. **A Century of Progress in the Natural Sciences: 1853-1953.** 1955

Leidy, Joseph. **The Extinct Mammalian Fauna of Dakota and Nebraska,** Including an Account of Some Allied Forms from Other Localities, Together with a Synopsis of the Mammalian Remains of North America. 1869

Lyon, Marcus Ward, Jr. **Mammals of Indiana.** 1936

Matthew, W[illiam] D[iller]. **Climate and Evolution.** 1915

Mayr, Ernst, editor. **The Species Problem.** 1957

Mearns, Edgar Alexander. **Mammals of the Mexican Boundary of the United States.** Part I: Families Didelphiidae to Muridae. 1907

Merriam, Clinton Hart. **The Mammals of the Adirondack Region,** Northeastern New York. 1884

Nuttall, Thomas. **A Manual of the Ornithology of the United States and of Canada.** Two vols. 1832-1834

Nuttall Ornithological Club. **Bulletin of the Nuttall Ornithological Club:** A Quarterly Journal of Ornithology. 1876-1883. Eight vols. in three.

[Pennant, Thomas]. **Arctic Zoology. 1784-1787.** Two vols. in one.

Richardson, John. **Fauna Boreali-Americana;** Or the Zoology of the Northern Parts of British America, Containing Descriptions of the Objects of Natural History Collected on the Late Northern Land Expeditions Under Command of Captain Sir John Franklin, R. N. Part I: Quadrupeds. 1829

Richardson, John and William Swainson. **Fauna Boreali-Americana:** Or the Zoology of the Northern Parts of British America, Containing Descriptions of the Objects of Natural History Collected by the Late Northern Land Expeditions Under Command of Captain Sir John Franklin, R. N. Part II: The Birds. 1831

Ridgway, Robert. **Ornithology.** 1877

Selected Works By Eighteenth-Century Naturalists and Travellers. 1974

Selected Works in Nineteenth-Century North American Paleontology. 1974

Selected Works of Clinton Hart Merriam. 1974

Selected Works of Joel Asaph Allen. 1974

Selections From the Literature of American Biogeography. 1974

Seton, Ernest Thompson. **Life-Histories of Northern Animals: An Account of the Mammals of Manitoba.** Two vols. 1909

Sterling, Keir Brooks. **Last of the Naturalists:** The Career of C. Hart Merriam. 1974

Vieillot, L. P. **Histoire Naturelle Des Oiseaux de L'Amerique Septentrionale,** Contenant Un Grand Nombre D'Especes Decrites ou Figurees Pour La Premiere Fois. 1807. Two vols. in one.

Wilson, Scott B., assisted by A. H. Evans. **Aves Hawaiienses:** The Birds of the Sandwich Islands. 1890-99

Wood, Casey A., editor. **An Introduction to the Literature of Vertebrate Zoology.** 1931

Zimmer, John Todd. **Catalogue of the Edward E. Ayer Ornithological Library.** 1926